IFAC

International Federation of Automatic Control

REAL TIME DIGITAL CONTROL APPLICATIONS

IFAC Proceedings Series, 1984. Number 1

IFAC Proceedings volumes, Published and Forthcoming

AKASHI: Control Science and Technology for the Progress of Society, 7 Volumes
ALONSO-CONCHEIRO: Real Time Digital Control Applications
ATHERTON: Multivariable Technological Systems
BABARY & LE LETTY: Control of Distributed Parameter Systems (1982)
BANKS & PRITCHARD: Control of Distributed Parameter Systems (1977)
BAYLIS: Safety of Computer Control Systems (1983)
BEKEY & SARIDIS: Identification and System Parameter Estimation (1982)
BINDER: Components and Instruments for Distributed Computer Control Systems
BULL: Real Time Programming (1983)
CAMPBELL: Control Aspects of Prosthetics and Orthotics
Van CAUWENBERGHE: Instrumentation and Automation in the Paper, Rubber, Plastics and Polymerisation Industries (1980) (1983)
CICHOCKI & STRASZAK: Systems Analysis Applications to Complex Programs
CRONHJORT: Real Time Programming (1978)
CUENOD: Computer Aided Design of Control Systems
De GIORGIO & ROVEDA: Criteria for Selecting Appropriate Technologies under Different Cultural, Technical and Social Conditions
DUBUISSON: Information and Systems
ELLIS: Control Problems and Devices in Manufacturing Technology (1980)
FERRATE & PUENTE: Software for Computer Control (1982)
FLEISSNER: Systems Approach to Appropriate Technology Transfer
GELLIE & TAVAST: Distributed Computer Control Systems (1982)
GHONAIMY: Systems Approach for Development (1977)
HAASE: Real Time Programming (1980)
HAIMES & KINDLER: Water and Related Land Resource Systems
HALME: Modelling and Control of Biotechnical Processes
HARDT: Information Control Problems in Manufacturing Technology (1982)
HARRISON: Distributed Computer Control Systems (1979)
HASEGAWA: Real Time Programming (1981)
HASEGAWA & INOUE: Urban, Regional and National Planning — Environmental Aspects
HERBST: Automatic Control in Power Generation Distribution and Protection
ISERMANN: Identification and System Parameter Estimation (1979)
ISERMANN & KALTENECKER: Digital Computer Applications to Process Control
JANSSEN, PAU & STRASZAK: Dynamic Modelling and Control of National Economies (1980)
JOHANNSEN & RIJNSDORP: Analysis, Design, and Evaluation of Man-Machine Systems

KLAMT & LAUBER: Control in Transportation Systems
LANDAU: Adaptive Systems in Control and Signal Processing
LAUBER: Safety of Computer Control Systems (1979)
LEININGER: Computer Aided Design of Multivariable Technological Systems
LEONHARD: Control in Power Electronics and Electrical Drives (1977)
LESKIEWICZ & ZAREMBA: Pneumatic and Hydraulic Components and Instruments in Automatic Control
MAHALANABIS: Theory and Application of Digital Control
MARTIN: Design of Work in Automated Manufacturing Systems
MILLER: Distributed Computer Control Systems (1981)
MUNDAY: Automatic Control in Space (1979)
NAJIM & ABDEL-FATTAH: Systems Approach for Development (1980)
NIEMI: A Link Between Science and Applications of Automatic Control
NOVAK: Software for Computer Control (1979)
O'SHEA & POLIS: Automation in Mining, Mineral and Metal Processing (1980)
OSHIMA: Information Control Problems in Manufacturing Technology (1977)
PAU & BASAR: Dynamic Modelling and Control of National Economies (1983)
PONOMARYOV: Artificial Intelligence
RAUCH: Applications of Nonlinear Programming to Optimization and Control
RAUCH: Control Applications of Nonlinear Programming
REMBOLD: Information Control Problems in Manufacturing Technology (1979)
RIJNSDORP: Case Studies in Automation related to Humanization of Work
RIJNSDORP & PLOMP: Training for Tomorrow — Educational Aspects of Computerised Automation
RODD: Distributed Computer Control Systems (1983)
SANCHEZ: Fuzzy Information, Knowledge Representation and Decision Analysis
SAWARAGI & AKASHI: Environmental Systems Planning, Design and Control
SINGH & TITLI: Control and Management of Integrated Industrial Complexes
SMEDEMA: Real Time Programming (1977)
STRASZAK: Large Scale Systems: Theory and Applications (1983)
SUBRAMANYAM: Computer Applications in Large Scale Power Systems
TITLI & SINGH: Large Scale Systems: Theory and Applications (1980)
WESTERLUND: Automation in Mining, Mineral and Metal Processing (1983)
Van WOERKOM: Automatic Control in Space (1982)
ZWICKY: Control in Power Electronics and Electrical Drives (1983)

NOTICE TO READERS

If your library is not already a standing/continuation order customer or subscriber to this series, may we recommend that you place a standing/continuation or subscription order to receive immediately upon publication all new volumes. Should you find that these volumes no longer serve your needs your order can be cancelled at any time without notice.
Copies of all previously published volumes are available. A fully descriptive catalogue will be gladly sent on request.

ROBERT MAXWELL
Publisher at Pergamon Press

IFAC Related Titles

BROADBENT & MASUBUCHI: Multilingual Glossary of Automatic Control Technology
EYKHOFF: Trends and Progress in System Identification
ISERMANN: System Identification Tutorials (*Automatica Special Issue*)

REAL TIME DIGITAL CONTROL APPLICATIONS

*Proceedings of the IFAC/IFIP Symposium
Guadalajara, Mexico, 17-19 January 1983*

Edited by

A. ALONSO-CONCHEIRO

*Universidad Nacional Autónoma de México
México, D. F., México*

Published for the

INTERNATIONAL FEDERATION OF AUTOMATIC CONTROL

by

PERGAMON PRESS

OXFORD · NEW YORK · TORONTO · SYDNEY · PARIS · FRANKFURT

U.K.	Pergamon Press Ltd., Headington Hill Hall, Oxford OX3 0BW, England
U.S.A.	Pergamon Press Inc., Maxwell House, Fairview Park, Elmsford, New York 10523, U.S.A.
CANADA	Pergamon Press Canada Ltd., Suite 104, 150 Consumers Road, Willowdale, Ontario M2J 1P9, Canada
AUSTRALIA	Pergamon Press (Aust.) Pty. Ltd., P.O. Box 544, Potts Point, N.S.W. 2011, Australia
FRANCE	Pergamon Press SARL, 24 rue des Ecoles, 75240 Paris, Cedex 05, France
FEDERAL REPUBLIC OF GERMANY	Pergamon Press GmbH, Hammerweg 6, D-6242 Kronberg-Taunus, Federal Republic of Germany

Copyright © 1984 IFAC

All Rights Reserved. No part of this publication may be reproduced, stored in a retrieval system or transmitted in any form or by any means: electronic, electrostatic, magnetic tape, mechanical, photocopying, recording or otherwise, without permission in writing from the copyright holders.

First edition 1984

British Library Cataloguing in Publication Data
IFAC/IFIP Symposium on Real Time Digital Control Applications *(1983: Guadalajara, Mexico)* Real time digital control applications. — (IFAC proceedings)
1. Digital control systems—Congresses 2. On-line data processing—Congresses
I. Title II. Alonso-Concheiro, A. III. Series
629.8'95   TJ213
ISBN 0-08-029980-6

These proceedings were reproduced by means of the photo-offset process using the manuscripts supplied by the authors of the different papers. The manuscripts have been typed using different typewriters and typefaces. The lay-out, figures and tables of some papers did not agree completely with the standard requirements; consequently the reproduction does not display complete uniformity. To ensure rapid publication this discrepancy could not be changed; nor could the English be checked completely. Therefore, the readers are asked to excuse any deficiencies of this publication which may be due to the above mentioned reasons.

The Editor

Printed in Great Britain by A. Wheaton & Co. Ltd., Exeter

IFAC/IFIP SYMPOSIUM ON REAL TIME DIGITAL CONTROL APPLICATIONS

Sponsored by:
International Federation of Automatic Control (IFAC)

Technical Committee on Applications (main sponsor)
Technical Committee on Computers
Technical Committee on Education
Technical Committee on Developing Countries

Co-sponsored by:
International Federation for Information Processing (IFIP)

Technical Committee 5

Organized by:
Centro Nacional de Enseñanza Técnica Industrial (CeNETI)
Centro Regional de Enseñanza Técnica Industrial (CeRETI) Guadalajara
Asociación de México de Control Automático (AMCA)

International Program Committee
R. Canales-Ruiz, Mexico (Chairman)
A. Alonso, Mexico
J. Alvarez, Mexico
P. R. Belanger, Canada
D. R. Bristol, U.S.A.
P. Castrucci, Brazil
N. Cohn, U.S.A.
M. Cuenod, Switzerland
C. M. Doolittle, U.S.A.
M. España, Mexico
J. Gertler, Hungary
R. Isermann, F.R.G.
L. Keviczky, Hungary
B. Kuo, U.S.A.
A. Weinmann, Austria
I. Landau, France
O. Lara, Cuba
P. M. Larsen, Denmark
M. Najim, Morocco
J. O'Shea, Canada
Y. Oshima, Japan
R. Padilla, Venezuela
D. Tabak, Israel
T. Takamatsu, Japan
M. H. Thoma, F.R.G.
P. Uronen, Finland
V. Vliestra, The Netherlands
H. J. Warnecke, F.R.G.
J. M. Wu, China

National Organizing Committee:
M. Rubio (Chairman)
R. Méndez (Co-Chairman)
J. Dueñas (Vice-Chairman)
J. Alvarez G
M. España V
G. García T
H. Guevara V
D. B. Hernández
A. Lemus V
J. Motolinía
B. Nakashima
G. Ruíz
J. Sabás
R. Silva

FOREWORD

Automatic control devices are at least as old as the Alexandrian engineering school (III century BC). Control theory is much more recent, starting with Maxwell's and Vishnegradskii's works (late XIX century AD). Further, the first applications of real-time digital control systems date only from 1960, about the time of IFAC's Foundation (1957). The number of technological advances and the speed of diffusion of their applications has been enormous by all standards during the last decade. Microcomputers, which appeared in the market in 1974, have probably been the main force behind this rapid and accelerated development. Real-time digital control systems can be considered the basis of a new industrial revolution which is already with us.

Up to now, developing countries have been mainly spectators of all this and have acted mainly as markets opened to indiscriminate imports of technological gadgets. As a first step towards the generation of their own technology in any field, and as an essential step towards a rational and intelligent selection and adaptation of certain technical advances, developing countries must learn what is being done elsewhere. This Symposium on Real-Time Digital Control Applications, organized in Guadalajara, Mexico, by the Asociación de México de Control Automático (AMCA) and the Centro Regional de Enseñanza Técnica Industrial de Guadalajara, with the sponsorship of the International Federation of Automatic Control (IFAC) and the International Federation for Information Processing (IFIP), is the first IFAC event to be organized in Latin America. Hopefully it will provide an opportunity for a fruitful and critical exchange of ideas and experiences among experts from many countries of the world, and in this manner contribute to the technical advancement of developing countries in general, and of Mexico in particular.

Many individuals and organizations have made this IFAC/IFIP Symposium possible. It is a pleasure to acknowledge their cooperation and to express our gratitude: to the Chairmen and Vice-chairmen of IFAC's and IFIP's Technical Committees who agreed to sponsor or co-sponsor the Symposium and helped us with valuable suggestions; to Professor T Vamos, IFAC's President, for his constant support; to our sister organization *Osterreichisches Zentrum fur Wirtschaftlichkeit und Produktivitat*, IFAC's Austrian National Member Organization, who kindly withdrew its application to organize an event on real-time control applications in Vienna later in 1983, in order that AMCA's application was considered favourably; to Fred Margulies, IFAC's Honorary Secretary, who was always helpful and understanding; to the members of the International Program Committee whose active participation was invaluable to obtain a technical program of quality; in particular we would like to thank Professor Nathan Cohn for his enthusiastic support and encouragement during all stages of the Symposium organization; to the Key-note Speakers who kindly accepted our invitation in spite of their heavy schedules; to those colleagues who spent some of their valuable time helping us by organizing invited sessions to complete the technical program; to the staff of Pergamon Press, who always responded promptly to our requests, including those made under time pressure; and finally, to all colleagues who contributed to the technical program of the Symposium and were patient with us throughout the selection procedure, making their best effort to comply with our modified deadlines.

The Symposium organizers are also very grateful to those organizations and institutions who kindly supported the event financially. Without them the Symposium would not have been possible. These non-profit and government institutions were: Centro Nacional de Enseñanza Técnica Industrial (CeNETI); Fondo de Estudios e Investigaciones Ricardo J Zevada; United Nations Educational, Scientific and Cultural Organization (UNESCO); and Instituto de Ingeniería, Universidad Nacional Autónoma de México.

A. Alonso-Concheiro
Symposium Editor

México, D.F., México
January 1983

CONTENTS

SESSION 1 - PLENARY PAPERS

The Evolution of Real Time Control Applications to Power Systems   1
N. Cohn

The Early Stages of Robotics   19
R.P. Paul

Real Time Control of Water Systems   33
R. Canales-Ruiz

SESSION 2 - METAL PROCESSING

Optimal Computer Combustion Control at the Soaking Pit   43
Y. Yamamoto, K. Mori, K. Fukuda, Y. Suzuki, K. Azumi and N. Saito

The Distributed Control System of Coil Annealing Furnaces   51
Y. Nariai, I. Yamazaki, M. Ono, T. Makino, S. Miki and R. Michioka

Combustion Control System of a Furnace for Hot Bloom   63
S. Tanifuji, Y. Morooka, J. Kumayama, K. Doi, T. Kawasumi and T. Shinmura

A New Adaptive Controller for Cold Rolling Mills   69
J. Hrušák, J. Mošna, E. Janeček and M. Simandl

DISCUSSION   75

SESSION 3 - MONITORING AND FAILURE DETECTION

A Specification of Real-Time Applications by Events and Links with Processes   77
M. Benmaiza and J.P. Thomesse

Distribution of Architecture and Allocation of Functions in an Integrated Microprocessor-based Substation Control and Protection System   85
M. Kezunović

Failure Detection and Prediction System by Using Adaptive Digital Filter   93
S. Oe, Y. Tomita and T. Soeda

Development of Nuclear Power Plant Automated Remote Patrol System   101
R. Nakayama, K. Kubo, K. Sato, J. Taguchi

DISCUSSION   107

SESSION 4 – ADAPTIVE CONTROL

Adaptive Control of Muscle Relaxation
C.S. Berger and W.A. Brown — 109

Microcomputer Implementation of an Adaptive Control Algorithm
R. Lozano and A. Noriega — 119

Adaptive Control of Discrete Multivariable Systems
R. Lozano and M. Bonilla — 125

Identification and Adaptive Control of a Sugar Factory Vacuum Pan
A. Aguado and A. Gómez — 133

Adaptive Control of a Steam Turbine
D.N. Oliva, E.L. Morris and M.T. Oliva — 139

DISCUSSION — 145

SESSION 5 – FUEL AND HEAT CONTROL

Model Reference Adaptive Control of an Industrial Phosphate Drying Furnace
B. Dahhou, K. Najim and M. M'saad — 147

Control Strategies for Multi-fuel Power Plants
U. Kortela, B. Salmelin, F. Wahlström and J. Joensuu — 159

Digital Decoupling of a 3-Zone Electrical Furnace by Means of Multivariable-PI Control
J. Gómez de Silva — 165

Combustion Stabilization and Improvement of the Efficiency in a Peat Power Plant
F. Wahlström and U. Kortela — 173

Digital Control of Furnaces in Ceramic Industry
H. El Hajjar, J.B. Pourciel and J.P. Babary — 183

The Mathematical Model of Computer Control of the Furnace Temperature of Mechanical Endurance Testing Machine
Jia-Sheng Wang and Yu-Fan Zheng — 191

DISCUSSION — 195

SESSION 6 – CEMENT INDUSTRY

Real Time Digital Control Systems for the Cement Industry
T. Ohta and K. Ishida — 197

Adaptive Control of a Ball Mill with Self-tuning Reference Model
R. Schulz — 203

A Microprocessor-based Adaptive Composition Control System
J. Hetthèssy, I. Vajk, R. Haber, M. Hilger and L. Keviczky — 209

Experiences from a Digital Quality Control System for Cement Kilns
T. Westerlund — 215

Computer Control of a Cement Plant
V.M. Dozortsev, E.L. Itskovich, I.V. Nikiforov and I.I. Perel'man — 221

SESSION 7 - ROBOTICS

The Diffusion of Industrial Robots in Sweden — 227
J. Carlsson and H. Selg

A Hierarchical Distributed Information Processing System for Forest Manipulation — 237
P. Kärkkäinen and M. Manninen

Direct Digital Robot Control Using a Force-torque Sensor — 243
G. Hirzinger

Tracking Control System for Complex Shape of Welding Groove Using Image Sensor — 257
M. Kawahara

A Multi-microcomputer-based Robot Control System — 265
K.W. Plessmann

DISCUSSION — 283

SESSION 8 - INDUSTRIAL APPLICATIONS I

On-line Scheduling for Transportation of Raw Materials — 285
K. Azumi, Y. Yamamoto, S. Ishikawa, Y. Maeda and Y. Ienaga

Optimization Control for Combustion Air in Refuse Incinerators — 293
M. Kawahara and K. Uosaki

Controlling a Distribution Conveyor by a Dedicated Microprocessor — 297
L.E.M. Boullart

Distributed Traffic Control System — 301
M. Nakai and M. Kasahara

DISCUSSION — 307

SESSION 9 - PULP AND PAPER

Trends in Digital Control Applications in Pulp and Paper Industry — 309
P. Uronen

SESSION 10 - POWER SYSTEMS I

Evolution of Digital Control in Energy Control Centers — 315
J.L. Carpentier

Area Control Performance Measurement and Corrective Control in Interconnected Systems — 325
N. Cohn

Emergency Control During Stability Crises by Tracking the Observation Decoupled Reference — 335
J. Zaborszky

Digital Control Applied to Power System Protection — 345
Y. Sekine and T. Matsushima

Digital Control in Nuclear Power Plants — 353
B. Bouzon

DISCUSSION — 357

SESSION 11 - CHEMICAL AND BIOCHEMICAL

Computer Control of Simple Variable Flow Processes 361
C.P. Jeffreson

Real Time Digital Multivariable Control for a Fermentation System 371
J. Carrillo, J. Alvarez and J.A. Gallegos

A Digital Approach to Monitoring and Controlling Fiberfill Plants 377
C. Gressel and A. Cohen

A Pilot-scale Distillation Facility for Digital Computer Control Research 385
J.L. Marchetti, A. Benallou, D.E. Seborg and D.A. Mellichamp

An Adaptive Feedforward Control Algorithm for Computer Control of Wastewater Neutralization 395
R.A. Balhoff and A.B. Corripio

Inferential Control Applied to Industrial Autoclaves 407
J.R. Parrish and C.B. Brosilow

Distributed Microcomputer Control in Real Time of the Process of Fermentation of Sugar Cane Derivatives (By-product) 415
S. Teijero Páez and J. Olivera Reyes

DISCUSSION 429

SESSION 12 - EDUCATION

A Model Program for Undergraduate Education in Real-time Computer Process Control 433
T. Olsen, R.H. Heist, H. Saltsburg and J.C. Friedly

A Cuban Experience in the Development of Courses in Microprocessor and Real Time Process Control with Microprocessors 441
J. Olivera Reyes and S. Teijero Páez

A Training About Real Time Digital Control in a French Engineer High School 447
J.P. Thomesse

DISCUSSION 453

SESSION 13 - MODELLING, IDENTIFICATION AND SOFTWARE

Local Optimisation for Correcting the Inputs in Non-linear Identification 455
M. de la Sen and M.B. Paz

A Moving Model of Discrete-data Systems and its Application in Control 461
Wen-Teng Wu, Yung-Chung Fang and J.R. Hopper

An Approach to the Design of Real-time Database Models 467
G. Rodriguez

Digital Image Coding by C-Matrix Transform 475
R. Srinivasan and K.R. Rao

Identification and Control Programs for Microprocessors 479
J. Enriquez, J.A. Hormaza and A.C. Campos

A Real Time Monitor, Its Representation by Petri Nets and an Application *A. Maldonado and F. Rivera*	485
DISCUSSION	491

SESSION 14 - INDUSTRIAL APPLICATIONS II

Microprocessor-based Control of Industrial Sewing Machines *B. Hertzanu and D. Tabak*	493
Microcomputer Direct Voltage Control of a PWM Inverter *G.S. Buja and D. Longo*	499
Industrial Applications of Vision Technology *M. Ejiri*	507
Direct Digital Control of Electrical Drive System Based on Improved Optimal Regulator Theory *T. Tsuchiya*	517
Stepping Motor Control *R. Canales-Ruiz and L. Alvarez-Icaza*	527
DISCUSSION	535

SESSION 15 - GLASS INDUSTRY

Digital Control Application on a Glass Gob Feeder *F. Saldaña and R. Solís*	537
A Digital Controller for a Glass Machine with Press-blow on Coated Molds Process *M.A. González*	541
On the Design of a Discrete Feedforward Control for Cooling Molds in the Glass Industry *F.L. Elizalde*	545
An Intelligent Digital Controller in the Formation Process of Glass Bottles *H. Rodríguez and S. Rodríguez*	551
Digital Control for a Pneumatic System *D. Figueroa and J. Heredia*	555
DISCUSSION	561

SESSION 16 - POWER SYSTEMS II

Real-time Computer System for the National Energy Control Center of Mexico *J.L. Calderon and M.A. Avila*	563
Management and Implementation of the Sictre Project *L. Rance*	577
Transmission of Digital Information via Satellite for the Real Time Control System of "Comision Federal de Electricidad", Mexico *D. Carrasco, G. Torres and A. Vazquez*	587
New Concepts for Automatic Generation Control in Electric Power Systems Using Parametric Quadratic Programming *J.L. Carpentier, G. Cotto and P.L. Niederlander*	595

On-line Observability Determination in Electric Power Network 601
P. Albertos, C. Alvarez and J.A. de la Puente

DISCUSSION 609

SESSION 17 - CONTROLLERS

A Self-tuning Controller with a PID Structure 613
F. Cameron and D.E. Seborg

Prediction of Optimal Direct Digital Control System for Process Industries 623
T. Moriyama, S. Fujii, H. Mitani, K. Achiba and T. Terada

Self-tuning Control of a Liquid-saturated Steam Heat Exchanger 631
S. Bittanti, D.W. Clarke, F. Romeo and R. Scattolini

DISCUSSION 639

Author Index 641

THE EVOLUTION OF REAL TIME CONTROL APPLICATIONS TO POWER SYSTEMS

N. Cohn

1457 Noble Road, Jenkintown, PA 19046, USA

Abstract. Paralleling the extensive growth and expansion of interconnected electric power systems in the United States and Canada during the past sixty years, has been the related need to regulate generation in the constituent areas, and the power flow between them, to achieve equitable, reliable and economic system and area operation. Many individuals and groups have made contributions to these objectives. These constitute the evolution of the system and area real time control art from modest, tentative beginnings to the comprehensive, broadly scoped and highly capable present day on-line digital control systems. This paper presents one individual's view, based largely on personal experience and observation, of significant steps in this evolutionary process. The paper deals primarily with the analog phases of these developments, many of the philosophies and techniques of which remain basic to current digital executions.

Keywords. Computer control, power system control, power station control, load dispatching, distributed control systems, large scale systems.

INTRODUCTION

The history of a discipline is probably best written, certainly most objectively written, by a non-participant in the events being recounted and evaluated. When reviewed by a participant, as in this instance, it is quite likely to have an autobiographical cast and perhaps even a bias. In any event, I hope that this review of earlier techniques, which I have been asked to undertake, and which recalls some of the steps and experiences that got us to where we are, may be helpful in providing a better basis for comprehending and appreciating the advanced technologies that are practiced in today's computer-directed world.

This paper will be based largely on the observations and experiences of my fifty-five years in this field, forty-five of these with Leeds & Northrup (Philadelphia, 1927-29 and 1955-72; San Francisco, 1929-36; Chicago, 1937-55) and ten years (1972-present) as a consultant. It will discuss activities in the United States and Canada, without at all diminishing the importance of work done elsewhere. It will describe developments that occurred primarily in the analog domain, many of the latter-day philosophies and techniques of which remain basic in current digital executions. Present day digital technology has of course moved far beyond the limitations of the analog domain, and introduced greatly expanded and valuable real time monitoring and control techniques for integrated power systems. A session including a presentation on the evolution of digital control for energy control centers, Carpentier (1983), is scheduled later in this symposium.

Over the years, many individuals and engineering and operations groups have contributed to the definition of system operating objectives and to the formulation, appraisal, revision and implementation of techniques for achieving them. To try to name all of them would be impossible. To name none would scarcely be reasonable. Despite its risks, I will identify some, primarily in this paper's bibliographic references. Others may see origins or events differently. In any event, with this paper I salute all individuals, whether specifically identified in it or not, who over the years have contributed to the development and application of real time power systems control.

Useful bibliographic tabulations are Preminger (1960) and IEEE (1977, 1981). Earlier historical reviews are Brandt (1953), Morehouse (1965) and McDaniel (1974).

Throughout the history of the power industry, dependable, real time automatic control to insure safe, reliable, responsive operation has been a necessary element of power system installations. That was true a hundred years ago at Thomas Edison's first central generating station at 255-257 Pearl Street in New York, placed into operation on Sept. 4, 1882, and generally regarded as marking the founding of the electric power industry, IEEE (1982). Each of the station's six 100 kw generators was equipped for control with speed governors, lineal descendants of James Watt's pioneering contribution to feedback control.

Control engineers are permitted a sympathetic and understanding chuckle when reading of Edison's early problems (later of course resolved) with the Pearl St. governors, encountered when first trying to run two generators in parallel.

An Edison biographer, Conat (1979), writes, "With one engine running, everything was fine. 'Then we started another engine', Edison reported, 'and threw them in parallel. Of all the circuses since Adam was born, we had the worst. One engine would stop, and the other would run up to 1000 revolutions. Then they see-sawed. The trouble was with the governors.'"

As in the very beginning, there are still speed governors. Such governors throughout the system, together with the frequency coefficient of connected customer load, and the variation of system stored energy as a function of frequency, serve as the basic self-regulating forces of the system. Jollyman (1927) describes the utilization of these effects for regulation of an isolated system. These governing effects are, however, singly-dimensioned and lack geographical discrimination in responding to load changes on an interconnected system, Cohn (1971b). They therefore require supplementary area controls to reallocate generation changes in order to satisfy individual interconnected area responsibilities and objectives, which include programmed bulk power transfers to other areas, and economic and secure operation within the area. Recounting the development of such supplementary controls is the prime objective of this paper.

POWER INDUSTRY -- 1927

To provide a background reference for the evolution and growth of on-line power systems controls, let's see what the domestic power industry was like at the time I embarked on my career. In the forty-five years since Pearl Street it had had what then would have been called great growth. The U.S. generating capability was about 25,000 MW. Thomas Edison and Nikola Tesla, the giant geniuses of the electric power field, were still alive, and working. The dc-ac battle between their respective technologies had been resolved in favor of the latter, though there were still many metropolitan areas that were distributing dc power. No one could have visualized then that dc would one day be back, as the preferred medium for long distance, extra high voltage transmission lines, and for asynchronous interconnections.

In contrast to the present capability of 600,000 MW, the 1927 capability seems small indeed. Also at that time, transmission voltages were lower, transmission distances shorter, and generating units smaller. And in the context of present day plant and system coordination, we can note that most fossil fueled power plants were at the time I now speak of, like Gaul, divided into three separate parts: a boiler room, a turbine-generator room, and an electrical switchboard room, with heavy walls separating them, lest, one can assume, there be communication between them.

Automatic Control

In 1927, supplementary automatic control was indeed in its infancy. Voltage control was, it is true, regularly used. Boiler feed water control was customary, and boiler combustion control was relatively new. System control depended primarily on generator speed governors, supplemented by manual control. In Preminger (1960), only three papers are listed for the period 1922-1928.

There was at that time, and for some time thereafter, relatively little control theory. Simulation as practiced in recent years was not available for control experimentation. It was not, however, especially missed. For the following two decades experimentation on the best of all simulators, power systems themselves, was feasible, and was practiced.

Telemetering was quite limited. A good watt transducer was not available. Such power telemetering as actually occurred was executed with a fairly complex and expensive transmitting unit, generally over dc telephone lines. Some telemetering was done from impulse generators on watthour meters, frequently over carrier, which tended to be limited and noisy. Analog computation, where executed, was generally done with servo driven slidewires. We had not yet entered the electronic age. Analog computers, as we later knew them, were yet to come, and digital computers were still far off in the distant future.

Interconnections

By 1927, the potential benefit of interconnections between adjacent areas, sharing generation and reserves, and in some cases plant construction, had been recognized and the practice started.

Comments by the late Samuel Insull (1921), whatever his faults may later have been, reflect considerable understanding of the value and the probable future extension of interconnections. Humphery (1927) provides a comprehensive summary of interconnections as they had then been developed in the northeast, in the mid-Atlantic states, and in the Chicago region. His paper outlines the potential benefits of interconnection. More particularly, however, it emphasizes prevailing operating problems, by no means then yet resolved, such as control of frequency, control of power flow and proper dispatching. It emphasizes that at that stage of the game, the existence of interconnections didn't mean their continued capability. Clearly, challenging control problems lie ahead.

Other interconnections already in service at that time include early ties of the Southern Company Pool and ties in California and the Pacific Northwest. The Pennsylvania-New Jersey Pool, later to be the Pennsylvania-New Jersey-Maryland Pool, was within a few months of being established.

The current full extent of interconnections, related sub-stations and plants in the United States and Canada, 230 kV and above, is shown in NERC (1981). A map of North American interconnected control areas appears in U. S. Dept. of Energy (1981).

I think it is clear that the extensive growth of interconnections, the corresponding increase in the number of generating stations, and the significant differences in their sizes and incremental efficiencies, introduced substantial hierarchical multivariable multi-level control problems. We will shortly see how these have been approached.

Measurement Developments

It is a self-evident maxim that what you cannot measure, directly or inferentially, you cannot control, or at least you ought not try to control. It will be clear from the discussion thus far, that two of the major parameters involved in power systems control are system frequency and megawatt load, the latter applying either to generators or transmission tie lines, or both. Apparatus for making such measurements prior to 1924 was of limited flexibility or precision, or of inadequate applicability to control systems, or far too costly.

Three developments, one initially unrelated to power systems activities and two that occurred virtually simultaneously but totally independently of each other, filled the measurement voids for power systems applications and were major factors in stimulating the early work in power systems real time control. These developments were:

1. The self balancing potentiometer high-torque servo recorder, invented by Leeds (1912).

2. The adaptation of the Leeds self-balancing recorder to a self-balancing ac Wien brige frequency recorder by Wunsch (1925).

3. The Lincoln thermal converter, introduced in 1924 by Lincoln Meter Company of Canada, as described in Lincoln (1929).

These have played so important a role in the development of power systems control that I should like to say a few words about each of them.

Leeds recorder. This instrument was originally developed for the automatic measurement of small dc potentials such as those encountered with thermocouples or resistance thermometer circuits. It was a revolutionary development and was a great stimulus to scientific and industrial measurement in many applications throughout the world. Its major characteristic was that in measuring very small electrical voltages, it did not draw power from or alter the measured voltage. In addition it possessed, from its own energy source, adequate power to drive a pen without restraint on a ten inch wide chart, to operate control contacts and to operate a number of retransmitting slide-wires in independent circuits in which were reproduced the measured voltage at high levels for analog computation and automatic control use. I dare say that when the instrument was developed in 1912 no one could have anticipated that it would become the cornerstone of frequency and load measurement and control, serving such functions widely through World War II and beyond.

Wunsch frequency recorder. Sometime in 1923 Nevin Funk (president of AIEE in 1943-44) then Chief Engineer of Philadelphia Electric Company, who had been using Leeds recorders for the measurement of generator and transformer temperatures was anxious to have an equally open scale recorder for a precise measurement of system frequency. He asked L&N if it would not be possible to build such a unit. The task was given to Felix Wunsch in the Company's Engineering Department. In due course he adapted a Leeds recorder to serve as a self balancing ac Wien bridge suitable for the precise measurement of system frequency, using a range of 58 to 62 cycles over a ten inch chart. The recorder was installed at Philadelphia Electric in 1924. Many followed elsewhere, and brought a whole new understanding of the nature of frequency variations on power systems.

Lincoln thermal converter. This unit invented, I believe, by Prof. Paul Lincoln of Cornell (AIEE president 1914-15) and developed for practical use jointly by Lincoln, Louis Paine of Lincoln Meter Company of Canada and Perry Borden of Hydro Electric Power Commission of Ontario was introduced in 1924 just about coincident with -- but totally unrelated to -- the development by Wunsch of the frequency recorder with which it was later to have so close and extensive an association.

The thermal converter was a most unusual device. It had no moving parts, developing a temperature difference between two self-contained heaters which was directly proportionate to ac power, independent of phase angle or frequency. It possessed high precision and stability. Self-contained thermocouples measured the temperature difference and turned out a dc milli-voltage proportionate to ac power, of sufficient magnitude to permit measurement on a self-balancing basis by the Leeds dc recorder referred to earlier. The dc outputs of a number of converters could be connected in series for reliable totalizing purposes, a new dimension for power system dispatching.

Thermal converters were in use extensively in Canada starting with the Hydro Electric Power Commission of Ontario in 1926, HEPC (1926). The Sangamo Meter Company had a relationship with Lincoln Meter Company and in 1931 an arrangement was made between L&N, Sangamo and Lincoln Meter Company for L&N to serve as the distributor of the units. Thereafter the converters were very intimately related in power systems applications to L&N recording and controlling assemblies. Their use in such applications multiplied and they were not withdrawn from sale until 1978, after more than fifty years of useful application.

LOAD FREQUENCY CONTROL -- EARLY TECHNIQUES

Dispatching for Manual Control

Conventional practice in power systems operations had been to depend on generator governors to respond to system load changes and to utilize manual adjustment of governor settings on one or more machines to achieve desired distribution of generation between alternative sources. An early, 1924, central dispatching installation to facilitate such operation was that of the Philadelphia Electric Company. Recorders showing the generation at each of their four stations, the total system generation and the first Wunsch recorder showing the system frequency were provided at the dispatching center. These were pre-Lincoln Thermal Converter days. The telemetering for the station load readings utilized Westinghouse Type R Kelvin Balance totalizing recorders. In each of the stations there was attached to the recorder a transmitting potentiometer slidewire, the output of which, connected to telephone lines, was measured potentiometrically with a Leeds recorder at the central office. Retransmitting slidewires on each of the receiving recorders provided a voltage summation of the individual station loads and was recorded as total system load by a fifth recorder. A Warren master clock provided the reference for periodic manual adjustment of the system speed to maintain time within limits considered appropriate. These were not close limits since it was not felt that close synchronous time was a service commitment to customers.

The information continually available at the central dispatching center minimized the need for communication with the individual plants, and the extent of manual adjustment required to fulfill operating objectives was not regarded as oppressive.

Other utilities had differing views on the need for close synchronous time.

Supplementary Automatic Control

One, New England Power, had embraced the policy of selling time to their customers, and assigned continuing manual adjustment to one of their stations for time regulation. This proved to be an arduous task, particularly considering the many other activities for which the operators were responsible. This led to the installation by New England Power in 1927 of what is regarded as the first use of automatic frequency control on a power system.

That step, to which reference will additionally be made shortly, was followed in the period 1928 to 1934 by comparable frequency control installations by other companies. For some, objectives went beyond close frequency for time regulation, and included additional objectives of simultaneous control of several generators within a station to achieve appropriate automatic division of loading between them, and regulating frequency to assist in control of tie line loading when interconnected.

I think it is appropriate to say that the pioneering work done in the east and middle west in this period provided the fundamental bases for moving on in subsequent years to the fully coordinated control of bulk power transfers between interconnected areas. Let's now see what some of these individual control developments were.

Frequency Control

As above noted, the first system to undertake automatic frequency regulation was New England Power. Two types of controllers were installed at Harriman Station. One was an adaptation by L&N of the Wunsch frequency recorder. Rather than regulate with simple "on" and "off" contacts, it was recognized that a relationship between control action and the extent of frequency deviation was desirable. The instrument was accordingly equipped to provide "lower" or "raise" control impulses proportional to the deviation of the instantaneous frequency from 60 Hz. Contact closure operated the governor synchronizing motor to lower or raise generation respectively. Sketch (a) of Fig. 1 shows the control characteristic of such a controller, drawn as the control balance points on a plot of frequency versus tie line flow. The controller would endeavor to hold scheduled frequency, F_o, regardless of tie line flow.

The other frequency controller was by Warren Telechron Company (later GE), and was a Warren Master clock with a mechanism arranged to provide contact closure related to the integration of frequency deviation over the previous two seconds. The results were apparently satisfactory for the both types of controllers as reported by Brandt (1929). A year's operation confirmed the validity of automatic frequency and time control, and additionally indicated that closely regulated frequency would contribute to regulation of bulk power transfers.

Shortly after construction of the New England recorder controller, L&N devised a simplified controller which had a fixed but adjustable balance point thereby eliminating the balancing slidewire of the recorder and the time required to operate it. Raise and lower impulses were proportionate to instantaneous frequency deviation and were made on a two-second cycle basis. The first of these units was installed at Wallenpaupack Station of Pennsylvania Power and Light in 1928. This type of unit remained standard for many years and was used on most of the installations in the United States and Canada.

A general discussion of frequency control, including descriptions of the L&N Southern California Edison Big Creek installation, and of the available Warren type equipment as well as comments on operating experiences and problems encountered are contained in Hunt (1930). Another report of that same general period which is of interest in reflecting the understanding of the problems and needs related particularly to interconnected systems, for which solutions had not yet been provided, and which were not to be available until several years later, is Fitch (1930).

Frequency Control with Time Correction

Regulation from "instantaneous" frequency might or might not result in a precise synchronous time, depending on the calibration of the controller and the overall effectiveness of the control system. This was recognized at an early date and for those who preferred the type of frequency regulation that was based on "instantaneous" values, an automatic time correction feature was added. One technique for achieving this was to have an automatic vernier adjustment of the frequency control set point from accumulated system time deviation, Heath (1929). An early execution was added by Southern California Edison to their Long Beach frequency controller. The arrangement worked very well, and in effect served as an overall corrective unit for both the calibration of the frequency controller and the integrated control responses of the system.

Commercial assemblies for such operation were supplied by L&N using a Warren clock as the master time standard. A number of installations of this type were made in the 1930-34 period, including the City of Vernon, Calif. Diesel Station, City of Seattle Diablo Station and Hoover Dam Station.

Multi-Unit Control

As systems grew, it was in many cases not feasible for a single unit in a multiple unit station to undertake the swings essential to regulate frequency. Further, better economy could be achieved by dividing total station load among units of the station in accordance with efficiency considerations. Three approaches were devised for such operation.

Proportionate loading. An early approach was to divide the load among participating units of the station in accordance with predetermined ratios, Doyle (1928, 1929). This technique originally proposed the use of shunts in the secondary circuits of generator current transformers, appropriately phased to be responsive to unit kw output, and distributing control pulses from the frequency regulator to maintain the desired ratio of outputs of the individual machines. I do not believe any installations were made using such shunts. Later, however, when Lincoln thermal converters became available, they were used in place of shunts to measure the outputs of the individual generators and the converter outputs were balanced through individual load controllers to maintain the preset ratios. This general technique has been very widely used.

Economic loading. This technique, intended for hydro stations, divided the load between the units of a station in accordance with their incremental input/output curves to match incremental water rates. Equipment for such application was jointly engineered by the I. P. Morris and De La Vergne Inc., and L&N and is described in Kerr (1930). Installations were made at Carolina Power & Light, Washington Water & Power and Montana Power & Light in 1930. The means of developing the incremental curves were cumbersome. They were comparable to the transmitters used for the Philadelphia Electric telemetering installation mentioned earlier, utilizing a Westinghouse Type R wattmeter driving an L&N slidewire rated to match the incremental characteristic.

Valve point loading. This technique seeks to take advantage of the fact that steam units have their best efficiencies when steam controlled inlet valves are not in a throttling position. Load division within the plant is accordingly based on programming all controlled units, save one, to operate to have fully opened valves. Description of a pioneering system for such load programming is described in Purcell and Powel (1931). This paper is of additional interest and importance because of other material it contains, to which reference will be made later. Valve point loading continues to be regarded by many operators as important, despite operating difficulties in precisely determining when a given valve is or is not in a throttling position.

INTERCONNECTED SYSTEMS CONTROL -- EARLY EFFORTS

By 1930-31 automatic frequency control was well established. It was in use in many locations, some equipped with time error correction, many with multiple units participating within the regulating station. Summaries of experiences with supplementary control up to this time are contained in Henry and others (1929) and Fitch (1931). Major questions then were, "now that we have frequency control, how do we spread the regulation and control power flow on interconnecting tie lines?"

It was clear that frequency control, by its very nature, meant absorbing load changes on the regulated unit or units regardless of where they originated. When the load changes were in a remote area, it was absorbed by the local frequency regulating station, resulting in undesirable changes in tie line flow. "How to avoid this?" That was the question.

There was full agreement that each company should endeavor to absorb its own load changes. Two separate techniques were explored in the 1930-31 period: (1) Parallel frequency control on interconnected companies, and (2) Constant tie line control on a company connected to a system already under frequency control. Both will be reviewed.

Parallel Frequency Controllers

A major program was undertaken in 1930-31 by West Penn Power and American Gas and Electric (now American Electric Power) utilizing concurrently operating frequency controllers at three different stations: at their jointly owned Windsor station, at the West Penn Springdale Station and at the AG&E Philo Station. This was a really "noble experiment" involving cooperative effort by the two power companies, their station operating personnel and two manufacturers. I say two manufacturers, because it was decided to conduct the experiment with two complete sets of control equipment, the one being the proportional step instantaneous type frequency control by L&N and the other being the Warren short period time deviation integration type by G.E.

The tests established that controllers could operate in parallel, but differences in calibration and sensitivities resulted in more deviation from schedule of power flow on interconnecting tie lines than was considered desirable or acceptable. Further work was indicated.

The experiment did establish, however, preference on the part of the operating people for the instantaneous type regulators, and thereafter these became the essential standard in the United States and Canada.

At about this same time, there were independent tests on parallel frequency controllers at Washington Water Power and Montana Power. Results, reported by McNair and others (1932), were comparable to those experienced in the Midwest, namely, too much variation of tie line flow.

Constant Tie Line Control

At approximately that same time, Duquesne Light installed a constant tie line controller at their Colfax Station to regulate power flow on their tie to Springdale of West Penn Power, the latter station already being under frequency control. Different types of regulators were used, one involving a solenoid assembly by Westinghouse, another utilizing a Westinghouse Kelvin balance unit, and the third an L&N proportional step controller, similar to the one used for frequency control except that instead of an ac Wien measuring bridge, it made a dc comparison of the output of a Lincoln thermal converter that metered the tie line power flow and a dial set to the desired tie line flow. Sketch (b) of Fig. 1 is the control characteristic of constant tie line control. The control would act to hold tie line at scheduled flow, T_o, regardless of frequency. Purcell and Powel (1932), previously referred to is a report of this installation. It notes that, at least from the point of view of Duquesne itself, the constant tie line controller provided a solution to the problem of a fixed interchange with a neighboring utility. As can be seen from the discussions of the paper, however, there were differing views by interconnected companies. They noted that in holding a constant tie line, Duquesne frequently contributed adversely to system frequency, by opposing its own self-regulating forces that responded to remote load changes. Thus it failed to provide assistance to remote areas in their time of need and indeed aggravated prevailing conditions. The discussions reflect a considerable difference of opinion as to the relative virtues of frequency control versus constant tie line control.

In retrospect, that paper and its discussions reflect a watershed period in the evolution of real time controls for the effective regulation of bulk power transfers on interconnected systems. It was the first to introduce constant tie line control as an operating technique. Its discussions include purposeful comments by Fitch, Sporn, Brandt, Hunt and Juncke, all then very active in the work being done on power systems control. Most had formulated in papers and presentations what they felt were the prevailing needs to insure effective control of scheduled bulk power transfers on interconnected systems.

Sporn's discussion, in reporting on the parallel frequency control experiment (which he described as "distributed control" -- probably the earliest use of a term and technique now frequently encountered) and on the work of others, correctly identified the limitations of flat frequency control and tie line control as being unable to distinguish as to whether the load changes to which they were responding occurred in or out of their respective areas. He also recognized the problems introduced when there were multiple ties and not just single ties between areas.

As a less than significant autobiographical note I might add that, being then stationed in San Francisco, I was present at the Lake Tahoe presentation of this paper. Electrical West (1931) in its report of the meeting, quotes me as commenting at the meeting on the merits of both frequency and tie line load control, and the probable uniqueness of each installation.

A pertinent reference of that period, Sporn and Marquis (1932), provides an excellent summary of frequency and tie line load responses to system load changes, a summary of experiences with Windsor Station regulating frequency for the entire interconnection, and the analysis that led to the parallel frequency control experiment at Windsor, Philo and Springdale. It outlines achievements, but also defines remaining unsolved problems as of that time.

The paper emphasized the need of proper overall coordination for economy purposes and plant loading. It reiterated the need for each area to absorb its own load changes. The importance of close frequency for more than time control was again stated. It noted most particularly the need for "proper coordination of tie line and frequency controllers so that the two function toward the same end with a minimum load swing and so that the functioning of one does not vitiate some prime function of the other or of the system at a time when such functioning is badly needed for proper system performance".

Such comments constituted a clarion call for significant additional steps in the evolution of power systems controls. That's how things were in 1933. There followed the development work of 1934-38 out of which came major significant steps in the regulation of bulk power transfers.

INTERCONNECTED SYSTEM CONTROL --
INITIAL FREQUENCY-TIE LINE
COMBINED CONTROLLERS

Steps taken in the Midwest and East to fulfill bulk power transfer objectives included combined frequency-tie line control techniques identified as "selective frequency", "tie line zoned" and "frequency zoned" controls. All were intended to be used in concert with a master frequency controlling area. Each will now be briefly examined.

Selective Frequency

The first development of a combined frequency and tie line controller was, I believe, at Crawford Station of Commonwealth Edison in 1934. Its control characteristic is shown in sketch (c) of Fig. 1. It is arranged so that control action is based on frequency deviation from schedule, but control is permitted to act only when it will also correct prevailing deviation of tie line flow, as in the first and third quadrants of the figure. There is no control action in the second and fourth quadrants since at that time control action needs to be taken in one or more remote areas to correct both system frequency and the tie line flow of those areas. This was an interesting development and was a move in the right direction. It added an element of stability when operated with master frequency control elsewhere, but there were still questions of differences in calibration and sensitivities between controllers, and the non-specific nature of uncontrolled deviation of tie line flow from schedule. The use of selective frequency control was quite limited, however, and it was superceded within a relatively short time by the introduction of the "tie line bias" controller, which will be discussed later, after a brief review of "zoned controls".

Zoned Controls

These controls were introduced shortly after the introduction of the selective frequency controller. I am uncertain as to their original formulation or extent of use. They were a standard option on L&N installations after the introduction of the tie line bias type of control. They were of two types. The one whose control characteristic is shown in sketch (d) of Fig. 1 was identified as "tie line zoned by frequency deviation". It would normally operate to maintain a constant tie line loading at scheduled flow T_o but would automatically shift to frequency control at preset high and low frequency levels when respective high and low frequency deviations occurred.

The other, designated "frequency zoned by tie line deviations", is illustrated in sketch (e) of Fig. 1. It would normally be operated as a constant frequency controller to maintain frequency at F_o, but at preset high and low tie line deviations, it would automatically shift to constant tie line control at the respective preset values.

These controls represented additional steps in combining frequency and tie line flow into a single controller to help overcome the limitations of constant frequency or constant tie line controls. They represented additional steps in the evolution to a fully coordinated combined frequency-tie line controller.

INTERCONNECTED SYSTEM CONTROL --
BIAS CONTROLS WITH CENTRALIZED
FREQUENCY CONTROL

A major contribution in the evolution to present-day interconnected system control practice was the development and introduction of "tie line bias control" devised by Williams and Morehouse (1935). In my view this was the genesis of current practice. It combined instantaneous frequency and prevailing tie line flow in manner illustrated by the solid line control characteristic of sketch (f), Fig. 1. In effect, the control characteristic is a linear curve of frequency versus tie line, passing through the point defined by scheduled frequency F_o and scheduled tie line T_o. The effective operating tie line schedule is shifted automatically in the controller along the control characteristic by the magnitude of frequency deviation from schedule. The amount of the shift is defined by the slope of the characteristic curve, the inverse of which is the "frequency bias", usually specified in MW per 0.1 Hz.

This technique has been optionally referred to as "tie line bias control" or "frequency biased tie line control" or just plain "bias control". It went through three phases before evolving into the design and applications practice that remains standard to this day. Each of these will be briefly summarized, noting that in all three of these phases one area was retained on constant frequency control to insure a constant frequency for the interconnection.

Tie-Line Bias Control With Bias Withdrawal

In the original design of bias control, the operating philosophy was that assistance should initially be provided -- but only temporarily -- to an area or to areas in need. The bias characteristic of the controller would provide for such assistance, since for a decrease in frequency, for example, the tie line schedule would automatically be shifted to a higher level of power "out" to provide the desired assistance. It was further felt, however, that the assistance should be rendered for only a limited time, by which time the offending area or areas should have adjusted their own generation to correct their deficiencies. If they did not do so, however, then a time dependent mechanism in the tie line bias controller would rotate the controller characteristic from the initial bias diagonal shown in sketch (f), Fig. 1 toward the vertical or constant tie line characteristic, shown as a broken line. This would withdraw the bias assistance that had therebefore been provided.

In other words, a constant tie line flow for the area was given higher priority than rendering sustained assistance to others.

The initial installation of this type of equipment was made in 1935 at Harriman Station of New England Power Co. followed by one at Carolina Power & Light in 1936, and one in early 1937 at Twin Branch Station of Indiana & Michigan Electric, a property of American Gas and Electric Co., now American Electric Power. All three units were of the bias withdrawal type.

Tie Line Bias Control With Sustained Bias

At the end of 1936 I was transferred from the L&N San Francisco Office to Chicago. It fell to my lot in 1937 to place into service the tie line bias control at Twin Branch Station. I was joined there by Clark Nichols from Company Headquarters, who was familiar with the two earlier installations at New England Power and Carolina Power & Light.

Though the controllers at New England and Carolina were presumably providing acceptable performance, getting satisfactory operation with the bias controller at Twin Branch proved to be very difficult, and in fact we didn't achieve it. What we noted after long hours of adjustment and observation was that at the start of a significant frequency change caused by remote load changes, the controller would cooperate nicely, would act to assist the remote areas, and would permit its own tie line to go off schedule in accordance with the bias curve. After a few minutes, however, depending on the setting of the controller, the assistance to the remote areas would be automatically withdrawn, creating a further departure in the already off-schedule frequency. Altogether, since remote load and generation changes were occurring all the time, this resulted in excessive tie line hunting between the Twin Branch area and other areas, a totally unacceptable situation.

It seemed to me on reflection that what was needed was to retain the bias response of the controller, but to delete the bias withdrawal action. This would permit the bias assistance to the remote area to continue uninterruptedly until the remote area or areas did what had to be done by generation changes within their areas. Twin Branch would not move back toward constant tie line control with all of its limitations, while frequency remained off normal. When the frequency did move back toward normal as remote areas made the requisite changes to their generation, the bias controller would, in accordance with its bias characteristic, smoothly remove assistance that was no longer required. The area would be back on normal tie line schedule when frequency had been returned by the remote areas to normal. It seemed to me that such cooperation with prevailing system needs should have higher priority than the area desire to maintain constant tie line transfer. Such an arrangement, I felt, would lead to stable and non-hunting operation for the area.

We accordingly arranged to rebuild the controller to remove the bias withdrawal function. With that done, it worked, as predicted. It was another example of using the system as our test simulator.

A sustained bias did thus prove to be the solution to the problem. Thereafter all of the tie line controllers we built, to this day, have been of the sustained bias type.

The control characteristic of this controller then became sketch (f) of Fig. 1 with the dash line and the time-dependent arrow removed. We were not yet at sketch (g) Fig. 1, however, since we were on a "cascade" not a "net interchange" basis, and a central frequency regulating station was still in use.

Net Interchange Versus Cascade

In relating the transition to sustained bias at Twin Branch, I think it is appropriate to identify one of the pioneers of that period who contributed significantly to the subsequent expanded use of bias control in the middle west. I refer to Jack Girard, at that time, and for many years thereafter, the Operations Chief at Indiana and Michigan Electric. Girard had a clear concept of the potential benefits of interconnected operation, and in preparing new contracts with neighbors he stipulated their use of tie line bias control to optimize reliability and economy.

For many years after our initial meeting at Twin Branch I met regularly with Girard at Twin Branch or Chicago to analyze operations and explore new and expanded needs. We sent many restaurant tablecloths to the laundry laden with exploratory and tutorial sketches.

In 1938 one of the I&M interconnected neighbors, Public Service Co. of Indiana, was to install tie line bias control at the company's Dresser Station. They had six "boundary" ties with I&M, Cincinnati G&E, Louisville G&E and Northern Indiana Public Service. The question arose as to which tie or ties should serve as the basis for the control. This was resolved by Girard and Joe Trainor, Operating Chief of PSC of I. They decided to take the net of all six ties. I was invited to lay out the control system and accordingly arranged for Dresser to operate on a basis of area net interchange with all neighbors. I believe this was the first use of net interchange bias control on a multiple tie area.

At this time I&M was regulating on the basis of their two ties to the frequency regulating area, Ohio Power. In effect, all areas beyond their own were, from the point of view of control at Twin Branch, effectively a part of the I&M area. The postulate defining this was simply to draw a line through whatever ties were serving the bias control and then complete the line as a "circle" back on itself without crossing any other ties. All of the system encompassed in such a "circle" constituted the area for which that particular company was regulating. Where companies within this enlarged "circle" were themselves regulating, such as PSC of I and Commonwealth Edison for the I&M case, they were in effect independent control areas within the overall larger I&M area. Any deficiency, however, in their fulfilling their regulating responsibilities would become a regulating burden for I&M. Further, the bias setting for I&M would have to include, in addition to a bias for itself, the summation of bias settings of all of these additional areas, a procedure referred to as "cascading".

A preferred technique for each area would be to net all of its own boundary ties and set its own bias independently of the bias settings of its neighbors. With this practice, each area would follow its own load swings. It would depart from its net interchange schedule when there were frequency deviations to provide assistance to areas in need in an amount related to its own responsibilities, and not the cumulative assistance due from itself and from its self-contained control areas.

Girard agreed that there would be potential improvement to I&M operations with such arrangements. In 1942 I&M arranged to have additional telemetering added and integrated with the bias controller to shift Twin Branch operation from "cascade" to net interchange control. As time went on, such operation became standard throughout the United States and Canada for multiple-tie areas.

INTERCONNECTED SYSTEM CONTROL -- FULLY DISTRIBUTED AREA CONTROL

Just as World War I had stimulated interconnections to serve the power needs of the time, so World War II, on a much larger scale, caused expansion and extension of the then existent interconnections. The expansions in the Southwest and their use of net interchange tie line bias control are described by Morehouse (1945).

The "Midwest Interconnection", later known as the Interconnected Systems Group (ISG), had also grown and expanded. Philo Station of Ohio Power continued to provide the master frequency regulating function. Its control area, in effect, was the entire interconnection. Generation variations on the station were great. In addition to picking up Ohio Power load variations, it would endeavor to absorb the regulating deficiencies of all of the other areas of the considerably expanded interconnection. While I had no direct contacts with Philo, my friends at I&M, a sister company of Ohio Power, reported that the normal regulating assignment at Philo was plus or minus 80,000 kW, and that frequently station generation would fluctuate beyond these high and low extremes and correspondingly interrupt frequency control.

We discussed this situation a lot at I&M. It seemed to me that the solution rested in removing frequency control from Ohio Power, and placing it like all of the other areas of the interconnection, on net interchange bias control. In other words, it seemed quite in order to use distributed bias control throughout the system rather than on all areas but one, and simply have no master frequency regulating station. My friends at I&M felt that this might someday come, but that first Ohio and others would have to be convinced that system frequency could be maintained at 60 Hz even if no station had the assignment to achieve this.

I had developed analyses to demonstrate, to my own satisfaction, that 60 cycles would automatically result if certain criteria were followed. For all participating areas, power flow on each boundary tie should be metered at a common point and telemetered to one of the two involved areas as power "out" and to the other as power "in", so that the algebraic sum of net interchanges would be zero. Similarly, schedules between each pair of areas should be of equal magnitude but of opposite sign, so that in summation the net interchange schedules for all areas would also be equal to zero.

With these criteria fulfilled, and with all areas regulating effectively, system frequency would automatically be maintained at 60 Hz and individual area net interchange schedules would be met.

Should one or more of the areas not regulate effectively, there would be a corresponding departure of frequency from normal, and all remaining areas, acting on their respective bias characteristics, would automatically shift their net interchange schedules and assist the area or areas then in need. Assistance from each of these areas would be a function of the area bias setting and the frequency deviation and would persist on a sustained basis until the area or areas at fault adjusted generation to restore frequency to normal.

I was located at that time in Chicago, and although there was skepticism on the part of some of my associates at Headquarters in Philadelphia that such an arrangement was practical, I felt that the technique of no master frequency control would work, and anxiously awaited an opportunity to try it out. An apparent opportunity arose in the Chicago area in 1944 when we were invited to recommend control arrangements for interchange between Carnegie-Illinois steel mills at Gary and South Chicago Works, but the control project did not materialize.

The following year, 1945, responding to an inquiry from Iowa Power & Light at Des Moines concerning a prospective tie with Iowa Electric Light and Power Co. at Cedar Rapids, a letter was prepared and sent by the L&N Chicago Office recommending tie line bias control at each end, outlining the benefits of having neither area operating on constant frequency control and defining the criteria that would make such operation satisfactory. That project was deferred.

The following year, 1946, however, a major new interconnection, identified as the United Pool, later designated Interchange Power Services, Inc., was planned to include Iowa Power & Light at Des Moines, Iowa-Illinois Gas & Electric Company, Kansas City Power & Light, and St. Joseph Light & Power Company. I was invited to meet with representatives of these companies and of United Light & Railway Service Company which was to do the engineering for the project, to recommend telemetering and control equipment for the interconnection, and of course did so. An opportunity to apply the new concept of no central frequency controlling station had arrived.

I recommended fully distributed frequency biased net interchange control for each of the four control areas. In discussions with representatives of the several participating companies, I described the way in which the proposed equipment would work and why no master frequency regulating area would be required. Equipment to operate in accordance with these concepts was ordered in 1947, and initial operation occurred in late 1948. Operations were fully in accord with expectations. We now had operation in accordance with sketch (g) of Fig. 1.

I would like to point out that in that same general time period Brandt (1947) described independently derived considerations for the same concept of eliminating the master frequency regulating area. The subsequent application of his ideas to the Northeast interconnection is described in McCormack and Metcalf (1949).

In Cohn (1950), I reviewed and analyzed inter-area control techniques that had until then been developed and used, summarizing their characteristics and limitations. Included were the combinations of a master frequency control in one area and constant tie line controls in other areas, a master frequency control in one area and frequency biased net interchange controls in other areas, and finally, distributed frequency biased net interchange control in all areas without a master frequency regulating area.

The fully distributed frequency biased net interchange control technique in all areas, without a central frequency regulating area, has, for close to 35 years, been the standard inter-area control practice on all US-Canada interconnected systems.

MAGNITUDE OF BIAS SETTINGS

With the basic technique of frequency biased net interchange tie line control in all areas well established, questions arose in various operating areas as to just what the proper magnitude of area bias settings should be. In Cohn (1950), I had suggested that the settings be equal to the area governing characteristic. Questions developed concerning methods of determining this characteristic. There were also questions related to its variability, and to the effects of the frequency coefficient of area load. It was reported that there were strong differences of opinion within ISG on the proper magnitude of bias settings, culminating in an unresolved debate at the group's 1955 annual meeting in Corpus Christi. To help resolve the disagreement, the Interconnected Systems Committee (ISC)

Test Committee headed by L. V. Leonard of PSC of Indiana was asked to investigate the matter and report to the ISG 1956 annual meeting scheduled for Des Moines.

The committee invited representatives from four manufacturing companies to present their views at a two-day meeting in Cincinnati. I was invited to speak for L&N and did so. At the end of the Cincinnati sessions I was asked by A. L. Richmond of Ohio Edison, Chairman of ISC, to "save my notes", and present the same analysis to the entire ISC group at Des Moines. This latter presentation is transcribed in Cohn (1956a) and was formalized and expanded into an AIEE paper, Cohn (1956b).

Equations developed in the analysis demonstrated that settings lower than the combined governor-load governing characteristics resulted in undesirable withdrawal of assistance to areas in need. Such withdrawal is appreciably greater in relative magnitude than additional assistance that would be provided if settings were above the combined governing characteristic. Clearly, a setting higher than the anticipated governor-load governing characteristic is much more preferable, and system cooperative, than a setting below the characteristic. The Mollman discussion of this AIEE paper included an analysis, based on the paper's equations, of a then recent large load drop. It showed that system bias settings averaged about one-half the system governing characteristic, accounting for the observed adverse response of the system to the load drop.

The Des Moines presentation provided a basis for the operating guideline adopted by ISC at that meeting. The guideline provided that area bias setting be at least as high as the area governing characteristic at peak load, and not less than 1%. Most of the time, therefore, bias settings would be higher than the area governing characteristic. This same guideline later became a part of the North American Power Systems Interconnection Committee (NAPSIC) Operating Manual, and is now a part of the successor North American Electric Reliability Council Operating Manual, NERC (1982).

INADVERTENT INTERCHANGE AND TIME DEVIATION

When one or more areas do not fulfill their regulating obligations, there will be frequency deviations and net interchange schedule deviations consistent with the bias characteristics of the assisting areas. There will be corresponding accumulations of inadvertent interchange for both the offending areas and the assisting areas, as well as an accumulation of system time deviation. Techniques for correcting for these deviations have been developed over the years and present practices are summarized in NERC (1982).

General practice insofar as bulk power transfers are concerned is to pay for scheduled transfers only, and pay back inadvertent accumulations in kind at "on peak" or "off peak" periods consistent with the period of their respective accumulations. Time deviations are corrected by appropriate frequency schedule offset, with participation by all areas.

Difficulties have been encountered at times in some aspects of the present payback procedures and interest has been expressed by some operators in a technique of payback in dollars. It has been felt that this would be an incentive for better area regulation.

Equitable payback in dollars would require a technique for determining how much of each area's inadvertent interchange had been caused by itself and how much by each of the other areas of the interconnection. Such a technique has not been available. In Cohn (1971a) there are developed equations to define the area-caused components of system time deviation and of each area's inadvertent interchange. These equations have not had practical use since they include non-measurable parameters. More recently, however, in Cohn (1982), these equations have been more fully developed and their understanding expanded. They now contain only known or measurable parameters. They can be used as a basis for a penalty/reward, credit/debit dollar payback technique for inadvertent interchange, and also for unilateral correction of area-caused time deviation and inadvertent interchange to replace present NERC corrective techniques. These new techniques, not yet supported by practical use, will not be further discussed in this presentation. They will be separately summarized in Cohn (1983) to be presented in the Carpentier invited session on Real Time Applications later in this symposium.

AREA CONTROL -- ECONOMY DISPATCH

Thus far, this presentation has concerned itself primarily with bulk power and energy transfers between areas of an interconnected system. Maintenance of bulk power transfer schedules while area load is varying, and simultaneously contributing to areas in need, as programmed by the bias characteristic when system frequency departs from normal, dictate the total generation requirement for an area. The next step is to distribute the total required generation among alternative available area sources to optimize economy and fulfill other operating criteria such as security and environmental obligations.

Techniques for computing desired generation distribution within a control area, taking into consideration incremental efficiencies of units, fuel costs and transmission losses, were developed over the years by a number of individuals, as were techniques for controlling plants and units to achieve these desired generation levels. I'll identify some of these contributions.

Transmission Loss Computation

References describing contributions to determining incremental cost of power delivered from alternative sources include Steinberg & Smith (1943); George (1943); Ward, Eaton & Hale (1950); Kirchmayer & Stagg (1951, 1952); Harder, Ferguson, Jacobs & Harker (1954); Early & Watson (1956); Watson & Stadlin (1959). These and related contributions provided a basis for achieving optimum economy by loading stations to equal costs of power delivered, i.e. equal "lambda".

In the years up to about 1950, general practice was for companies to prepare charts or tables or their equivalent to guide operators in achieving, generally with manual control, the desired generation allocations. On-line operation of multiple-plant control soon followed.

Multiple Station Control With One Station Serving As The Reference

The earliest technique used for multiple station control was based on proportioning the load distribution between stations in much the manner that had previously been used for proportioning load between units. One station served as the "master" in receiving control pulses from the central dispatching office and served additionally as the reference against which other plants were loaded. Other stations received control pulses to follow the master in pre-set patterns. The central load dispatcher had manually set adjustments for "base load" and "ratio" for each source and used these adjustments to establish the loading relationship of the "follow" plants to the master plant. Initial installations, about 1950, included Virginia Electric Power and Detroit Edison. The latter installation is described in Campbell (1952).

Multiple Station Control With Area Required Generation As The Reference

In 1951, Union Electric, Illinois Power, and Central Illinois Public Service Company contracted to supply large amounts of electrical energy to the atomic energy plant in the Kentucky area. I was invited to meet with representatives of their engineering and operating departments to lay out control to satisfy the new operating needs. It was planned to make a transition, in a single step at each company, from relatively little control (although Union Electric already had extensive telemetering) to full scale on-line control for virtually all of each area's units. Using a master station within the area as a reference as on previous installations had, I felt, limitations. Allocation assignment ranges were limited, and interaction between stations was probable.

A reference of total area required generation, developed with feedforward from the parameter area control requirement (later renamed area control error, or ACE) and a feedback from total area generation would represent total required generation and would be a superior reference. With base point and participation setters it would provide full and completely flexible ranges for the loading of each plant. Furthermore, the feedforward/feedback reference would eliminate interaction between plants regardless of their respective rates of response to control assignments.

This technique, identified as Area-Wide Generation Control and described in Cohn (1953), was installed at all three companies in 1952 and performed very well. Many similar systems have been installed elsewhere over the years. An interesting installation for the regulation of the Columbia River hydro system is described in Benson, Johannson & McNair (1963).

From its feedforward-feedback reference circuit, there is derived a unique feature, later used for both analog and digital desired generation computer controls. Computed area control error can be automatically decomposed into source control errors whose sum is equal to the area control error. The economy dispatch computer then searches for source control errors whose algebraic sum matches prevailing area control error. These source control errors are then used for non-interactive source control to achieve zero area control errors and economy dispatch for prevailing area generation.

A summary of control techniques prevalent in this period is contained in Nichols (1953).

The Early Bird

In 1953, Don Early of Southern Services devised an on-line computer capable of providing continuous real-time data on the cost of power delivered from various operating sources. An operating unit is described in Early, Phillips & Shreve (1955). It was an important forward step in economy dispatch.

Desired Generation Computer

The next step in the advance to complete automatic economy dispatch was proposed by Miller (1954), who suggested the combination of the on-line cost of power delivered computations of Early with the feedforward-feedback allocation computations of Cohn. He thereby provided an on-line multi-station desired generation control system serving the purposes of both area control and economy dispatch complete with real-time incremental transmission loss computation for operating sources. Many such units were installed, one of which installed at the Central Illinois Public Service Co. is described in Derks and Preston (1965).

Lambda Dispatch

In 1961 the Pennsylvania-New Jersey-Maryland (PJM) 12-company system, that had operated until then as an isolated single area with free flowing ties between companies, planned to establish permanent ties north to the New York, New England, and Canada system and west to the Interconnected Systems Group.

PJM had not used, nor from their point of view needed, automatic frequency control during its many years of operation as an independent single area system. Operating practice had been to manually dispatch a desired incremental cost of power delivered (system lambda) to each of its several operating entities who in turn would adjust the generation sources of their respective systems to operate at this common lambda value, and thereby achieve optimum operating economy for the pool.

From time to time, on instructions from the control center at Philadelphia, manual adjustments would be made to system frequency to keep system time within desired limits considered adequate.

Automatic area control was now to be installed which would utilize the by then well developed frequency biased net interchange technique for inter-area bulk power transfers. In addition, it was desired to supplement such control with a system lambda dispatch which would closely emulate, automatically, the previously prevailing manually operated dispatch system. I was invited to suggest suitable control arrangements.

Data on total generation at each of the companies was available at the control center. There it was arranged to be combined with feedforward from prevailing area control error to provide the parameter total pool required generation. An adjustable function generator was pre-programmed to simulate, from available pool data, the relationship between total required generation and approximately the necessary lambda to achieve it. This initial value of lambda, itself closely matching the real need, was automatically refined to the precisely requisite value by control action of the prevailing area control error. The computed lambda was continuously broadcast to all participating companies which, it was intended, would install distributed station and unit control of their own choices to maintain their own respective generation outputs at the broadcast common lambda value. Area control error was forced to zero by lambda signal initiated station generation responses, with feedback from the boundary tie lines and system frequency.

In this way, the pool would fulfill its obligations to its interconnected neighbors, while achieving optimum economy for its own internal operations. This basic lambda dispatch system is described in Cohn (1962).

It performed well over the years. Though now replaced with digital equipment, it has been retained for standby use at the PJM Valley Forge Control Center.

Advances in Telemetering

I should note that paralleling the progress in area control systems, there were advances in telemetering, permitting ready data monitoring and collection over the greater distances involved. My own experiences were largely with impulse, impulse duration, and frequency techniques developed primarily by E. D. Doyle, W. E. Phillips and J. B. Carolus, and utilizing microwave transmission.

Control Execution

Before going on to the last of the pre-direct digital power system control techniques, let me say a few words about control execution. Thus far I have limited my remarks in this review of evolution to commenting on the levels of total generation that are required within an area, and on the desired generation levels at stations or at units within the area. That's of course only part of the story. Achieving the desired limits by application of control signals is the other and equally important part of the control requirement. I will comment on three aspects of this, namely, processing of the area control signal, supplementing the desired generation control action with ancillary action when the rate of response of the desired generation assignment is not adequate to satisfy the area control demand, and coordinating generation control signals sent to fossil fuel units with related boiler controls.

Processing the control signal. Power system control engineers and operators are familiar with the rapid random variations in system frequency and comparably rapid "synchronizing swings" in area net interchange. These are the two parameters that make up area control error, which correspondingly itself has related short-term swings, in the order of seconds, in addition to longer term "sustained" variations. The rapid swings are not controllable, certainly not at the present state of the control art. An objective in control operation is to use suitable filtering or adaptive means in processing the area control signal to secure a truly effective signal for the initiation of control action.

One such technique has found appreciable use. It is the "error adaptive computer controller" (EACC) described in Ross (1966).

Assist action. When the units currently receiving area control signals lack the ability to respond at a rate adequate to fulfill prevailing area control requirement, assistance from other operating units is desirable.

One early technique for securing such assistance was designated "base load backup". It was first used at I&M Twin Branch Station in 1944, and at Des Moines Station of Iowa Power & Light in 1948. In each case, a new efficient generating unit having capability of rapid response was available, but the understandable operating desire was to keep it as close as possible to a selected base load. This apparent contradiction was resolved by permitting the new unit to act as the basic regulating source. In addition, its load recorder carried contacts which, whenever the unit moved away from its base load setting in response to regulation demands, initiated supplementary control action to the slower units which it was intended should ultimately absorb the regulation. These slower units were moved up and down, as required, at their most rapid rate, to provide generation changes that would cause the area controller, recognizing these changes, to move the new unit back toward its base load setting. This was a sort of "hare and turtle" arrangement, which served its purposes very well.

A later technique, still in use, is identified as "normal assist action". Assistance from units not otherwise at that time assigned a regulating function, is triggered by deviation of area control error from zero, either at preset deviation limits or in direct preset proportion to area control error deviation. Descriptions of such techniques are included in Cohn (1966).

Coordinated control. It was recognized early that it was one thing for the control to demand generation changes from a fossil fuel fired unit. It was another to achieve required energy conversions to fulfill the control demand. Early steps included coordinating boiler steam pressure and elements of boiler combustion control with generation control signals. A fully comprehensive coordination technique including tie-ins of the generation control signal with all significant controlled variables of the boiler and its auxiliaries is described in Bristol (1956). It has found extensive use, particularly on once-through units.

Digitally Directed Analog Control

All of the control techniques I have thus far discussed have been of the analog type. With the advent of digital technology and digital computers, interest developed in adapting their flexibility and capability to power system control. Reliability and programming technology had not yet, however, been fully developed and a first step was to replace the analog desired generation computations with digital computation, and use the results automatically to set analog allocation equipment.

A pioneering installation of this type, utilizing an L&N 3000 computer was made at Detroit Edison in 1961. It is described in Blodgett, Hissey, Falk and Schultz (1962).

Another early installation was made at Canton, Ohio for American Electric Power, utilizing for that application an IBM 1710 computer. The latter was utilized at AEP's request to provide capability for later adaptation to Direct Digital Control. This installation is described in Kinghorn, McDaniel and Zimmerman (1965) and Morgan and others (1965). The step to DDC was indeed later taken by AEP, as described in Stagg and others (1967).

Direct Digital Control

I have noted above the adaptation by AEP of the Canton Digitally Directed Analog Control to Direct Digital Control. As planned in the introduction to this paper, that is where I will stop with evolutionary history. Certainly what has occurred since the introduction of digital computers in electric power systems controls represent modern advances of mammoth magnitude. The many great direct digital control center installations that have been made in recent years and the additional ones currently being made, with their expanded capabilities for load flow, contingency studies, security determinations and many other new applications add great cubits to the reliability and economy of system operation. As earlier noted, we will hear more of those important and advanced facets of real time evolution in power systems controls later at this conference.

Scheduling Practices

Discussions of inter-area practices that had been developed and applied for short and long term scheduling of various types of bulk power transfers are contained in Mochon and others (1972).

CONCLUSION

I have always regarded the interconnected electric power system, with its hierarchical, multi-variable, multi-level parameters, its geographically wide ranging extent, its parallel operation of literally hundreds of energy sources, its extensive and non-linear self-regulating forces, and its many objectives of operating economy, security and environmental influences, as a most challenging systems control problem, perhaps one of the most challenging of all large scale systems control problems. Much has been done in providing operable solutions during the past sixty years, at paces that at times may seem very slow in retrospect, though with increasing rapidity in later periods, and with highly accelerated activity in current digital and microprocessor environments. Although a lot has indeed been done, there is much yet to be done. In Hunt (1931), my old friend from Southern California Edison Company said, "Load frequency control is in a state of evolution". Now, more than fifty years later, I think it very proper to borrow and repeat his phrase with just a slight variation namely, "Power systems control remains in a state of evolution".

There is the ongoing need to challenge the continuing validity of present day practices of distributed frequency biased net interchange control, already in its 35th year, there remains the need for a dynamic real-time on-line computation of generating unit incremental heat rates for better concurrent economy dispatch, there is a need for more effective adaptive and predictive technology for greater effectiveness in control execution. And there remains the need of the education of system operators in digital and control technology, and the comparable need of educating computer and software specialists in the technology of power system operation. Who will do these things, and the many others that obviously I can't begin to identify? The late Nobel Laureate, Harold Urey, said it very well, in another context, but equally applicable to our field. "We must leave something for the young people to solve. It would be most disappointing if we older people solved all the problems, which of course none of us will ever do", (Urey, 1976).

To youth, then -- with confidence -- my best wishes for success in the continued evolution of these and related fields. And to age, this admonition: Let them do it.

ACKNOWLEDGEMENT

I should like to express appreciation to L&N Management who authorized company personnel to assist in the preparation of this paper. I extend particular thanks to S. L. Peirce, in whose area the paper was processed, to Eileen Robinson who typed it, and to Adina Zupanick for her assistance with the references.

REFERENCES

NOTE: Patent references carry the filing date.

Benson, A. R., Johannson, D. E. and McNair, H. D. (1963). Centralized load-frequency control for the United States Columbia River power system. IEEE Winter Meeting, paper CP63-230.

Blodgett, D. G.; Hissey, T. W.; Falk, A. K. and Schultz, W. B. (1962). Application of an on-line digital computer for dispatch and control of the Detroit Edison system. IEEE Winter Meeting, N.Y., Paper CP62-247.

Brandt, Robert (1929). Automatic frequency control. Electrical World, 93, No. 8, 385-388.

Brandt, Robert (1947). Theoretical approach to speed and tie line control. AIEE Trans., 66, 24-29.

Brandt, Robert (1953). Historical approach to speed and tie line control. AIEE Transactions Pt. III, 72, 7-9.

Bristol, E. S. (1956). Control systems for electrical generating units. U. S. Patent 2,861,194.

Campbell, W. J. (1952). New system control allocates changing load among 29 generating units. Electrical World, 137, No. 1, 32-33.

Carpentier, Jacques (1983). Digital automatic generation control. IFAC Symposium on Real Time Digital Control Applications, Guadalajara, Mexico, January.

Cohn, Nathan (1950). Control of power flow on interconnected systems. Proceedings of the American Power Conference, Chicago, 12, 159-175. Electric Light and Power, 28, Nos. 8, 9.

Cohn, Nathan (1953). A new method of area-wide generation control for interconnected systems. Proceedings of the American Power Conference, Chicago, 15, 316-344. Electric Light and Power, 31, Nos. 7, 8, 9.

Cohn, Nathan (1956a). A step-by-step analysis of load-frequency control showing the system regulating responses associated with frequency bias. Minutes of the Annual Meeting of the Interconnected Systems Committee, Des Moines, Iowa.

Cohn, Nathan (1956b). Some aspects of tie-line bias control on interconnected systems. AIEE Trans., 75 Pt. III, 1415-1428.

Cohn, Nathan (1962). Economic Loading of Power Systems. U. S. Patent 3,270,209.

Cohn, Nathan (1966). Control of generation and power flow on interconnected systems. John Wiley and Sons, New York. Second edition 1971.

Cohn, Nathan (1971a). Techniques for improving the control of bulk power transfers on interconnected systems. IEEE Trans. on Power Apparatus and Systems, PAS-90, 2409-2419.

Cohn, Nathan (1971b). Power system control practice. Proceedings of the Ninth Annual Allerton Conference on Circuit and System Theory, 719-727.

Cohn, Nathan (1982). Decomposition of time deviation and inadvertent interchange on interconnected systems. I: Identification, separation and measurement of components. II: Utilization of components for performance evaluation and corrective control. IEEE Trans. on Power Apparatus and Systems, PAS-101, No. 5, 1144-1169; No. 8, 2711-2720.

Cohn, Nathan (1983). Area control performance measurement and corrective control in interconnected systems. IFAC Symposium on Real Time Digital Control Applications, Guadalajara, Mexico, January.

Conot, Robert (1979). A Stroke of Luck. Seaview Books, N.Y.

Derks, R. A. and Preston, E. H. (1965). Unique operating features of the Central Illinois Public Service Co. dispatch computer system. Proceedings of the American Power Conference, Chicago, 27, 1111-1118.

Doyle, E. D. (1928). System of distribution. U. S. Patent 2,054,121.

Doyle, E. D. (1929). Apparatus for controlling load distribution. U. S. Patent Re. 20,548.

Early, E. D., Phillips, W. E., and Shreve, W. T. (1955). An incremental cost of power delivered computer. AIEE Trans. Pt. III, 74, 529-535.

Early, E. D. and Watson, R. E. (1956). A new method of determining constants for the general transmission loss equation. AIEE Transactions Pt. III, 74, 1417-1423.

Electrical West (1931). Report of session on transmission and distribution, AIEE Pacific Coast Meeting, Lake Tahoe, 67, No. 4, 194-195.

Fitch, H. S. (1931). The Pennsylvania-Ohio-West Virginia interconnection. Presented at the AIEE Summer Convention, Ashville, N.C., June.

George, E. E. (1943). Intrasystem transmission losses. AIEE Trans., 62, 153-158.

Harder, E. L. and others (1954). Loss evaluation, II - current-power-form loss formulae. AIEE Trans., 73, 716-731.

Heath, L. O. (1929). Apparatus for speed control. U. S. Patent Re. 19,157.

Henry, R. T. and others (1929). Symposium on frequency control. Annual meeting of the Electrical Section, Empire State Gas and Electric Association, Briarcliff Manor, N.Y., December 6.

Humphrey, G. S. (1927). The interconnection of power systems surrounding the Pittsburgh area. The Electric Journal, XXIV, 251-258.

Hunt, L. F. (1930). Automatic control in hydro-electric plants. Electrical West, 64, No. 6, 337-354.

Hydro-Electric Power Commission of Ontario (1926). The Toronto load totalizer. The company Bulletin, October. Courtesy the late P. A. Borden.

IEEE Systems Controls Subcommittee (1977). Bibliography on automatic generation control and boiler-turbine-governor controls. 1971-74. 77CH1211-2-PWR

IEEE Working Group (1981). Description and bibliography of major economy-security functions. IEEE Trans. on Power Apparatus and Systems, PAS-100, 211-235.

IEEE Power Engineering Society (1982). Pearl Street. IEEE Center for the History of Electrical Engineering.

Insull, Samuel (1921). Speech at Peoria, Illinois. Purdue University Collection. Courtesy Prof. M. E. Van Valkenburg, U. of Illinois.

Jollyman, J. P. (1927). The speed control of a transmission system. The Electric Journal, XXIV, 250-251.

Kerr, S. L. (1930). Automatic operator for economy control. Presented at the AIEE Middle Eastern District Meeting No. 2. Philadelphia, October.

Kinghorn, J. H.; McDaniel, G. H. and Zimmerman, C. P. (1965). Development of coordination and control of generation and power flow on American Electric Power System. Proceedings of the American Power Conference, 27, 1068-1078.

Kirchmayer, L. K., and Stagg, G. W. (1951). Analysis of total and incremental losses in transmission systems. AIEE Trans., 70, 1197-1204.

Kirchmayer, L. K., and Stagg, G. W. (1952). Evaluation of methods of coordinating incremental fuel costs and incremental transmission losses. AIEE Trans. Pt. III, 71, 513-521.

Leeds, M. E. (1912). Electrical recorder. U. S. Patent 1,125,699.

Lincoln, P. M. (1929). Totalizing of electric system loads. AIEE Trans., Pt. III, 48, 775-782.

McCormack, J. E. and Metcalf, C. N. (1949). Load-frequency control of the Northeast interconnection. Electric Light and Power, 24, February, 70-73.

McDaniel, G. H. (1974). Evolution of dispatch techniques and facilities. Presented to a meeting of the IEEE Power Systems Engineering Committee, Scottsdale, Az., Nov. 13.

McNair, J. S. and others (1932). Operating experiences with the "Automatic Operator" in the Northwest. Presented to the Engineering Section Meeting of the Northwest Electric Light and Power Association, Portland, Oregon, April 13-15.

Miller, W. G., Jr. (1954). Generation control system. U. S. Patent 2,836,731.

Mochon and others (1972). Symposium on scheduling and billing of bulk power transfers. Proceedings of the American Power Conference, Chicago, 34, 904-967.

Morehouse, S. B. (1945). Inter-system power coordination in the Southwest region. Electric Light and Power, 23, December, 62-68, 72, 105.

Morehouse, S. B. (1965). Some historical highlights in the operation of electric utility systems on an interconnected and coordinated basis. Presented at the Annual Joint Meeting of the Northeast Regional Committee of the Interconnected Systems Group and the System Operation Committee of the Pa. Electric Association, Pittsburgh, Pa.

Morgan, W. S. and others (1965). Facilities for the American Electric Power System production and control center. Proceedings of the American Power Conference, 27, 1079-1085.

NERC (1981). Map of 230 kV and above transmission lines, power stations and substations. North American Electric Reliability Council, Princeton, N.J.

NERC (1982). Operating manual. North American Electric Reliability Council, Princeton, N.J.

Nichols, C. (1953). Techniques in handling load-regulating problems on interconnected power systems. AIEE Transactions Part III, 72, 447-460.

Preminger, J. (1960). Bibliography of load and frequency control literature 1922-1957. AIEE Winter General Meeting, Paper CP 60-489.

Purcell, T. E. and Powel, C. A. (1931). Tie-line control of interconnected networks. AIEE Trans. Pt. III, 51, 40-50.

Ross, C. W. (1966). Error adaptive control computer for interconnected power systems. *IEEE Trans. on Power Apparatus and Systems*, PAS-85, 742-749.

Sporn, Philip and Marquis, V. M. (1932). Frequency, time and load control on interconnected systems. *Electrical World*, March 12, 495-500; April 2, 618-624.

Stagg, G. W. and others (1967). On-line computer optimizes loading of 38 generators. *Instrumentation Technology*, 14, No. 1, 31-34.

Steinberg, M. J. and Smith, T. H. (1943). *Economy loading of power plants and electric systems*. Wiley, New York.

Urey, H. C. (1976). *Geochim, Cosmochim. Acta*, 40, 570.

U. S. Dept. of Energy (1981). Map, North American interconnected control areas.

Ward, J. B. and others (1950). Total and incremental losses in power transmission networks. *AIEE Trans.*, 69, 626-632.

Watson, R. E. and Stadlin, W. O. (1959). Calculation of incremental transmission losses and general transmission loss equation. *AIEE Trans.*, 78, Pt. III, 12-18.

Williams, A. J., Jr. and Morehouse, S. B. (1935). Electrical generating system. *U. S. Patent 2,124,725*.

Wunsch, Felix (1925). System of frequency or speed measurement and control. *U. S. Patents 1,751,538 and 1,751,539*.

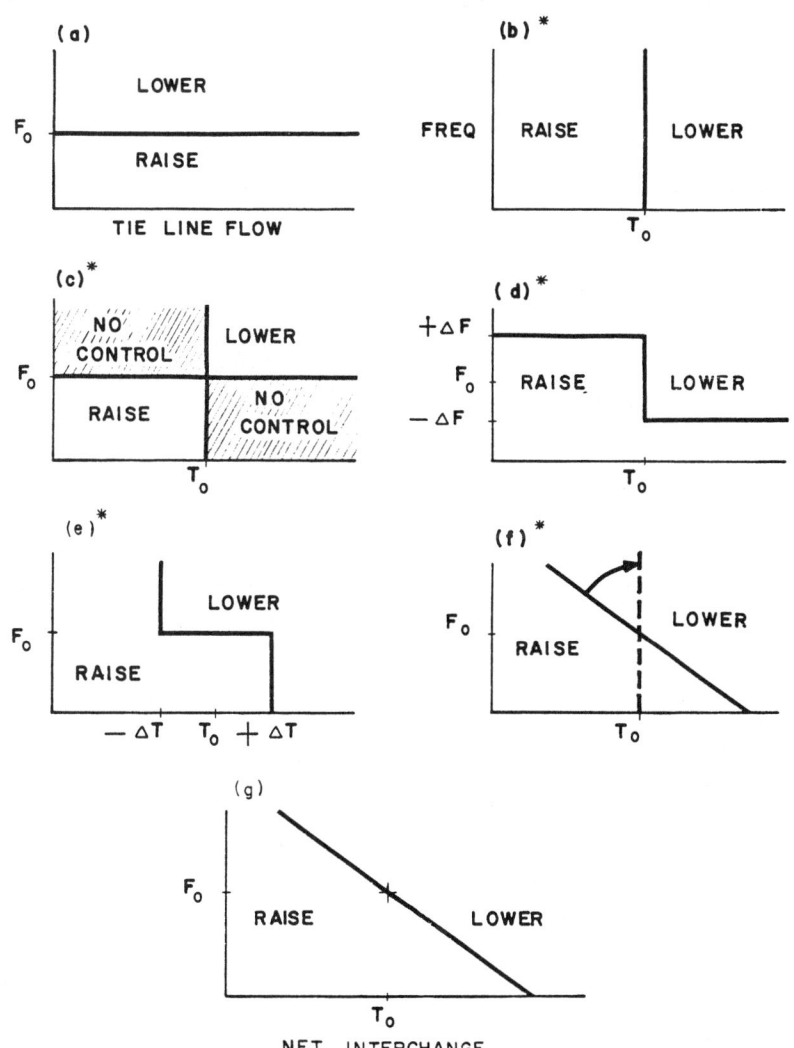

Fig. 1 Evolution of real time control characteristics. Curves show balance points of controller on plots of frequency versus tie line flow. Points not on characteristic result in "raise" or "lower" control action.

F_o is scheduled frequency. T_o is scheduled tie line flow. Increase in outgoing power is to the right of the tie line coordinate. Individual sketches apply to the following types of control:
(a) Constant frequency   (b) Constant tie line
(c) Selective frequency   (d) Tie line zoned
(e) Frequency zoned   (f) Tie line bias -- withdrawal
    (g) Frequency biased net interchange -- sustained. In all areas.

* Control types (b), (c), (d), (e), and (f) used in concert with a master frequency controlling area, (a).

THE EARLY STAGES OF ROBOTICS

R. P. Paul

School of Electrical Engineering, Purdue University, West Lafayette, IN 47907, USA

ABSTRACT: The early stages of robotics are recalled by tracing the evolution of robot manipulator programming languages. Initial manipulator control was by means of procedures imbedded in a high level language. This evolved into a geometric based planning and execution system where planning was time independent and execution was time efficient. With the addition of sensor feedback and the need to develop the ad hoc procedures involved in assembly an interpretative system evolved. Finally we have come full circle with manipulator control once again imbedded in a high level language. This is now possible because of the development of high level languages, the simplification of manipulator control eliminating of the need to plan trajectories and to pre calculate dynamics, and the increasing computer power available to provide for the economic real time control of the manipulator.

INTRODUCTION

The present day industrial robot has its origins in both teleoperators and numerically controlled machine tools. The teleoperator, is a device to allow an operator to perform a manual task from a distance. The numerically controlled machine tools shapes metal automatically, based on digitally encoded cutting data.

The teleoperator was developed during the second world war to handle radioactive materials.[1] An operator was separated from a radioactive task by a concrete wall with one or more viewing ports through which the task could be observed. The teleoperator was a substitute for the operator's hands; it consisted of a pair of tongs on the inside (the slave), and two handles on the outside (the master). Both tongs and handles were connected by linkage mechanisms to provide for arbitrary positioning and orientation of the master and slave. The mechanical linkage caused the slave to replicate the motion of the master.

In 1947, the first servoed electric-powered teleoperator was developed. The slave was servo-controlled to follow the position of the master. Force information was, however, no longer available to the operator, and tasks requiring parts to be brought into contact were difficult to perform. To quote Goertz, "The general-purpose manipulator may be used for moving objects, moving levers or knobs, assembling parts, and manipulating wrenches. In all these operations the manipulator must come into physical contact with the object before the desired force and movement can be made on it. A collision occurs when the manipulator makes this contact. General-purpose manipulation consists essentially of a series of collisions with unwanted forces, the application of wanted forces, and the application of desired motions. The collision forces should be low, and any other unwanted forces should also be small."[2] In 1948, a servo system was introduced in which the force exerted by the tongs could be relayed to the operator by back driving the master; the operator could once again feel what was going on.

In 1948, faced with the need to procure advanced planes whose parts were designed to be machined rather than riveted, the Air Force sponsored research in the development of a numerically controlled milling machine.[3] This research was to combine sophisticated servo system expertise with the new, developing digital computer techniques. The pattern to be cut was

---

Richard Paul is the Ransburg Professor of Robotics. This material is also based upon work supported by the National Science Foundation under Grant No. MEA-8119884. Any opinions, findings, conclusions, or recommendations expressed in this publication are those of the authors and do not necessarily reflect the views of the National Science Foundation.

stored in digital form on a punched tape and then a servo-controlled milling machine, using the tape as input, would cut the metal. The MIT Radiation Laboratory was awarded a subcontract and demonstrated such a machine in 1953.

THE INDUSTRIAL ROBOT

In the 1960's, George Devol demonstrated what was to become the first Unimate industrial robot, a device combining the articulated linkage of the teleoperator with the servoed axis of the numerically controlled milling machine. The industrial robot could be taught to perform any simple job by driving it by hand through a sequence of task positions which were recorded in digital memory. Task execution consisted in replaying these positions by servoing the individual joint axes. Task interaction was limited to opening and closing the tongs or end effector and to either signaling external equipment or waiting for a synchronizing signal. The industrial robot was ideal for pick and place jobs such as unloading a diecasting machine (see Figure 1). The part would appear in a precise position, defined with respect to the robot; it would be grasped, moved out of the die, and dropped on a conveyor. The success of the industrial robot, like the NC milling machine, relied on precise, repeatable digital servo loops. There was no interaction between the robot and its work. If the diecast machine were moved, the robot could in no way adapt to the new position, any more than an NC milling machine could successfully cut a part if the stock were arbitrarily relocated during cutting. If the diecasting machine were moved, the robot could, however, be retaught.

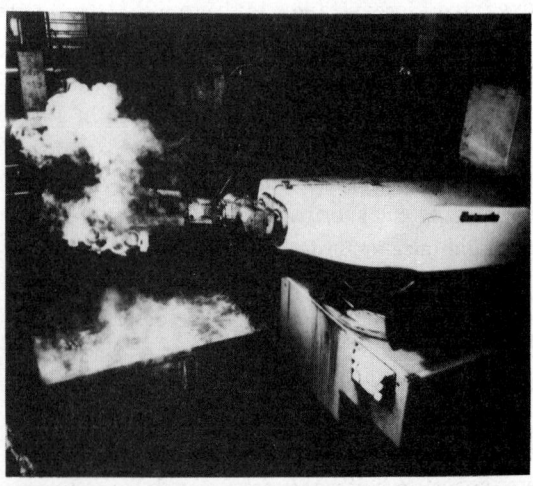

Fig. 1. The Unimate Robot at Work Diecasting

The success of the industrial robot lay in its application to jobs in which task positions were absolutely defined and in its reliability and positioning repeatability in lieu of adaptation. The industrial robot was a piece of automation like a transfer line component; however, the position sequence could be taught to it directly - there were no cams or gears to cut.

Automation in the form of an industrial robot was different from all previous forms of automation. In previous models, some machine was introduced which performed the task faster or differently from the existing process; in the case of the industrial robot, a worker performing a job was replaced by a machine having an anthropomorphically identifiable arm which performed the job in much the same way as the replaced operator. Both forms of automation increase productivity.

THE SENSOR-CONTROLLED ROBOT

Simultaneously with the development of the industrial robot an attempt was made to automate the teleoperator, an attempt made possible only by the development of digital computers. A device, which we shall refer to as a sensor controlled robot, was proposed by Shannon and Minsky in 1958. The sensor controlled robot was to consist of a teleoperator equipped with all forms of sensors connected to a computer. The computer was to be informed of a goal and the robot, by means of its sensors, would size up the environment and decide on the actions necessary to accomplish the required goal. Although this device was never built, Ernst[4] built a robot with touch sensors located in the hand which could be programmed to perform tasks such as locating and picking up blocks and putting them in a box (see Figure 2). Programming was in the form of instructions such as *"move in direction x with speed v until sense element s indicates a "or" if sense element s indicates 1, go to the next instruction otherwise continue the same action"*. A program of 600 lines of code, made up of instructions of this form, resembling an assembly language program, was required for the block program. The lack of any global idea of the position of objects limited this robot as much as the complete lack of task information limited the position-controlled industrial robot.

VISION

In 1963, Roberts demonstrated the feasibility of processing a digitized halftone picture of a

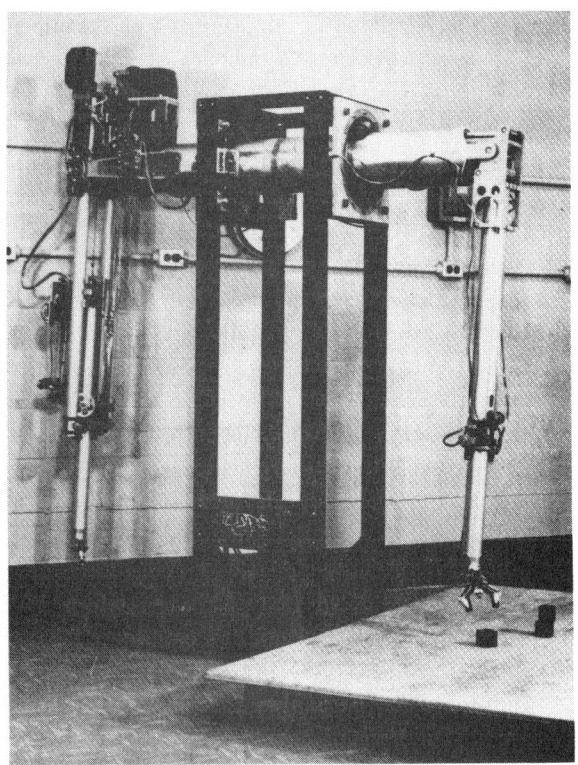

Fig. 2. The First Sensor Controlled Robot.

scene to obtain a mathematical description of the block-like objects in the scene, expressing their location and orientation by homogeneous transformation[5] (see Appendix A). This work was important for two reasons: it demonstrated that objects could be identified and located in a digitized halftone image, and it introduced homogeneous transformations as a suitable data structure for the description of the relative position and orientation between objects. If the relative position and orientation between objects is represented by homogeneous transformations, the operation of matrix multiplication of homogeneous transformations can establish the overall relationship between any two objects.[6,7]

HAND-EYE SYSTEM

Touch feedback, because of its slow, groping nature, was dropped in favor of vision as an input mechanism. By 1967, a computer equipped with a television camera as an input mechanism could, in real time, identify objects and their location.[8] Scenes normally contained more than one object; the vision processing would first locate edges, then vertices, and finally identify all the objects in the scene. These objects were represented in a world model. The world model frequently contained prototypes of the possible objects which could comprise the scene, and after vision processing the world model would contain a set of instances of these objects whose positions and orientation were described by homogeneous transformations. The manipulator, stripped of its touch sensors, would then rely on position-servoed joint axes. Homogeneous transformations, however, expressed the position and orientation of the end effector in Cartesian coordinates, not as the angles between a series of unorthogonal joints. Pieper was able to apply the theories of kinematics relating to closed-link chains to obtain a solution to this problem, and the manipulator could then be commanded to move to Cartesian positions in its workspace.[9]

The manipulator was controlled by means of a small number of high level language subroutines and functions similar in concept to the scientific library which provides such familiar routines as SIN, COS and SQRT. In the case of the manipulator there was some form of initialization routine, and further routines to move the manipulator and to open and close its gripper. The manipulator was treated as an output device, much like a line printer, information was transmitted to it and the required actions were performed. The routines shown in table 1 provided the above functions.

Table 1

NAME	TYPE	DESCRIPTION
INIT	SUBROUTINE	initialize the arm
SERVO (J(6))	SUBROUTINE	move the manipulator until all the joints are as specified in the array J.
OPEN (DIST)	SUBROUTINE	open the gripper to DIST
CLOSE (DIST)	FUNCTION	close to gripper as far as it will close and return FALSE if the opening is less than DIST.

When a routine was called, program execution was suspended until the action was completed. A manipulator program took the form of a sequence of subroutine calls, typically with constant arguments. Of the four subroutines listed, only CLOSE provided for any interaction with the environment in which the manipulator was working. If an object of a certain known size was to be picked up, the size could be provided to CLOSE in order to verify that the object was correctly acquired.

The arguments to SERVO were the six joint coordinates of the desired position of the arm. These were typically integer variables specifying the non-orthogonal coordinates of the manipulator joints. The form of motion whether joint by joint, all starting off together, coordinated, etc. was not specified. This servo routine, by itself, was not very useful, but when coupled with two solution sub-programs it provided the necessary support for the eye system. HANDPOS (T(4,4)) a subroutine, returned a 4 by 4 homogeneous transformation expressing the Cartesian coordinate position and orientation of the gripper. SOLVE (J(6), T(4,4)) a function, returned the appropriate joint coordinates J, to position the gripper in the Cartesian coordinate position and orientation specified by a transform T. The routine returned FALSE if it was not possible to position the manipulator at T.

The basic concept of the system was that the manipulator was to be used as an output device for the vision system. The vision system would analyze a scene and locate objects. The necessary manipulator position would be calculated in order to grasp each object and SOLVE called to obtain the joint coordinates. The manipulator would then be moved by calling SERVO and the object grasped by calling CLOSE. Typical scheduling of the computation is shown in figure 3. There were three levels of control: at the top level calls were executed, at the next level SOLVE converted positions into joint coordinates, and at the lowest level the manipulator was servoed.

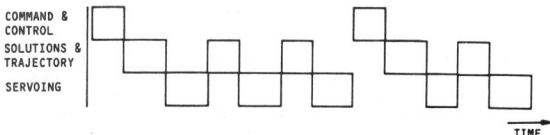

Figure 3. Hand Eye System Scheduling.

In order to move an object a series of manipulator positions is required: a safe approach position, the grasping position, the lift off position, the approach position for the object, etc. In the hand eye system the manipulator moved and came to rest at each of these positions while the necessary joint angles for the next position were computed. While this was of little consequence in the hand eye system, it is extremely wasteful of time and energy in general.

The hand eye system differed from a conventional industrial robot in two ways. The motion and gripper primitives were embedded in a high level computer language and the manipulator was programmed symbolically in Cartesian coordinates. As the manipulator was merely an output device, the embedding of the manipulator primitives in a high level language did not produce much benefit over conventional systems, there were no tests and thus no use for conditional statements. The use of Cartesian coordinates was, however, an important step and was to form the basis of a world model which would allow for decision making, planning, and task verification.

This form of manipulator programming is awkward as basic data types representing positions and orientations do not exist in the high level language, the program is executed step by step requiring that the manipulator be stopped at each program step. The overhead of a compiler is not warranted as manipulator programs are executed at a very low statement rate and an interpreter would provide more flexibility and better debugging interaction.

BLOCKS WORLD SYSTEM

By 1970, a camera- and arm-equipped compute could play real-world games and the "instant insanity" puzzle was successfully solved at Stanford University.[10] In this puzzle, four cubes with different-colored faces must be stacked so that no two similar colors appear on any side. At MIT a block structure could be observed and copied. In Japan, research led to a hand-eye system which could assemble block structures when presented with an assembly drawing.[11] In this system a world model of fixed objects and of objects located by the vision system was represented as instances of prototype objects whose orientation was described by homogeneous transformations[12]. The manipulator system was told what object to move, and the position to which it should be moved. The manipulator system then determined stable grasping position. It also determined whether the object needed to be set down and regrasped in order to be moved to the specified destination[13]. A collision avoider (never implemented) was to determine a safe path for the arm through the world model, describing the path as a sequence of homogeneous transformations. In order to avoid stopping at every point making up such a path or trajectory, a continuous curve was fitted through the sequence of joint angles corresponding to the sequence of transformations making up the path.

A manipulator trajectory was specified as six sequences of joint angles through which the six joints were to pass in a time coordinated manner. At the first and last points of each of the six trajectories, zero velocity and acceleration constraints were imposed. At all the intermediate points continuity of velocity and acceleration was required. A sequence of polynomial splines was calculated to meet these requirements. Unfortunately, the spline fit was a lengthy procedure and could not be performed as the manipulator moved. However, in the blocks world system, as all the positions were known, the lengthy spline fit could be computed before the manipulator moved. This introduced the concept of a planning phase and a runtime phase, see figure 4.

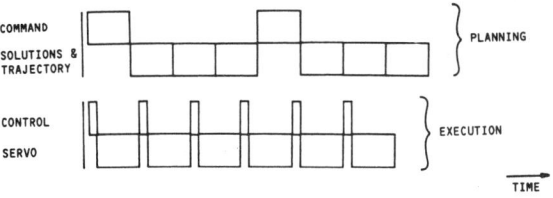

Figure 4. WAVE Scheduling.

By dividing the task in this manner, the planning could be performed off-line, with no time constraints. Additional calculations relating to dynamics, servo gains and offsets could also be calculated. At runtime, no delays were incurred while the solutions for the next positions were obtained, and the resulting motion was smooth and continuous.

The use of trajectories had an important safety side effect. In order to fit the spline through the trajectory points, the manipulator joint accelerations were specified as continuous, bounded, functions of time. During execution, actual joint accelerations were compared to the specified values by the servo and the motion aborted if the difference became excessive. This simple check resulted in years of accident free experimental operation.

The new manipulator system consisted of just one high level procedure, MOVE_INSTANCE, with two parameters, the name of the instance of a prototype to be moved and the desired final transformation of the instance. Based on this information, trajectories and gripper commands were assembled into a file which, when executed by the arm, resulted in the desired action. Interaction with the actual environment was limited to the calls on the gripper CLOSE command which verified the correct object width at each grasp. By the addition of a planning phase the step by step execution of a program could be turned into a continuous motion during execution. Programs still had to be written in a high level language which lacked appropriate data types and operations.

While this system was excellent for generating graphics displays of an ideal arm moving ideal objects, problems occurred in a real environment. No interaction was specified between the arm and the environment; the arm simply moved through space opening and closing its gripper. At every interaction with the environment, the forces and torques generated were ignored. Consider, for example, the task of placing an object on a surface. The arm is commanded to move the object to zero height and the gripper opened. The position tolerance of this move is zero, for if the arm stops above the surface and opens its gripper, then the object is dropped, not placed. If the arm tries to move below the surface, it is stopped by the object while infinite forces are exerted on the object. The placement task should be specified to move the object towards the surface until an appropriate contact force is detected and the gripper then opened. Similarly, in grasping an object, the hand should be centered over the object using touch feedback; the fingers should not simply be closed, possibly displacing the object.

Force and touch feedback were added to the arm, in a rather crude manner, to perform the above functions. With the addition of this feedback came a great deal of sensory information from the environment. While the position of objects could be modelled using homogeneous transformations and problem solvers could function with such a model, there was no model for the interactive forces, torques, and touches. This information was not used by the problem solver, other than to call vision to locate an unknown object. The manipulator programming became ad hoc and experimental. In order to meet the needs of this type of programming, the WAVE system was developed.

THE WAVE SYSTEM

In the WAVE system[14] the procedure calls previously embedded in a high level language could be typed in directly by name with parameters or read from a file. A macro facility was added to build simple sequences of manipulator primitives to make up higher level commands such as MOVE_INSTANCE had become. An on-line macro editor provided a quick interactive way of developing these macros. Motion force and touch commands became the primitives of the language. The WAVE system functioned in two modes: a planning mode in which instructions read in were assembled into a file for later execution, and a direct mode in which each instruction was executed as it was typed in.

Provision was made for defining transformations by placing the gripper in the appropriate location and typing HERE, at which point the arm position was read and converted to a homogeneous transformation. These positions could then be modified and verified by executing a direct move to the named transformation.

Initially the sensor information was simply used to verify task execution as CLOSE had been used. By adding tests and jump instructions, however, a first level of error recovery was built in to the execution program, adding to the robustness and reliability of the programs.

The force and touch primitives in WAVE were modifiers of motion statements; for example, the WAVE instruction to insert a pin along the z axis of station coordinates is as follows:

```
FREE 2 X,Y   ; COMPLY WITH FORCES IN X AND Y
SPIN 2 X,Y   ; COMPLY WITH TORQUES IN X AND Y
STOP Z 100   ; STOP WHEN THE Z FORCE EXCEEDS 100
MOVE IN      ; WHILE MOVING TO POSITION 'IN'
SKIPN 23     ; IF THE STOPPING CONDITION WAS MET
             ; SKIP THE NEXT INSTRUCTION
JUMP EROC    ; JUMP TO ERROR RECOVERY ROUTINE
```

At the time the motion is terminated by the STOP instruction, indicating that the pin is fully inserted, the position and orientation of the arm are not specified by the transform IN as the motion has been allowed to translate in X, Y and Z and to rotate about the X and Y axis. The position and orientation of the arm, however, accurately represent the pin in the hole location. Provision was made to save this information in the form of a homogeneous transformation in case the arm needed to return to this position or to a position nearby. In assembling objects, quite complicated hierarchical saving algorithms were developed resulting in remarkably adaptive behavior of the manipulator in performing assemblies of parts of low tolerance.

The WAVE system still made use of a planning phase but eliminated the need to write programs in an inappropriate high level language. The form of programming was similar to assembly language programming of computers and lacked structure.

The WAVE system was finally extended to run two manipulators, but in a very simple manner, chiefly through synchronization, with very limited interaction. The use of two manipulators, however, greatly simplified the task of programming assemblies as one hand could function as a jig, or fixture, for the other as needed.

CARTESIAN MOTION

The WAVE system specified positions and orientations in Cartesian coordinates but moved the manipulator in joint coordinates. In order to provide a system capable of working on moving objects of known position and velocity one must also move the manipulator in Cartesian coordinates. The position and orientation of the object, described by a homogeneous transformation, are simply post-multiplied by a second transformation representing the desired position and orientation of the gripper with respect to the object. The product yields a transformation representing the required base coordinate position and orientation of the manipulator. From this transformation the joint coordinates can be obtained. These transformations must be performed at a rate sufficiently high to provide for continuous tracking motion of the manipulator. A system designed to function in this manner was developed at SRI[15] and performed these transformations at a 20 hertz rate in order to control a UNIMATE manipulator. The pre-planned spline fit trajectories of WAVE were discarded in favor of a simple on-line method which calculated trajectories segment by segment. The resultant Cartesian motion although elegant, was simply a result of the necessity of combining Cartesian positions of objects when one object was in motion.

The programming style was also changed in that a move through a series of positions was programmed as a move to each individual position. During execution, the moves through a series of positions were turned into a continuous motion by the run-time trajectory calculator. During each move segment a series of functions could be performed such as opening and closing the gripper. This form of programming was very similar to the original Unimate style of programming. The main differences were the following: 1) Positions were specified in terms of homogeneous transformation products, one of which could represent a moving coordinate system such as a conveyor. 2) Motions were made in a coordinated well controlled manner; 3) Motions did not stop at each intermediate point but transitioned smoothly through intermediate points. This system eliminated the need for a planning phase.

INTEGRATING ROBOT CONTROL INTO HIGH-LEVEL LANGUAGES

With the elimination of the need for a planning phase, robot manipulator control could be represented as a sequence of program statements to move the manipulator from one position to another. A manipulator control process, much like an input/output device driver, would actually move the manipulator. The manipulator control process would also provide for smooth path motion transitions and for bringing the manipulator to rest when no further move statements were pending. The move statement would provide for synchronization between program execution and manipulator motion.

With the development of new high level programming languages such as: Algol 60, c, and Pascal; it became possible to represent the data structures necessary for manipulation directly in the high level language. The integration of sensors with a manipulator could then be achieved by imbedding the manipulator control directly into one of the above languages. Sensors would be treated as input devices and all the well understood control and data types of the language would be available to the robot programmer. We will describe the imbedding of manipulator control into PASCAL[16,17] but any of the other languages would do as well.

INTEGRATING MANIPULATOR CONTROL INTO PASCAL

We will first describe data structures to define transformations and transform expressions. We will then show how our data representation can be used to solve the necessary transform equations. Motion primitives will then be introduced which correspond to joint motion, Cartesian motion, and a new form of functional motion. Program execution will be considered in terms of a co-processor structure: one processor for the program and one for the manipulator. A synchronization structure will be developed to coordinate the two processors. These processors may, of course, be implemented in one physical processor or in a multi-processor configuration.

Task Description

Consider the following task. The task consists in picking up pins, and inserting them into holes in a subassembly. By defining a series of manipulator end effector positions pn (see Figure 5), we can describe the task as a sequence of manipulator moves and actions referring to these numbered positions.

    MOVE   p1   Approach pin
    MOVE   p2   Move over pin
    GRASP   Grasp the pin
    MOVE   p3   Lift it vertically
    MOVE   p4   Approach hole at an angle
    MOVE   p5   Stop on contact with hole
    MOVE   p6   Stand the pin up
    MOVE   p7   Insert the pin
    RELEASE
    MOVE   p8   Move away

Such a program could be executed by any one of a number of commercially available industrial robots, and it also exhibits all the limitations of such robots. There is no provision for compliance, such as is required during the pin insertion or the contact between the pin and the hole. No provision is made for storing information related to the actual position of any objects. After the pin is inserted in the first hole, information relating to the position of the hole needs to be retained in order to simplify the insertion of the second pin.

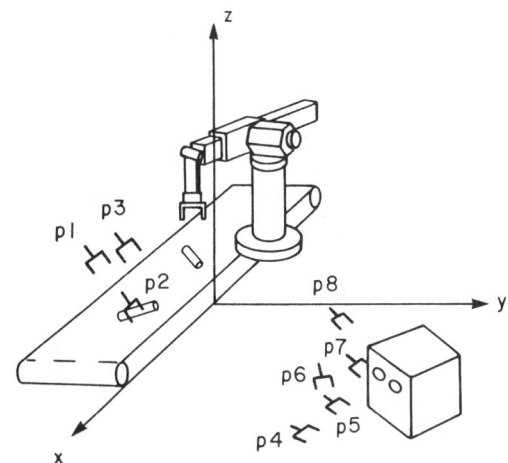

Figure 5. Task Positions

If the manipulator is moved, the entire program must be retaught. To insert the second pin, the entire program must be repeated, but with slightly different positions relating to the second hole. What is missing is the structure of the task.

Let us begin to define the structure of the task by defining the structure of the manipulator. We will describe the manipulator by the product of three transformations such that the positions in the task description are replaced by

$$\text{MOVE pn} = \text{MOVE } Z\ T_6\ E$$

where
- Z represents the position of the manipulator with respect to the base coordinate system;
- T_6 represent the end of the manipulator with respect to its base. T_6 is a computable function of joint coordinates;
- E represents a tool or end effector at the end of the manipulator.

With such a description, the calibration of a manipulator to the work station is represented by Z. If the task is to be performed with a change of tool, only E must be changed.

We will now represent the structure of the task in terms of the following transforms:
- P the position of the pin in base coordinates;
- H the position of the block with the two holes;
- $^H HR_i$ the position of the ith hole in the block with respect to the H coordinate system;
- $^P PG$ the position of the gripper holding the pin with respect to the pin;
- $^P PA$ the gripper approaching the pin;
- $^P PD$ the gripper departing with the pin;
- ^{HR}PHA the pin approaching the ith hole;
- ^{HR}PCH the pin at contact with the hole;
- ^{HR}PAL the pin at the beginning of insertion;
- ^{HR}PN the pin inserted.

The task can now be represented as a series of transform equations solvable for T6, the manipulator control input, as follows:

p1: $Z\ T_6\ E = P\ PA$

p2: $Z\ T_6\ E = P\ PG$

  GRASP

p3: $Z\ T_6\ E = P\ PD\ PG$

p4: $Z\ T_6\ E = H\ HR_i\ PHA\ PG$

p5: $Z\ T_6\ E = H\ HR_i\ PCH\ PG$

p6: $Z\ T_6\ E = H\ HR_i\ PAL\ PG$

p7: $Z\ T_6\ E = H\ HR_i\ PN\ PG$

  RELEASE

p8: $Z\ T_6\ E = H\ HR_i\ PN\ PA$

While this may appear complicated, it represents the essential structure of the task, and each transformation represents a separate piece of information.

Declarations and Data Structures

We will begin this section by defining two data types: vectors and transforms. A vector presents no real challenge and is represented in PASCAL simply as

type vector = record x, y, z: real end;

A transform is then defined as four vectors

type transform = record n,o,a,p: vector end;

Transforms also have inverses, which we will store in another record and link together

```
type transpointer = ↑transform;
     transform = record n,o,a,p: vector;
         inverse: transpointer;
     end;
```

The field inverse is a pointer to another record of type transform. Some transforms are functionally defined, a transform describing the instantaneous position of a moving conveyor for example. We will enumerate all current functions by a scalar type functionname and include it in the transform description

```
type transform = record n,o,a,p: vector;
     inverse: transpointer;
     fn: functionname;
end; {transform}
```

The scalar type functionname is a list of all transform functions. For each function included in functionname a function must be declared.

Transform Equation Data Structure

We are now ready to represent the transform equations defining a manipulator task. We will do this in two stages. Initially, we will represent the left hand side of an equation, then the right hand side, and finally we will link them together to represent the equation. We will represent each term of a transform equation by a term record which links the transform into the equation.

```
type transform = ↑term;
  term = record
    nxt, inv:termpointer;
    trans:transpointer
  end; {term}
```

The first (right most) element of either the left or right hand side of the equation is formed by the function atom

Additional elements of the equation (reading right to left) are added to the front of the data structure by the function cons

In order to construct the data structure corresponding to p1 of the task described above

$$p1: Z\ T_6\ E = P\ PA$$

we would first define a series of functions which could be used to create data structure lists corresponding to equations which contain 2,3,...,n transformations: list2, list3, ..., listn, respectively. For example

```
function list3 (a,b,c:transpointer):termpointer;
  begin {list3}
    list3:=cons(a,cons(b,atom(c)))
  end; {list3}
```

and then execute the following calls

```
var z, t6, e, p, pa; transpointer;
  lhs, rhs:termpointer;

lhs:=list3(z,t6,e);
rhs:=list2(p,pa);
```

The resulting data structure is shown in Figure 6.

The full transform equation is represented by forming two circular data structures to represent
the closed link chain in both directions (see Figure 7). This is accomplished by a position function makeposition, which returns a position record. This record points to the equation data structure in terms of T_6, and to the tool coordinate frame. Information relating to the manipulator configuration might also be stored in this record, thus defining position as follows

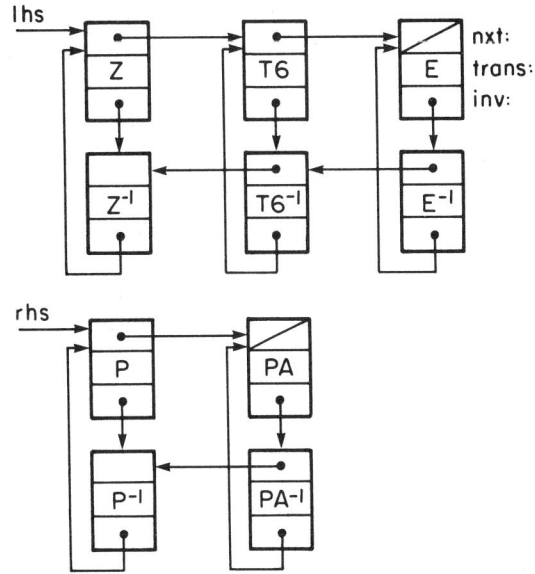

Figure 6. Transform Equation Data Structure

```
type positionpointer = ↑position;
  position = record
    t6ptr, tolptr:termpointer;
    rightly, flip:boolean
  end
```

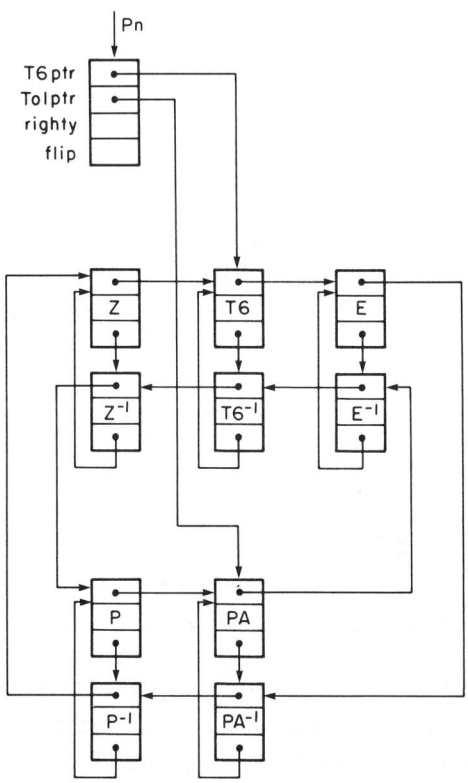

Figure 7. Complete Transform Equation Data Structure

We are now in a position to define all of the manipulator positions in a task and to solve the resulting transform equations for all the information necessary to execute the task. How is this to be done? We will define a motion procedure move with two arguments: a position equation and a mode record. The position equation describes the next position to which the manipulator is to be moved. The mode record describes how the manipulator is to be moved and is defined partially as follows

```
type modepointer = ↑mode;
  mode record
  typeofmotion:(joint,Cartesian);
    {joint motion or Cartesian}
  tacc, tsegment:integer;
    {acceleration time and segment time}
  mass:real;{mass of load}
  end;
```

Times are specified in milliseconds The procedure move is aware of the minimum possible acceleration and segment times and will use these if smaller times are specified. Thus times of 0 inform move to compute the times.

Let us define the positions and modes for the task we have described

```
p1:=make position (list3(z,t6,e),
  list3(conv,p,pa),e,true,true);
with m1↑do begin {set up mode}
  typeofmotion:=joint;
  tacc:=0; tsegment:=0; mass:=0 end;
p2:=makeposition(list3(z,t6,e),
  list3(conv,p,pg),e,true,true);
m2↑:=m1↑; with m2↑do typeofmotion:=Cartesian;
p3:=makeposition(list3(z,t6,e),
  list4(conv,p,pd,pg),pg,true,true);
m3↑:=m2↑; with m3↑do begin
  tsegment:=500; mass:=10 end;
p4:=makeposition(list3(z,t6,e),
  list4(h,ht,pha,pgJ),pg,true,false);
m4↑:=m3↑; with m4↑do begin
  typeofmotion:=joint; tsegment:=0 end;
p5:=makeposition(list3(z,t6,e),
  list4(h,ht,pch,pg),pg,true,false);
m5↑:=m4↑; with m5↑do begin
  typeofmotion:=Cartesian; tsegment:=1000 end;
p6:=makeposition(list3(z,t6,e),
  list4(h,ht,pac,pg),pg,true,false);
m6↑:=m5↑; with m6↑do tsegment:=0;
p7:=makeposition(list3(z,t6,e),
  list3(h,ht,pn),pg,true,false);
m7↑:=m6↑; with m7↑do tsegment:=500;
p8:=makeposition(list3(z,t6,e),
  list4(h,ht,pn,pa)e,true,false);
m8↑:=m2↑{an unloaded move}
```

The program then becomes:

```
for i:=1,2 do begin
  read(camera, pc);{Read in position of pin}
  matrixmultiply(p,cam,pc);{Set p}
  move(p1,m1);{approach pin}
  move(p2,m2);{over pin}
  move wait;
  close;
  move (p3,m3);{departure position}
  ht↑:=hr[i]↑;{Temp position of hole}
  move(p4,m4);{Hole approach}
  move(p5,m5);{Contact hole}
  move(p6,m6);{Stand up}
  move(p7,m7);{Insert pin}
  move wait;
  open(10)
  move(p8,m8){Depart from pin}
end
```

Software Organization

The procedure move communicates with another process which, driven by a real time interrupt, actually runs the manipulator. The move procedure simply passes to the interrupt process two pointers defining the position and mode (more will be said about the interrupt process below). If, at the time of the call, the manipulator is already in motion, then program execution waits until the manipulator approaches the currently specified position. At this time, a transition to the next position commences without the manipulator stopping. Program execution then continues. If the manipulator reaches the transition point without a move statement pending, then the manipulator is brought to rest at the specified position. Sometimes it is necessary to bring the manipulator to rest, as in the case at positions p2 and p7, when the object is grasped and released. This is accomplished by a procedure movewait, which holds program execution until the manipulator is brought to rest at its current final position.

The camera and end-effector are also asynchronous processes and we will define some procedures by which to communicate requests for image processing.

camerafindpin;
    {scan for a pin and record conveyor position}

and for the communication of results

readcamera(p:transpointer; conveyorposition:integer);
    {set the transform pointed to by transpointer equal
    to the image transformation at the time the
    image was processed.
    set conveyorposition to the position of the
    conveyor when the image was processed}

If no image has been found then readcamera will hold program execution until a valid image is processed. Finally, we define the next procedure to hold program execution while the end effector is operating

operatewait;
    {wait until end effector operation finishes}

Armed with these new procedures, we can rewrite the program in terms of three processes: the manipulator process, the end effector process, and the camera process

```
camerafindpin;{find first pin};
for i:=1,2 do begin
  ht↑:=hr[i]↑;{Temp position of hole}
  readcamera(pc,sc);{wait here until pin found}
  matrixmultiply(p,cam,pc);{Set p}
  move(p1,m1);{approach pin}
  move(p2,m2);{over pin}
  movewait;
  close;
  operatewait;{wait until grasped}
  move(p3,m3);{departure position}
  move(p4,m4);{Hole approach}
  camerafindpin;{start looking for next pin}
  move(p5,m5);{Contact hole}
  move(p6,m6);{Stand up}
  move(p7,m7);{Insert pin}
  movewait;
  open(10);{no need to wait here}
  move(p8,m8){Depart from pin}
end
```

Specifying Compliance

We will specify the compliance of the manipulator in terms of a servo mode for each of the six Cartesian coordinates. The servo mode is itself a record, which is defined as follows

```
type servomode = record
  case servo:(position,force,stopforce,goforce) of
    position:(tolerance:real);
      {the position tolerance if the manipulator
      is to stop}
    force:(value:real);
      {the force to be exerted in a compliance mode}
    stopforce:(limit,distance:real);
      {monitor the force along this axis and change to
      a force command when the force condition is met
      terminate the current motion when this condition
      is met or distance is exceeded}
    goforce:(value,limit,distance:real)
      {exert this force until position error in this
      direction exceeds limit or the motion exceeds
      distance, then change to a position servo mode.
      Terminate the current motion when this condition
      is met}
end;{servomode}
```

The manipulator process is aware of the minimum values of tolerance, force, stopping force, and goforce displacement. If these are specified as zero, then the minimum values are employed instead. Mode is redefined to be

```
type mode = record
  typeofmotion:(joint,Cartesian);
  tacc,tsegment:integer;
  mass:real;{mass of load}
  dx,dy,dz,rx,ry,rz:servomode
    {dx,dy, and dz refer to translations or forces
    along the principal axes of the TOOL frame.
    rx,ry, and rz refer to rotations or torques about
    the axes}
end;
```

For example the mode for the pin insertion m7 is expanded to:

```
with m7↑do begin
  tsegment:=500;
  with dx do begin servo:=force; value:=0 end;
  with dy do begin servo:=force; value:=0 end;
  with dz do begin servo:=stopforce; limit:=50;
    distance:=5 end;
  with rx do begin servo:=force; value:=0 end;
  with ry do begin servo:=force; value:=0 end;
end;
```

Functionally Defined Motion

Whenever it is waiting for a move statement, the manipulator enters the Cartesian servo mode and continues to evaluate the current transform expression set point. This feature enables us to provide functional defined motions. For example, we might wish to describe a circle with the end effector or trace out some curve defined by a polynomial. But before we discuss functionally defined motions we first need to discuss the time variable.

The time variable, time, simply increments at a millisecond rate. It is however, assignable, so it may be reset to zero at any time. After it is assigned it simply continues to increment at a millisecond rate.

We will now employ the Cartesian servo mode and the time variable to illustrate functionally defined motion. Consider the crank shown in Figure 8. The transform equation

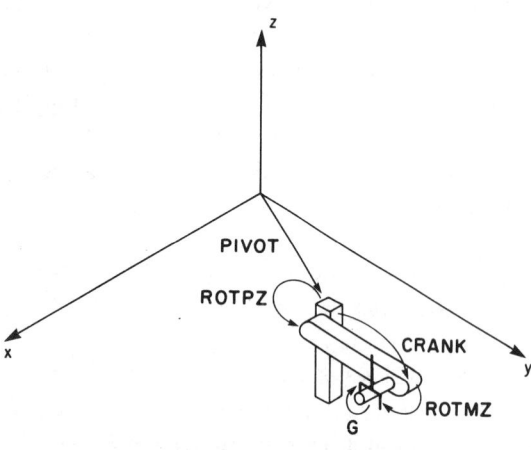

Figure 8. Crank Turning Example

describing the manipulator holding the crank handle is

$$Z\ T_6\ E = PIVOT\ ROTPZ\ CRANK\ ROTMZ\ G$$

where ROTPZ is a function transform representing a positive rotation about the z axis of theta degrees. ROTMZ is another function transform, representing a negative rotation about the z axis of theta degrees. The following program will move the manipulator to the crank handle and then turn the crank around twice

```
p1:=makeposition(list3(z,t6,e),
    list5(pivot,rotpz,crank,rotmz,g),
    rotmz,pivot,true,true);
theta:=0;

    etc.
```

```
move(p1,m0);{move to the crank handle}
movewait;{and stay there}
close;
time:=0;
{reset time}
while time<5000 do theta:=720 * (time / 5000);
```

The manipulator moves to the crank handle and grasps it. The while statement takes five seconds to execute, and the manipulator remains in Cartesian servo mode, evaluating the transform equation which is functionally dependent on theta. This causes the crank to be rotated twice.

CONCLUSIONS

We have traced the development of robotics through the evolution of robot manipulator programming languages. Initially manipulator control was by means of procedures imbedded in a high level language. This evolved into geometric based planning and execution systems where planning was time independent and execution was time efficient. With the addition of sensor feedback and the need to develop the ad hoc procedures involved in assembly an interpretative system evolved. Finally we have come full circle with manipulator control once again imbedded in a high level language. This has become possible because of the development of high level languages, the simplification of manipulator control eliminating of the need to plan trajectories and to pre calculate dynamics, and the increasing compute power available to provide for the economic real time control of the manipulator.

The sensor input, computer controlled robot is an extremely rich and interesting research area which is beginning to develop as an independent discipline. In the past it was seen as an interesting application of various research results and this has resulted in a fragmented and chaotic development of robotics in general. While work in the area of sensors, actuators, kinematics, dynamics, structures, and end-effectors is of great importance it is the area of language that distinguishes robots from all other devices. It is a tremendous undertaking to integrate the description of tasks, models, data reduction from sensors, and control of actuators into a unified system.

Perhaps the greatest area of importance in language development is the identification of language primitives, that small set of instructions which will allow us to describe the infinite number of tasks a robot can perform. A closely related area is that of modelling; at present we rely almost entirely on the geometric models of computer graphics. What kinds of models are necessary to relate force, touch, and vision? How would we model the randomly distributed parts of a clock such that the robot could deduce the correct assembly sequence? How is a bucket of paint modelled such that the robot could correctly stir the paint in order to mix it? How is the language to include these capabilities, yet be simple enough to be processed by a small computer? Robotics will need to develop as a strong independent area in order to answer these questions, to define needed developments in other areas, and to incorporate new developments in an intelligent and consistent manner.

REFERENCES

1. R. C. Goertz, "Manipulators Used for Handling Radioactive Materials," Chap. 27 of *Human Factors in Technology*, E. M. Bennett, et al., eds., McGraw-Hill, New York, 1963, pp. 425-443.
2. *ibid*
3. J. Rosenberg, *A History of Numerical Control 1949-1972: The Technical Development, Transfer to Industry, and Assimilation*, U.S.C. Information Sciences Institute, Marina del Rey, Calif., ISI Report ISI-RR-72-3, 1972.
4. H. A. Ernst, *A Computer-Operated Mechanical Hand*, Sc.D. Thesis, M.I.T., Cambridge, Mass., 1961.
5. L. G. Roberts, *Machine Perception of Three-Dimensional Solids*, M.I.T. Lincoln Lab., Cambridge, Mass., Report No. 315, 1963.
6. L. G. Roberts, *Homogeneous Matrix Representation and Manipulation of N-Dimensional* Constructs, M.I.T. Lincoln Lab., Cambridge, Mass., Document MS1045, 1965.
7. R. L. Paul, "The Mathematics of Computer Controlled Manipulators," *Proc. 1977 Joint Automatic Control Conference*, San Francisco, June 1977, pp. 124-131.
8. M. W. Wichman, *Use of Optical Feedback in Computer Control of an Arm*, Stanford Artificial Intelligence Lab., Stanford, Calif., Memo No. 56, 1967.
9. D. L. Pieper, *The Kinematics of Manipulators Under Computer Control*, Stanford Artificial Intelligence Lab., Stanford, Calif., Memo AIM-72, 1968, p. 144.
10. J. Feldman, et al., "The Use of Vision and Manipulation to Solve the Puzzle," *Proc. Second Int'l. Joint Conf. on Artificial Intelligence*, London, Sept. 1971, pp. 359-364.
11. M. Ejiri, T. Uno, H. Yoda, T. Goto, and K. Takeyasu, "A Prototype Intelligent Robot That Assembles Objects From Plane Drawings," *IEEE Trans. Computers*, Vol. C-21, No. 2, Feb. 1972, pp. 161-170.
12. Paul, R. P., "The Computer Representation of Simply Described Scenes," in *Pertinent Concepts in Computer Graphics*, ed. by M. Faiman and J. Nievergelt, University of Illinois Press, pp. 87-103, 1969.
13. Paul, R. L., *Modeling, Trajectory Calculation and Servoing of a Computer Controlled Arm*, Stanford Artificial Intelligence Laboratory Memo AIM-177, Stanford University, 1972.
14. Paul, R. P. C., "Wave: A Model-Based Language for Manipulator Control," The Industrial Robot, March 1977.
15. Paul, R. P., "Manipulator Cartesian Path Control," *IEEE Trans. on Systems, Man and Cybernetics*, SMC-9, 11, 702-711, Nov. 1979.
16. Jensen, K. and Wirth, N., *PASCAL User Manual and Report*, Springer-Verlag, 1974.
17. Paul, Richard P., "Robot Manipulators: Mathematics, Programming, and Control," MIT Press, 1981.

APPENDIX A

Homogeneous transformations are used to express the relative position and orientation between two coordinate systems. The components of a vector, from the origin of the first coordinate frame to the origin of the second frame, describe the relative position. The components of each of three unit vectors along the principal axes of the second coordinate frame describe the relative orientation. The components of all vectors are in terms of the first frame (see Figure 9). If we have two objects described in terms of coordinate systems fixed to each object then, given a homogeneous transformation expressing the relationship between coordinate frames, we can fully describe the position and orientation of one object relative to the other. At the end of the position vector we

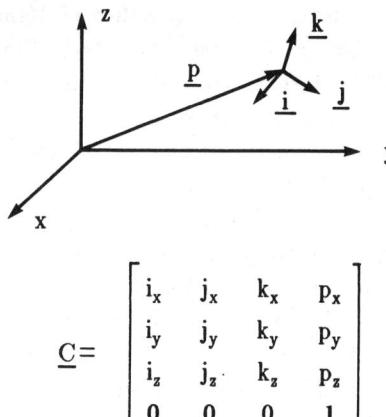

$$\underline{C} = \begin{bmatrix} i_x & j_x & k_x & p_x \\ i_y & j_y & k_y & p_y \\ i_z & j_z & k_z & p_z \\ 0 & 0 & 0 & 1 \end{bmatrix}$$

Fig. 9. Coordinate Systems

draw the three unit vectors in terms of their components in the reference object coordinate frame. If we know the relationship of the described object to its coordinate frame, we can then reconstruct both objects.

We will arrange the components of the three unit vectors as the three columns of a rotation matrix. The x vector is the first column, the y vector is the second column and the z vector is the third. The position of any vector in the described object, in terms of the reference object, can then be obtained by considering the vector as a column matrix and premultiplying by the rotation matrix. Why does this work? Consider vectors along each of the principal directions of the described object, these vectors consist of two zeros and a number representing the length of the vector. When multiplied by the rotation matrix, this vector will simply scale the appropriate column of the rotation matrix representing the vector to which it is parallel. We must, however, add the vector describing the position of the second object in terms of the first object to the result of the matrix multiplication. We may perform the addition by adding the position vector as a fourth column to the rotation matrix and by adding a "one" as a fourth element to the vector we wish to transform from the second object to the first. In terms of matrix multiplication, this has the effect of adding the components of the position vector as required. Finally, by adding a fourth row to the 3x4 matrix, consisting of three zeros and a "one", we can produce a transformed vector consisting of four elements, the x, y, and z components, and another one in the last row. This 4x4 matrix is known as a homogeneous transform. If we had a third object described in terms of the second by a homogeneous transformation $\underline{T}_2$ and a vector in terms of the third object $\underline{v}_3$ then we would have:

$$\underline{v}_2 = \underline{T}_2 \, \underline{v}_3 \qquad (1)$$

Further, if we had described the second object in terms of the first by a transform $\underline{T}_1$ then we would have the vector $\underline{v}_1$ given by:

$$\underline{v}_1 = \underline{T}_1 \, \underline{v}_2 \qquad (2)$$

and combining equations 1 and 2:

$$\underline{v}_1 = \underline{T}_1 \, \underline{T}_2 \, \underline{v}_3 \qquad (3)$$

but, if we had described the third object in terms of the first by a transform $\underline{T}_3$, we would have:

$$\underline{v}_1 = \underline{T}_3 \, \underline{v}_3 \qquad (4)$$

as this is true for all vectors $\underline{v}_1$ and $\underline{v}_3$, we have by combining equations 3 and 4

$$\underline{T}_3 = \underline{T}_1 \, \underline{T}_2 \qquad (5)$$

This means that if we have a description of one object in terms of a second, and a description of the second object in terms of a third, then the matrix product of the two transformations is the transformation which describes the first object in terms of the third. Further, if we have a description $\underline{T}_1$ of one object in terms of a second object, and also a description of the second object in terms of the first, $\underline{T}_2$, then we must have:

$$\underline{T}_1 + \underline{T}_2 = I \qquad (6)$$

where $\underline{I}$ is the identity transform, and thus:

$$\underline{T}_2 = \underline{T}_1^{-1} \qquad (7)$$

where the superscript -1 refers to the operation of matrix inversion, a simple operation for homogeneous transformation.

REAL TIME CONTROL OF WATER SYSTEMS

R. Canales-Ruiz

Instituto de Ingeniería, UNAM, México, D.F., México and
Instituto de Investigaciones Eléctricas, Cuernavaca, Morelos, México

Abstract. The evolution of digital computer control of water supply, distribution and disposal is presented. Explanations of the difference in intensity of the use of computer control in water and electrical systems are given. The main control problems of water distribution are presented in mathematical and non-mathematical terms. Early attemps of the real time solution of the network equations and actual installation of digital control of water supply systems are described. The use of self-tuning control technique as an alternative to the computer aided control of water systems, commonly employed in actual installations, in suggested.

The water disposal control problem and several actual installations with different degrees of automation are briefly treated.

Keywords. Water distribution. Sewer system. Computer control. Large scale systems. Data acquisition.

INTRODUCTION

As it occurs in many services today, in the water supply, distribution, treatment and waste water disposal services, the digital computer presence is been felt. However, in comparison with other utilities, such as the electrical utilities, the use of the digital computer as a direct control device is rather limited. While the number of digital control centers of electrical systems have grown enormously in recent years (Dy Liacco, 1978), there are only a few of such center devoted to water works controls.

Even though there has been a considerable activity in the field of real time forecasting and control of water resources (Wood, 1980), this paper will be devoted only to the urban water systems.

New technologies are adopted because a) are attractive from the benefit-cost point of view and for b) give the possibility of attacking unsolved problems with them.

Only recently both reason are becoming clear in the water industry as it will be detailed later.

The digital computers entered the urban water scene through the billing, accounting and administrative doors. Soon, they were being used in helping in the design of system expansions. Later, extended to simulation of operations and only recently, as a control tool (Caves and Earl, 1979).

Water supply systems in medium to large cities were operating rather satisfactorily in the fifties. System expansions and enhancementes were made through civil engineerings works and in agreement with the times, without energy saving in mind. The systems were designed and built for easy operation.

To increase the reliability of the water distribution networks "circuits" were built in such a way as to guarantee water supply in the event of a single pipe failure. In some cities this decision have reduced the controllability of the networks.

The energy crisis of the seventies changed the criteria. Energy bill savings became the targets and economical operations were sought. Cities grew out of hand and distribution, all of a sudden, became a complex problem. Underground water was needed and its integrity was closely watched (Alexander 1976). Then, and only then, the digital computer control for water supply and distribution began.

The operation of sewer systems were handled with conventional controls, and even manually, until some twelve years ago. When antipollution laws throughout the world required the maintanance of a water quality it was needed that the sewerage had to be treated before discharging it into rivers, lakes or sea. It was demostrated that with digital computer control the installed water treatment capacity can be more effectively used and major works investments can be reduced (Field, 1982).

In México, a real time digital computer control program for its drainage system has been started and the first phase, the data acquisition system, is fully operational (Guerrero-Villalobos, 1982).

In Fig. 1, the main components of the water cycle in a block diagram form are presented.

The rest of the paper is divided in two parts; in the first the water supply and distribution system computer control problems and practices will be presented. In the second, the sewer systems will be discussed.

WATER SUPPLY AND DISTRIBUTION

There are many similarities between the supply of water and the supply of electrical energy, but there is a basic difference that explains why the digital control had been used so much in the latter and so little in the former: water can be stored, either in large city reservoirs or in individual home storages, for later use.

The time scale of the two problems are widely different; while in the electrical industry the Automatic Generation Control function is performed every 4 to 10 seconds (de Mello, 1973) the valve manipulation is done every 10 minutes to several hours, making in many instances the manual or conventional control appropiate.

In places where there is enough water the inmediate goal is to provide domestic water, under normal conditions, with enough presure (typically from 4 to 9 atm.). In some countries like the USA there should be enough water for emergencies such as fires.

When there is a water shortage, like it occurs in many fast growing cities like in Mexico, the aim is to evenly distributed the water among the population.

For the operation it should be decided what primary sources (springs, wells, ponds, lakes and rivers) should be exploited; what pumps should be turned on: what should be the valves openings, etc, to guarantee the appropiate quality of the service guarding the integrity of the ground water and reducing the costs.

In the decade of the sixties, when computers were batch processing oriented, there were several ambitious attempts to use them in the control loop. Many of these efforts felt short of their expectation and produced a not very positive reactions among potential users. (Shamir, 1981).

On the other hand, there has been very little communication between the developers of similar systems. Also, the number of technical papers dealing with actual installations has been very small. These factors have prevented the diffusion of digital systems applications to water distribution systems.

Evolution of network control

In the Fig. 2 three degrees of automation in water networks are presented. Only in very rare cases systems have evolved in a different fashion. At first, the supervisory control is installed and digital technology is used only to monitor the network. All decisions are made by the operator. In this stage most of the transducers and actuators are installed making simpler the development of later stages.

At the second stage, the computer is used as an aid in the decision making process. Data is transfered to it automatically by the SCADA (Supervisory Control and Data Adquisition System) and the operator request simulations and/or evaluation of tentative control strategies.

Only at the third stage is where the computer is a real "real time" device.

The city of Dallas is a typical example of the evolution of automation in water supply systems, this evolution is shown in table 1.

YEAR	FUNCTION
1950	All pumps were manned. Communication was made by telephone. Every hour the operators reported to a central station.
1960	Telemetry was introduced, but operations were manned.
1970	Telecontrol started.
1975	Digital Computers were used to process data and presented it in a meaningful way to the central station operator.
1980	The digital computer was used as an aid to evaluate alternative operations.
1985?	Close loop digital control

TABLE 1. Automation Function Evolution.

Network analysis and simulation

The heart of the problem in water distribution control is the solution of the network equations. These are a set of equations that relate flows and pressures at different nodes. Typical systems have hundreds of pipes and nodes, and solving the equations might be a formidable task.

The equations that relate the variables Q_{ij}, the flow in pipe joining nodes i and j, h_i, the pressure at node i and D_i, the flow outside the network are: (Canales, 1979).

Continuity at nodes

$$\sum_{j \varepsilon \Gamma(i)} Q_{ij} + D_i = 0 \quad i = 1,\ldots n_n \quad (1)$$

Losses in pipes

$$K_{ij} Q_{ij} |Q_{ij}|^{\alpha_{ij}-1} = h_i - h_j \quad (2)$$
$$i \varepsilon \Gamma(j) \; \forall j$$

Terminal elements

$$F(h_i, D_i, x_i) = 0 \quad i = 1,\ldots n_n \quad (3)$$

where $\Gamma(j)$ are the set of nodes connected to node j by a pipe, n_n the number of nodes of the network

x_i is a state variable associated with node i

α_{ij} and K_{ij} parameters associated to the i, j pipe.

There are several types of terminal elements. Some of them are:

Constant node demand

$$D_i = D_o$$

Constant head

$$h_i = H_o$$

Free discharge

$$D_i = \begin{cases} k_i(h_i - h_{ei})^{1/2} & \text{if } h_i > h_{ei} \\ 0 & \text{otherwise} \end{cases}$$

Pumps

$$\beta_{i1} D_i^2 + \beta_{i2} D_i + \beta_{i3} - h_i = 0$$

Storage tank

$$D_i = f(x_i - h_i)$$

$$\frac{dx}{dt} = g(D_i - D_o)$$

Associated with the equations (1) - (3) there are several problems that have to be solved in real time to have digital close loop control.

D-H Given the demand D_i at every node and the head at a reference node, find h_i in all other nodes

$H_p - H$ Given h_i at some nodes and supplementary hypotesis (e.i., proportional load at nodes, that is $D_i = k_i D$), find h_i at the rest of the nodes.

$C - H_D$ Given an initial state, find control settings (valve positions, on pumps) such that a desired head profile H_D is obtained.

An early result

In the early seventies, one of the first studies to look into the applications of real time digital control techniques to typical water distribution networks was sponsored (De moyer et al, 1975).

The posed problem consisted in the following: using the measured pressure at some nodes of a network, take the pressure of a specific node to a fixed value by manipulating pumps and valves.

In order to test the feasibility of the results, a mathematical model to simulate the network was developed. Due to the limited core memory, a relaxation method (Hardy-Cross) was used to solve the D-H problem. The inverse problem was solved using the algorithm despicted in Fig. 3. This way, the estimation of the demand was made.

To solve the control problem very rudimentary search techniques were used.

By the middle of the seventies several efficient computer codes to simulate water distribution systems were available. (Cesario, 1980).

Most of the algorithms used Newton like iterative methods. However, the difference in acceptability was due to the capacity of effective man-machine interface.

Actual installations

Honolulu

A digital computer automated system was put into opperation in the first half of the last decade in Honolulu. Demographic growth of Oahu island after World War II made it necessary to use additional water sources. Local wells were put into operation and were inexpensive, but overexploitation might lead to sea water instrussion.

Storage and main pipe systems did not grow according to demand. The least expensive solution was a more intelligent use of avalaible resources (Alexander, 1976).

The main automatic control functions were:

a) Supply water to customers within prescribed pressure limits.
b) Monitoring system operation and alarm detection
c) Management water resources
d) Minimize pumping costs.

The automatic control actions are based on dynamic simulations. Every half hour the computing system, based on partial information (pressure at few points), makes an estimation of the actual demand and analyzes the effect of 8 possible control actions. The operator chooses the best.

Denver

(Colorado USA), a system operates around a simulation program WATSIM developed at Systems Control (Rao and Seitle, 1975). As in Honolulu the operator simulates several actions and chooses the one he considers the best.

St. Gallen and Zurich.

(Switzerland), Computer based automated systems with the major functions oriented toward the control of the water supply were installed last decade. Under normal conditions pumping stations are controlled automatically taking into account levels, future demands and electricity costs.

Many more cities, that are lenghtly to mention, do have digital supervisory control.

Self Tuning Control for a Water-Distribution System

Most of the water used in Mexico City is brought from remote wells and rivers. Because of undersupply, those sources are exploited to capacity and for practical pourposes they can be considered of constant flow. Water is stored in large tanks located around the city. Distribution is done by gravity and can be changed only through very complex valve manipulations. Because of water shortage, domiciliary tanks are used to store water whenever the pressure is high enough.

Schemes, either for simulation or control, used in other cities do not apply because of the different type of end terminal elements. New techniques had to be found. A simulation model was built.

In Fig. 4 the evolution of municipal tank levels and in the Fig. 5 the domiciliary tank levels are shown.

For control purposes a self tuning technique was applied to the network to evenly distribuite water among users. The controller had the task of identification and control.

In the Fig. 6 a network with similar structure of Mexico's is presented. The idea is to operate the valves to raise the pressure of individual nodes, one at the time, and at the same time identify the parameters of the terminal devices. A linearize relation between input (valve positions) and outputs (node pressures) are obtained in every observation period. (Hernández, 1981).

The results are shown in Fig. 7.

WATER DISPOSAL

The combined sewer operation was used for the fisrst time in England after the well known Broads Street epidemy in 1854. Sewers systems until then were meant for storm water run-off only. People were ordered to discharge the sanitary waters into the sewers systems, in order to secure public health.

In the 1960 a great concern about the pollution of the waters became a big issue. Laws and international aggreements were signed and a intense compaign to reduce pollution started. The municipal water dispossal offices were asked to treat the water and discharge it into lakes, rivers and sea only when a certain quality was obtained.

It was believed that the answer was to undo what the british started more than a centeny before: to separate the combined sewer into storm-run off and sanitary and process only the latter.

Soon it was found that storm run-off water was in itself a source of stream pollution and that it had to be treated also, so the initial investment to separate the sewers services was not enough.

When the combined sewers systems were decided upon, then for peak loads the capacity of the treatment plant had to be large and too expensive. However, if the load could be leveled significant cost reductions could be realized.

The storage of the storm run-off coudn't be in only one place, it had to be distribuited, either the piping system could be used or ad-hoc storage had to be built.

Some system in operation.

In table 2 some of the computerized sewer systems in the United States are presented, and they will be commented briefly (Brueck et.al, 1981).

	DATA LOGGING	SCADA	AUTOMATIC
Seatle, Washington	1971	1971	1974
Detroit, Mich.	1968	(1968) ANALOG	--
San Francisco, Ca.	1970	--	--
Cleveland, Ohio	1972	--	1974
Minneapolis, MN	1968	1968-1976	--
Lima, Ohio	1979	1979	1980

TABLE 2. Computerized sewer systems in the U.S.A.

Seatle, Washington.

This was the first system with a real time digital control. The project CATAD (Computer Augmented Treatment and Disposal System) was started in 1966, and the purpose was to use the piping system as storage in order to maximize the water treated before dumping it into the sea.

The run-off water is pumped from some parts to other in the piping system, depending on the storage capacity. In Fig. 8 a schematic diagram of the system is shown.

The data acquisition system was operational in 1971 after having two contractors to develop the software. The problem then was that at time the project started there were not available event driven computers.

After three years of collecting data, then the system was run on a completed automated fashion, and currently is going through some modernization of equipment.

In Detroit, San Francisco, Victoria (B. C. Canada) and Mineapolis have a data acquisition systems, and are operated manually with the same ideas as in Seatle.

Lima, Ohio. (Brueck, 1981)

This is one of the most modern systems in operation. The reasons that move the city authorities to automated operation were economical. Separated sewers systems were too expensive.

In Fig. 9 there's an schematic of the sewer system. Water is stored temporarity and released gradually to the treatment plant, with the following priority system:

a) Avoid in system flooding
b) Avoid river overflow:

   b.1 Dewater system prior to rain fall event
   b.2 Increase treatment at plant
   b.3 Maxime storage systems capacity

c) Reduce gate operations.

The main difficulties found in the development of the automated system were: lack of dynamic performance data, the non linear relationship between sewer levels and system storage capacity and tuning.

A simulation model was used to supply the overall dynamic behaviour. The implementation phase was initially manual and only after sufficient data was collected, was put into operation in the automatic mode.

A computer was needed because appropriate decisions could not be made with only local information and even if it were centralized it was too difficult to process by humans. With the successful operation of the system it was confirmed that sewer systems real time digital computer control is feasible and practical.

Mexico.

Mexico city is located in a closed valley, no rivers for discharge are avalaible. Few years ago, the city authorities ask national research institutions to examine how computers can be used to control water flooding, one of the major problems of the city. At this point in time, water quality release is not an issue. Looking at the problem it was early learned that there was a complete lack of data. Obviously, the real dynamic behavior of the system could not be assesed. The logical step was to develop an automatic data gathering system. Special effort was made to built such a system with the own country technology. The automated rain gage network and alarm system was built and is in operation (Guerrero-Villalobos, 1982).

The next two steps are:

a) To predict sewer levels
b) To control them

CONCLUSIONS

It is clear, from previous descriptions, that most of the schemes used in Real Time have been developed for specific systems and there is a lack of unifying concepts and practices.

Very few experts in real time digital control have participated in the system development and there is a lot to be done in this field; there are robust and proved ways to control some large scale systems from where operation of water networks might benefit. Is up to control engineers to make it happen.

Acknowledgments.

The author is grateful to Luis Alvarez and Rafael Carmona for their comments.

REFERENCES

Alexander S. M. (1976). Multipurpose Computer for Honolulu Board of Water Supply. *American Water Works Association Journal*, 68, 147-155.

Brueck, T. M., Knudsen D. I. and Peterson D. F. (1981), Automatic Computer-Based Control of a Combined Sewer System. *Water Science and Technology*, 13, 103-109.

Canales R. R. (1979). *Real-Time Control of the Mexico City water supply system.* (in Spanish), Instituto de Ingeniería, Internal Report.

Caves J. L. and Earl C. T. (1979). Computer Applications: A tool for Water Distribution Engineering. *American Water Works Association Journal*, 71, 230-235.

Cesario, A. L., (1980) Computer Modelling Programs: Tools for Model Operations *American Water Works Association Journal.*, 72, 508-513.

Davis A. L. and Jeppson R. W. (1979) Developing a Computer Program for Distribution System Analysis. *American Water Works Association Journal.*, 71, 236-242.

De Mello, F. P. (1973). Automatic Generation Control. *IEEE Transactions on Power Aparatus and Systems*, 92, 710-720.

De Moyer R. Jr. and Horwitz L. R. (1975) Automatic Control of Water Distribution. *Transactions of the ASME.*

Dyliacco T. M. (1978). Incorporating security functions in system control, present status *IEEE Spectrum*, 15, 43-50.

Field R. (1982). Storm and Combined Sewers; Part of the Treatment Process. Water Engineering and Managment, 129, 34-39.

Guerrero Villalobos G. (1982) (Ed.), El Sistema Hidraulico del Distrito Federal, Departamento del Distrito Federal, Mexico.

Hernández A. A. (1981). Self-Tunning Control of a Model of the Mexico City water Supply System. (in Spanish). M. S. thesis. Mexico.

Nemetz, P. (1973). A Dispatching Center for Gas and Water Supply in St. Gallen. Brown Bovery Rev, 10/11, 443-449.

Rao H. S. and Seitle R. A. (1975), Computer Applications in Urban Water Distribution System Control. ASME Journal of Dynamic Systems, Measurement and Control, 117-119.

Russel J. and Wahl L. (1976) Automation of the Public Water Supply System of the City of Zurich, Brown Bovery Rev, 9, 579-584.

Shamir U. (1981). Real Time Control of Water Supply Systems. Proceedings of International Symposium on Real-Time Operations of Hydrosystems. University of Waterloo, Waterloo, Ontario, Canada, 550-562.

Wood E.F. (1980) (Ed.), Real Time Forecasting/Control of Water Resources Systems, Pergamon Press, Oxford, New York.

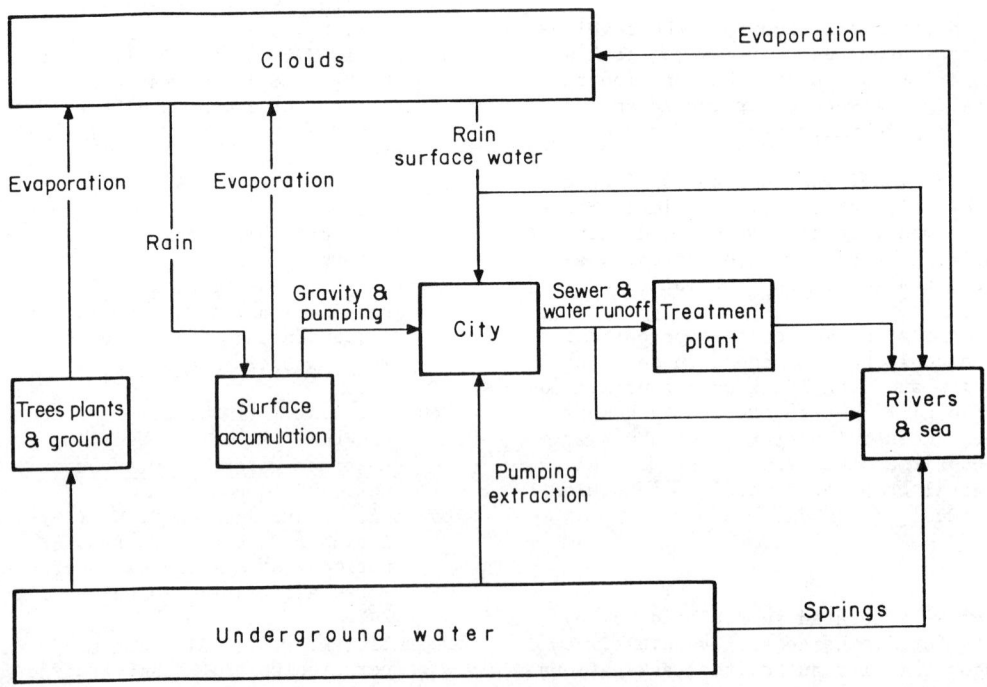

Fig 1 Block diagram of water cycles

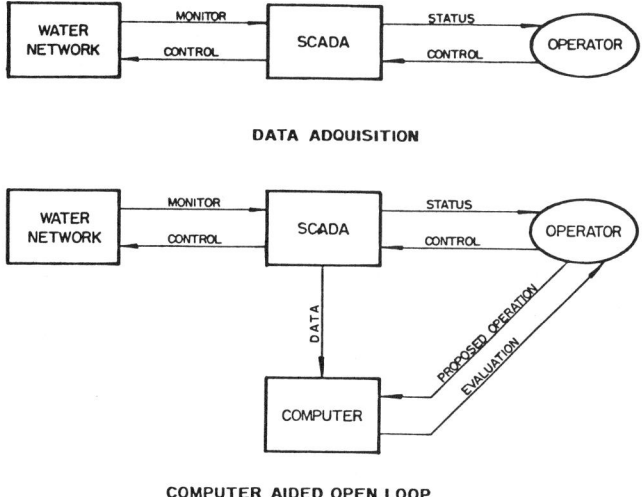

DATA ACQUISITION

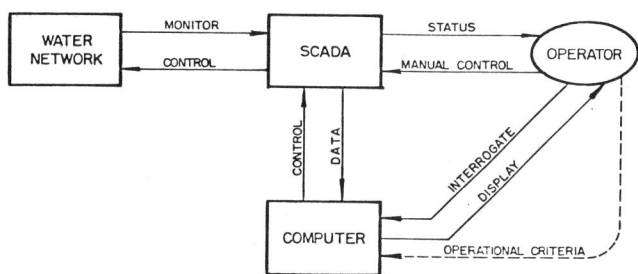

**COMPUTER AIDED OPEN LOOP
CONTROL EVALUATION**

FIG 2 CLOSED LOOP COMPUTER CONTROL

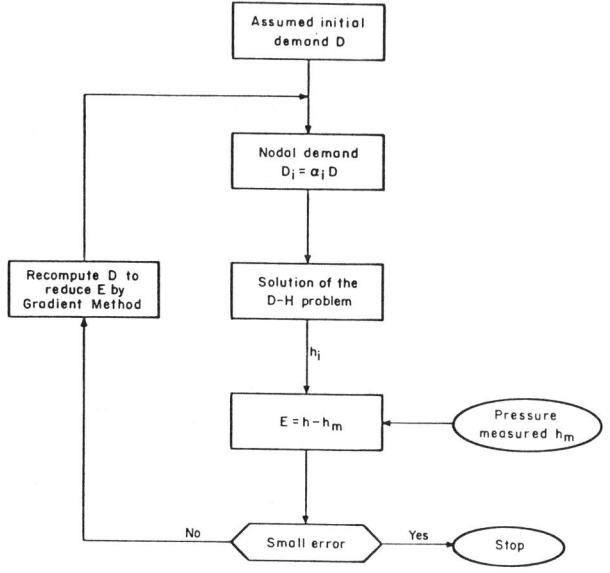

Fig 3 Solution of the inverse problem

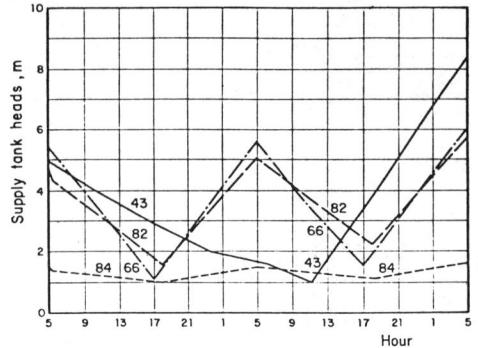

Fig 4 Water stored variation in supply tanks

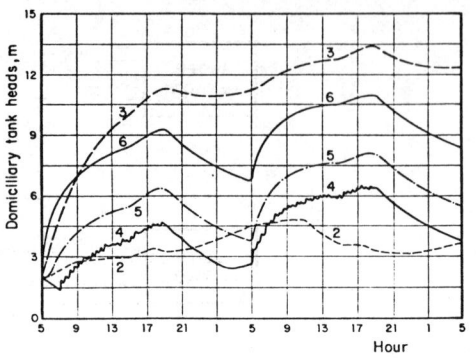

Fig 5 Water stored variation in domiciliary tanks

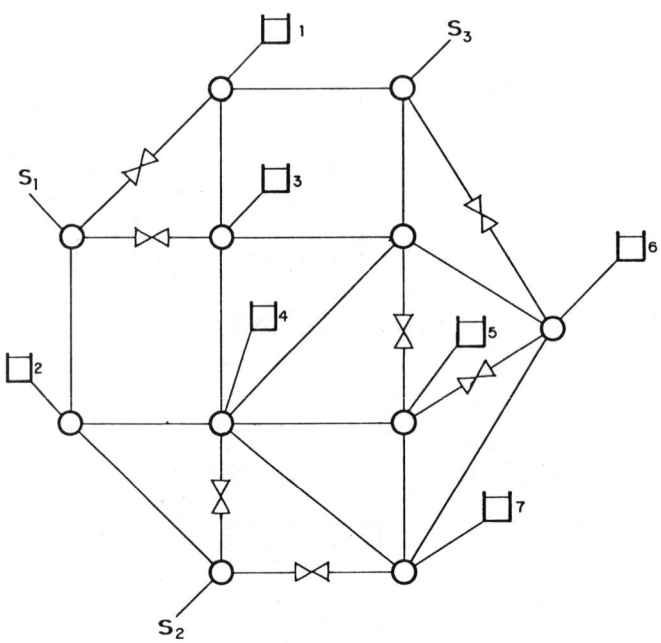

FIG 6 NETWORK FOR SELF TUNING CONTROL

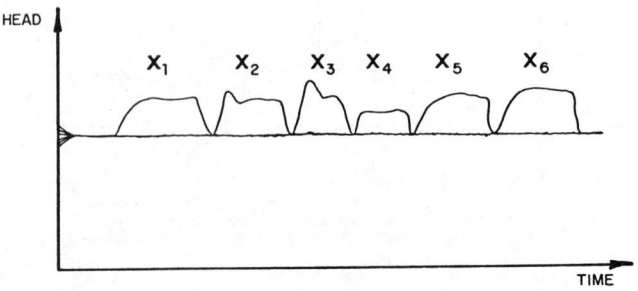

FIG 7 SELF TUNING CONTROL RESULTS

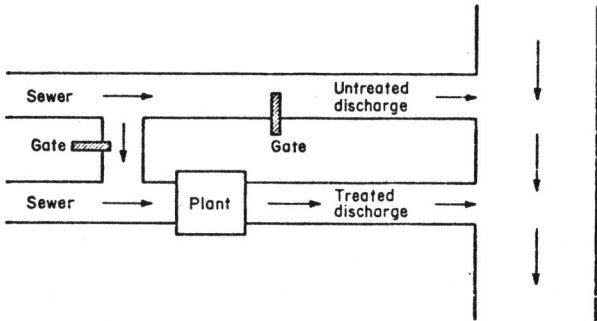

Fig 8 Discharge control

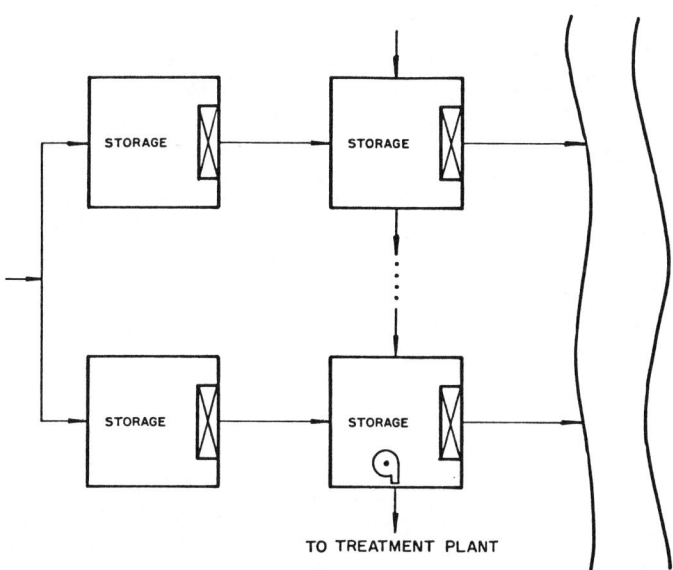

FIG 9 LIMA, OHIO STORM WATER MANAGEMENT SYSTEM

OPTIMAL COMPUTER COMBUSTION CONTROL AT THE SOAKING PIT

Y. Yamamoto*, K. Mori*, K. Fukuda*, Y. Suzuki**, K. Azumi***, and N. Saito***

Wakayama Steel Works, Sumitomo Metal Industries Ltd., Japan
**Hasaki Research Center, Sumitomo Metal Industries Ltd., Japan*
***Sumikin System Development Ltd., Japan*

Abstract. The fuel consumption in the soaking pit represent 60% of all energy used by the blooming mill plant. Therefore, energy savings at the soaking pit is most important factor for energy savings in the blooming mill plant as a whole. From the point of view of saving energy, there are two especially important factors of combustion control in the soaking pit. First, the fuel-air ratio control and second the heat pattern control, we developed a new computer system, and succeeded in optimizing those important controls. First, we optimized the fuel-air ratio gradually adapting the change to the combustion environment by observing either the pit temperature or the gas flow for a few minutes. Regarding the heat pattern control we first determine the most desirable heat pattern and consequently input the set points to the pit temperature control in accordance with the necessary time. We have developed the optimal heat pattern control with respect to the given conditions. We compose the heat pattern as an 8-dimensional vector $(=\chi)$ and represent a heat state of the ingot as a function of vector $X(=f(X))$ using the heat conduction model, and represent a fuel consumption as a function of vector $X(=g(\chi))$ using the heat balance model of the soaking pit. We optimize the vector X so as to minimize $g(X)$ under the given conditions about the heat state $f(X)$ by applying the "Gradient Method".

Keywords. Digital computer applications; energy control; nonlinear programing; mathematical programing; modeling; optimal control; steel industry.

INTRODUCTION

The main function of the blooming mill is to roll ingots which have been casted at the steel making plant. The main function of a soaking pit is to heat up and send out ingots to the blooming mill.

Recently, saving energy is the most important factor of all production activities. Consequently, operations of the soaking pits have been changed to emphasize saving energy rather than large scale production. To facilitate these changes of operation, knowing the accurate temperature state of the ingots which have been hated, and being able to easily adjust the delivery time of ingots from the soaking pits, have become important necessary considerations for combustion control at the soaking pits. Fuel consumption at combustion has begun to be used as an estimation of the control performance, recently.

From the point of view of saving energy, control of the fuel-air ratio and the heat pattern are important factors of fuel consumption in the soaking pits.

In this paper, we describe two controls, which were developed for the computer systems for the No. 1 Blooming Mill of Wakayama Steel Works.

BACKGROUND

Soaking Pit Operation

Use of a soaking pit is the first process of a blooming mill and is used to supply ingots which have been heated to the appropriate temperature for rolling. Therefore, in general, there are two main governing factors with regard to the soaking pits as follows;

(1) the schedule for charging ingots which have been received from the steel making plant into the soaking pits and further, delivering ingots from the soaking pits to the blooming mill.

(2) the combustion control of the soaking pits to minimize fuel consumption.

Fundamental Combustion Control in The Soaking Pits

In the soaking pits of No. 1 Blooming Mill, a mixture of BFG (Blast Furnace Gas) and LPG (Liquid petroleum gas) which has a gas calorie of about 1850kcal/m^3 is used. Fig. 1 shows the fundamental combustion control system and Fig. 2 shows the heat balance of the soaking pits. With respect to the heat balance in Fig. 2, to increase thermal efficiency, we have to decrease the amount of heat in the exhausted gas and loss of heat from the walls of the soaking pits.

To realize the these demands, it is important to optimize the fuel-air ratio as well as to optimize input heat by controlling the heat pattern as shown in Fig. 1. These two methods are typical energy saving controls used in the soaking pits, for which we have developed a computer system as shown in Fig. 3. We will describe the fuel-air ratio control briefly and the heat pattern control in detail.

FUEL AIR RATIO CONTROL

Hitherto, various techniques for fuel-air ratio control, such as a control method based on continuous analysis of the O_2 value in exhaust gas or a constant fuel-air ratio presetting method, etc., have been used. However these techniques are not always the most effective since the optimum fuel-air ratio varies under the influence of such factors as variation of fuel gas calorie level, temperature of the soaking pit, etc.. Moreover, O_2 analyzer equipment is expensive and hard to maintain.

Principle

The optimum fuel-air ratio coincides closely with the maximum flame temperature, i.e., suitability of the fuel-air ratio appears to change with the pit temperature. On the other hand, we control the pit temperature using a heat pattern, therefore the most suitable fuel-air ratio can be found by looking at the fuel flow.

Control Method

Fig. 4 shows the control method. We express the fuel flow curve with equation (1).

$$y = R_0 + R_1 \exp(-\alpha_1 t) + k_2 \exp(-\alpha_2 t) \quad (1)$$

where
  y: fuel flow
  t: time
  $k_0, k_1, k_2, \alpha_1, \alpha_2$: unknown parameters

In order to forecast the future fuel flow at t_2, we estimate the unknown parameters k_0, k_1, k_2, α_1, α_2 by using sampling data $Y_1, \ldots Y_n$ during period T_1 and the method of least squares. If we use the fuel air ratio R during the period T_1 and $R+\Delta R$ during the period T_2, at time t_2 we can find the actual fuel flow, Y_A, using the fuel-air ratio $R+\Delta R$ and an estimation of fuel flow, Y_E, using the fuel-air ratio R. By comparing the estimation with the actual value, we can judge whether the fuel-air ratio is suitable or not. This process is repeated until reaching the optimum fuel-air ratio. Fig.10 shows actual results of application at the soaking pits.

OPTIMAL HEAT PATTERN CONTROL

Outline

The requirements of the soaking pits are as follows:

1. to heat ingots so as to satisfy the following 4 conditions, which makes them suitable for rolling.

solidification ratio of ingot $\geq C_{sol}$ (2)
average temperature of ingot $\geq C_{ave}$ (3)
difference between max. and min. temperature of ingot $\leq C_{dif}$ (4)
minimum temperature of ingot $\geq C_{min}$.   (5)

2. to heat the ingots to meet the target delivery time.

target delivery time = time at which conditions (2)~(5) need to be met.   (6)

At the same time, suppose we input the heat pattern to the set point of the pit temperature controller such as in Fig. 1. Then we define the optimum heat pattern as follows:

"The optimum heat pattern is that which minimizes fuel consumption subject to condition (2)~(6)".

To solve this problem, we introduce two mathematical models of the soaking pits and one optimization method.

Heat conduction model. This model forecasts the ingot temperature according to each heat pattern.

Heat balance model. This model forecasts fuel consumption according to each heat pattern.

Optimization method. This method optimizes heat pattern with respect to fuel consumption using the heat conduction model and the heat balance model at the same time.

Fig. 5 shows this relation.

Mathematical Model of The Soaking Pit

As we must calculate the temperature of ingots within an appropriate calculating time

(2~3 seconds), we adopt the difference equations as follows.

Heat conduction model. Fig. 6 shows heat conduction model.

1. Heat transfer. Between t_i and t_{i-1}, heat transfer is as follows:

$$Q_{i,i-1} = \lambda \cdot F \cdot \tau \cdot (t_i - t_{i-1})/b \quad (7)$$
$$(i=2\sim 6)$$

where
 λ; thermal conductivity
 τ; time
 b; distance between t_i and t_{i-1}
 F; area of heat transfer.

2. Heat emission. Between the surface of ingot, t_6 and pit temperature, t_F

$$Q_{6,F} = 4.88 \cdot \phi \cdot \left(\frac{t_F+273}{100}\right)^4 - \left(\frac{t_6+273}{100}\right)^4 \quad (8)$$

where
 ϕ; constant

Elements of heating condition.

Solidification rate.

$$P_{sol} = \sum_i QLi / TQL \quad (9)$$

Average temperature.

$$T_{ave} = \sum_i t_i V_i / \sum V_i \quad (10)$$

Minimum temperature.

$$T_{Min} = \underset{i}{Min.} \ T_i \quad (11)$$

Maximum temperature.

$$T_{max.} = \underset{i}{Max.} \ T_i \quad (12)$$

Difference between max. and min. temperature

$$T_{dif} = \underset{i}{Max.} \ T_i - \underset{i}{Min.} \ T_i \quad (13)$$

where i = 1~6
 V_i : volume of mesh
 QL_i : latent heat
 TQL: total latent heat

Suppose that S is a temperature distribution before heating, when we heat the ingots with heat pattern $\mathbf{X}$, we formally describe each element of the heating conditions as follows.

$$P_{sol} = f_1(S, \mathbf{X}) \quad (14)$$
$$T_{ave} = f_2(S, \mathbf{X}) \quad (15)$$
$$T_{dif} = f_3(S, \mathbf{X}) \quad (16)$$
$$T_{min} = f_4(S, \mathbf{X}) \quad (17)$$

Heat balance model. Fig. 7 shows the heat balance of the soaking pit.

As incoming heat into soaking pit is equal to outgoing heat from it, so during Δt(=sufficiently small)

$$V_f = \frac{\frac{\Delta Q_{iron}}{\Delta t} + \frac{\Delta Q_{wall}}{\Delta t} + \frac{\Delta Q_{loss}}{\Delta t}}{H_\ell + q_a - q_w} \quad (18)$$

$\Delta Q_{iron}/\Delta t$ is calculated from heat conduction model of ingot and $\Delta Q_{wall}/\Delta t$ is calculated from heat conduction model of pit walls which is similer to ingot. Suppose that S is the temperature distribution of ingots and that F is the temperature distribution of the walls, we can describe formally the fuel volume V_f being consumed using heat pattern $\mathbf{X}$ as follows;

$$V_f = g(S, F, \mathbf{X}). \quad (19)$$

Optimization Method

Mathematic expression of optimum heat pattern. We compose a heat pattern as an 8-dimensional vector as shown in Fig. 8.

$$\mathbf{X} = (x_0, x_1, x_2, x_3, x_4, x_5, x_6, x_7) \quad (20)$$

1st, 2nd, 3rd temperature maintenance period is parallel to the time axis.

Since x_0 is pit temperature at the time when we calculate optimum heat pattern, we can eliminate x_0 from variables $(x_0,\ldots,x_7)$ which we must optimize.
We express finally a heat pattern as vector $\mathbf{X}$.

$$\mathbf{X} = (x_1, x_2, x_3, x_4, x_5, x_6, x_7) \quad (21)$$

We define the optimum heat pattern $\mathbf{X}^{OPT}$ as follows.

Conditions of ingot temperature.

$$P_{sol} = f_1(S, \mathbf{X}) \geq C_{sol} \quad (22)$$
$$T_{ave} = f_2(S, \mathbf{X}) \geq C_{ave} \quad (23)$$
$$T_{dif} = f_3(S, \mathbf{X}) \leq C_{dif} \quad (24)$$
$$T_{Min} = f_4(S, \mathbf{X}) \geq C_{Min} \quad (25)$$

Conditions of operation at soaking pit and heat pattern, etc. which are realistic conditions.

$$x_1 - x_0 \geq 0 \quad (26)$$
$$x_2 - x_1 \geq 0 \quad (27)$$
$$\frac{x_1 - x_0}{x_3 - x_5} \leq c_1 \quad (28)$$
$$\frac{x_2 - x_1}{x_4 - (x_3 + x_6)} \leq c_2 \quad (29)$$

where c_1, c_2 are maximum temperature grade on 1st and 2nd heating period.

Conditions of heating time.

$$Cout - \sigma \leq x_4 + x_7 \leq Cout + \sigma \quad (30)$$

where Cout = delivery time - present time
  σ : small time

Under the above conditions.

$$\text{Min. } V_f = g(S, F, \mathbf{X})$$
$$\mathbf{X} \in I\{\mathbf{X}\}$$

where $I\{\mathbf{X}\} = \{\mathbf{X} \mid$ a set of $\mathbf{X}$ satisfying Eqs. (22)~(30)$\}$

Linearizing and Gradient Method

suppose $\mathbf{X}^0 \in I\{\mathbf{X}\}$ and $\mathbf{X}^1 = \mathbf{X}^0 + \Delta\mathbf{X}$ (where $|\Delta\mathbf{X}| \leq \epsilon$), then $\Delta\mathbf{X}$ have to satisfy following conditions so that $\mathbf{X}_1$ belongs to $I\{\mathbf{X}\}$.

$$\sum_{ij} a_{ij}(\mathbf{X}^0)\Delta x_j \leq b_j$$
$$i = 1, 2, \ldots 10$$
$$j = 1, 2, \ldots 7$$

where
$$[a_{ij}(\mathbf{X}^0)]$$

$$= \begin{pmatrix}
\frac{-\partial f_1}{\partial x_1}, \frac{-\partial f_1}{\partial x_2}, \frac{-\partial f_1}{\partial x_3}, \frac{-\partial f_1}{\partial x_4}, \frac{-\partial f_1}{\partial x_5}, \frac{-\partial f_1}{\partial x_6}, \frac{-\partial f_1}{\partial x_7} \\
\frac{-\partial f_2}{\partial x_1}, \frac{-\partial f_2}{\partial x_2}, \frac{-\partial f_2}{\partial x_3}, \frac{-\partial f_2}{\partial x_4}, \frac{-\partial f_2}{\partial x_5}, \frac{-\partial f_2}{\partial x_6}, \frac{-\partial f_2}{\partial x_7} \\
\frac{-\partial f_3}{\partial x_1}, \frac{-\partial f_3}{\partial x_2}, \frac{-\partial f_3}{\partial x_3}, \frac{-\partial f_3}{\partial x_4}, \frac{-\partial f_3}{\partial x_5}, \frac{-\partial f_3}{\partial x_6}, \frac{-\partial f_3}{\partial x_7} \\
\frac{-\partial f_4}{\partial x_1}, \frac{-\partial f_4}{\partial x_2}, \frac{-\partial f_4}{\partial x_3}, \frac{-\partial f_4}{\partial x_4}, \frac{-\partial f_4}{\partial x_5}, \frac{-\partial f_4}{\partial x_6}, \frac{-\partial f_4}{\partial x_7} \\
-1, 0, 0, 0, 0, 0, 0 \\
1, -1, 0, 0, 0, 0, 0 \\
1, 0, -C_1, 0, C_1, 0, 0 \\
-1, 1, C_2, -C_2, 0, C_2, 0 \\
0, 0, 0, -1, 0, 0, -1 \\
0, 0, 0, 1, 0, 0, 1
\end{pmatrix} \quad (32)$$

where $\frac{\partial f_i}{\partial x_j}$ (i=1,~4) (j=1,~7) stands for

$$\frac{\partial f_i(S, \mathbf{X}^0)}{\partial x_j}.$$

$$[b_j] = \begin{pmatrix}
f_1(S, \mathbf{X}^0) - C_{sol} \\
f_2(S, \mathbf{X}^0) - C_{ave} \\
C_{dif} - f_3(S, \mathbf{X}^0) \\
f_4(S, \mathbf{X}^0) - C_{min} \\
x_1^0 - x_0^0 \\
x_2^0 - x_1^0 \\
C_1(x_3^0 - x_5^0) + x_0^0 - x_1^0 \\
C_2(x_4^0 - x_3^0 - x_6^0) - x_1^0 - x_2^0 \\
x_4^0 + x_7^0 - Cout + \sigma \\
Cout + \sigma - x_4^0 - x_7^0
\end{pmatrix} \quad (33)$$

Suppose that $\Delta\mathbf{X}$ is variable, we define the optimum heat pattern $\mathbf{X}^{0,OPT}$ at neighborhood of $\mathbf{X}^0$ as follows:

Optimum heat pattern $\mathbf{X}^{0,OPT}$

Minimize; $V_f = g(S, F, \mathbf{X}^0 + \Delta\mathbf{X})$

subject to; $\sum_{ij} a_{ij}(\Delta\mathbf{X}^0)\Delta x_j \leq b_{ij}$.

We have solved this problem by using the Gradient Method under conditions of linear inequalities.
Fig. 9 shows this algorithm.

Applied Results for No. 1 Blooming Mill Computer System

We show typical applied results at Fig. 11 and Fig. 12.
Fig. 11 shows that the conventional heat pattern is excessive and Fig. 12 shows that keeping a low temperature at the 1st temperature maintenance period is not an effective way for prolongation of delivery time.

The time which computer system takes to calculate the optimum heat pattern, is 2~3 minutes.

CONCLUSION

In this paper, we have described the control of both the fuel-air ratio and the heat pattern as energy saving controls with respect to combustion control in the soaking pits. We used the method of least squares in the fuel-air ratio control and the Gradient Method in the heat pattern control to minimize the fuel consumption.

These controls have been developed at No. 1 Blooming Mill of Wakayama Steel Works and have achieved significant results.

First, since we can now control the temperature of ingots by changing the heating conditions, we always have a adequate supply of ingots ready for the rolling mill and we have achieved smooth soaking pit operations.

Second, fuel consumption has been decreased by 15%.

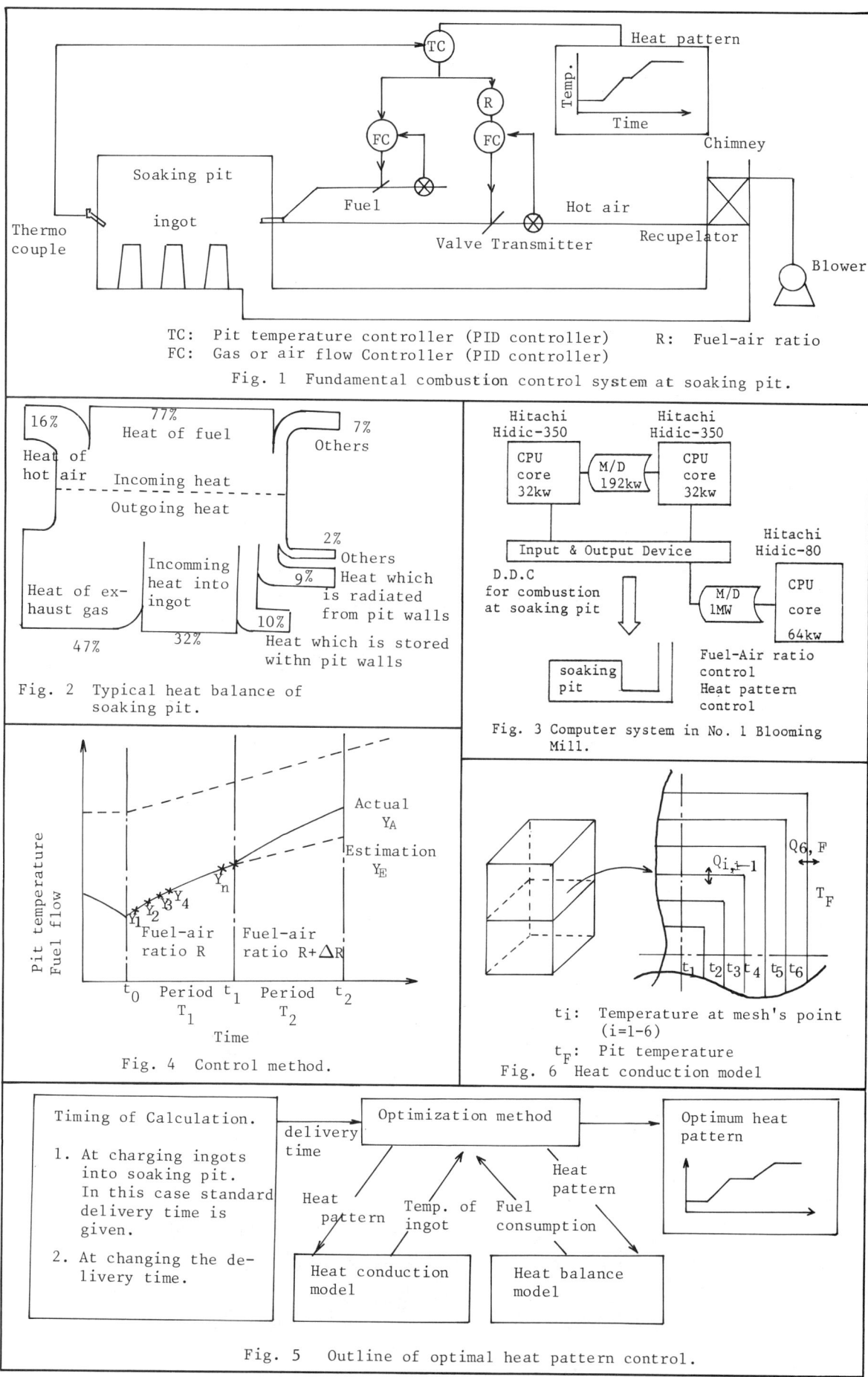

Fig. 1 Fundamental combustion control system at soaking pit.
TC: Pit temperature controller (PID controller)　　R: Fuel-air ratio
FC: Gas or air flow Controller (PID controller)

Fig. 2 Typical heat balance of soaking pit.

Fig. 3 Computer system in No. 1 Blooming Mill.

Fig. 4 Control method.

Fig. 6 Heat conduction model
t_i: Temperature at mesh's point ($i=1$–6)
t_F: Pit temperature

Fig. 5 Outline of optimal heat pattern control.

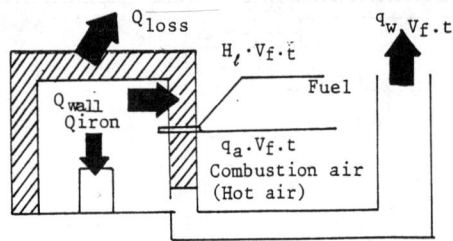

H_ℓ: Calorie of fuel
V_f: Fuel volume
q_a: Heat of hot air for combustion
q_w: Heat of exhaust gas
Q_{iron}: Incoming heat to ingot
Q_{wall}: Heat which is stored within pit walls.
Q_{loss}: Heat which is radiated from pit walls.
t : time

Fig. 7 Heat balance model

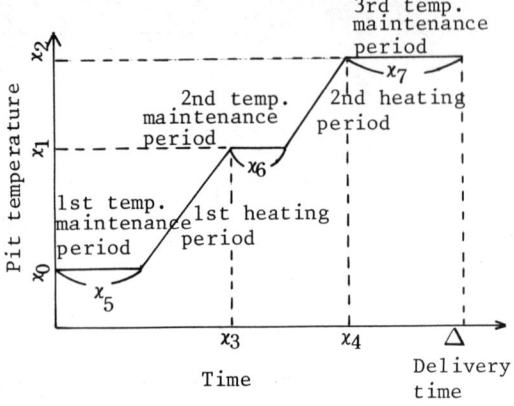

Fig. 8 Mathematical expression of heat pattern

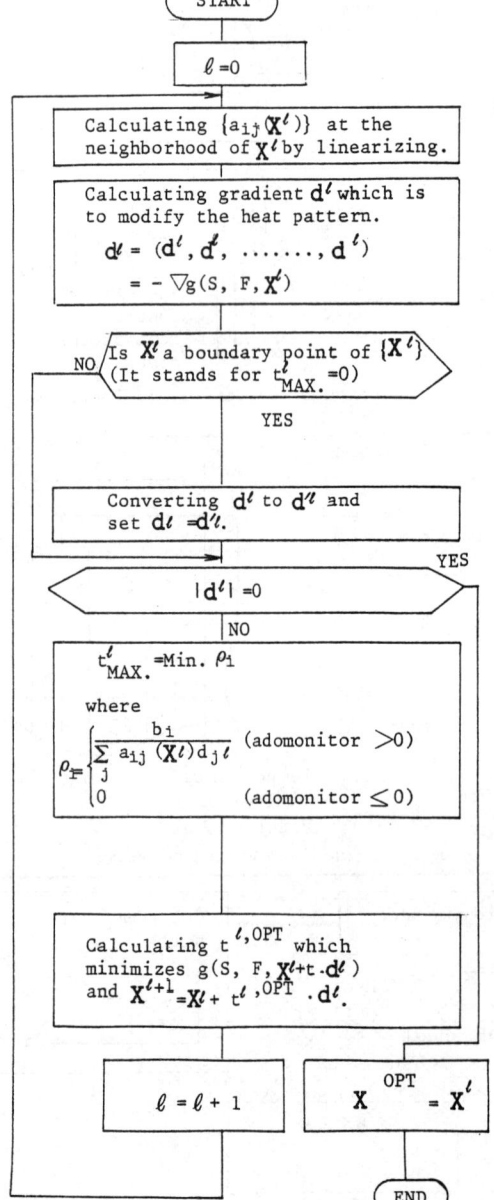

Appendix of flow chart

$I\{X\} = \{X \mid$ a set of X satisfying Eqs.(22)$\sim$(30)$\}$

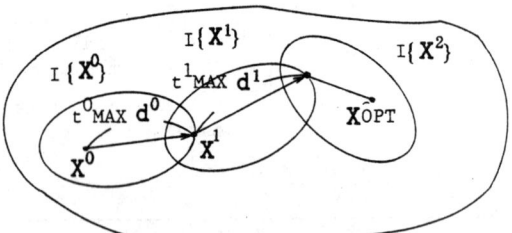

where $t^\ell_{MAX}\cdot d^\ell$; the longest vector $t\cdot d^\ell$ in $I\{X^\ell\}$

The algorithm at the boundary point.

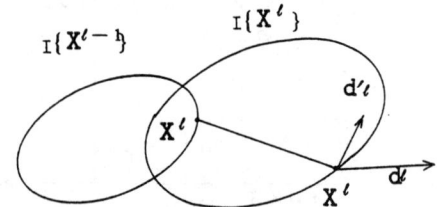

If the direction of d^ℓ is the outside of $I\{X^\ell\}$, we convert the vector d^ℓ to d'^ℓ follows;
for i which $\sum_j a_{ij}(X^\ell)d^\ell_j = 0$ is satisfied

$$d'^\ell_j \leftarrow \begin{cases} d^\ell_j : a_{ij}(X^\ell)d^\ell_j \leq 0 \\ 0 : a_{ij}(X^\ell)d^\ell_j > 0. \end{cases}$$

that is,

$g(S, F, X^\ell + d'^\ell)$
$= g(S, F, X^\ell) + \sum_j \frac{\partial g(S, F, X^\ell)}{\partial x_j} \cdot d'^\ell_j + \cdots\cdots$
$= g(S, F, X^\ell) - \sum_j d^\ell_j \cdot d'^\ell_j + \cdots\cdots$
$\leq g(S, F, X^\ell)$
    if $|d'^\ell|$ is sufficiently small

Fig. 9 Flow chart of algorithm

Computer Combustion Control at the Soaking Pit

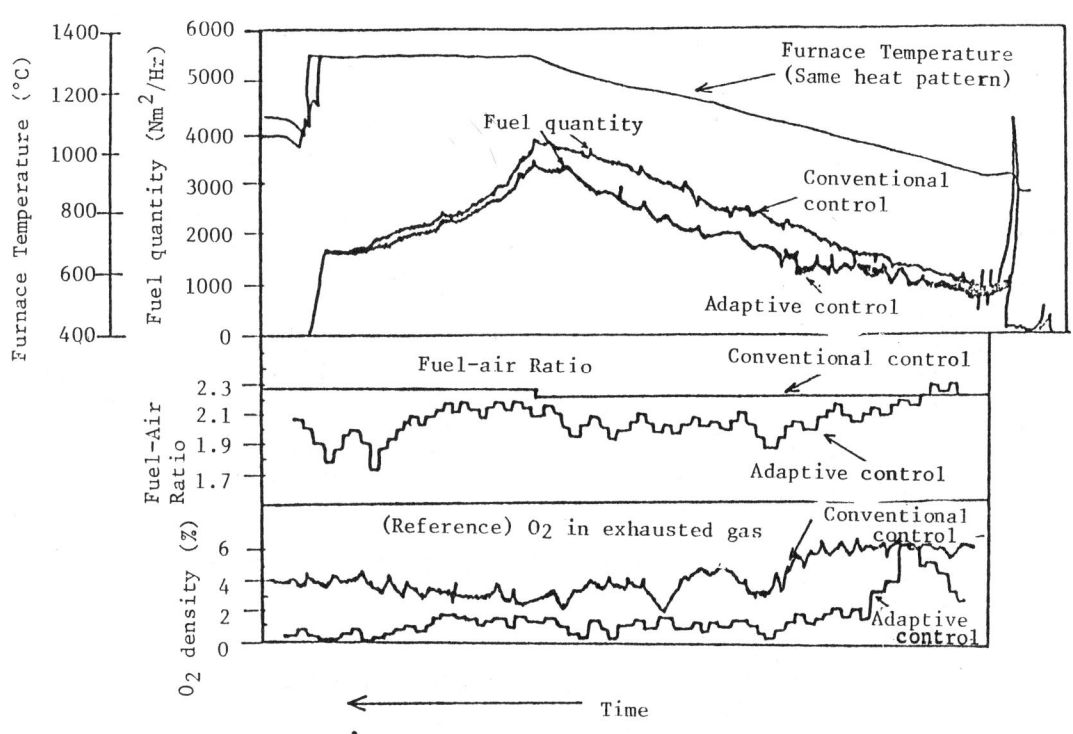

Fig. 10 Comparison of control Ability between Conventional Control and Adaptive Control

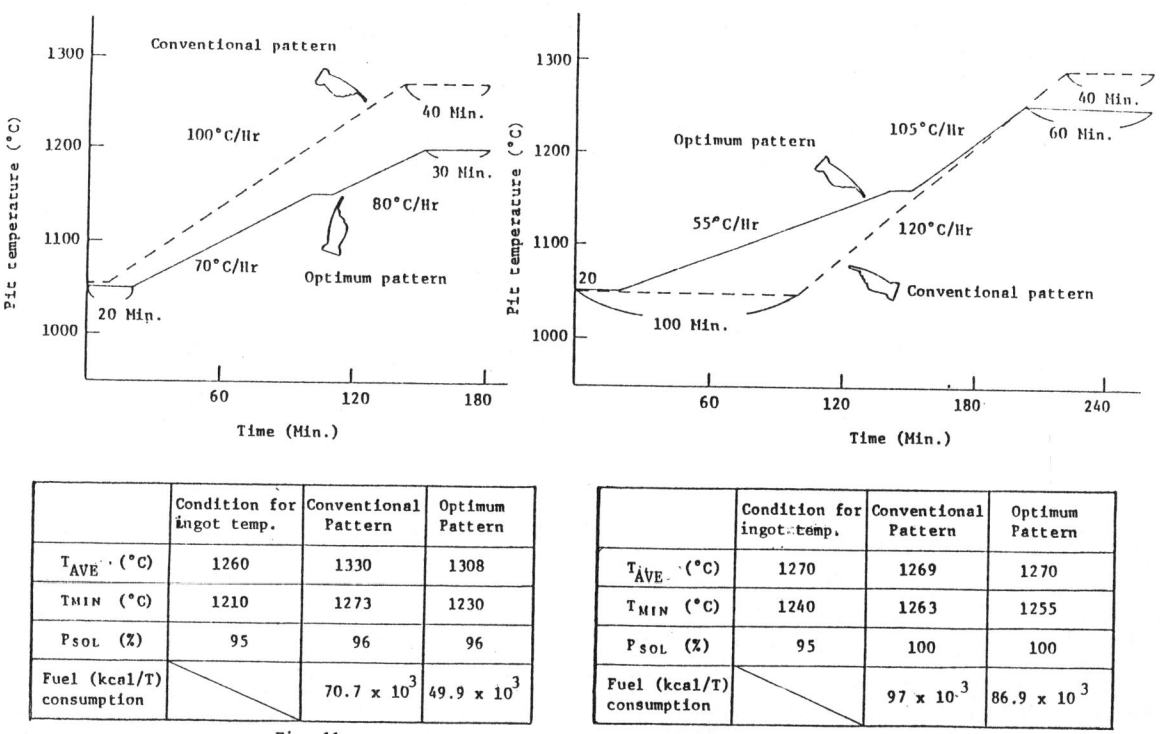

	Condition for ingot temp.	Conventional Pattern	Optimum Pattern
T_{AVE} (°C)	1260	1330	1308
T_{MIN} (°C)	1210	1273	1230
P_{SOL} (%)	95	96	96
Fuel (kcal/T) consumption		70.7×10^3	49.9×10^3

Fig. 11

	Condition for ingot temp.	Conventional Pattern	Optimum Pattern
T_{AVE} (°C)	1270	1269	1270
T_{MIN} (°C)	1240	1263	1255
P_{SOL} (%)	95	100	100
Fuel (kcal/T) consumption		97×10^3	86.9×10^3

Fig. 12

Fig. 11, Fig. 12, Applied Result on rimmed Steel at Wakayama

THE DISTRIBUTED CONTROL SYSTEM OF COIL ANNEALING FURNACES

Y. Nariai*, I. Yamazaki*, M. Ono**, T. Makino**, S. Miki*, and R. Michioka*

*Kashima Steel Works, Sumitomo Metal Industries Ltd., Ibaraki, 314, Japan
**Central Research Laboratories, Sumitomo Metal Industries Ltd., Osaka, 660, Japan

Abstract. This paper introduces a distributed micro-computer system applied to the combustion control of batch-type annealing furnaces for cold strip coils. This system consists of 21 micro-computers : 4 supervisory, 2 interface, and 15 DDC controler. Each DDC computer has an advanced mathematical model which optimizes heat-up by predicting the internal temperatures of piled coils.

By means of this system, fuel gas consumption has been reduced and the capability of process supervision and reliability have improved.

Keywords. microprocessors, combustion, control, annealing furnace, distributed system, cold spot, energy saving, hierarchical systems.

INTRODUCTION

In the cold strip mills in the iron and steel industries, there are many batch-type annealing furnaces for tight coils. The purpose of annealing is to remove the residual stress of strip produced during coil rolling and to give good forming property to coils. Consequently, it is a very important process in the production of cold coils.

A typical single stack bell-type annealing furnace is shown in Fig. 1. Coils are heated indirectly from outside the inner cover by the burners. Inert gas, which is circulated with the base fan, activates the heat transmission to the coils.

For annealing coils, it is necessary to heat all parts of the coils to a prescribed temperature. The coils are heated by radiation, conduction, and by the convection of the inert gas. However, the internal temperature cannot be measured during operation, so the conventional control method is to raise the temperature of the inert gas and the surface of the coils to the prescribed value or higher until the cold spot temperature goes up to the desired value. At the same time, the temperature of the surface of the coils is kept lower than the upper limit.

In the Kashima Steel Works, an analog-type combustion control system based on the conventional control method had been installed for 31 furnaces (Single-stack Coil AF : 16, Multi-stack Coil AF : 15). In this system, however, there were a few problems as follows :

(1) Inefficient consumption of fuel gas in the heat-up control of the cold spot.

(2) Shortage or excess of soaking because of inexact determination of the cold spot temperature.

(3) Too many troubles in the control equipment such as contact relays, analog-controlers, etc.

To cope with these problems, two countermeasures were developed and applied at the Kashima Steel Works. In order to save fuel gas (coke oven gas), a new mathematical control model was introduced. A microcomputer, using the mathematical model, calculates the internal temperatures utilizing the heat transfer equation in unsteady state, and it controls the increasing rate of the cold spot temperature in accordance with an optimally pre-set pattern which is effective in saving fuel gas. This also allows exact determination of when soaking is finished.

To deal with the hardware problems, a distributed control system which consists of 21 micro-computers was applied to improve reliability and maintainability.

MICRO-COMPUTERS AND ANNEALING CONTROL SYSTEM

Configuration

Fig. 2 illustrates the new annealing control system in use at the Kashima Steel Works, which is a hierarchical system that includes distributed subsystems.

The process control level is in charge of controling combustion and monitoring the cooling stage, and it consists of 15 DDC controllers. There are 4 controllers for the SCA, and 10 for the MCA. The extra one is a spare for the SCA or the MCA.

The essence the supervisory control level is the operator's console, and operations are centralized on the two colored graphic CRT displays. The CRT displays make the supervision of control easier.

Basic information needed for annealing control, such as coil dimensions and specifications of heat treatment, are sent from the on-line production control system to the supervisory controller ①. These data are not only transmitted to each DDC controller, but are also stored in the mini drums. Furthermore, the results of an annealing operation are reported to the on-line system.

The process data gathered by DDC controllers are stored in the mini durms which form data base of this system. Data from annealing operations performed within the last 7 days can be monitored on the CRT at any time. An example of a CRT display is shown in Fig. 3.

The process data gathered during an annealing cycle (from ignition to completion of cooling) are preserved by copying both the characters and the graphic images on the CRT onto a hard copy.

In this system, if a problem occurs in one controler it does not affect all the other controlers. A disorder occurring in a controller or in a peripheral device is detected when another controller tries to communicate with the malfunctioning controller and receives an invalid response or no response at all. These response cause an alarm to be sent to an operator, and the malfunctioning controller is isolated from the system and repaired.

The general specifications of a DDC controller are described in Table 1, and the function of each controller in Table 2 and Fig. 4.

Calculation of Internal Temperature of Coils

Differential equation (1) describes unsteady heat transmission in an isotropic and homogenious cylindrical body assuming no tangential temperature gradient.

$$c\rho \frac{\partial T}{\partial t} = \lambda \Delta^2 T$$

$$= \lambda_r \frac{\partial^2 T}{\partial r^2} + \lambda_r \frac{1}{r} \frac{\partial T}{\partial r} + \lambda_z \frac{\partial^2 T}{\partial z^2} \quad (1)$$

where
- T = temperature
- t = time
- ρ = density
- c = specific heat
- r = radius of cylinder
- z = height of cylinder
- λ_r, λ_z = thermal conductivity of r, z

To calculate the temerature in the coil with a digital computer, equation (1) must be expressed in a finite difference. Hence, a cross section of a coil is divided as shown in Fig. 5 and the temperature of the area around each point represented by a small circle is calculated.

In this way, equation (1) can be rewritten as equation (2).

$$c\rho \frac{T_{i,j}(t + \Delta t) - T_{i,j}(t)}{\Delta t} =$$

$$\lambda_r \frac{T_{i+1,j}(t) + T_{i-1,j}(t) - 2T_{i,j}(t)}{\Delta r^2}$$

$$+ \frac{\lambda_r}{r_i} \frac{T_{i+1,j}(t) - T_{i,j}(t)}{\Delta r}$$

$$+ \lambda_z \frac{T_{i,j+1}(t) + T_{i,j-1}(t) - 2T_{i,j}(t)}{\Delta z^2} \quad (2)$$

where
- $T_{i,j}$: temperature at (i,j)
- r_i : r-coordinate at (i,j)

In the DDC controlers, it was decided that m = 7 and n = 5, due to the limitations of the arithmetic capability of the micro-computers.

Specific heat and thermal conductivity vary with temperature, as shown in Fig. 6 and Fig. 7. As to thermal conductivity, λ_r is determined to be 1/10 of λ_z based on the results of experiments.

In evaluating the heat-up process in coils, heat transmission in the convector plate has not been considered in many previous studies. But absorption of heat energy through the edges is the dominant heating factor in the heat-up of coils and the heat flow from the inert gas to the coils through the convector plates influences precision of simulation, particulary in the starting period of combustion. Consequently, this model calculates the temperatures not only in the coils but also in the convector plates. Regarding the convector plates, the temperature gradient in r-direction and z-direction is ignored. Fig. 8 shows the effect of the calculation of the temperatures of the convector plates.

Fig. 9 illustrates the results of the calculation, which shows good agreement with the measured value by thermo-couples.

The real-time calculation of the cold spot temperature have made it possible for an operator to know when soaking is completed. The cold spot temperature is displayed on the CRT in real-time, and shortage or excess of annealing can be prevented. This contributes to the reduction of fuel gas consumption and the reinforcement of quality control. Moreover, prediction of cold spot temperature is helpful in the scheduling of the rotation of outer covers and cooling covers.

Combustion Control Model

Fig. 10 shows the conventional control method. Furnace temperature is kept constant below the value at which coils will be damaged by overheating, and heat absorption of coils is accelerated. After the base temperature reaches the prescribed value and the soaking period begins, it is kept constant instead of the furnace temperature.

In this method, fuel gas consumption decreases slowly with time and then suddenly decreases when the coils enter into the soaking period. Thus, since the greatest amount of fuel is used in the starting period, the beginning of the annealing process offers the greatest potential for saving fuel.

In addition, a great deal of heat is lost in the exhaust gas, as shown in Fig. 11. Another way to save fuel gas is to reduce this heat loss.

It has been said that heating slowly by keeping the furnace temperature lower than the conventional value leads to an improvement in heat efficiency, but it has not been used because the heating time gets longer and the throughput goes down.

But many numerical analysis and experiments have made it clear that the heating time does not increase if the furnace temperature is raised to the conventional value at the proper time. This is because the heat loss in exhaust gas increases in proportion to fuel gas consumption, while an increase of furnace temperature or an increase of fuel gas supply does not contribute much to the amount of absorbed heat energy when the furnace temperature is relatively high.

Based on these experimental results, a new control method was developed. The main points are as follows.

(1) The heat-up speed of the cold spot is determined and controled so that the combustion control will finish approximately at the same time as in the conventional method.
(2) The cold spot is heated up to the prescribed temperature, keeping the furnace temperature below the tolerable limit in the starting period of annealing.

To realize (1) and (2) requires control of the heat-up of the cold spot temperature. As shown in Fig. 12, the heat-up rate of the cold spot or (ta,Ta) is determined by the dimensions of coils, specifications of heat treatment, and throughput of annealing. No other conditions need to be considered in this control. Energy saving and annealing throughput can be adjusted by changing ta.

Methematical Model

In this new control model, the relationship between the furnace temperature and the cold spot temperature is approximated by the differential equation (3).

$$a \frac{\partial^2 T_C}{\partial t^2} + b \frac{\partial T_C}{\partial t} + T_C = T_{in}(t - t_L) \quad (3)$$

where
- a,b : constant
- T_C : cold spot temperature
- t_L : lag time
- T_{in} : inside wall temperature of inner cover

However, the temperature of the inner cover cannot be measured during operation, so it is substituted by the modified furnace temperature T_f. Then equation (3) is rewritten as (4).

$$a \frac{\partial^2 T_c}{\partial t^2} + b \frac{\partial T_c}{\partial t} + T_c = T_f \quad (4)$$

In order to calculate the equation with a digital computer, differential terms are expressed in a finate difference.

$$\frac{\partial^2 T_c}{\partial t^2} = \ddot{T}_c(t_i) = \frac{T_c(t_i+\Delta t) + T_c(t_i-\Delta t) - 2T_c(t_i)}{\Delta t^2} \quad (5)$$

$$\frac{\partial T_c}{\partial t} = \dot{T}_c(t_i) = \frac{T_c(t_i+\Delta t) - T_c(t_i-\Delta t)}{2\Delta t} \quad (6)$$

where
  t : time increment
  $T_c(t_i)$: cold spot temperature at time t_i

when $t_i = t_1, t_2$, equation (4) gives (7).

$$a\ddot{T}_c(t_i) + b\dot{T}_c(t_i) + T_c(t_i) = T_f(t_i) \quad (7)$$
$$(i = 1,2)$$

than a, b are calculated by soving linear equation (7).

In the starting period, the furnace temperature is controled at the optimal value which is determined by the algorithm shown in Fig. 13. This calculation is executed every 20 minutes till the cold spot is heated up to the prescribed temperature. In this alogorithm, ΔT_f is determined based on the calculated a, b using (7).

Fig. 14 compares the new control method with the conventional control method. Control of the heat-up of the cold spot leads to the capability of setting the furnace temperature lower than in the conventional method. An energy saving of about 9 percent is achieved under average annealing weights and coil numbers. Since the introduction of the system at the Kashima Steel Works, overall specific fuel consumption (kcal/ton) has decreased, as shown in Fig. 15.

Automatic Calibration of Analog Input Modules

In this system, temperature is the most important process input and analog input modules in the DDC controlers are equipped only for 213 CA thermo-couples. Zero and span of analog input modules can be automatically checked once a day by scanning the standard voltage, and the result is printed out.

This function has made periodical calibration unecessary, and contributes to a saving of labor time in hardware maintenance.

Real-time Control of Heat Cycle

Each DDC controler monitors the process from start of combustion to completion of cooling the coils for each base, and has the following functions :

(1) Notification of an operator on the CRT display when it is time for the next job :
  extinguishment of burners, exchange of inert gas, opening valves for water cooling etc.
(2) Validity check of combustion and cooling process.
(3) Real-time display of concumption amount of coke oven gas (Nm^3).
(4) Self diagnosis :
  on-line : RAM, ROM
  off-line : PI/O(DI,DO,AI), Arithmetic Operation Unit, MODEM Interface etc.

CONCLUSION

This new combustion control system controls the heat-up speed of the cold spot efficiently and thus decreases fuel gas consumption. Real-time calculation of the cold spot temperature informs the operator of the completion of soaking, and it is helpful in the reinforcement of quality control.

The distributed control system also brings about an improvement in the capability of process supervision and reliability, and it has contributed to the rationalization of the Kashima Steel Works.

REFERENCES

Mizikar, E.A., Bresky, N.P., and Veitch, R.A. (1972). An improved Method for Calculating Soak Times in Batch Annealing. Iron and Steel Engineer. July, 53-59.
Kunioka, K., Kurihara, K., and TADA, T. (1971). Heat Transmission Analysis of Tight Coils in Single-Stack Bell-type Annealing Furnace. Proceedings ICSTS, Supple. Trans. ISIJ. Vol. 11. 796-800.

Distributed Control System of Coil Annealing Furnaces

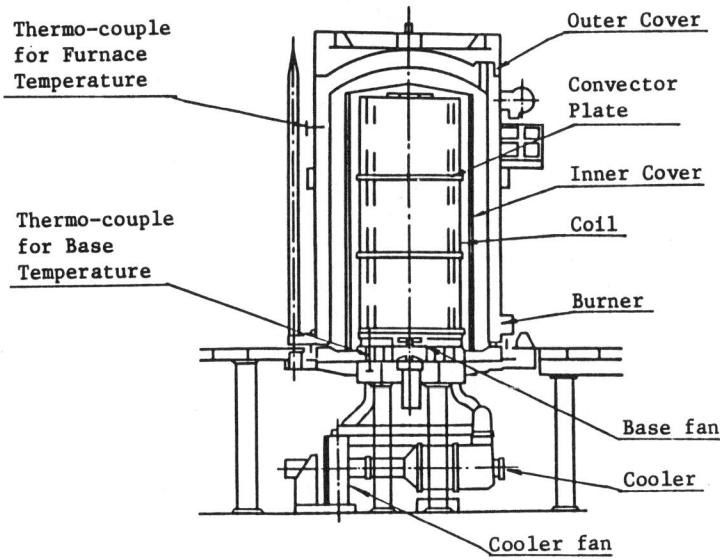

Fig. 1 Bell-type single-stack annealing furnace.

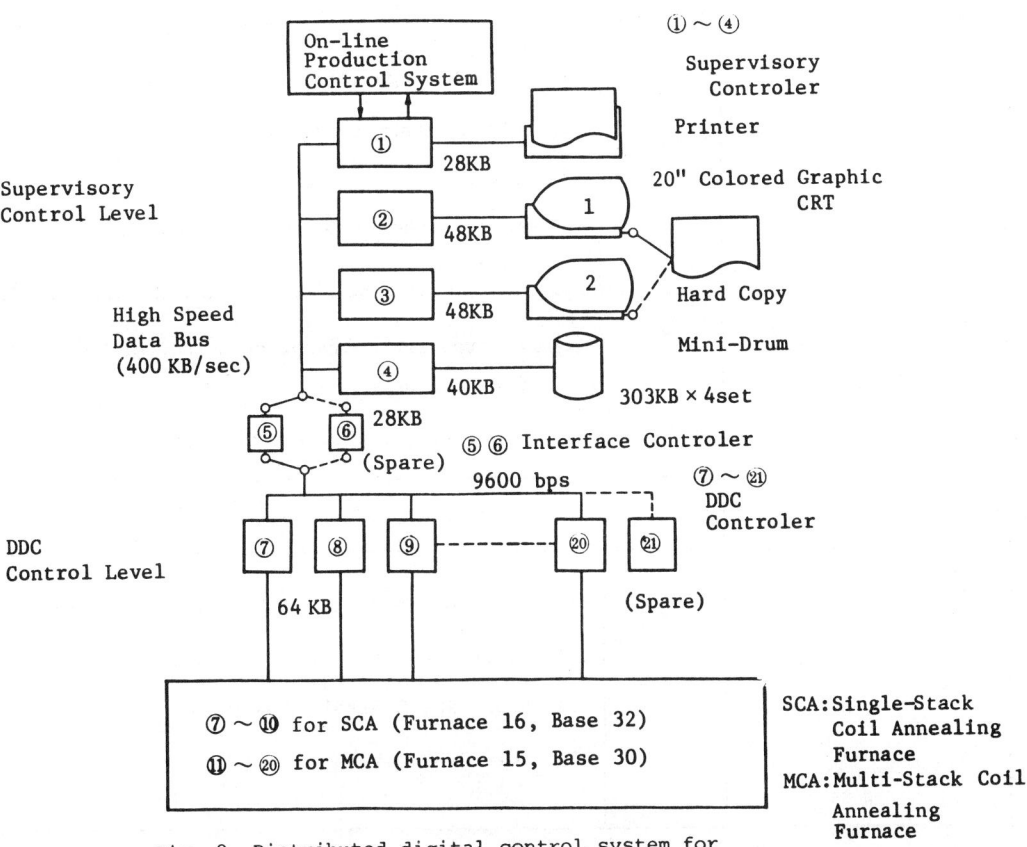

Fig. 2 Distributed digital control system for bell-type annealing furnaces.

TABLE 1 Specification of DDC Controler

Item	Contents
CPU	Intel 8085 (8bits)
Memory	ROM 40 KB
	RAM 24 KB
Process I/O	DI 32 points
	DO 48
	AI 24

TABLE 2 Functions of Supervisory & DDC Controlers

Level	Controler	Functions
Supervisory Control	①	(1) Data Linkage with On-line System (2) Alarm output to Printer (3) System Maintenance Console (Read/Write of Memory, diagnosis of Data Bus etc.)
	② ③	(1) Centralized Operator's console of DDC Controlers (2) Monitoring of Process Condition
	④	Control of Data Base (Mini Drum 1.2MB)
DDC Control	⑦～㉑	(1) Calculation of the Cold Spot Temperature (2) Combustion Control of SCA & MCA (3) Real-time Control of Heat Cycle (4) Self Diagnosis

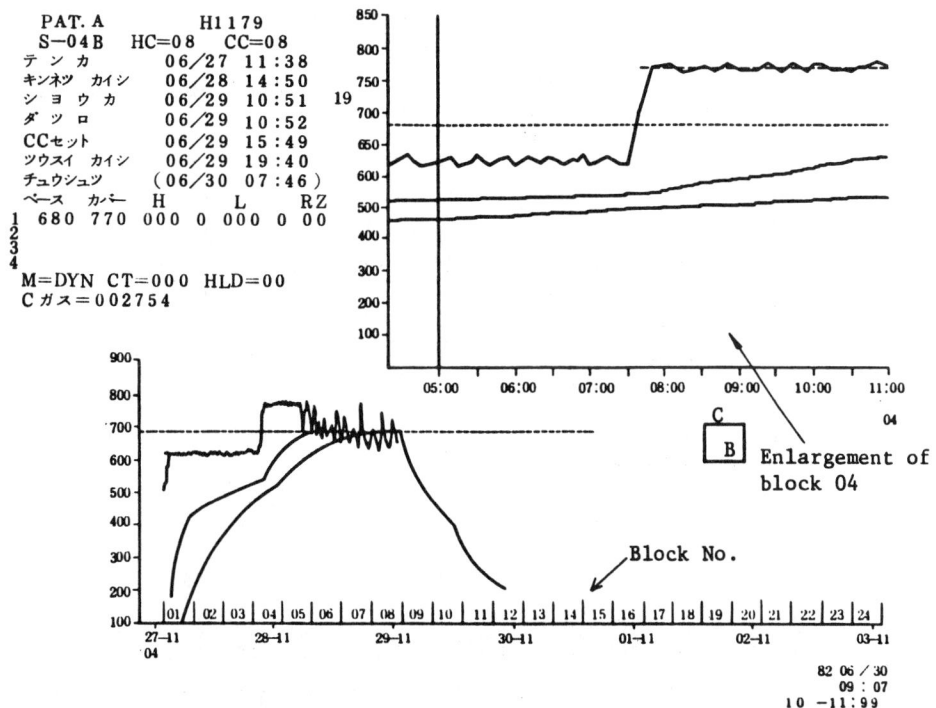

Fig. 3 Example of CRT display (hard copy).

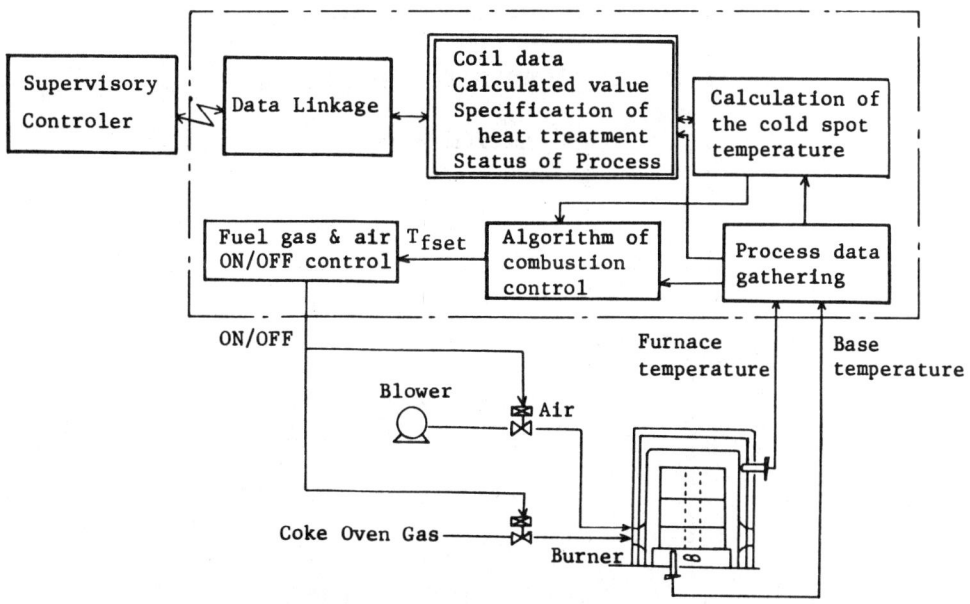

Fig. 4 Function of DDC controler.

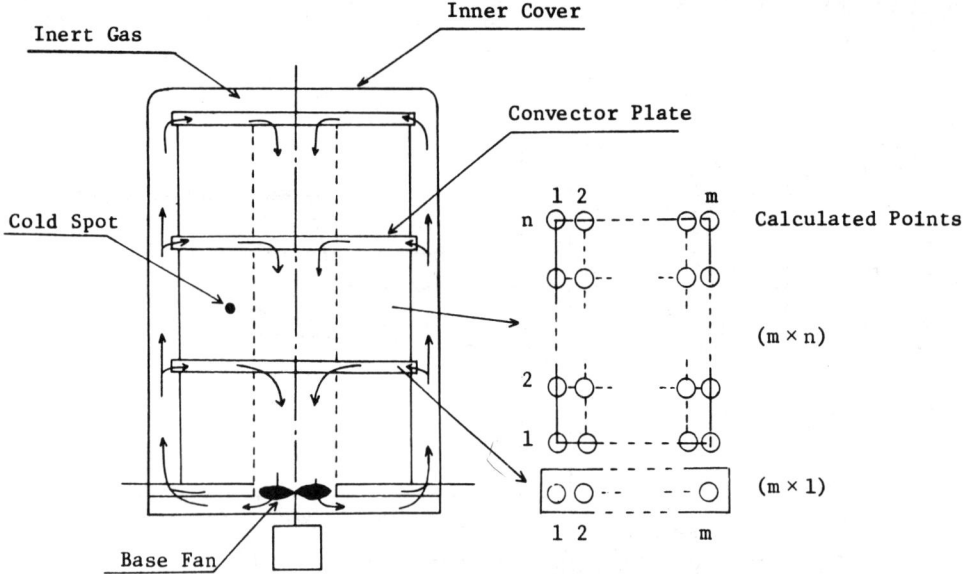

Fig. 5 Calculation of internal temperature.

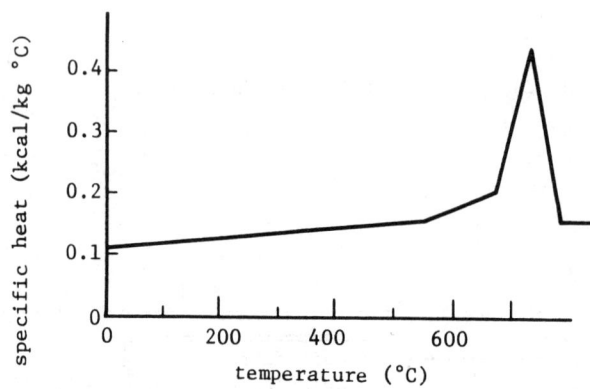

Fig. 6 Specific heat.

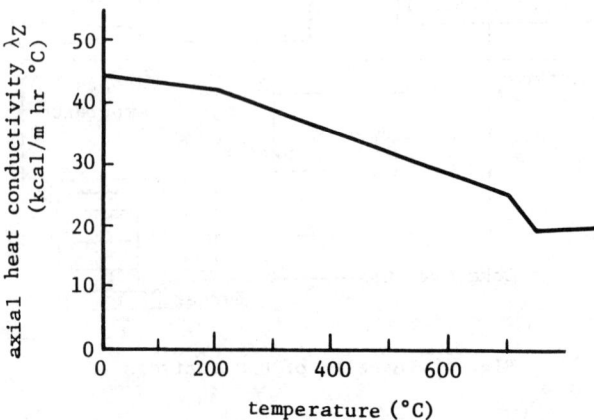

Fig. 7 Axial heat conductivity λ_Z.

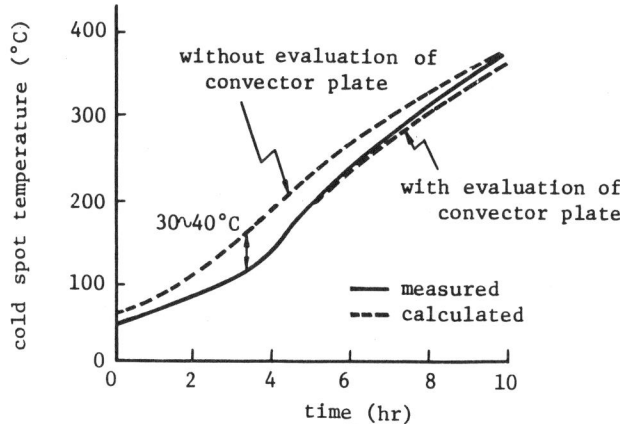

Fig. 8 Effect of evaluation of convector plate.

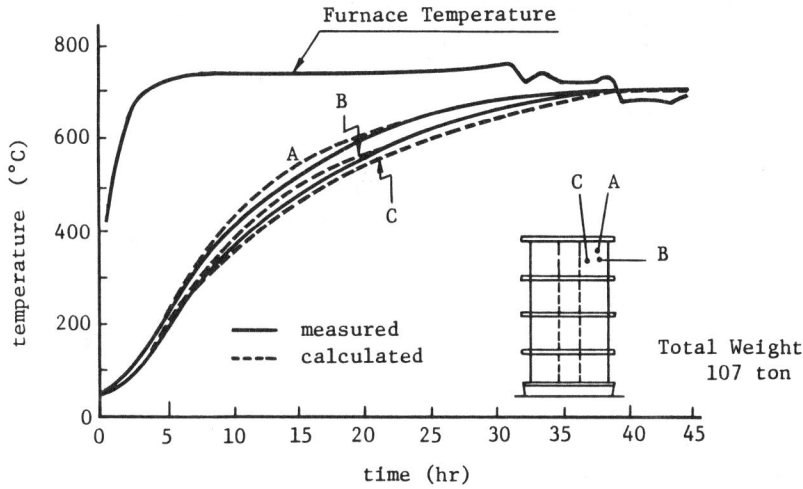

Fig. 9 Result of calculation of internal temperature.

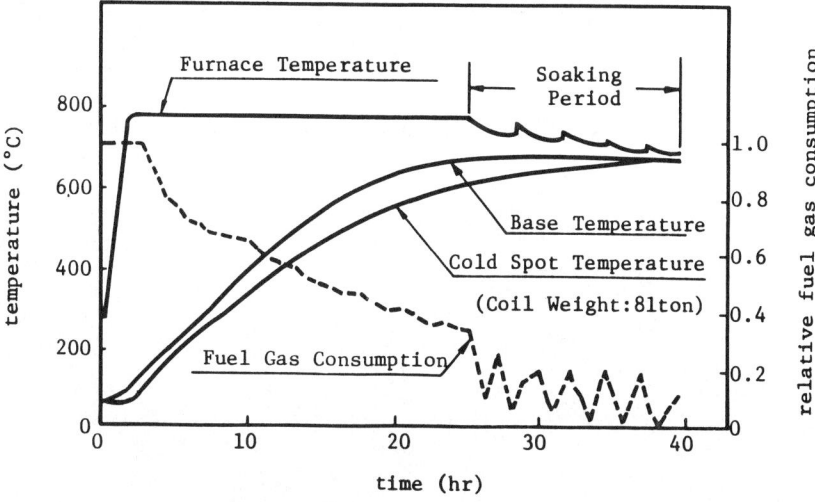

Fig. 10 Conventional combustion control.

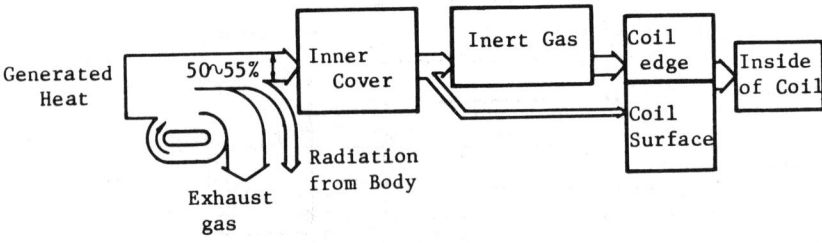

Fig. 11 Heat transmission in a furnace.

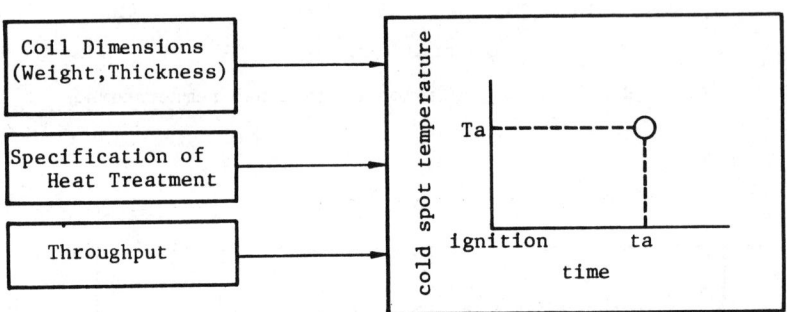

Fig. 12 Decision of heat-up speed of the cold spot temperature.

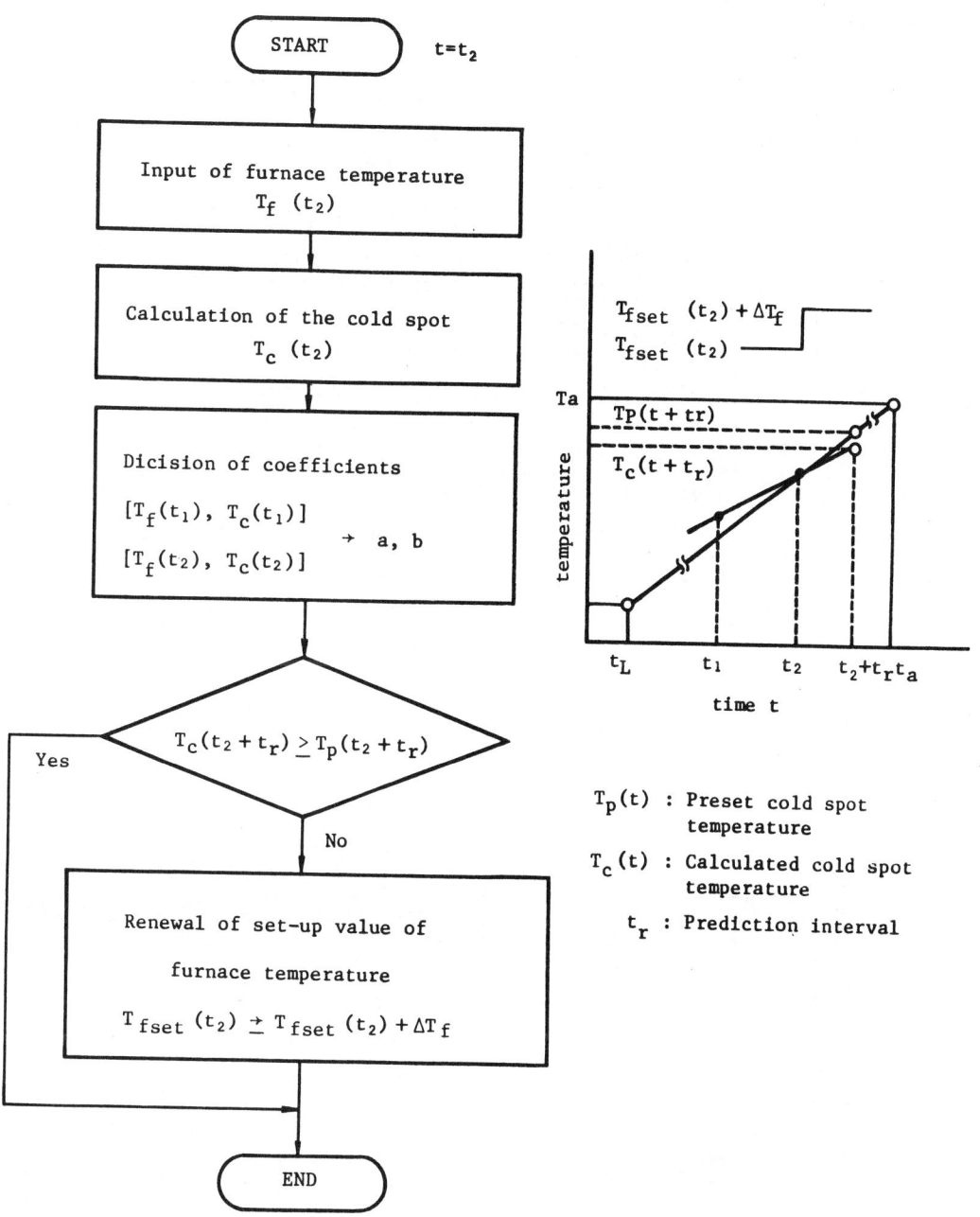

Fig. 13 Algorithm of combustion control.

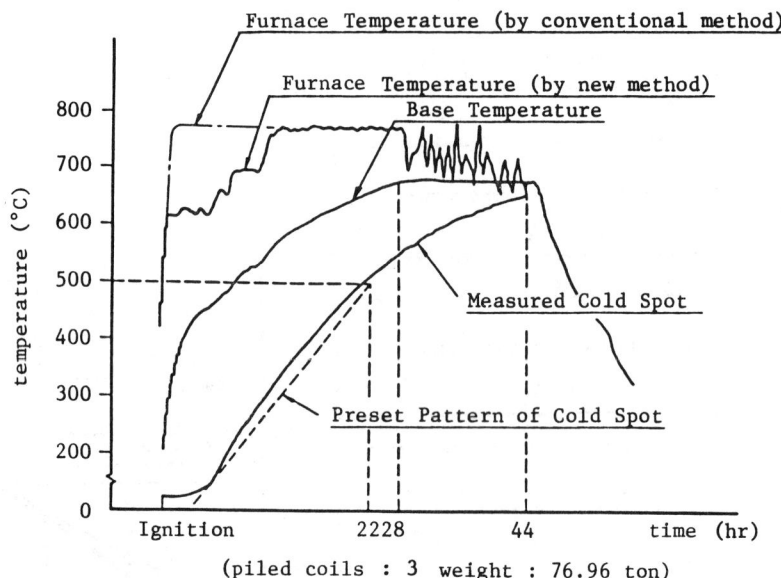

Fig. 14 Heat-up control of cold spot.

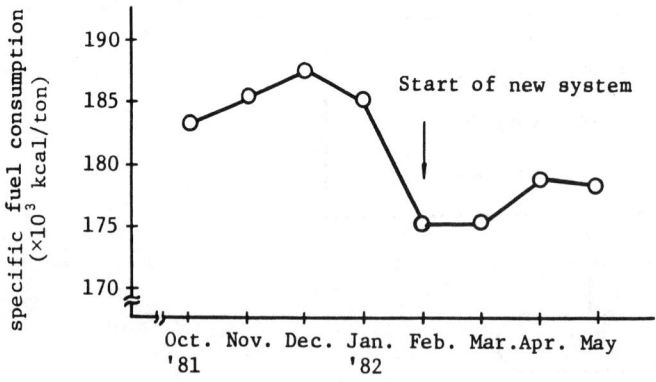

Fig. 15 Effect of new control method.

Copyright © IFAC Real Time Digital Control Applications
Guadalajara, Mexico 1983

COMBUSTION CONTROL SYSTEM OF A FURNACE FOR HOT BLOOM

S. Tanifuji*, Y. Morooka*, J. Kumayama**, K. Doi***, T. Kawasumi*** and T. Shinmura***

*Hitachi Research Laboratory, Hitachi Ltd., Hitachi-shi, Ibaraki-ken, Japan
**Ohmika Works, Hitachi Ltd., Hitachi-shi, Ibaraki-ken, Japan
***Kobe Works, Kobe Steel Ltd., Nada-ku, Kobe-shi, Japan

Abstract. A computer control system has been developed to save energy in the operation of multi-zone furnace, which heats hot continuously casting blooms. The system determines an optimal temperature trajectory of the bloom so as to minimize total fuel consumption under the actual tracking schedule. The total fuel consumption is predicted by using a heat exchange model. The system regulates the zone temperature so that the bloom temperature coincides with the trajectory. From the actual operation with this system it is found that the temperature deviation of the extracted blooms is less than 15 °C from the desired value and that the avarage fuel consumption is reduced by 10 % from that of manual operation.

Keywards. Computer control; reheating furnace; energy savings; optimization; linear programming.

INTRODUCTION

A Digital computer has been widely used in a combustion control of a reheating furnace for slabs, in order to save energy and improve a control accuracy of extracted bloom temperature. The slabs are mostly heated so as to follow along a predetermined trajectory corresponding to a typical tracking schedule of the slab. However, the schedule is often disturbed by unexpected troubles of rolling mills and alternations of production plan. Therefore, the predetermined trajectory is not always adequate under the actual operations.

Recently, another control algorithm has been proposed. The distribution of the zone temperature is controlled so as to lower the temperature of the charging zone or to raise the temperature of the discharging zone. Although this idea is based upon experiences of manual operation, the fuel consumption is not evaluated directly. Therefore it is not clear whether the optimal combustion is realized under various conditions of the furnace.

This paper proposes a new optimization algorithm to minimize total fuel consumption. With this algorithm, an optimal temperature trajectory of the bloom can be determined through online calculation under the actual tracking conditions. The computer control system with this algorithm has been successfully used at Kobe Works of Kobe Steel, Ltd.. In the first, section a process model of the furnace is explained and the optimization algorithm is presented in the second section with the use of the process model. Final section is devoted to the results of the actual operation of this computer controlled system.

MATHEMATICAL MODEL OF THE FURNACE

Profile of The Furnace

Figure 1 shows a profile of a multi-zone furnace for continuously casting bloom at Kobe Works. Since the furnace is equipped with independent walking beams at each zone, a group of hot charged blooms can be transported by some of the walking beams towards the discharging zone with high thermal efficiency, without changing the positions of other blooms. High-temperature exhaust gas, produced by burning fuel at each zone, flows toward a chimney in the charging zone.

Process Model

The process model is composed of the heat exchange model and the bloom temperature model.

Heat exchange model. Gas flow and heat destribution in the furnace are very complicated and change with time. For the online use of the process model, we assume that the temperature in each zone is uniform, since the heat unbalance in each zone can be considered to be temporal and can be smoothed out within the response time of the zone temperature. A heat transfer in the I-th zone can be described in a heat balance model by

$$Q_F + Q_A + Q_{IN} = Q_S + Q_L + Q_{OUT} \quad (1)$$

See Fig. 2 in detail.
Here
$$Q_F = v_I \cdot H_f \quad (2)$$
: heat content of the fuel,
$$Q_A = v_I A_0 A_R (T_{PRE} - T_{RM}) C_A \quad (3)$$
: sensitive heat of preheating air,
$$Q_{IN} = V_G (\xi_{I+1} T_{I+1} - T_{RM}) C_G \quad (4)$$
: sensitive heat of the exhaust gas flowing-in,
$$Q_{OUT} = (V_G + v_I (G_0 - A_0 + A_0 A_R)) \cdot (\xi_I T_I - T_{RM}) C_G \quad (5)$$
: sensitive heat of the gas flowing-out,
$$Q_L = \eta_1 T_I + \eta_2 \quad (6)$$
: heat-loss escaping from skids of this zone,
Q_S : sum of heat flow into the blooms in the i-th zone,
$$V_G = \sum_{K=i+1}^{N} v_K (t + D_{KI})$$
: volume of the exaust gas,
N : the index number of the discharging zone
v_I : fuel flow-rate of the I-th zone,
T_I : zone temperature of the I-th zone,
T_{RM} : room temperature,
T_{PRE} : temperature of the preheated air,
H_f, A_0, G_0 : calolific value, required volume of air and volume of the generated gas per unit fuel flow-rate, respectively,
C_A, C_G : specific heats of the air and the gas, respectively,
A_R : adjustable parameter of air flow-rate,
D_{KI} : delay time of the influence of the combustion in the K-th zone to reach the I-th zone.

The fuel flow-rate v_I is derived from eqs. (1) to (6).

$$v_I = \frac{Q_S + Q_L}{\psi_1 - \psi_2 T_I} - \frac{\psi_3 T_{I+1} - \psi_2 T_I}{\psi_1 - \psi_2 T_I} V_G \quad (7)$$

Since ψ_1, ψ_2 and ψ_3 are nearly constant, the value for v_I can be obtained if Q_S, Q_L, T_I, T_{I+1} and V_G are known. The estimation of values for Q_S and Q_L will be explained later. The quantities T_{I+1} and V_G represent the influence of the combustion in other zones.

Bloom temperature model. The heat conduction in the direction of the length of the bloom is negligibly small compared with those in the direction of thickness and width. Then two-dimensional heat conduction equation can be applied.

$$c(\theta) \rho \frac{\partial \theta}{\partial t} = \kappa(\theta) \left(\frac{\partial^2 \theta}{\partial x^2} + \frac{\partial^2 \theta}{\partial y^2} \right) \quad (8)$$

The quantity q_U is the heat flow density at the top and the bottom of the bloom and q_S is the heat density at the surface:

$$q_U = \kappa(\theta) \frac{\partial \theta}{\partial x} = \phi \left(\left(\frac{T+273}{100} \right)^4 - \left(\frac{\theta_S + 273}{100} \right)^4 \right) \quad (9)$$

$$q_S = \kappa(\theta) \frac{\partial \theta}{\partial y} = \phi \left(\frac{1}{2}(1-\cos Z_1) \left(\frac{T_U+273}{100} \right)^4 \right.$$
$$+ \frac{1}{2}(1-\cos Z_2) \left(\frac{T_L+273}{100} \right)^4$$
$$+ \frac{1}{2}(\cos Z_1 + \cos Z_2) \left(\frac{\bar{\theta}_N+273}{100} \right)^4$$
$$\left. - \left(\frac{\theta_S+273}{100} \right)^4 \right) \quad (10)$$

Here, (x,y) : position coordinates in the bloom,
θ : temperature profile of the bloom at (x,y),
θ_S : surface temperature of the bloom,
$c(\theta), \kappa(\theta), \rho$: specific heat, thermal conductivity and density of the bloom, respectively,
T_U, T_L : temperature of the upper part and lower part of the zone, respectively,
$T = (T_U \text{ or } T_L)$, $\bar{\theta}_N$: surface temperature of an adjacent bloom,
ϕ : effective emissivity,
Z_1, Z_2 : angles shown in Fig. 3.

The above equations can be solved numerically, once the value for the zone temperature is given, whether it is a measured value or an assumed value. An average temperature $\bar{\theta}$ and a temperature difference $\Delta \theta$ between the surface and the center of the bloom can be easily determined, and $\bar{\theta}$ and $\Delta \theta$ are used to judge the heating condition of the bloom. In addition, the heat flow Q_S defined in the heat exchange model can be calculated from q_U and q_S.

OPTIMAL COMBUSTION PROBLEM

Formulation of The Optimization

The optimization in the present paper in defined by the following.
(i) Minimization of the total fuel consumption not for an individual bloom but for a group of the blooms in a period τ from the present time to the extraction.
(ii) Minimization of the deviation of the extracted bloom temperature from the desired temperature.

In order to satisfy the above conditions, the zone temperature and the fuel flow-rate must be within allowable values. The fuel flow-rate and the bloom temperature are parameters which represent future state of the furnace. They can be estimated by using the process model, if any heat schedule is given. We define a group heated in the M-th zone as 'group M' and devide the period into n intervals. The heat schedule of the group M is defined as (T(1), T(2), ····,

T(n)), where T(i) indicates the zone temperature of the i-th time interval.

The condition (i) on the total fuel consumption J_M can be represented as

$$J_M = \sum_{i=1}^{n} v(i) \Delta\tau_i \longrightarrow \min \qquad (11)$$

where v(i) is the fuel flow-rate in a zone to which the group M belongs at the i-th time interval, and $\Delta\tau_i$ is the period of the i-th interval. The other restrictions are defined by the following inequalities,

$$\bar{\theta}^L \leq \bar{\theta}_{OUT} \leq \bar{\theta}^U \qquad (12)$$

$$\Delta\theta^L \leq \Delta\theta_{OUT} \leq \Delta\theta^U \qquad (13)$$

$$0 \leq v(i) \leq v^U \quad (i = 1,n) \qquad (14)$$

$$T^L \leq T(i) \leq T^U \quad (i = 1,n) \qquad (15)$$

$$\delta^L \leq (T(i+1)-T(i))/\Delta\tau_{i+1} \leq \delta^U$$
$$(i = 1, n-1) \qquad (16)$$

where the superscripts L and U indicate the lower limit and the upper limit, respectively. The quantities $\bar{\theta}_{OUT}$ and $\Delta\theta_{OUT}$ represent the mean temperature and the temperature difference at the time of extraction, respectively. Equation (16) represents the restriction on the changing rate of the temperature.

The optimum heat pattern can be obtained from satisfying the inequalites (12)-(16) and from making the total fuel J_M minimum. However it is not practical to solve this problem in online procedure because of the following reasons.
(i) The nonlinearity in the process model forces to repeat complex calculations until the solution converges in a numerical analysis. The repetition usually takes quite a long time resulting in delay of the control.
(ii) Since the equations to obtain optimization conditions for a group of blooms, all the groups must be considered simultaneouly. This makes it very difficult to perform in a short time.
To overcome these difficulties, a new algorithm for the online calculation is developed.

Algorithm for Online Optimization

Order of the optimization. Since exhaust gas flows toward the charging zone only, the group of blooms at the discharging zone is not affected by the combustion in other zones. We can solve the optimization problem of group of blooms at the discharging zone independently of other groups. This gives the optimal heat pattern as well as the fuel flow-rates at the discharging zone. It is assumed that the group in the adjacent zone is affected by the determined fuel flow-rates from the combustion of the discharging zone. Then the optimization problem of the adjacent group can be solved. In this way, all groups can be optimized easily.

Linear representation of the problem. In order to avoide the repetition in solving the nonlinear problem, we linearize $\bar{\theta}_{OUT}$, $\Delta\theta_{OUT}$ and v(i) in the neighborhood of a particular heat pattern ($T_0(1)$, $T_0(2)$, ..., $T_0(n)$) shown in Fig.4 (This pattern is discussed at the end of this section).

$$\bar{\theta}_{OUT} = \bar{\theta}_{OUT,0} + \sum_{k=1}^{n} \alpha_k x_k \qquad (17)$$

$$\Delta\theta_{OUT} = \Delta\theta_{OUT,0} + \sum_{k=1}^{n} \beta_k x_k \qquad (18)$$

$$v(i) = v_0(i) + \sum_{k=1}^{n} \gamma_k(i) x_k \quad (i = 1,n) \quad (19)$$

where $\bar{\theta}_{OUT,0}$, $\Delta\theta_{OUT,0}$ and $v_0(i)$ are the values of $\bar{\theta}_{OUT}$, $\Delta\theta_{OUT}$ and v(i), respectively, when the bloom is heated under this particular heat pattern, and x_k is the small change of the zone temperature in the k-th time interval. The quantities α_k, β_k and $\gamma_k(i)$ are partial derivatives of $\bar{\theta}_{OUT}$, $\Delta\theta_{OUT}$ and v(i) with respect to x_k. The values of $\bar{\theta}_{OUT,0}$, $\Delta\theta_{OUT,0}$ and $v_0(i)$ can be calculated by applying the heat pattern $\{T_0(i)\}$ to the process model. The temprature change in the k-th time interval causes the changes not only of the bloom temperature at the extraction but also of the fuel flow-rates in other time intervals. The simulation under new heat pattern ($T_0(1)$, ..., $T_0(I) + \delta x$, ..., $T_0(n)$) gives new values $\bar{\theta}_{OUT,k}$, $\Delta\theta_{OUT,k}$ and $v_k(i)$ (i=1,n), where δx represents the small change parameter in the k-th interval. The derivatives α_k, β_k and $\gamma_k(i)$ can be determined as follows.

$$\alpha_k = (\bar{\theta}_{OUT,k} - \bar{\theta}_{OUT,0})/\delta x \qquad (20)$$

$$\beta_k = (\Delta\theta_{OUT,k} - \Delta\theta_{OUT,0})/\delta x \qquad (21)$$

$$\gamma_k(i) = (v_k(i) - v_0(i))/\delta x$$
$$(i = 1,n) \qquad (22)$$

The following linear equations can be obtained by substituting the eqs. (17)-(19) into (11)-(16).

$$J_M = c_0 + \mathbf{C X} \longrightarrow \min \qquad (23)$$

$$\mathbf{b}^L \leq \mathbf{H X} \leq \mathbf{b}^U \qquad (24)$$

where

$$X = (x_1\ x_2\ \cdots\ x_n)^T$$
$$C = (c_1\ c_2\ \cdots\ c_n)^T$$
$$b^L = (b_{L1}\ b_{L2}\ \cdots\ b_{Lm})^T$$
$$b^U = (b_{U1}\ b_{U2}\ \cdots\ b_{Um})^T$$

$$H = \begin{pmatrix} \alpha_1 & \alpha_2 & \alpha_3 & \cdots & \alpha n \\ \beta_1 & \beta_2 & \beta_3 & \cdots & \beta n \\ \gamma_1(1) & \gamma_2(1) & \gamma_3(1) & \cdots & \gamma n(1) \\ \vdots & \vdots & \vdots & & \vdots \\ \gamma_1(n) & \gamma_2(n) & \gamma_3(n) & \cdots & \gamma n(n) \\ 1 & 0 & 0 & \cdots & 0 \\ 0 & 1 & 0 & \cdots & 0 \\ \vdots & \vdots & \vdots & & \vdots \\ 0 & 0 & 0 & \cdots & 1 \\ -1 & -1 & 0 & \cdots & 0 \\ 0 & -1 & 1 & \cdots & 0 \\ \vdots & \vdots & \vdots & & \vdots \\ 0 & 0 & 0 & \cdots & 1 \end{pmatrix}$$

In above equations, m is the number of the inequalities, and c_i (i=1,n), b_{Lj} and b_{Uj} (j=1,m) are constants determined by using the simulation results and the lower and upper limits of eqs.(12)-(16). Equations (23) and (24) are regarded as a typical linear programming problem and their solution can be obtained by the simplex method in a short time. Optimal temperature $T_{opt}(i)$ (i=1,n) is equal to sum of $T_0(i)$ and $x_{opt,i}$ which is the optimal small change given by the simplex method. It has been confirmed that the total calculation time by the linearization technique is about one-tenth of that with nonlinear model.

It is necessary to choose the heat pattern $\{T_0(i)\}$ adequately in order to prevent approximation error by the linearization process.
(i) If the group M is heated in the M-th zone in the i-th time interval, currently measured temperature will be used as $T_0(i)$.
(ii) If not, we use the results of the optimization of the (M+1)-th group as the temperature $T_0(i)$. In other words the zone temperature of any zone into which the M-th group moves is assumed to be equal to that obtained beforehand by optimizing the (M+1)-th group.

CONTROL SYSTEM AND ITS EFFECTS

Computer Control System

A computer control System is shown in Fig.5. The combustion condition is controlled by a cotrol computer HIDIC 80E with 512K-byte main memory. Main functions of this system are as follows.

Prediction of tracking schedule.
Tracking schedule of the bloom is predicted from the schedule of the rolling mills and the casting machines. With the use of the predicted tracking schedule, the replacement of the optimal heat trajectory of each group is ordered in the following cases.
(i) When the difference between a newly predicted schedule and old ones exceeds a predetermined value.
(ii) When a group is tracked into the charging zone or is moved to an adjacent zone.
(iii) When the optimal trajectory does not change for 20 to 30 minutes.

Determination of the optimal temperature trajectory.
With this function optimal temperatures $\{T_{opt}(i)\}$ for every group is calculated at the first step, then the temperature trajectory is determined by applying $\{T_{opt}(i)\}$. The feedforward controller of this system adjusts the actual zone temperature so that the bloom temperature coincide with this trajectory.

Feedforward control.
The feedforward control consists of the following three steps.
Step 1 ; Estimation of the bloom temperature after a particular period D_T from the present time on the assumption that the zone temperature is kept at the current level (See Fig.6).
Step 2 ; Evaluation of the following cost function J' by using the results for all blooms in a certain zone of interest.

$$J' = \sum_j w_j(\bar{\theta}_{Pj} - \bar{\theta}_j) \quad (25)$$

where $\bar{\theta}_j$ is the average value of the estimated temperature of the bloom and $\bar{\theta}_{Pj}$ is the temperature on the trajectory shown in Fig.6. The quantity w_j is the weighting factor.
Step 3 ; Determination of the set value T'_S for the zone temperature. T'_S is given by

$$T'_S = T + c_T \cdot J' \quad (26)$$

where T is the measured value for the zone temperature and c_T is a given constant. The output value for T_S is used as an input to the PI controllers shown in Fig.1.

Results

Identification of the heat-loss model.
Total heat-loss Q_L of the furnace can be determined by using the following equation, derived from the total heat balance of the furnace.

$$\sum_T Q_L = (\sum_T \tilde{v}_I)(H_f + A_0 A_R (\tilde{T}_{PRE} - T_{RM})C_A$$
$$- (G_0 - A_0 + A_0 A_R)(\tilde{T}_{gas} - T_{RM})C_G)$$
$$- \sum_T Q_S \quad (27)$$

where T_{gas} : measured temperature of the exhaust gas in the chimney, shown in Fig.1,
T_{PRE} : measured temperature of the preheated air,
$\sum_{T} V_I$: sum of the fuel flow-rate measured at each zone,
$\sum_{T} Q_S$: sum of the heat flow into each bloom.

We estimate the heat-loss Q_L of each zone by distributing the total heat-loss in proportion to the number of the skid in each zone. The solid line in Fig.7 represents the estimated heat-loss Q_L and the dotted line represents the zone temperature measured in the charging zone. It is clear that the heat-loss Q_L depends on the zone temperature. The parameters n_1 and n_2 in eq.(6) are identified recursively by the online least square method [3].

Calculation accuracy and control accuracy of the bloom temperature. In order to cofirm the accuracy of the calculated bloom temperature, the real temperature was measured by thermo-couples burried in the bloom. Figure 8 compares the measured temperatures and the calculated temperatures. They agree with each other very well at all heating period. From repeated experiments, it is shown that the deviation of the discharged bloom temperature from the desired value is less than 15 °C.

Energy savings. Figure 9 shows the behavior of the fuel consumption of newly constructed furnace in Kobe Works over one year period. The furnace had been controlled by a manual operation for the first half year. The specific fuel of the furnace had been improved as the production increased. The computer control has been introduced gradually for the latter half year, and the specific fuel has been reduced to 170000 Kcal/ton. It has been confirmed that the fuel consumption is improved by about 10 % by using this computer control system compared with the manual operation.

CONCLUSION

A combustion control of a hot bloom furnace has been developed. This system determines an optimal temperature trajectory of the bloom through online calculation under the actual tracking conditions. The computer control system with this trajectory has been successfully used at Kobe Works of Kobe Steel, Ltd.

REFERENCES

1) Iwahashi, Y. and others (1981). Compute Control System For Continuous Reheatin Furnaces, Preprint of IFAC 8th Triennia World Congress at Kyoto.
2) Hollander, F. and R. L. Haisman, (1970) Online Computer Control For Five Zon Reheating Furnace In A Modern Hot Stri Mill, Internationale EisenHuttentagun;
3) YOUNG, P. C. (1969). Applying Paramet Estimation To Dynamic System, Contr Engineer, Oct., P P. 119-125.

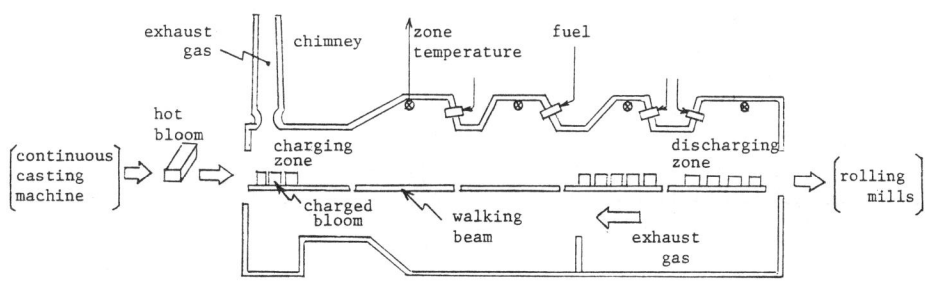

Fig. 1 Profile of the furnace for continuously cast bloom.

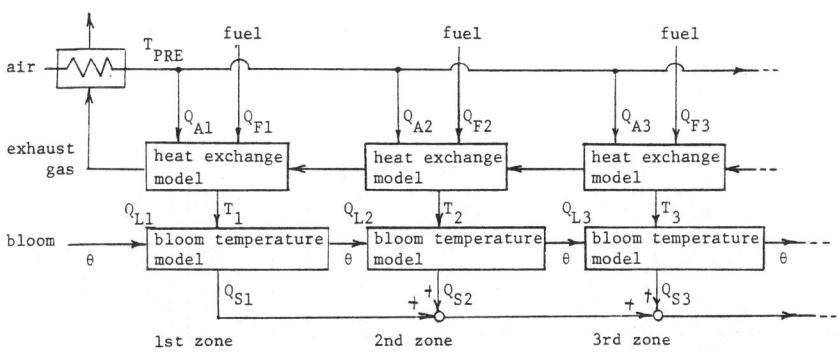

Fig. 2 Process model of the multi-zone furnace

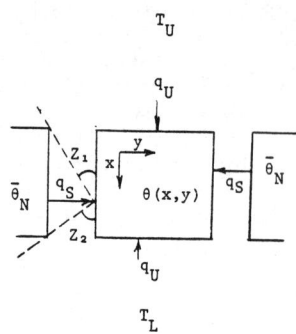

Fig. 3 Temperature model of the bloom.

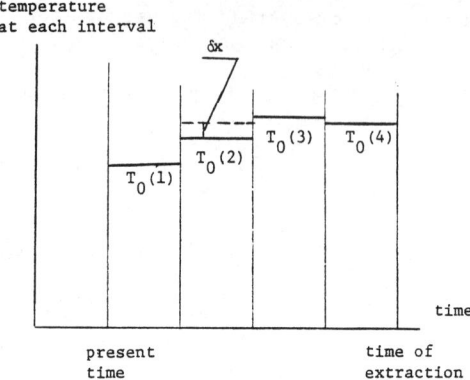

Fig. 4 Computation algorithm of linear model.

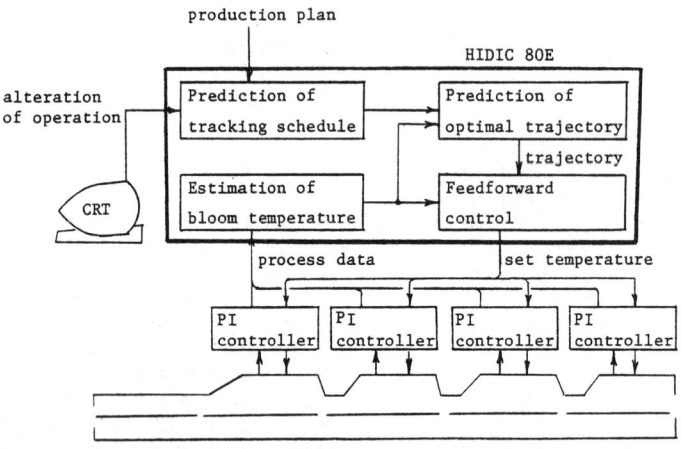

Fig. 5 Computer control system of the furnace.

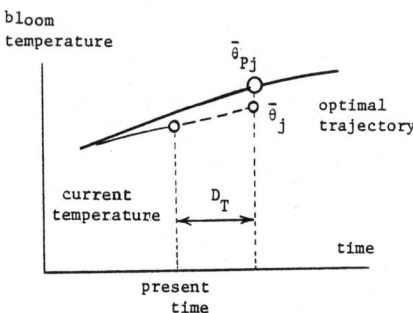

Fig. 6 Feedforward control of the bloom temperature.

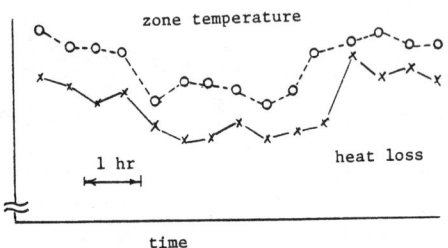

Fig. 7 Dependence of the heat loss Q_L on the zone temperature.

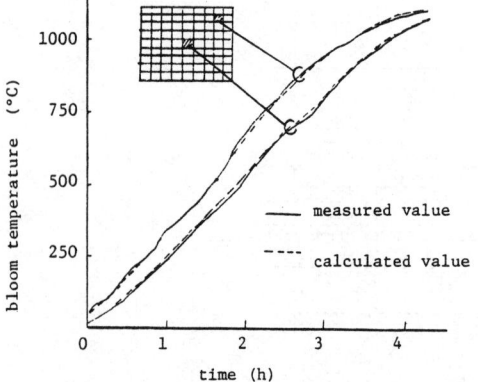

Fig. 8 Accuracy of the calculated bloom temperature.

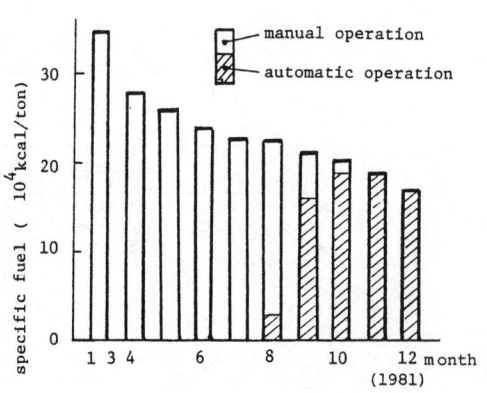

Fig. 9 Improvement of specific fuel.

A NEW ADAPTIVE CONTROLLER FOR COLD ROLLING MILLS

J. Hrušák, J. Mošna, E. Janeček and M. Šimandl

Department of Technical Cybernetics, Institute of Technology, Plzeň, Czechoslovakia

Abstract. The paper deals with the problems concerning the design of thickness-control subsystem at cold rolling. A strip thickness-control algorithm respecting technical equipments of SKODA-mills is designed. A mathematical model of controlled process is obtained by considering force-balance equation in the roll gap. Derivated model respects roll eccentricity of back-up rolls. Thickness-control is carried out by an adaptive input error controller. The designed algorithms were tested in real conditions and their effectivity is illustrated.

Keywords. Adaptive systems; digital control; optimal filtering; optimal control; rolling mills.

INTRODUCTION

In designing of thickness-control algorithm we have found which fundamental physical relationships between physical variables of controlled process must be respected. The formal derivation of the mathematical model of controlled process is based on the force-balance model in the roll gap. This model respects roll eccentricity of back-up rolls which can arise during of a operation of some cold-rolling-mills. Stochastic processes with independent increments are used as an uncertainty model. Possible approaches to the synthesis of open-loop feedback adaptive controller are discussed. An adaptive controller minimizing variance of the defined "input error" is suggested. This approach removes "turn-off" and/or "escape" effects that appear when regulators minimizing variance of "output error" are used. The closing part of the paper is devoted to the verification of the designed algorithm in real conditions.

CONTROLLED PROCES MODEL

A two level structure of thickness-control system was drafted for SKODA cold-rolling mills. The roll position control is performed by an analog regulator on the first control level while the ste-ups of the analog regulator can be adapted by a digital controller from the second control level. The control objective of the controller is to compensate for disturbances influencing output-strip thickness fluctuations. Controlled process properties are bound up with the position of the strip and therefore, the bulk of production can be considered as a natural measure of time, instead of real time. This fact will be respected by the definition of "rolling-mill-time" l

$$l(t) \triangleq \frac{l_2(t)}{\lambda} \triangleq \frac{1}{\lambda}\int_{t_0}^{t} v_2(\tau)\, d\tau \quad (1)$$

where
$v_2(t)$ is output-strip speed at the time t,
$l_2(t)$ output-strip position at the time t,
λ normalizing constant defined by the step-length of the position-gauges,
t_0 initial time.

Control system structure is given in Fig. 1.

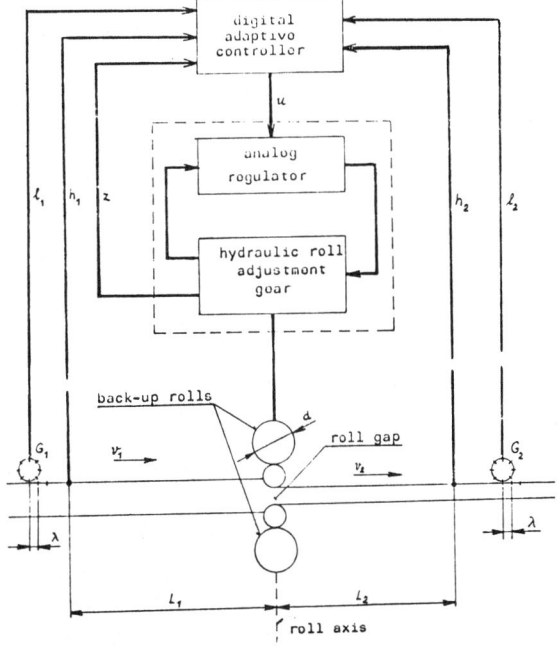

Fig. 1. Control system structure

where
- h_1 is input thickness error (diference between actual input thickness and nominal thickness H_1^*),
- h_2 output thickness error (diference between actual output thickness and required thickness H_2^*),
- l_1, l_2 input and output-strip positions (measured by input and output position-gauges G_1, G_2 respectively),
- v_1, v_2 input and output velocities of the strip,
- L_1, L_2 distances of input and output thickness-gauges from the roll axis respectively,
- z measured value of roll-position,
- u designed value of roll-position (control variable).

From the control point of view the following force-balance model seems to be acceptable for describing situation in the roll gap, see Kubík, and co-workers (1979),

$$P_V(1) = R(1) \cdot \sqrt{\Delta + h_1(1-L_1) - h_2(1-L_2)} \quad (2)$$

where
- $P_V(1)$ is roll-force at the time 1,
- $R(1)$ a parameter representing mechanical constants of the stand and involving some mechanical properties of the strip (and to a certain extent other circumstances). Dynamic properties of the parameter will be modelled by means of a stochastic process with independent increments for which it holds

$$\{E\ dR(1)\} = 0 \quad E\{dR^2(1)\} = \sigma_R^2 dl \quad (3)$$

- σ_R^2 reflects the measure of unhomogenity along the strip during stabilized technological conditions,
- Δ is known constant representing basic strip thickness reduction

$$\Delta = H_1^* - H_2^* . \quad (4)$$

- $\bar{L}_1$ is the rolling-mill-time interval determined by the time instant when the input thickness-gauge gets information about the value of input-thickness error h_1 and the time instant when this h_1 comes into the roll gap. According to (1), and Fig. 1., $\bar{L}_1(1)$ is defined implicitly by

$$\frac{1}{\lambda} L_1 = \int_{1-\bar{L}_1(1)}^{1} \frac{v_1(\tau)}{v_2(\tau)} d\tau \quad (5)$$

- $\bar{L}_2$ is the rolling-mill-time delay determined by the time instant when h_2 leaves the roll gap and the time instant when the output thickness gauge gets information about this value of output thickness error h_2.

According to Fig. 1 it holds $\bar{L}_2(1) = \frac{1}{\lambda} \cdot L_2$.

The roll-force $P_V(1)$ is not measurable. If the stand is replaced by a model of linear spring in the vicinity of an operating point the global roll-force can be described with sufficient accuracy, see Kubík, and co-workers (1979), as

$$P_V(1) = k_V [H_2(1) - Z_0(1)] \quad (6)$$

- k_V is constant characterizing elasticity of the stand,
- $Z_0(1)$ roll openning without stress,
- $H_2(1)$ strip-thickness in the roll gap.

The roll openning $Z_0(1)$ is measured by means of the roll position gauge.
It holds

$$Z_0(1) = z(1) - f(e,1,l_0) + \eta(1) + \psi(1) \quad (7)$$

- $z(1)$ is measured value of roll-position,
- $f(e,1,l_0)$ periodic function reflecting roll eccentricity of back-up rolls,
- e unknown parameter depending on the roll eccentricity magnitude,
- l_0 unknown parameter reflecting the roll eccentricity phase shift,
- z_0 unknown pre-setting of the roll-position gauge,
- $\eta(1)$ white stochastic process modelling roll position error,
- $\psi(1)$ stochastic process with independent increments modelling reproducibility error of working-rolls adjustment in the roll gap.

Processes $\{\eta(1)\}$ and $\{\psi(1)\}$ are characterized by

$$E\{\eta(1)\} = 0 \quad E\{\eta^2(1)\} = \alpha^2 \quad (8)$$

$$E\{d\psi(1)\} = 0 \quad E\{[d\psi(1)]^2\} = \delta^2 dl \quad (9)$$

$H_2(1)$ can be expressed by

$$H_2(1) = h_2(1+\bar{L}_2) + H_2^* . \quad (10)$$

Due to practical reasons force-balance model (2) is linearized in the mill stand operating point determined by H_1^*, H_2^*.

When linearizing the model the square-root is aproximated by

$$\sqrt{\Delta + h_1(1-\bar{L}_1) - h_2(1+\bar{L}_2)} \doteq$$

$$\sqrt{\Delta} + \frac{1}{2\sqrt{\Delta}} [h_1(1-\bar{L}_1) - h_2(1+\bar{L}_2)] . \quad (11)$$

Substituting (6), (7), (10), (11) into (2) we get

$$h_2(1+\bar{L}_2) - z(1) = p_1(1)[h_1(1-\bar{L}_1) - h_2(1+\bar{L}_2)] +$$

$$p_2(1) + f(e,1,l_0) + \eta(1) \quad (12)$$

where $p(l) \triangleq [p_1(l), p_2(l)]^T$ is a process with independent increments, characterized by the following relations

$$E\{dp(l)\} = 0 \quad E\{dp(l) \cdot dp^T(l)\} = Qdl \quad (13)$$

$$Q = \frac{\sigma_R^2}{k_v^2} \begin{bmatrix} \frac{1}{4\Delta} & , & \frac{1}{2} \\ \frac{1}{2} & , & \Delta + \frac{\partial^2 \cdot k_v^2}{\sigma_R^2} \end{bmatrix} \quad (14)$$

The periodic function $f(e, l, l_0)$ can be approximated by Fourier expansion

$$f(e, l, l_0) \sim a + e \cdot \sin\left[\frac{4\lambda}{d}(l+l_0)\right] \quad (15)$$

where
a   is unknown constant,
d   back-up roll diameter.
Substituting (15) into (12) and including a into $p_2(l)$ we get

$$h_2(l+\bar{L}_2) - z(l) = p_1(l)\left[h_1(l-\bar{L}_1) - h_2(l+\bar{L}_2)\right] +$$

$$p_2(l) + e \cdot \sin\left[\frac{4\lambda}{d}(l+l_0)\right] + \eta(l). \quad (16)$$

The linearized force-balance model will be used as the underlying model for adaptive control design.

A NEW ADAPTIVE CONTROLLER FOR COLD-ROLLING MILLS

A strip thickness-control algorithm for cold-rolling mills without roll eccentricity elimination control is considered in Hrušák, and co-workers (1981, 1982). Fourier analyzer of roll eccentricity detector and control system for elimination of roll eccentricity is presented in Imai, and Susuki (1973).

In this section an unified approach to synthesis of adaptive input error controller including roll eccentricity elimination control will be considered.

The main aim of input disturbances compensation consists in generating such control variables u that would make it possible to achieve the minimum of output thickness error. The control problem can be formalized as the problem of output error minimum variance, i.e. we are facing the problem of minimization of the following performance index

$$J)u) = E\left\{\int_{1-\bar{L}_2}^{\frac{1}{\lambda}L-\bar{L}_2} h_2^2(\tau) \, d\tau\right\} \quad (17)$$

where L is an effective length of the strip. Optimal output error regulator (belonging into the class of closed-loop policies) is not efficiently utilizable because of high numerical demands. However, a feedback type control, where learning is "passive" see Bar-Shalom, and Tse (1974), can be used. It is well known that this type of strategies being used in real conditions (or simulations) can be threatened by "turn-off" and/or "escape" effects see Åström, and Wittenmark (1971) and Kubík, and co-workers (1980).

We will try to avoid some of the above mentioned problems via formulation and solution of input error minimum variance problem (IEMV). When formulating IEMV problem it is suitable to start with the concept of "an ideal control". The ideal control $u^{id}(\tau)$, $\tau \in (1, \frac{1}{\lambda}L>$ is such an action that will quarantee ideal fulfilment of the compensation objective, i.e.

$$h_2(\tau + \bar{L}_2) = 0.$$

For the linearized force-balance model (16) the ideal control is given by

$$u^{id}(\tau) = -p_1(\tau)h_1(\tau - \bar{L}_1) - p_2(\tau) -$$

$$e \cdot \sin\left[\frac{4\lambda}{d} \cdot (\tau + l_0)\right] - \eta(\tau). \quad (18)$$

An ideal controller (18) $u^{id}(\tau)$ is unrealizable due to incomplete information about the controlled proces. The fact that plant relation is continuous on natural topologic spaces of stimulations and responses implies that if any realized control $u(\tau)$ is close to the ideal control $u^{id}(\tau)$ then the output strip thickness h_2 will be close to the required h_2^*. Thus, we can accept performance index in the following form

$$J = \min_{S} E\left\{\int_1^{\frac{1}{\lambda}L} \tilde{u}^2(\tau) d\tau\right\} \quad (19)$$

where S is a set all open-loop feedback strategies, and $\tilde{u}(\tau)$ is the system input error defined by

$$\tilde{u}(\tau) \triangleq u(\tau) - u^{id}(\tau). \quad (20)$$

Let us denote all the information that has been gained up to the time l as $\mathcal{X}(l)$. At the time l we do not suppose that any information will come in future (open-loop) but if it comes it will be used in an optimal way (open-loop feedback). The solution of the problem consists in the solution of open-loop problem specified by a priori data $\mathcal{X}(l)$ and performance index

$$J(u) = \int_l^{\frac{1}{\lambda}L} E\left\{[u(\tau) - u^{id}(\tau)]^2 / \mathcal{X}(l)\right\} d\tau \quad (21)$$

Supposing non-randomized control strategy, performance index (21) can be (after some rearrangements) rewritten as

$$J(u) = \int_1^{\frac{1}{\lambda}L} \left[u^2(\tau) - 2u(\tau) \cdot E\{u^{id}(\tau)/\mathcal{X}(1)\} + E\{[u^{id}(\tau)]^2 / \mathcal{X}(1)\} \right] d\tau . \quad (22)$$

For the linearized force-balance model (16) the corresponding optimal open-loop policy is described by

$$\forall \tau > 1: u^*(\tau/\mathcal{X}(1)) = -E\{p_1(\tau)/\mathcal{X}(1)\} .$$

$$h_1(\tau - \bar{L}_1) - E\{p_2(\tau)/\mathcal{X}(1)\} -$$

$$E\left\{ e \cdot \sin\left[\frac{4\lambda}{d} (\tau + l_0) \right] / \mathcal{X}(1) \right\} \quad (23)$$

From (23) it follows that the optimal open-loop feedback controller is given by

$$u^*(1) = -k_1^*(1) \cdot h_1(1-\bar{L}_1) - k_2^*(1) - k_3^*(1) \quad (24)$$

where $k_1^*(1)$, $k_2^*(1)$, $k_3^*(1)$ are the best linear estimates of $p_1(1)$, $p_2(1)$, $e \cdot \sin\left[\frac{4\lambda}{d}(1+l_0)\right]$ obtained under the condition that $\mathcal{X}(1)$ is known, i.e.

$$k^*(1) = \begin{bmatrix} k_1^*(1) \\ k_2^*(1) \\ k_3^*(1) \end{bmatrix} = E\left\{ \begin{bmatrix} p_1(1)/\mathcal{X}(1) \\ p_2(1)/\mathcal{X}(1) \\ e \cdot \sin\frac{4\lambda}{d}(1+l_0) /\mathcal{X}(1) \end{bmatrix} \right\} \quad (25)$$

Let us suppose that measured data are equidistantly synchronized with output position of the strip and the period is determined by the step-length of the output position gauge. It means that $\mathcal{X}(1)$ is specified by

$$\mathcal{X}(1) \triangleq \mathcal{X}(i) \quad i \leq 1 < i+1$$
$$\mathcal{X}(i) \triangleq \{h_1(0), z(0), h_2(0), \ldots, h_1(i-1), z(i-1), h_2(i-1)\} \quad i=0, 1, 2, \ldots \quad (26)$$

It seems to be hard task to obtain the estimates $k^*(1)$ considering the structure $k_3^*(i) = E\{e \cdot \sin[\frac{4\lambda}{d}(i+l_0)] / \mathcal{X}(1)\}$ with unknown parameters e, l_0. Nevertheless a solution is possible, if $k_3^*(i)$ is considered as state estimate of a system generating periodic function $e \cdot \sin\left[\frac{4\lambda}{d}(i+l_0)\right]$. Filtering model in question has following structure

$$x(i+1) = A x(i) + \xi(i)$$
$$y(i) = C(i)x(i) + \eta(i) \quad (27)$$

where

$$x(i) \triangleq [p_1(i-\bar{L}_2), p_2(i-\bar{L}_2), e \cdot \sin[\frac{4\lambda}{d}(i+l_0-\bar{L}_2)], e \cdot \cos[\frac{4\lambda}{d}(i+l_0-\bar{L}_2)]]^T$$

$$y(i) \triangleq h_2(i) - z(i-\bar{L}_2)$$

$$A \triangleq \begin{bmatrix} 1 & 0 & 0 & 0 \\ 0 & 1 & 0 & 0 \\ 0 & 0 & \cos\frac{4\lambda}{d} & \sin\frac{4\lambda}{d} \\ 0 & 0 & -\sin\frac{4\lambda}{d} & \cos\frac{4\lambda}{d} \end{bmatrix}$$

$$C(i) \triangleq [h_1(i-\bar{L}_1-\bar{L}_2) - h_2(i), 1, 1, 0]$$

$$\xi(i) \triangleq [\xi_1(i), \xi_2(i), \xi_3(i), \xi_4(i)]^T$$

$\xi(i)$ is white noise, characterized by the following relations

$$E\{\xi(i)\} = 0 \quad E\{\xi(i) \cdot \xi^T(i)\} = R = \begin{bmatrix} Q & \vdots & 0 \\ \cdots & \vdots & \cdots \\ 0 & \vdots & r_1 & 0 \\ & \vdots & 0 & r_2 \end{bmatrix}$$

where Q is given by (14) and r_1, r_2 reflect drift of eccentricity phase caused by strip of back-up rolls and by changes of technologic conditions. For the model (27) the best linear estimate can be obtained by means of linear filtering theory in the following form

$$\hat{x}(i+1) = A\hat{x}(i) + \varkappa(i)[y(i) - C(i)\hat{x}(i)] \quad (28)$$

$$\varkappa(i) = \frac{AP(i)C^T(i)}{\alpha^2 + C(i)P(i)C^T(i)} \quad (29)$$

$$P(i+1) = AP(i)A^T - \varkappa(i)C(i)P(i)A^T + R \quad (30)$$

where

$$\hat{x}(i) = E\{x(i)/\mathcal{X}(i)\}$$
$$P(i) = E\{[x(i) - \hat{x}(i)][x(i) - \hat{x}(i)]^T / \mathcal{X}(i)\}$$

Using (25) and (27) we obtain

$$k_1^*(i) = \hat{x}_1(i), \quad k_2^*(i) = \hat{x}_2(i),$$

$$k_3^*(i) = \hat{x}_3(i)\cos(\frac{4\lambda}{d}\bar{L}_2) + \hat{x}_4(i)\sin(\frac{4\lambda}{d}\bar{L}_2)$$

thus the optimal control (24) is realizable.

CONTROL SYSTEM REALIZATION

The important special case of the considered algorithm arises if the roll eccentricity elimination control is not necessary. In such a case the IEMV controller is described by the following equation, see Hrušák, and co-workers (1982)

$$u^*(1) = -k_1^* \cdot h_1(1-\bar{L}_1) - k_2^*(1). \quad (31)$$

The best linear estimate of parameters $p_1(1)$ and $p_2(1)$ can be obtained from (25)-(30), where

$$x(i) \triangleq [p_1(1), p_2(1)]^T$$

$$y(i) \triangleq h_2(i) - z(i-\bar{L}_2)$$

$$A \triangleq I$$

$$C(i) \triangleq [h_1(i-\bar{L}_1-\bar{L}_2) - h_2(i), 1]$$

$$\xi(i) = [\xi_1(i), \xi_2(i)]^T$$

$$E\{\xi(i)\} = 0 \quad E\{\xi(i)\xi^T(i)\} = Q$$

During the long-term testing operation IEMV controller proved that required quality of production could be assured with considerable reserve if either previous force-balance model of its linearized version had been accepted for 20-roll mills. The effectivity of the chosen algorithm is demonstrated in Fig. 2. Besides the versions that make use of Bayes approach (in the form of Kalman filter) controllers based on gradient adaptive algorithm and recursive least square algorithm with exponential forgetting were tested in Kubík, and co-workers (1980) and Hrušák, and co-workers (1981). The gradient adaptive algorithms used to fail in real conditions because no "universal" setting-up of weighing-matrices, assuring continuously required quality of control, was found. The least-square adaptive algorithm proved to be utilizable. Nevertheless, the obtained results were worse in spite of the enormous effort that had been exerted when seeking the proper value of forgetting factor. That was obviously caused by the fact that the algorithm was not able to respect the stochastic dependence between dynamic evolution of parameters in the linearized force-balance model.

The whole thickness control algorithm together with all the necessary software (program package) has been implemented on a minicomputer with the storage capacity of 8k with 16 bit-word. There is a real possibility of implementing it on a microcomputer with a 8 bit microprocessor, e.g. INTEL 8080 if double arithmetic is used.

The general form of control algorithm was tested both on the 20-roll mills and four-high rolling mills. From the experimental results it follows that the advantage of the general algorithm is more evident when used on four-high rolling mill. This is due to the fact that the roll eccentricity doesn't play substantial role at 20-roll mills.

CONCLUSION

An adaptive control algorithm for cold rolling-mills respecting the roll eccentricity has been developed. The proposed IEMV algorithm belongs into the class of open-loop feedback strategies. The suggested IEMV controller was tested both on 20-roll mills and four-high rolling mills. From experimental results it follows that the suggested algorithm in its general form is preferable for four-high rolling mills, where the roll eccentricity play a very substantial role. On the other hand the special case of the algorithm (31) see also Hrušák, and co-workers (1982) has guaranteed approximately the same quality of production as the general form of algorithm applied to 20-roll mills.

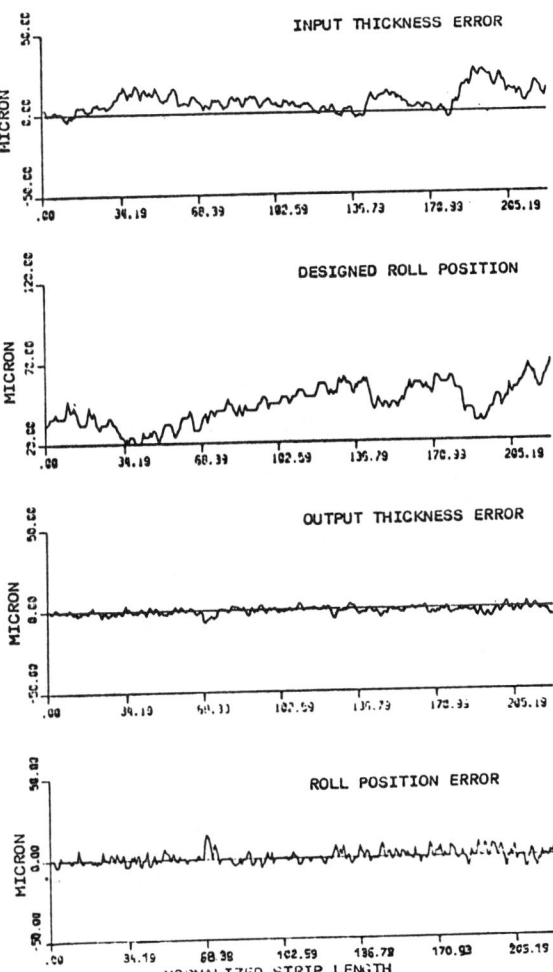

Fig. 2. Illustrative results of experiments in real conditions

REFERENCES

Kubík S. and co-workers (1979): Adaptivní algoritmus řízení tlouštky pasu pro válcovací stolice ŠKODA. Report No. 209-02-79, Institute of Technology, Plzeň

Bar-Shalom, Y., and Tse E. (1974). Concepts and Methods in Stochastic Control. Leondes (Ed.) Control and Dynamic Systems Advances in Theory and Applications, Vol. 12, Academic Press

Aström, K.J., and Wittenmark, B.(1971). Problems of Identification and Control, J. Math. Anal. Appl., Vol. 34, No. 1, pp. 90-113

Kubík, S. and co-workers (1980). Ověřovací provoz adaptivního algoritmu řízení tlouštky pasu na dvacetiválcové válcovací stolici ŠKODA, Report No. 209-06-80, Institute of Technology, Plzeň

Hrušák, J., Mošna, J., Janeček, E., and Šimandl, M. (1981). Provozní ověření adaptivního algoritmu řízení tlouštky na dvacetiválcové válcovací stolici, Sborník COMEMOP´81, Žilina

Hrušák, J., Mošna, J., Janeček, E., and Šimandl, M. (1982). Strip-thickness Adaptive control for Cold-Rolling Mills, Problems of Control and Information Theory, to be published

Imai, I., and Susuki, T. (1973). Fare Detector and Control System for Elimination of Roll Eccentricity, Ishikawajima-Harima Heavy Industries Co., Ltd

Aström, K.S. (1970). Introduction to Stochastic Control Theory, Academic Press

DISCUSSION

SESSION MA1: METAL PROCESSING

Paper: THE DISTRIBUTED CONTROL SYSTEMS OF COIL ANNEALING FURNACES

Authors: Y Hariai, I Yamazaki, M Ono, T Makino, S Miki, R Michioca (Sumitomo Metal Industries Ltd, Kashima, Japan)

Discusser: *Jan Carlsson,*
Computers and Electronics Commission, Drottningattan 20
111 51 Stockholm
Sweeden

Questions or Comments:

1) Could you say something about the total development cost (including hardware) for the system?

2) What is the return of investment, for example in terms of the pay-off time?

3) Could you give an estimate of how much the savings in fuel consumption are in terms of money?

Author's reply:

1) Total development cost are 144×10^6 yen ($\sim$ 626,000 US dollars)

2) Pay-off time is about 1.3 years

3) The estimated value of savings is about 112,500 US dollars per month, calculated as follows: The unit price of gas in Japain is 2.5 cents per kCal. The savings in fuel consumption due to our system improvements are estimated to be 30,000 kCal per each tonnage of products in the case of the blooming mill. As a result, we can save 75 cents per each tonnage of products. The monthly production of the blooming mill reaches 150,000 tons per month. Consequently we have saved 112,500 US dollars per month, which corresponds to the saving of 1,350,000 US dollars per year.

A SPECIFICATION OF REAL-TIME APPLICATIONS BY EVENTS AND LINKS WITH PROCESSES

M. Benmaiza* and J. P. Thomesse**

Centre d'Etudes et de Recherche en Informatique, Alger, Algérie
Centre de Recherche en Informatique de Nancy, France
Centre de Recherche en Automatique de Nancy, France
**Centre de Recherche en Informatique de Nancy, France*
Centre de Recherche en Automatique de Nancy, France
2 rue de la Citadelle, BP 850, 54011, Nancy Cedex, France

Abstract : A specification method, for RT applications, based on events and links is presented. An analysis of event concept is made and attributes are defined. Taking into account the occurences of events, links between events and tasks are then defined. Full integration of exception-handling is made by means of events. Fault-tolerance problem is also taken into account.

Key-words : Process-control application, events, occurences, assignment, evolutivity, fault-tolerance, implementation independance.

- INTRODUCTION -

The paper presented here deals with the specification of process-control applications (noted P.C.A.). What is meant by specification is to provide the designer with a good tool which allows him to precisely define his application in a "natural" way. To reach this aim, the specification must then, be based on a natural concept : the concept of event. This concept seems to be natural because most situations (natural or artificial) can be described in terms of events and reactions to these events. So we think that this concept should enable the specification to be at the same time near the user requirements and the implementation mechanisms.

Throughout the undermentioned chapters, no hypothesis is made about the physical system. Nevertheless, our specification is chiefly based on an asynchronous view of the system. Control could be, however, implemented in a synchronous way. This possibility allows us to add or to suppress some sensors in a fault-tolerance goal. Logical events are assigned to the sensors and processing tasks are attached to these events. This assignment operation allows us to dynamically take into account some failures.

Another important point for our specification is to be static with a full integration of exceptions-handling. We wish express in a static way and with the same tool the normal behaviour of a P.C.A. as well as the abnormal one. This static expression allows some verifications without any references to implementation problem.

The heart of our method is obviously the concept of event. Events may be external to cope with the environment (physical system) or internal for a synchronization purpose. The reader will successively find in the paper :

. a discussion about the event concept :
    - events and occurences
    - life of occurences
    - relation between external and internal events
. a definition of composed events :
    - definition
    - life of such events
    - use
. a duality between events and conditions
. the links between the events and the processing tasks
. and finally an example to illustrate the previous concepts.

Whenever it is necessary, a discussion about fault-tolerance and securities problems are developped.

We do not end this introduction without mentioning the taking into account of the evolutivity problem. Indeed, our asynchronous specification allows a very modular description of the applications and a dynamic integration of events and associated actions without modifications.

- I - CONCEPT OF EVENT -

Following (1) and (2), a distinction is made between an occurrence and an event. This distinction will have a great importance when we will specify how these events are treated : one must precisely specify what occurrence is to be treated.

Definition : An occurrence is a state-change of a variable at a given instant (this variable may be a numeric input, an analogic input compared to a threshold ...). An event is a set of time-ordered occurrences of a same state-change.
In the following chapters, we shall try to give some precisions about the definition and the utilization of the events (e.g. their nature, their life and their sharing).

1.1. External and internal events :

As said in the above introduction, a P.C.A. is divided into two parts : the environment which represents the physical process, and the control system. At the control-system level, the environment is perceived as external events. To coordinate actions in the control-system, internal events may be used. In a ch cking and prove purpose, internal events cannot be defined by the designer ; they are automatically generated by the system and only concern the begin and the end of the tasks (a task represents the processing of an occurrence).
External events are also called physical events so far as they represent physical phenomena. The physical events are then closely related to the real world.
Logical events are, merely, representation of physical events at the control-system level. They are defined by the designer and have several attributes described afterwards. An independance between a physical level and a logical level is consequently allowed. It will be fairly usefull to take into account the fault-tolerance problem.
Now, it is necessary to define an operation which establishes a link between a physical event and a logical event : the assignment operation.

1.2. Assignment operation :

Syntax : <u>assign</u> (logical event ident) <u>to</u> (physical event definition)

where (physical event definition) could be "Interruption n°j on" or "True-False transition of numeric input n°k"...

Generally used at a statical level, it is however interesting to allow the operation to use it : in case of an equipment failure, the operator can dynamically change a link without affecting the structure of the application.
We shall see below that we also can assign several logical events to a same physical event.

1.3. Life of occurences :

Only physical events will be concerned. Three instants can be considered during the life of an occurrence :

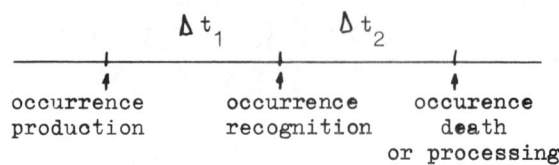

occurrence production — occurrence recognition — occurence death or processing

If Δt_1 only depends on detection system, Δt_2 is more related to the event nature and to the processing task. Δt_2 is called the life of an occurrence.
We can divide this life into two parts : intrinsic life (e.g. passing of a product through a manufacturing station) and additional life which directly concern memorization of occurrences. The additional life is typically variable, depending on processing tasks. Then, the memorization appears to make the processing of an occurrence less critical ; by this way it is directly related to the priority processing. However a problem arises : how could the designer define this life ?

1.4. Events and time-out :

The interesting for the designer at this stage is probably not how to define the event life but chiefly what he will have to do when an occurrence is missing. Here, it is very usefull to define a time-out linked to the event with the following meaning : it is conceptually related to each occurrence and indicates a fault in a processing task or in the tasks scheduling. By the same way a time-out may be define in accordance with the tasks which warns an occurrence-producer failure.

1.5. Events, tasks and priority :

A first type of priority is attached to tasks and defines a processing urgency. A second type of priority is also considered : priority attached to the events and indicating a detection urgency. This distinction is also made in (3).

1.6. Sharing of events :

When several tasks will be attached to the same event, it will be called shared. In that paragraph, we want firstly to precise what occurrences are concerned with the sharing : the same occurrences or not ? It is easy to understand that some tasks are only concerned with the latest occurrence while others are concerned with all occurrences for example. The occurrences are then considered as resources and may be consumed or consulted. There is an obvious problem if several tasks have to process the same occurrence. To solve it, an additional operation is introduced : the duplication operation. This operation allows us, whenever it is necessary, to duplicate an occurrence (for a logical event) with the respect to time-ordering of occurrences. As illustrated by the following schema, two cases of sharing may be considered :

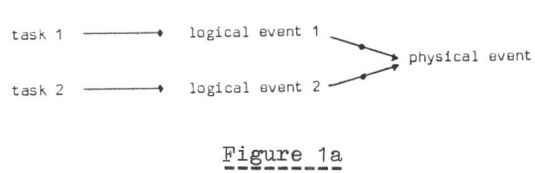

Figure 1a

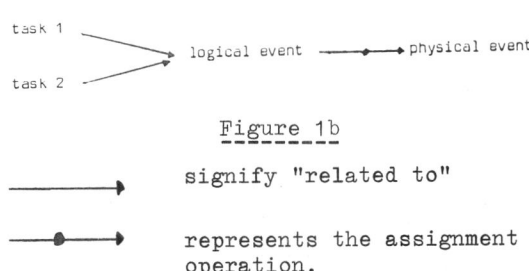

Figure 1b

———▶ signify "related to"

———•▶ represents the assignment operation.

Fig. 1.a states that the assign operation is a duplication operation. Tasks are only concerned with logical events and by this way, the designer may define various attributes for the same physical event.

Fig. 1.b is not fondamentally different from fig. 1.a, but states that duplication is an explicit operation.

First solution may appear more elegant than the second one. It hides the duplication operation to the users.

We will then adopt this solution.

1.7. Elementary events. Composed events :

Hitherto, only elementary events are treated. This is not enough if we want to consider all potential situations. For example, it will often occur that an action takes place if two events appear simultaneously. To get rid of these situations, composed events will be considered.

- II - COMPOSED EVENTS -

We introduce here operators which seem to have the ability to well supply the needs of the P.C.A. (they are really quite natural !) : AND, ØR, NØT, BEFØRE, AFTER, delay after, number of occurrences of.

As it has been said above, an event is a set of time-ordered occurrences. To compose two events or more, it is necessary to precise what are the occurrences the operators applied on : are they the earlier occurrences or all ? (let us drawn your attention to the definition of "earlier" and "all". This problem will be discussed in § 4).

At this step, we suppose that we are able to define an initial instant. Other topics will be examined : the life of composed occurrence.

2.1. Occurrences concerned :

The monadic operators NØT, number of occurrences of, does not cause any problem. For the other operators a case analysis which takes into account all occurrences and recent occurrences is presented below.

2.1.1. AND operator

a) All occurrences are concerned :

Let us consider the expression e : evt 1 <u>and</u> evt 2, and the schemas :

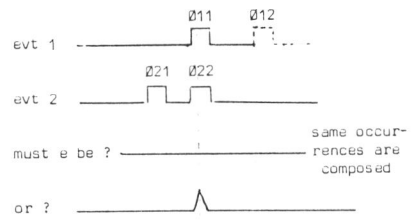

Figure 2a

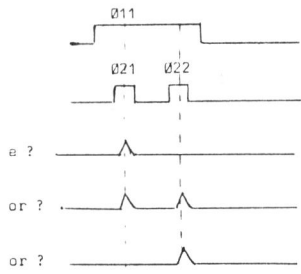

Figure 2b

where $\emptyset ij$ is the abreviation for "occurrence ij".

The question is : do we compose the same occurrences or not ? (if we suppose that we are able to number the occurrences). In that case it is necessary to point out that the nature of the occurrences has not any importance any more, and that we will only consider occurrences occuring at the same time.

If E1 is defined as : vat n°1 full
    E2 is defined as : tap n°2 open
and E : E1 and E2,

an occurrence of E happens each time vat n° 1 is full when tap n° 2 is open. No matter with first or n^{th} opening of the tap !
In fact, we think it is essentially a problem of detection.
As an illustration let us examine a second case :

    b) The earliest occurrences of evt1 and evt2 will be here concerned :

We have the scheme :

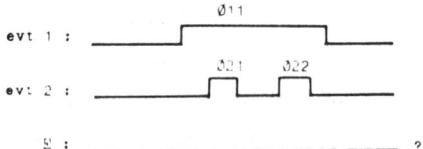

How can we produce E ? There is an ambiguity in considering the earlier occurrence of evt2. Indeed, Ø11 is still alive when Ø22 occurs. But we have already considered Ø21 as an earlier occurrence and an occurrence of E has been already produced !

In order to avoid this problem we have again to consider all occurrences without numbering.

But it remains another point to solve : how to do the evaluation of an events-expression ? There will be, for example, an equivalence between fig. 3.a and fig. 3.b concerning the detection of the occurrences :

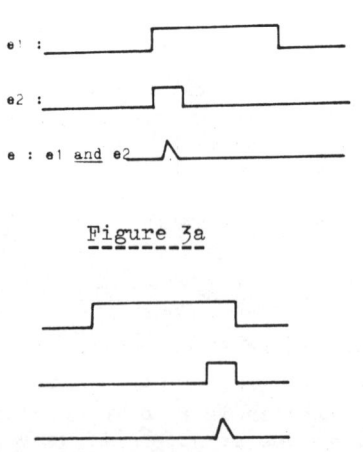

Figure 3a

Figure 3b

(i.e. e1 and e2 = e2 and e1)

To assure the detection of all occurrences of e (e : e1 and e2) the evaluation must be done for all occurrences of e1 and e2. This evaluation must also begin with one occurrence of e1 or e2 and go on as long as this occurrence is alive without occurrence missing (we assume that it exists an evaluator which is sufficiently performant to take into account all occurrences). Indeed, occurrences cannot be lost at the evaluation stage. The idea of an evaluator is also retained for the other operators.

2.1.2. ØR operator (inclusive ØR)

Let be the chronogram of fig. 4 where e : e1 OR e2

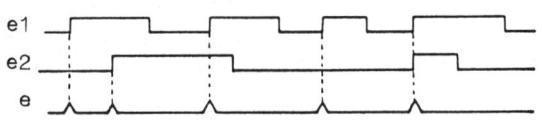

Figure 4

As with the AND, an evaluation must be made for each occurrence. It can also be seen in (2) where the authors define the event E1 + E2 as a set of all time-ordered occurrences of E1 and E2.

2.1.3. BEFØRE operator

In this case, it is essential to consider the time-ordering of the occurrences. If the ordering of occurrences is always possible in a centralized system, this is not obvious on the other hand in a decentralized system (which constitutes our working hypothesis) because of clocks asynchronism. Nevertheless we can refer to Lamport's work (6) for this topics. Furthermore, we may suppose that it is always possible to make a time-distinction between occurrences of e1 and e2 (e : e1 before e2). So, the before operator is always defined.

Let us consider the chronogram of fig. 5 :

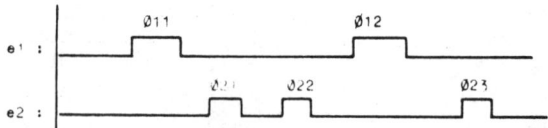

Figure 5

and e defined as e1 before e2.

We suppose that the initial instant is fixed (depicted by the vertical line).

Here, Ø11 is surely before Ø21, but what can we say about Ø22 ?
- It can be compared to the last occurrence of e1 and then all e1 occurrences are before e2 occurrences depending only on the initial instant.

- One can use the order of occurrences to compare the like-numbered occurrences. As the second solution seems us to be the most natural, we will adopt it.

In fig. 5, for example :

- Ø21 is compared to Ø11
- Ø22 is compared to Ø12
- Ø23 will be compared to Ø13.

We have then only one occurrence of e.

The After operator can be deduced from the before operator.

Nota : There can be no occurrence missing. The occurrences are received by the evaluator which constitues the central (but not necessarily centralized !) part of our system.

2.1.4. Delay after event operator

This operator is obviously applied to all occurrences and produce delayed occurrences.
Till now we only talked about the production of occurrences but not about their use. This will be seen in § 4 (links between events and tasks).

2.2. Composed occurrences and intrinsic life

Two cases are to be examined :

* It does not exist dependancy between the relation operator and the composed life : operators before, after, number of occurrences, delay after. In that case which is not furthermore detailed, a certain freedom is given to the designer.

* It exists a dependancy between relation operator and composed life : AND, ØR, NØT.
We shall discuss the AND and the ØR afterwards, the NØT being obvious.

2.2.1. AND case

Consider a chronogram (with e : e1 and e2) :

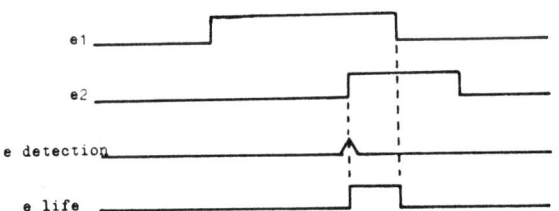

e intrinsic life appears to be the "intersection" between e1 and e2 intrinsic lifes. This life is variable and restricted by the shortest life of e1 or e2. Notice that the composed occurrences can be in pendantly memorized and processed.

2.2.2. ØR case

We begin to take an example to illustrate this case. Let e : e1 or e2 be a composed event. We have the chronogram:

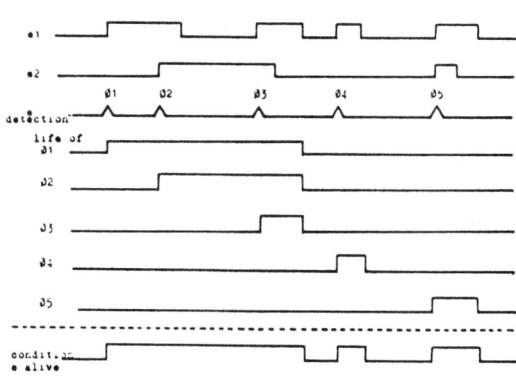

So as to obtain a composed life, you only have to "add" together the different lifes of time-interleaved occurrences.
If we consider the condition "e alive", we may notice that it exactly corresponds to this life (of course !). We must also notice the existing difference between the occurrence detection and its life.
We think, all the cases are explained in the above schema.

As a summary, we can say that in all cases the designer must define, derived from the life, a time-out for each event. He cannot directly define a time-out for all events. A good tool based on simulation may eventually help him in this task.

- III - LOGICAL EVENTS ATTRIBUTES -

After the preceeding analysis, we give here the logical events attributes. We use a BNF notation. Some explanation will also be given.

(logical event definition)::= event (event-id), (nature):(attributes list)

(attributes list)::= (priority),(time-out), (memorization),(sharing)

(priority)::= (PRI, integer)

(time-out)::= (TØ, delay [,(ident of exception event)]

(memorization)::= [(MEM, integer ,(ident of exception event)])

(sharing)::= SHARED | NOSHARED

(nature)::= ELEMENTARY | COMPOSED(expression of events)

- The priority, as explained above, is the detection priority.

- To the time-out is attached an event with the following meaning : if, for a given occurrence, the delay is elapsed then one occurrence of this event happens. By a link of this event to a task, one can easily take into account the exceptions relating to the missing of occurrences.

- The memorization ident of exception event has the same meaning with the difference that it is related to a number of occurrences.

- Sharing is only used in a statically checking aim.

Note that we have no implementation attributes at this level ; they appear in the physical event definition as in the example below :

<u>assign</u> ev1 to <u>true-false transition of numeric input n° 3</u> on (implementation site).

- IV - ABOUT THE DUALITY BETWEEN EVENTS AND CONDITIONS -

In P.C.A. an occurrence is generally processed under some conditions (representing the system state at this moment). We then may have a syntax of the form :

(1) <u>ON</u> event <u>if</u> condition <u>do</u>...

where condition is a boolean condition (may be <u>true</u> or <u>false</u>).

An event can be easily associated with this condition : a false-true transition of the condition.
Dually we can associate a condition to an event : condition event alive.
We can therefore express (1) as :

(2) <u>On</u> FT-Transition of cond <u>if</u> event alive <u>do</u>...

This is very interesting because conditions and events can be processed in the same way. The event concept is now defined. It remains us to precise the links between events and tasks.

- V - LINKS BETWEEN EVENTS AND TASKS -

What we have to consider in this chapter at first, are the features of these links:

- may be statically defined in a checking and expression easiness aim.
- may be dynamically established in a fault-tolerance purpose.
- must allow to take into account various occurrences of concerned events.

These links can be expressed as :

(link id) : <u>on</u> evt... <u>do</u>...

<u>and</u>

<u>link</u> (link id)::= (date) (event id)

The dynamic aspect of this expression is represented by the event. It means that the link is established on an occurrence happening of this event or at the indicated date.
Here again we have the problem of the concerned occurrence and during of link.

5.1. Instant of link :

We suppose that the link is established at the first occurrence. The link instant is noted 1o.
Io does not necessarily correspond to the starting of the application. So, we have to consider the <u>past</u> occurrences and the <u>future</u> occurrence with respect to lo.
This may be very usefull in case of task-failure (there are many non-processed occurrences) if we are dealing with the recovery problem.

5.2. Link expression (1) :

We give to the designer the possibility to take into account :
- all the past occurrences
- all the future occurrences
- all the occurrences (past and future)
- the latest occurrence
- the future occurrence.(7)

We must also precise how to use these occurrences : consuming, consulting. We think that these possibilities allow us to process most of the situations.

Syntax :

$$\langle link\ id \rangle : \underline{on} \left\{ \begin{array}{l} \underline{each}\ \langle evt\text{-}id \rangle \left[\left\{ \begin{array}{l} \underline{past} \\ \underline{future} \end{array} \right\} \right] \\ \underline{last}\ \langle evt\text{-}id \rangle\ \underline{past} \\ \underline{next}\ \langle evt\text{-}id \rangle\ \underline{future} \end{array} \right\} \left\{ \begin{array}{l} \underline{with} \\ \underline{without} \end{array} \right\} \underline{consuming}$$

$$\underline{if}\quad cond\ \underline{do}\ \langle task\ id \rangle\ \underline{with}\ priority = number$$

Condition can be defined in the same manner than event.

5.3. During of the link :

Several cases are to be considered :

(a) li : <u>on</u> <u>last</u>... <u>do</u> T1
              <u>next</u>...

Only one occurrence is to be processed. This link ends wtith the occurrence of event "End of T1". Nevertheless, li may be "reactivated" if it is defined as : evt k <u>or</u> "end of T1".

(b) lk : <u>on each</u> evt <u>past</u> ... <u>do</u> T2

All the past occurrences are to be processed. The link ends with the processing of all past occurrences. It may be reactived with the event "end of T2", with T2 defined as cyclic task.

A problem may arise if lk is defined as :

lk : evt k <u>or</u> "end of T2"

and there are no occurrences at the link

instant. To prevent this, we introduce a <u>null execution</u> which produces the end-o task event.

(c) ln : <u>on each</u> evt <u>future</u>... <u>do</u>T3
lm : <u>on each</u> evt <u>do</u> T4

Here, we must have an explicit destroy operation to break the link.

To end this chapter, we may notice that the occurrences can be processed or not depending ont he conditions value. In other words, conditions may be used to make "deaf" one task for a given time-interval.

- VI - EXAMPLE -

We consider a manufacturing industry example with a transport system, a working station and a sampling station.

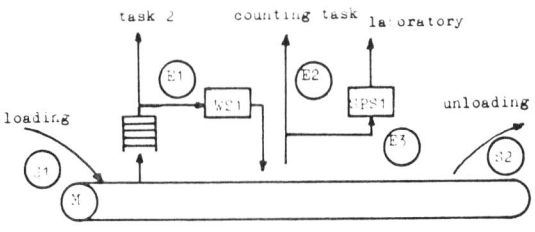

S1, S2, E1, E2, E3 represent logical events
WS1 : Working station
SPS1 : Sampling station.

Task 1 states for "another processing". There is a waiting queue before WS1.
We want to specify here the normal behaviour as well as the abnormal one of this system according to the previous definitions. Abnormal behaviour is essentially related to motor breakdown, WS1 et SPS1 failures.

Some comments are enclosed by "%" symbol. "Init" states for "starting of the application".
For a readability purpose, links, assignments and events definition relating to the same processing are grouped together.

We then have :
(a) WS1 processing
<u>link</u> l1 : init ;

l1 : <u>on each</u> E1 <u>with consuming if</u> cond 1
    <u>do</u> WS1 <u>with</u> priority = pl1 ;
%E1 is a logical event shareg between WS1 and task 1%

<u>event</u> E1, <u>Elem</u> : (PRI,p1),(TØ, delay 1, ex 1),(MEM,n1,EX2),SHARED ;
    <u>link</u> 12 : ex2 ; % exception event defines WS1 saturation %
12 : <u>on each</u> E1 <u>past</u> <u>with consuming if</u>
    cond2 <u>do</u> task 2 <u>with</u> priority = p12 ;
    <u>link</u> 13 : ex1 ; %WS1 failure %
L3 : <u>on each</u> E <u>with consuming if</u> cond 3
    <u>do</u> task 2 <u>with</u> priority = p13

Notice, here, that cond 1 may be used to take into account WS1 failure and then to make "deaf" WS1-task :

cond 1 : "Normal working of WS1".
The value of cond 1 may be changed by task 2.

As the assign operation, we have, for example :
<u>assign</u> E1 <u>to</u> False-true Transition of Numeric Input n°1 <u>on</u> site 1.

(b) Sampling and counting processing
<u>link</u> 14 : init ;
<u>link</u> 15 : init ;
<u>event</u> E2,Elem :(PRI,p2)(TØ,delai2,ex3),
                    (MEM,O),NOSHARED
<u>event</u> E3,Elem :(PRI,p3),(TØ,delay3,ex4),
                    (MEM,O),NOSHARED
%E2 and E3 assigned to the same physical event : there is an implicit duplication %
<u>Assign</u> E2 <u>to</u> false-true transition of NI n°2 <u>on</u> site 1
<u>Assign</u> E3 <u>to</u> false-true transition of NI n° 2 <u>on</u> site 1

ex3 and ex4 represents counting and sampling anomalies and are not detailed furthermore.

14 : <u>on each</u> E2 <u>with consuming if</u> cond 4
    <u>do</u> counting <u>with</u> priority = pl4
15 : <u>on next</u> E3 <u>future with consuming</u>
    <u>if</u> cond 5 <u>do</u> SPS1 <u>with</u> priority = pl5

(c) Exceptions handling
% motor breakdown %
<u>event</u> E4, COMP(delay5 <u>after</u> S1 <u>and not</u> E1):
    (PRI,p4),(TØ, delay4),(MEM,O),NOSHARED
% unloading failure%
<u>event</u> E5, COMP(delay6 <u>after</u> E2 <u>and</u> notS2):
    (PRI,p5),(TØ,delay5),(MEM,O),NOSHARED
No events-exception are defined for the events-exception E4 and E5. We suppose that after delay 4 or delay 5 is elapsed, a standard action is carried out.

<u>link</u> 17 : init
<u>link</u> 18 : init

I7 : <u>on next</u> E4 <u>future if</u> cond 7 <u>do</u> 17
    <u>with</u> priority = pl7
    where T7 is a task which warns the operator for example.
18 : same definition as 17.

Now S1 and S2 have to be defined :
<u>event</u> S1,Elem : (PRI,PS1),(TØ,OO),(MEM,O),
                    NOSHARED
<u>event</u> S2,Elem : (PRI,PS2),(TØ,OO),(MEM,O),
                    NOSHARED

We have (T∅,00) because there is no processing task for these events.
We have also :
Assign S1 to false-true transition of NI n°4 on site 1.
Assign S2 to false-true transition of NI n°5 on site 1.

- VII - CONCLUSION -

In this paper, the main features of currently called events have been investibated. An important distinction has been underlined between occurrences and events. Many attributes has been identified. They are related to the events, to their occurrences or finally to the tasks.

This approach allows to take into account the normal behaviour as well as the abnormal one of an industrial process. No hypothesis has been made about implementation. The use of these concepts is essentially usefull at the specification step. Nevertheless this description is not very far of the programming step.

This study is now applied to specify a flexible workshop for a manufacturing industry.

Such industries are typically sequential systems with a great deal of events and are then an important application field for such a method.

However, it must be noticed that this proposal applies to continous process as well as to discontinous ones.

- BIBLIOGRAPHY -

(1) P. LADET
Outils de structuration temps réel dans la commande des procédés industriels.
Thèse de 3° cycle, INPG, 1977.

(2) P. CASPI, N. HALBWACHS
Algebra of events : a model for parallel and real time systems.
R.R. n°285, january 1982, IMAG, France.

(3) H.G. MENDELBAUM
Structuration dans la conception et la réalisation des systèmes en temps réel.
Thèse de Doctorat es Sciences, IPP, PARIS VI, janvier 1976.

(4) J.P. THOMESSE
SYGARE : une structuration pour la conception d'applications en temps réel et réparties.
Thèse de Doctorat es Sciences, INPL, Nancy, mai 1980.

(5) LE CALVEZ-LISCH
Définition d'un langage de description globale des applications en temps réel.
Thèse de 3° cycle, PARIS VI, janvier 1979.

(6) L. LAMPORT
Time, clocks and the ordering of events in a distributed system.
CACM, july 1978, vol. 21 n° 7, pp. 558-565.

(7) P. LADET
Contribution à l'étude des applications informatiques réparties pour la commande des procédés industriels
Thèse d'Etat INPG Grenoble - 1981.

DISTRIBUTION OF ARCHITECTURE AND ALLOCATION OF FUNCTIONS IN AN INTEGRATED MICROPROCESSOR-BASED SUBSTATION CONTROL AND PROTECTION SYSTEM

M. Kezunović

Energoinvest-Institute for Control and Computer Science, Sarajevo, Yugoslavia

Abstract. This paper is concerned with two critical problems which are appearing during a design procedure of an Integrated Microprocessor-Based Control and Protection System to be applied in H.V. Electric Power Substations. Those problems are related to strategies for selection of distributed architecture and allocation of functions for the computer system beeing designed. Topics discussed in this paper are: Functional characteristics of Electric Power Substation, Functional organization of an Integrated System, Distribution of computer system architecture, Allocation of functions to elements of a distributed microprocessor-based computer system. Five different approaches for distribution of architecture and allocation of functions are proposed as a result of the analysis. A brief discussion of advantages and disadvantages for each of the approaches is outlined having in mind real time digital control application enviroment related to control and protection of Electric Power Substations.

Keywords. Power system control; Computer control; Integrated substation control and protection; Multiprocessing systems; Computer architecture; Microprocessors; Hierarchical systems.

INTRODUCTION

Application of microprocessors in Electric Power Substations was initiated in the mid-seventies. A number of microprocessor-based devices are developed and tested so far in various research organizations, companies, and universities in Europe, U.S.A., Canada, Japan, Australia, India and elswere (Kezunović, 1981a).

The concept of Integrated Control and Protection Systems for Electric Power Substations was approached in the late-seventies. Microprocessor-Based Integrated Control and Protection Systems are beeing developed in the U.S.A. under EPRI/Westinghouse and EPRI/G.E. research projects (EPRI Workshop, 1979) as well as at the American Electric Power Service Corporation (Phadke, Horowitz, 1979). Similar systems are beeing designed in Europe at the LABORELEC Institute in Belgium (Miegroet and others,1981) and at the Electricite de France Research Laboratories in Clamart (Pavard and others, 1981). A proposal is also published by the University of Calgary in Canada (Malik, Hope, 1981). The only commercially available Integrated System can be purchased from Mitsubishi Company of Japan (Sugiyama and others, 1982).

Design of an Integrated Microprocessor-Based Control and Protection System represents quite a complex problem and requires a system approach design methodology (Kezunović, 1981b). A detailed analysis of functional requirements reveals several design criteria which are pretty much inherent in most of the real time digital control applications. This application enviroment requires specific solutions related to computer system architecture (Kezunović, 1982) and hardware, software and communications (Kezunović, 1981c).

This paper is concerned with two critical problems which are appearing during a design procedure of an Integrated Microprocessor-Based Control and Protection System. Those problems are related to strategies for selection of distributed architectures and allocation of functions for the computer system beeing designed.

FUNCTIONAL CHARACTERISTICS

The first step of a design procedure for the Integrated System is to analize characteristics of the control and protection functions in a substation. Having in mind a microprocessor system, the following functions are briefly discussed:

- protection functions,
- control functions,
- data acquisition functions,
- operator interface functions,
- computer system functions.

The most relevant issues such as function time response, I/O characteristics and operational requirements are considered for each of the functions listed.

Protection Functions

In this group of functions are included protection of transmission lines, transformers, buses and some auxiliary functions.

Time response. Protection functions have the most stringent requirements for time response comparing to other functions. For most digital relaying algorithms in use today it is common that the sampling rate for input data is in order of 12 to 16 samples per cycle of the power signal. It is also a common requirement that the fastest time for protection algorithm to react to a fault condition should be 1/4 to 1/2 of the power signal cycle. Therefore, it is needed that two conditions for time response are satisfied: each iteration of the protective relaying algorithm calculations should be completed in the time frame of approximately 1 ms; protective relaying algorithm should recognize a fault and initiate the tripping in the time frame of 4 to 10 ms.

I/O characteristics. It is interesting to note that there are overlapping zones for different protection functions in terms of I/O signal connections. As it is shown in Figure 1, data needed for a line protection come from breaker which is providing data for the bus protection as well. Also, tripping signal outputs for two different protection functions are connected to the same breaker. This situation should be carefully analyzed when an Integrated System interfaces are designed, becouse there is a possibility to provide a common interface for the mentioned signals.

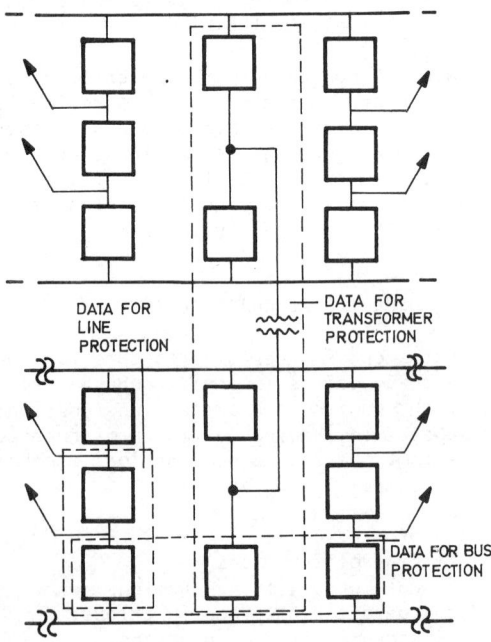

Fig. 1. One line diagram of a substation

Operational requirements. Protection functions operate as direct control functions which means that there is a closed loop which enables comparison of input signal values with the preset values, and tripping signals are issued if there is a defined mismatch. Each of the protection functions is autonomous and requres separate wirings for all I/O signals that are generated, and there is no need for data exchange among various protection functions.

Control Functions

Those functions include: autoreclosing, automatic switching sequences, load shedding, LTC control, auxiliary breaker and switch operations.

Time response. This parameter for the control functions requires sampling time to be in order of tens of milliseconds. Algorithm operation response time may vary and typically ranges from several milliseconds for autoreclosing to several hundred of milliseconds for circuit restoration and several seconds for LTC control.

I/O characteristics. Control functions I/O signals are connected in parallel with the protection signals to circuit breakers and switches. There is a hardwired arbitration logic performing a task of determining which breaker and/or switch should be operated by either control or protection function. Some of the control functions require status information from all of the breakers and switches in a substation, and are capable of operating all of them.

Operational requirements. Some of the control functions need a coordination at the substation level since actions taken by those functions are related to all of the apparatus in substation. Typical functions of this nature are automatic switching sequences, load shedding and LTC control for parallel transformers.

Data Acquisition Functions

The following functions are considered as typical: sequence of events recording, alarming, operator and revenue metering, oscillography.

Time response. Parameter of interest is the resolution with which the data are captured. There is a fearly wide range of requirements in this respect. SOE recording requires resolution in order of 1 ms, which is comparable to the resolution for protective relaying and high speed control functions. Oscillography requires sampling rate of at least 1KHz, which is requirement for revenue metering as well. Alarming and operator metering have relaxed requirements asking for sampling rate of several hundred of milliseconds.

I/O characteristics. Data acquisition functions require signal connections with data flow directions from field apparatus to data acquisition subsystem. Signal sources for analog inputs are instrument transformers with different clases of accuracy for protective relaying, operator metering and revenue metering signals.

Operational requirements. Data acquisition functions have to present a large ammount of collected data to several users. There is a problem associated with data storage requirements. Most demanding data acquisition situations are associated with disturbances in the substation operation, which then requires high speed data collection and presentation to the operator.

Operator Interface Functions

Most commonly provided outputs to the operator come from mimic boards, CRT displys, printing and recording devices and signal panels. Operator initiated inputs are generated using switches, pushbutons and typewriter-like keyboards.

Time response. The most critical requirements in this respect are placed on the CRT interfaces. It is needed that each operator action creates a response which will appear on the screen after 2-3 seconds at most. If completion of action requires several seconds, the CRT interface should generate responses which keep operator informed about the progress of the execution.

I/O characteristics. In this case I/O characteristics are related to formats for representing the data to both local and remote operators. There is a need to use several I/O devices for representation of different formats including charts, lists, tables and one-line diagrams. Also, the interface should enable operator to access data related to power apparatus, to change parameters of the control and protection functions, and to operate breakers and switches.

Operational requirements. It should be noted that operator interface functions are very critical in accepting the Integrated System concept. It should be possible for operator to examine the status of the computer system as well as the substation apparatus. Also, there should be a provision for system testing, maintenance diagnostics, system reconfiguration and system parameter changes.

Computer System Functions

There is a number of functions aimed towards operation of the computer system. Those functions are: automatic restart, self-checking, programs initiation, power failure recovery, system synchronization, tests and diagnostics, maintenance and fail-over switching.

Time response. This requirement is very much dependent on the specific strategy used for system operation, failure recovery and maintenance. However, the most critical time response requirements are in the order of several milliseconds. This situation is related to fail-over switching and to system synchronization function.

I/O characteristics. It is interesting to realize that most of the I/O signals for these functions are generated within the computer system itself. Operation of hardware and execution of software creates various events which are then used by the computer system functions to create outputs which can be directed towards hardware or software modules in the system.

Operational requirements. Those requirements depend on the overall organization of the computer system architecture, hardware, software, and communications. Generally speaking, there are two extremes of computer system operation: loosely coupled and tightly coupled processor operations.

ORGANIZATION OF AN INTEGRATED SYSTEM

Taking into account functional characteristics discussed in the previous section, it is possible to view functional organization of an Integrated System as a hierarchical structure (Kezunović, 1981b). This is shown in Figure 2. It is relevant to emphasize the following characteristics of the functional organization:

- functional levels,
- data flow.

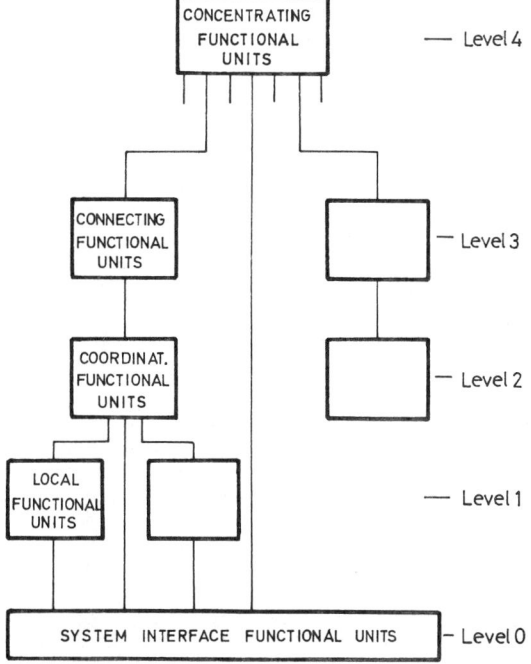

Fig. 2. Hierarchically structured functional organization of an Integrated System

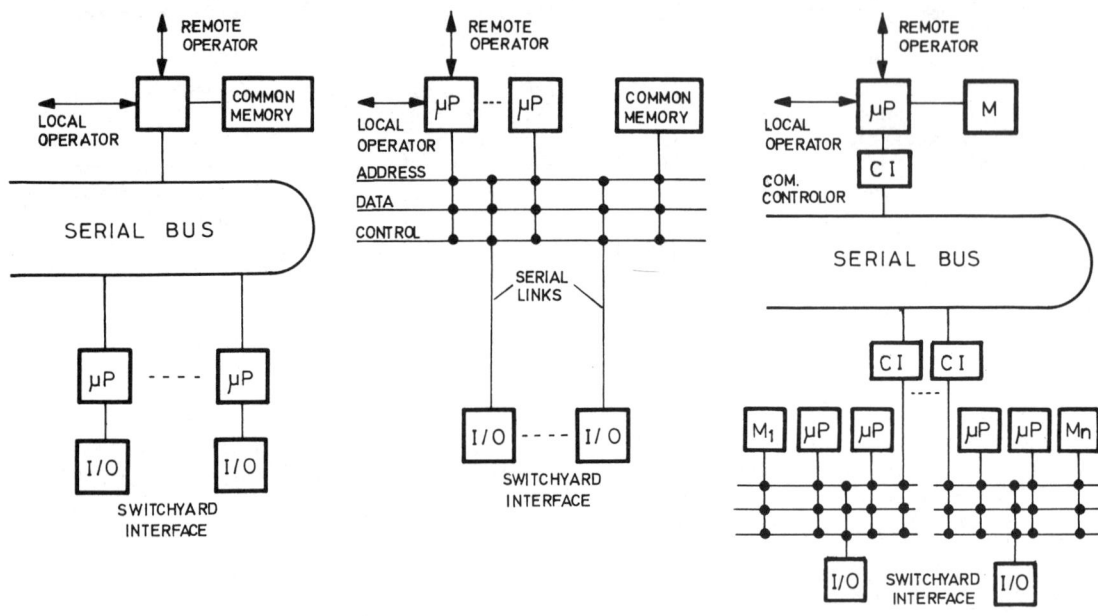

Fig. 3-a. Serial inter-
connections

Fig. 3-b. Parallel inter-
connections

Fig. 3-c. Parallel and serial
interconnections

cated at a centralized place. In the case of a centralized common memory, its location should be at a position which is easelly accessed by the operator interface.

I/O subsystem. This subsystem consists of interfaces towards the switchyard equipment and interfaces towards the local and remote operators.

Switchyard equipment interfaces can be organized in conventional way or as unconventional interface if a multiplexed data link is used. In this case several signals from a portion of the substation is multiplexed on one physical data link and switchyard data are shared by several application functions.

Operator interface can be either centralized or decentralized. Centralized interface assumes that all of the MMI devices are connected a one physical location while decentralized interface assumes that each microprocessor has its own operator interface.

Possible Architecture Configurations

There are three basic types of architecture configurations which are characterized by the physical location of the distributed computer system components. Those locations are: control house, substation switchyard, control house and substation switchyard.

Control house. The three most commonly proposed architectures are given in Figure 3-a, 3-b, and 3-c. Switchyard interfaces can be located in the control house where the signal wires are terminated, or those interfaces can be located in the switchyard.

Architecture with serial interconnections enables separate microprocessors to be directly connected to the switchyard interfaces. Specific input data can be processed at this level and some of this data can be retransmitted to the centralized microprocessor which provides operator interfaces. There is a common system memory provided at the centralized location for storing the overall system data. It is important to note that microprocessors at the lower level have direct access to the output interfaces going to switchyard.

Architecture with parallel interconnections provides common parallel interconnection for all of the microprocessors in the system. Switchyard interfaces as well as the system common memory are also connected to the parallel bus. This architecture enables all of the processors to operate in a similar maner with an arbitration logic implemented for solving the contention problems on the parallel bus.

Architecture with parallel and serial interconnections is implemented using clusters of microprocessors which are connected using parallel buses. Those clusters are then connected

Functional Levels

As it can be seen from Figure 2, there are 5 functional levels within an Integrated System.

Level 0. This level is related to the system interface functions. It includes interfaces to the switchyard equipment and to the local and remote operators.

Level 1. All of the direct control functions are associated with this functional level. It accomodates a number of local functional units which are related to specific portions of the substation switchyard. Typical example are various protection functions which can be viewed as local functional units.

Levels 2 and 3. Those levels enable exchange of data between local units and concentrating functional units at the highest level in the hierarchy. Functions at those levels enable exchange of information among various control, protection and data acquisition functions as well as information exchange between system operators and switching and measuring equipment in the substation switchyard.

Level 4. This level accomodates functions which are related to the operation of the overall substation. It provides interfaces for local and remote operators and some control functions that need information from all of the switching equipment are located at this level.

Data Flow

A brief discussion of the most important data flows within an Integrated System reveals three most prominent data sources.

Substation switchyard. The largest ammount of data for Integrated System comes from the substation switchyard equipment. This data has digital or analog form and is processed by system functions in order to make an operational decision, or in order to make a record of this input data. Most of the data is processed by the control and protection functions contained in the local functional units. However, some substation switchyard data is retransmitted to either some local units that do not have access to this specific data, or to the centralized level in an Integrated System.

Operator interfaces. There is significant ammount of data that is beeing presented to an Integrated System by local and remote operators. This data contains either controls for the switchyard equipment, or requests related to operation of the computer system.

System functions. Each of the system functions performs processing of input data comming from the outside. As a result, there are generated tripping signals, in the case of control and protection functions, or data exchange signals in the case of data acquisition and computer system functions. These sources of data can be located at any of the functional levels within an Integrated System, and data can be routed to system interfaces located either at level 0 or at level 4.

DISTRIBUTION OF COMPUTER SYSTEM ARCHITECTURE

This section presents strategies that can be applied in distributing computer system architecture. Different configurations sugested in this section are selected after a carefull analysis of functional characteristic and system organization (Kezunović, 1982). The following topics are discussed:

- elements of a distributed architecture
- possible architecture configurations

Elements of a Distributed Architecture

The main elements of a distributed system architecture are: microprocessors, communication subsystem, system memory space, I/O subsystem.

Microprocessors. The system considered is a distributed processing computer system. This implies that system consists of a number of microprocessors that are interconnected to form some sort of local area network of microprocessors. It allows for various application functions, allocated to different processors within the system, to be executed simultaneously. The overall system response time is increased as well as the processing power. It is expected that standard 16-bit microprocessors can be utilized to form the distributed processing computer system.

Communication subsystem. It is a vital part of the system since it enables required data exchanges among microprocessors and makes distributed system work. Communication subsystem can be designed using various performance measures such as: speed of data transmission, data transmission protocols, communication procedures for data exchange, physical media for data transmission.

Required speed of data transmission varies between several kbit/sec to several Mbit/sec. Data transmission protocols include the physical layer, data link control layer and the network layer. Communication procedures are related to master slave and peer-to-peer data exchange situations. Physical media used are processor multibus parallel lines and serial data links implemented using one of the following media: twisted pair, coaxial cable, fiber optic links.

System memory space. Each microprocessor in a distributed system has its own private memory. However, there is a large ammount of input data that is collected from the switchyard and processed during each sampling time. There is also a need to represent this data to local and/or remote operators. Therefore, there should be also a common memory which can be either distributed throughout the system, or lo-

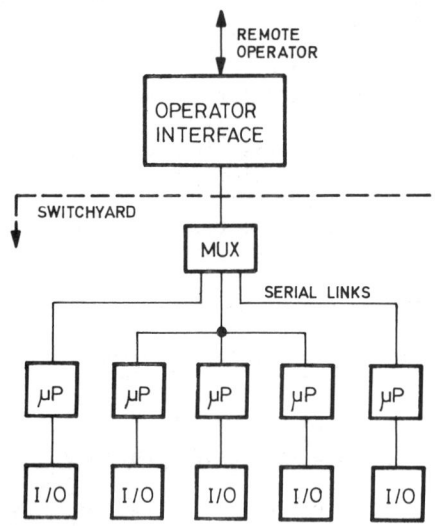

Fig.4. Distributed architecture located in the substation switch yard

with a serial link to the centralized microprocessor which provides coordination for the overall substation system. Common system memory can be distributed to each of the clusters and to the centralized location. This architecture provides a direct connection between microprocessor clusters.

Substation switchyard. This type of architecture is given in Figure 4. It consists of number of microprocessors housed in a cabinet located in the bays of the switchyard close to the switching equipment cabinets. The microprocessor cabinets are connected via serial links either to other cabinets or directly to the centralized operator panel. This operator interface panel can be located in the control house or at a cabinet in the switchyard. System memory is associated with each of the microprocessors.

Control house and switchyard. As it can be seen from Figure 5, one part of distributed system is located in the switchyard and other part in the control house. Switchyard-based microprocessors are located close to the switching equipment. Those microprocessors are interconnected using serial links and are also connected to the control house equipment. Microprocessors in the control house are connected using parallel data buses. Operator interfaces and common memory space are provided at this hierarchical level as well.

ALLOCATION OF FUNCTIONS

This section is concerned with strategies for allocating application functions to processors within a distributed computer system. Some of the most attractive strategies are discussed making a relation to tradeoffs in system performance. Discussion includes the following topics:

- grouping of functions,
- allocation strategies for given architectures.

Grouping of Functions

It is extremely important to note that digital algorithm implementation of control, protection and data acquisition functions shows that there is a number of common algorithm processing modules that can be shared among the mentioned functions. Therefore, there are at least three approaches for grouping the functions: traditional, unconventional, compromising.

Traditional approach. This approach assumes strict separation among control, protection and data acquisition functions. It is not only that software modules have to be separated, but also the hardware has to be uniquely allocated. This means that each of the mentioned functions should be implemented using separate microprocessors, which use a separate memory space as well. It is also assumed that

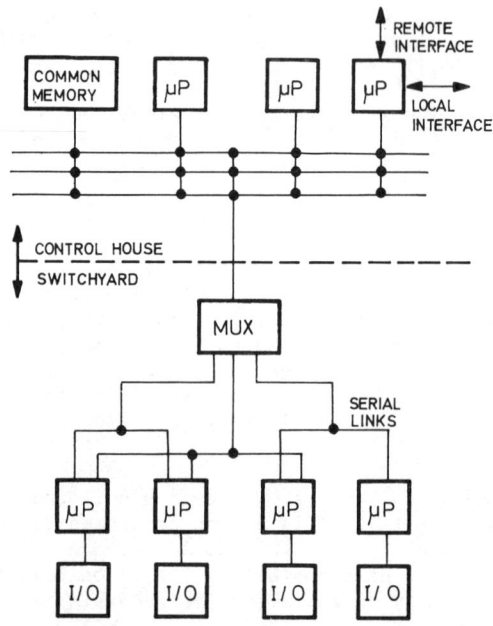

Fig.5. Distributed architecture located in control house and the switchyard.

conventional signal wiring is used. As it can be seen, this approach is very much alike the approach taken using electromechanical and solide state technologies.

Unconventional approach. A full utilization of flexibilities introduced by the LSI technology is applied in this approach. Application functions are assumed to consist of a number of software modules which perform specific data processing. If there is a modul which in neede in several protection functions, then this modul is used once and its capabilities are shared among related application functions. A one step further is taken by allowing sharing of input data as well. It has been shown that some of the input data is shared by at least two application functions. Therefore, a multiplexed data link can be used to bring this data to a location in distributed computer system where this data can be shared by a number of software modules.

All of the mentioned possibilities show that unconventional approach enables optimization of computer system resources by grouping the functions in the most siutable way as far as data processing is concerned.

Compromising approach. As it always has been, there is an approach that compromises the two oposite approaches. In our case this approach assumes a strict separation between control and protection functions, but data acquisition functions are implemented using software modules of the control and protection functions. In some instances there is only input data sharing among those functions. This implies that there are microprocessors performing either control or protection functions. Those processors perform required data processing and then this data is passed on to data acquisition processors. Input data is brought to protection functions using dedicated signal wires, and there are dedicated signal wires for tripping and blocking signals as well.

Allocation Strategies for Given Architectures

The selected architectures given in Figures 3-a, 3-b, 3-c and Figures 3 and 4 are particularly suitable for specific allocation strategies. Those strategies for each of the architectures are discussed emphasizing advantages and disadvantages.

Figure 3-a. This distributed architecture is suitable for conventional function allocation (Phadke, Horowitz, 1979). Each of the protection functions such as line, transformer and bus protection are allocated to a separate microprocessor next to the switchyard interfaces. Control functions are also allocated to a processor at this level. Data acquisition and operator interface functions are performed by a processor at the upper level in the hierarchy. Serial bus for data exchange among processors in the system is a low speed standard serial data bus. Signal wiring to the switchyard is of the conventional type.

Advantages of such a system are: it resembles all of the design criteria of conventional systems, it can be built in a modular way by adding each new function as needed, organization of software is straight forward, communication system is quite simple, system testing is simplified. Disadvantages are the system cost as well as the difficulty in implementing new application functions which can be developed using this system architecture.

Figure 3-b. This proposal (Malik, Hope, 1981) assumes that separation of functions is retained by allocating functions to separate microprocessors on the bus. There is an advantage in providing a common data base accessable by all of the processors. It is also easy to implement the proposed architecture by using a standard Multibus structure. However, this system requires quite a complex arbitration scheme for accessing the parallel bus and this can be a major bottle-neck for implementing high speed protection functions. This architecture can be promising if use of array processors is considered for this application.

Figure 3-c. This approach is suitable for accommodating new design phylosophies as far as the unconventional functional grouping goes (EPRI Workshop, 1979). It is convinient to accommodate protection functions and some of the high speed control and data acquisition functions in the microprocessor clusters at the lower level. Each of the clusters can be related to a bay in the substation switchyard. All of the needed data can be brought through a common multiplexed data link to the cluster. Application functions can be allocated in the optimal way to the processors in the cluster and input data can be shared as needed. After data is processed, it is passed on to the microprocessor which provides operator interface functions. This processor also performs some of the control functions that are related to the overall substation.

This architecture optimizes needed hardware and software and therefore reduces system cost. There is a possibility of reducing the signal wiring cost as well. A great advantage comes from possibilities to develop new application functions by providing complex strategies for control, protection and data acquisition associated with specific power apparatus. On the other hand this architecture and allocation strategy are quite involved for design since they are a breakaway from traditional design phylosophy. There is also a need to ese the most advanced hardware, software and communication concepts which affect the development risk, which is also associated with the commercial affect. An example of such a situation shows that markets for this type of systems are yet to be developed (Sigiyama and others, 1982).

Figure 4. This situation enables strict separation among application functions by having each of the control, protection and data acquisition functions allocated to a separate microprocessor. System signal wiring is conventional and function response time is optimized. This system is straight forward to build, it is easy to test and easy to expand and/or modify. However, there is a problem associated with bringing microprocessors close to the sources of severe EMI which is radiated from circuit breaker arcs and transmission line transients. Neverthelless, this concept seems to be quite attractive (Pavard and others).

Figure 5.This architecture enables conventional allocation of protection and high speed control functions by providing a microprocessor, located in the switchyard, to perform either protection or control function. Signal wiring is conventional using separate interfaces for separate functions. On the other hand, there are several microprocessors located in the control house. Those processors perform control, data acquisition and operator interface functions that require data related to the overall substation This architecture is therefore a good compromise between the traditional solutions and new concepts and it would be interesting to develop and test such a system (Miegroet and others, 1981). Possible disadvantages of such a system may come from the system cost considerations.

CONCLUSIONS

It is quite clear that microprocessor applications in Electric Power Substations have opened new possibilities for designing control, protection and data acquisition systems. Distributed processing approach has enabled concept of integration of all of the substation functions to be implemented. However, there are various strategies for architecture distribution and functional allocation in an Integrated system which have to be carefully investigated in order to optimize system performance criteria. A number of different approaches for Integrated system design. that are beeing investigated today, show that there is no unique approach that is widely accepted. Therefore, it is needed to investigate all of the technical, commercial and operational practice issues in order to make the final decision. This paper has given basic considerations in that respect.

REFERENCES

EPRI Workshop. (1979). Control and Protection of Transmission Class Substations. EPRI, Publ. No WS 79-184. U.S.A.

Kezunović, M. (1981a). Digital Protective Relaying Algorithms and Systems-an Overview. Electric Power Systems Research Journal, 4, 167-180, Switzerland.

Kezunović, M. (1981b). A System Approach to the Design of An Integrated Microprocessor Based Control and Protection System. 8th IFAC World Congress, XX.60-XX.65. Japan.

Kezunović, M. (1981c). Hardware, Software and Communication Requirements of an Integrated Substation Control and Protection System. A.I.M. Conf. on Data Processing for H.V. Power Systems, 10.1.-10.7., Belgium.

Kezunović, M. (1982). Distributed Architectures for an Integrated Microprocessor Based Substation Control and Protection System. 17th Universities Power Engineerig Conference. paper No 7.1. England.

Malik, O.P., Hope, G.S., (1981). Design Concepts for A Distributed Microprocessor-Based Transmission Line Control and Monitoring System. 8th IFAC World Congress, XI.143-XI.147., Japan.

Miegroet, P., Goemiune, P., Monseu, P. (1981). ALPES: Une Famille D' Equipements Programmes pour Les Postes des Reseaux Electriques. A.I.M Conf. on Data Processing for H.V. Power Systems, 9.1-9.11., Belgium.

Pavard, M.M., Bornard, P., Boussin, J.L., Tesseron, J.M. (1981). Digitisation of Line-Control Functions in EHV Substations. 8th IFAC World Congress, XX.54-XX.59, Japan.

Phadke, A.G., Horowitz. S.H. (1979). Report on Computer Applications for Integrated Substations on the American Electric Power System. Colloquium of CIGRE Study Committee No. 34, Australia.

Sigiyama, T., Kameoka,S., Maeda, K., Kaneda, A., Goda, T. (1982).Development and Field Experience of Digital Protection and Control Equipment in Power Systems, IEEE PES Winter Meeting, Paper No 82 WM 173-3, U.S.A.

Copyright © IFAC Real Time Digital Control Applications
Guadalajara, Mexico 1983

FAILURE DETECTION AND PREDICTION SYSTEM BY USING ADAPTIVE DIGITAL FILTER

S. Oe*, Y. Tomita* and T. Soeda**

Department of Information Science and Systems Engineering, Faculty of Engineering, Tokushima University, Tokushima, 770, Japan
**President of Tokushima University, Tokushima, 770, Japan*

Abstract. This paper deals with the detection of the catastrophic failure and the detection and prediction of the deteriorative failure by using adaptive digital filter. Assuming that the change of states of system can be measured as the change of statistical characteristics of observed randam signals, the autoregressive(AR) model is fitted to the signals, and the deteriorative performance index is calculated to detect and predict the deteriorative failure continuously in time. The index is computed to measure the statistical difference between normal and other states quantitatively. The three kinds of indices, that is, quadratic distance of AR parameter differencies, variance of the residuals for prediction scheme, and distance of the time series by the Kullback information, are introduced. Furthermore, the present method is able to eliminate the inferior signals contained in the observed signals and to improve the detection and prediction accuracies by an on-line algorithm. Finally, the effectiveness of the present algorithm is shown by numerical simulations.

Keywords. Failure detection; Failure prediction; Signal processing; Modelling; Time-varying system; Identification.

INTRODUCTION

To maintain the safety and reliability of systems, it is a very important problem to detect and predict failures. There are two types of failures, that is, one is such catastrophic failure as breaking of wire which occures suddenly and the other is such deteriorative failure as the bearing wear which results in a failure state slowly from normal state.
For the former failure, various method for detecting failures in automatic systems using signal processing techniques have been introduced up to now (Nakamizo and others,1979; Willsky,1976),but less consideration for predicting this failure has been made since it seems to be impossible to predict the failure because of suddenly change of state. For the latter failure, as the change from normal state to failure one progresses slowly, it seems to be able to predict failures. But few paper for predicting this failure has been reported (Sata,1978). If the catastrophic failure occurs during the deterioration process, then it is difficult to detect the failure rapidly by using the detection methods of deteriorative failure reported up to now. Furthermore, the inferior signals such as noises are contained in the obserbed signals occasionary, and then the detection and prediction accuracy deteriorates.
In this paper, we propose a new algorithm to detect and predict both catastrophic and deteriorative failure by using an adaptive digital filter(ADF). We assume that the change of states of the system can be measured as the change of statistical characteristics of observed random signals. The presented method enables us to eliminate inferior signales contained in the observed signals and to improve the detection and prediction accuracies by an on-line algorithm. Namely, by using the adaptive digital filter we fit an autoregressive(AR) model to the time series of observed signals, and calculate the deteriorative performance index which shows the degree of change of state at all times. Then we use it as an index to detect and predict the deteriorative failure. Further, we detect the catastrophic failure and inferior signals by the test of whiteness for the error sequences which are the output of adaptive digital filter and eliminate the inferior signals by the on-line algorithm. The deteriorative performance index is calculated to measure the statistical difference berween the normal state and the other one quantitatively. The difference between the normal and the other state is obtained at all times by computing the various indices based on the parameters of the AR models and prediction error sequences. We introduce the three kinds of indices, that is, quadratic distance of the AR parameters, variance of the residuals for the prediction schemes, and distance of Kullback information for the time series. By comparison with these indices, we prove that the index using the Kullback information is better than

the others.Finally,we show the effectiveness of the presented algorithm by numerical computations.

PRINCIPLE OF ADF AND IDENTIFICATION OF TIME SERIES

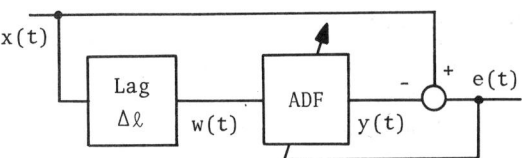

Fig. 1. Identification system by using ADF.

The system to model the time series by using ADF is shown in Fig. 1. Let $w(t)$ and $y(t)$ be the input and output of ADF,respectively,$x(t)$ be the input of this system,$e(t)$ be the error between $x(t)$ and $y(t)$,and M be the order of ADF. Assuming that the relation of input and output of ADF is given by

$$y(t)=\Phi_t^T W(t) \qquad (1)$$

where, $\Phi_t = \text{col}[\phi_{t1},\phi_{t2}, \cdots ,\phi_{tM}]$
$W_t = \text{col}[w(t),w(t-1), \cdots ,w(t-M+1)]$,

ϕ_{tm},$m=1,2,\cdots,M$ are weights of the filter at time t and "T" denotes transpose.
As the construction method of ADF,noisy least mean square(LMS) algorithm which does not need a priori information of the time series such as autocorrelation has been developed(Widrow and co-workers,1975).The algorithm is shown by the following recursive formula

$$\Phi_{t+1} = \Phi_t + 2\nu W_t e(t) \qquad (2)$$

Where ν is a constant value which satisfies the relation

$$0 < \nu < 1/\lambda_{max} \qquad (3)$$

and λ_{max} denotes the maximum eigenvalue of $E[W_t W_t^T]$.

When ADF converges for the stationary time series$\{x(t)\}$,it is proved that the prediction error sequence$\{e(t)\}$ is a Gaussian white noise process with zero mean and constant variance σ_e^2. For lag $\Delta \ell = 1$,the pulse transfer function of this system shown in Fig. 1 is written as

$$\frac{x(t)}{e(t)} = \frac{1}{1-\phi_1 z^{-1} -\phi_2 z^{-2} \cdots -\phi_M z^{-M}} \qquad (4)$$

The equation concludes that the time series $\{x(t)\}$ is identified by follwing Mth order autoregressive model:

$$x(t) = \phi_1 x(t-1) + \phi_2 x(t-2) + \cdots + \phi_M x(t-M) + e(t) \qquad (5)$$

THE DETECTION OF CATASTROPHIC FAILURE AND THE DETECTION AND ELIMINATION OF INFERIOR DATA

After the ADF converges to the stationary time series$\{x(t)\}$,when the abnormal data by catastrophic failure or the inferior one by additive noise in measurement system enters to the system shown in Fig. 1 as input data,the prediction error changes suddenly and the whiteness of $\{e(t)\}$ is destroyed.Thus,it is possible to detect the abnormal or inferior data automatically by using the on-line test of whiteness. We compute the statistic

$$Q(k) = \frac{n}{S_e^2} \sum_{\tau=1}^{p} r_{ee}^2(\tau) \qquad (6)$$

as the detection index of whiteness from error sequencies$\{e(j),j=k-n+1,k-n+2,\cdots,k-1,k\}$,and the test the hypothesis that this statistic $Q(k)$ is approximately distributed as χ^2 with freedom p,where n is the number of $\{e(j)\}$, S_e and $r_{ee}(\tau)$ are variance and autocovariance of $\{e(t)\}$,respectively,and p is free but is less than n/4.The detection method for abnormal data and inferior data is the same processing method as before,but we can distinguish between the inferior data and the occurrence of catastrophic failure by calculating the number of inferior or abnormal data which are detected successively.In the case that there is a number of these data,we decide that the catastrophic failure occures in the system. In other case we decide that the inferior data exist in the acquisition data.Fig. 2 is a flowchart of detecting and eliminating inferior data while modelling.

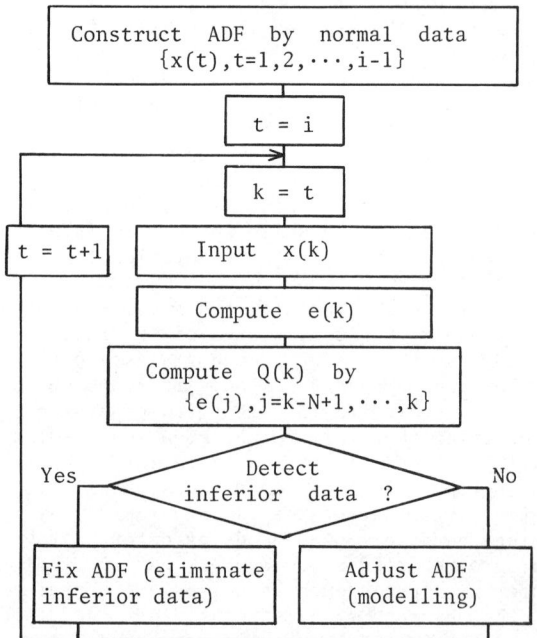

Fig. 2. Flow-chart for detecting and eliminating inferior data.

An explanation of this procedure is as follows:
① ADF converges to the stationary time series $\{x(t), t=1,2,\cdots,i-1\}$ without inferior data.
② Error $e(k)$ is computed by new input data $x(k)$ at time k.
③ Detection index $Q(k)$ is computed from n error sequence $\{e(j), j=k-n+1, k-n+2, \cdots, k\}$.
④ The existence possibility of inferior data is judged from χ^2-test about $Q(k)$.
⑤ If the inferior data do not exist, the weights of ADF are modified by using equation (2), and then go to ② to get next new data and so on. If the inferior data exist, the weights are not modefied before the whiteness of $\{e(j)\}$ to be assured.

Thus, the method does not use the inferior data for modelling, that is, this means the elimination of inferior data. (Oe and co-workers, 1980)

DETERIORATIVE PERFORMANCE INDEX

As mentioned in the preceding section, ADF has the excellent ability to detect the suddenly change of state, so, there are many papers (Widrow and co-workers, 1975; Mehra and co-workers, 1971; Griffiths, 1969; and Kikuch and co-workers, 1977) in which ADF is applied to failure detection such as the catastrophic failure.

In the case that the state changes slowly such as deteriorative failure, little change in statistical characteristic of the output error sequence will occur, because ADF converges to the time series with slow change of the system state. In this case we recognize the change of statistical characteristic of time series as the variation of the model fitted to the input time series, and then evaluate the degree of change of state quantitatively by computing the deteriorative performance index using the model. For the model building procedure based on ADF, special attention is not paid for setting data length used for computing the index, and it is assumed to be N. Since the state changes from normal state to failure one slowly, the sequence of N data may be regarded as stationary. Let us describe some states during the deterioration by i, where $i=0$ denotes the normal state and it is assumed that i increases according to the progress of deterioration.

Using ADF, the observed time series $\{x^i(t), t=1, 2, \cdots, N\}$ for the state i is fitted to the following AR(M) model

$$x^i(t) = \sum_{m=0}^{M} \phi_{i,m} x^i(t-m) + e^i(t) \quad (7)$$

where $\{e^i(t)\}$ is a white Gaussian process with zero mean and variance σ_i^2.

We describe the AR parameter in normal state and another state i by

$$\Phi_0 = [\phi_{01}, \phi_{02}, \cdots, \phi_{0M}]^T$$

$$\Phi_i = [\phi_{i1}, \phi_{i2}, \cdots, \phi_{iM}]^T,$$

respectively. Let us denote a deteriorative performance index which measures the difference between the normal and the other state i

of system by P(i). Assume that the relation between P(i) and state i is shown in Fig. 3.

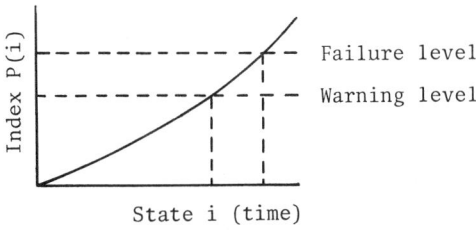

Fig. 3. Relation between state i and index P(i)

Since i corresponds to the degree of deterioration, P(i) increases with elapsed time. Therefor, if we set $P(i)=P_1$ or P_2 as shown in Fig. 3, we can detect the warning or failure time and predict the failure of system at the same time. In this paper, we consider the following three types of deteriorative performance indices which measure the difference between the normal state and the other one.

Derivation of Deteriorative Performance Index

(a) <u>Quadratic distance of AR parameters of difference models</u>
Taking into consideration that the observed data are characterized by AR parameters and the variance of residual sequences, we adopt the following distance $P_a(i)$;

$$P_a(i) = [\Phi_0 - \Phi_i]^T [\Phi_0 - \Phi_i] + \left(\frac{\sigma_0^2 - \sigma_i^2}{\sigma_0^2}\right)^2 \quad (8)$$

(b) <u>Variance of the residual sequences</u>
It is reasonable to measure the goodness of fitted models by the variance of the prediction error. From equation (7), the models fitted to the series $\{x^0(t)\}$ for normal state and the series $\{x^i(t)\}$ for the other state i are given, respectively by

$$x^0(t) = \phi_{01} x^0(t-1) + \cdots + \phi_{0M} x^0(t-M) + e^0(t) \quad (9)$$

$$x^i(t) = \phi_{i1} x^i(t-1) + \cdots + \phi_{iM} x^i(t-M) + e^i(t) \quad (10)$$

If we predict $\{x^i(t)\}$ by using equation (9), then the variance σ_a^2 of the residual sequence $\{a(t)\}$ becomes as follows;

$$\sigma_a^2 = E[a(t)^2]$$
$$= E[\{x^i(t) - \phi_{01} x^i(t-1) \cdots \phi_{0M} x^i(t-M)\}^2]$$
$$= E[\{e^i(t) + (\phi_{i1} - \phi_{01}) x^i(t-1) + \cdots$$
$$+ (\phi_{iM} - \phi_{0M}) x^i(t-M)\}^2]$$
$$= \sigma_i^2 + [\Phi_0 - \Phi_i]^T R_i [\Phi_0 - \Phi_i] \quad (11)$$

where
$$R_i = \begin{pmatrix} r_i(0) & r_i(1) & \cdots & r_i(M-1) \\ r_i(1) & r_i(0) & \cdots & r_i(M-2) \\ \vdots & \vdots & & \vdots \\ r_i(M-1) & r_i(M-2) & \cdots & r_i(0) \end{pmatrix} \quad (12)$$
$$r_i(k) = E[x^i(t)x^i(t-k)]$$

we calculate the ratio of σ_a^2 to σ_0^2 and adopt it as the deteriorative performance index $P_b(i)$;

$$P_b(i) = \frac{\sigma_a^2 - \sigma_0^2}{\sigma_0^2} = \frac{\sigma_i^2}{\sigma_0^2} + \frac{1}{\sigma_0^2}[\Phi_0 - \Phi_i]^T R_i [\Phi_0 - \Phi_i] - 1 \quad (13)$$

Conversely, if we predict $\{x^0(t)\}$ by using equation (10), then the variance σ_b^2 of the residual sequence $\{b(t)\}$ becomes

$$\sigma_b^2 = \sigma_0^2 + [\Phi_0 - \Phi_i]^T R_0 [\Phi_0 - \Phi_i] \quad (14)$$

where
$$R_0 = \begin{pmatrix} r_0(0) & r_0(1) & \cdots & r_0(M-1) \\ r_0(1) & r_0(0) & \cdots & r_0(M-2) \\ \vdots & \vdots & & \vdots \\ r_0(M-1) & r_0(M-2) & \cdots & r_0(0) \end{pmatrix} \quad (15)$$
$$r_0(k) = E[x^0(t)x^0(t-k)].$$

Similarly, we have the index $P_c(i)$ as follows;

$$P_c(i) = \frac{\sigma_b^2 - \sigma_0^2}{\sigma_0^2} = \frac{1}{\sigma_0^2}[\Phi_0 - \Phi_i]^T R_0 [\Phi_0 - \Phi_i]. \quad (16)$$

(c) <u>Kullback information</u>
We introduce the Kullback information (Kullback, 1959) that provides various benifits in the fields of pattern recognition or time series analysis as the deteriorative performance index (Oe and co-workers, 1980).
Let the model of normal state be described by (9) and let the joint probability density function of N random variables, $x_1, x_2, \cdots, x_N$ be $f_0(x_1, x_2, \cdots, x_N)$. Similarly, assume that the model of the other state i is described by (10) and the joint probability density function of N random variables is $f_i(x_1, x_2, \cdots, x_N)$.
The Kullback information shows the mean information for discrimination of the distribution with $f_0(\cdot)$ against the distribution with $f_i(\cdot)$.
Then the Kullback information $I(f_0, f_i)$ per sample of $f_i(x_1, x_2, \cdots, x_N)$ averaged by $f_0(x_1, x_2, \cdots, x_N)$ is given by

$$I(f_0, f_i)$$
$$= \frac{1}{N} \int f_0(x_1, \cdots, x_N) \ln \frac{f_0(x_1, \cdots, x_N)}{f_i(x_1, \cdots, x_N)} dx_1 \cdots dx_N \quad (17)$$

In general $N \gg M$, and using the following relation

$$f_i(x_1, x_2, \cdots, x_N)$$
$$= f_i(x_N | x_{N-1}, x_{N-2}, \cdots, x_1) f_i(x_{N-1}, x_{N-2}, \cdots, x_1)$$
$$= f_i(e_N) f_i(x_{N-1}, x_{N-2}, \cdots, x_1) = \cdots$$
$$= f_i(e_N) f_i(e_{N-1}) \cdots f_i(e_1),$$

equation (17) is rewritten as follows;

$$I(f_0, f_i) = \frac{1}{N} E_0 \left[\ln \left\{ \prod_{t=1}^{N} \frac{\frac{1}{\sqrt{2\pi}\sigma_0} \exp\left(-\frac{(x(t) - \phi_{01}x(t-1) - \cdots - \phi_{0M}x(t-M))^2}{2\sigma_0^2}\right)}{\frac{1}{\sqrt{2\pi}\sigma_i} \exp\left(-\frac{(x(t) - \phi_{i1}x(t-1) - \cdots - \phi_{iM}x(t-M))^2}{2\sigma_i^2}\right)} \right\} \right]$$

$$= \ln \frac{\sigma_i}{\sigma_0} - \frac{1}{2\sigma_0^2 N} \cdot \sum_{t=1}^{N} E_0 [(x(t) - \phi_{01}x(t-1) - \cdots - \phi_{0M}x(t-M))^2] + \frac{1}{2\sigma_i^2 N}$$
$$\cdot \sum_{t=1}^{N} E_0 [(x(t) - \phi_{i1}x(t-1) - \cdots - \phi_{iM}x(t-M))^2] \quad (18)$$

where $E_0[\cdot]$ means that the expectation is taken about the time series with p.d.f. $f_0(\cdot)$.
Therefore, the second and third terms of right hand side in equation (18) are written as follows;

$$E_0[(x(t) - \phi_{01}x(t-1) - \cdots - \phi_{0M}x(t-M))^2]$$
$$= E_0[(x^0(t) - \phi_{01}x^0(t-1) - \cdots - \phi_{0M}x^0(t-M))^2] = \sigma_0^2 \quad (19)$$
$$E_0[(x(t) - \phi_{i1}x(t-1) - \cdots - \phi_{iM}x(t-M))^2]$$
$$= E_0[(x^0(t) - \phi_{i1}x^0(t-1) - \cdots - \phi_{iM}x^0(t-M))^2]$$
$$= E_0[(e^0(t) + (\phi_{01} - \phi_{i1})x^0(t-1) + \cdots (\phi_{0M} - \phi_{iM})x^0(t-M))^2] = \sigma_0^2 + [\Phi_0 - \Phi_i]^T R_0 [\Phi_0 - \Phi_i]. \quad (20)$$

We adopt the $I(f_0, f_i)$ which is obtained by substituting (19) and (20) to (18) as the deteriorative performance index $P_d(i)$.

$$P_d(i) = I(f_0, f_i)$$
$$= \frac{1}{2}\{\ln \frac{\sigma_i^2}{\sigma_0^2} - 1 + \frac{1}{\sigma_i^2}(\sigma_0^2 + [\Phi_0 - \Phi_i]^T R_0 [\Phi_0 - \Phi_i])\} \quad (21)$$

Similarly, we can get the Kullback information $I(f_i, f_0)$ of $f_0(x_1, x_2, \cdots, x_N)$ by the average of $f_i(x_1, x_2, \cdots, x_N)$ which is adopted as a deteriorative performance index $P_e(i)$.

$$P_e(i) = I(f_i, f_0)$$
$$= \frac{1}{2}\{\ln \frac{\sigma_0^2}{\sigma_i^2} - 1 + \frac{1}{\sigma_0^2}(\sigma_i^2 + [\Phi_0 - \Phi_i]^T R_i [\Phi_0 - \Phi_i])\} \quad (22)$$

where R_i is the autocovariance function defined by equation (12).

The Kullback information $I(f_0, f_i)$ means the distance between two distributions $f_0(x_1, x_2, \cdots, x_N)$ and $f_i(x_1, x_2, \cdots, x_N)$, but it does not possess the symmetric property, that is, $I(f_0, f_i) \neq I(f_i, f_0)$. So, we consider the following distance $J(f_0, f_i)$ which is called the divergence measure, and adopt it as the index $P_f(i)$.

$$P_f(i) = J(f_0, f_i) = I(f_0, f_i) + I(f_i, f_0)$$

$$= \frac{1}{2\sigma_i^2} \{\sigma_0^2 + [\Phi_0 - \Phi_i]^T R_0 [\Phi_0 - \Phi_i]\}$$

$$+ \frac{1}{2\sigma_0^2} \{\sigma_i^2 + [\Phi_0 - \Phi_i]^T R_i [\Phi_0 - \Phi_i]\} - 1 \quad (23)$$

Comparison of Deteriorative Performance Index

As the deteriorative performance index is used to detect the statistical difference between the time series of normal state and of the other state, it should possess the properties which the index is zero only for the case that the statistical characteristics of the two time series are equal and it is positive for the other case. The conditions with the same characteristics, that is, $P(i)=0$, holds only when $\Phi_0=\Phi_i$ and $\sigma_0^2=\sigma_i^2$. Considering these conditions, we investigate those six indices.

(a) $\underline{P_a(i)}$ This index has the characteristic such that it becomes zero only for the case of $\Phi_0=\Phi_i$ and $\sigma_0^2=\sigma_i^2$, but it has a defect of lackness of the theoretical strictness because of intuitive index. However, for the simplicity of computation, this index is preferable when the fast computational speed is need.

(b) $\underline{P_b(i)}$ This index has been used formerly for the detection of catastrophic failure. The index is zero when $\Phi_0=\Phi_i$ and $\sigma_0=\sigma_i$, but if $\sigma_i^2 < \sigma_0^2$ there is a case that the index happens to be zero in spite of $\Phi_0 \neq \Phi_i$. In other words, this index has a serious defect that it may not defect the difference of two time series even if the two time series have different statistical characteristics.

(c) $\underline{P_c(i)}$ For this index to be zero, it is necessary that $\Phi_0=\Phi_i$. But, $\sigma_0^2=\sigma_i^2$ is not necessary. Since the index has not any information about the variance of time series, it has a defect that this index cannot detect the difference between two time series for the case that the two time series have the same autocorrelation but different variance.

(d) $\underline{P_d(i), P_e(i), P_f(i)}$ The relations between $P_d(i)$ and $P_c(i)$ and between $P_e(i)$ and $P_b(i)$ are obtained by

$$P_d(i) = \frac{1}{2}\{\ln\frac{\sigma_i^2}{\sigma_0^2} - 1 + \frac{\sigma_0^2}{\sigma_i^2} + \frac{\sigma_0^2}{\sigma_i^2} P_c(i)\} \quad (24)$$

$$P_e(i) = \frac{1}{2}\{\ln\frac{\sigma_0^2}{\sigma_i^2} + P_b(i)\} \quad (25)$$

Therefore, it is evident that $P_d(i)$ and $P_e(i)$ have more information than $P_b(i)$ and $P_c(i)$, respectively. Using the following inequality;

$$[\Phi_0 - \Phi_i]^T R_0 [\Phi_0 - \Phi_i] \geq 0 \; , \; [\Phi_0 - \Phi_i]^T R_i [\Phi_0 - \Phi_i] \geq 0 \; ,$$

we have

$$P_d(i) \geq \frac{1}{2}\{\ln\frac{\sigma_i^2}{\sigma_0^2} - 1 + \frac{\sigma_0^2}{\sigma_i^2}\} \geq 0 \quad (26)$$

$$P_e(i) \geq \frac{1}{2}\{\ln\frac{\sigma_0^2}{\sigma_i^2} - 1 + \frac{\sigma_i^2}{\sigma_0^2}\} \geq 0 \quad (27)$$

$$P_f(i) \geq \frac{1}{2}\{\ln\frac{\sigma_0^2}{\sigma_i^2} + \frac{\sigma_i^2}{\sigma_0^2}\} - 1 \geq 0 \quad (28)$$

In these equations, the equality holds only when $\Phi_0=\Phi_i$ and $\sigma_0^2=\sigma_i^2$. Thus, these indices have the characteristics that they become zero only when the time series have the same statistical characteristics and become positive for the other case. Furthermore, these indices denote the difference of amplitudes and dynamic characteristics of two time series.
In consequence, they are very excellent indices compared with the other indices.
$P_c(i)$ and $P_d(i)$ need compute the autocovariance function R_0 in normal state only one time and need not compute R_i in state i which changes in time. Consequently, $P_c(i)$ and $P_d(i)$ have the high speed processing characteristic compared with the other indices.

FAILURE DETECTION AND PREDICTION SYSTEM

The failure detection and prediction system by using ADF is shown in Fig. 4. The methods of pre-processing the inferior data and modelling the acquisition data are shown in Fig. 2.
In this system, the function which counts the number of abnormal data and alarms to detect the catastrophic failure is added to the system shown in Fig. 2. In addition to it, this system has a subsystem which predicts and detects the deteriorative failure by using the deteriorative performance index computed by the proposed method. This system enables us to compute the detection index $Q(k)$ for abnormal data and the deteriorative performance index $P(k)$ every time getting one input data. Since the parameter Φ_k is a random vector, the index $P(k)$ is a random variables. Thus, we can have the smoothed

performance index by calculating the moving average of the index.

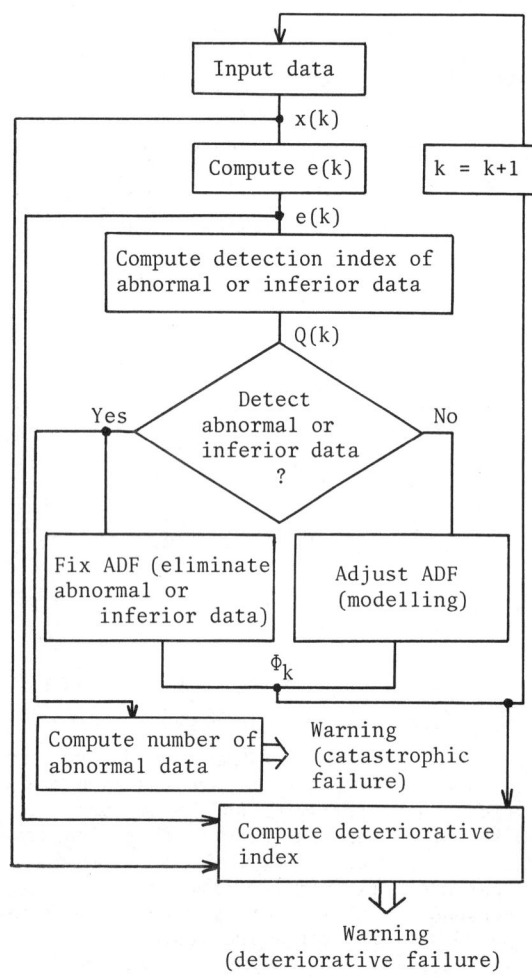

Fig. 4. Failure detection and prediction system by using ADF.

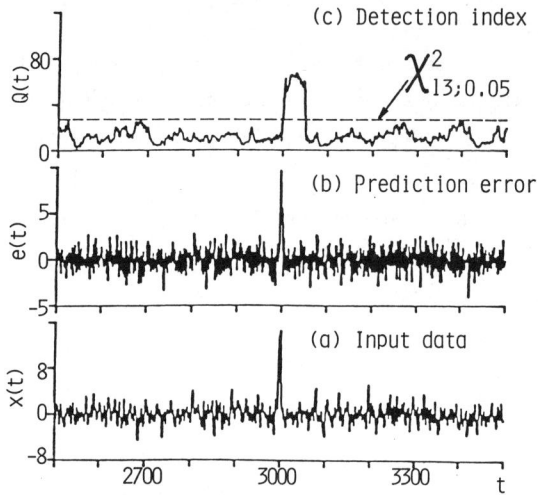

Fig. 5. Detection of inferior data contained in stationary time series.

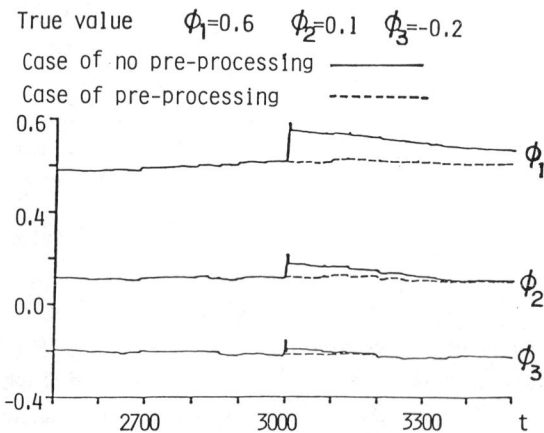

Fig. 6. Time history of estimated parameters.

NUMERICAL EXAMPLES

Detection and Elimination of Inferior data

We use the simulation data generated by the stationary third order autoregressive model as follows;

$$x(t)=\phi_1 x(t-1)+\phi_2 x(t-2)+\phi_3 x(t-3)+\varepsilon(t).$$

Fig. 5(a) shows the simulation data which include the inferior data between t=3000 and 3005 in the stationary time series generated from the model with $\phi_1=0.6, \phi_2=0.1$, and $\phi_3=-0.2$.

Fig. 5(b) shows the residual sequence $\{e(t)\}$, and (c) shows the time history of detection index Q(t) calculated with n=52 and p=13. In the Figure, Q(t) is larger than dotted line denoting $\chi^2_{13;0.05}$ between t=3000 and 3057, which shows that the number of inferior data is about (3057-3000)-52=5. Using this method, we can count the number of inferior data.
Fig. 6 shows the time history of three parameters about two cases where one is pre-processed by the method presented here and the other is not pre-processed. After the inferior data disappeared, the estimated parameters show the violent variation for a while in the case of no pre-processing, but the exact model is get in the case of pre-processing. The effectiveness of this method is confirmed from these results.

Computation of Deteriorative Performance Index

Assuming the deteriorative failure, we generate the simulation data by using the second order AR model with time variant parameters as follows;

$$x(k)=\phi_{k1} x(k-1)+\phi_{k2} x(k-2)+\varepsilon(k)$$

$$\phi_{k1}=0.3+8\times 10^{-7} k, \quad \phi_{k2}=0.3+4\times 10^{-7} k$$

where $\{\varepsilon(k)\}$ is a Gaussian white noise process with zero mean and unit variance.

TABLE 1 Estimated Parameters and Deteriorative Performance Index

k	True values ϕ_k		Estimated values $\hat{\phi}_k$		Estimated variance of $\{e(k)\}$ $\hat{\sigma}_k^2$	$P_a(k)$	$P_d(k)$	$P_e(k)$	$P_f(k)$
5000	ϕ_{k1}	0.304	$\hat{\phi}_{k1}$	0.309	1.0412	—	—	—	—
	ϕ_{k2}	0.302	$\hat{\phi}_{k2}$	0.290					
54000	ϕ_{k1}	0.343	$\hat{\phi}_{k1}$	0.347	1.0413	2.009×10^{-3}	1.98×10^{-3}	2.22×10^{-3}	4.19×10^{-3}
	ϕ_{k2}	0.311	$\hat{\phi}_{k2}$	0.321					
103000	ϕ_{k1}	0.382	$\hat{\phi}_{k1}$	0.386	1.0415	7.941×10^{-3}	7.63×10^{-3}	10.10×10^{-3}	17.73×10^{-3}
	ϕ_{k2}	0.334	$\hat{\phi}_{k2}$	0.341					
152000	ϕ_{k1}	0.422	$\hat{\phi}_{k1}$	0.423	1.0419	17.612×10^{-3}	16.89×10^{-3}	27.39×10^{-3}	44.28×10^{-3}
	ϕ_{k2}	0.361	$\hat{\phi}_{k2}$	0.356					
201000	ϕ_{k1}	0.461	$\hat{\phi}_{k1}$	0.460	1.0427	30.914×10^{-3}	29.64×10^{-3}	63.30×10^{-3}	92.84×10^{-3}
	ϕ_{k2}	0.380	$\hat{\phi}_{k2}$	0.379					
250000	ϕ_{k1}	0.500	$\hat{\phi}_{k1}$	0.496	1.0444	47.802×10^{-3}	45.80×10^{-3}	148.76×10^{-3}	194.56×10^{-3}
	ϕ_{k2}	0.400	$\hat{\phi}_{k2}$	0.402					

Assuming that the time series until k=5000 indicates the normal state, we show the estimated value $\hat{\phi}_{k1}$ and $\hat{\phi}_{k2}$ of parameter ϕ_{k1} and ϕ_{k2}, estimated variance of error sequence $\hat{\sigma}_k^2$, and deteriorative performance indices $P_a(k)$, $P_d(k)$, $P_e(k)$ and $P_f(k)$ at time k=5000, 54000, 103000, 152000, 201000 and 250000 in Table 1. This Table shows that the parameters are estimated accurately and the deteriorative performance indices detect the change of state precisely. From these two numerical examples, it is verified that the failure detection and prediction system proposed in this paper can not only detect the catastrophic failure but also detect and predict the deteriorative failure.

CONCLUSION

In this paper, first we proposed a method to build the AR model while detecting and eliminating inferior data by using ADF and second we introduced new deteriorative performance indices. As a result of comparing these indices, it has been made clear that the indices based on the Kullback information are the most superior. Furthermore, we proposed a new failure detection and prediction system which was able to detect the catastrophic failure and to detect and predict the deteriorative failure on line. Finally, we showed the usefulness of the present method by simulations.

ACKNOWLEDGEMENT

We would like to express our cordinal thanks to Dr.S. Omatu and Dr.H. Sakai for their helpful consultation and valuable assistance.

REFERENCES

Nakamizo, T., K. Akizuki and T. Soeda (1979). Statistical failure detection methods of system. Trans. Soc. Instrum & Control Eng., 18, 471-480 (in Japanese).

Willsky, A.S. (1976). A survey of design method for failure detection in dynamic systems. Automatica, 12, 601-611.

Sata, T. (1978). About malfunation detection and prediction techniques in mechanical system. Symposium on malfunction detection and prediction techniques, 1-8 (in Japanese).

Widrow, B., J. Glover, J. Mccool, J. Kaunitz, C. Williums, R. Hearn, J. Zeidler, E. Dong and R. Gooldlin (1975). Adaptive noise cancelling principles and applications, Proc. IEEE, 63, 12, 1692-1716.

Mehra, R.K. and J. Peschon (1971). An innovations approach to fault detection and diagonosis in dynamic systems. Automatica, 7, 637-640

Griffiths, L.J. (1969). A simple adaptive algorithm for real time processing in antenna array. Proc. IEEE, 57, 1696-1704.

Kikuch, A., S. Omatu, T. Soeda (1978). Detections of signal contained in Noise and extraordinary level of signal by using adaptive digital filter. IECE Trans., J61-A, 7, 657-664 (in Japanese).

Kullback, S. (1959). Information Theory and Statistics, John Wiley & Sons, New York, Chap.1, PP. 3-7.

Oe, S., T. Soeda and T. Nakamizo (1980). A method for predicting the failure or life by using AR models. Int. J. Syst. Sci., 11, 10, 1177-1188.

Oe, S., Y. Shinohara and T. Soeda (1980). Pre-processing and modelling for time series containing extraordinary data by using ADF. IECE Trans., J63-A, 9, 643-644 (in Japanese).

Copyright © IFAC Real Time Digital Control Applications
Guadalajara, Mexico 1983

DEVELOPMENT OF NUCLEAR POWER PLANT AUTOMATED REMOTE PATROL SYSTEM

R. Nakayama, K. Kubo, K. Sato and J. Taguchi

Nuclear Engineering Laboratory, Toshiba Nuclear Energy Group, Kawasaki, Japan

Abstract. An Automated Remote Patrol System was developed for a remote inspection, observation and monitoring of nuclear power plant's components. This automated remote patrol system consists of; a vehicle moving along a monorail; three rails mounted in a monorail for data transmission and for power supply; an image fiber connected to a TV camera; an arm type mechanism (manipulator) for moving image fiber; a computer for control and data processing and operator's console. Special features of this Automated Remote Patrol System are as follows:
1. The inspection vehicle runs along horizontal and vertical (up/down) monorails.
2. The arm type mechanism (manipulator) on the vehicle is used to move image fiber.
3. Slide type electoric collectors are used for data transmission and power supply.
4. Time-division multiplexing is adapted for data transmission.
5. Voice communication is used for controlling mechanisms.
6. Pattern recognition is used for data processing.
The experience that has been obtained from a series of various tests is summarized.

Keywords. Nuclear Power Plant, Patrol, Automation, Monorail, Manipulator Vehicle, Computer Control, Time-divission multiplexing.

INTRODUCTION

Nuclear power plants are patrolled to inspect and monitor many parts of them in various ways.

The patrolling job is now performed by workers. An idea occurred to develop a monitor robot to perform the patrolling job now undertaken by operators. The introduction of such a robot would reduce human exposure to radiation and lighten the burdens imposed on the operators. A pilot system incorporating such a robot must be able to move and recognize patterns as it is supposed to take the place of man. The patrol system introduced herein employs a drive motor mounted on an inspection vehicle which runs on a monorail; and recognizes patterns by image processing and voice input and output.

The system is described in detail in the following pages.

SYSTEM DESCRIPTION

The patrol system consists of the following.
(1) Monorail
(2) Vehicle
(3) Manipulator mounted on the vehicle
(4) Signal transmitters
(5) Control units (vehicle/manipulator)
(6) Pattern recognition units (image processor, voice input/output units)
(7) Display units and control console

Here is a brief description of the patrol system (see figure 1).

The vehicle moves on the monorail laid in a nuclear power plant. The vehicle stops at an inspection position that is input from the audio input unit, where the manipulator is driven by signals from the control unit and an image fiver/TV camera picks up an image for inspection. The video signals thus obtained are routed via rails within the monorail to the display unit which is installed at a fixed position. The image processing unit recognizes the pattern, and a voice output is sent to notify the operator whether there is anything abnormal (see figure 2). It is possible to do some light work with the fingers of the manipulator, for example, operating a switch to the on or off position, by controlling the manipulator.

(1) Monorail
The monorail is made of aluminum. It is light in weight, and can be easily laid inside a nuclear power plant. The monorail has a cross section of the shape ⊏⊐ (see figure 3). There are 3 rails laid along the bottom of the monorail. Two of the 3 rails are for supplying electric power and transmitting manipulator control signals. The other rail is used for transmitting video signals.

The transmitter for vehicle control is mounted on the side of the monorail near a stop position. There is also a reflector located near a vehicle stop position to improve the accuracy of the vehicle stopping at the stop position.

The monorail which moves up and down has a row of racks along the bottom to improve running performance. The course of the monorail used for the last tests is shown in figure 1.

(2) Vehicle

Figure 4 shows external views of the vehicle and manipulator. The vehicle can be roughly divided into two parts as follows:
(a) Vehicle drive unit
(b) Carrier

(a) Vehicle drive unit

A DC voltage of 24 V is supplied from the power rail in the monorail to the DC motor via a slide type electoric collectors.

The output power of the DC motor is transmitted via a gear to the drive wheel. The motor output shaft has a detector for measuring running vehicle speed so that vehicle speed can be monitored at all times. An approximate location of the vehicle on the monorail can be determined by integrating the value given by the detector. The motor has an electromagnetic brake to brake the vehicle to a stop.

(b) Carrier

The carrier mounts a TV camera, manipulator, control circuits (for the vehicle and manipulator), signal transmitter, etc.

The carrier is built of FRP (fiber reinforced plastics) on the outside and aluminum structural members to reduce wieght.

(3) Manipulator

Figure 4 shows an external view of the manipulator, and figure 5 its degree of freedom.

The manipulator performs two functions. One of the two functions is to move the image fiber to any of various positions against the object to be inspected; and the other is to do light work by using the finger mechanism at the tip of the manipulator.

A manipulator which satisfies the needs of these two functions is shown in figure 5.

Each axis of the manipulator is provided with a DC motor, reduction gear, rotary encoder, and limit switch. Each part of the manipulator is driven by the DC motor.

Manipulator operations are described in reference to figure 3. Motor 1 swings the manipulator relative to the vehicle, and motor 2 changes the root angle of the manipulator. (Motors 1 and 2 are installed inside the vehicle.)

Motor 3 changes the angle of the manipulator part which corresponds to a human elbow.

Motor 4 rotates the image fiber axis; motor 5 turns the manipulator fingers; and motor 6 moves the fingers forward and back, that is, to press the pushbutton switch on and off.

A manipulator which could satisfy the needs of the abovementioned two functions was designed and manufactured by incorporating these 6 degrees of freedom.

The structural members are made of aluminum, the gear is built of steel and the external covering of FRP. This made it possible to reduce the manipulator weight.

(4) Signal Transmitters

Figure 6 shows a system block diagram. The signal transmitters are used for two purposes, that is, for controlling the vehicle and for controlling the operation of the manipulator.

The signal transmitter for controlling the vehicle is fixedly installed on the monorail. The control unit on the ground transmits control signals one way to the vehicle about its address and route.

An optical transmission system using photoelectric elements and infrared rays is employed. The optical transmission signal transmitter has a capacity of 8 bits. (The capacity must be increased if there are many inspection points on the monorail.)

The signal transmitter for manipulator control is provided for the vehicle and the control unit on the ground respectively. After the vehicle stops at an inspection position, the control unit sends a predetermine angle, converted from an analog signal into a digital signal, to the vehicle by multiplex. The vehicle sends an angle signal from the manipulator shaft angle detector, processed in a similar way, to the stationary control unit. An FM transmission system is used for sending these signals through the power rail in the monorail at a multiplex transmission speed of 2400 bits/sec.

(5) Control Units (for Vehicle and Manipulator)

As shown in the system block diagram (figure 6), there are two types of control units, one for controlling the vehicle and the other for controlling the operation of the manipulator.

Vehicle control involves two kinds of control, that is, running speed control and stop position control. These two kinds of control are performed to move the vehicle accurately to a preset inspection position.

The running speed control unit is mounted inside the inspection vehicle, and regulates DC motor rpm with a pulse width control system (PWM), which selects either low speed or high speed.

The stop position control unit uses marks attached to the side of the monorail and a positioning sensor mounted aboard the vehicle to accurately stop the vehicle at an inspection position. As the vehicle detects a mark, the running speed control unit slows down the vehicle to low speed. When the vehicle detects the next mark, the DC motor is switched off and the vehicle is braked to a stop.

The manipulator control unit is mounted inside the vehicle, and controls by accurately transmitting the manipulator's each axis angle from the signal transmitter to the DC motors. A teaching playback system for teaching the rotary angle of each axis corresponding to a manipulator posture is employed. All commands for manipulator postures are compared with the values given by the rotary angle detector attached to the DC motor for each axis, and the difference between taught and actual posture values is reduced by controlling the difference.

(6) Pattern Recognition Units

As shown in the system block diagram (figure 6), two kinds of pattern recognition units are used. One is a voice recognition unit, and the other an image recognition unit. (Both are installed in the control panel.)

The voice recognition unit uses a microphone on the operator console to receive input data on vehicle destination and manipulator posture. It also outputs image pattern recognition data, which is mentioned later, and data on whether the above-mentioned voice inputs are correct.

Combinations of simple sentences are put to spectral analysis for recognition.

The image recognition unit analyzes video signals from the TV camera mounted aboard the vehicle to find whether anything is abnormal. It can recognize colors and the presence or absence of objects by using such data as luminance signal and color difference signal levels in video signals.

(7) Display Units and Operator Console

The display units and operator console appear as shown in figure 1. Three color CRTs are used as display units. One is for the TV camera mounted aboard the vehicle; another is for displaying the location of the vehicle on the monorail and the posture of the manipulator; and the last is for displaying input and output data.

The operator console has a voice input microphone, voice output speaker, auto/manual select switch, and various select switches to be used in manual operation. The display units and the operator console can be operated by a single operator.

TESTS

The following tests were conducted on the main components of the system.
(1) Vehicle performance test
(2) Vehicle stop accuracy test
(3) Test on signal transmission through rails
(4) Voice input and output test
(5) Image recognition test on image fiber
(6) Overall system test

(1) Vehicle Performance Test

Using a test course as shown in figure 2, the vehicle performance test was conducted on a vehicle carrying a 20-kg dummy weight on the following items.
(a) Running speed control test
(b) Vertical running test
(c) Running noise level measurement

The vehicle was found to have no performance problems as a result of the above tests.
(a) Running speed control test

Using the pulse width modulation system, running speed variation with DC motor rpm was measured.

Vehicle speed changed from 5 m/min to 31 m/min when the pulse width ratio was changed from 13% to 75%, indicating an approximately proportional change.
(b) Vertical running test

The vehicle tends to run slower in moving upward on the vertical section of the monorail than in running level on the horizontal section of the monorail, and faster in moving downward. This was corroborated by the test as follows:
Horizontal running speed:
    31 m/min (with 20-kg load)
Vertical upward running speed:
    23 m/min (with 20-kg load)
Vertical downward running speed:
    40 m/min (with 20-kg load)
It was found that vehicle speed changed within the range of 0.7 to 1.3 times.
(c) Noise level measurement

The noise produced by a running vehicle was measured. Though a slightly high noise level was indicated by a descending vehicle, the noise generated by the vehicle was negligible on the whole.

(2) Vehicle Stop Accuracy Test

The reflector located near a stop position was used as a mark in testing vehicle stop accuracy. The test revealed that the vehicle would stop within an accuracy of ±2 mm or less on the horizontal section of the monorail at a maximum speed of 31 m/min and a minimum speed of 5 m/min.

(3) Test on Signal Transmission through Rails

A signal transmission test was conducted by using the rails laid in the monorail. A video signal of P-K 0.75 V, susceptible to noise effect, was used for the test.

The video signals generated by the TV camera mounted on the vehicle were transmitted via the collector to the TV monitor on the ground. The results were as follows:
(i) Video transmission was satisfactory when the vehicle was not running.
(ii) Noise from the power rail was mixed with the video signal when the vehicle was running.

By studying these rsults, it was decided to employ a line noise filter for the vehicle and the ground control system. The noise was virtually eliminated by the use of the line noise filters.

(4) Voice Input and Output Test

First, a voice input and output was conducted with a combination of a spectral analyzer, voice synthesizer and microcomputer. Voice inputs, such as "coffee" and "beer", from an ordinary person were reproduced in numerals corresponding to the input words on the CRT with an accuracy of about 90% or more. Voice outputs also showed approximately the same accuracy. On the basis of these findings, 4 destination addresses were assigned to the vehicle and a voice command test was conducted. The results were approximately the same as those of the previous test. Because speakers were not specified, recognition accuracy was sometimes lower than normal. There will be little problem in actual cases because a specific speaker will be used.

(5) Image Recognition Test on Image Fiber

If an image fiber is connected to a TV camera, its monitor images differ slightly from general TV images, that is, appear a little darker than the latter. It was found, however, that even such images were good for the above-mentioned pattern recognition.

(6) Overall System Test

The overall system test has just been started subsequent to the individual function tests mentioned in the preceding pages. It is not possible to present the detailed results of the overall system test as yet. They will be explained in detail at the coming general meeting.

CONCLUSIONS

This monorail type Automated Remote Patrol System was developed for a remote inspection and monitor of nuclear power plant's components.

Several tests were conducted on the main components of this system. The results of these tests were completely good. And now overall system test has just been started. Main results of system tests satisty the major purpose of the patrolling job.

After these tests, this Automated Remote Patrol System will be applied to nuclear power plant in near future.

Development of Automated Remote Patrol System

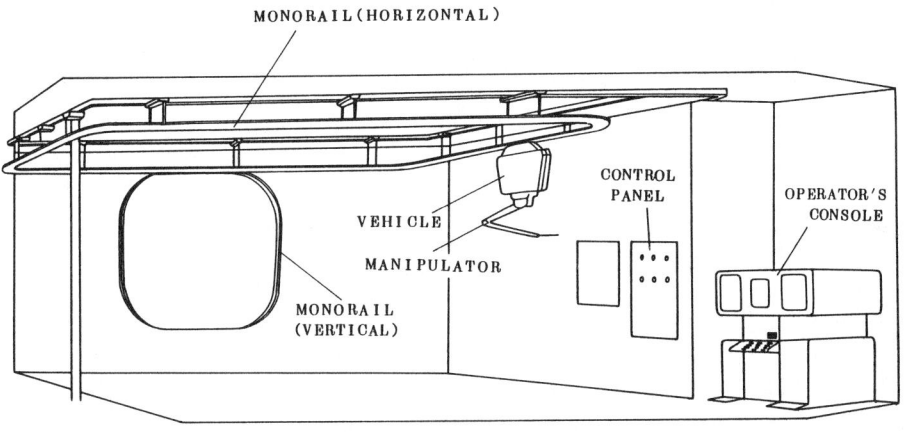

Fig. 1 Automated Remote Patrol System

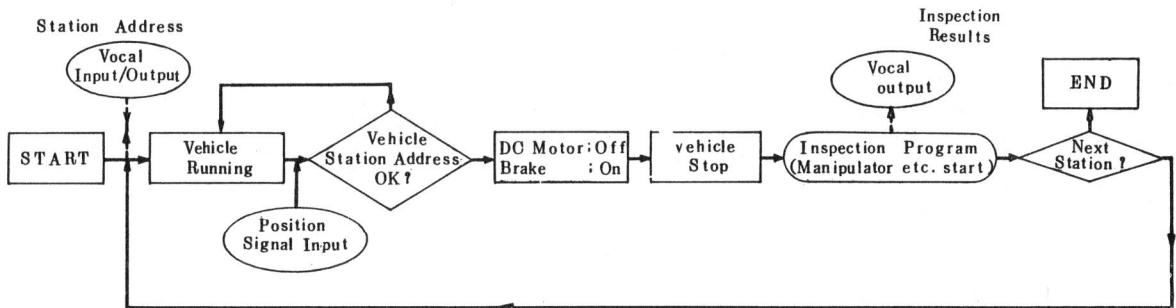

Fig. 2 Inspection Flow Diagram

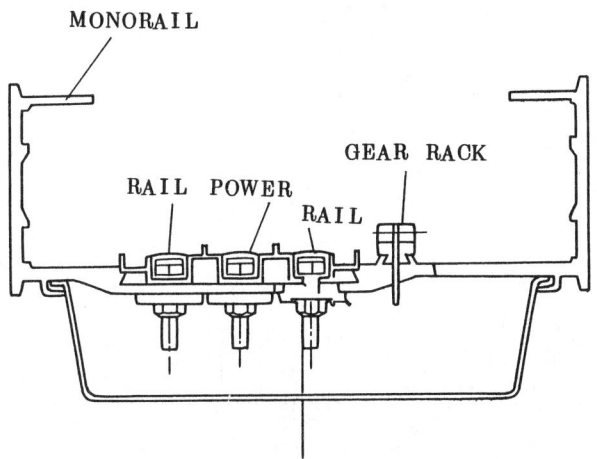

Fig. 3 Section of Monorail

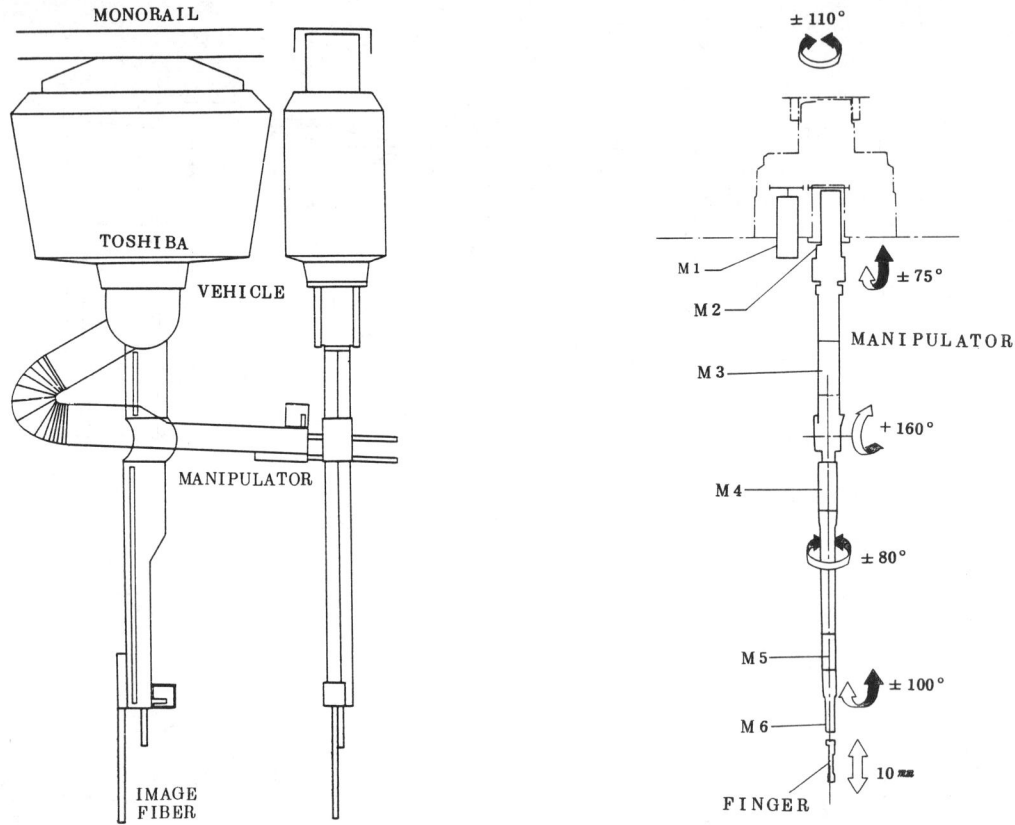

Fig. 4 Inspection Vehicle (Monorail Type)

Fig. 5 Manipulator Functional Diagram

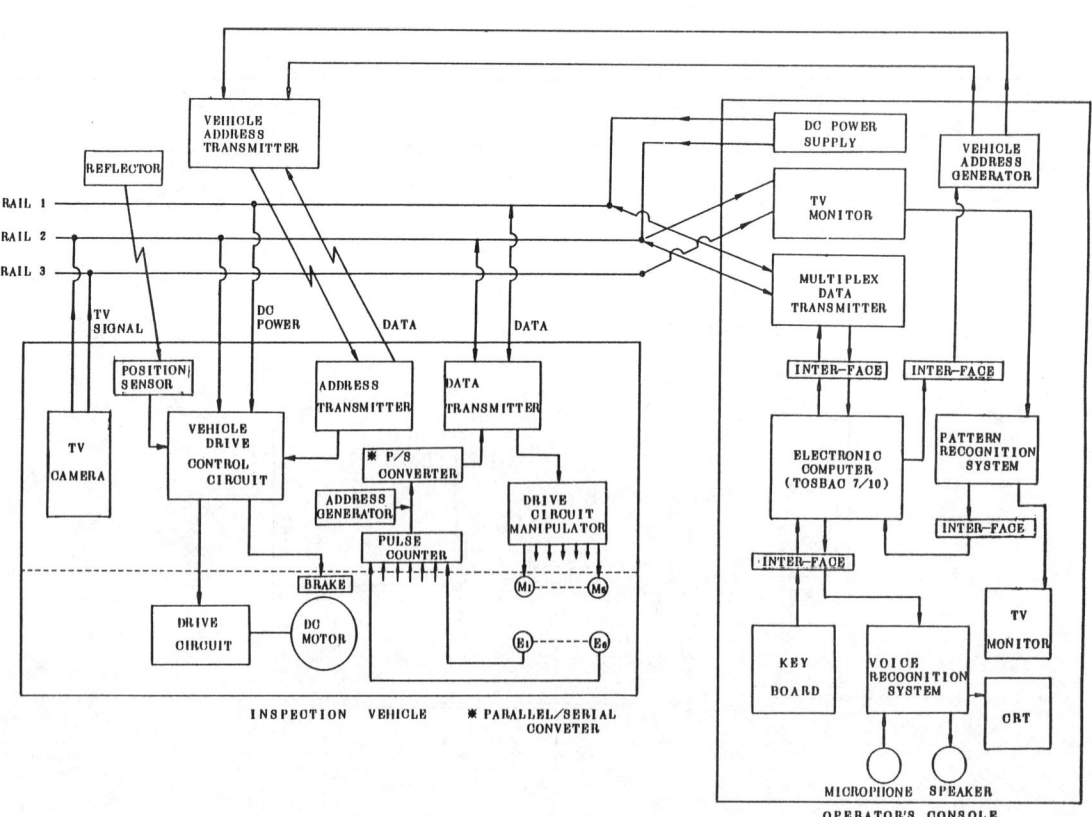

Fig. 6 Automated Remote Patrol System Block Diagram

DISCUSSION

SESSION MA2: MONITORING AND FAILURE DETECTION

Paper: A SPECIFICATION OF REAL TIME APPLICATIONS BY EVENTS AND LINKS WITH PROCESSES

Authors: *M Ben Maiza, JP Thomese* (Centres de Recherches en Informatique et Automatique de Nancy, Nancy France)

Discusser: *José Luis Farah*
IIMAS
Apdo Postal 20-726
México 01000, DF
México

Questions or Comments:

1) Could you please comment on how you would implement the "AFTER" operator, and how would you limit both time and memory?

2) How would you recognize if a concurrent event has ocurred if you only monitor elementary events?

Authors' reply:

1) It is essential to consider the time-ordering of the ocurrences. The ocurrences may be time-stamped and then ordered. Such a solution is then available even in a distributed system.

2) All the events (elementary and composed) are recognized or evaluated. It's independent from the sharing (or not) of events between several tasks.

ADAPTIVE CONTROL OF MUSCLE RELAXATION

C. S. Berger and W. A. Brown

Department of Electrical Engineering, Monash University, Clayton, Victoria 3165, Australia

Abstract. The implementation of feedback control of muscle relaxation is discussed. A nonlinear model of the response of the muscle relaxation of a sheep to a dosage of curare is derived and used to test different adaptive control strategies. The adaptive controller is successfully applied to the on-line control of the muscle relaxation of a sheep.

Keywords. Adaptive control; biomedical control; computer control; identification.

INTRODUCTION

The feedback control of muscle relaxation has been achieved both in the laboratory (Cass and colleagues, 1976) and in the operating theatre (Brown and colleagues, 1981). The technique ensures a more uniform level of paralysis and leads to a reduced total dosage of relaxant drug for procedures of any significant duration. Furthermore, it automates the administration of relaxation, thus relieving the anaesthetist of the task.

Whilst a fixed coefficient controller can work satisfactorily with careful attention, there are several reasons for considering the use of adaptive control. Subjects vary in their susceptibility to relaxant drugs, depending upon age, circulation, renal function etc. Furthermore, drug response at any time is dependent upon the preceding pattern of drug infusion. Hence adaptive control offers the possibility of good quality control for a wide range of biological conditions.

Simulations and on-line experiments have shown that the time-varying nonlinear behaviour of the process can, for typical perturbations about a set-point, be adequately represented by a linear time-varying model. This has enabled the use of the pole assignment controller (Wellstead, Prager and Zanker, 1979) but various precautions were found necessary to avoid the occasional anomalous control action.

MUSCLE RELAXATION AND IEMG MONITORING

Muscle relaxation is achieved by the intravenous injection of a small amount of a paralysing drug such as curare. The drug blocks neuromuscular transmission in skeletal muscle. In order to achieve an adequate level of paralysis the anaesthetist may inject several boluses of the drug or may follow an induction bolus with a steady drug infusion using a motor driven syringe. The duration of effect for blockers such as curare is of the order of 20 minutes.

The degree of muscle relaxation is measured using the evoked electromyogram (Lam and colleagues, 1981). In sheep we stimulate an intercostal nerve once every ten seconds and measure the corresponding surface electromyogram (EMG) from a rectus muscle. The EMG is gated and integrated to give a steady measure of neuromuscular transmission. In human subjects the EMG is evoked using transcutaneous stimulation of the median nerve and measured from the abductor pollicis brevis of the thumb.

For control purposes the relaxant drug is infused using a motor driven syringe. It has proven to be convenient to realize variations in requirements by using an on-off mode of operation, whereby the drive is turned on for a controlled time within a periodic interval. In our experiments the period used has been either 40 or 60 seconds. Thus the control signal is a number less than unity representing the fraction of time for which the drive is ON within any period. A control signal of unity corresponds to a drug dosage of 37.5 cc/hr.

The normal procedure during an experiment or operation is to make connections to the monitoring apparatus, inject a bolus of drug to induce paralysis, and then to switch over to automatic feedback control

THE ADAPTIVE CONTROL SCHEME

The self-tuning adaptive control scheme, shown in Fig. 1, was used. At each sample time the current input and output data were used to update a model of the relaxation response to curare; the model was then used to redesign the controller.

The structure of the model was determined off-line. The response of the relaxation of muscle to curare is both nonlinear (see below) and time-varying. Experience has shown, however, that a linear time-varying model, for fluctuations about the set-point, is adequate for the design of a control system for the usual clinical procedures. This enables the use of efficient identification and control design algorithms.

A particular problem with the control of drug administration is that negative inputs do not always exist (unless antidotes to the drug are available). Hence, if the set-point is over-shot, the system will be in open-loop until the drug is eliminated by natural processes. A number of precautions have thus been taken to decrease the probability of overshoots occurring.

The Identification Algorithm

The following model was used

$$(y_n - r_n) = b_1(y_{n-1} - r_n) + \ldots b_p(y_{n-p} - r_n) + a_1 u_{n-1} + \ldots a_m u_{n-m} + M \quad (1)$$

where y, u and r are the relaxation response, the drug dosage and the desired set-point. The process parameters a, b and M were treated as time-varying by the identification algorithm. The dimension of the input parameters, m, was made large enough to model both the transport lag and the order of the system.

The above model may be more compactly written as

$$w_n = \underline{b}^T \underline{w}_{n-1} + \underline{a}^T \underline{u}_{n-1} + M$$
$$= \underline{c}^T \underline{v}_{n-1} \quad (2)$$

where

$$\underline{c}^T = [\underline{b}^T \ \underline{a}^T \ M] = [b_1 \ b_2 \ldots b_p \ a_1 \ldots a_m \ M]$$
$$\underline{v}_{n-1}^T = [\underline{w}_{n-1}^T \ \underline{u}_{n-1}^T \ 1]$$
$$= [y_{n-1} - r_n \ldots y_{n-p} - r_n \ u_{n-1} \ldots u_{n-m} \ 1]$$

and $w_n = y_n - r_n$

Two methods of tracking time-varying parameters have been proposed

(a) The parameter vector $\underline{c}$ may be regarded as being generated by a Markov process

$$\underline{c}_n = \underline{c}_{n-1} + \underline{\zeta}_{n-1}$$

where $\underline{\zeta}_{n-1}$ is a random white noise sequence with covariance matrix

$$Q = E\{\underline{\zeta} \ \underline{\zeta}^T\}$$

The least squares estimate of $\underline{c}$ is then given by
$$\hat{\underline{c}}_{n+1} = \hat{\underline{c}}_n + P_n \underline{v}_n^T (w_{n+1} - \hat{\underline{c}}_n^T \underline{v}_n)/f_n \quad (3)$$
where
$$P_{n+1} = \frac{1}{\lambda}\left(P_n - \frac{P_n \underline{v}_n \underline{v}_n^T P_n}{f_n}\right) + Q; \quad f_n = \lambda + \underline{v}_n^T P_n \underline{v}_n$$
and $\lambda = 1$.

(b) Past data may be weighted by an exponentially decreasing "forget factor" λ. In this method λ in the above equations is usually chosen to be between .99-.995 and Q is zero. Simulations showed, however, that the best control performance was obtained with $\lambda = .99$ and Q retained as a diagonal matrix with .001 on the diagonal.

Precautions should be taken to ensure that P remains bounded (Astrom, 1980). The implemented algorithm set Q=0 and $\lambda=1$ whenever the input vector $\underline{u}_{n-1}$ was zero. This ensured that large changes in the estimated parameters only occurred when sufficient data was available.

The Control Design Algorithm

The pole assignment design algorithm (Wellstead, Prager and Zanker, 1979) was used. This method has the advantage over others (see Berger, 82 for list of references) in that the closed loop system remains stable even if the model is non-minimum phase; the closed loop poles converge to the required values even if the least squares estimate is biased; the implementation can be made numerically efficient.

Defining z^{-1} as a delay operator (2) may be represented by the transfer function description

$$w_n = \frac{A(z)}{1 - B(z)} u_n + \frac{M}{1 - B(z)} \quad (4)$$

where $A(z) = a_1 z^{-1} + \ldots a_m z^{-m}$; $B(z) = b_1 z^{-1} + \ldots b_p z^{-p}$

The required closed loop poles for the system may be achieved by the controller

$$u_n = \frac{-G(z)}{1 - F(z)} y_n + \frac{d}{1 - F(z)} \quad (5)$$

where G(z) and F(z) are polynomials in z of appropriate dimension which are designed to place the resulting closed loop poles in desired positions and d is a constant designed to drive the output, y_n, to some interim set-point s. Note that s may not always be equal to the desired set-point r. The closed loop equation thus becomes

$$w_n = \frac{A(z)d}{[1-B(z)][1-F(z)] + A(z)G(z)}$$
$$+ \frac{[1-F(z)]M}{[1-B(z)][1-F(z)] + A(z)G(z)}$$
$$= \frac{A(z)d}{1-T(z)} + \frac{[1-F(z)]M}{1-T(z)} \quad (6)$$

where

$$1-T(z) = 1 - t_1 z^{-1} - t_2 z^{-2} \ldots - t_{m+p-1} z^{-(m+p-1)}$$

is the desired closed loop characteristic equation. The control parameters in the F and G polynomials are thus obtained by solving the equation

$$[1-B(z)][1-F(z)] + A(z)G(z) = 1-T(z) \quad (7)$$

An efficient algorithm is given in the appendix. The variable d may be designed using the final value theorem.

$$w_n \Big|_{n \to \infty} = s = \left.\frac{A(z)d}{1-T(z)}\right|_{z=1} + \left.\frac{[1-F(z)]}{1-T(z)}\right|_{z=1}$$

Hence for a chosen interim set-point s

$$d = \left.\frac{[1-T(z)]}{A(z)}s\right|_{z=1} - \left.\frac{[1-F(z)]M}{A(z)}\right|_{z=1} \quad (8)$$

Alternatively the output can be maintained at a set-point by placing an integrator in the loop. (see Fig. 2).

$$u_n = u_{n-1} + x_n$$

The open loop transfer function relating x_n to w_n is thus

$$w_n = \frac{z}{z-1} \frac{A(z)}{1-B(z)} x_n \quad (9)$$

The parameters for the above open loop system may be obtained by regarding $x_n = u_n - u_{n-1}$ as the input to the system. The pole placement algorithm may be similarly modified to treat the open loop transfer function as w_n/x_n.

It was noted above (below equation 5) that the interim set-point s was different from the desired set-point r. This was introduced to enable a more gradual approach to the desired set-point r such that the chance of an overshoot, due to inaccurate parameter estimates during the transition between set-points, was decreased. The prevention of overshoots is important because, as noted above, negative inputs are not available. The interim set-point was determined by the equation

$$s_n = y_n + \rho(r_n - y_n) \quad (10)$$

where ρ determines the rate at which $s \to r$. Simulation showed that $\rho = .7$ provided the best compromise between overshooting the reference and the rate of convergence.

Practical Precautions

The precautions taken with the identification algorithm were described above.

(a) The solution of the pole placement algorithm requires the inversion of a matrix (see appendix). Simulation studies showed that the matrix could become sufficiently ill-conditioned to produce invalid controllers. A condition number for the matrix A

$$C = \frac{|\det A|}{\sum_i |a_{ii}|} \quad (11)$$

was therefore used to check the validity of the control calculation. It was observed that sudden changes in C, as well as its magnitude, indicated the likelyhood of an invalid control calculation. The exponentially averaged number

$$K_n = .7 K_{n-1} + C_n \quad (12)$$

was therefore calculated and the latest control parameters discarded if

$$C_n < K_n/15$$

(Note that for a constant C, $K \to 3.33C$). The various parameters were determined by simulation.

(b) The validity of the parameter estimate was checked by monitoring the prediction error $e_n = w_n - \hat{w}_n$, where $\hat{w}_n$ is the model prediction of the output. If the prediction error exceeded 5% the conservatively designed conventional controller

$$u_n = .5 u_{n-1} + .6(r - y_n) \quad (13)$$

was used.

(c) An artificial limit L was placed upon the dosage to prevent the injection of a large spurious dose when $y_n \approx r$, where

$$L = 3 + r/20 \quad (14)$$

The constants for these precautionary equations were found from simulation and real experiments. They were only seldom active but did prevent the administration of anomalous dosages.

SIMULATION STUDIES

A Nonlinear Model of the Response of Muscle Relaxation to Curare

Results from previous experiments were used to determine the nonlinear model. Figure 4 compares a typical response of muscle relaxation to a step dosage of curare with that of the model. The model contains two nonlinearities

(a) A threshold
(b) Saturation

The magnitude of the threshold was obtained by extrapolating the initial response backwards until it intersected zero time, as shown in Fig. 4.

The "Law of Mass Action", used in receptor kinetics, can be used to provide a physiological device for saturation. It failed, however, to provide a good model for the experimental results. The model response shown in Fig. 4 was obtained using the heuristically derived relationship.

$$y = K[1 - \exp(-x a)]$$

where x is the input, y the output and K and "a" experimentally found constants.

The dynamic component of the model was found by first taking out the nonlinear effects, in the order shown in Figure 3, and then finding the extended least squares estimate of a autoregressive - moving average model for the modified data.

The best constants for the saturation equation were found by trial and error to be

$$K = 100, \quad a = 1/50$$

The threshold was found to be -83.33, whereas the extended least squares estimate for the dynamic model was

$$z_n = 1.25 \, z_{n-1} - .268 \, z_{n-2} + .895 \, u_{n-1} + .18 \, u_{n-2}$$

Control Studies

The adaptive control algorithms described above were tested on the nonlinear model. A small amount of discrete white noise was added to the output to simulate measurement noise (see Fig. 5).

The simulation was used to compare the performance of the system for variations in the following factors:-

(a) Start up procedures
(b) The model orders (m and p)
(c) The forget factor (λ, see eq. 3)
(d) The diagonal value of the covariance matrix (q, see eq. 3)
(e) The set-point factor (ρ, eq. 10)
(f) pole placements

The effect of integral control was also evaluated.

The Start Up Procedure

The simulation was initiated by applying the maximum dosage (37.5 cc/hr.) until the threshold was reached and the % relaxation began to rise. When the relaxation exceeded 5% the conventional controller (eq. 13) took over and model identification commenced. The adaptive controller was introduced after 20 discrete time intervals of conventional control provided the prediction error of the identification algorithm was less than 5%. The initial set-point was set at 50% relaxation. The subsequent set-point schedule is shown in Fig. 5.

The above procedure was modified in the on-line sheep experiment by initiating the experiment with a bolus of curare. This caused a much more rapid onset of relaxation.

Tuning the Parameters

A third order linear model was found to be adequate for the simulation. The on-line sheep experiments showed, however, that the input order, m, should be increased to 5 to allow for variations in the transport lag.

Two measures of performance were used for tuning the remaining parameters: a prediction error cost function

$$E_1 = \sum_{i=40}^{250} (\hat{w}_i - w_i)^2 / 210$$

and a control error cost function

$$E_2 = \sum_{i=40}^{250} (y_i - r_i)^2 / 210$$

The control error was dominated by the error during the interval 120-140 and 170-190 when the output was decaying to lower set-points. This sometimes invalidated the measure because a sluggish response, which failed to follow increases in the references, was able to more rapidly match a subsequent decrease in the reference. The measure E_2 is therefore omitted where this was seen to occur.

Preliminary experiments showed that the integral controller did not perform as well as the controller given by eq. 5 and thus it was not used in the tuning experiments.

Methods of choosing the desired closed loop poles are still being considered. As an interim measure it was decided to contract the open loop poles by a contraction factor α. This enabled an easy parametrisation of the closed loop system and ensured negative feedback.

The tables below show the effect of varying the four parameters

α - the pole contraction factor
ρ - the set-point factor
λ - the forget factor
q - the covariance factor (diagonal value of Q)

TABLE 1 The Effect of Varying the Pole Contraction Factor

α	0	.3	.5	.7	.9	
E_1	1.0	1.03	.98	1.0	.92	$\rho = .7$
E_2	4.84	4.92	4.92	5.13	5.58	$\lambda = .99$
						$q = .001$

The pole contraction factor is seen from the above table to have a small effect on the prediction error but did change the control performance significantly. Small α's (large contractions) required large feedback and feedforward signals (d) which made the system sensitive to identification errors, whereas large α's resulted in sluggish responses (see Fig. 5 where the response with α=.9, ρ=.7, λ=.99 and q=.001 is shown). The prediction errors decrease as the system response is slowed down.

TABLE 2 The Effect of Varying the Set-Point Factor

ρ	.02	.2	.5	.7	1.0	α=.5
E_1	.96	.96	.96	.98	1.03	λ=.99
E_2	–	5.06	4.83	4.34	4.88	q=.001

The set-point factor has a similar effect on the prediction error as the pole contraction factor : the slower the resulting response the more accurate the prediction. The control performance is seen to improve the more rapidly the interim set-point approaches the desired set-point provided the the model accuracy is not significantly degraded. At ρ = 1, however, the model errors resulted in the set-point being overshot.

TABLE 3 The Effect of Varying the Forget Factor

λ	.9	.95	.97	.99	.995	α=.5
E_1	7.1	5.	1.19	1.06	8.5	ρ=.7
E_2	–	–	–	4.7	–	q=.0001

The prediction errors are seen in the above table to be very sensitive to variations in λ. Most of the E_2 values were omitted as they were judged invalid for the reasons given above. The conventional controller was often activated by the check on the prediction error (see 3.3).

TABLE 4 The Effect of Varying the Covariance Factor

q	0	.0001	.001	.01	α=.5
E_1	1.05	1.05	.98	1.13	ρ=.7
E_2	6.5	4.96	4.34	4.84	λ=.99

The prediction errors are seen from the above table to be relatively insensitive to the change made in q. The control performance was, nevertheless, markedly affected. Figure 5 shows the deterioration in the performance obtained by changing q from .001 to zero.

The values of the parameters chosen for the on-line experiment were α=.5, ρ=.7, λ=.99, q=.001 which gave values of E_1=.98 and E_2=4.34.

The integral controller was tested with the same parameters but resulted in a more oscillatory response with E_1=2.03 and E_2=5.93.

THE ON-LINE SHEEP EXPERIMENT

A sample time of 40 seconds was used. The set-point schedule for r and the resulting response is shown in Fig. 6. The experiment was initiated by applying a bolus of curare (an impulsive input). This was given offline and was thus not recorded by the computer. The controller was programmed to inject a dose of 5cc/hr. until the relaxation rose above 5%. This occurred at n=20. The conventional controller then took over for the next 20 sample times; the adaptive control thus commenced at n=40.

The adaptive controller caused oscillatory behaviour during the first set-point at r=60 (for n=50 → 106) and at n=95 injected an anomalously large dose. The prediction errors are seen to be large during this initial period (Fig. 6): this explains the unsatisfactory performance.

At n=106 the set-point was changed and from this point on the controller behaved reasonably well, even when the set-point r was subsequently returned to 60. Overshoots are all less than 4% and the prediction errors are small except at times following a change in the set-point. These occurred at

n = 106, 130, 160, 227, 287 and 333.

The transport lag at these set-point changes was measured from the trajectory to be 3 for positive changes and 4 for negative changes. This finding is also reflected in the relatively large magnitudes of a_3 and a_4 shown in Fig. 7 for most of the experiment.

The output parameters **b**, shown in Fig. 7, gave a marginally unstable pole, p ≈ 1, over large periods of the experiment. The instability of the model poles and the negative values for a_1 suggest that a better choice of sample time and model structure could have been made.

CONCLUSION

Preliminary experiments have shown that adaptive control is a promising method of maintaining muscle relaxation at desired levels. Various constraints on the adaptive controller have been found which improve its reliability. Areas where improvements are still necessary have been located.

After an initial learning period the relaxation followed set-point changes with very little steady-state error. Since the steady-state level depends on the accuracy of model parameter estimates, via the calculation of the feed-forward signal d, this is a confirmation of the validity of the model.

The transient performance after 70 minutes was also satisfactory but it should be possible to reduce overshoot by using a more sophisticated method of choosing the closed loop poles.

REFERENCES

Astrom, K.J. (1980). <u>Design Principles for Self-Tuning Regulators.</u> In H. Unbehauen (Ed.), <u>Methods and Applications in Adaptive Control</u>; Proceedings of an International Symposium, Bochum, 1980. Springer-Verlag, Berlin, Heidelberg, New York.

Berger, C.S. (1982). New Pole Placement Method for Adaptive Controllers. <u>IEE Proc.</u>, Vol. 129, Pt. D, No. 1, Jan., 13-14.

Brown, W.A., D.G. Lampard, K.C. Ng, H.S. Lam, M. Haysom, N.M. Cass, and J.R.W. Allan (1981). Microprocessor control of muscle relaxation. Reprints of VIIIth, IFAC Congress, Kyoto. Vol. XXI, 150-154.

Cass, N.M., D.G. Lampard, W.A. Brown, and J.R. Coles (1976). Computer controlled muscle relaxation - A comparison of four relaxants in the sheep. <u>Anaesth. Intens. Care</u>, 4, 36-40.

Lam, H.S., N.W. Cass, and K.C. Ng (1981). Electromyographic monitoring of neuromuscular block. <u>Br. J. Anaesth.</u>,53.

Wellstead, P.E., D. Prager, and P. Zanker (1979). Pole assignment self-tuning regulator. <u>Proc. IEE</u>, Vol. 126, 781-787.

ACKNOWLEDGEMENTS

The authors wish to thank Dr. K.C. Ng for setting up the IEMG monitoring equipment and Mr. R.H. Mitchell for assisting with the programming.

APPENDIX

Equating coefficients of like powers of z equation 7 may be represented in the following matrix form

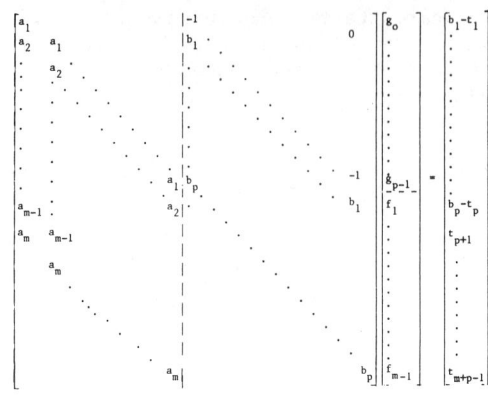

or

$$A \left[\frac{\underline{g}}{\underline{f}} \right] = \underline{y}$$

The -1's in the second partitioned matrix may be used, by elementary row-transformations, to eliminate all the elements below the 1's. Thus

$$TA \left[\frac{\underline{g}}{\underline{f}} \right] = \left[\begin{array}{c|c} A_{11} & -I \\ \hline A_{21} & 0 \end{array} \right] \left[\frac{\underline{g}}{\underline{f}} \right]$$

$$= T \underline{y} = \left[\frac{\underline{x}_1}{\underline{x}_2} \right]$$

where

A_{11} has dimension $(m-1) \times p$

A_{21} has dimension $p \times p$

I is a $(m-1) \times (m-1)$ identity matrix

and

$\underline{x}_1$ and $\underline{x}_2$ are p and (m-1) dimensional vectors. The solution to the above equation is therefore given by

$$\underline{g} = A_{21}^{-1} \underline{x}_1$$

$$\underline{f} = A_{11} \underline{g} - \underline{x}_2$$

Hence only a $(p \times p)$ matrix needs to be inverted for the solution of the $\underline{g}$ and $\underline{f}$ control vectors.

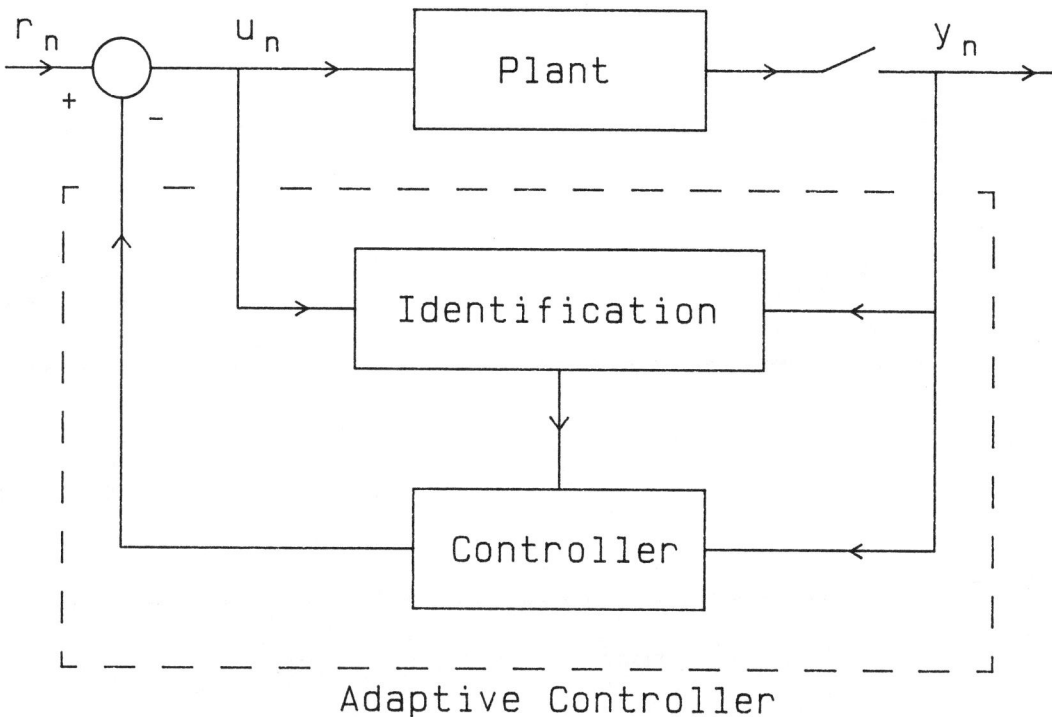

Fig. 1 Self-tuning Adaptive Control

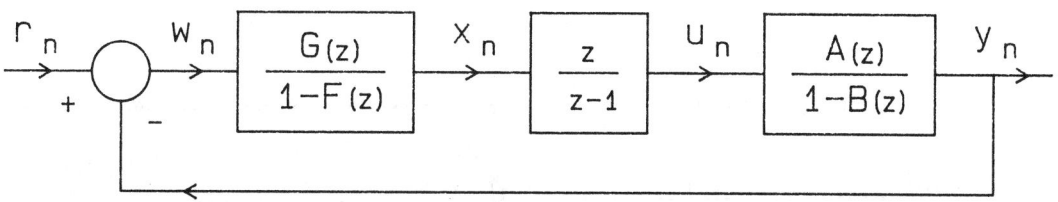

Fig. 2 Integral Control

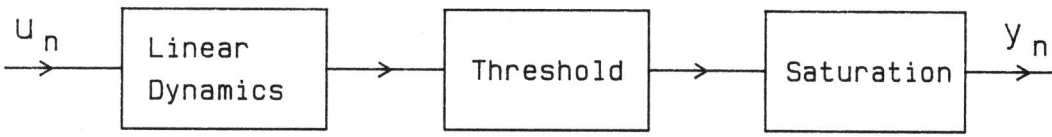

Fig. 3 Non-linear Model

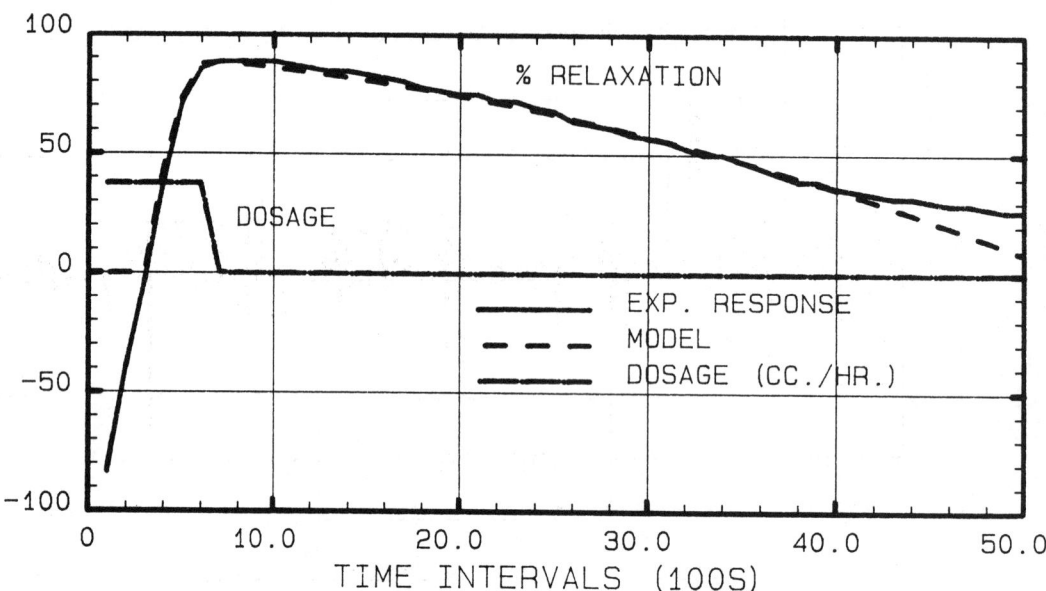

Fig. 4 Comparison of Model and Experimental Results.

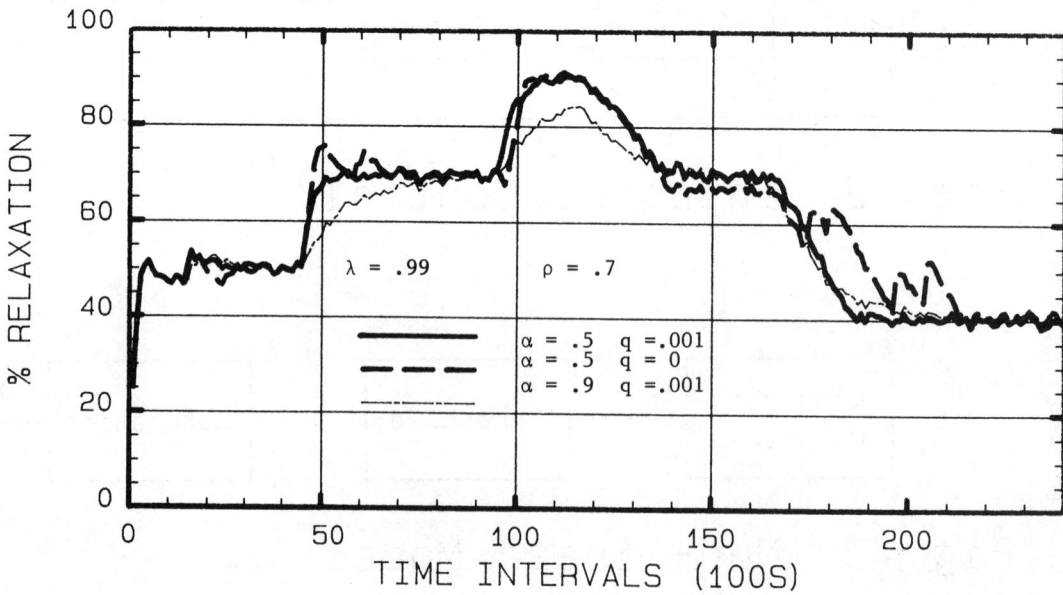

Fig. 5 Simulation Experiments.

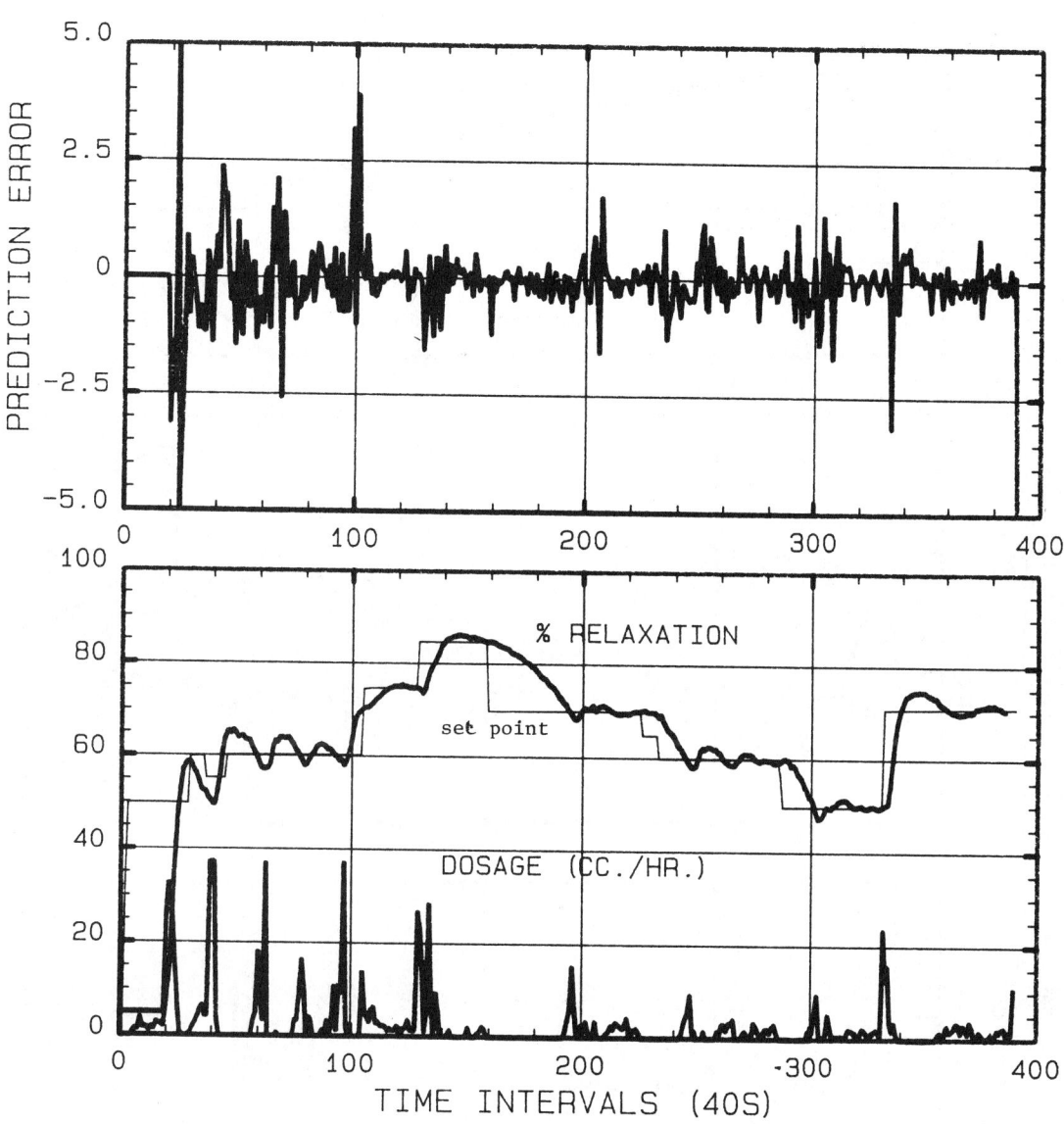

Fig. 6 Performance of Adaptive Controller.

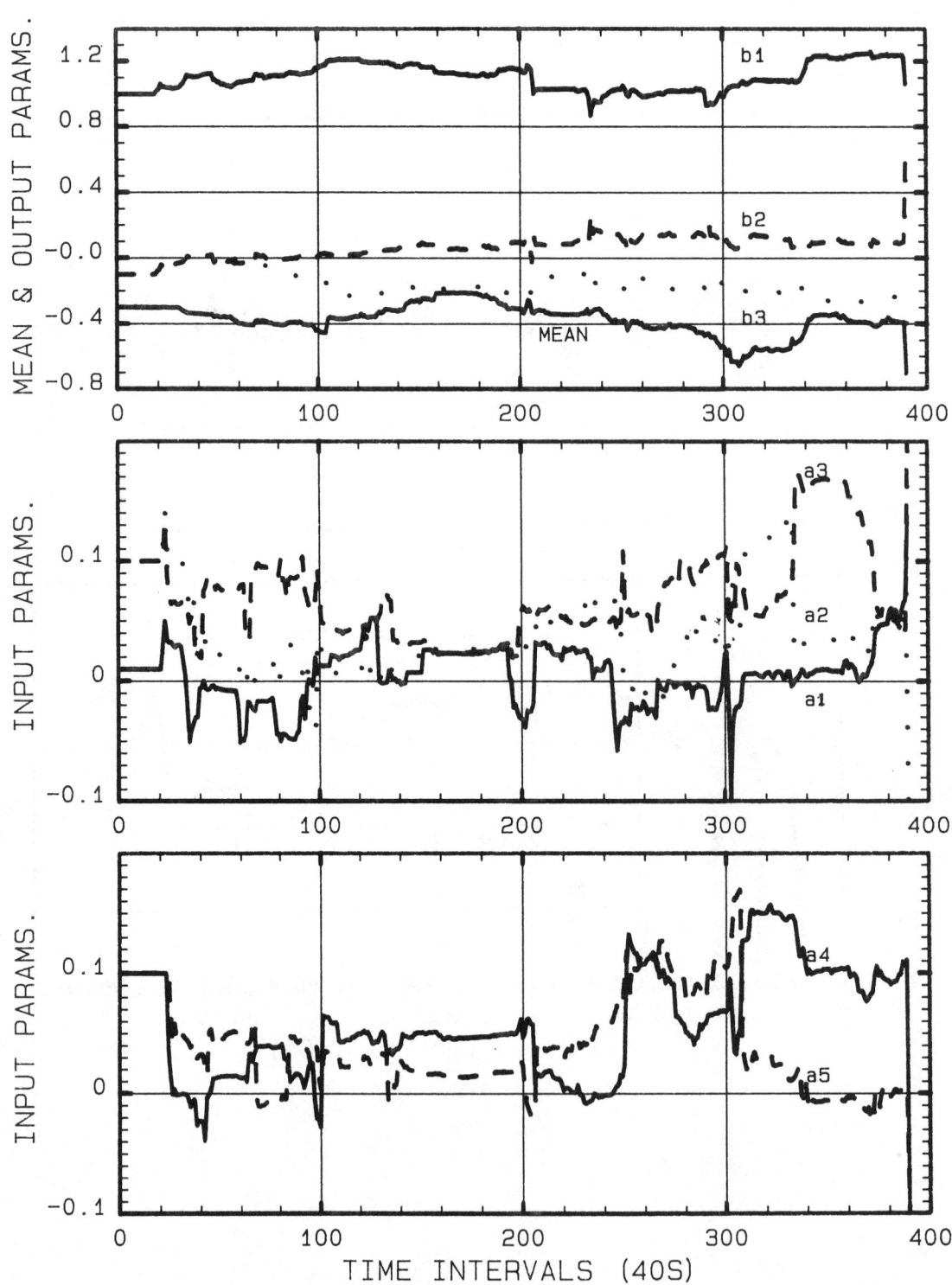

Fig. 7 Variation of Model Parameters.

MICROCOMPUTER IMPLEMENTATION OF AN ADAPTIVE CONTROL ALGORITHM

R. Lozano* and A. Noriega**

*Depto. de Ing. Eléctrica, CIEA-IPN Ap. Postal 14-740, 07000 D.F., México
**Instituto Tecnológico de Querétaro, Querétaro, Qro., México

ABSTRACT. This paper describes the implementation of an adaptive control algorithm in a MEK 6800-D2 microcomputer for speed control of a DC motor. The adaptation algorithm uses a forgetting factor and asymptotically achieves independent tracking and regulation objectives. Real time experimental results are presented.

Keywords. Adaptive Control, Microprocessors, Motor Control, Recursive Identification

INTRODUCTION

In the latest decade different adaptation algorithms have appeared in the literature. These approaches have been focused mainly on the analysis at convergence. In contrast to theoretical studies, few real time applications have been reported.

In this paper the adaptive algorithm with forgetting factor of [2] is used for speed control of a DC motor. We use a constant trace adaptation algorithm, see [5]. One of its main properties is that the controller parameters are continuously updated with no need to reset the adaptation gains.

The control algorithm was programmed in assembly language in order to reduce the computation time and was implemented on a MEK 6800-D2 microcomputer with only 2K bytes of RAM.

The paper is organized as follows: In section II the design of a control law for linear plants with known parameters is presented. Section III describes the corresponding adaptive control algorithm. Finally, the experimental results are presented in section IV.

II CONTROL OF DISCRETE LINEAR PLANTS WITH KNOWN PARAMETERS

Consider a SISO discrete linear time-invariant plant described by:

$$A(q^{-1})y(k) = q^{-d} B(q^{-1})u(k) \quad (2.1)$$

where

$$A(q^{-1}) = 1 + a_1 q^{-1} + \ldots + a_{n_A} q^{-n_A}$$

$$B(q^{-1}) = b_0 + b_1 q^{-1} + \ldots b_{n_B} q^{-n_B} \quad ; \quad b_0 \neq 0$$

$\{q^{-1}\}$ is the backward shift operator, $\{d\}$ represents the plant time delay, $\{u(k)\}$ and $\{y(k)\}$ are the plant input and output, respectively. We assume the zeros of $B(q^{-1})$ are all in $|q|<1$; therefore they can be cancelled without leading to an unbounded control-input.

The objectives of the control which we are considering are the following:

i) Regulation

The control should be such that an initial disturbance $\{y(0) \neq 0\}$ is eliminated thus resulting.

$$c_r(q^{-1}) y(k+d) = 0 \quad ; \quad k > 0 \quad (2.2)$$

where

$$c_r(q^{-1}) = 1 + c_1 q^{-1} + \ldots + c_{n_c} q^{-n_c} \quad (2.3)$$

is an asymptotically stable polynomial.

ii) Tracking

The control should be such that the plant output could asymptotically track a given reference sequence $\{y^M(k)\}$ i.e.:

$$\lim_{k \to \infty} \varepsilon(k) = 0 \quad (2.4)$$

where

$$\varepsilon(k) = y(k) - y^M(k) \quad (2.5)$$

is the tracking error.

The reference sequence can be defined for instance as:

$$c_t(q^{-1}) y^M(k) = q^{-d} D(q^{-1}) u^M(k) \quad (2.6)$$

where

$$c_t(q^{-1}) = 1 + c_1^t q^{-1} + \ldots + c_{n_t}^t q^{-n_t} \quad (2.7)$$

$$D(q^{-1}) = d_0 + d_1 q^{-1} + \ldots + d_{n_D} q^{-n_D} \quad (2.8)$$

The equation (2.6) defines a reference model. The $c_t(q^{-1})$ polynomial is asymptotically stable and $u^M(k)$ is the bounded reference model input.
The $c_r(q^{-1})$ polynomial plays also a filtering role and the control performance will depend on its choise. In fact, this polynomial smooths the adaption process and its zeroes are the closed loop poles of the system.

It is well known [Lozano, Landau (1981)] that there exist unique minimum degree polynomials.

$$S(q^{-1}) = 1 + s_1 q^{-1} + \ldots + s_{n_S} q^{-n_S} \quad (2.9)$$

$$R(q^{-1}) = r_0 + r_1 q^{-1} + \ldots + r_{n_R} q^{-n_R} \quad (2.10)$$

with (2.11)

$$n_S = d-1 \quad ; \quad n_R = \max(n_A - 1, n_c - d)$$

such that the following polynomial identity holds:

$$c_r(q^{-1}) = A(q^{-1}) S(q^{-1}) + q^{-d} R(q^{-1}) \quad (2.12)$$

Using the polynomial identity (2.12) and equations (2.1) and (2.3) we can write:

$$c_r(q^{-1}) \varepsilon(k+d) = c_r(q^{-1}) \{ y(k+d) - y^M(k+d) \}$$

$$= A(q^{-1}) S(q^{-1}) y(k+d) +$$

$$+ q^{-d} R(q^{-1}) y(k+d) -$$

$$- c_r(q^{-1}) y^M(k+d)$$

$$= B(q^{-1}) S(q^{-1}) u(k) +$$

$$+ R(q^{-1}) y(k) -$$

$$- c_r(q^{-1}) y^M(k+d) \quad (2.13)$$

The above equation can also be written as:

$$c_r(q^{-1}) \varepsilon(k+d) = b_0 u(k) + \Theta_0^T \phi_0(k) -$$

$$- c_r(q^{-1}) y^M(k+d)$$

$$= \Theta^T \phi(k) - c_r(q^{-1}) y^M(k+d) \quad (2.14)$$

where

$$\phi_0^T(k) = \{ u(k-1), \ldots, u(k-d-n_B+1), y(k),$$

$$, \ldots, y(k-n_R) \} \quad (2.15)$$

$$\Theta_0^T = \{ b_0 s_1 + b_1, \, b_0 s_2 + b_1 s_1 + b_2,$$

$$, \ldots b_{n_B} s_{d-1}, \, r_0, \ldots, r_{n_R} \} \quad (2.16)$$

$$\phi^T(k) = \left[u(k) \; ; \; \phi_0^T(k) \right] \quad (2.17)$$

$$\Theta^T = \left[b_0 \; ; \; \Theta_0^T \right] \quad (2.18)$$

The control input is computed such that the right hand side of equation (2.13) or equation (2.14) is equal to zero. i.e.

$$u(k) = \frac{1}{b_0} \{ c_r(q^{-1}) y^M(k+d) - R(q^{-1}) y(k) -$$

$$- B_s(q^{-1}) u(k-1) \} \quad (2.19)$$

where

$$B_s(q^{-1}) = \{ B(q^{-1}) S(q^{-1}) - b_0 \} q \quad (2.20)$$

or equivalently:

$$u(k) = \frac{1}{b_0} \{ c_r(q^{-1}) y^M(k+d) - \Theta_0^T \phi_0(k) \} \quad (2.21)$$

The control scheme for plants with known parameters is shown in figure 2.1.

Given that $y^M(k)$ is a bounded sequence, we can see from equation (2.11) that the plant output is also bounded, and assuming that we will work with minimum phase plants, the fact that the plant output is bounded garanties that the plant input is bounded too.

Note that if $c_r(q^{-1})=1$, all the closed loop poles are placed at the origin, and then $\varepsilon(k+d)=0$, which means that the tracking error is canceled d-steps after the application of the control signal.

III DESIGN OF THE ADAPTIVE CONTROL LAW

In this section, we present an adaptive controller applicable to minimum - phase plants described by equation (2.1), We assume that:

i) The time delay $\{d\}$ is known

ii) Upper bounds for $\{n_A\}$ and $\{n_B\}$ are known.

In this case the unknown parameters b_0 and ϕ_0 in equation (2.11) can be replaced by adjustable parameters $\hat{b}_0(k)$ and $\hat{\Theta}_0(k)$ which are updated by an adaptation mechanism. Therefore the control input in the adaptive case is given by:

$$u(k) = \hat{b}_0^{-1}(k) \{ c_r(q^{-1}) y^M(k+d) - \hat{\Theta}_0^T(k) \phi_0(k) \} \quad (3.1)$$

or equivalently:

$$\hat{\underline{\Theta}}^T(k) \underline{\phi}(k) = c_r(q^{-1}) y^M(k+d) \quad (3.2)$$

where

$$\hat{\Theta}^T(k) = \{ \hat{b}_0(k) ; \hat{\Theta}_0^T(k) \} \quad (3.3)$$

These parameters estimates $\hat{\Theta}(k)$ are obtained by using the following identification algorithm with forgetting factor:

$$\hat{\Theta}(k) = \hat{\Theta}(k-1) + \frac{F_k \underline{\phi}(k-d) \varepsilon(k)}{1 + \underline{\phi}^T(k-d) F_k \underline{\phi}(k-d)} \quad (3.4)$$

where F_k is the adaptive gain matrix given by:

$$F_{k+1} = \frac{1}{\lambda(k)} \left(F_k - \frac{F_k \phi(k-d) \phi^T(k-d) F_k}{1 + \phi^T(k-d) F_k \phi(k-d)} \right) \quad (3.5)$$

with:

$$F_1^{-1} > 0 \; ; \; 0 < \delta < \lambda(k) < 1 - \varepsilon \; ;$$
$$; \; \delta > 0 \; ; \; 0 < \varepsilon < 1 \quad (3.6)$$

and $\lambda(k)$ such that:

$$0 < \lambda \max \left(F_k^{-1} \right) < \infty \quad (3.7)$$

The figure 4.1 shows the adaptive control law, where F_k is computed by equation (3.5)

IV EXPERIMENTAL RESULTS

The experimental control system was built around a MEK 6800-D2 microcomputer. A schematic block diagram is ilustrated in figure 4.1. The following is a description of the various system components:

i) DC motor. The plant used in the experiment is a DC motor whose characteristic are: 45 watts and 6000 rpm max.

The transfer function from the armature voltage to the angular velocity of the motor is:

$$\frac{W(s)}{V_a(s)} = \frac{K_s}{1 + \zeta_m s} \quad (4.1)$$

where ζ_m is the motor time constant, K_s is the speed constant, and "s" is the Laplace operator. It should be noted that the value of ζ_m is a combination of electrical and mechanical parameters (Motor armature resistance, Moment of inertia, etc.)

ii) The microcomputer system. The microcomputer MEK 6800-D2 is based on the MC6800 Microprocessing Unit (MPU) and its family of associated memory and I/O devices. This microcomputer has an 8 bit microprocessor with an average execution speed of 2 μ sec.- per instruction.

iii) Interface circuit. This circuit uses two MC3410 (ten bits D/A converter) with 0.25 μ sec. conversion time and relative accuracy of ±0.05% maximum error. This devices are used for succesive approximation A/D converters with a maximum conversion time of -- 475 μ sec.

The reference model, which specifies the tracking objective is defined by:

$$\frac{q^{-d}D(q^{-1})}{c_t(q^{-1})} = \frac{q^{-1}(0.0902+0.0646\ q^{-1})}{1-1.213q^{-1}+0.3679q^{-2}} \quad (4.2)$$

The DC motor-tachogenerator set is initially characterized by:

$$G_M(s) = \frac{34.3}{1+0.5s} \rightarrow \frac{q^{-d}B(q^{-1})}{A(q^{-1})} = \frac{6.18\ q^{-1}}{1-0.8187\ q^{-1}} \quad (4.3)$$

sampling period = 1 sec

At time $k=t_1$, the motor load is changed and the transfer function is then:

$$G_M(s) = \frac{61.3}{1+0.25s} \rightarrow \frac{q^{-d}B'(q^{-1})}{A'(q^{-1})} = \frac{20.2\ q^{-1}}{1-0.6703\ q^{-1}} \quad (4.4)$$

The desired regulation behavior will be given by the polynomial:

$$c_r(q^{-1}) = (1-0.5q^{-1})^2 \quad (4.5)$$

A constant trace adaptive algorithm ($F_0 = 2\ I_5$) was used: i.e. $\lambda(k)$ in eq. (3.5) was such the trace of F_{k+1} was constant and equal to 10.

Figure 4.2 ilustrates the tracking reference sequence, $y^M(k)$ when $u^M(k)$ is a square wave. Figures 4.3 and 4.4 show the control signal $u(k)$ and the output $y(k)$ of the DC motor respectively when the linear control law for known parameters in section II is -- used. Experimental results showed that the linear control law depicted in figure 2.1 verifies the tracking and regulation objectives. But the control performances may change drastically if the plant's parameters used in the design change.

Figure 4.5 shows the plant output and figure 4.6 the corresponding adaptive control input when the parameter's change occurs at t_1. It can be observed that the tracking error converges to zero in 4 sampling periods.

The regulation behavior can also be observed from figure 4.5. The c_r-polynomial in equation 4.5 causes the plant output to be regulated around 1000 r.p.m. as depicted in figure 4.5. The transient period could be reduced by changing c_r at the expense of more input energy.

CONCLUSIONS

An adaptive control algorithm has been implemented in a microcomputer to control the speed of a DC Motor.
The computation time has been reduced by programming the control algorithm in assembly language with fixed point arithmetic.
This allows to control fast system like DM Motors with small time constants.

A recursive least squares identification algorithm with forgetting factor was used. This algorithm allows to update the parameter estimates continuously. It has been shown that the adaptive controller asymptotically achieves the tracking and regulation objectives..
Some of the problems encountered in this application were lack of precision in the A/D and D/A convertions and non-linearities of the plant.

It has been observed that when a constant trace adaptation algorithm is used the convergence speed depends strongly on the trace value. An adequate trace value has been empirically obtained improving the control performance.

REFERENCES

[1] Lozano L. R. (1981). Adaptive Control with Forgetting Factor. VIII World Congress. Kyoto, 1981. To be published in Automatica.

[2] Lozano L. R., I. D. Landau (1981). Redesign of Explicit and Implicit Discrete Time Model Reference Adaptive Control Schemes. Int. Journal of Control Vol. 33. pp. 247-268.

[3] Noriega A. P., R. L. Lozano (1982). Implantación de un Algoritmo de Con-

trol Adaptable en una Microcomputadora.
VII Congreso de la Academia Nacional de Ingeniería. Torreon, Coah. 1982.

[4] Egardt B. (1980). Stability Analysis of Discrete Time Adaptive Control Schemes. IEEE Trans on - Aut. Control, AL-25, 710.

[5] Irving E. Private Communication.

[6] Jacobs, Donaghey (1977) Microcomputer Implementation of Direct - Digital Control Algorithm for -- Thermal Process Control Applications. Journal of Dynamical Systems, Measurement, and Control. pp. 233-240.

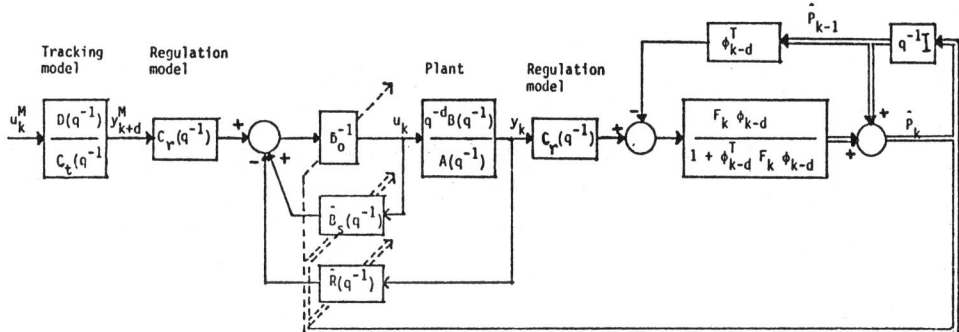

FIGURE 4.1

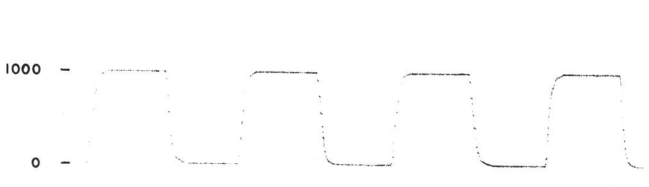

Fig. 4.2 The desired output. $y^M(K)$

Fig. 4.3 The control input with known parameters. u(K)

Fig. 4.4 The DC. Motor output. y(K)

Fig. 4.5 The adaptive control signal. u(K)

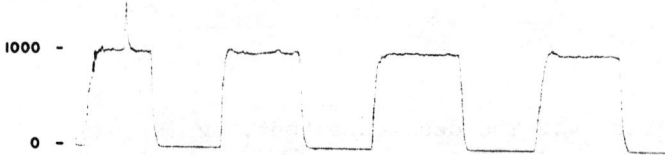

Fig. 4.6 The DC. motor output. y(K)

ADAPTIVE CONTROL OF DISCRETE MULTIVARIABLE SYSTEMS

R. Lozano and M. Bonilla

Depto. de Ing. Eléctrica, Centro de Investigacion y de Estudios, Avanzados. Ap. Postal 14-740, 07000 México, D.F., México

Keywords. Adaptive Control, Multivariable Systems, Pole Placement, Hydraulic Systems, Discrete Systems.

ABSTRACT. This paper presents a multivariable control algorithm for linear systems and its corresponding adaptive control scheme. The control algorithm is applicable to a class of multivariable systems including some non-stably invertible plants. The controller achieves tracking and regulation objectives independently and the adaptation algorithm uses a forgetting factor. This algorithm is applied to a two inputs-two outputs linear model simulated in an analogue computer. The linear model represents a laboratory hydraulic process consisting of three interconected tanks. Experimental results are presented.

INTRODUCTION

In the past few years there has been an increasing interest in obtaining simple, globally convergent, adaptive control algorithms.

In the single input-single output case Goodwin, Ramadge and caines (1980) proposed an algorithm that achieves a tracking objective with bounded input for minimum phase systems. They proved global convergence for their algorithm. The introduction of independent tracking and regulation objectives and adaptive algorithms with forgetting factor were studied by Lozano, Landau, (1981)

Adaptive algorithms for the general linear scalar case, i.e. either minimum or non minimum phase systems have been proposed recently. Goodwin and Sain (1981) stablished local convergence for an indirect scheme without resort to sufficiently rich input signals.

In contract to the single input-single output case, few multivariable adaptive controllers have been reported in the literature. The basic self-tuning controller has been extended to the multivariable case by Borrisson (1979). Koivo (1980) carried out an extension of the algorithm of Clarke and Gawthrop (1975) for the multivariable case. The controller is obtained by penalizing the control input in the cost function. Goodwin, Ramadge and Caines (1980) also proposed an extension of their own algorithm to the multivariable case.

In this paper we present a multivariable adaptive control applicable to a class of multivariable systems including some non stably invertible plants. One of its features is that tracking and Regulation objectives can be achieved independently. The control algorithm is indirect in the sense that the controller's parameters are not estimated directly, but computed from the plant parameter's estimates. The plant parameters are estimated by using the least squares recursive identification algorithm with forgetting factor studied in [Lozano (1982)].

In our experimental study, we considered a linearized model of a Hydraulic set up consisting of three interconnected tanks. The equations were simulated on an analogue computer and the controller was implemented on a microcomputer.

The paper is organized as follows: In section II the multivariable controller structure achieving both regulation and tracking objectives is presented. The corresponding indirect adaptive control scheme is presented in section III. section IV is devoted to show experimental results.

II. CONTROL OF MULTIVARIABLE LINEAR PLANTS OF MINIMUM PHASE WITH KNOWN PARAMETERS.

Consider a multi-input multi-output (MIMO) discrete linear time invariant plant described by:

$$A(q^{-1})y(t) = q^{-d}B(q^{-1})u(t)+w(t) \quad (2.1)$$

where $y \in R^m$ is the output, $u \in R^m$ is the input, $w \in R^m$ is a bounded disturbance and q^{-1} is the backward shift operator. The polynomial matrices A and B are given by:

$$A(q^{-1}) = I + A_1 q^{-1} + \ldots + A_{n_A} q^{-n_A} \quad (2.2)$$

$$B(q^{-1}) = B_0 + B_1 q^{-1} + \ldots + B_{n_B} q^{-n_B} \quad (2.3)$$

with $B_0 \neq 0$ and all the zeroes of $\det\{B(q^{-1})\}$ strictly inside the unit disc.
We assume also that $B(q^{-1})$ has full rank.
If we define an input $\bar{u}(t) \in R^m$ such that

$$u(t) = \text{Adj}\{B(q^{-1})\}\bar{u}(t) \quad (2.4)$$

where $\text{Adj}\{B(q^{-1})\}$ represents the Adjugate of $B(q^{-1})$.

Then the eq. (2.1) takes the following form:

$$A(q^{-1})y(t) = q^{-d} b(q^{-1})\bar{u}(t) + w(t) \quad (2.5)$$

where

$$b(q^{-1}) = \det\{B(q^{-1})\} = b_0 + b_1 q^{-1} + \ldots + b_{m\,n_B} q^{-m\,n_B};$$

$$; b_0 \neq 0 \quad (2.6)$$

We wish to accomplish the following objectives in regulation and tracking:

- Tracking. The control input should be such that the plant output $\{y(t)\}$ asymptotically follows a desired sequence $y^M(t) \in R^m$.

- Regulation. In regulation the matrix transfer function from the disturbance to the plant output must be asymptotically stable, i. e.

$$c(q^{-1})y(t) = S(q^{-1})w(t) \quad (2.7)$$

where

$$c(q^{-1}) = 1 + \ldots + c_n q^{-n} \quad (2.8)$$

is an asymptotically stable scalar polynomial matrix that depends on the system.

It readily follows that both objectives will be achieved if the following equation holds:

$$c(q^{-1})\left[y(t) - y^M(t)\right] = S(q^{-1})w(t) \quad (2.9)$$

It can be easily shown [Lozano, Landau (1981)] that there exist unique polynomial mxm matrices $S(q^{-1})$ and $R(q^{-1})$ of order $(d-1)$, $r = \max(n-d, n_s - 1)$ respectively, such that:

$$c(q^{-1})I_m = S(q^{-1})A(q^{-1}) + q^{-d}R(q^{-1}) \quad (2.10)$$

Using the above polynomial identity and eqs. (2.9) and (2.5) we obtain:

$$c(q^{-1})\left[y(t) - y^M(t)\right] =$$

$$= q^{-d}\left[b(q^{-1})S(q^{-1})\bar{u}(t) - c(q^{-1})y^M(t+d) + R(q^{-1})y(t)\right] + S(q^{-1})w(t) \quad (2.11)$$

then the control objectives given by eq. (2.9) will be achieved if the control law is:

$$b(q^{-1})S(q^{-1})\bar{u}(t) = c(q^{-1})y^M(t+d) - R(q^{-1})y(t) \quad (2.12)$$

From eq. (2.9) and given that $c(q^{-1})$ is an asymptotically stable scalar polynomial, $y(t)$ is bounded.

Since $B(q^{-1})$ has full rank and the zeroes of $b(q^{-1})$ are strictly inside the unit disc, it follows that $\bar{u}(t)$ is bounded and from (2.4) $u(t)$ is also bounded.

The solution of the polynomial identity, given by eq. (2.10), is very simple to obtain. Equating the terms multiplying q^{-1} $i=0,\ldots,r+d$, the solution can be expressed by the following set of equations.

$$S_0 = I_m$$

$$S_i = c_i I_m - \sum_{j=0}^{i-1} S_j A_{i-j} \; ; \; i=1,\ldots,d-1$$

$$R_i = c_i I_m - \sum_{j=0}^{i+d-1} S_i A_{d+i-j} \; ; \; i=0,\ldots,r$$

with:

$$S_i = 0_m \; ; \; i \geq d$$

$$A_i = 0_m \; ; \; i > n_A$$

$$c_i = 0 \; ; \; i > n \quad (2.13)$$

Particular case.

If we have a system in a state representation with measurable states (i.e. $n_A = 1$, $n_B = 0$, $d = 1$) and if we choose $n \triangleq 1$, then the control law, defined in eqs. (2.12),(2.13) and (2.4), takes the following form:

$$u(t) = B_0^{-1} \left[y^M(t+1) + A_1 y(t) - c(q^{-1})(y(t) - y^M(t)) \right] \quad (2.14)$$

Remark.

If $b(q^{-1}) = \det B(q^{-1})$ has zeroes outside the unit disc, but $b(q^{-1}) I_m$ and $A(q^{-1})$ have no common factors, then the following control law can be used [Lozano, (1982)].

$$S(q^{-1})\bar{u}(t) = \frac{c(q^{-1}) y^M(t)}{b(1)} - R(q^{-1}) y(t) \quad (2.15)$$

with

$$u(t) = \text{Adj } B(q^{-1}) \bar{u}(t) \quad (2.16)$$

where S and R are polynomials of order (r-1) with $r = \max(n_A, m \cdot n_B)$, solution of the polynomial identity

$$S(q^{-1})A(q^{-1}) + R(q^{-1})b(q^{-1}) = c(q^{-1}) I_m \quad (2.17)$$

It can be easily proved that the above control algorithm achieves the objective expressed in the following equation:

$$c(q^{-1}) \left[y(t) - \frac{b(q^{-1})}{b(1)} y^M(t) \right] =$$

$$= S(q^{-1}) w(t) \quad (2.18)$$

III. INDIRECT ADAPTIVE CONTROL.

We use an indirect adaptive control scheme for multivariable systems. In this scheme the controller parameters are up dated in two steps; the algorithm first identifies the plant parameters and second the polynomial identity (2.13) is solved for the plant estimates.

We will employ a least squares algorithm with forgetting factor to identify the plant parameters.

The convergence properties of this identifier are studied in [Lozano, (1981)].

Identification Algorithm.

The polynomial matrices $A(q^{-1})$ and $B(q^{-1})$ are formed by $m \times m$ scalar polynomials, $a^{ij}(q^{-1})$ and $b^{ij}(q^{-1})$ respectively, which are given by:

$$a^{ij}(q^{-1}) = \delta^{ij} + \sum_{k=1}^{n_A} a_k^{ij} q^{-k} \; ;$$

$$; \; i,j = 1,\ldots,m$$

$$b^{ij}(q^{-1}) = \sum_{k=0}^{n_B} b_k^{ij} q^{-(k+d)} \; ;$$

$$; \; i,j = 1,\ldots,m \quad (3.1)$$

where:

$$\delta^{ij} = \begin{cases} 1 \text{ if } i=j \\ 0 \text{ otherwise} \end{cases} \quad (3.2)$$

Then, the plant in eq. (2.1) can be expressed by the following set of equations ($w(t)=0$)

$$y_i(t) = \Theta_i^T \phi(t-1) \; ; \; i=1,\ldots,m \quad (3.3)$$

where

$$\Theta_i^T = \left[-a_1^{i1},\ldots,-a_{n_A}^{i1}, -a_1^{i2},\ldots,-a_{n_A}^{im}, \right.$$
$$\left. b_0^{i1},\ldots,b_{n_B}^{i1}, b_0^{i2},\ldots,b_{n_B}^{im} \right] \quad (3.4)$$

$$\phi^T(t-1) = \left[y_1(t-1),\ldots,y_1(t-n_A), \right.$$
$$, y_2(t-1),\ldots,y_m(t-n_A),$$
$$, u_1(t-d),\ldots,u_1(t-d-n_B),$$
$$\left. , u_2(t-d),\ldots,u_m(t-d-n_B) \right] \quad (3.5)$$

The plant parameters Θ_i are estimated by an adaptive algorithm with forgetting factor. This algorithm is given by the following set of equations:

$$\hat{\Theta}_i(t) = \hat{\Theta}_i(t-1) +$$
$$+ \frac{F_t \phi(t-1) \left[y_i - \hat{\Theta}_i^T(t-1)\phi(t-1) \right]}{1 + \phi^T(t-1) F_t \phi(t-1)} \;,\;$$

$$; \; i=1,\ldots,m \quad (3.6)$$

where

$$F_{t+1}^1 = F_t - \frac{F_t \phi(t-1) \phi^T(t-1) F_t}{1+\phi^T(t-1) F_t \phi(t-1)} \quad (3.7)$$

$$F_{t+1} = \frac{1}{\lambda(t)} F_{t+1}^1 \quad (3.8)$$

$$F_t > 0 \; ; \; \delta < \lambda(t) < 1-\varepsilon \; ; \; 0 < \delta, \varepsilon < 1 \quad (3.9)$$

this algorithm has the following properties

i) $\hat{\Theta}_i(t)$ are bounded and converge.

ii) $\lim_{t \to \infty} \left[y_i(t) - \hat{\Theta}_i(t)\phi(t-1) \right] = 0$ \quad (3.10)

A special particular case of the above algorithm that has been successfully used in practical applications is the so called constant trace algorithm [Irving, 1979]. In this case $\lambda(t)$ is computed such that the trace of the gain matrix $\{F_t\}$ remains constant; i. e.:

$$\lambda(t) = \text{trace}\{F_{t+1}^1\}/K \; ; \; K = \text{cte.} > 0 \quad (3.11)$$

Adaptive Control Law.

The adaptive control law is obtained by replacing in the control law (2.12) the controller's parameters by adjustable parameters, i.e.:

$$b(t,q^{-1}) S(t,q^{-1}) \bar{u}(t) = c(q^{-1}) y^M(t+d) -$$
$$- R(t,q^{-1}) y(t) \quad (3.12)$$

where

$$u(t) = \text{Adj}\{B(t,q^{-1})\} \; \bar{u}(t) \quad (3.13)$$

$$b(t,q^{-1}) I_m = B(t,q^{-1}) \text{Adj}\{B(t,q^{-1})\} \quad (3.14)$$

The polynomial matrices, $S(t,q^{-1})$ and $R(t,q^{-1})$, are obtained from the following eq.

$$S(t,q^{-1}) A(t,q^{-1}) + q^{-d} R(t,q^{-1}) = c(q^{-1}) I_m \quad (3.15)$$

where $A(t,q^{-1})$ and $B(t,q^{-1})$ are the estimated polynomials of A and B in eqs. (2.2) and (2.3) respectively.

This identity must be solved at each sampling step. The solution can be -

obtained by using eq. (2.13).

IV SIMULATION RESULTS.

We have considered a set of three interconnected tanks shown in fig. 4.1. A linear model of the system was simulated in an analogue computer EAI-1000. The adaptive controller was implemented in a minicomputer cromemco based on the Z-80 microprocessor to control the simulated system.

The system is described by the following set of eqs.:

$$A\dot{N}_1(t) = \alpha_0 D^* - \alpha_1 k_1 \sqrt{N_1 + k_2} \quad (4.1)$$

$$A\dot{N}_2(t) = \alpha_1 k_2 \sqrt{N_1 + k_2} - \alpha_2 k_1 \sqrt{N_2 + k_2} \quad (4.2)$$

Where K_1 and k_2 are constants depending on the dimensions of the various system's components and D^* is the maximum entrance flow.

Let us define

$$\alpha_i = \bar{\alpha}_i + \Delta\alpha_i \quad (4.3)$$

$$N_i = \bar{N}_i + n_i \quad (4.4)$$

where $\Delta\alpha_i$ and n_i represent variations arround the nominal values $\bar{\alpha}_i$ and $\bar{N}_i$.

The linearized system is:

$$\begin{pmatrix} \dot{n}_1 \\ \dot{n}_2 \end{pmatrix} = \begin{pmatrix} a_{11} & 0 \\ a_{21} & a_{22} \end{pmatrix} \begin{pmatrix} n_1 \\ n_2 \end{pmatrix} +$$

$$+ \begin{pmatrix} b_{11} & b_{12} \\ 0 & b_{22} \end{pmatrix} \begin{pmatrix} \Delta\alpha_0 \\ \Delta\alpha_1 \end{pmatrix} \quad (4.5)$$

where:

$$a_{11} = -a_{21} = -\frac{\bar{\alpha}_1 k_1}{2 A \sqrt{\bar{N}_1 + k_2}} \quad (4.6)$$

$$a_{22} = -\frac{\bar{\alpha}_2 k_1}{2 A \sqrt{\bar{N}_2 + k_2}} \quad (4.7)$$

$$b_{11} = D^*/A \quad (4.8)$$

$$b_{12} = -b_{22} = -\frac{k_1}{A} \sqrt{\bar{N}_1 + k_2} \quad (4.9)$$

We have chosen two operating points in which the linearized system is represented by:

P_1:

$$\begin{pmatrix} \dot{y}_1(t) \\ \dot{y}_2(t) \end{pmatrix} = \begin{pmatrix} -0.014 & 0 \\ 0.014 & -0.016 \end{pmatrix} \begin{pmatrix} y_1(t) \\ y_2(t) \end{pmatrix} +$$

$$+ \begin{pmatrix} 0.049 & -0.051 \\ 0 & 0.051 \end{pmatrix} \begin{pmatrix} u_1(t) \\ u_2(t) \end{pmatrix} \quad (4.10)$$

P_2:

$$\begin{pmatrix} \dot{y}_1(t) \\ \dot{y}_2(t) \end{pmatrix} = \begin{pmatrix} -0.05 & 0 \\ 0.05 & 0.05 \end{pmatrix} \begin{pmatrix} y_1(t) \\ y_2(t) \end{pmatrix} +$$

$$\begin{pmatrix} 0.049 & -0.01 \\ 0 & 0.01 \end{pmatrix} \begin{pmatrix} u_1(t) \\ u_2(t) \end{pmatrix} \quad (4.11)$$

The reference model which specifies the tracking objective was chosen as:

$$\begin{pmatrix} \dot{y}_1^M(t) \\ \dot{y}_2^M(t) \end{pmatrix} = \begin{pmatrix} -0.016 & 0 \\ 0 & -0.016 \end{pmatrix} \begin{pmatrix} y_1^M(t) \\ y_2^M(t) \end{pmatrix} +$$

$$+ \begin{bmatrix} 0.016 & 0 \\ 0 & 0.016 \end{bmatrix} \begin{bmatrix} u_1^M(t) \\ u_2^M(t) \end{bmatrix} \quad (4.12)$$

The scalar polynomial defining the regulation dynamics was chosen as:

$$c(q^{-1}) = 1 - 0.1q^{-1} \quad (4.13)$$

Identification.

We first identified on line the system of eq. (4.10) simulated in an analogue computer using the algorithm described in section III. The system was excited using two 1023 long PRBS displaced each other by 511 time periods.

Figure 4.2 shows the performance of the identification algorithm using different constant trace (k) values (see eq. (3.11)). In all cases we have assumed null initial conditions. It can be observed that there exist a tradeoff between the convergence speed and the noise level in the parametric distance.

Adaptive Control.

Figure 4.3 shows the performance of the adaptive control algorithm when applied to the plant of eqs. (4.10) and (4.11). We used the adaptive control law of equation (2.14) and the controller parameters, A_1 and B_0, were estimated using eqs. (3.6) to (3.8) with a constant trace k=10 (see eq. 3.11). From time 0 to 170 the plant was that of eq. (4.10) and from time 170 to 590 the plant was described by eq. (4.11). It can be observed that 200 periods of time later both output errors are very close to zero. At time 590 the plant was again changed to that in eq. (4.10) and it can be seen that after a transition period the output errors are again zero. It has been observed that the smaller the trace values the longer the transition periods with a corresponding reduction of the transient errors. Therefore there is a tradeoff between adaptation period duration and abrupt changes during the transient in terms of the trace value.

V CONCLUSIONS.

A new indirect adaptive control for a class of multivariable systems has been proposed. This class of systems includes some non-stably invertible systems.

The controller has being designed by using pole placement techniques and allows to achieve tracking and regulation objectives independently.

The control algorithm has been applied to a linearized model of a hydraulic set up simulated on an analogue computer. It has been observed that in spite of the lack of precision of the A/D and D/A interfaces the output error converges to zero.

The identifier involved in the control algorithm was of constant trace type. It was pointed out that there exist a tradeoff between adaptation period duration and abrupt changes during the transient in terms of the trace value. We are presently studying algorithms with variable trace in order to improve the control performance during the adaptation period.

References.

Åström K. J., Borrisson U., Ljung L. and Wittenmark B., (1977). Theory and applications of self-tuning regulators. Automatica 13, pp. 457-476

Borrison, U. (1979). Self-tuning regulators for a class of multivariable systems. Automatica 15, 209-215

Clarke D. W. and P. J. Gawthrop, (1975). Self-tuning controller Proc. IEE 122, 929-934,

Elliot H., Wolovich W. A. (1979). Parameter Identification and Control. IEEE Trans. Automat. Contr. Vol. AC-24 pp 592-599

Goodwin G. C., Ramadge P. J. and Caines P. E. (1980) Discrete time multivariable adaptive control. IEEE Trans. Automat. Contr. Vol. AC-25 pp 449-456

Goodwin G. C. and K. S. Sin (1981). Adaptive control of non minimum phase plants. IEEE Trans. Automat. Contr., 26, 2.

Koivo H. N. (1980). A multivariable self-tuning controller. Automática 16, 351-366.

Kuora V. (1979). Discrete linear control - the polynomial equation approach. John Willey and Sons.

Landau I. D., Lozano R. L. (1981).
Unification and evaluation of discrete time explicit model adaptive control designs. Automática 17, No. 4 pp 595-611

Lozano R. L. (1981). Adaptive control with forgetting factor. 8th IFAC World Congr. Kyoto. Japan. To be published in Automática.

Lozano R. L. Landau I. D. (1981). Redesign of explicit and implicit -- discrete time model reference adaptive control schemes. Int. J. Control. Vol. 33, No. 2, 247-268.

Lozano R. L. (1982). Stochastic Adaptive Control for M.I.M.O. Discrete Systems. Technical Report IECA 01/82 CIEA México D. F.

Wolovich W. A. (1974). Linear multivariable systems. New York, Springer-Verlag.

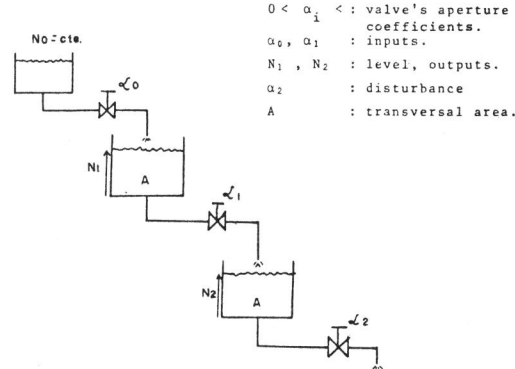

Fig. 4.1 Laboratory hydraulic set-up.

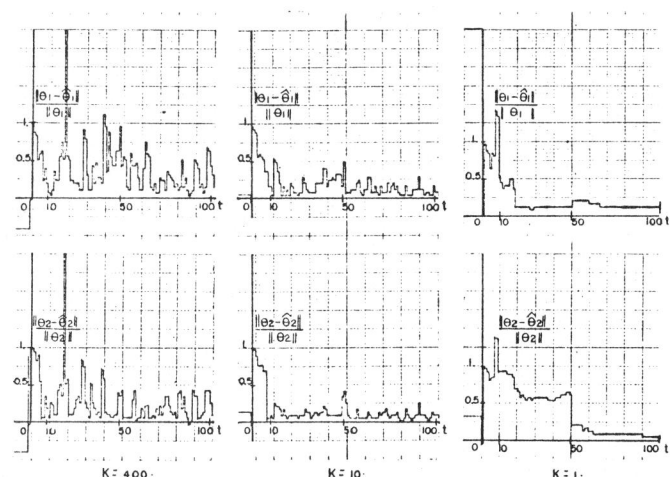

Figure 4.2

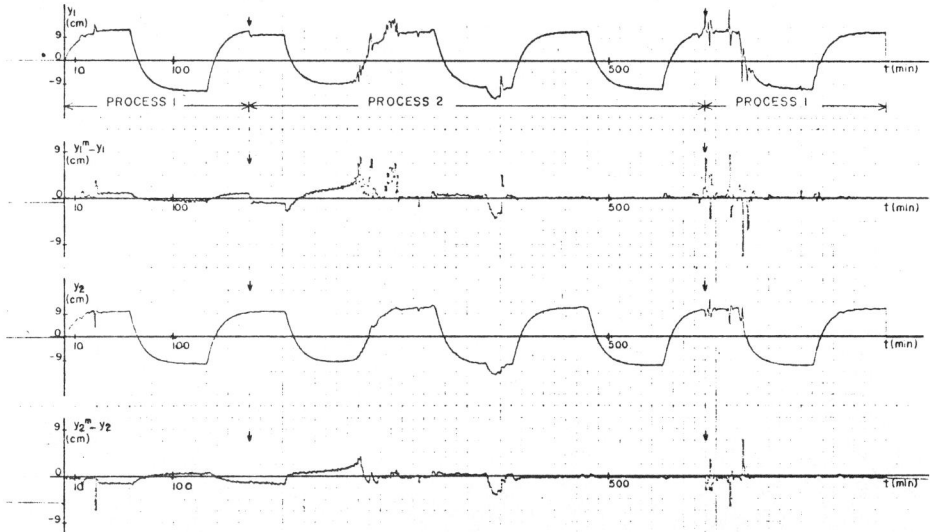

Figure 4.3

IDENTIFICATION AND ADAPTIVE CONTROL OF A SUGAR FACTORY VACUUM PAN

A. Aguado and A. Gómez

IMACC— Academia de Ciencias de Cuba, Cuba

Abstract. In the paper, the results obtained by means of digital simulation of a self-tuning simplified regulator for the conductivity control-loop of vacuum-pans are discussed. The control algorithm that was used is presented in details, remarking its computational advantages. The simulations are based on a varying parameters model that was identified on the basis of experimental data. Some special characteristics of the application are discussed, that is: the possible use of two control variables, the slow-changing of the reference value and the variation of the dynamic characteristics of the process with time. Some words are dedicated at the end of the paper to the on-line implementation of the described self-tuning regulator and the preliminar results obtained until now.

Keywords. Self-tuning regulators, on-line identification, vacuum-pans, conductivity control.

INTRODUCTION

As it is known, the cristalization and cristal growing process that take place in the vacuum-pans are probably the more complex and sensitive in all the sugar production process. At the same time, it is in this area where a near-optimal control may have a more strong influence over the efficiency of the whole process and on the quality of the final product.

The conventional control schemes of vacuum-pans includes as a fundamental loop, the regulation of the condcutivity of the mass in the pan using as control variable, the flow of syrup or molasses, according to the type of process. Other important controlled variables are the vacuum and the steam pressure.

The conductivity of the mass is an indirect measurement of the supersaturation, that is, the variable which determines the speed of cristal growing. The conductivity is also related with the level of the mass in the pan according to a decreasing curve as it is shown in figure 1. In order to ensure that supersaturation is kept constant along the batch process that takes place in the vacuum pan, the conductivity set-point must follow a curve similar to that shown in fig. 1. This goal is hardly accomplished by conventional control schemes, specially taking into account that the precise form of the curve may change depending of some characteristics of the mass that is being processed as its purity and others.

In order to get a deeper knowledge of the dynamic characteristics of the vacuum-pan process, some identification experiments were conducted in the CENTRAL ESPARTACO, a sugar factory near Cienfuegos, Cuba. The results obtained were used in a simulation study of several self-tuning regulators for this process which concluded in an experimental on-line application which seems very promising. The rest of the paper is devoted to describe the identification and simulation results, as well as the control algorithm of the self-tuning type that was implemented by means of a microcomputer developed in IMACC and installed in the Central Pablo Noriega, near Havana.

IDENTIFICATION EXPERIMENTS

The identification experiments consisted in the application of pseudo-random binary signals in the main inputs of the vacuum-pans and the simultaneous measurement of the conductivity of the mass which is the most important controlled variable. Sequences of 63 binary values were applied several times until a number of approximatelly 300 measurements of inputs and the output were obtained. The sample period was choosen empirically equal to 0.5 minutes which seemed to be adequate according to the time-constants of the process.

The input and output variables are represented as follows:

u_1 - air pressure over the valve which regulates the syrup or molasses flow

u_2 - air pressure over the valve which regulates the steam flow.

y – conductivity of the mass

y_r – reference value or set-point for the conductivity

The identified model was the conventional second-order discrete regression model, with the structure:

$$y(t) = a_1 y(t-1) + b_{11} u_1(t-1) + b_{12} u_2(t-1)$$
$$a_2 y(t-2) + b_{21} u_1(t-2) + b_{22} u_2(t-2) + e(t) \qquad (1)$$

where $e(t)$ represents the stochastic term of the model. The order was determined experimentally after testing models from first to fourth order, according to the prediction error criterium.

The identification method used was the least-squares in the square root reccursive version due to Peterka(1975) and called by the author REFIL. The REFIL algorithm has proved to have very good numerical properties both in its off-line and on-line implementations.

In table 1 a typical result of parameters identification after 20, 100, 200 and 300 sample periods is given. In this case an exponential forgeting coefficient of 0.999 was used.

The % of prediction error at the output was calculated according to the formula:

$$\% \text{ error} = \frac{y(t) - \hat{y}(t)}{y(t)} \times 100 \qquad (2)$$

being $\hat{y}(t)$ the prediction of the output at time t calculated through the parameters identified at that time.

The analysis of the results shown in table 1 and of others obtained with different exponential forgeting values permitted to arrive to some conclusions. First, the prediction error obtained was sufficiently small. That is a confirmation of the adequacy of the model order and sample-period choosen. Second, the values of some identified parameters change considerably with time. This fact reflects a physical characteristic of the process, that is: At the beginning of the strike, the quantity of mass in the pan is small and for that reason its conductivity is more sensitive to the changes in the input flows that at the end of the strike when the level of the mass reached the maximum level.

The varying parameters characteristics of the model and the slow change of the conductivity set-point served as a basis for the idea that an adaptive regulator could solve the control problem of the vacuum-pan in a more effective way that the conventional regulators of PID type which are normally used in most of the control schemes for this process.

SIMULATION STUDY

The identified second-order, two inputs and one output model described in equation (1), served, for the purposes of the simulation study, as the "plant" to be controlled. In figure 2 is shown schematically the structure of the simulation program. The stochastic term $e(t)$ is generated by means of a subroutine that computes a discrete white noise with adjustable variance.

As the control algorithm that was used is a simplified minimum-variance monovariable regulator, a fictitious control variable $v(t)$ is introduced, so that:

$$u_1(t) = Kv(t) \qquad (3)$$
$$u_2(t) = (1-K)v(t) \qquad (4)$$

In this form the original two-inputs system is identified by means of a one-input model which serves as a basis for the monovariable control algorithm. Of course, this approach has the restriction that a constant relation is imposed over the two real inputs $u_1(t)$ and $u_2(t)$, by means of the coefficient K. This coefficient was varied between 0 and 1 and it represents the relative participation taken by every input variable in the control action.

In the simulations, the varying parameters of the original second-order model, denoted by means of $\hat{P}(t)$, were stored in magnetic-tape, and they are recovered every sample-time. Only the identified parameters after sample-period No. 40 are used for the simulation. In this way the initial oscillations of the parameters caused by the convergence of the identification algorithm are avoided and only those changes which really reflect the variation in the dynamic characteristics of the process are maintained.

The simplified model used in the self-tuning regulator, has the structure:

$$\varepsilon(t) = \hat{b}_1 \Delta v(t-1) + \hat{a}_1 \varepsilon(t-1) + \hat{b}_2 \Delta v(t-2) + \hat{a}_2 \varepsilon(t-2) + e(t) \qquad (5)$$

where:

$$\varepsilon(t) = y(t) - y_r \qquad (6)$$
$$\Delta v(t) = v(t) - v(t-1) \qquad (7)$$

In figure 1, $\hat{P}_1(t)$ is the simplified-model parameter vector, that is:

$$\hat{P}_1(t) = [\hat{b}_1, \hat{a}_1, \hat{b}_2, \hat{a}_2] \qquad (8)$$

The order and the increments introduced in the model given in (5) were determined by the requirements of the control algorithm, as will be seen in the next section.

CONTROL ALGORITHM

The minimum output-variance stable regulator for a system described by the model given in (5), can be calculated by solving the general polynomial equation (Peterka, 1972):

$$B^* = (1-A)(1+R) + q^{-1}BS \quad (9)$$

where:

$$\frac{\mathcal{E}(t)}{\Delta v(t)} = \frac{q^{-1}B(q^{-1})}{1-A(q^{-1})} \quad (10)$$

is the discrete transfer function of the process. In our case it has the next form for polinomies $B(q^{-1})$ and $A(q^{-1})$:

$$B(q^{-1}) = \hat{b}_1 + \hat{b}_2 q^{-1} \quad (11)$$

$$A(q^{-1}) = \hat{a}_1 q^{-1} + \hat{a}_2 q^{-2} \quad (12)$$

S and R are the polinomies of the regulator transfer function, that is:

$$\frac{\Delta v(t)}{\mathcal{E}(t)} = \frac{S(q^{-1})}{1+R(q^{-1})} \quad (13)$$

q^{-i} is the shift operator, that is $q^{-i}x(t) = x(t-i)$.

The polinomy $B^*(q^{-1})$ which appears in equation (9) is defined by

$$B^* = B^+ \tilde{B}^- \quad (14)$$

where:

B^+ is a factor of B that contains the stable roots of the polinomy (over and in the outside of the unitary circle)

B^- contains the unstable roots of B.

$\tilde{B}^-$ is the reciprocal of polinomy B^- in that way that if B^- is a polinomy of degree n:

$$\tilde{B}^- = \frac{1}{b_n} q^{-n} B^-(q^{-1}) \quad (15)$$

In our case polinomy B is of first order and B^* can be determined easily in the following way:

$$B^* = 1 + \alpha q^{-1} \quad (16)$$

where:

$\alpha = b_2/b_1$ if $|b_1| \geq |b_2|$ (minimum-phase system)

$\alpha = b_1/b_2$ if $|b_2| > |b_1|$ (non minimum-phase system)

Equation (9) can be written then as follows:

$$(1-a_1 q^{-1}-a_2 q^{-2})(1+r_1 q^{-1}) + q^{-1}(b_1+b_2 q^{-1})$$
$$(s_0+s_1 q^{-1}) = 1+\alpha q^{-1} \quad (17)$$

and solving equation (17) with respect to the regulator parameters r_1, s_0 and s_1, the next equations are obtained:

$$D = b_1^2(a_2-a_1\alpha) - b_2^2 \quad (18)$$

$$s_0 = ((\alpha+a_1)(a_2 b_1 - a_1 b_1) - a_2 b_2)/D \quad (19)$$

$$s_1 = ((a_2 b_1 - b_2(\alpha+a_1)))/D \quad (20)$$

$$r_1 = b_2 s_1/a_2 \quad (21)$$

The following is the equation of the regulator:

$$\Delta v(t) = -s_0 \mathcal{E}(t) - r_1 \Delta v(t-1) - s_1 \mathcal{E}(t-1) \quad (22)$$

Notice that the use of the increments $v(t)$ instead of the absolute value of the control variable $v(t)$ ensures the inclusion of an integral mode in the control algorithm, because it is required to calculate:

$$v(t) = \sum_{i=1}^{t} v(i) = v(t-1) + \Delta v(t) \quad (23)$$

The inclusion of an integral mode in the control action, helps to mantain the astatism of the system, specially with respect to the changes of the controlled variable set-point.

The self-tuning algorithm explained above was tested in the simulations and compared with other algorithms as the conventional PID regulator, second-order minimum variance, etc., giving a very good performance in spite of its simplicity as will be illustrated in the next section. The main advantages are the small number of operations that requires (only 15 multiplications and divisions) and its adaptive nature in the sense that it works well in spite of the minimum-phase or non minimum phase characteristic of the system model.

The final version of the algorithm includes the limit checking for the control variable and the substitution of the coefficient α for a prefixed number when parameters $\hat{b}_1$ and $\hat{b}_2$ are very close each other. In this last case, the poles of the closed loop transfer function can be very near to the unitary circle and it can cause undesirable oscillations in the control variable.

SOME SIMULATIONS RESULTS

In the tables 2 and 3, some results of the simulations made following the scheme given in the figure 2 are presented. They correspond to exponential forgetting coefficients of 0.995 and 1 respectively. The values of K in both tables is equal to 1, that is, only the flow of input syrup is used as control variable. The results of several simulations showed that there is not any special advantage in using the two possible control variables for the regulation of the conductivity and this coincide with the experiences obtained from many conventional schemes for the control of vacuum pans.

In tables 2 and 3 are given the results from sample time 40 to 140. Simulations continued until t=300 but there was not any appreciable change and for that reason the last part was omitted.

As it is known, the exponential forgetting coefficient is used in connection with the on-line identification methods to avoid the "saturation" of the identification algorithm and to follow possible slow changes in the parameters of the model (Aguado,et. al.,1980). As the vacuum-pan is a batch process, where the number of sample periods from the beginning to the end of the strike is relatively small, it seems that no special advantages are obtained with the use of a coefficient different from one. This point seems to be confirmed by the results shown in tables 2 and 3, although it must be confirmed by the on-line experiments. In the simulations the change of the refference value for the output was introduced considering it as a time function parabola.

As it can be seen in tables 2 and 3, the error and the variance of the output variable are extremely small, with the exception of the first steps, where they are much bigger. This effect is due to the fact that relativelly arbitrary initial values were choosen for the parameters and for the output variable. In the on-line application this problem can be solved because after a number of experiences, very tight values could be choosen for the initial parameters.

SOME CONCLUDING REMARKS

The adaptive regulator for the conductivity of the mass in the vacuum-pan presented in this paper, has been partially tested until now in an on-line experiment conducted in Central Pablo Noriega, near Havana. The micro-computer developement system MICRO-IMACC 8101, based on INTEL-8080 microprocesor, was used for the test. The memory requirements for the identification and control algorithm, as well as a floating-point arithmetic package and some other general purpose programs was of 3.8 Kbytes. Even when some very promising results were obtained, the experiments must proceed in the next campaign starting on December. The on-line experiments made until now permitted to visualize some problems concerning the selection of an adequate sample-period. It seems that this period must change, in that way that at the beginning of the strike, when the process is more rapid, an smaller sample-period must be used that at the end, when the process is considerably slower. Values between 10 and 30 seconds could be used.

The final version of the microprocessor controller for the vacuum-pan, must include not only the conductivity control loop described here, which is the more complex and important, but also vacuum and steam pressure loops as is usual in conventional control schemes for this process, as well as logical sequentional operations that are required at the start and end of the strike.

REFERENCES

Aguado,A., Martínez,J., Enríquez,J.(1980) Identificación y control adaptivo de procesos tecnológicos, Revista CCA, No. 4.

Aguado,A.(1982). Reguladores de autoajuste simplificados, Revista CCA, No. 2

Peterka,V.(1972). On steady state minimum variance control strategy, Kybernetika, Vol. 8, No. 3.

Peterka,V.(1975). A square root filter for real time multivariate regression. Kybernetika, Vol. 11, No. 1.

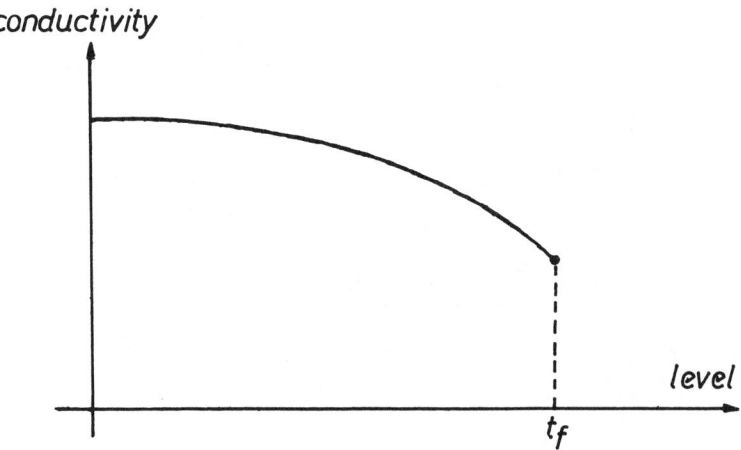

fig. 1. Form of the conductivity vs level dependence.

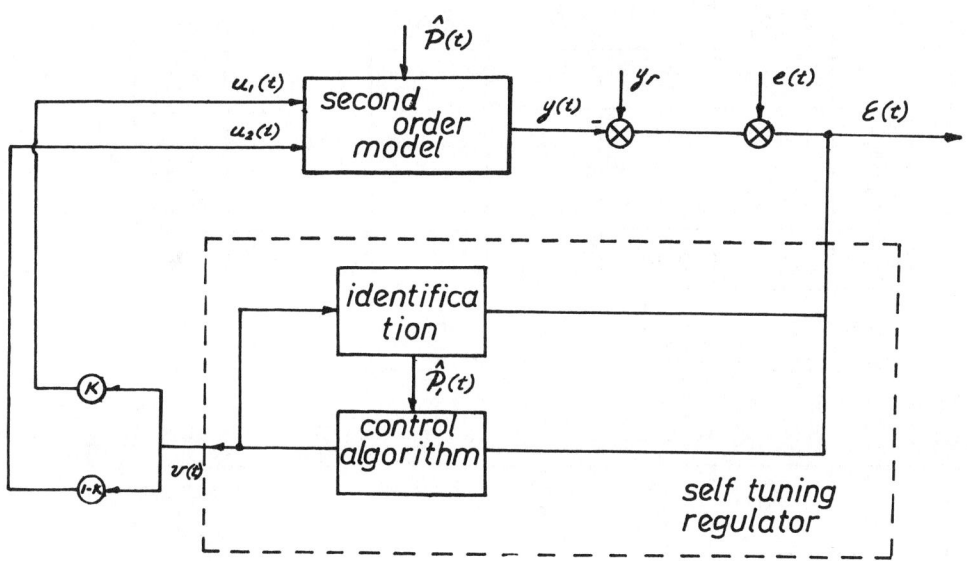

fig. 2. Scheme of the simulations.

sample period n°.	b_{11}	b_{21}	a_1	b_{12}	b_{22}	a_2	% error at the output
20	7.297	1.759	0.920	-2.755	-1.968	0.049	6
100	6.518	0.831	0.921	-3.1518	-0.865	0.054	1.78
200	6.087	-0.493	0.925	-4.018	0.381	0.060	2.24
300	5.125	-0.058	0.902	-3.108	-0.163	0.084	1.12

table 1. Results of the identification for exponential forgetting equal 0.999.

t	$\hat{b}_1$	$\hat{a}_1$	$\hat{b}_2$	$\hat{a}_2$	$y-y_r$	output variance
40	4	2.000	-3.000	-1.000	-6.470	41.98
60	1.653	0.713	-1.634	-0.261	-0.170	6.67
80	1.659	0.715	-1.631	-0.261	0.015	0.0001
100	1.800	0.746	-1.566	-0.256	-0.181	0.0165
120	1.802	0.746	-1.553	-0.254	-0.041	0.0172
140	1.804	0.747	-1.550	-0.253	0.020	0.0010

table 2. Results of the simulation of the self tuning regulator. Exponential forgetting equal 1.

t	$\hat{b}_1$	$\hat{a}_1$	$\hat{b}_2$	$\hat{a}_2$	$y-y_r$	output variance
40	4.000	2.000	-3.000	-1.000	-6.470	41.98
60	0.949	0.563	-1.837	0.246	1.183	8.16
80	0.975	0.570	-1.804	-0.243	-1.397	1.1558
100	0.997	0.579	-1.794	-0.238	0.600	0.3606
120	1.000	0.581	-1.794	-0.235	-0.082	0.0104
140	1.006	0.583	-1.794	-0.235	-0.142	0.0134

table 3. Results of the simulation of the self tuning regulator. Exponential forgetting equal 0.995

ADAPTIVE CONTROL OF A STEAM TURBINE

D. N. Oliva*, E. L. Morris** and M. T. Oliva***

*Fac. de Ciencias, Universidad Central de Venezuela
**Department of Engineering Science, University of Exeter, Exeter, UK
***Fac. de Ingeniería, Universidad Central de Venezuela, Venezuela

Abstract. The paper describes the development of a modular adaptive controller based on linear least squares identification, pole assignment and state variable feedback. The controller is applied to a small steam turbine, and is compared with other types of controllers.

Keywords. Adaptive control; identification; pole assignment; process control; digital control.

INTRODUCTION

During the last decade, as first mini- then micro-computers have become readily available there has been a swing from analog to digital controllers in all areas of applications. This has been followed by a slower shift from classical to more modern controller and design techniques. This is exemplified in the evolution of control systems for steam turbines, where analog electro-hydraulic controllers replaced mechanical-hydraulic systems in the mid 60's, only to be replaced by computer controlled electro-hydraulic systems within the last few years. (Podolsky, 1972)

Associated with this development of computer controlled systems has been a desire to produce a general purpose controller sufficiently flexible to be used on a wide variety of different systems. This has been made possible by the development of adaptive control algorithms which allow the designer a degree of latitude in this choice of the controller parameters in the expectation that these values would soon be modified by the control computer.

There were three principal objectives in the design of the controller described in this paper: to produce a controller for a system expected to be non-linear but whose parameters were known only approximately; to devise a system which could be broken down into separate modules which could be easily implemented on one or more microprocessors; and to use only relatively simple building blocks, each of which could be easily understood and modified as required.

The algorithms chosen were tested by controlling a small steam turbine system with a PDP 11-03 microprocessor. Following the work of Morris and Abaza (1976) the system was assumed to be of second order, and the sampling time to be 0.5 sec. These algorithms were a state variable feedback controller with integral action, whose parameters are set by using the Gopinath pole assignment technique. The knowledge of the system required to use this method was obtained on-line by using the adaptive linear least squares identification technique. The three sections of the controller are completely independent and were implemented as separate subroutines in the control computer, but could just as easily be run on separate but interconnected processors. Only the controller is vital to the operation of the system; failure in the other components would lead to degradation of performance but not complete loss of control.

DESCRIPTION OF ALGORITHMS

A system may be described by a set of linearized difference equations:

$$x_k = A\, x_{k-1} + B\, u_{k-1}$$

and $y_k = C\, x_k$

where the coefficients A and B will depend on the current operating condition of the system, and are in general neither independent of the state variables, x, nor independent of time. However, it is assumed that the model is valid over the range of the normal perturbations of the system about its operating point and over a period many times longer than the sampling time used on the system.

The control law (Young and Willems, 1972) which has been implemented is given by:

$$u_k = -K_p\, x_k - K_i\, w_k$$

with $w_k = w_{k-1} + T\,(y_{ref} - y_{k-1})$

where T is the sampling time and y_{ref} is the input reference value. Augmenting the state difference equation by this equation for w gives:

$$\begin{vmatrix} x \\ w \end{vmatrix}_k = \begin{vmatrix} A & 0 \\ -TC & 1 \end{vmatrix} \begin{vmatrix} x \\ w \end{vmatrix}_{k-1} + \begin{vmatrix} B \\ 0 \end{vmatrix} u_{k-1} + \begin{vmatrix} 0 \\ T \end{vmatrix} y_{ref}$$

The Gopinath (1968) algorithm was used to select the controller gains. If the augmented state equations are written:

$$x_k = \phi x_{k-1} + D u_{k-1}$$

then the characteristic polynomial of the closed loop system can be written:

$$p + K Q D = 0 = r$$

where p is the characteristic polynomial of the original open loop system, K is the feedback gain matrix, (K_p, K_i), and

$$Q = \text{adjoint } (zI - \phi).$$

If p is written as $p = (p_1 \ldots p_n)^t$ and $r = (r_1 \ldots r_n)^t$ then it can be shown that

$$\begin{vmatrix} 1 & 0 & 0 \\ p_1 & 1 & 0 \\ p_2 & p_1 & 1 \end{vmatrix} \begin{vmatrix} D^t \\ D^t \phi \\ D^t (\phi^2)^t \end{vmatrix} K^t = (r-p)$$

Note that the r_i are the coefficients of the desired characteristic polynomial, that ϕ and D are 'known' and that the p_i may be obtained from ϕ relatively simply. The major problem in computing the feedback matrix K, is the inversion of the square matrix required. For the control of the turbine, the inversion of the 3 x 3 matrix is done by direct knowledge of the equations.

To evaluate the transition and driving matrices of the turbine, the adaptive linear least squares technique was used. Assuming that the state equations can be written (Young, Shellswell and Neethling, 1971)

$$x_k = \phi x_{k-1} + D u_{k-1} + e_k$$

where e is the residual error in the computation together with any noise in the system. Form a vector $z_k = (x_{k-1}^t, u_{k-1})^t$, then

$$x_k = (\phi \mid D) z_k + e_k$$

now form matrices

$$X_k = \begin{vmatrix} \sqrt{\gamma} \; X_{k-1} \\ x_k^t \end{vmatrix} ; \; Z_k = \begin{vmatrix} \sqrt{\gamma} \; Z_{k-1} \\ z_k^t \end{vmatrix}$$

and $\Theta = (\phi \mid D)^t$.

Now if we define

$$V_k = \sum_{i=1}^{k} e_i^t e_i \gamma^{k-i}$$

$$= \sum_{i=1}^{k} \gamma^{k-i} (x_i^t - z_i^t \Theta)(x_i - \Theta^t z_i)$$

and minimize with respect to Θ, we obtain

$$\sum_{i=1}^{k} \gamma^{k-i} z_i x_i^t = \sum_{i=1}^{k} \gamma^{k-i} z_i z_i^t \hat{\Theta}_k$$

which can be written

$$\hat{\Theta}_k = (Z_k^t Z_k)^{-1} Z_k^t X_k$$

This relationship can be made recursive if we define $P_k^{-1} = Z_k^t Z_k$:

$$P_k = \frac{1}{\gamma}(P_{k-1} - P_{k-1} z_k (\gamma + z_k^t P_{k-1} z_k)^{-1} z_k^t P_{k-1})$$

$$\hat{\Theta}_k = \hat{\Theta}_{k-1} - P_k z_k (z_k^t \hat{\Theta}_{k-1} - x_k)$$

For computational purposes it is convenient and more reliable to use a square root algorithm (Peterka, 1975) to caculate P, as P is positive definite. We can write $P_k = S_k S_k^t$, and S may be updated instead of P. This guarantees that P remains positive definite despite any numerical problems with particular sets of data.

DESCRIPTION OF SYSTEM

The plant to be controlled was a single stage impulse type de Laval steam turbine designed to attain maximum efficiency at 22.4 KW load and 20.000 revolutions per minute. The load was provided by an alternator driven through a reduction gear and delivering power to a resistance bank. The steam valve was operated by an electro-hydraulic actuator controlled by a voltage controlled 4-20 mA current source The turbine also possessed a mechanical overspeed trip for safety reasons.

The torque produced by the turbine and the output speed measured after the reduction gear were used as the state variables of the system. Both these signals were filtered using low pass filters and amplified to be within the range +/- 10 Volts. The voltage input to the current source acted as the reference input to the system. The steam pressure was maintained as constant as possible during the experimental work.

Following the work of Morris and Abaza (1976) a sampling time of 0.5 seconds was used throughout, although tests were made to verify that this value was suitable for the modified configuration of the turbine and the new controller algorithms.

For the purposes of the identification, the input signals were offset with DC voltages and then filtered with digital high pass filters to remove unwanted steady state voltages. The control input signal had a 63 bit Pseudo Random Binary Sequence added to it on occasions when identification and adaptive control routines were being used. The amplitude of the PRBS was adjusted to produce a 5 percent speed fluctuation. Reduction of this amplitude caused the convergence of the identification routines to be delayed. In cases where there was a significant and persistant fluctuation of the load, this signal could be removed, but in the artificial environment of these experiments, this was not in general possible.

The control computer used was a Digital Equipment Corporation PDP 11-03, with 24 K-bytes of memory and the floating instruction set (FIS). The computer possessed no backing store or peripherals other than a visual display unit (VDU) which served as the console. All the software was written in Assembly language, in as modular a fashion as possible, on the Faculty of Engineering PDP 11-40, assembled and subsequently downline loaded into the PDP 11-03. The control computer possessed a 16 channel analog multiplexer and a 12 bit analog to digital converter with conversion time of 20 micro seconds. In addition there were two 12 bit digital to analog converters and a line clock.

The console VDU could be used to set the constants and initial values of the variables of the system, and to control the program flow by using a number of software switches.

EXPERIMENTAL RESULTS

A preliminary series of identification tests were performed, both on- and off-line. A variety of identifications techniques were used, including Linear Least Squares, the adaptive linear least squares, and the instrumental variable methods. The first two of these are well known to produce biased results. However, as the noise levels in the system were found to be very low, it was hoped that they would prove satisfactory in practice. Indeed the results were found to compare quite well with the instrumental variable results. The adaptive L.L.S. method proved to be the only technique capable of satisfactorily tracking the variation in the turbine parameters as the load was changed. This was therefore the method chosen to be used in the adaptive controller. The 'forgetting factor', γ, was chosen to be 0.995. Reduction of this parameter makes the tracking faster, but the fluctuations in the identified parameters more violent. Increase of this parameter reduces tracking to unsatisfactory levels.

It proved possible to identify the steam turbine system both with and without an enclosing proportional feedback loop. Again this was assisted by the low level of noise in the system. The perturbing PRBS signal was adjusted to give a 5 percent speed fluctuation, which proved difficult in the open loop case.

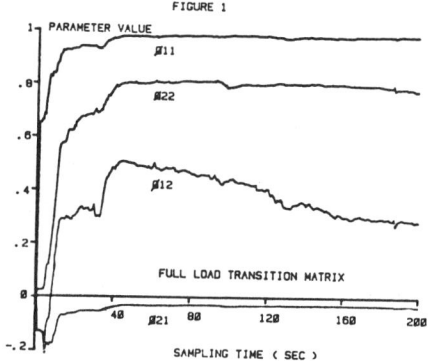

Figures 1 and 2 show the identification matrix of the turbine, as a function of time, using the instrumental variable technique, under full and half load conditions. These calculations were done off-line on the

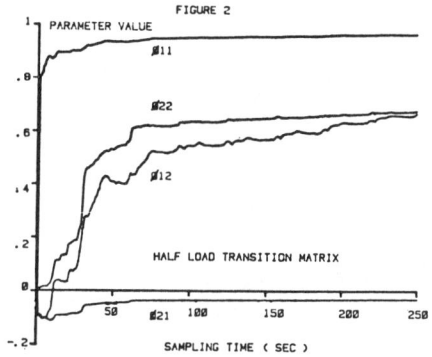

faculty PDP 11-40. It can be seen that the convergence to stable values is reasonably rapid, and that at least one of the parameters varies considerably with load. (Oliva M. T., 1981) Table 1 shows values for the transition and driving matrices for full and half load conditions. Again it can be seen that ϕ_{12} varies most with load, although the other parameters are affected.

TABLE 1   Steam Turbine Parameters

	ϕ		D
Full	0.970	0.389	0.105
Load	-0.022	0.855	0.013
Half	0.988	0.943	0.148
Load	-0.018	0.790	0.004

These results show clearly that the turbine system was non-linear and that an adaptive controller should be of some benefit. However, a variety of different control algorithms were implemented in an attempt to show the differences between them. The controllers included proportional, proportional plus integral, state variable feedback, state variable feedback with integral action, the self tuning controller (Clarke and Gawthrop, 1975) and the pole assignment self tuning controller (Wellstead and Allidina, 1979) as well as the adaptive controller described in this paper. It was clear that none of the controllers without integral action were satisfactory. This was largely due to the fact that the gain of the turbine system was dependant on load. The performance of the state variable feedback with integral action controller was clearly better than the proportional plus integral controller and compared favourably with the adaptive controller, as shown in Figures 3 and 4. Figure 5 shows the step response of the system, on startup for the state variable controller. This can be compared with Figure 6. The

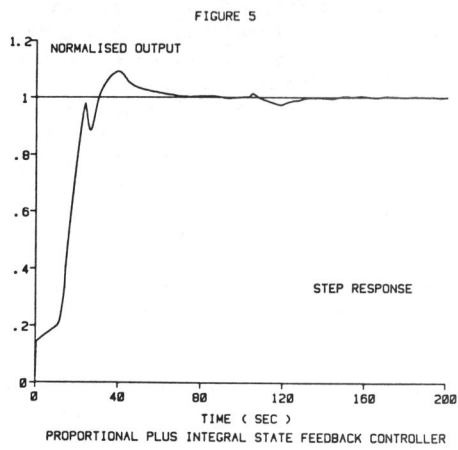

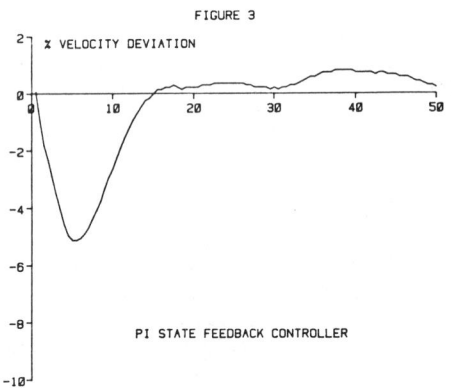

adaptive controller had a slightly better performance for small speed changes a different loads. At full load, the adaptive controller and the state variable controller were essentially the same. The adaptive controller, of course, suffered from the disadvantage of the injected PRBS perturbation signal. This might well be unacceptable in some cases; equally it might prove unneccessary in cases where there were other naturally occuring disturbances. (Oliva D. N., 1981)

A more unexpected finding was that neither of the self tuning controllers could cope adequately with the variation in the system parameters that was experienced with the turbine. Admittedly these controllers were not designed for this situation, but the result was still dissappointing.

Figures 6 and 7 give a good idea of the abilities of the adaptive controller as the system was started up under full load, using the initial parameters of the state variable

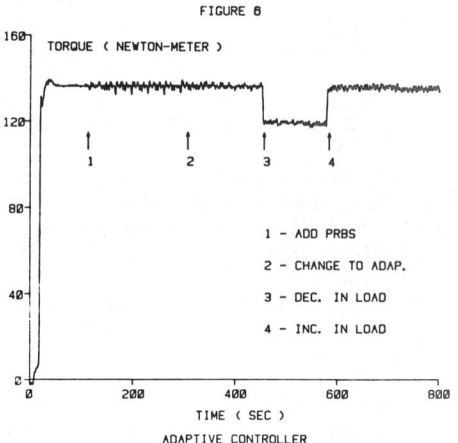

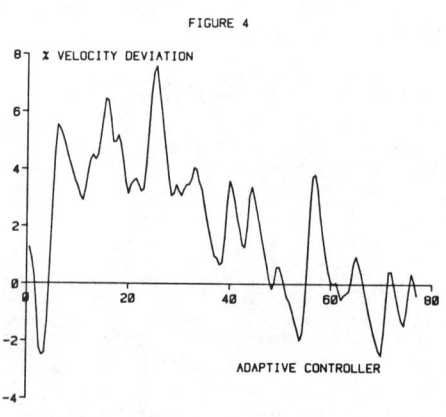

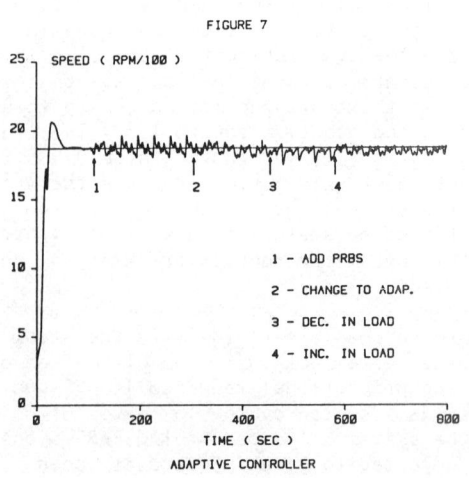

controller. Identification was commenced when the PRBS was injected, and the adaptive controller was included after enough samples for the identifier to converge to reasonably stable values. Subsequent decrease and then increase in the load causes virtually no effect on the speed, as can be seen clearly.

CONCLUSIONS

It is clear that the combination of algorithms proposed provide a satisfactory adaptive control. The use of the Gopinath algorithm, involving a matrix inversion as it does, consumes the major part of the time available for the computations. The identification algorithm, being recursive, is relatively efficient in terms of processor time. As the various algorithms are virtually independant of one another, it would be possible to operate them in parallel. Indeed, further development in this area is under way.

The adaptive algorithm provided good control under all load conditions, the only defect being the size of the perturbations caused by the injected PRBS. It compared very favourably with all the other controllers tested. Of these, the best was the state variable feedback with integral action. Under full load conditions, these two were designed to be the same.

ACKNOWLEDGEMENTS

The authors would like to acknowledge the support and encouragement they received in the Department of Engineering Science of the University of Exeter, and their gratitude to the Science Research Council for a grant to carry out this work. Two of the authors would also like to thank the government of Venezuela for grants to continue their studies.

REFERENCES

Clarke, D. W. and Gawthrop P. J. (1975). Self-tuning Controller, Proc. IEE, 122, 929.

Gopinath, B. (1968). On the Control of Linear Multiple Input Output Systems. Conf. Rec. 2nd Asilomar Conf. on Circuits and Systems, New York, IEEE, p 21.

Morris, E. L. and Abaza B. A. (1976). Adaptive Digital Control of a Steam Turbine, Proc. IEE, 123, 549.

Oliva, D. N. (1981). Application of Adaptive Control Techniques to a Steam Turbine, Ph. D. Thesis, University of Exeter.

Oliva, M. T. (1981). On-line Identification Techniques for State Variable Systems, Ph. D. Thesis, University of Exeter.

Peterka, V. (1975). A Square Root Filter for Real Time Multivariable Regression, Kybernetica, 11, 53.

Podolsky, (1972). Digital Electro-hydraulic Control for Large Turbine Generators, Proc. of the 14th International ISA Power Instrumentation Symp., p 40.

Wellstead, P. E. and Allidina (1979). Pole Assignment Self-tuning Regulators, Proc. IEE, 126, 781.

Young, P. C., Shellswell S. H., and C. G. Neethling (1971). A Recursive Approach to Time Series Analysis, CUED/B Control TR 16, Department of Engineering, University of Cambridge.

Young, P. C. and J. Willems (1972). An Approach to the Linear Multivariable Servomechanism Problem, Int. J. Control, 15, 961.

DISCUSSION

SESSION MA4: ADAPTIVE CONTROL

Paper: ADAPTIVE CONTROL OF MUSCLE RELAXATION

Authors: CS *Berger, WA Brown* (Dept of Electrical Engineering Monash University, Victoria, Australia)

Discusser: *Rogelio Lozano*
Depto de Ingeniería
Eléctrica, CIEA-IPN
Apdo Postal 14-740
México 14, D.F.
México

Questions or Comments:

1) How fast do the plant parameters change with respect to the convergence speed of the identification algorithm?

Authors' reply:

1) Under steady state operation the parameters' change is slow. The linearised parameters change instantly with changes in the operating conditions. The interim set point was therefore introduced to ensure that the operating conditions approach the desired set-point sufficiently slowly for the identification algorithm to work. The choice of closed loop poles was also governed by the rate of convergence of the identification algorithm. Note also that a relatively low forget factor was used.

Paper: MICROCOMPUTER IMPLEMENTATION OF AN ADAPTIVE CONTROL ALGORITHM

Authors: *R Lozano, A Noriega* (Depto de Ing Eléctrica, Centro de Investigación y Estudios Avanzados, IPN, Apdo Postal 14-740, 07000 México DF, México)

Discusser: *Alberto Aguado*
Carmen No 170, C
Habana, Cuba

Questions or Comments:

1) Which forgetting factor did you use?

Author's reply:

1) We have chosen the forgetting factor such that the gain matrix trace is equal to 10.

Paper: ADAPTIVE CONTROL OF DISCRETE MULTIVARIABLE SYSTEMS

Authors: *R Lozano, M Bonilla* (Depto de Ing Eléctrica, Centro de Investigación y de Estudios Avanzados, IPN, Apdo Postal 14-740, 07000 México, DF, México

Discusser: *G Gilles*
Laboratoire d'Automatique
Université de Lyon 1
43 Boulevard du 11 novembre 1918
69622 Villeurbanne Cedex, France

Questions or Comments:

1) I am interested in some comments on numerical values: What are the values of:
   - The different time constants involved in the plant?
   - The sampling period?
   - The computation time of the control

algorithm on the z-80 system?

2) Did you processed experimentations on the real plant after using simulation on an analog computer? If so, which are the conclusions?

Authors' reply:

1) We can consider that the plant time constant is around 10 sec. The sampling period was 1 sec. The computing time of the control algorithm did't matter because the control input U_t is not a function of y_t, but only a function of Y_{t-1} and past data

2) No

Discusser: José Luis Farah
Apdo Postal 20-726
México 01000, DF
México

Questions or Comments:

1) Could you please comment on the choice of the forgetting factor and of its sensitivity?

2) If possible, can you relate its choice to the system's structure?

Authors' reply:

1) We choose the trace of the gain matrix stablishing a compromise between speed of convergence of the identification algorithm and variance of the parametric distance (see fig 42)

2) See answer to Prof. Berger's questions

Discusser: CS Berger
19 Sheringham Dr,
Wheelers Hill,
Vic, Austria

Questions or Comments:

1) In your work you chose the structure *a priori*. Have you had experience with using adaptive control to determine the structure?

Authors' reply:

1) In one of our experiments we used the identification algorithm to identify the structure, but it takes too long to identify the system.

SESSION 5 — FUEL AND HEAT CONTROL

MODEL REFERENCE ADAPTIVE CONTROL OF AN INDUSTRIAL PHOSPHATE DRYING FURNACE

B. Dahhou, K. Najim and M. M'saad

Laboratoire d'Electronique et d'Etude des Systèmes Automatiques, Faculte des Sciences, B.P. 1014, Rabat, Morocco

Abstract. This paper presents experimental result of a model reference adaptive control algorithm with independent tracking and regulation objectives presented in (Landau, Lozano, 1981) to the control of a phosphate drying process at the Beni-Idir Factory of the OCP (Office Chérifien des Phosphates - Maroc).
The main control objective is to keep the moisture content of dried phosphate at a prescribed value (1,5%), independently of external perturbations acting on the drying process.
The plant dynamic characteristics vary under the effect of variations of the input material characteristics such as the phosphate nature and humidity that vary from one layer to another.
The implementation of the adaptive algorithm was based on a reduced order plant model previousely checked and uses a small size minicomputer.
An energy saving close to 4,5% and ten times reduction of the variance of the output humidity error with respect to the desired one were obtained. This led to the motivation of introducing an advanced computer control in Moroccan Phosphate Industry.

Keywords. Adaptive Control - Model Reference - Energy Saving - Phosphate Processing - Drying Furnace.

INTRODUCTION

During the past few years different approaches to adaptive control have been suggested, studied and applied. Among these approaches, the Model Reference Adaptive System and the Self-tuning Regulator seem to be the most attractive ones.

This paper deals with the application of a Model Reference Adaptive Control Algorithm, presented in (Landau, Lozano 1981), to the control of a Phosphate Drying Process at the Beni-Idir Factory of OCP.

The phosphate, independently of its way of extraction has about 17% humidity. Before being sold, this high humidity has to be reduced to around 1,5% in Rotary Drying Furnaces.

The drying process in one of industrial operations that requires a great consumption of energy, hence an increase in the price of the produced dried material.

The objective of this study is to keep the humidity of the dried phosphate close to the prescribed value (1,5%), independently of raw material humidity variation (7∼20%); feed flow rate variations (100∼240 t/h) and other perturbations that may effect the drying process.

There is, invariably, some uncertainty in the characretistics of the processed phosphate that can be attributed to variable moisture content and the nature of the damp product. The phosphate drying process is therefore non-linear and non-stationnary in its nature. The change in dynamic characteristics with operating conditions is such that a fixed parameter controller is inedequate to achieves satisfactory performances in the entire range over which the characteristics of the process may vary. An adaptive control holds obvious attractiveness in such situation because controller parameters are adjusted operation to maintain specified dynamic performances.

A Model Reference Adaptive Control Scheme, developed by I.D. Landau and R. Lozano and based on reduced order plant model, previously checked was implemented using smale size minicomputer.

The main motivations of such control scheme are the following :

- It is simple :i.e. it can be implemented even on microcomputer.
- It ensures the asymptotic convergence of the plant output (the humidity of the dried phosphate) to the reference sequence and the boundness of the control applied to the plant.
- It allows to solve the problem of independent specification of tracking and regulation objectives.

This paper is organized as follows. In section II, we provide physical description of the used drying process. In section III, a mathematical model of the drying furnace is formulated. In section IV, the adaptive control scheme used to control the phosphate drying furnace is presented while in section V, the hardware and software facilities are described and the furnace control performances using the Model Reference Adaptive Control Algorithm are reported.

PROCESS DESCRIPTION

The phosphate drying furnace is mainly constituted of the following components (fig.1)
- Feeding system
- Combustion chamber
- Drying tube
- Dusting chamber
- Ventilator and chimney

These elements are described in the following.

Feeding system

The main part of the feeding system is a contant speed moving belt that carries the raw phosphate into the furnace. A large container spreads the phosphate over the belt at regulated rate by controlling the opening of the container to the belt. This will allow the phosphate to be fed into the furnace at the rate needed for production.

Combustion chamber

The combustion chamber produces the hot gas needed for the drying process. The heavy fuel is initially heated to 100°C by steam. To facilate its mixing with the air, the fuel is pulverised by the aid of auxiliary jet of steam. The necessary oxygen for the combustion is produced by the primary air injected under low pressure by a ventilator in the combustion chamber. The heat produced is transfered into the drying tube by secondary air current.

Drying tube

This is a horizontal tube of 25 m length, its rotation velocity is constant; its production capacity is in the order of 150 ton/hr. The tube has cascades in its inner side arranged helically, to facilate the thermal exchange between the hot gas and the phosphate, and also they help in driving the phosphate to the output of the tube. Contrary to cement furnaces the movement of the phosphate and the hot gas occurs in the same direction in the drying furnace, from the combustion chamber to the dusting chamber.

Dusting chamber

The dusting chamber is made up mainly of schelved tubes whose primary function is to slow and recapture the phosphate fine particles which are carried into the dusting chamber by the hot gas. These fine particles make up about 30% of the dried phosphate.

Ventilator and chimney

The main role of the ventilator is to create a reduction in the pressure at the head of the drying tube to induce a secondary air current and to prevent trapping of the phosphate in the drying tube. The chimney action will serve as evacuator of the hot gas out of the furnace.

The final product is received at the exit of the dust chamber by the main conveyor.

The existing conventional control loops on the phosphate dry process are shown in figure 1.

The flows of primary air and steam are adjusted with respect to the fuel flow in order to ensure a complete combustion.

PROCESS MODEL

Several models have been developed in (K. Najim and all 1976,1977,1978,1979) to describe the dynamic behaviour of the phosphate drying furnace. We have chosen a single input single output one, by letting the product feed rate to be kept constant (e.g. maximum production). The fuel flow (the control variable) and the humidity of the dried phosphate (the output variable) are the key variables for suitable single input - single output model of the furnace. A simple representation of the simplified model can be written as :

$$A(q^{-1})y(t) = q^{-d} B(q^{-1})u(t) + w(t) \quad (1)$$

with

$$A(q^{-1}) = 1 + a_1 q^{-1} + a_2 q^{-1} + \ldots + a_{n_A} q^{-n_A}$$

$$B(q^{-1}) = b_0 + b_1 q^{-1} + \ldots + b_{n_B} q^{-n_B}; b_0 \neq 0 \quad (2)$$

where

$\{q^{-1}\}$ is the backward shift operator, $\{d\}$ represents the process time delay, $\{u(t)\}$ and $\{y(t)\}$ are the process input (the fuel flow) and output (the humidity of the dried phosphate) respectively, and $w(t)$ is a bounded disturbance.

This model is most adaptable to adaptive control system which we have adopted. Moreover, it uses the variables to which the operating of the furnace is the most sensitive. The sampling period T and the process time delay have been determined from an a priori caracterisation study of the process, while the process model order has been chosen to allow satisfactory performances of adaptive control system. The obtained values are :

$$T = 45s$$
$$d = 2$$
$$n_B = 1$$
and $$n_A = 3$$

PRESENTATION OF ADAPTIVE CONTROL SCHEME

We will use the notation of (Landau, Lozano 1981) and give only a brief outline of the basic theory of the control scheme adopted. The theory and design of this scheme is widely discussed in the above reference.

The main objective of the control system is to find a control law so that an initial error between the plant output (described by the equations (1) and (2) and assumed to be a minimum phase plant) and a reference sequence $y^M(k)$ or an initial output disturbance converge to zero with the dynamics of the C_R-polynomial, i.e.,

$$C_R(q^{-1})(y(k+d) - y^M(k+d)) = S(q^{-1})w(t) \quad (3)$$

where

$$C_R(q^{-1}) = 1 + C_1^R q^{-1} + \ldots + C_{n_{C_R}} q^{-n_{C_R}}$$

is an asymptotically stable polynomial and the polynomial S is so that :

$$S(q^{-1}) w(t) = 0 \quad \text{for} \quad k \geq k^* \quad (5)$$

The reference sequence can be realized by the output of a reference model described by :

$$C_T(q^{-1}) y^M(k) = q^{-d} D(q^{-1}) u^M(k) \quad (6)$$

where

$$C_T(q^{-1}) = 1 + C_1^T q^{-1} + \ldots + C_{n_{C_T}} q^{-n_{C_T}} \quad (7)$$

is an asymptotically stable polynomial and

$$D(q^{-1}) = d_0 + d_1 q^{-1} + \ldots + d_{n_D} q^{-n_D} \quad (8)$$

An appropriate control configuration used for the case of known plant parameters to realise the objectif (3) is given by

$$u(k) = \frac{C_R(q^{-1})y^M(k+d) - R(q^{-1})y(k)}{B(q^{-1})S(q^{-1})} \quad (9)$$

where the polynomials $S(q^{-1})$ and $R(q^{-1})$ verify the following identity.

$$C_R(q^{-1}) = A(q^{-1})S(q^{-1}) + q^{-d}R(q^{-1}) \quad (10)$$

where

$$S(q^{-1}) = 1 + s_1 q^{-1} + \ldots + s_{n_s} q^{-n_s} \quad (11)$$

$$R(q^{-1}) = r_0 + r_1 q^{-1} + \ldots + r_{n_R} q^{-n_R} \quad (12)$$

which has a unique solution for the polynomials $S(q^{-1})$ and $R(q^{-1})$ for a given $C_R(q^{-1})$ if one chooses :

$$n_s = d-1$$
$$\text{and} \quad n_R = \text{Max}(n_A - 1, n_{C_R} - d) \quad (13)$$

The control law (9) can be written :

$$p^T \phi(k) = C_R(q^{-1}) y^M(k+d) \quad (14)$$

where

$$\phi^T(t) = [u(t), u(t-1), \ldots, u(t-d-n_B+1), y(t), \ldots, y(t-n_R)] \quad (15)$$

$$p^T = [b_0, b_0 s_1 + b_1, \ldots, b_{n_B} s_{d-1}, r_0, \ldots r_{n_R}] \quad (16)$$

When the plant parameters are unknown the parameter vector p of the control law (14) given by Eq (16) can not be computed. Landau and Lozano have developed an extension of the linear controller design given by Eq (14) which is applicable to minimum phase plants and for which only the time delay {d} and upperbounds of the degrees of polynomials $A(q^{-1})$ and $B(q^{-1})$ denoted n_A and n_B are known. The parameter p in Eq (14) is replaced by adjustable parameter vector $\hat{p}(k)$ which will be updated by the adaptation mechanism. Therefore the control law is given by :

$$\hat{p}^T(k)\phi(k) = C_R(q^{-1})y^M(k+d) \quad (17)$$

and the design objectif (3) will be asymptotically achieved if :

$$w(k) = 0 \quad (18)$$

and if the following adaptation algorithm is used :

$$p(k) = p(k-1) + F(k)\phi(k-d)\gamma^*(k) \quad (19)$$

with

$$F(k+1) = \frac{1}{\lambda_1(k)} \left[F(k) - \frac{F(k)\phi(k-d)\phi^T(k-d)F(k)}{\frac{\lambda_1(k)}{\lambda_2(k)} + \phi^T(k-d)F(k)\phi(k-d)} \right]$$

where

$$0 < \lambda_1(k) \leq 1 \quad ; \quad 0 < \lambda_2(k) \leq 2 \quad ; \quad F(1) > 0 \quad (21)$$

and $\gamma^*(k)$ is the adaptation error defined as:

$$\gamma^*(k) = \frac{H_1(q^{-1})}{H_2(q^{-1})} \varepsilon^*(k) \quad (22)$$

where $H_1(q^{-1})$ and $H_2(q^{-1})$ are asymptotically stable monic polynomials and should be chosen such that the transfer function

$$H(z^{-1}) = H(z^{-1}) - \frac{1}{2} \quad (23)$$

is strictly positive real function with

$$2 > \lambda \geq \text{Max}(\lambda_2(k)) \quad \text{for} \quad k_0 < k < \infty \quad (24)$$

and $\varepsilon^*(k)$ is the augmented error defined as :

$$\varepsilon^*(k) = [p - \hat{p}(k-d)]^T \phi(k-d) \quad (25)$$

The adaptive control algorithm which has been adopted for the control of the phosphate drying furnace is derived from the previous one for :

$$H_1(q^{-1}) = H_2(q^{-1}) = 1 \quad (26)$$

The positivity condition in Eq (23) is automatically verified and the expression for the adaptation error in Eq (22) becomes :

$$\gamma^*(k) = \frac{C_R(q^{-1})y(k) - \hat{p}(k-1)\phi(k-d)}{1 + \phi^T(k-d)F(k)\phi(k-d)}$$

Figure 2 shown the block diagram of the adaptive control scheme.

PRACTICAL ASPECTS OF THE CONTROL SYSTEM

Computer hardware and software facilities

The DDC computer hardware used for implementing the controller algorithm was based on a D.E.C LSI-11 microcomputer. The configuration involves a 16 bit microprocessor with the minimum hardware arithmetic facilities, i.e.

all integer and floating point multiplication and division performed by software, 64K memory dual floppy disc mass storage, console terminal and teletype printer.

The experimental data interface consisted of a 16 channel multiplexed successive approximation A/D converter, 4 D/A converters all with 12 bit resolution and programmable real-time clock counter.

The standard DEC real-time operating system RT-11 was used to develop the programme and to control its execution, using the real-11 Fortran software facilities.

The flowchart of the real-time algorithm with the interface between the process and computer is shown in Fig.3 .

The choice of the C_R polynomial

The choice of the polynomial $C_R(q^{-1})$ results from a compromise between the tracking error and the control value. Indeed, we have observed that when the tracking error decrases quickly after any perturbation, the control becomes more energetic. In the case of our experiment, the following polynomial

$$C_R(q^{-1})=1-0.85q^{-1}+0.25q^{-2}-0.0585q^{-3}$$

has been chosen in order to avoid abrupt changes in the plant output.

"Start-up" of the control system

The initialisation of the control system has been done as follows

$$P^T(o) = [0,..,0]$$
$$\phi^T(o) = [UN,..,UN,HN,..,HN]$$

where UN and HN represents the fuel flow and the humidity of the dried phosphate respectively at the operating point

$$F(1) = 1000 \text{ I}$$
$$\lambda_1(k) = \lambda_2(k)=0.95$$

The use of such initial values lead to a control too important for the process this induces us to fix the control to its nominal value UN until the computed control is close to an interval around its nominal value UN ,

and this in constant way (the control may remains in the prescribed interval for about ten iterations). This being done, the control system operated with the "descreasing gain" algorithm ($\lambda_1(k)= \lambda_2(k)=0.95$) as long as the trace of the adaptive gain matrix is greater than a prescribed value. If not so, the control system operated with "constant trace" algorithm ($\lambda_1(k)=\lambda_2(k)$ and $\lambda_1(k)$ is such trace $(F(k))$=constant).

Results

In order to compare the performances of the adaptive control scheme with those achieved when using conventional PID controllers, the following experiments have been carried out.

- The PID controllers are used to control the phosphate drying furnace, its parameters are adjusted by an operator in order to provide acceptable performances. The microcomputer is used only to supervise the furnace operating and for production management.
- The adaptive control system presented above is used to control the phosphate drying furnace. The microcomputer is then used to control and supervise the furnace operating and for production management.

The operating conditions of the dryer for both adaptive control system and conventional PID controllers were the most common ones : at the input, the product feed rate was close to 220 t/h and its moisture content was subject to random variations. The range of these variations is between 10 and 15%.

The recorded curves of the humidity of the damp and dried phosphate and the fuel flow obtained by the two experiments are shown in figures 4 and 5.

Table 1 summaries statistical results that allows to appreciate the performances by using the two control systems.

Records	Stat.Char.	Conv.Cont.	Adap. Cont.
Damp phos humi.	Espe.	12,3%	12.70%
	Vari.	0.18	0.71
Dried pho humi.	Espe.	1.78%	1.42%
	Vari.	0.57	0.15
Fuel Flow	Mean Cons.	11.38	10.9
	Vari.	0.16	0.039

Tableau 1. Recorder statistical characteristics.

phos. : phosphate
humi. : humidity
Espe. : Esperance
Vari. : Variance
Stat. : Statistical
Char. : Characteristics
Conv. : Conventional
Cont. : Controller
Adap. : Adaptive
Cons. : Consumption

CONCLUSION

The control studies reported in this paper demonstrate a successful application of model reference adaptive controller to an industrial phosphate dryer.

The results of the experimentation illustrate the key features of the model reference adaptive controller, especially its potentiality to ensure suitable performances when changes of the plant dynamic characteristics occur.

On the other hand, the adaptive control system presented above allows, an energy saving of 4,5% and satisfactory quality of regulation wich involves the material saving, because of less thermic solicitations leading to a longer period between revisions.

Acknowledgement : The authors gratefully acknowledge the financial and material support of the OCP of Morocco.

REFERENCE

Landau, I.D. (1981). Unification of discrete time explicit model reference adaptive control designs. Automatica, vol-17, n°4 pp. 453-611

Najim, K.;Najim, M.;Koehret,B. and Ouazani,T. Modelisation and simulation of a phosphate drying furnace. 7th Annual Pittsburgh Conf. on Modeling and Simulation. April 1976 Pittsburgh. USA.

Najim, K. and Jouhari, D. (1977) Identification of a multivariable industrial system : A phosphate drying furnace. 20th Mid west symposium on circuits and systems Lubbock, Texas, August 15-17 USA

Najim, K.;Najim,M. and Jouhari, D.(1978) Identification of a phosphate drying furnace, JACC 18-20 Oct. Philadelphia.

Najim, K. (1979). Commande des systèmes complexes par apprentissage stochastique These de Docteur-ès-Sciences, Université Paul Sabatier, Toulouse, Mai 1979.

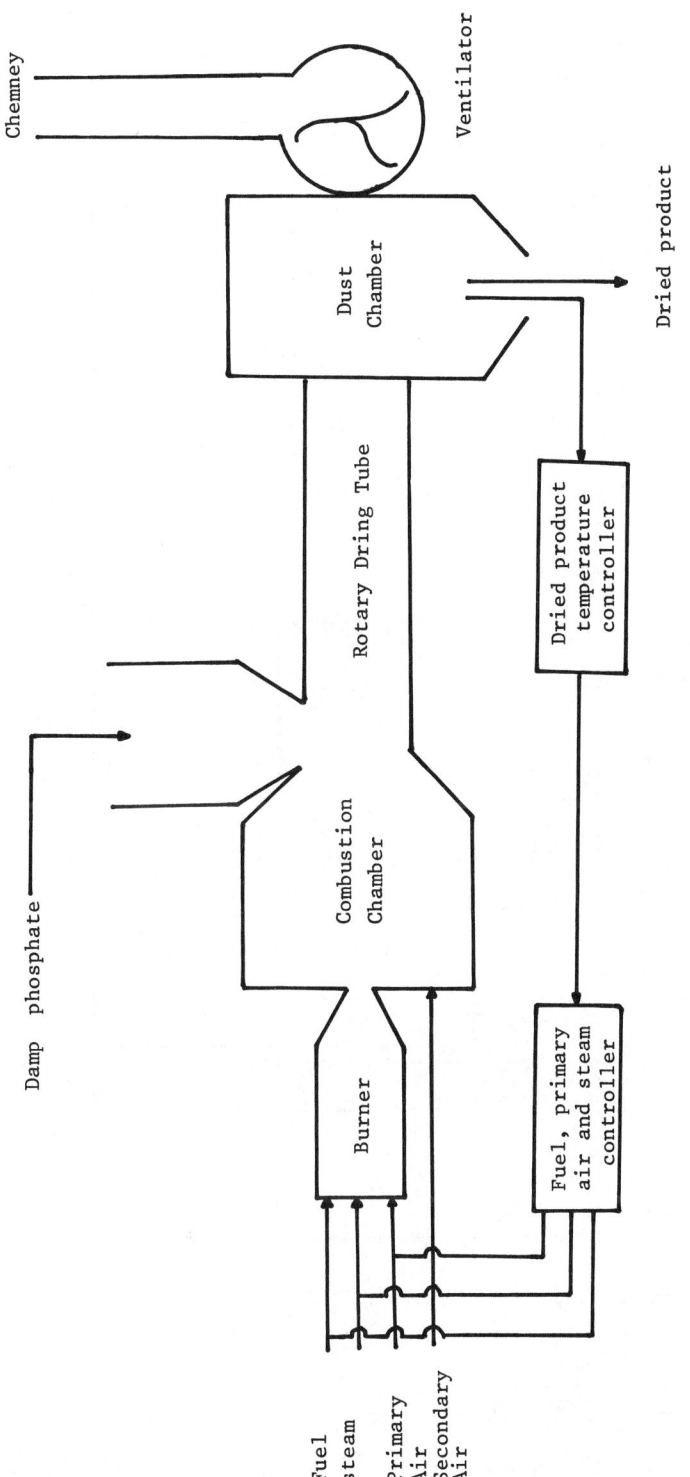

Fig.1. Drying furnace

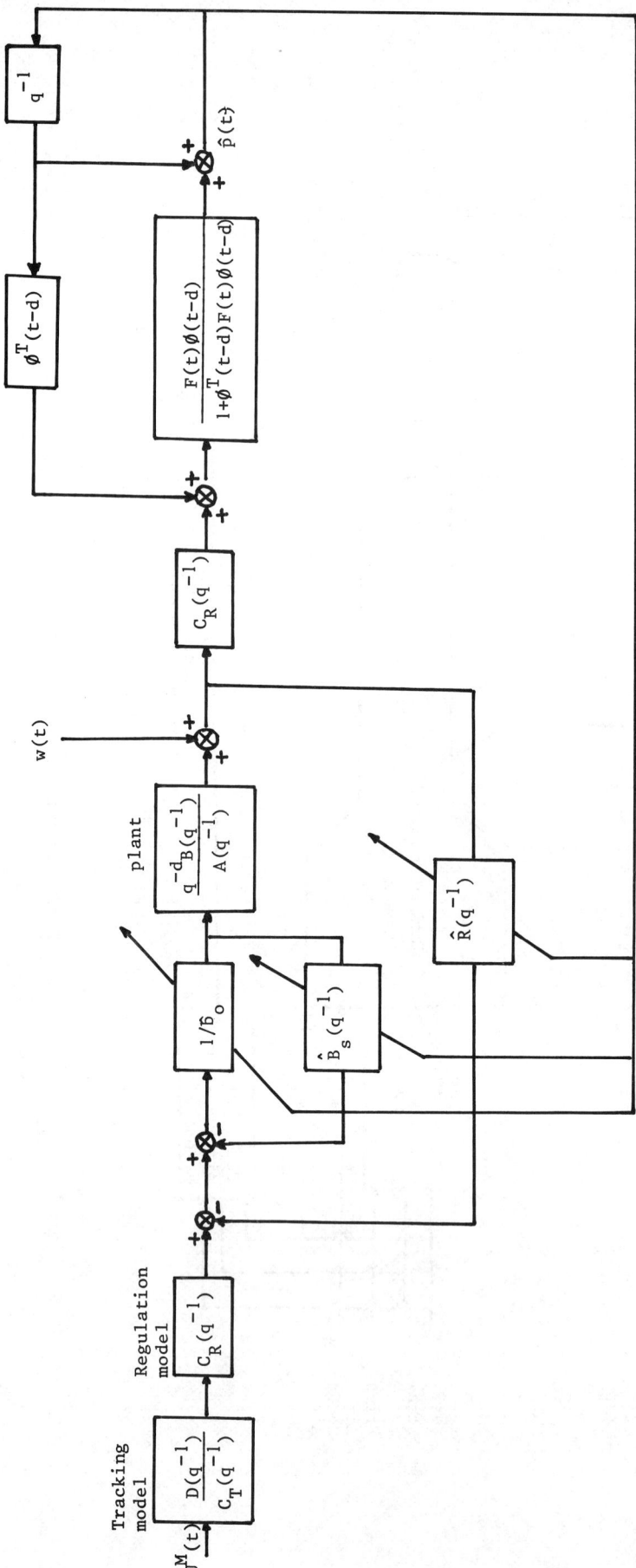

Fig.2. Block diagram of the adaptive control scheme

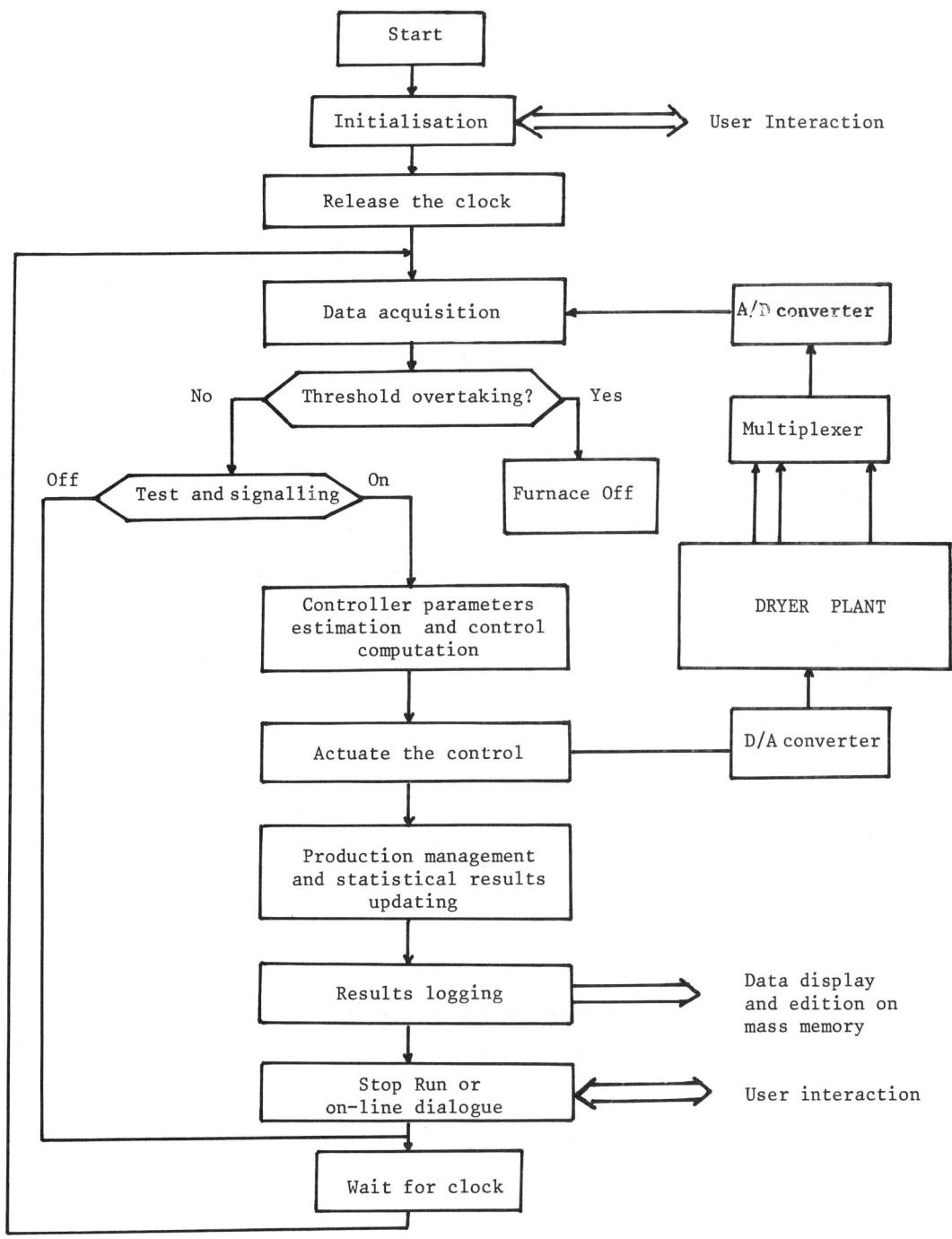

Fig.3. Flowchart of the real-time algorithm

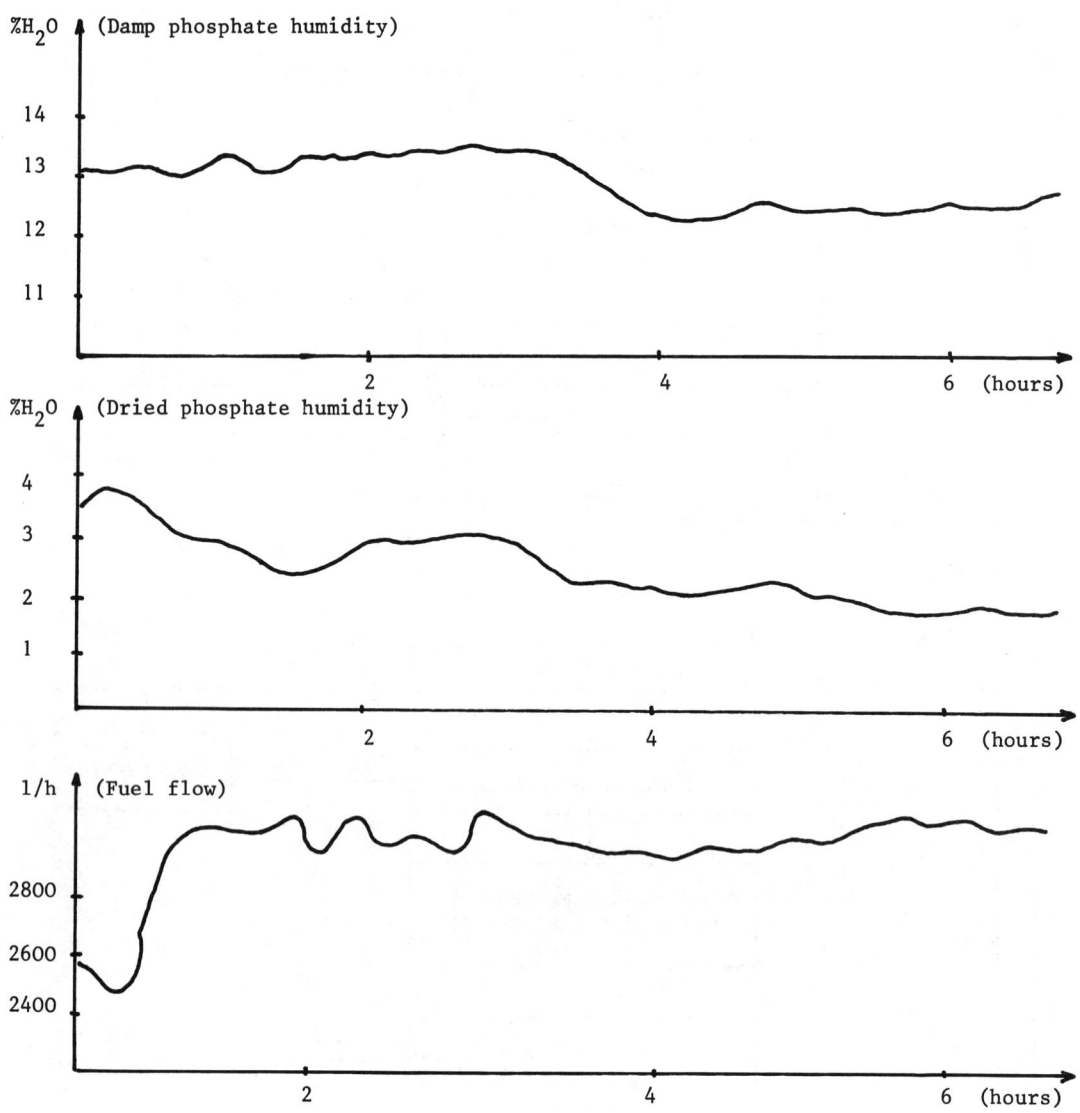

Fig.4. Typical conventional control recordings

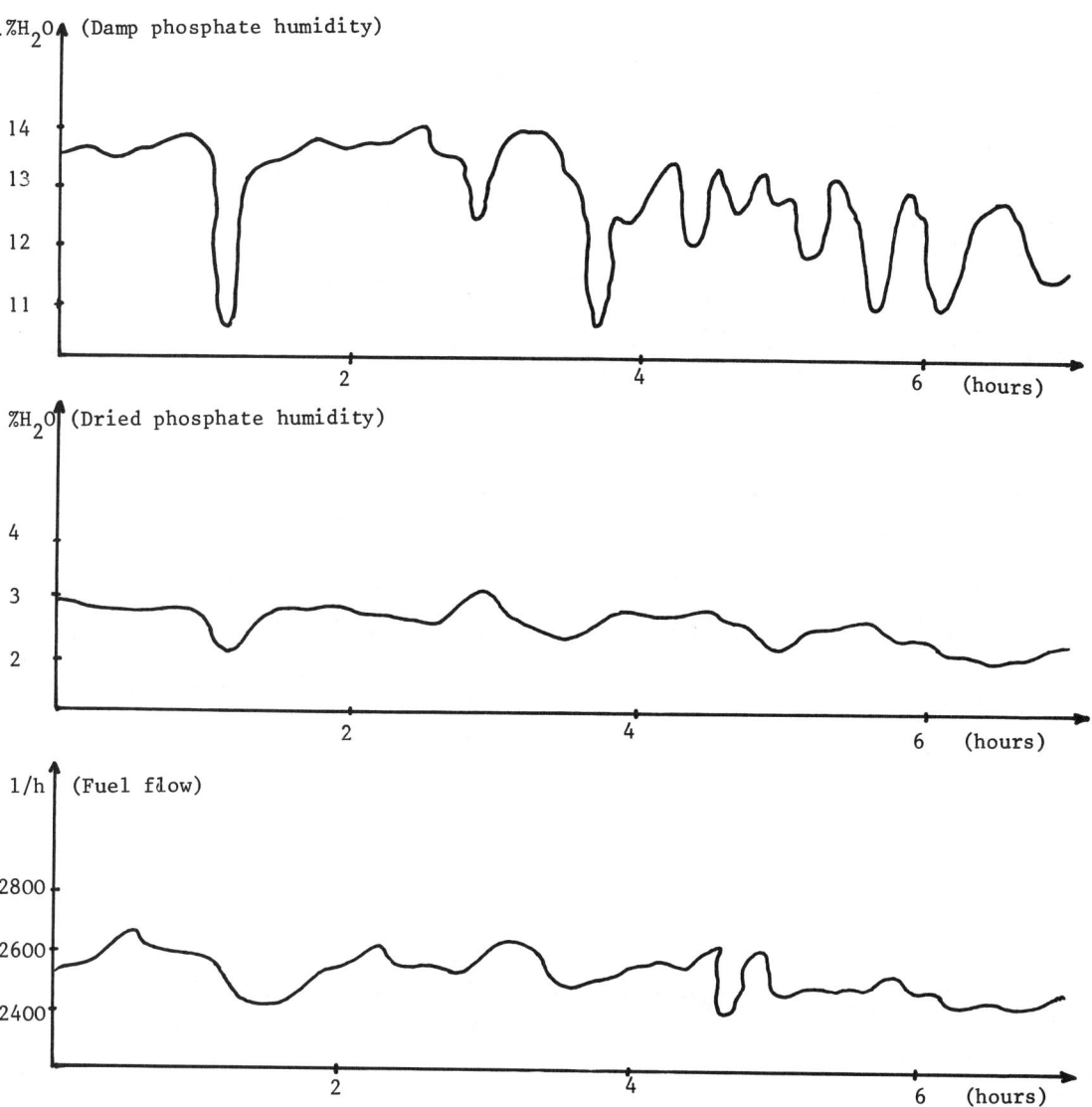

Fig.5. Typical adaptive control recordings

CONTROL STRATEGIES FOR MULTI-FUEL POWER PLANTS

U. Kortela*, B. Salmelin*, F. Wahlström** and J. Joensuu*

*Control Engineering Laboratory, Helsinki University of Technology, Finland
**Control Engineering Laboratory, Tampere University of Technology, Finland

Abstract. Much work has been done to develope the energy production and load allocation systems for steam boilers in pulp and paper industry. In multi-fuel power plants several problems exist, as load allocation, steam power smoothing and fuel allocation including the fuel power control. The aims will be energy savings so that the boiler has a stable behaviour even in load change situations and also when the feed of the fuel is disturbed.

In our case the problem is to use two different size multi-fuel boilers optimally. Both boilers use peat, coal, natural gas and oil as fuels. On the load part the main disturbances are caused by the start-ups of the paper machine and the failures in the paper production line. On the other hand also the fuel supply especially the feeding of the peat can be disturbed for many reasons. The moisture and the heat value of the peat may vary as well as the volumetric fuel feed in the feeder. To optimize the use of the boilers a special network as function of load and fuels should be generated. This has been done by using the real measurements and on the basis of the expectations concerning the energy needs of the paper machines.

The steam pressure and the steam power disturbances on the steam side have been smoothed by using the feed water tank as a storage for the system. Moreover by using the furnace as an active actuator the effect of these disturbances can be compensated also on the supply side of the fuel. To carry this out the fuel power estimation and control methods developed in our earlier studies should be used.

The paper deals with the real-time computing system for load allocation calculations in a paper mill where two boilers are used as steam generation source for several paper machines. The fast load disturbances are smoothed by the feedwater control system. The rest of the disturbances will be compensated on the combustion part of the boilers. The system includes one minicomputer and three microprocessors for compensation and allocation purposes. The man-machine communication will be carried out by the minicomputer.

Keywords. Power plant control; multi-fuel boiler, hierarchical control; combustion control.

INTRODUCTION

The aim of this study is to optimize the burning process in an industrial multi-fuel power plant. The fuels are peat, coal, oil and/or earth gas. The main disturbances are caused by non-homogenous fuels and load changes, such as breakdowns of paper machines etc. The load changes can be up to 50% of the whole nominal load in a certain operating point. The control strategy of the system stabilizes the burning conditions by compensating the feed disturbances, both in the quality and in the quantity.

The load changes are taken into account as increments to the beforehand calculated production prognose.

The System

The power plant consists of two different size boilers, the one is 8-33 kg/s and the other 19-56 kg/s. The power plant produces steam for a relatively big paper mill, in which there are e.g. papermachines which consume 8 kg/s each. The system is connected to a waste-liquour boiler (recovery boiler), which also causes some disturbances in the whole steam production. In Fig. 1 the steam production system is presented.

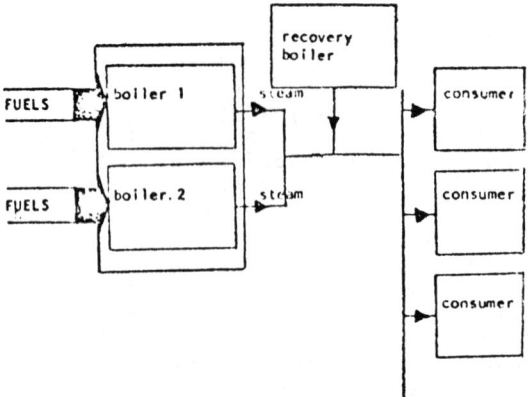

Fig.1 The steam production system

The purpose of the controller can be divided into three sections, which all are of major economical importance:

1. The system divides the load optimally to the two boilers and determines how much of each fuel should be feeded in at each operating point.

2. The system makes possible to use the balancing drive in minor load disturbances.

3. The system stabilizes the burning conditions and optimizes the use of the fuels.

Because of the hierarchical structure of the production system, the structure of the control and optimization strategy is also hierarchical.

The production prognose, i.e. the amount of each paper quality produced pr day etc. determines the basic loads of each boiler and also each fuel. It determines whether the both boilers are in use or not, so it is a very rough "decision block".

When more details are taken into account we get deeper in the hierarchy. The next step is to measure and calculate how near to the prognose we are, and make corrections according to the difference. Now we have a relatively close approximation of the real, correct feed of each fuel.

If the fuel would be homogenous, this would be a reasonable strategy, but because it is not, the quality of the fuel, that is the heat value have to be estimated in real-time. Because it is not possible to calculate beforehand the heat value of the fuel accurately enough, the O_2-analysis of the exhaust gases have been used.

In single-fuel boilers this estimation procedure has been in use for a couple of years, and it has been also reported in several conferences, /1,2,3/.

The compensation module has to take into account the states of the feeders, because in real process environment they all are often not working. The limitations of the states of the boilers due to reparations etc. have also to be taken into the concideration when making decisions in the higher hierarchy. In Fig.2 the structure of the control system is presented.

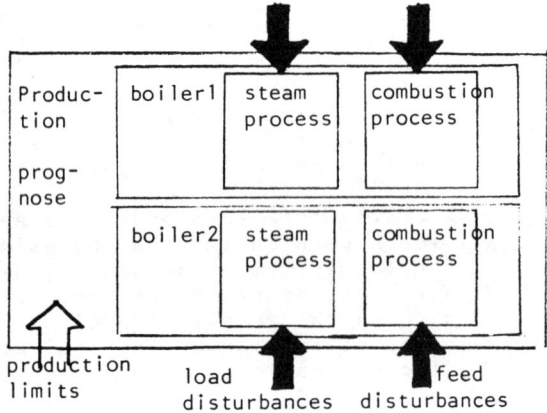

Fig.2 The control system hierarchies

Because the steam production process and the combustion process are independent subprocesses which have few functional interconnections, the fuel compensation and the steam production load disturbance compensating can be done in separate blocks.

The different fuels have to be compared with each other according to their steam production capability (kg/s of steam). In higher hierarchical levels are the actual fuel flows not calculated or used in control, but at the compensation level the actual feeder set points are used.

The system is also hierarchical due to the time: the load disturbances are very fast compared with the fuel quality variations.

In practice almost all disturbances are so big that the balancing drive is not sufficient to compensate the load changes thus also the capacity of the boilers should be used. Although the balancing drive belongs to the steam process and the fuel distribution to the burning process, they both are hierarchically at the same level.

The control strategy

The purpose of this control system is to use optimally the two boilers with minimum fuel costs, when the restrictions are taken into consideration. Because the boilers are of different size (steam production, pressure etc.) the bigger one has a somewhat higher grade of efficiency when driven near the nominal power. When the total load is small, the smaller boiler is more efficient than the bigger one. During the winter time, or when the load is big, both the boilers have to be in use. However, the smaller one is less efficient than the big one which leads to the strategy, where the big one covers the basic load and it operates at a relatively stable operating point while the smaller one follows the load changes. The strategy is described graphically in fig.3.

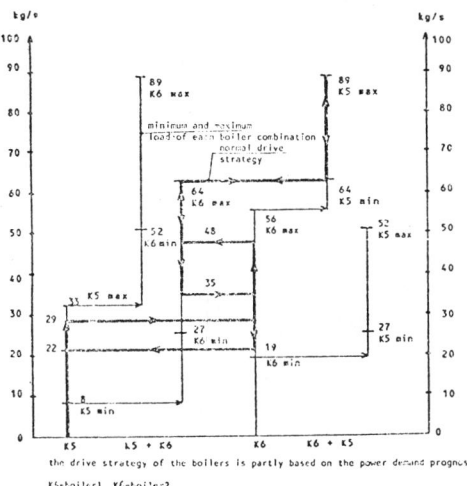

Fig.3 The control strategy of the boilers

The control strategy of the power division is based on the production prognose and is therefore quite rough. The actual steam demand is superposed to this basic load and the marginal to absolute load limitations have to be big enough to compensate the difference between the actual and calculated load. The switching curve is plotted in fig. 3 and shows the load points where the boilers should be switched on and/or off. The hysteresis shown in the switching curve is needed to minimize the number of switchings because the life-time of boilers is heavily reduced when they are rapidly driven up and down. Usually the actual steam demand is not the same as the estimated steam demand but the marginals cover the difference. The switchings of boilers are done manually.

The fuel allocation

All the fuels are in principle comparable to each other when the heat value is calculated. The preference order is determined by several factors such as
- limitations in accessibility
- the states of feeders
- the relative prices
- manual lockings.

The strategy is based on two main ideas
- the cheapest fuel is used maximally within limitations
- the earth gas compensates the disturbances of load (short period) and the coal and peat compensate the changes of fuel properties (long period).

After disturbances the gas burning level is reduced to its minimum as rapidly as possible because of the high price of the gas relative to the other fuels.

Peat and coal are the main fuels and changes in basic load levels cause changes in the feed of these fuels. Minor disturbances can be compensated with the (balancing drive) system which tells the main program the capability of balancing drive. When balancing drive has been completed the main program begins to recover to bring the system ready to accept new disturbances. The whole system is described in fig.4.

The compustion power compensation

The compensation principle has been reported in earlier reports and papers, see e.g.

The idea is to calculate from all air flows the real combustion power using a real time O_2-analyser. The result tells the heat value of each fuel which in turn can be used to calculate the feeder set points at each operating point of the boiler.

The automatic compensation and estimation units are independent for each boiler and are realized by microprocessor-based computing units. These low-level-hierarchy-units communicate with the higher-level control unit but not directly with the other low-level units. A scheme of the functional blocks in the combustion power compensation unit is presented in fig.5.

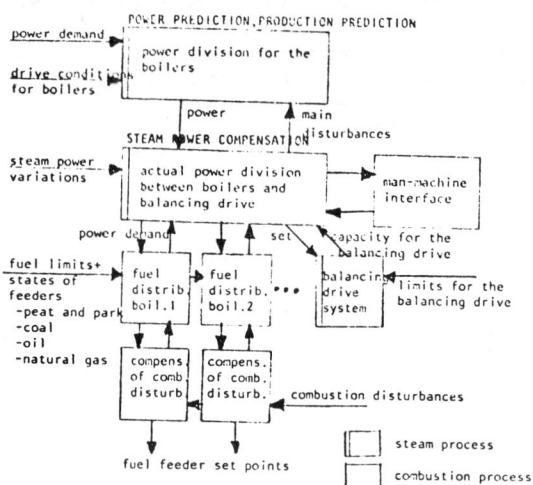

Fig.4 The Control and Compensation System

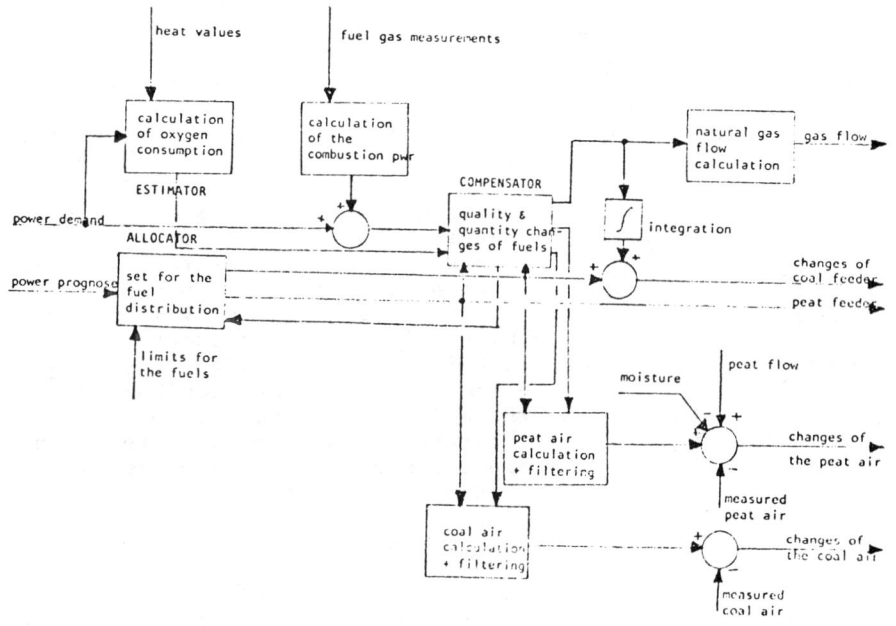

Fig.5 The combustion power compensator and estimator

The compensator calculates the quality changes in the different fuels and makes corrections to the set point of each feeder separately, within limitations. Because the quality changes usually are relative slow the computing time of the heat values doesn't cause any problems.

The balancing drive system

The balancing drive is hierarchically close to the fuel allocation problem shown in fig.4, because the balancing drive provides a short-term power source or storage which can be used to compensate identified disturbances in the load of the boilers.

As a matter of fact, the possibility to use the balancing drive also in major and long term disturbances helps to save resourses but the relative savings are then smaller. The essential saving can be achieved in small load disturbances around the operating point.

The system consists of a three state substrategy which drives the feedwater tanks optimally. The states and the transitions are shown in fig. 6.

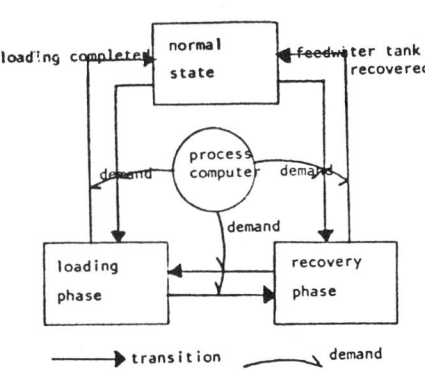

Fig.6 Transition possibilities of the balancing drive

When the boilers are in a stable state at the operating point the balancing drive system is at the so called "nominal state", which means that the feed-water tank levels are stable, and do not cause any changes in the mass-energy balance of the boiler.

The system is in "the loading phase" when two requirements are fulfilled. They are
- the system power produced is greater than the power consumed, due to e.g. failures in process equipment (paper-machines etc.)
- the balancing drive system has capacity to load.

The system observes all the time its state and controls itself to maintain the optimal levels in all feed water tanks and so the optimal energy flow from or to the tanks, too. The balancing drive system is also able to report it's state to the operators, whenever wanted. Manual lockings and manual drive is also possible, which is very essential in industrial power plants. The system tries always to recover as soon as possible to be able to compensate new, unwanted disturbances. The strategy of the balancing drive can be described graphically, as in fig.7.

The figure explains how the balancing drive operates in actual disturbances such as line failures of the paper-machines.

The diagram shows the actual steam consumption, which drops due to the disturbance -20 kg/s (hypothethical value). The duration of the disturbance can, of course not generally be known beforehand but the tanks can ahead be driven to suitable levels for a new start-up (at ~ 15 min in the figure).

When the disturbance starts the system moves into the loading phase. At the same time also the control fuels are

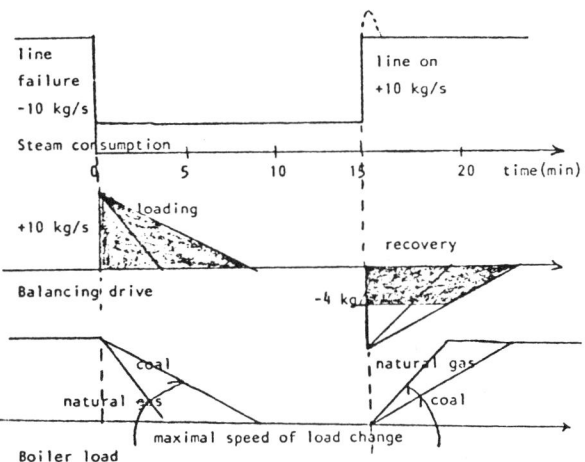

Fig.7 The functions of the balancing drive

driven down as rapidly as possible, because the balancing drive capacity is relatively small. When all the loading capacity is used, the possible load change still needed is done by fuel feed changes only. The recovery of the boiler is begun when the line is on, when the duration of the disturbance is relatively short, but the recovery has to be done earlier, when the disturbance happens over a long period, because there can exist new failures in the papermachines and the system has to be able to meet the new disturbances also.

The fuels are driven to wards a new level as soon as the disturbance is over.

The process computer in the balancing drive system superposes all the existing disturbances, so that the actual disturbance can e.g. look as follows:

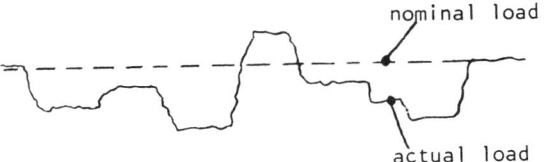

Fig.8 The superposed steam demand diagram

The balancing drive is then controlled by this resultant load demand, which takes into account all the load changes. The changes in fuel quality and the states and disturbances in fuel feeders is taken into consideration in the actual fuel compensation program of this system.

Test results and conclusions

The system methodology has been tested in single fuel (peat) power plants, which has shown that the new estimation method gives an improvement in the boiler efficiency compared with conventional control methods. The table below tells the major features of the new system

	CONVENTIONAL CONTROL SYSTEM	COMPENSATED CONTROL SYSTEM
ESTIMATION OF HEAT VALUE	—	REASONABLE ACCURACY
OIL BURNER FAILURES	?	GOOD COMPENSATION
FEED DISTURBANCES	AMPLITUDE, SETTLING TIME (6%, BIG)	AMPLITUDE, SETTLING TIME (2%, ~10 MIN)
LOAD DISTURBANCES	AMPLITUDE, SETTLING TIME (~7%, BIG)	AMPLITUDE, SETTLING TIME (2,5%, ~15 MIN)
AIR CONTROL EXCESS O_2	3 → 3,5%	~2,4%

Table 1 The new system compared with conventional control methods

The system presented in this paper has not yet been installed in a practical power plant, but it has been simulated in a MINC process computer.

The simulation results show, that although the control strategy of the boilers in principle is as simple as shown in fig.3, it is practical to implement some more decision blocks in the strategy to minimize the switchings of the boilers. The system now suggests to the user only those switchins, which are absolutely necessary. The strategy calculations are mainly based on future drive conditions so that the system can economically be driven to each operating point.

The combustion control and compensation units are tested in real process environment and the system has shown its ability to control the boilers optimally.

References

Kortela U.,Lautala P.:A New Control Concept for a Coal Power Plant,8th IFAC World Congress, Kyoto,Japan,1981

Kortela U.:Estimation and Compensation of Feed Disturbances in Some Problematic Industrial Processes,Dr.Techn. thesis,Helsinki Univ. of Technology,Helsinki,1980

Lehtomäki K.,Kortela U.,Luukkanen J.:New Estimation and Control Methods for Fuel Power in Peat Power Plants,8th IFAC World Congress, Kyoto,Japan,1981

Leffler N.:The Economics of Process Control in Energy Management,4th IFAC Conference PRP Automation,Gent,Belgium,1980

Copyright © IFAC Real Time Digital Control Applications
Guadalajara, Mexico 1983

DIGITAL DECOUPLING OF A 3-ZONE ELECTRICAL FURNACE BY MEANS OF MULTIVARIABLE-PI CONTROL

J. Gómez de Silva

Instituto de Ingeniería, Automatización, UNAM, México

Abstract. The real-time digital control of a 50kW 3-zone metallurgical furnace is presented, using decoupling techniques and a discrete multivariable PI-structured controller. The plant is described by a 3-input/3-output z-transformed rational transfer function matrix revealing a highly coupled dynamical structure. Particularly suited for real-time digital control, Multivariable Nyquist Array methods yield a simple PI-structured precompensator achieving efficient decoupling at all frequencies. A classical approach, comprising three monovariable PI-controllers, is presented for comparision purposes. Experimental results performed on the furnace confirm the superiority of the multivariable solution.

Keywords. Digital control; multivariable control systems; temperature control; decoupling; real-time processing; Multivariable Nyquist Array.

INTRODUCTION

The need to cope with increasing system complexity provided strong motivation for the development of multivariable control theories since the second half of the century. Application of the different methodologies arising from this activity wouldn't be possible without the availability of modern digital computing facilities.

Real-time digital control applications to multivariable systems were initially performed using State-Space techniques. Although they proved to be extremely valuable for numerous aero-space applications, and enabled the gain of substantial insight in certain important aspects of automatic control, these techniques presented certain drawbacks when implemented for many industrial problems. The difficulty to define appropriate performance indexes, the necessity of dealing with an accurate plant description and the obtention of , generally, complex controllers, made these methods, in occasions, hard to apply in real-time, where a major concern is *race against time*. Therefore, research attention, particularly in Great Britain, turned to the development of frequency-domain multivariable methods seeking for a generalization of the well grasped classical frequency-response approaches (Mac Farlane, 1978; Gómez de Silva, 1982). One of these methods, the Multivariable Nyquist Array (MNA) method (Rosenbrock, 1974; Leininger, 1979), was applied in this paper to derive a simple PI-structured multivariable controller particularly suited for real-time control of interactive systems.

DIGITAL CONTROL LOOP DESCRIPTION

The plant under consideration is a metallurgical 50kW electrical furnace with 3 independent heating zones. Temperature ranges from ambient to approximately 925°C. Each zone comprises 3 cylindrical resistive elements (one for each phase of a 3-phase network), as shown schematically in fig. 1 for phase i. The figure shows the different blocks integrating the digital control loop:

- a measurement block, comprising a set of pyrometers, a scanner/multiplexer and a digital voltmeter (DVM).

- an industrial type HP 2100S minicomputer, provided with a disc operating system (DOS), an alphanumeric terminal, a line printer, a recording unit, and input-output interfaces for the measured temperatures, safety system status, control strategy commands, TV monitor display, etc.

- a triac heating-power control block.

Temperature in each zone is sensed by a set of 3 pyrometric rods containing NiCr/CrAl thermocouples. Zone temperature is obtained be averaging the 3 corresponding pyrometers. A computer controlled scanner/multiplexer sequentially samples the temperature at the measure points. The read out is performed by a computer controlled DVM giving 10μV resolution (corresponding to about 0.25°C). The non-linear characteristic of the thermocouples is compensated by means of a Chebychev polinomial interpolation and a least squares parameter identification.

Control and monitoring of the system is accomplished by the minicomputer, initializing and starting the system, updating the

$$G(z) = \begin{pmatrix} \dfrac{.0002z^{-1}+.00113z^{-2}+.00121z^{-3}}{1-1.79z^{-1}+.868z^{-2}+.0637z^{-3}-.138z^{-4}} & \dfrac{.00053z^{-1}+.00072z^{-2}+.00038z^{-3}}{1-1.246z^{-1}-.115z^{-2}+.55z^{-3}-.19z^{-4}} & \dfrac{.00038z^{-1}+.00033z^{-2}+.00033z^{-3}}{1-.76z^{-1}-.195z^{-2}-.227z^{-3}+.19z^{-4}} \\[1em] \dfrac{.00052z^{-1}+.0007z^{-3}}{1-1.365z^{-1}+.032z^{-2}+.48z^{-3}-.147z^{-4}} & \dfrac{.00014z^{-1}+.00121z^{-2}+.00115z^{-3}}{1-1.59z^{-1}+.45z^{-2}+.36z^{-3}-.223z^{-4}} & \dfrac{.00057z^{-1}+.0009z^{-2}+.0005z^{-3}}{1-1.077z^{-1}-.266z^{-2}+.434z^{-3}-.0874z^{-4}} \\[1em] \dfrac{.00006z^{-1}+.00027z^{-2}+.00029z^{-3}}{1-1.05z^{-1}-.275z^{-2}+.189z^{-3}+.138z^{-4}} & \dfrac{.00006z^{-1}+.00043z^{-2}+.00065z^{-3}}{1-1.546z^{-1}+.32z^{-2}+.409z^{-3}-.182z^{-4}} & \dfrac{.0002z^{-1}+.00119z^{-2}+.00145z^{-3}}{1-1.765z^{-1}+.836z^{-2}+.065z^{-3}-.133z^{-4}} \end{pmatrix} \quad (1)$$

parameters of the programmable units (DVM, scanner/multiplexer, power command block, etc.), fetching information (temperature measurements, state of alarms and securities, etc.) and establishing the communication between the operator and the system. All these functions are performed in real-time by means of a disc resident monitor achieving appropriate management of the CPU so as to allow the execution of several programs simultaneously. An Assembler and a Fortran compiler enables the creation, modification and simulation of source programs.

The power control block incorporates a set of 9 triacs and their control circuitry. Power reference input, for each heater, is contained in a 10 bit register allowing 1024 control levels. Power command may be assigned according to three modes:

1) Manual, in which the power is defined by a set of thumb wheels.

2) External, in which power specification is assigned by a remote unit such as a supervisor or an external regulator.

3) Automatic, in which the power delivered to the heaters is specified by the computer who loads the command registers at each sampling period.

The 10 bit word, specifying the power, is converted into a sequence of pulses, uniformly distributed in time over a fixed period, to be applied to the gates of the corresponding triacs. The number of pulses per period is directly proportional to the specified power and the pulse width is determined so as to maintain the triac in conduction for 20 ms (one period of our 50Hz power line). In this manner we obtain a fixed cycle period of 1023 x 0.02 s = 20.46 s containing from 0 to 1023 pulses allowing a linear control of the heating power for each zone.

THE MODEL

To describe the system, a 3-input/3-output z-transformed rational transfer function matrix (TFM) was derived by real-time identification using standard correlation techniques (Boillot, 1979; Gómez de Silva, 1982). With a set point of 700°C, a pseudorandom binary noise of length 63 was applied sequentially to each of the furnace inputs with a sampling period of 20 s. Cross-correlation sequences were obtained and used in a parametric model in the form of difference equations. Parameter estimation for each element was performed by a least squares algorithm yielding the transfer function matrix of eq. 1, which reveals a highly coupled dynamical structure. Non diagonal dominance may be observed in Fig. 2 where the Gershgorin band associated with the element $g_{22}(z)$, of the Multivariable Nyquist Array (MNA) of the furnace, contain the origin of the G-Plane.

OUTLINE OF THE METHOD

Multivariable Nyquist Array (MNA) methods (Rosenbrock, 1969, 1974; Leininger, 1979) arise as a generalization of classical frequency-response theories used exhaustively in monovariable control approaches. They provide a stability theorem, analogous to the single-loop Nyquist theorem, which allows multivariable systems to be designed using an intuitive understanding of transfer functions as in classical theory.

Let $G(z)$ be a mxm transfer-function matrix (TFM) representing the multivariable system to be controlled, and $K(z)$ a mxm TFM representing the controller. Defining the open-loop TFM as $Q(z)$ we have

$$Q(z) = G(z) K(z) \quad (2)$$

Considering unity feedback, the closed-loop TFM is

$$H(z) = \left[I_m + Q(z) \right]^{-1} Q(z) \quad (3)$$

Two important theorems serve to prove the stability of multivariable systems. The first one uses the *fundamental equation*, relating open and closed-loop behaviour in multiple-loop control systems (Mac Farlane, 1970; Rosenbrock, 1969),

$$|I_m + Q(z)| \triangleq |R(z)| = \frac{\text{closed-loop characteristic polynomial}}{\text{open-loop characteristic polynomial}} \quad (4)$$

to show that the system is closed-loop stable if, and only if, Γ does not enclose the origin of the $R(z)$-plane, i.e. no zero of the closed-loop characteristic polynomial is contained in the closed right-half complex plane. Since we may write

$$|R(z)| = \prod_{i=1}^{m} r_i(z) \quad (5)$$

the stability of the system may be assured if none of the loci Γ_i, $i = 1, m$ enclose the origin of the R(z)-plane.

Using the characteristic-value-shift theorem, stating that the characteristic values of $R(z) = I_m + Q(z)$ are $\{r_i(z) = 1 + q_i(z)\}_{i=1,m}$, the previous theorem may be expressed as: A necessary and sufficient condition for a linear multivariable system (LMS) to be stable is that none of the loci traced by $q_i(z)$, as ω takes values on D, enclose the critical point $(-1, j0)$.

This theorem shows the importance of the characteristic values $q_i(z)$ of the open-loop TFM in describing the behaviour of LMS. Nevertheless it is inconvenient for practical applications: characteristic values of rational polynomial matrices may not, in general, lie in the field of rational functions. On the other hand, the avoidance of characteristic value determination requires computing the determinant of a polynomial matrix. A more rewarding approach is to consider Gershgorin theorem along with the diagonal dominance criterion (Rosenbrock, 1974; Mac Farlane, 1970).

Gershgorin theorem states that all m x m complex matrices R(z) have their characteristic values contained in the union of the circles defined by

$$|z - r_{ii}(z)| \leq \sum_{\substack{j=1 \\ j \neq i}}^{m} |r_{ij}(z)| \quad i=1,m \quad (6)$$

as well as in the union of the circles

$$|z - r_{ii}(z)| \leq \sum_{\substack{j=1 \\ j \neq i}}^{m} |r_{ji}(z)| \quad i=1,m \quad (7)$$

Defining (8)

$$\rho_i(z) = \sum_{\substack{j=1 \\ j \neq i}}^{m} |r_{ij}(z)| \text{ and } \rho'_i(z) = \sum_{\substack{j=1 \\ j \neq i}}^{m} |r_{ji}(z)|$$

a LMS is said to be diagonal dominant on D if the elements $\{r_{ij}\}_{i,j=1,m}$ of its TFM satisfy

$|r_{ii}(z)| - \rho_i(z) > 0$ (row dominance)

and (9)

$|r_{ii}(z)| - \rho'_i(z) > 0$ (column dominance),

for all $z \in D$

It may be shown (Rosenbrock, 1969; Mac Farlane, 1970) that a multivariable system satisfying condition (9) is stable, with all feedback loops closed, if none of the loci traced by $r_{ii}(z)$, as z takes values on D, encloses the critical point $(-1, j0)$.

A useful graphical procedure for checking diagonal dominance, and the stability of LMS, was introduced by H. H. Rosenbrock, (1974), as follows:

Let $r_{ii}(z)$, $i=1,m$, map D into the R(z)-plane.

For each point $\tilde{z} \in D$ on $r_{ii}(\tilde{z})$, as center, trace a circle of radius $\rho_i(z)$ (for checking row dominance) or $\rho'_i(z)$ (for checking column dominance) as in (8). As $\tilde{z}$ varies on D these circles sweep a band which has been designated Gershgorin band. If, for all i, the bands exclude the origin of the complex plane, the system is said to be (row or column) dominant. As in the monovariable case, it is convenient to analyse the open-loop TFM Q(z) instead of R(z). In this case the loci is traced by $q_{ii}(z)$, the circles are defined by

$$\rho_i(z) = \sum_{\substack{j=1 \\ j \neq i}}^{m} |q_{ij}(z)| \quad \rho'_i(z) = \sum_{\substack{j=1 \\ j \neq i}}^{m} |q_{ji}(z)| \quad (10)$$

and the condition is that the Gershgorin bands must exclude the critical point $(-1, j0)$, for all i. Fig. 2 shows the loci traced by the diagonal elements $\{g_{ii}(z)\}$ of G(z), as z varies on D, with its Gershgorin bands superimposed. The analysis is made by column. Large bands show high interaction terms.

The importance of the characteristic values $q_i(z)$ on the dynamic behaviour of LMS was mentioned previously. An objective of multivariable controller design would then be the modification of the *characteristic loci* traced by the characteristic values $q_i(z)$ so as to obtain a desired closed-loop dynamic behaviour, i.e. choosing the controller matrix K(z) so that the characteristic values of the open-loop TFM(2) possesses certain prescribed properties. This is a difficult task since little is known about how the characteristic values of the product of two matrices are related to the characteristic values of the matrices taken independently. A solution could be to diagonalize the system and deal with its diagonal elements considered as a set of m single-loop systems on which classical monovariable theory may be applied. Nevertheless the resulting diagonalizing controller is, in general, too complicate for practical utilization and in occasions unstable. A much practical approach, leading to very simple controllers, is to search among the class of *constant real matrices* the one which renders the system the more diagonal as possible, for a given angular frequency $\tilde{\omega}$, and to operate on the diagonal terms of the resulting open loop TFM so as to approach the prescribed closed-loop behaviour, with the condition of preserving diagonal dominance at all operating frequencies. This has led to the pseudodiagonalization technique (Rosenbrock, 1974) which minimizes

$$\sum_{\substack{i=1 \\ i \neq j}}^{m} |qij(\tilde{z})|^2 \quad j = 1,m, \; \tilde{z} \in D \quad (11)$$

subject to the constraint

$$\sum_{k=1}^{m} k_{kj}^2 = 1 \quad (12)$$

where $\{k_{kj}\}_{k=j=1,m}$ are the elements of the constant matrix K sought.

Letting

$$q_{ij}(\tilde{z}) = \sum_{k=1}^{m} g_{ik}(\tilde{z}) \cdot k_{kj}$$
$$= \sum_{k=1}^{m} (\alpha_{ik} + j\beta_{ik}) \cdot k_{kj} \qquad (13)$$

and using Lagrange multipliers, the function to minimize becomes

$$\phi_j = \sum_{\substack{i=1 \\ i\neq j}}^{m} |\sum_{k=1}^{m} (\alpha_{ik} + j\beta_{ik}) k_{kj}|^2 + \lambda\left[1 - \sum_{k=1}^{m} k_{kj}^2\right]$$

$$= \sum_{\substack{i=1 \\ i=j}}^{m} \left[(\sum_{k=1}^{m} \alpha_{ik} k_{kj})^2 + (\sum_{k=1}^{m} \beta_{ik} k_{kj})^2 \right] +$$

$$+ \lambda\left[1 - \sum_{k=1}^{m} k_{kj}^2\right] \qquad (14)$$

leading, after performing $\frac{\partial \phi_j}{\partial k_{\ell j}} = 0$, $\ell = 1, m$, to the standard characteristic vector problem

$$A_j k_j - \lambda k_j = 0 \qquad (15)$$

where we have defined $k_j \triangleq k_{\ell j}$, the j-th column of the matrix K sought; and

$$A_j = (a_{k\ell}^{(j)}) \triangleq \left(\sum_{\substack{i=1 \\ i\neq j}}^{m} (\alpha_{ik} \cdot \alpha_{i\ell} + \beta_{ik} \cdot \beta_{i\ell}) \right) \qquad (16)$$

which is, at least, positive semi-definite so its characteristic values are real and non-negative. To minimize (11) it is necessary to choose k_j corresponding to the minimum λ. Solving (15) for each j gives the constant matrix K achieving maximal diagonal dominance at $\tilde{z} = e^{j\tilde{\omega}}$. If the elements of the open-loop TFM Q(z) are continuous functions of $\tilde{\omega}$, the frame vectors k_j obtained are continuous functions of $\tilde{\omega}$, and it is reasonable to assume that the decoupling properties of the compensation matrix K obtained, persist over a range of frequencies around $\tilde{\omega}$.

CONTROLLER DESIGN

Pseudodiagonalization techniques leads to very simple controllers particularly appropriate for real-time digital control applications. For the furnace, we have separated the problem into low frequency and high frequency compensation leading to a controller structure of the form

$$K(z) = k\left[\frac{\alpha T}{1 - z^{-1}} K_L + K_H\right] \qquad (17)$$

where K_L is a low frequency constant precompensation matrix derived by pseudodiagonalization using $\tilde{z} = e^{j\omega_L}$, ω_L being an appropriately choosen low angular frequency; K_H is a high frequency constant precompensation matrix obtained by pseudodiagonalization at $\tilde{z} = e^{j\omega_H}$, ω_H being an adequate high angular frequency; k a gain factor and α a real constant allowing a suitable transition from low to high frequency compensation so as to have

$$K(z) \simeq \frac{\alpha kT}{1-e^{-j\omega}} K_L \quad \text{at low frequencies} \qquad (18)$$

$$K(z) \simeq k K_H \quad \text{at high frequencies} \qquad (19)$$

The controller (17) is of the discrete multivariable proportional-plus-integral (PI) type which have proved to give well decoupled responses at all frequencies and good overall dynamic characteristics in the case of the furnace.

The precompensation matrices, K_L and K_H, were obtained with $\omega_L = 10^{-4}$ rad/s and $\omega_H = 0.025$ rad/s, respectively, giving

$$K_L = \begin{pmatrix} 0.865 & -0.362 & 0.13 \\ -0.499 & 0.88 & -0.557 \\ 0.043 & -0.307 & 0.82 \end{pmatrix} \qquad (20.a)$$

$$K_H = \begin{pmatrix} 0.93 & -0.306 & 0.077 \\ -0.363 & 0.909 & -0.428 \\ 0.034 & -0.281 & 0.9 \end{pmatrix} \qquad (20.b)$$

The gain k was choosed equal to 34 so as to have a gain margin of approximately 8 dB, α was choosed equal to 1.5×10^{-4} and the sampling period T equal to 20s. Fig. 3 shows the effect of high frequency compensation after post multiplying G(z) by K_H. Note the contraction of the Gershgorin bands on the high frequency part of the Nyquist diagrams. Fig. 4 shows the effect of low frequency compensation after postmultiplying G(z) by K_L. The previous data, substituted in equation (17) yields the global controller

$$k(z) = \frac{1}{1-z^{-1}} \begin{pmatrix} 0.088 & -0.037 & 0.013 \\ -0.051 & 0.09 & -0.056 \\ 0.004 & -0.031 & 0.084 \end{pmatrix} +$$

$$+ \begin{pmatrix} 31.62 & -10.4 & 2.62 \\ -12.34 & 30.91 & -14.55 \\ 1.16 & -9.55 & 30.6 \end{pmatrix} \qquad (21)$$

Considering

$$u(z) = k(z)\, \varepsilon(z) \qquad (22)$$

we may derive the following difference equations, governing the control law for the furnace:

$$u_i(k) = \sum_{j=1}^{3} \left[a_{ij}\, \varepsilon_j(k) - b_{ij}\, \varepsilon_j(k-1) \right] +$$

$$+ u_i(k-1) \qquad i = 1,3 \qquad (23)$$

where (24)

$$(a_{ij}) = \begin{pmatrix} 31.71 & -10.44 & 2.63 \\ -12.39 & 31 & -14.61 \\ 1.16 & -9.58 & 30.68 \end{pmatrix} \text{ and}$$

$$(b_{ij}) = \begin{pmatrix} 31.62 & -10.4 & 2.62 \\ -12.34 & 30.91 & -14.55 \\ 1.16 & -9.55 & 30.6 \end{pmatrix}$$

These equations were programmed in fortran II on a HP2100 industrial type minicomputer endowed with real-time executive facilities.

EXPERIMENTAL RESULTS

For the sake of comparision, the same experimental sequence was applied to the regulation loop driven, first, by a set of 3 discrete monovariable PI controllers, characterizing a classical approach, and next, by the discrete multivariable PI controller obtained by MNA methods. The experimental sequence consisted in the application of a temperature step of 10°C at the central zone, to analyse decoupling, and the cut-off of one of the heaters (phase 2 heater) of zone 1 (door) to analyse robustness of the algorithm. Each experiment was performed, on the same environmental conditions, with the furnace stabilized at a set point of 700°C.

Fig. 6a shows the responses of the 3 zone of the furnace, when a step of 10°C is applied in the central zone, for the two approaches adopted. Fig. 6b display the control histories generated by the controllers. Ordinates are expressed in command units (from 0 to 1023). Note the abrupt change in the commands of the lateral zones, "anticipating" the correction, when MNA approach is used.

Temperature outputs of the 3 zones, when the furnace is perturbed by disconnecting the heater corresponding to phase 2 of the first zone (the door of the furnace), are shown in fig. 5.

REFERENCES

Boillot, E. and Tanguy Y. (1979) Identification of a Heat Treating Furnace Application to Digital Control Proc. 5-th IFAC Symposium, Darmstadt, FRG.

Gómez de Silva, J. (1982) Commande Multivariable: Approche Fréquentielle. Application a un four Multizone Docteur Ingenieur thesis. Université de Paris-Sud. Ecole Supéricure d'Electricité.

Leininger, G.G. (1979) Diagonal Dominance for Multivariable Nyquist Array Methods Using Function Minimization Automatica, Vol. 15, pp. 339-345.

MacFarlane, A. G. J. (1970) Return-Difference and Return-Ratio Matrices and their Use in Analysis and Design of Multivariable Feedback Control Systems Proc. IEE. Vol. 117, No. 10.

MacFarlane, A. G. J. (1978) Frequency-Response Methods in Control Systems IEEE Press. Cambridge.

Rosenbrock, H.H. (1969) Design of Multivariable Control Systems Using the Inverse Nyquist Array Proc. IEE. Vol. 116. No. 11.

Rosenbrock, H.H. (1974) Computer Aided Control System Design, Academic Press. London.

NOMENCLATURE

- z : complex variable defined as e^{sT}. $s = \sigma + j\omega$
- T : sampling period
- ω : angular frequency
- $G(z), K(z), Q(z), H(z), R(z)$: mxm discrete-time rational polynomial matrices representing, respectively, the plant, controller, open-loop transfer function matrix (TFM), closed-loop TFM and return-difference matrix.
- $r(z), \varepsilon(z), u(z), y(z)$: mx1 vectors representing, respectively, the reference input, error signal, command and output signal.
- $|Q(z)|$: determinant of matrix $Q(z)$
- $q_i(z)$: i-th characteristic value of $Q(z)$
- $q_{ij}(z)$: ij-th element of matrix $Q(z)$
- D: complex plane closed contour consisting of a unit-modulus circle and a circle of radius α joined by a double path along the real axis (α choosen large enough to ensure that every finite zero and pole of all functions involved lies within D).
- Γ: map of D by $|R(z)|$
- Γ_i: map of D by $r_i(z)$

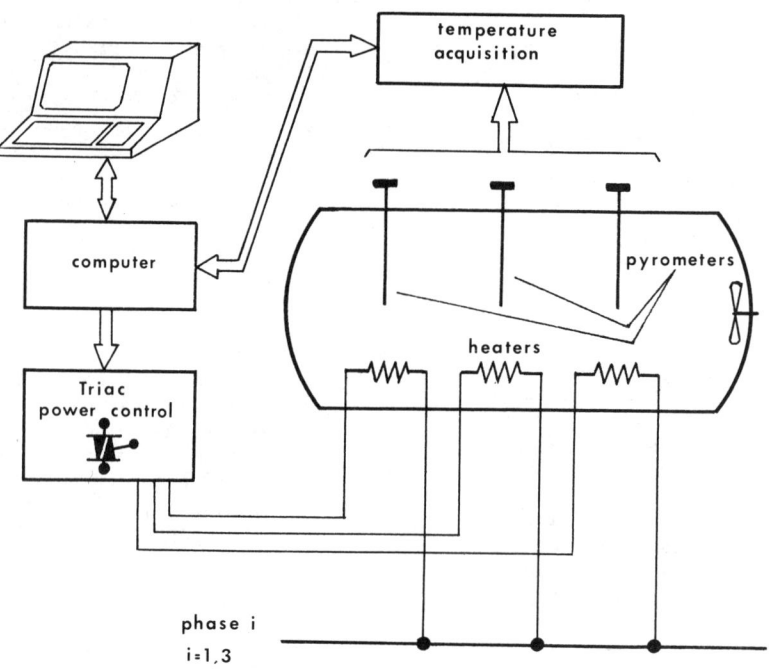

Fig. 1 The furnace in its digital control-loop

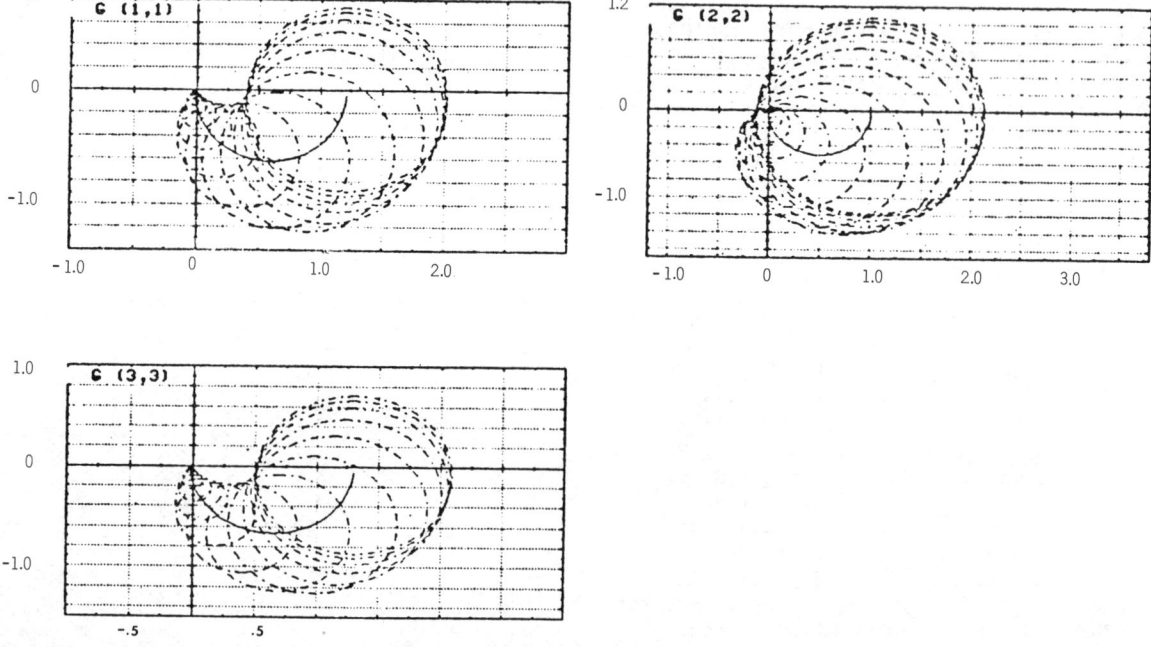

Fig.2 Multivariable Nyquist Array of the plant with Gershgorin bands superimposed.

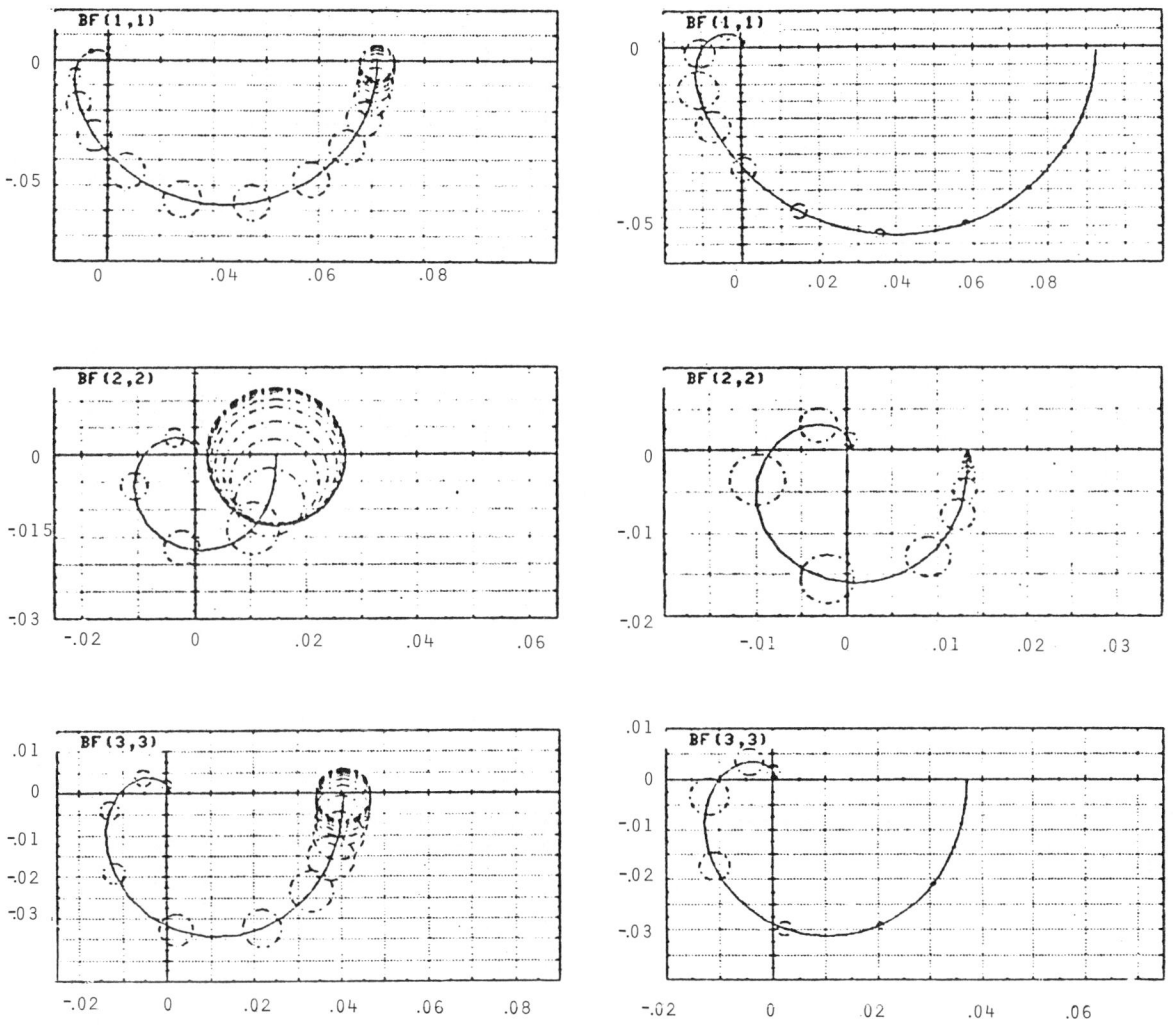

Fig.3 High frequency compensation.   Fig.4 Low frequency compensation.

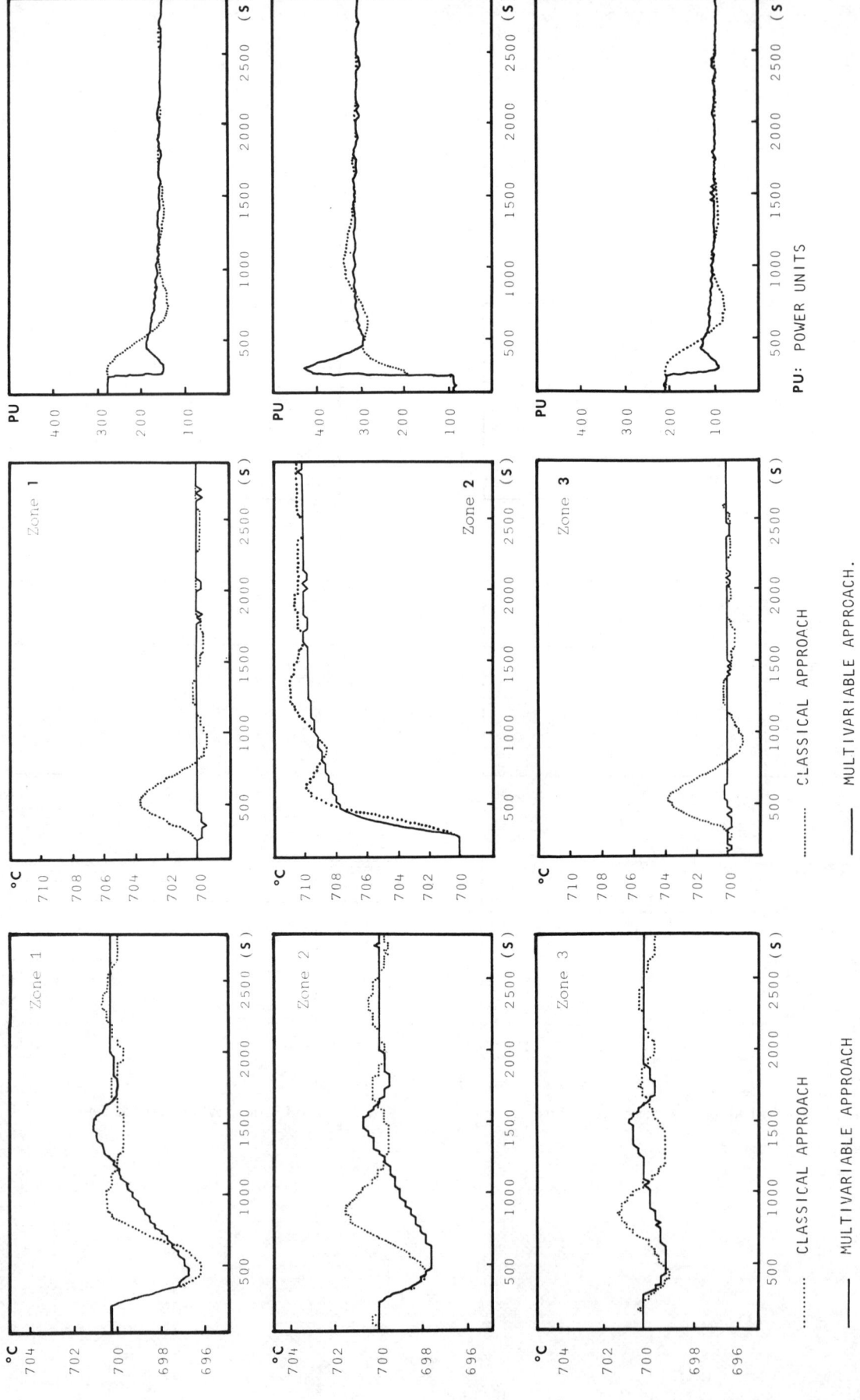

Fig. 5 Effects of perturbations (phase 2 heater cut-off).

Fig. 6a Step responses.

Fig. 6b Control strategies.

Copyright © IFAC Real Time Digital Control Applications
Guadalajara, Mexico 1983

COMBUSTION STABILIZATION AND IMPROVEMENT OF THE EFFICIENCY IN A PEAT POWER PLANT

F. Wahlström* and U. Kortela**

Department of Electrical Engineering, Tampere University of Technology, Tampere, Finland
**Department of Electrical Engineering, Helsinki University of Technology, Espoo, Finland*

Abstract. Improved technology in the instrumentation of power plants permits new control designs in the plants. Big benefits can be achieved using measurements from new flue gas analyzers combined with the cheap computing capacity of small microprocessor based control systems. The burning process in the furnace is the most difficult subprocess of the plant to control because of the very few measurements. Peat is an inhomogeneous fuel and the stability of the combustion process controlled by conventional methods is weak. For efficient combustion control the fuel power in the supply and/or the heat power released in the furnace should be measured or estimated.

In our study it has been shown that the heat power can be estimated using a Kalman filter when the air supply and the excess oxygen in the flue gas are measured. To improve the accuracy of the estimation the support oil supply, the speed of the peat feeder and the peat moisture can be used. For parameter fine tuning the power of the boiler might be calculated.

The paper presents a microprocessor based estimation and compensation system, which compensates the disturbances in the fuel and air supplies. The estimation and compensation algorithms as well as the connections to an analog instrumentation system are presented. The system has been used at the biggest Finnish peat power plant in Oulu during several months. The stability of the plant improved remarkably and the oxygen content in the flue gas can be kept within a very narrow band. The load following capability of the plant improved also clearly.

In multifuel power plants the control of the air supply is difficult. Using measurements of both O_2- and CO-contents in the flue gas and calculating the heat losses of the flue gas the total losses as a function of the excess oxygen can be obtained through a recursive identification process. Knowing the loss function the optimal excess oxygen can be determined. This type of total air supply control has been tested through simulation studies.

INTRODUCTION

During the last years the usage of peat for energy production has increased extensively. Especially the district heating boilers of many larger cities are peat fuelled. Most of the power plants use oil as a supporting fuel for security reasons. In the design phase of several peat power plants there have been used methods, which had been proved to work in coal and brown coal fuelled plants. The biggest peat power plants burn pulverized peat (Fig. 1). The peat is fed into the drying pipe, where hot flue gases (700... 1000°C) from the upper part of the furnace carry the peat into a hammer mill. The peat dries from initially 40...65 % moisture to about 5...25 % moisture. The pulverized peat is carried from the mill through the blower into the piping, which splits the peat flow to different burners.

Because of the big quality variations of peat the peat usage has caused trouble e.g. in the control of the burning process. The effective heat value of peat can vary between 1200...4600 MJ/m^3 due to the varying moisture, density and age of the peat. The peat feeders are usually volume feeders and thus the actual fuel power can vary highly although the peat feed is constant. The situation could be helped by using weighing systems and moisture analysators in the peat feed system. The plants use however usually several feed lines and this solution would be a very expensive one. Besides the moisture analysators on the market have been shown to be badly applicable to measurement of peat moisture as a consequence of the fluctuating physical and chemical characteristics of the peat. The quality variations of the peat appear in the operation of the plant as variations in the steam pressure and temperature, which the regulation system cannot eliminate.

This paper presents a method to estimate the heat power released in the burning process (combustion power), which will give a measure of the actual amount of burned peat considerably earlier than the heat transferred to the steam would give. The method is based on the calculation of oxygen consumption using measurements of the burning air flow, the flue gas flow and the flue gas oxygen content. Since the flue gas flow not often is measured, it will be calculated using data from the fuel flows, the fuel mean compositions and the burning air flow. The combustion power can be calculated with static equations but this will lead to some disturbances in the estimate of the combustion power. For this reason an optimal filter for the combustion power was built. A linear dynamic model for the feed part of a pulverized peat fuelled plant was constructed and the Kalman equations were applied. In addition to the combustion power the filter calculates the effective heat value of the peat. The calculation method of the combustion power has been proved reliable and a control method has been developed for the new measurement. The control method is designed to be easy to add to old control systems. The method was tested in a big peat power plant in Finland (Oulu, Toppila; 60 MW_e, 120 MW_t). The tests were successful and the design of a microcomputer based combustion power estimation and compensation system was begun. The system has been in use in the peat power plant in Oulu, Toppila and the experiences have been very positive. Besides the compensation of the combustion power the system controls the burning air flow according to the oxygen content in the flue gas and taking the boiler load into consideration.

In power plants, which burn mainly one fuel, the burning air can be controlled using the oxygen content measurement as the only feedback information, because in this case the optimal set point of the oxygen content depends approximately only on the boiler load. On the contrary if the plant is fuelled with several fuels the oxygen content set point is extremely difficult to determine, because it depends strongly also on the fuel proportions. In this case it is necessary to use some other measurement to an additional feedback. For this aim the carbon monoxide content in the flue gas can be used. The new CO-monitors, which measures across the stack absorption of infrared light beams, provide sufficient sensitivity and reliability for the purpose. The paper presents an optimal control strategy for the burning air flow, which attempts to minimize the expectation value of the flue gas losses. The method is tested in simulation studies and proved to be possible to apply. A disadvantage is however a heavy calculation burden. In an application is a minicomputer or a very efficient microcomputer to be used.

MEASUREMENT OF THE COMBUSTION POWER

An immediate way to measure the combustion power does not exist. The only measure of the released heat power has earlier been obtained via the boiler from the measurements of the steam pressure, temperature and flow. In our research we have however shown, that there is several ways to measure combustion power indirectly dynamically earlier than measurement of the power of the boiler. These measurement methods are e.g. measurement of the temperature of the flame, the heat power of the radiation superheater and the oxygen consumption of the burning process. The flame temperature is however not to be used because of the difficulties and poor reliability in the measurement. In the initial phase of the research there were made extensive process test runs in two big peat power plants (Tampere, Naistenlahti and Oulu, Toppila; both 60 MW_e, 120 MW_t). The collected process data was used for modelling and correlation analysis. A conclusion from these analyses was that the oxygen consumption is a fast and reliable measure of the combustion power. This is in fact a natural conclusion because the burning is a reaction mainly between carbon and oxygen. The theoretical oxygen consumption does very little depend on the composition of the dry peat. The normal ranges of the elements of dry milled peat are:

```
carbon   : 50    ...60 weight-%
hydrogen:  5    ... 7      "
oxygen   : 30   ...40      "
nitrogen: 0.5... 2.5       "
sulphur : 0.1... 0.4       "
ash      :  2   ...12      "
```

From this the theoretical oxygen need of different qualities of milled peat can be calculated (moisture constant 50 %)

```
normal      : 55.39 dm³O₂/MJ
lower limit: 53.86  -"-  ; Δ = -2.8 %
upper limit: 56.44  -"-  ; Δ = +1.9 %
```
(Δ = deviation from normal conditions)

Even the moisture of the peat varying in the normal range 40...65 weight-% does not influence too much on the oxygen need. With average composition of the peat we can obtain the theoretical values:

```
normal (moisture = 50 %)     :
   55.39 dm³O₂/MJ
lower limit (moisture = 40 %):
   53.28 dm³O₂/MJ ; Δ = -3.8 %
upper limit (moisture = 65 %):
   61.68 dm³O₂/MJ ; Δ = +11.4 %
```

Thus if the oxygen consumption can be measured reliably and with a small time constant the combustion power can be calculated from

$$P_c = c(M) \cdot F_{O_2} \qquad (1)$$

where

P_c ≜ the combustion power, MW,
M ≜ the moisture of the peat, weight-%,
F_{O2} ≜ the oxygen consumption, m³/s,
$c(M)$ ≜ a parameter depending slightly on the peat moisture, MJ/m³.

The oxygen consumption can in turn be calculated from

$$F_{O2} = 0.01 \cdot (21 \cdot F_i - X_{O2} F_s) \qquad (2)$$

where

F_i ≜ the burning air flow, m³/s,
X_{O2} ≜ the flue gas oxygen content, vol.-%,
F_s ≜ the flue gas flow, m³/s.

The oxygen content of the flue gas is to be measured by an in-situ zirconium oxide sensor providing a highly reliable measurement. Usually the flue gas flow is not measured and it is thus to be calculated from other quantities. If the power plant uses two different fuels (peat and oil) the flue gas flow can be calculated from four measurements (the oxygen content, the air flow, the oil flow and the peat moisture) using the following equations [1]

$$m_s = \frac{21 F_i - (100 c_5 + \frac{21 c_4 X_{O2}}{21 - X_{O2}}) m_o}{\frac{21(c_1 M + c_2) X_{O2}}{21 - X_{O2}} + 100 c_3 (100 - M)}, \qquad (3)$$

$$F_s = \frac{(c_1 M + c_2) m_s + c_4 m_o}{1 - \frac{X_{O2}}{21}}, \qquad (4)$$

where

m_o ≜ the oil flow, kg/s,
$c_1, c_2 \ldots c_5$ ≜ parameters depending on the compositions of the fuels,
m_s ≜ the burned peat flow as a fictive "standard peat" (composition constant), kg/s.

The calculation of the flue gas flow requires a reasonably accurate information of the peat moisture. A ten per cent error in the moisture measurement will cause an error in the calculated flue gas flow of about 7...10 % depending on the moisture value.

OPTIMAL FILTERING

The calculation method of combustion power presented above contain several noise sources. The quantities used in the equations (the air flow, the O_2-content, the oil feed and the peat moisture) have to be correctly phased with each other. The dynamics of the process between different quantities can simply be described by a model with one time constant and a delay but this will cause some errors. Also the measurements in themselves contain noise. Thus the calculated combustion power has to be filtered. In the following the optimal filtering of the combustion power calculated in a pulverized peat fuelled boiler will be presented.

The Kalman equations can be used for the optimal filter and thus we have to obtain a linear dynamic model of the peat feed and the burning. The process of Fig. 1 can be parted into subprocesses according to Fig. 2. In practice there is always a small amount of peat stored in the beater mill. Changing the speed of rotation of the peat blower will affect the peat storage in the mill. The mill is described as a mixer of first order in regard to the effective heat value of the peat and the peat flow. In addition the derivated speed of rotation of the blower affect the peat flow leaving the mill. Thus

$$\dot{P}_M = \frac{1}{\tau_m}(h_{um} m_s - P_M) + k_1 h_{um} \dot{N}_k \qquad (5)$$

and

$$\dot{h}_{um} = \frac{1}{\tau_m}(h_u - h_{um}), \qquad (6)$$

where

m_s ≜ the peat fed into the mill, m³/s,
P_M ≜ the fuel power leaving the mill, MW,
N_k ≜ the speed of rotation of the blower, 1/s,
h_u ≜ the heat value of the peat feed. MJ/m³,
h_{um} ≜ the heat value of the peat leaving the mill, MJ/m³,
τ_m ≜ time constant, s,
k_1 ≜ constant, m³s.

The piping and the burners can be modelled as a mixer regarding the fuel power.

$$\dot{P}_c = \frac{1}{\tau_p}(P_M - P_c), \qquad (7)$$

where

P_c ≜ the combustion power, MW,
τ_p ≜ time constant, s.

The streaming of the flue gas in the furnace and measuring the O_2-content will also be modelled as a first order model (assumption: the oxygen consumption in the flames of the burners = $k_2 P_c$)

$$\dot{F}_{O2} = \frac{1}{\tau_f}(k_2 P_c - F_{O2}), \qquad (8)$$

where

F_{O2} ≜ measured oxygen consumption, m³/s,
τ_f ≜ time constant, s,
k_2 ≜ constant, m³/MJ.

For a complete model there is yet to model the dynamics of the heat value of the peat before the peat feeders. It will be assumed that the heat value of the peat brought to the plant behaves as white Gaussian noise and the peat treatment system of the plant (silos, conveyors, crushers etc.) can be thought as a colouring system to the noise. Thus

$$\dot{h}_u = -\frac{1}{\tau_u} h_u + w_1, \quad (9)$$

where

$\tau_u \triangleq$ time constant, s,
$w_1 \triangleq$ white Gaussian noise.

When the equations (5)...(9) are collected and linearized and noting that $\bar{h}_u = \bar{h}_{um}$ and $\dot{\bar{N}}_k = 0$ (steady-state values) a four-dimensional linear state-space model of the feed part of the plant is obtained. The discretizing of the model gives (Δ-notation $\triangleq$ deviation from steady-state values; ¯ -notation $\triangleq$ steady-state values)

$$\underline{x}_{k+1} = \underline{A}\,\underline{x}_k + \underline{B}\,\underline{u}_k + \underline{w} \quad (10a)$$

$$y_{k+1} = \underline{C}\,\underline{x}_k + v \quad (10b)$$

where

$$\underline{x} = \begin{bmatrix} \Delta h_u \\ \Delta P_M \\ \Delta P_C \\ \Delta F_{O2} \end{bmatrix}, \quad \underline{u} = \begin{bmatrix} \Delta m_s \\ \Delta \dot{N}_k \end{bmatrix},$$

$$\underline{A} = \begin{bmatrix} 1-\frac{T}{\tau_u} & 0 & 0 & 0 \\ \bar{m}_s \frac{T}{\tau_m} & 1-\frac{T}{\tau_m} & 0 & 0 \\ 0 & \frac{T}{\tau_p} & 1-\frac{T}{\tau_p} & 0 \\ 0 & 0 & k_2 \frac{T}{\tau_f} & 1-\frac{T}{\tau_f} \end{bmatrix},$$

$$\underline{B} = \begin{bmatrix} 0 & 0 \\ \frac{T}{\tau_m}\bar{h}_u & k_1 T \bar{h}_u \\ 0 & 0 \\ 0 & 0 \end{bmatrix}, \quad \underline{C} = [0\ 0\ 0\ 1].$$

$\underline{w}, v \triangleq$ white Gaussian noise.

For this system can optimal filtering equations, which use all a priori knowledge, be written (the Kalman filter)

$$\underline{V}'_{k+1} = \underline{A}\,\underline{V}_k \underline{A}^T + \underline{V}_w \quad (11a)$$

$$\underline{K}_{k+1} = \underline{V}'_{k+1}\underline{C}^T(\underline{C}\,\underline{V}'_{k+1}\underline{C}^T + V_v)^{-1} \quad (11b)$$

$$\underline{V}_{k+1} = (\underline{I} - \underline{K}_{k+1}\underline{C})\underline{V}'_{k+1} \quad (11c)$$

$$\hat{\underline{x}}_{k+1} = \hat{\underline{x}}_k + \underline{K}_{k+1}[y_{k+1} - \underline{C}(\underline{A}\hat{\underline{x}}_k + \underline{B}\underline{u}_k)] \quad (11d)$$

where

$\underline{V}_w \triangleq$ the variance matrix of $\underline{w}$,
$V_v \triangleq$ the variance of v,
$\hat{\underline{x}}_k \triangleq$ the optimal filtered value of $\underline{x}$,
$\underline{V}_k \triangleq$ the variance matrix of the estimate error,
$\underline{V}'_k \triangleq$ the variance matrix of the prediction error.

The optimal filter was tuned and tested using real process data from test runs in the Oulu plant. The parameters τ_u, τ_p, τ_f, k_1 and k_2 were estimated in advance from the time behaviour of the different process quantities and their interdependences. The filter was tuned experimenting with different values of V_v and $\underline{V}_w$ ($\underline{V}_w$ was held as a diagonal matrix). The results of an optimal filtering run is presented in Fig. 3. Note that the filter also is able to efficiently calculate the effective heat value of the peat. The figure also shows for comparing purpose the heat value calculated by a simple static equation from the measured peat flow and the boiler load. The inaccuracy in the static method is clearly observed. The estimate of the combustion power is reliable and a more detailed examination has shown that the changes in the estimated combustion power will occur even 2...3 minutes before corresponding changes in the power of the boiler.

CONTROL OF THE COMBUSTION POWER

Earlier there were no measurement available of the power of the burning process. For this reason the control of the fuel feed was carried out using measurements of the produced steam (Fig. 4). The advantages with the steam side measurements are simplicity and reasonably good accuracy. However a disadvantage is that the process to be controlled is characterized by big time constants and long delays. Thus the energy balance regulator has to be tuned slow to maintain stability in the circuit. Slow control means that the disturbances in the fuel feed and the boiler load have a strong effect upon the controlled quantity (usually steam pressure) and the disturbances settle very slowly.

Using the new combustion power measurement method it is possible to obtain data from the burning process dynamically before the steam measurements. (This will require the computing abilities of a digital system.) With the help of the new measurement the fuel feed control can be realized in a cascade type manner (Fig. 5). A fast fuel feed control is done according to the combustion power measurement and only the inaccuracies in this control loop will be corrected by the steam pressure control.

Further from Fig. 5 divergent control connections can be designed. The combustion power regulator can be cascade connected with the steam pressure regulator in a normal way or the connection can be realized according to Fig. 6. However reliability questions may prevent the use of a normal cascade circuit. The circuit of Fig. 6 is suited for larger digital systems, where the needed demand signals already are in digital form.

In the process tests we have used the circuit of Fig. 5 since this was easy to connect to the existing analog automation system.

The combustion power regulator of Fig. 5 or Fig. 6 cannot be a normal PI-regulator because this would lead to a situation where the steam pressure regulator and the combustion power regulator do not have fixed equilibrium positions. Thus one (or both) of the regulators would finally drift to the edge of the working range. The control characteristics of a P-regulator are limited and thus it is justified to use a special control algorithm. This is easy to program into the same digital system, which calculates the combustion power. A suitable control algorithm is e.g.

$$u_k = U_o + s[u_{k-1} + K(e_k - e_{k-1} + \frac{T}{T_i} e_k) - U_o] \quad (12)$$

where

- u_k ≙ the output of the regulator at time k,
- U_o ≙ the equilibrium value of the regulator,
- s ≙ filtering constant, $0 \ll s < 1$,
- e_k ≙ the input of the regulator at time k (error),
- K ≙ the regulator gain,
- T_i ≙ the regulator reset,
- T ≙ the sampling time.

The short time characteristics of the algorithm remind of those of a PI-regulator but if the error nearly equals zero u_k will converge towards U_o.

The errors to the two regulators (steam pressure and combustion power) are formed from completely different quantities. The calculation and measurement inaccuracies can thus result in that one of the regulator inputs does not equal zero in the equilibrium state of the system. In this situation the regulators resist each other. To avoid this an integrator can be added to the system, which drives the error of the combustion power regulator slowly towards zero.

In the microcomputer based combustion power estimation and compensation system presented in the following text are the connections of Fig. 5 and the control algorithm presented above used.

A DIGITAL ESTIMATION AND COMPENSATION SYSTEM

The control of the combustion power was tested in an early stage of the research work at the peat power plant in Oulu using a connection of analog components. The oxygen consumption was calculated with one multiplication unit and two summing units according to a heavily simplified equation. A PI-regulator was used as the combustion power regulator. In spite of inaccuracies in the calculations the test runs gave very encouraging results and the connections were in fact for some time in continuous usage.

After the successful tests we decided to design a digital estimation and compensation system so that the combustion power calculations could be made more accurate and the control algorithm of the combustion power regulator could be chosen freely. The hardware chosen was the DP2000 microcomputer system delivered by Digiproduct Oy. This system uses the Motorola MC6803 as CPU. The software is stored in 24 kB EPROM memory and the 3 kB RAM memory used as data storage is provided with a battery back-up. The process interface uses one eight-channel A/D-converter and one four-channel D/A-converter. The operation and parametrization is carried out through a keyboard and display card mounted at the front. Ten seven-segments and several LEDs work as display elements and the keyboard has 27 keys. For special purposes the system can be equipped with serial data transfer interface and digital inputs and outputs. The hardware is mounted in a 19" rack (Fig. 7).

In addition to the combustion power control the system is able to control the total burning air flow. The control is carried out as control of the flue gas oxygen content with the set point dependent on the boiler load. The fuel feed and air flow control strategies as well as the connections of the system to an external analog control system is presented in Fig. 8. The output of the pressure regulator is proportional to the power need and thus proportional to the oxygen consumption need. The output of the regulator is therefore a suitable signal for the total air flow demand. To achieve a feedback the air flow is corrected by the oxygen content regulator through a multiplication unit (correction of the air/fuel ratio).

The two regulators of the system (fuel and air) can be operated through the keyboard (manual/automatic, manual up and down). The 7-segments can display any of the measured and calculated quantities of the system. Two analog outputs are reserved for recorders and the quantities to be recorded can be chosen by parametrizing. Valuable quantities for the supervision are e.g. the combustion power, the heat value of the fuel, the flue gas flow etc.

The parameters of the system can be protected with a lock switch. After removal of the key only the display and the regulators can be operated. The parameters are also protected against power fails with a battery back-up (maximum 500 hours).

THE PROCESS TESTS

When the development work of the estimation and compensation system was finished there were planned process test runs in the Oulu plant. The tests were carried out in November 1981. Intentional disturbances were made in the fuel feed, the air feed and the boiler load and the performance of the compensation system was estimated by comparing to corresponding situations where the compensation system not was used. Process data was collected in the tests from 31 measurement points for over 14 hours using a process computer system.

In Figs. 9, 10, 11 and 12 are some test run results presented. Figs. 9 and 10 present disturbance situations in the peat feed. The disturbances are made with the second peat feeder, which was run manually. In Fig. 9 the compensation system is not connected. The steam pressure and temperature oscillate considerably after the disturbances. In the case of Fig. 10, where the compensation system is in usage, the oscillating tendency has completely disappeared and the deviations are clearly smaller.

Figs. 11 and 12 present boiler load disturbances. Both of the peat feed lines are connected for automatic control. In Fig. 11 the compensation system is not used and the load disturbances cause strong oscillations. Using the combustion power compensation (Fig. 12) the tolerance against load disturbances has improved remarkably and no oscillations occur.

The process tests proved that the compensation system was able to efficiently stabilize the burning process. The system has been in continuous use in the plant for several months after the tests. The good stability of the plant has made it possible for the total air flow control to keep the oxygen content in the flue gas extremely close to the set point. The deviations have been about $\pm 0.1 \% \: O_2$.

OPTIMIZING THE AIR FLOW

In the compensation system presented above the control strategy of the burning air flow was to control the flue gas oxygen content with the set point being a function of the boiler load. If the plant uses multiple fuels the function is difficult to determine because the optimal value of the oxygen content set point depends on the boiler load, the fuel proportions and the burner conditions. In this case the set point function would have to be tabulated in regard to several quantities. The table would require considerably memory and the parametrizing would be extremely troublesome. Thus it would be preferrable if the optimal set point of the oxygen content could be calculated from other quantities. This is possible if a sensitive carbon monoxide analysator is available.

The flue gas losses can be divided into losses as unburned matters (mainly CO) and the heat losses of the flue gas. The CO-losses decrease rapidly when the air flow is increased but the heat losses in turn increase nearly linearly. The loss curve is usually drawn as a function of the air/fuel ratio but since the oxygen content in the flue gas is nearly linearly dependent on the air/fuel ratio at small O_2-content values the loss curve can also be drawn as a function of oxygen content (Fig. 13). The optimum oxygen content set point can now be obtained from the minimum value of the loss function.

Thus the problem is to continuously identify a changing loss function. This can be done by identifying separately the CO-loss function and the heat loss function. The CO-losses are assumed to follow an exponent function ($L_{CO} = a_1 e^{b_1 X_{O2}}$) and the heat losses are assumed to be linear ($L_t = a_2 X_{O2} + b_2$). The CO-losses are calculated from the equation

$$L_{CO} = 0.01 \: h_{co} X_{co} \:, \qquad (13)$$

where

L_{co} ≙ the CO-losses per flue gas volume unit, MJ/m^3,
h_{co} ≙ the heat value of carbon monoxide ≈ $12.6 \: MJ/m^3$,
X_{co} ≙ the carbon monoxide content of the flue gas, vol.%.

The heat losses can be calculated from the equation

$$L_t = c_f(t_{out} - t_o) \:, \qquad (14)$$

where

L_t ≙ the heat losses per flue gas volume unit, MJ/m^3,
c_f ≙ the average heat capacity of the flue gas, $MJ/°C \: m^3$,
t_{out} ≙ the end temperature of the flue gas, °C,
t_o ≙ the outside air temperature, °C.

The CO-loss function is linearized for the recursive identification by taking the logarithm of the CO-losses ($\ln(L_{co}) = a_3 X_{O2} + b_3$) and thus both of the loss functions can be handled as linear functions. After each sampling period is hence to be calculated a new estimate for the parameters a_2, b_2, a_3 and b_3 from the measurement pairs $\ln(L_{co})$, X_{O2} and L_t, X_{O2}. For this purpose the recursive least squares method (RLS) [2] can be used. Since the parameters are time varying has the algorithm to forget old data. The forgetting can be realized using an exponential or a rectangular window. In the simulation studies was shown that the normal exponential window forgetting scheme works extremely badly in fast boiler load changes. On the contrary the algorithm with changing forgetting factor of Fortescue et al. [3] gave clearly better results. Likewise the rectangular window algorithm [2] worked satisfactoryly although the choose of the right window length was important. The actual RLS-algorithm can be realized in different ways. The standard form is advantageous from the calculation burden view but the algorithm can produce numerical problems. Numerically more stable algorithms are the square root algorithm [4] and the UD-algorithms [5]. The algorithm presented here is the Fortescue algorithm in standard form. The algorithm is used separately for the CO-loss function and the heat loss function. The algorithm is

$$y_k = \underline{x}_k^T \underline{\theta}_k \qquad (15a)$$

$$\underline{K}_{k+1} = \underline{P}_k \underline{x}_k / (1 + \underline{x}_k^T \underline{P}_k \underline{x}_k) \qquad (15b)$$

$$\hat{\underline{\theta}}_{k+1} = \hat{\underline{\theta}}_k + \underline{K}_{k+1}(y_k - \hat{y}_k) \qquad (15c)$$

$$\lambda_k = 1 - (1 - \underline{x}_k^T \underline{K}_{k+1})(y_k - \hat{y}_k)/\Sigma_o \qquad (15d)$$

if $\lambda_k < \lambda_{min} \Rightarrow \lambda = \lambda_{min}$

$$\underline{P}_{k+1} = (\underline{I} - \underline{K}_{k+1} \underline{x}_k^T)\underline{P}_k / \lambda_k \qquad (15e)$$

where

$$\underline{x} = \begin{bmatrix} X_{O2} \\ 1 \end{bmatrix}, \quad y = L_t \text{ or } y = \ln(L_{co}),$$

$$\hat{\underline{\theta}} = \begin{bmatrix} \hat{a}_i \\ \hat{b}_i \end{bmatrix} \triangleq \text{ the parameter estimates } (i = 2 \text{ or } 3),$$

$\lambda_{min}, \Sigma_o \triangleq$ tuning parameters.

The simulation study shows that the algorithm could be used in practice if a sufficient calculation capacity is available.

CONCLUSIONS

It was shown that the combustion power can be estimated if the oxygen consumption of the burning process is calculated. The combustion power control strategy was able to stabilize the conditions in the furnace. The stability provide savings mainly in two ways. The control of the burning air flow is eased, when the variations in the oxygen consumption are eliminated. The control can thus reduce the standard deviation of the flue gas oxygen content and the air flow can be lowered closer to the optimal flow. As a consequence of this the flue gas losses are reduced by a notable amount even in middle sized boilers. Secondarily the stabilized steam temperatures save the boiler constructions. These savings are difficult to estimate but they are probably quite large.

In an industrial boiler there are often required fast load changes. In this case the improved load following capability of the plant achieved due to the combustion power control is very valuable.

The combustion power measurement and control has in this research project been implemented only in peat power plants. The method is however usable in general and the same advantages can be achieved in other plants regardless of the fuels and burning method. The greatest benefits can however be attained in plants fuelled with inhomogeneous fuels (peat, coal, bark and waste) and in industrial boilers.

For the burning air flow optimization was an algorithm presented. The method requires however an efficient digital calculation system and a carbon monoxide monitor, which is able to measure small quantities of carbon monoxide in the flue gas.

REFERENCES

[1] Wahlström, F., Development of a Fuel Power Estimation and Compensation System for a Peat Power Plant, Turvop-report 15, Tampere University of Technology, Tampere 1981. (in Finnish)

[2] Goodwin, G.G., Payne, R.L., Dynamic System Identification: Experiment Design and Data Analysis. Academic Press, London.

[3] Fortescue, T.R., Kershenbaum, L.S., Ydstie, B.E., Implementation of Self-tuning Regulators with Variable Forgetting Factors. Automatica 17.6, 1981.

[4] Clarke, D.W., Some Implementation Considerations of Self-tuning Controllers. Numerical Techniques for Stochastic Systems, North-Holland Publishing Company, 1980.

[5] Thornton, C.L., Bierman, G.J., Filtering and Error Analysis via the UDU^T Covariance Factorization. IEEE Trans., 1978, Vol. AC-23, pp. 901-907.

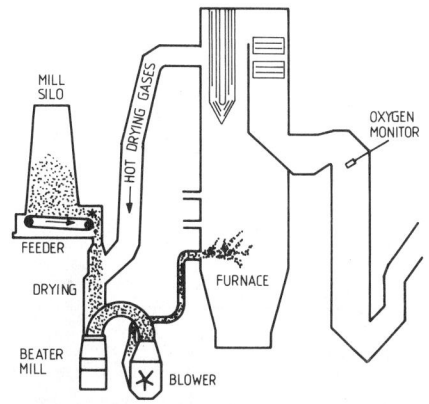

Fig. 1. Mill-drying system.

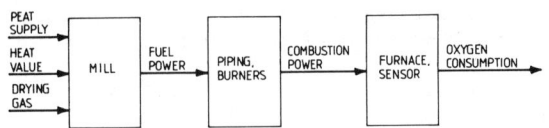

Fig. 2. The feed part of the pulverized peat fuelled power plant.

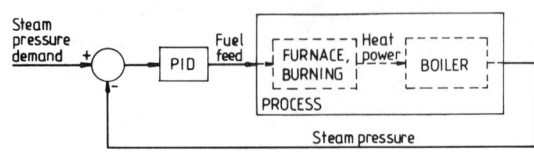

Fig. 4. Conventional fuel feed control.

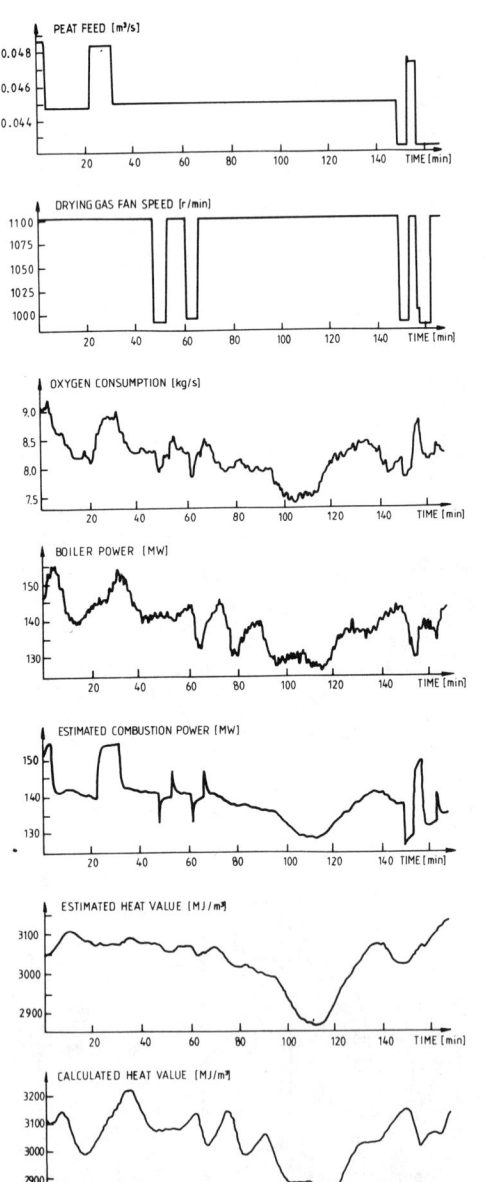

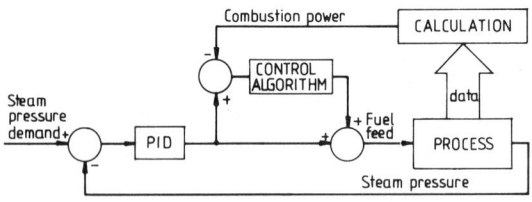

Fig. 5. Improved fuel feed control.

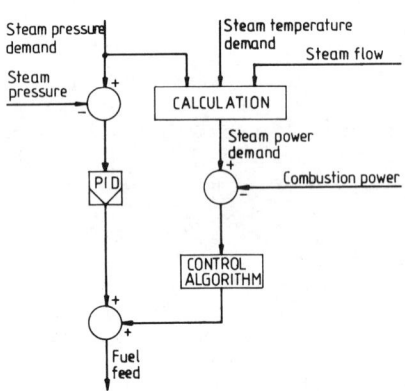

Fig. 6. Alternative combustion power control scheme.

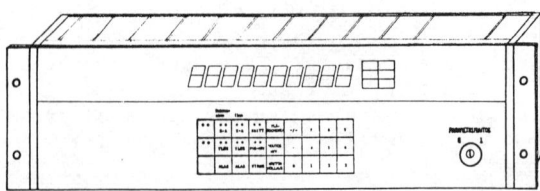

Fig. 3. Results of an optimal filtering run.

Fig. 7. The combustion power estimation and compensation system.

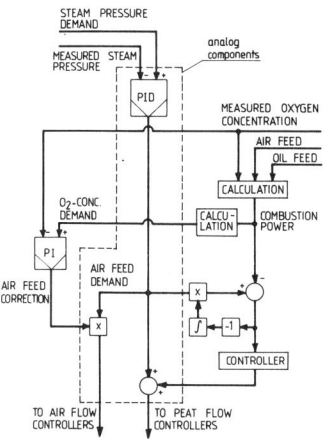

Fig. 8. The connections of the compensation system to an external control system.

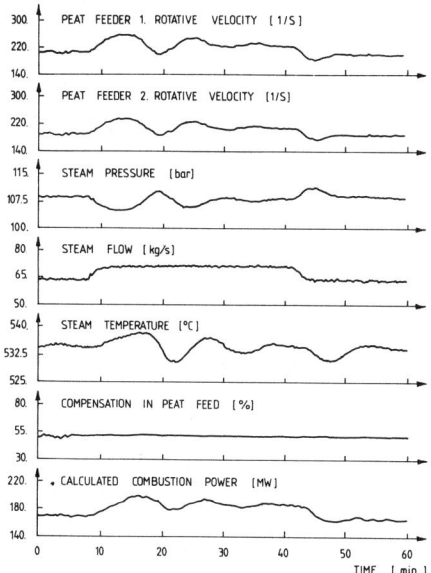

Fig. 11. Load disturbances. Conventional system.

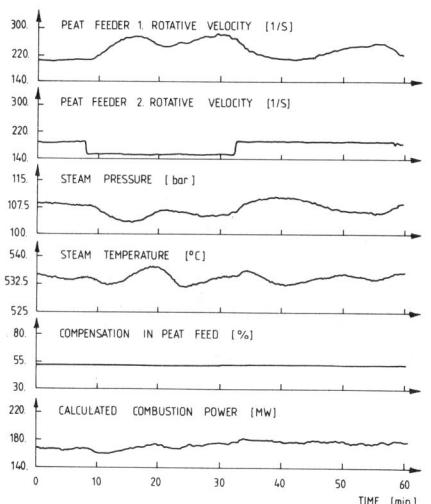

Fig. 9. Disturbances in peat feed. Conventional system.

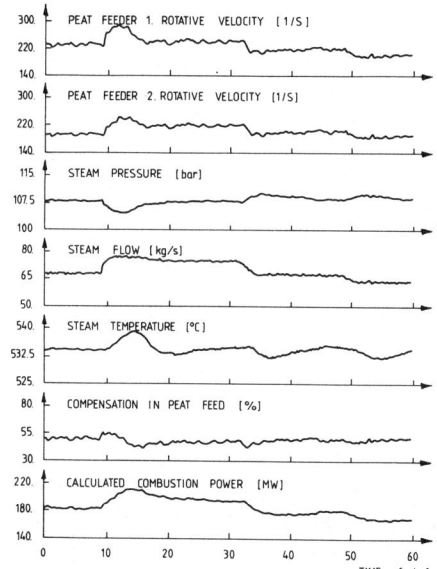

Fig. 12. Load disturbances. Stabilized system.

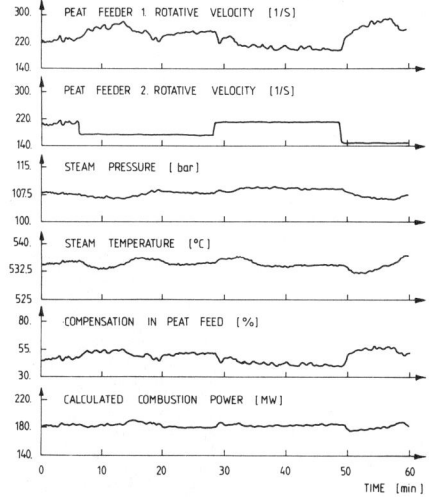

Fig. 10. Disturbances in peat feed. Stabilized system.

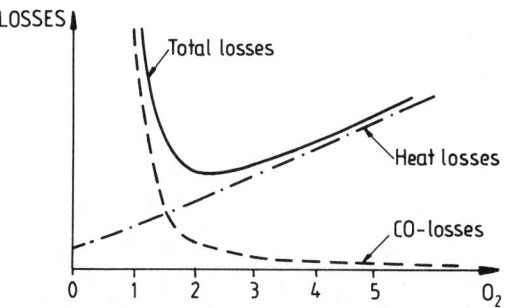

Fig. 13. The flue gas losses.

DIGITAL CONTROL OF FURNACES IN CERAMIC INDUSTRY

H. El Hajjar, J. B. Pourciel and J. P. Babary

Laboratoire d'Automatique et d'Analyse des Systèmes du C.N.R.S., 7 avenue du Colonel Roche, 31400 Toulouse, France

Abstract. In the field of ceramic industry, burning furnaces for products like bricks, tiles or pots, are characterized by an important energy consumption and need performing thermal profile control of the products to be burnt. In this paper, we show in a concrete example, how these two aspects are solved, emphasizing the role played by a two microcomputer structure, for off-line or on-line studies. In the first case, this structure allows variable measurements on the plant, required for its modelling in order to develop a performing control law. In the second case, we get a flexible and performing device for a closed loop control structure.

Keywords. Digital control; distributed parameter systems; ceramics

INTRODUCTION

In the field of ceramic industry (Krause, Plaul, Zollner), the drying and burning operations of products like bricks, tiles or pots, are relatively slow processes and need a large energy consumption (gas or fuel). Moreover, for the burning operation, there exists a critical temperature, which must be reached carefully during heating and cooling phases, and a high temperature phase which must be maintained for a given time interval.

The burning operation of industrial ceramic products is done through large tunnel furnaces which operate as counter flow heat exchangers. These furnaces are heated by means of groups of burners, whose number and position affect directly their energy consumption and their productivity. An optimization study can be made from a precise modelling of the thermodynamic balance.

This paper shows, considering one particular burning ceramic furnace, how the real-time digital control problem was solved using a two microcomputer structure. This structure was, at first used for the system analysis and the model parameter identification, and then for the real-time furnace operation control.

FURNACE DESCRIPTION

The furnace used as an experimental support for this study is a tunnel furnace for burning of ceramic pots (Fig. 1). It is 50 m long and has (1,6 m x 1,6 m) square cross section. The heating is provided by 20 burners (gas) split into four groups. Each group is controlled by one motorised valve. The charge (carriages) moves in discrete displacements of 50 cm. One carriage goes through the furnace in about 40 hours. A counter flow of air allows the carriages to be cooled in the cooling zone and to be heated in the heating zone. The gas temperature inside the furnace is measured by means of 12 thermocouples located under the ceiling, along the tunnel.

The main difficulties of the burning operation in such a tunnel furnace are :
- the longitudinal heterogeneity of useful charge (products to be burnt)
- the large time-constant of heat transfer through the charge
- the difficulty of measurement of the charge temperature.

OFF-LINE PRELIMINARY STUDY

An optimisation study needs an elaboration of a mathematical model of the thermal behaviour, for the determination of the best distribution of burners as well as for the designing of a closed loop control algorithm.

Modelling

The mathematical modelling of the thermal behaviour has been previously published, (Kassinopoulos, 1980; Barreteau, Laguerie, Babary, Kassinopoulos, 1980). We summarize here the basic hypotheses and the obtained model structure.
The model was obtained by writing the heat transfer balance equations which involve con-

vection, conduction and radiation, between the charge, the gas and the furnace walls, and mass transfer (charge, air and gas flows).

We assume that :
- the charge is homogeneous along the tunnel
- the gas flow is laminar and one-dimensional
- the conduction is negligible with respect to the convection and radiation phenomena.
- the wall surface temperature dynamics is such that its derivative with respect to the time variable is formally equal to zero.
- the temperatures are defined in a section of the tunnel by their average.

The mathematical model equations are the following :

Temperature T (x,t) of the charge :

$$\frac{\partial}{\partial t}[\rho_T S_T C_T T + \rho_{CH} S_{CH} C_{CH} T] + \frac{\partial}{\partial x}[U_T S_T \rho_T C_T T + U_T S_{CH} \rho_{CH} C_{CH} T] =$$
$$= A_{TG}(G-T) + B_{TG}(G^4-T^4) + B_{TP}(P^4-T^4) \quad (1)$$

Temperature G (x,t) of the gas :

$$\frac{\partial}{\partial t}[\rho_G S_G C_G G] - \frac{\partial}{\partial x}[U_G S_G \rho_G C_G G] = A_{TG}(T-G) + A_{GP}(P-G) +$$
$$+ A_{GE}(E_e-G) + B_{TG}(T^4-G^4) + B_{GP}(P^4-G^4) +$$
$$+ Q_R \cdot \sum_{i=1}^{N_B} D_{x_i} \delta(x-x_i) \quad (2)$$

Temperature P (x,t) of the wall :

$$A_{GP}(G-P) + A_{PE}(E-P) + B_{GP}(G^4-P^4) + B_{TP}(T^4-P^4) = 0 \quad (3)$$

Mass balance :

$$\frac{\partial D_{MG}}{\partial x} = \frac{\partial [U_G \rho_G S_G]}{\partial x} = -\sum_{i=1}^{N_B} \delta(x-x_i) D_{x_i}(17.9 + 16.9\, \gamma) \quad (4)$$

Boundary conditions :

$$T(x=0,t) = 433°K; \quad G(x=L,t) = 310°K \quad (5)$$

with $x \in]0,L[$ (L : lenght of the furnace)

Notation :

ρ : density; C : specific heat; $\mathcal{U}$: velocity; S : cross section; A : conductivity; B : radiation coefficient.
The indices T, G, P and CH are respectively related to the charge, the gas, the wall and the carriages.

E_e : air temperature under carriages
E : ambiant temperature
γ : "primary air" excess coefficient
D_{x_i} : gas flowrate per burner (i=1,2,...,N_B)
D_{MG} : "secondary air" flowrate
Q_R : combustion coefficient of gas
N_B : number of burners.

The utilization of the two microcomputer system (see below) for the acquisition of temperature and flowrate measurements, enables the off-line identification of unknown parameters such as loss coefficient, mean surface of pots, etc ...

Optimization (Meirelles-Ribeiro, 1981; Babary, Meirelles-Ribeiro, 1981)

The optimal burning profile of ceramic products is not well known by ceramic specialists. However, taking into account some of the physico-chemical properties of the products to be burnt, it is possible to write down a set of inequality constraints.

a) In the pre-heating zone, the charge heating from 773°K to 923°K must be done in at least 3.5 hours.

b) In the burning zone, the charge temperature must be more than 1223°K for 1.5 hour at least.

c) In the cooling zone, the charge cooling from 923°K to 773°K must be done in at least 3.5 hours.

The a) and c) constraints are related to the existence of a critical temperature (830°K) which must be slowly reached.
The b) constraint enables good mechanical characteristics of burnt products.

If at least one of these constraints is not satisfied, a relatively important product percentage cannot be sold and cannot even be used in any way. So, it is interesting to takle the optimal control problem of the furnace, taking into account the set of above constraints. We can also examine the problem in a wider sense by reconsidering the distribution of burners (number and location).

Static optimization of the furnace. A detailed study is published in Babary, Meirelles-Ribeiro (1981). We recall here the choosen criterion and the obtained results. The choosen (6a) is of economy type, taking into account the productivity and the energy consumption; the term "Pen" in (6a) acts as a penalty function when constraints are not satisfied

$$J = B - Pen \quad (6a)$$

with

$$B = Prod\ (P_v - C_p) - Cons.\ P_c - C_F \quad (6b)$$

B : benefit per hour
Prod : Production per hour of burnt products (proportional to the carriage velocity)
P_v : Mean sale fare of burnt products (per kg)
C_p : Mean variable cost of burnt products (per kg)
Cons : gas consumption
P_c : gas cost (per kg)
C_F : Fixed money charge per hour
Pen : $P_1 + P_2 + P_3$

with :

P_1 : penalty relative to the constraint in the pre-heating zone

P_2 : penalty relative to the constraint in the burning zone,
P_3 : penalty relative to the constraint in the cooling zone.

This criterion can be maximized with respect to the following parameters :
- the mean velocity of carriages
- the mass flowrate of "secondary air"
- excess coefficient of "primary air"
- the number and the position of burners
- the mean mass flowrate of burners

Results

The obtained solution consists in implementing four burners in the pre-heating zone, ten burners in the burning zone and four burners in the cooling zone (Fig. 2). The productivity is then increased by 20% with respect to the initial configuration.

Dynamic optimization. (El Hajjar). The temperature profile corresponding to static optimization results is a mean profile. A regulation algorithm around this profile is determined by minimizing a quadratic performance index (El Hajjar's thesis) with a higher weighing term on the difference $T_v(t_1)$ between the charge temperature and the static optimal profile at the final time of carriage displacement.

$$J = \underline{T}'_v(t_1).F.\underline{T}_v(t_1) + \int_0^t \left[\underline{T}'_v(t).Q.\underline{T}_v(t) + \underline{U}'(t).R.\underline{U}(t) \right] dt \qquad (7)$$

The optimal control $\underline{U}(t)$ is given by :

$$U(t) = - R^{-1} B^T P(t). \underline{a}(t) \qquad (8)$$

where $\underline{a}(t) = \phi . \underline{T}_v(t)$

$T_v(t)$: Difference between the charge temperature and the optimal profile.
ϕ, B : Constant matrices

P(t) is the matrix solution of the Riccati equation.

For the control computation, we need the charge temperature, which cannot be directly measured. A state observer is performed to resolve this problem. With the knowledge of the gas temperature, we can reconstitute the charge temperature profile. In Fig. 3, we show the evolution of the difference between the charge temperature and the optimal profile in four measurement points over six displacements.

REAL TIME DIGITAL CONSIDERATIONS

Hardware device and basic software

As the furnace is far from the laboratory (100 km) we have developed a two microcomputer asynchronous architecture, linked (Fig. 4) by a telephone line over a switched network, through two modems. The microcomputer A, installed near the furnace, is built around Motorola 6800 type microprocessor. It is composed

of :
- a central unit module asynchronous interface (ACIA) and parallel interface (PIA) (M68 ADS1).
- a module of 16 k bytes of read only memory (ROM)
- a module of 8 k bytes of random access memory (RAM)
- a real time module with an automatic restart system
- a digital input/output module with 32 input lines and 32 output lines
- a module of 16 analog channel inputs (acquisition time : 40 µs)

The microcomputer B, installed at the laboratory, is also built around Motorola 6800 type microprocessor. It is composed with :
- a central unit module (M68 ADS1)
- a read only memory module (16 k bytes)
- a random access memory module (32 k bytes)
- an alphanumeric module and one graphic module to manage a C.R.T. display
- a C.R.T. display and an ASCII keyboard
- two asynchronous lines (one for modem another for a minicomputer PDP 8A).

Basic software. In each microcomputer, software for general purpose is implemented, to decompose the whole set of operations into specific tasks and to organize then in real time. These tasks are the following.

The development monitor (1 k bytes ROM) (TDS User's Guide Motorola) which enables the program development in R.A.M., the memory area manipulation, the system initialization and the starting of the real time monitor.

Real time monitor. The work is divided in periodical tasks scheduled by a real time monitor (1 k bytes) developed at the laboratory. This monitor schedules 21 periodical tasks (this number can be increased). Each period of task can be choosen between 20 ms and 256 minutes. The monitor has two priority levels. The lower level (level 0) can be interrupted by level one tasks or external interrupts. The higer level (level 1) can be interrupted only by the non maskable interrupt (NMI). The priority in each level is determined by the order of the tasks in the task table. This monitor is implemented in each microcomputer.

Floating point library. This library contains all floating point subroutines (sin, cos ...) the digital input and the digital output subroutines and the analog input subroutines (4 k bytes) we have also developed a matrix calculation package (1 k bytes).

The program of the connection management between the two microcomputer, which are connected by an asynchronous line through two modems linked by a telephone line. The transmission baud rate is 300 (30 characters per second) the program enables either dialogue between both microcomputers, or operation of microcomputer B in transparent mode (in this case the microcomputer B works as a simple terminal of the microcomputer A).

Off-line aspects

The modelling of the furnace led to a distributed parameter hyperbolic system. The model parameters are very complex, and in this phase both microcomputers are used to make the model valid and to identify the unknown parameters. The measurements are stocked (on line) on a mass memory (PDP 8 A floppy disk), and are used off-line for parameter identification.

The knowledge of the model enables a good analysis of the thermal behaviour and makes easier the determination of the optimal distribution of burners by minimizing an economy type criterion. A control law was deduced to regulate the system around the optimal steady-state. In this phase, the two microcomputer structure is used to test the characteristics of the control law (convergence, implementation,...)

On-line aspects

The distance between laboratory and furnace led us to utilize the microcomputer A as an interface in charge to make data acquisition, to control the motor valves and to supervise the furnace operation. The microcomputer B is used to calculate the control of motor valves, to display all measurements and to manage the telephone line. With this organization, it is possible to change the control law from the laboratory, without stopping checking and supervision programs. The set of operations is divided in specialized tasks which are managed by the real time monitor implemented on each microcomputer. We detail below the main tasks performed by each microcomputer :

a) Microcomputer A tasks (Fig. 5)

1) Temperature acquisition : the temperature profile in the furnace is calculated by linear interpolation of 12 measurement points. This task has a 1 second period and does only one measurement each period. Each measurement point is sampled every 16 seconds.

2) Flow acquisition : each group of burners is characterized by the gas flow, the air flow and the motorized valve position. One multiplexer card, using three digital outputs and three analog inputs provides the acquisition for every variable. This task has 4 seconds period and each variable is sampled every 16 seconds.

3) Charge acquisition : in order to know the charge inside the furnace, each carriage has an optical transponder. The acquired signal from the transponder gives the number of the carriage. The carriages are also weighted at the entrance of the furnace.

4) Motor valves control : a set point for the primary air flowrate is given at each control computation and for each group of burners using the last received measurements. A local loop controls the motorized valves and a flow transducer measures the air flowrate. A simple bang-off-bang controller was used to implement this local control loop. The gas flowrate is regulated in an analog way with respect to the primary air flowrate at each group of burners. As the time constant of the flow transducer is relatively important (10 seconds), a predictor is implemented, in order to know the air flowrate after each modification.
A proportionnal integral loop is also implemented. This loop is used in the case when the telephone line breaks down. (It is the digital reconstitution of the analog control loop existing before this study).

5) Synoptical display (Fig. 6): A special synoptical display card controlled by 12 digital lines (PIA) enables the display of the temperature profile inside the furnace, and the gas flowrate, using luminous and coloured points. Moreover, this device shows the charge distribution inside the furnace and the time variable.
A set of coloured points shows the difference between measured temperature and the optimal profile. The synoptical module enables us to supervise the whole furnace operation. (This module is installed in the furnace control room).

b) Microcomputer B tasks

We display in Fig. 7 the microcomputer B area, its main tasks are :

1) Temperature profile display : the temperature profile is obtained by linear interpolation of 12 measurement points and displayed in a C.R.T.

2) Charge display : this task displays the number and the weights of carriages inside the furnace.

3) Transmission time display : this task displays the time (HH.MM) of the last transfer between the two microcomputers. This display helps us to check the operation of microcomputer A and the telephone line.

4) Valve position display : A "0 To 100%" bargraph displays the valve position on the C.R.T.

5) Gas and air flowrates : the gas and air flowrates are also numerically displayed on the C.R.T..

The period of each of these tasks is 10 seconds.

6) Telephone line management and control algorithm : A transmission baud rate 300 is used and the duration of this task is 32 seconds; this task can be divided in three phases :

- the microcomputer B suspends its tasks, put the microcomputer A in punch and put itself in load. The microcomputer A sends then its measurements without suspending its tasks, and the microcomputer receives them.

- with received measurements, microcomputer B computes the motor valves controls according to an off-line algorithm.

- the microcomputer B sends an instruction to the microcomputer A which suspends then its tasks and stays in load. The microcomputer B stays in punch and sends the calculated motor valves controls. At the end of this phase each microcomputer resumes its tasks.

The transmission is supervised and any transmission error is detected. The telephone line break-down is also detected. The period of this task is one minute.
The information exchange is done in ASCII format. This format (S format) corresponds to the memory contents with periodical verification of the check sum indication.

CONCLUSION

The development of microcomputer technology enables the control of complex systems, with efficient algorithms (short computation time and good accuracy).
A two-interconnected microcomputer structure has been applied to the control of a ceramic furnace. Such a structure allows an extension of these possibilities to the case where the experimental laboratory is far away from the industrial plant. Moreover centralization of supervision for several industrial processes is possible.

REFERENCES

Babary, J.P. and M. Meirelles Ribeiro (1981). Sur l'optimisation du fonctionnement d'un four d'industrie céramique. RAIRO Automatique, vol. 15, n° 4, pp. 303-317.

Barreteau, D., C. Laguerie, J.P. Babary, M. Kassinopoulos (1980). Modelling of a tunnel furnace in the ceramic industry. 3rd IFAC Conference on "System Approach for Development. Rabat.

El Hajjar, H. Thesis in preparation.

Kassinopoulos M. (1980). Modélisation et commande sous-optimale d'un four d'industrie céramique. Thèse de 3ème cycle, Universsité Paul Sabatier, Toulouse.

Krause, Plaul, Zollner (1973). Principes et Techniques de cuisson et de construction de fours céramiques. Ed. SEPTIMA, Paris.

Meirelles Ribeiro M. (1981) Régulation d'un four d'industrie céramique autour d'un régime optimal de production. Thèse de Docteur-Ingénieur, Université Paul Sabatier, Toulouse.

T.D.S. User's Guide - MOTOROLA Semiconductor Products Inc.

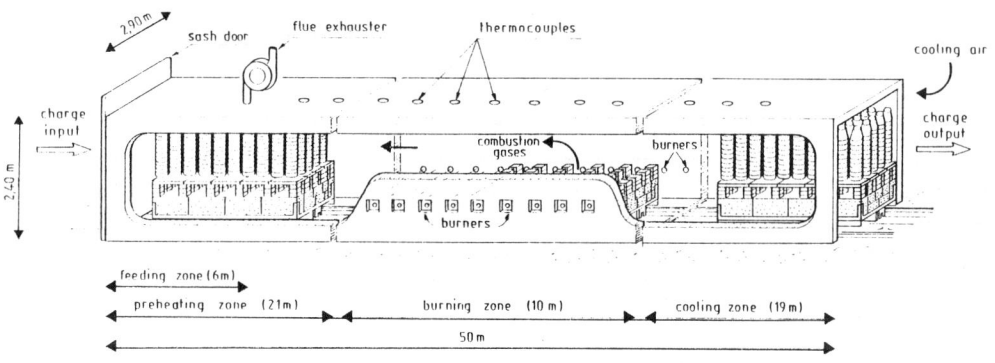

Figure 1. Scheme of the furnace

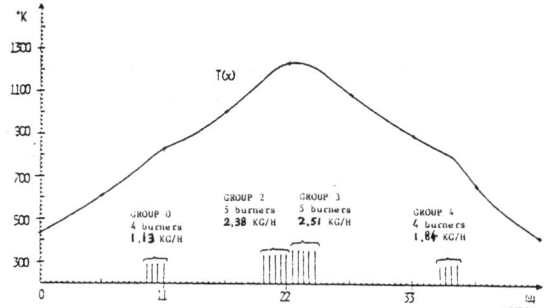

Fig. 2. Temperature profile and burners distribution obtained by the static optimization.

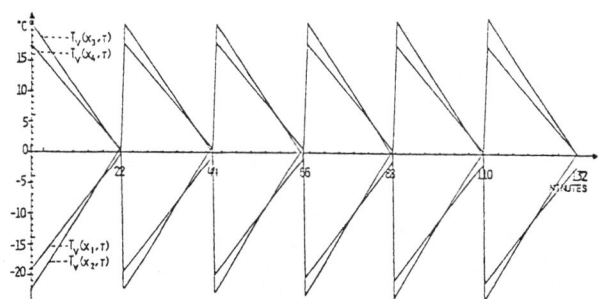

Fig. 3. Simulation using state observer

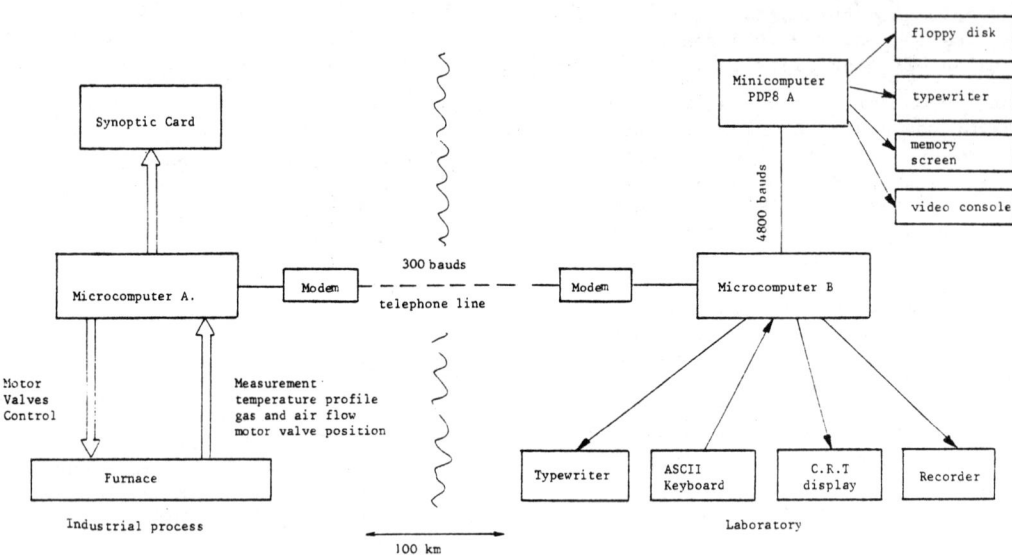

Fig. 4. Two microcomputer structure

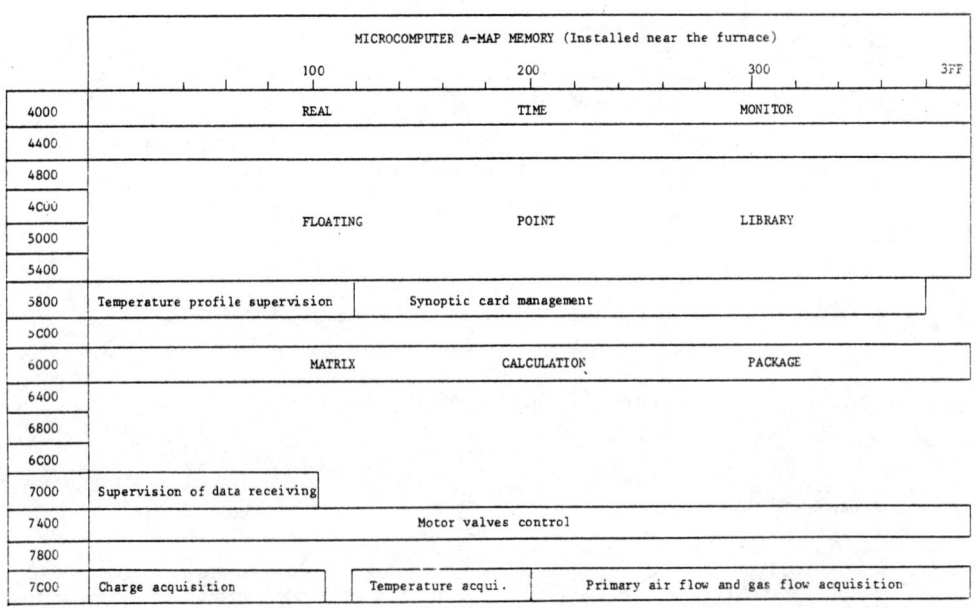

Figure 5.

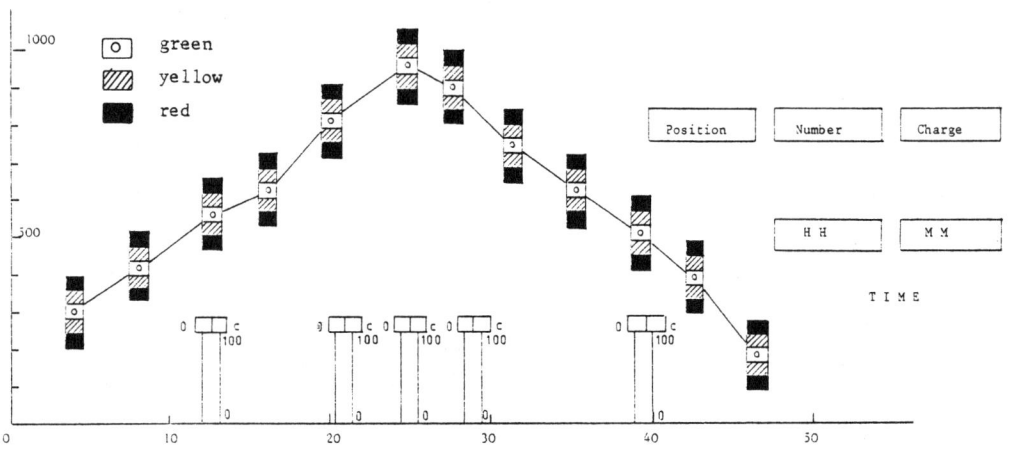

Figure 6.

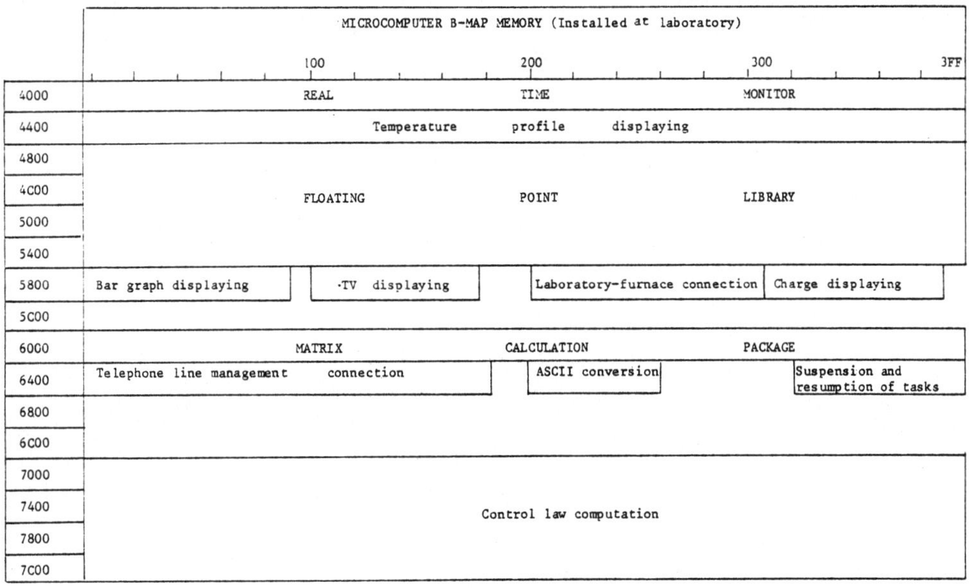

	MICROCOMPUTER B-MAP MEMORY (Installed at laboratory)			
	100	200	300	3FF
4000	REAL	TIME	MONITOR	
4400	Temperature profile displaying			
4800				
4C00	FLOATING	POINT	LIBRARY	
5000				
5400				
5800	Bar graph displaying	TV displaying	Laboratory-furnace connection	Charge displaying
5C00				
6000	MATRIX	CALCULATION	PACKAGE	
6400	Telephone line management connection	ASCII conversion	Suspension and resumption of tasks	
6800				
6C00				
7000	Control law computation			
7400				
7800				
7C00				

Figure 7.

THE MATHEMATICAL MODEL OF COMPUTER CONTROL OF THE FURNACE TEMPERATURE OF MECHANICAL ENDURANCE TESTING MACHINE

Jia-Sheng Wang and Yu-Fan Zheng

Department of Mathematics, The Teachers' College of Shanghai, China

Abstract.- An example of utilizing the modern control theory in discussed in this paper. The controlled object is an electrical furnace with two independent heating windings. Two thermocouples are used to measure the furnace temperature at two different points. The system demands that the temperature of upper and lower parts of the furnace must meet the technical requeriments. The controlled object is treated as a LQG system with two variables. The mathematical model includes the dynamic mathematical model, Kalman filter, optimal state feedback and the self-adapting following of set value of input.

INTRODUCTION

The mechanical endurance testing machine is used for performing the mechanical tests of different kinds of steels. During testing the steel specimen is put into the heating furnace. Under certain high temperature (e.g. 800°C), a tension force (e.g. 500 kg) is applied at the two ends of the specimen for certain time or until it ruptures.

Fig. 1 is the schematic diagram fo the interior part of the heating furnace of mechanical endurance testing machine. There are two sets of electrical resistance wire — (μ_1, μ_2) in the upper and lower parts of the furnace for heating. At the two ends of the specimen, two thermo-couples are used to measure temperature (Z_1, Z_2). This system demands taht the input voltage must be regulated according to the temperature measured, so that the difference between the output temperature (upper and lower parts) and set value must be within ±4°C. The difference of temperature between the upper and lower parts must be also within ±4°C.

In order to obtain the optimal control of the furnace temperature and realize the central control of more than 100 furnaces, we adopt computer system, so that we can get the real-time digital optimal control. The study of mathematical model includes the — identification of dynamic mathematical model, the design of Kalman filter, the design of optimal state feedback and the design of self adapting following of set value.

IDENTIFICATION

From above system, the output temperature of upper and lower ends (Z_1, Z_2) are affected by the input voltage of two ends (u_1, u_2) so it is a "two imput - two output" system. According to experimental data the object is treated as a LQG system of "two input - two output" of time invariant. This system is expressed by the following difference equation of two order:

$$Z(k) - A_1 Z(k-1) + A_2 Z(k-1) = B_1 u(k-1) + B_2 u(k-2) + \varepsilon(k) \quad (1)$$

where

$$Z(k) = [Z_1(k), Z_2(k)]^T$$

is output vector,

$$u(k) = [u_1(k), u_2(k)]^T$$

is input vector.

where the value of input and output are referred to difference between the actual measuring value and set value. $\varepsilon(k) = [\varepsilon_1(k), \varepsilon_2(k)]^T$ is Gaussian white noise sequence of zero expected value. A1, A2, B1, and B2 are matrices of parameters.

In order to estimate the parameter matrices in equation (1), we first obtain an impulse response matrix by experiment.

The relation between the input and output of the system is expressed in Fig 2. The impulse response matrix of system is as follows:

$$H(t) = \begin{bmatrix} h_{11}(t) & h_{12}(t) \\ h_{21}(t) & h_{22}(t) \end{bmatrix} \quad (2)$$

The Wiener-Hopf equation of "two input – two output" system is as follows:

$$R_{u_1 z_1}(\tau) = \int_0^{T_m} h_{11}(\theta) R_{u_1 u_1}(\tau-\theta)d\theta + \int_0^{T_m} h_{21}(\theta) R_{u_1 u_2}(\tau-\theta)d\theta \quad (3)$$

$$R_{u_1 z_2}(\tau) = \int_0^{T_m} h_{12}(\theta) R_{u_1 u_1}(\tau-\theta)d\theta + \int_0^{T_m} h_{22}(\theta) R_{u_1 u_2}(\tau-\theta)d\theta \quad (4)$$

$$R_{u_2 z_1}(\tau) = \int_0^{T_m} h_{21}(\theta) R_{u_2 u_2}(\tau-\theta)d\theta + \int_0^{T_m} h_{11}(\theta) R_{u_2 u_1}(\tau-\theta)d\theta \quad (5)$$

$$R_{u_2 z_2}(\tau) = \int_0^{T_m} h_{22}(\theta) R_{u_2 u_2}(\tau-\theta)d\theta + \int_0^{T_m} h_{12}(\theta) R_{u_2 u_1}(\tau-\theta)d\theta \quad (6)$$

Where T_m is the time of a control period, T_s is the regulation time of system. $T_m > T_s$, $R_{u_2 z_j}(\tau)$ is the mutual correlation function of U_i and Z_j. $R_{u_i u_j}(\tau)$ is the self correlation function of U_i and U_j, $i,j = 1,2$. If the input vector $U(t)$ is composed of two independent white noise, $R_{u_1 u_1}(\tau)$ and $R_{u_2 u_2}(\tau)$ will be impulse function and $R_{u_i u_j}(\tau) = 0$, $i \neq j$.

Thus (3) — (6) may be written as:

$$R_{u_\ell z_m}(\tau) = k\, h_{\ell m}(\tau) \quad \ell, m = 1, 2. \quad (7)$$

Therefore, when we have calculated the mutual correlation function of input and output, we may obtain the impulse response matrix (2).

We use the binary pseudorandom sequence as signal source to substitute white noise approximately. According to the system characteristic we adopt that the voltage amplitude variation of pseudorandom code is 5V, time width $\Delta t = 5$ minutes. If we take pseudorandom signal of 63 codes as one period, then the length of one period is beyond the regulation time of the system (4 hours).

We apply "individual input method" and "simulataneous input method" to carrying out tests respectively; the same result is obtained approximately.

(1) Individual Input Method. First stabilize the input end U2 of lower part at "zero position" (set power input value), we input a pseudorandom signal of 63 codes to the input end U1 of upper part to substitute the white noise approximately (adding to stabilizing input). Then we take sample every five minutes and take correlation analysis in on-line. Begining at the second period, the calculated mutual correlation function is as follows:

$$R_{u_1 z_m}(\tau) = \frac{1}{63} \sum_{i=1}^{63} U_1(i-\tau) z_m(i) \quad m = 1, 2$$

Then let $U_1(t) = 0$, we input the same pseudorandom signal to the input end U2, similialy we calculate

$$R_{u_2 z_m}(\tau) = \frac{1}{63} \sum_{i=1}^{63} U_2(i-\tau) z_m(i) \quad m = 1, 2$$

According (7), we can obtain impulse matrix $H(t)$ from $R_{u_\ell z_m}(\tau)$.

Because of the difference between the self correlation function of the pseudorandom signal and the ideal impulse function, it is required to take following calculation:

$$h_{\ell m}(\tau) = \left(R_{u_\ell z_m}(\tau) - \alpha_{\ell m} \right) \cdot \frac{N}{N+1} \cdot M$$

Where $\alpha_{\ell m}$ is the stabilizing value of $R_{u_\ell z_m}(\tau)$. The theoretical calculation of $\alpha_{\ell m}$ is given in {1}, where M is the proportional factor of unit.

(2) Simultaneous Input Method. We use two pseudorandom signal whose period are both 127 codes and whose mutual displacement is 63 codes to input to upper and lower ends simultaneously, and calculate the mutual correlation function $R_{u_\ell z_m}(\tau)$, similarly from (3)–(6). When $0 \leq \tau \leq 63$, we can obtain impulse response matrix.

We apply $U_1(k) = \delta(k)$; $U_2(k) = \delta(k-63)$ as the input data and apply
$z_1(k) = h_{11}(k) + h_{21}(k-63)$; $z_2(k) = h_{12}(k) + h_{22}(k-63)$
as output data (when $\tau < 0$ or $\tau > 63$, $h_{\ell m}(\tau) = 0$). By using the least square method, the 16 parameters in parameter matrices of model (1) are estimated. Then after derivation from model (1) we obtain "state equation" and "observation equation"

$$\underline{X}(k+1) = \Phi\, \underline{X}(k) + \Gamma[\overline{U}(k) + W(k)] \quad (8)$$

$$Z(k) = H\, \underline{X}(k) + V(k) \quad (9)$$

where $\underline{X}(k) = \begin{pmatrix} Z(k) - V(k) \\ Z(k+1) - V(k+1) - B_1 \overline{U}(k) - B_1 W(k) \end{pmatrix}$

$\Phi = \begin{bmatrix} 0 & I_2 \\ -A_2 & -A_1 \end{bmatrix}$; $\Gamma = \begin{bmatrix} B_1 \\ B_2 - A_1 B_1 \end{bmatrix}$; $H = (I_2 \quad 0)$

where $W(K)$ is dynamic noise of the system, $V(K)$ is the measuring noise, $W(K)$ and $V(K)$ are independent Gaussian white noise.

THE KALMAN FILTER AND OPTIMAL STATE FEEDBACK

We use Kalman filter to estimate the state of system.

$$\hat{\underline{X}}(k) = \hat{\underline{X}}(k|k-1) + K\, \Delta Z(k) \quad (10)$$

where $\hat{\underline{X}}(k|k-1) = \Phi\, \hat{\underline{X}}(k-1) + \Gamma\, \overline{U}(k-1)$

$\Delta Z(k) = Z(k) - H\, \hat{\underline{X}}(k|k-1)$

The recursion formula of Kalman filter is as follows

$$K_k = P_k'\, H^\tau\, (H P' H^\tau + R)^{-1}$$

$$P'_k = \Phi P_{k-1} \Phi^\tau + \Gamma Q \Gamma^\tau$$

$$P_k = (I - K_k H) P'_k$$

K is the steady matrix obtained by iteration, where Q,R are respectively the dynamic noise covariance matrix and the measuring noise covariance matrix. In consideration of the value of Q and R is difficult to estimate, we calculate some different values of K by selecting different proportional values of Q,R of several sets. Thus these values of K may be used selectively in real-time control.

In order to select the control sequence U(K) so that the system can optimize the transfer from initial state to zero state, we select the rule that the time is shortest and the energy consumption is minimum to establish the performance index of quadratic form:

$$J = \sum_{k=1}^{N} \left(\underline{X}(k)^\tau Q \underline{X}(k) + \overline{U}(k-1)^\tau \overline{U}(k-1) \right)$$

where $Q = \begin{bmatrix} 2q & -q & 0 & 0 \\ -q & 2q & 0 & 0 \\ 0 & 0 & 0 & 0 \\ 0 & 0 & 0 & 0 \end{bmatrix}$

N is the times of sample in one period of controllling. We take larger value of q, so that it can still transfer to zero state in case of smaller value of Z(K).

According to the separation theorem, when W is white noise, the problem of the above optimal control is equivalent to optimal control of the deterministic system adding Kalman filter. Under the condition of the state equation

$$\hat{\underline{X}}(k+1) = \Phi \hat{\underline{X}}(k) + \Gamma U(k)$$

the calculation formula for U(K) which minimizes J is as follows:

$$\overline{U}_{N-K} = -\Lambda_{N-K+1} \Phi \hat{\underline{X}}_{N-K}$$

$$\Lambda_{N-K+1} = (I + \Gamma^\tau Q^o_{N-K+1} \Gamma)^{-1} \Gamma^\tau Q^o_{N-K+1}$$

$$Q^o_{N-K+1} = Q + \Phi^\tau \tilde{Q}_{N-K+2} \Phi$$

$$\tilde{Q}_{N-K+2} = Q^o_{N-K+2} - Q^o_{N-K+2} \Gamma \Lambda_{N-K+2}$$

where K=1,2,...,N, take $\tilde{Q}_n = 0$. This set of recursive formula is referring to the time in reverse direction. When N is large enough, $\Lambda_{N-K+1} \to \Lambda$. Constant matrix Λ is the coefficient matrix in optimal feedback

$$U^* = -\Lambda \Phi \hat{\underline{X}}(k) \qquad (11)$$

It is the "optimal feedback control".

The stable matrix K of Kalman filter and the coefficient matrix Λ of state feedback are all calculated in off-line.

THE SELF-ADAPTING FOLLOWING OF SET VALUE OF INPUT

Referring to each heating furnace, when the controlled temperature is stabilized at set value y_0, the input voltage is also stabilized at a certain value V_0, which is called a set value of input. In general the V_0 is differential for different furnace. At time K, the actual input voltage of each furnace is $V(k) = V_0(k) + U^*(k)$, where U*(K) is determined by (11).

Because of the variation of external environment and the system itself during the operation of the system, it demands the self-adaptive ability of mathematical model. Referring to the control of fixed value, we adopt the method of corrected set values of input, so that the system may have self-adaptive ability.

Assuming that the error of set value of input is $\Delta U(K)$ and the change is slow, it may be considered that the error is also $\Delta U(K)$ at time K-1, and K-2. The prediction output is

$$\hat{Z}(k) = -A_1 Z(k-1) - A_2 Z(k-2) + B_1 \overline{U}(k-1) + B_2 \overline{U}(k-2)$$

and the actual output is

$$Z(k) = -A_1 Z(k-1) - A_2 Z(k-2) + B_1 [\overline{U}(k-1) + \Delta \overline{U}(k)] + B_2 [\overline{U}(k-2) + \Delta \overline{U}(k)]$$

therefore

$$\Delta Z(k) = Z(k) - \hat{Z}(k) = (B_1 + B_2) \Delta \overline{U}(k)$$

i.e.

$$\Delta \overline{U}(k) = (B_1 + B_2)^{-1} \Delta Z(k)$$

Before applying this formula, $\Delta Z(K)$ is processed by filteration using "index smooth method"

$$\Delta \tilde{Z}(k) = (1-\alpha) \Delta \tilde{Z}(k-1) + \alpha \Delta Z(k)$$

(take $\alpha = 0.75$)

then $\Delta \tilde{U}(k) = (B_1 + B_2)^{-1} \Delta \tilde{Z}(k)$

Thus the set value is $V_0(k) = V_0(k-1) - \Delta \tilde{U}(k)$

the actual input value is

$$V(k) = -\Lambda \Phi \hat{\underline{X}}(k) + V_0(k-1) - \Delta \tilde{U}(k)$$

where

$$\hat{\underline{X}}(k) = \Phi \hat{\underline{X}}(k-1) + \Gamma [\overline{U}(k-1) - \Delta \tilde{U}(k)]$$

THE EFFECT OF APPLICATION

A china-made minor computer is adopted in this system. After realizing or real-time digital control, good results are achieved. It advanced the accuracy of temperature control (within ±2°C), shortened the testing-period, realized group control of more than one hundred furnaces, decreased the manpower and material expenses.

Fig 3 shows the comparison of temperature graphs between the real-time digital control by computer and the manual control.

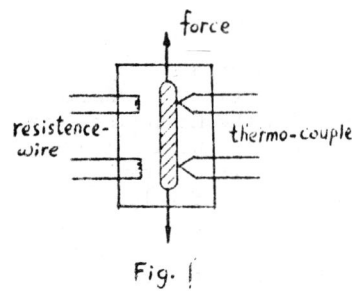

Fig. 1

---

* All experimental data in this article are supplied by engineers Tian-Chi Xia; Mr Ya-Ping Ma do a lot calculation work.

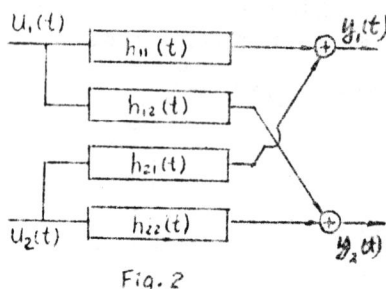

Fig. 2

REFERENCES

1. W.D.T.Davies "System Identification for Self-adaptive Control". John Wiley, - 1970.

2. J. Aström. "Introduction to Stochastic Control Theory". Academic Press. New York, San Francisco, London, 1970.

3. "Mathematical Method of Discrete Time Filter". By Probability Section of Mathematical Research Institute, China Science Institute, 1975.

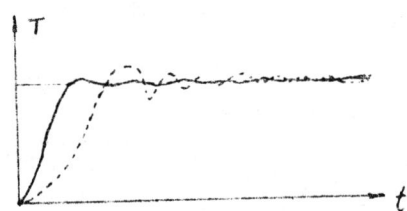

Fig. 3 Solid line indicates computer control.
Dot line indicates manual control.

DISCUSSION

SESSION MP2: FUEL AND HEAT CONTROL

Paper: DIGITAL DECOUPLING OF A 3-ZONE ELEC-
TRICAL FURNACE BY MEANS OF MULTI-
VARIABLE-PI CONTROL

Authors: *J Gómez de Silva* (Instituto de In-
genieria, UNAM, Apdo Postal 70-
472, 04510 México, DF, Mexico)

Discusser: *C.S. Berger*
19 Sheringhdam
Vic, 3150
Australia

Questions or Comments:

1) Wouldn't sub-optimal control have been easier?

2) Do you have methodical methods of chaging loci?

Author's reply:

1) I don't believe that in our case sub - optimal control would have been easier. Our search of constant precompensators requires the computation of a simple least squares algorithm and the solu - tion of a standard characteristic vec - tor problem.

2) Only when the matrices $K(z)$ and $G(z)$ conmute. In this case the characteris— tic functions $q_i(z)$ $i=1,m$ are simply

$$q_i(z) = k_i(z)\, g_i(z), \quad \forall\, i=1,m$$

Discusser: *C Jeffreson*
Chemical ENA'A Dept
University of Nevada Reno
Reno
Nevada 89557
USA

Questions or Comments:

1) Why should it be necessary to avoid in- teraction between zones in this case? or, do you really wish to vary the set point of each zone independently of the others? If so, what is the metallurgical reason for this emphasis on independence?

2) How did system performance vary with load on the furnace?

Author's reply:

1) The avoidance of interaction effects disminishes perturbation propagation between zones (for instance when you open the door of the furnace)

2) System performance (from the decoupling point of view) was greatly decreased when changing load conditions

Discusser: *Cristina Verde*
Universität Duisburg
Kommandantn Straße 60
4100 Duisburg 1
West Germany

Questions or Comments:

1) It is possible to find very good results using linear optimal regulators. There are many papers on how to choose the Q and R matrices.

Author's reply:

1) It is true! In general the state space approach leads to more complex control- ler structures.

Discusser: *Alberto Aguado*
Cuban Academy of Sciences
Habana-Cuba

Questions or Comments:

1) Which advantages does your frequency domain approach have in comparison with the state-space approach?

Author's reply:

1) Mainly the possibility to extend well known frequency response techniques (Nyquist, Bode,...) to the multivariable case. On the other hand the Multivariable Nyquist Array approach, in conjunction with pseudo diagonalization techniques, leads to a simple multivariable - PI controller.

REAL TIME DIGITAL CONTROL SYSTEMS FOR THE CEMENT INDUSTRY

T. Ohta* and K. Ishida**

*Fuji Facom Corporation, 1 Fuji-Machi, Hino-Shi, Tokyo 191, Japan
**Fuji Electric Co. Ltd., 1 Fuji-Machi, Hino-Shi, Tokyo 191, Japan

Abstract. Digital control systems has been applied to the Cement Industry and several kinds of new control technique has been creating successful results. There is wide variety of systems from simple single-loop controller to hierarchical complex system. In many cases, real time computer is connected with the off-line large scale computer, and process data and instructions for the operation are communicated. Therefore, man-machine communication system has the important role in the real time system. Actual examples of Kiln control, raw material blending control, electric power demand control, and mill feed control are described. For those, the following new control methods were applied.
  Auto-regressive model
  Physical models with long integration time constants
  Predictive model by the least mean square error
  Auto-tuning of PID control parameters
The results of the new control scheme are shown, as well as the system configuration in figures.

Keywords. Cement industry; digital control; kiln control; blending control; power demand control; auto-regressive model; auto-tuning.

INTRODUCTION

The cost for energy is very large in the product cost of cement industry and has grown up to 50%, according to a Japanese statistics. Therefore, energy saving is expected. On the other hand, the trend to coal has been promoted in order to make the cement produce settled.

For energy saving, fine control is needed. For the treatment of coal, difficulty of control increases. Under these circumstances, real time digital control systems have been developed and the new control methods have been researched in order to keep the stable automatic operation for a long time.

Among those new methods, this paper describes auto-regreassive kiln control as a probabilistic method, raw material blending control as deterministic method, electric power demand control as predictive method, and auto-tuning as control parameter optimization method.

Auto-regressive model realizes automatic kiln control which make the deviation from the reference smaller than manual control. Raw material blending control decreases the variation of powder composition remarkably. By power demand control, total plant can be operated in the contract limit and the limit can be decreased occasionally. Auto-tuning gives the optimal PID control parameters without apriori control information.

Man-machine communication by using CRT is introduced, because it is very important for the actual operation and many kinds of patterns have been applied. Because of the convenience for operation and the savings of space, graphic panels have been replaced to CRT display systems of digital control.

A new scheme of mill control is designed by using sound level for avoiding the overcharge.

The results of digital control are compared with correspondings of conventional control, which mainly relies upon human operation.

KILN CONTROL

Kiln is the most important component in the cement plant. However, the characteristics of a kiln is difficult to represent by the physical model based on the physical or chemical laws. Therefore, auto-regressive model (abbreviated as AR-model) is applied.

In order to obtain the practical AR-model, suitable period of the objective process is selected among the actual operation, transient behaviors of the process variables and the manipulated variables are sampled and recorded, the model of the process, i.e. variables and order of the model, is decided, and the coefficients of the model are calculated by using a off-line large scale computer. Finally, the model is implemented in the on-line digital control system, together with the interface elements between the model and the conventional control units.

Description of Kiln Control

Figure 1 is the diagram of kiln control. As this figure shows, the control system contains many variables and the variables have inter-relation to each other complicatedly. Therefore, automatic control of kiln is very difficult and it is said that only skillful operator can control it by manual empirical operation standard.

For the design of kiln control system, physical models had been studied. However, their parameters cannot be estimated on the design stage and cannot be fixed even after the plant operation starts. Accordingly, we applied the regressive model to this process.

Kiln control is consisted of the following two kinds of control mode.
$$\begin{cases} \text{Stationary control} \\ \text{Unstationary control} \end{cases}$$
In the large part of kiln operation, kiln is controlled stably interior of the proper range and the process variables vary a little according to the small disturbance of clinker product amount change or quality change of raw materials. This range of operation corresponds to stationary control.

On the otherhand, kiln sometimes receives large disturbances and the operation condition varies to a big extent. In such a case, unstationary control works and let the condition return into the range of stationary control by the special method for the individual process or the setpoint change in the simple situation.

For the stationary control mode, we apply the optimal control by AR-model.

AR-Model of Kiln control

In order to construct the AR-model, the objective kiln should be operated for some period so that the process variables vary large enough to get the model parameters.

For this purpose, the manipulated variables are changed intentionally in the random form of Maximum Period Sequence. Their amplitude is selected as small as possible and the disturbance to the actual operation is estimated previously by the computer simulation with the model of similar kiln.

The discrete AR-model for output variables is expressed as

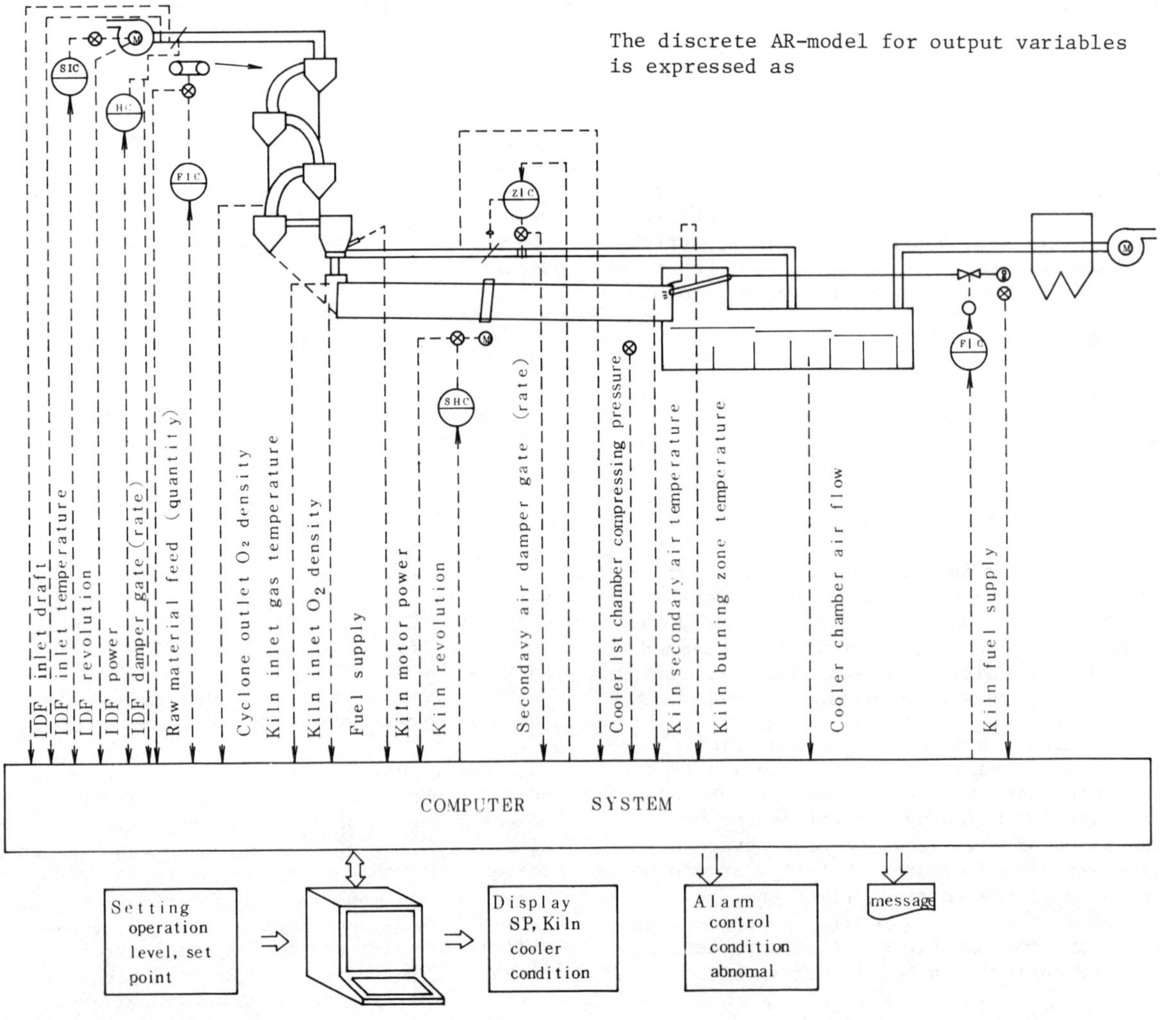

Fig.1 Kiln multi-variable control

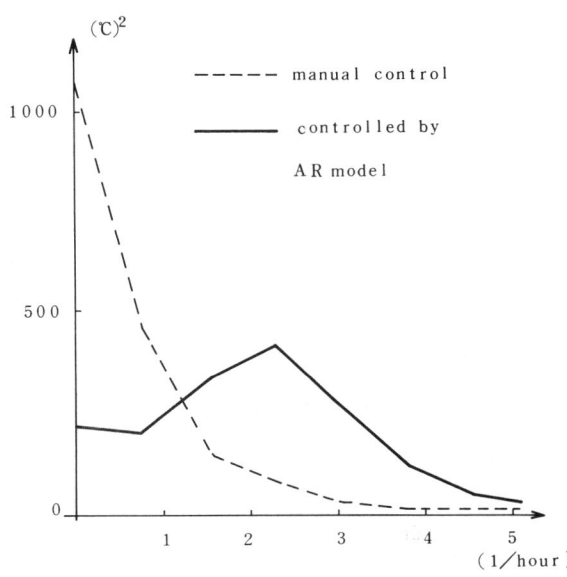

Fig. 2 Power spectrum of controlled secondary air temperature

$$X(k) = \sum_{m=1}^{M} \{A(m)X(k-m) + B(m)U(k-m)\} + W(k) \quad (1)$$

where $X = \text{col}(x_1, x_2, x_3)$, $U = \text{col}(u_1, u_2, u_3)$
$W = \text{col}(w_1, w_2, w_3)$

x_1, w_1; kiln inlet raw material temperature
x_2, w_2; kiln secondary air temperature
x_3, w_3; kiln moter power
u_1; raw material feed
u_2; kiln revolution
u_3; kiln fuel supply
M ; order of AR-model
A,B; coefficient matrices of the model

The order of the model is determined by the Akaike's method, which defines the most appropriate order M as minimizes the following value of MFPE(M). (Akaike, 1974)

$$\text{MFPE}(M) = \left(\frac{N + M \cdot K + 1}{N - M \cdot K + 1}\right)^K \cdot \det(d_M) \quad (2)$$

where N ; number of discrete data used
K ; dimension of X
d_M; covariance matrix of W

Coefficient matrices A and B are obtained by Yule-Walker's equation constructed with the recorded process data. The controller output is calculated by

$$u(k) = -G \cdot \text{col}\{x(k), \ldots, x(k-M+1), u(k-1), \ldots, u(k-M+1)\} \quad (3)$$

The optimal feedback gain G is obtained by using Dynamic Programming so as to minimize the performance index:

$$J_I = \sum_{k=1}^{I} \{x'(k) \cdot Q \cdot x(k) + u'(k) \cdot R \cdot u(k-1)\} \quad (4)$$

where I; interval for evaluation
Q,R; weighting matrices

Effect of AR-model of Kiln Control

In Fig. 2, the power spectrum of secondary air temperature controlled by AR-model is compared with that controlled manually through computer simulation. In low frequency domain, the effect of AR-model appears. In high frequency domain, AR-model varies more largely than the manual control which manipulate very little and let the kiln free.

The control by AR-model realizes automatic control, decreases the amplitude of process variable deviation from the expected value, and keeps the extreme operation away. As the result, the operational margin can be made small, two or three per cent of fuel is saved, and kiln can be operated stably not only by the special skillful man who have much experience and know-how.

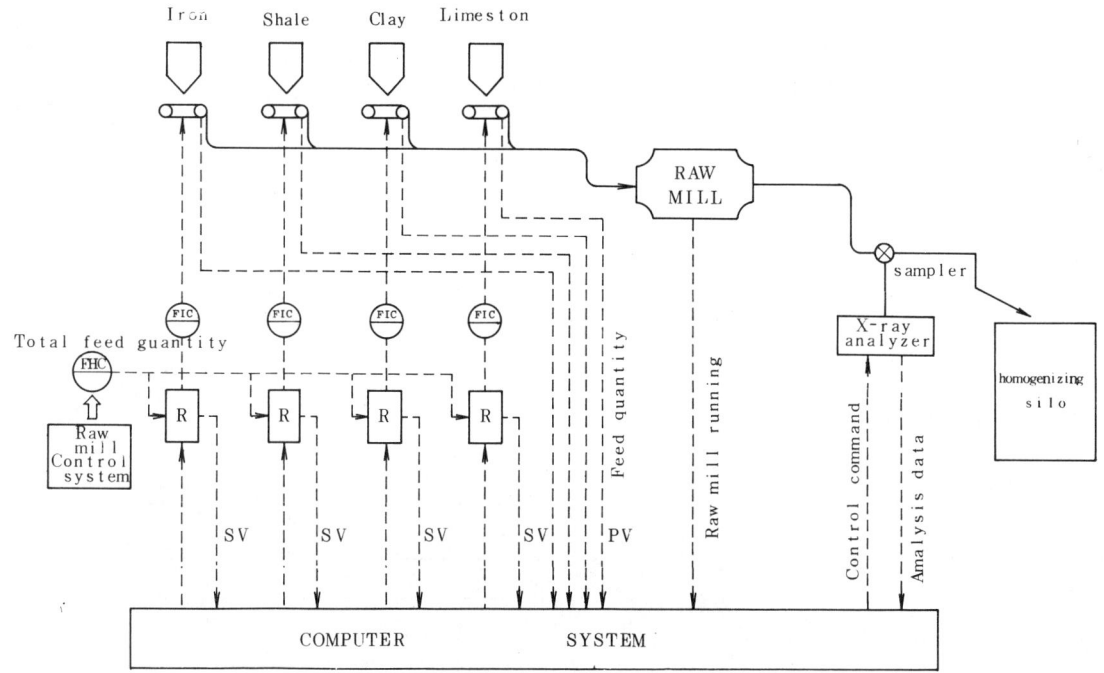

Fig.3 Raw material blending control

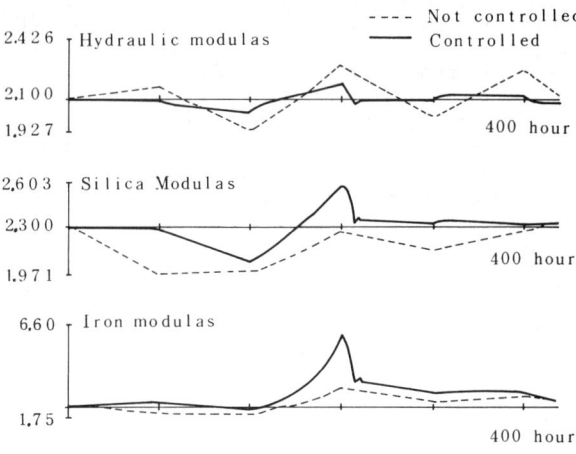

Fig. 4 Simulated modulas transients

RAW MATERIAL BLENDING CONTROL

Raw material blending control is the most effective of the computer controls in cement industry. The quality of blended raw material affects kiln process and product quality. Nonlinear calculation for modulus or composition and integration for long time range are implemented easily in digital computer but very difficult in conventional analog control system.

Description of Raw Material Blending Control

Figure 3 is the diagram of raw material blending control. Raw materials are fed according to the set values by the individual local feed controller and the total amount is controlled so as to agree with the quantity given from the raw mill control system by digital computer. Quality of the powder from raw mill is analysed by X-ray analyzer and controlled by changing the rates of raw materials to the total feed. This rates are calculated through digital computer.

Raw materials with different composition come from the place of production. They are mixed in raw mill for the first stage and blended in homogenizing silo again.

It is the purpose of the blending control to make the output of homogenizing silo agree with the reference quality. For this purpose, settings of raw materials feed are calculated in the computer under the consideration of mixing effect and dead time of mill and silo.

The important indeces of material composition are hydraulic modulus(HM), silica modulus(SM) and iron modulus which are defined as

$$HM = \frac{CaO}{SiO_2 + Al_2O_3 + Fe_2O_3} \quad (5)$$

$$SM = \frac{SiO_2}{Al_2O_3 + Fe_2O_3} \quad (6)$$

$$IM = \frac{Al_2O_3}{Fe_2O_3} \quad (7)$$

Effect of Raw Material Blending Control

The results of blending control by digital computer is compared with those of no control regarding to the three moduli. In this example, three of four kinds of raw materials are adjustable. Therefore, only two moduli, that is HM and SM, are controllable and IM is left in no control. However, the deviation of IM does not go out of its limit, because the larger change of IM is allowable than HM or SM.

As Fig. 4 shows, the deviation of raw material blending control is smaller than that of no control. The blending control uses real time signal of material composition by means of only one X-ray analyzer and uses statical constants of raw material composition proper to their root.

In this control, arithmetic calculation and memorization by digital computer create big effect. From the history of powder composition stored in the computer, the reference values for the feed controllers are calculated on the base of stored data for long interval.

The moduli do not coincide to the reference exactly, but the deviation becomes smaller than half of no control. Thus, the operation on the following stages is stabilized.

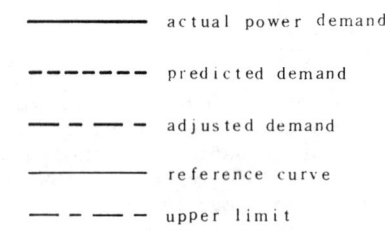

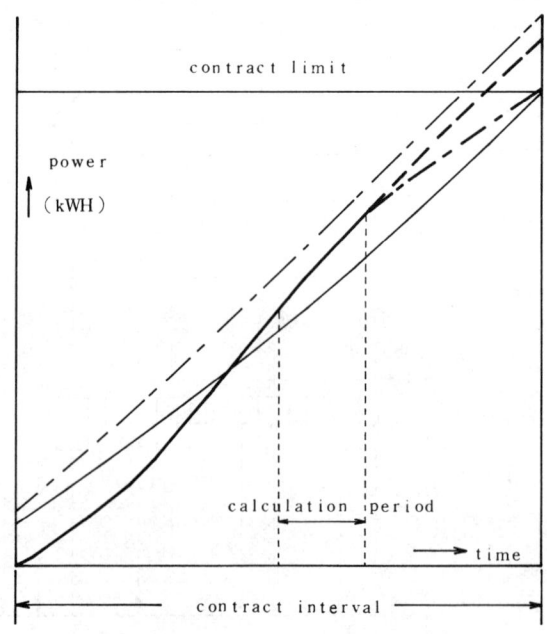

Fig.5 porwer demand control scheme

ELECTRIC POWER DEMAND CONTROL

The electric power should be consumed in the contract limit and the overcharge is required if the actual consume exceeds the limit. Therefore the power demand is controlled so as to be maintained in this limit.

The limit is contracted not for the instantaneous power but for the accumulated power between a certain interval. Accordingly, the demand control by digital computer is carried out by the following scheme corresponding to Fig. 5.

The contract interval (one hour for example) is devided into several periods (ten minutes) for calculation. In this period, the instantaneous power is sampled (at each one minute), the accumulated value is calculated, and the final value of the contract interval is predicted by least mean square error method. The accumulated power and the predicted power are compared with the reference curve or the contract limit, respectively.

In the case when the power demand prediction exceeds over the contract limit, the message is printed on typewriter, the buzzer sounds and the lamp is turned on, in order to inform operators. Operators communicate with computer by using CRT display which shows the curves like Fig. 5 and they decide whether some machines should be cut off or not.

In some application examples, the contract power could be decreased to 90 or 95 per cent of the power without demand control.

MILL CONTROL

Data processing for mill control is shown in Fig. 6. It is the purpose of the control to keep constant ratio of mill hold up for the stable operation. In new systems, grinding sound is used as the signal of the ratio of mill hold up and it can be used to adjust the reference value of feed.

Mill control by computer is executed in the normal operation and not in the special condition such as start up or shut down. Reference value of feed is decided through separator return or power of bucket elevator.

By the application of advanced control, mill over charge is avoided and the stop operation decreases as the result.

CRT DISPLAY APPLICATIONS

Man-machine communication is very important in a large scale control system. Operation is stabilized by human adjustment which is guided by the concentrated information through computer processing of plant data.

In one example of large systems, following kinds of display were applied with the indicated numbers of the different patterns.
Display of machine operation condition.... 3
Display of transportation line condition..12
Supervising display of controlled state... 8
Display for setting control parameters ... 8
Supervising display of instrumentation ...32
Trend display of process operation 48
Display of conditions after alarm 2

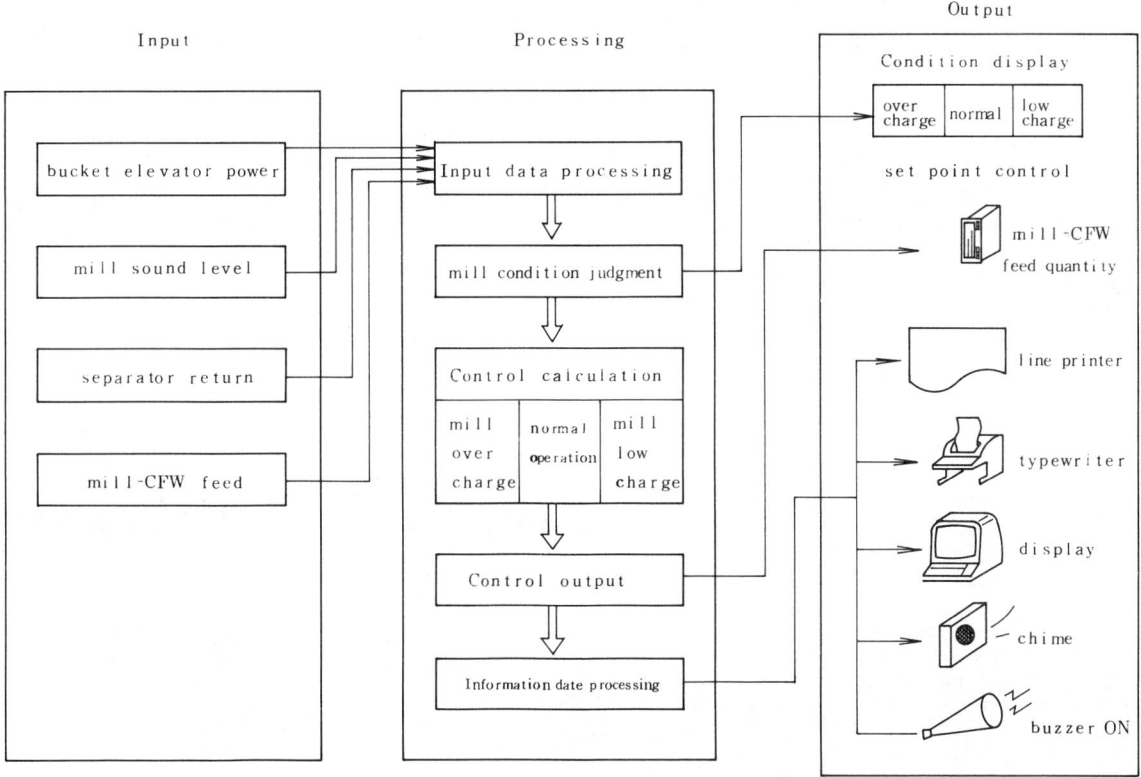

Fig.6 Mill feed control

AUTO-TUNING OF CLINKER FEED CONTROL

Auto-tuning method of PID control parameters is developed and implemented in several kinds of digital control systems such as multiple loop DDC unit, process cumputer system, and supervising unit of hierarchical DDC system. The procedure is cosisted of (1) pulsewise test signal output for process parameter estimation, (2)data sampling of the process response to the test signal, (3)determination of the process self-regulation quality, (4)calculation of characteristic areas of the process step response which are newly defined as the process parameter, and (5)calculation of optimal PID control parameters by using simple polynomials of the approximated relation between estimated process parmeters and optimal settings. (Nishikawa, Sannomiya, Ohta, and others,1981)

Auto-tuning means automatic estimation of process parameters and calculation of optimal PID settings. It is started by the operator's command, because process operation is disturbed a little. The obtained optimal settings are used for the actual control after the operator's confirmation, because process response may be affected by the large noise and estimation error may be not small.

In order to get optimal settings, We adopt a new form of performance index, i.e. the weighted ISE:

$$J(\beta) = \int_0^\infty \{\Delta x(t) \cdot e^{\beta t}\}^2 dt \qquad \beta > 0 \qquad (8)$$

β is given from the period of ultimate oscillation of the loop. This weighted ISE is easily calculated and this optimization algorithm can be applied to wide variation of the transfer functions, even to the process with long dead time.

A clinker feed process as shown in Fig. 7(a) is one of self regulating processes. To this process, auto-tuning was applied in the setting up period of the control system. The test signal and the process response recorded are shown in Fig. 7(b). From the two auto-tuning tests of this figure, almost same PID parameters are obtained without no apriori information about the process dynamic characteristics. One of them is set and the process was controlled well as shown in Fig. 7(c).

CONCLUSION

New digital control systems are explained. Auto-regressive model is described in detail, but others are introduced briefly, because the space is limited.

For the application of a new method, hardware and software of the system must be rich enough. At the same time, the characteristics of the process should be understood.

Process model and computer simulation are useful not only for the analysis or synthesis of control systems, but also for the test or estimation of new digital control schemes.

REFERENCES

Akaike, H.(1974) A New-Look at the statistical Model Identification. <u>IEEE Trans, AC-19-16</u>, 716-723.

Nishikawa, Y., N. Sannomiya, T. Ohta, H. Tanaka, K. Tanaka (1981) A Method for Auto-tuning of PID-control Parameters. <u>Pref. of 8th IFAC Congress Session 33, VII</u>, 65-70.

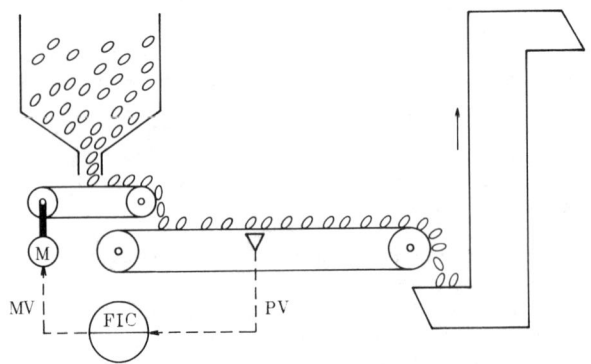

(a) clinker feed control loop

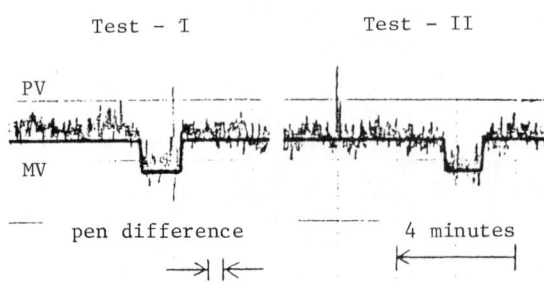

(b) Test signal and process response

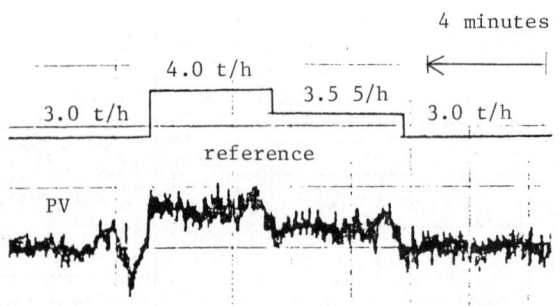

(c) Result of the DDC by the settings of the auto-tuning, where P_b=150% T_i=30 sec., T_d=7 sec., T_ℓ=10 sec.

Fig. 7 Auto-tuning of feed control loop

ADAPTIVE CONTROL OF A BALL MILL WITH SELF-TUNING REFERENCE MODEL

R. Schulz

Department of Electrical Engineering, Fachhochschule Bielefeld, Bielefeld, Federal Republic of Germany

Abstract. The transfer behavior of a ball mill is both very slow and characterized by time variant behavior and considerable nonlinearities. For on-line process identification a self-tuning reference model without additional test signal is used. The parameters of the real time grinding power cascade controller are optimized in a fictitious control loop using the identified reference model. The control speed in the fictitious control loop is the multiple of the control speed in real process.
The time constants and the nonlinearities of the ball mill 17o t/h were determined in plant measurements. An 8 bit microcomputer was used as digital controller.

Keywords. Adaptive control; cascade control; cement industry; grinding; nonlinear systems.

INTRODUCTION

In the implementation of the control algorithm of a ball mill, the importance of the operating conditions in the respective cement plant is often neglected. Whenever the control principles are not accepted by the operating personnel, they are likely to be changed again as soon as the commissioning engineer has left. There is also the question of expenses, since any new installation of a digital controller has to be supervised by a scientist due to many restrictions not obvious to the operating engineer. One must also take into consideration that control concepts with additional test signals must effect modifications of the manipulated variable also at steady-state process behavior. Thus expensive weight-belt feeder equipments are constantly strained and worn out. [7]

This paper reports on an adaptive control capable of acting without external signals. For that reason it is accepted so far that the finished product flow cannot be maximized continually day and night under constant process behavior.
On principle the advantages of the auxiliary controlled variable of the degree-of-fill of a ball mill are used in the cascade control of the throughput control. This can be effected independently of the particular control algorithm of the master controller.

The objective - optimal adaptation of a closed circuit without additional test signals to the actual process behavior - can be obtained by continuous minimization of the control error e(t) Intentional changes of controller parameters lead to a minimum of a quality criterion - mostly the integral of the control error. Such integration takes place within the sampling interval. Therefore the results of the parameter changes of the controller, the process and the disturbance values are recognized also in a sample time. The optimization of controller and process parameters by the gradient method is the base for adaptive control algorithms described in the following.

PROCESS IDENTIFICATION

The identification of dynamic behavior is based on an a priori knowledge of th nonlinear process. The necessary approx mation for that purpose takes into consideration the nonlinear relations between tailings flow/degree-of-fill and finished product flow/degree-of-fill according to Keviczky [6]. Correspondin to the signal flow chart (Fig. 1) it is added a reference model 3 to the proces 2. The input signal: $Y(s)$ - fresh feed material flow plus $X(s)$ -

tailings flow is at the same time the input signal of the model. The difference E(s) of the output signals L(s) - degree-of-fill and M(s) - reference model is the degree for the approximation of the transfer behavior of process and model. Fig. 1.

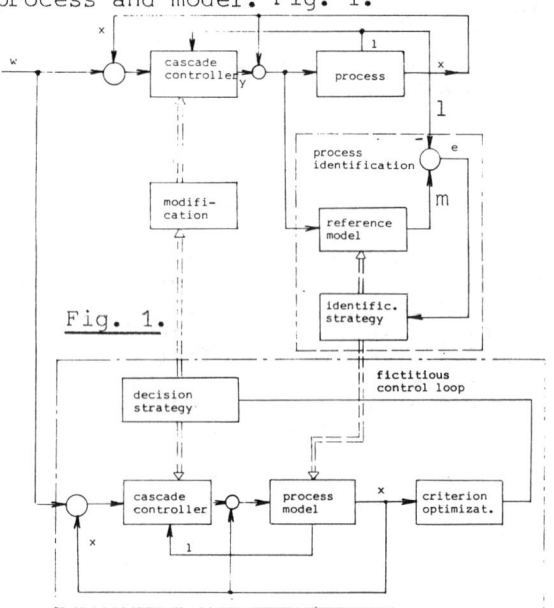

Fig. 1.

The model parameters a_i are as long varied as a minimum is obtained of the average value of the even function.

$$e(t, a_i) = m(t, a_i) - l(t) \quad (1)$$

$$\overline{f[e(t, a_i)]} = \overline{e^2(t, a_i)} \rightarrow \text{Min} \quad (2)$$

The operating experience [4,8,10] was used here that above all the adaptation of the slave control loop is decisive during disturbances in grindability for the cascade control of a ball mill. T was considered as model parameter because the reference model contains the above mentioned nonlinearities of the process. This renders possible a realization without problems in the assigned small 8 bit digital controller.

$$E(s,T) = (Y(s) + X(s)) \cdot$$

$$\cdot \left(\frac{1}{(1+Ts)(FB+Ts)} - \frac{1}{(1+T_1s)(FB+T_1s)} \right) \quad (3)$$

$$\frac{\partial}{\partial T} M(s,T) = (Y(s) + X(s))$$

$$\left(\frac{-s}{(1+Ts)^2(FB+Ts)} - \frac{s}{(FB+Ts)^2(1+Ts)} \right) =$$

$$= -M(s,T) \left(\frac{s}{1+Ts} \right) + \frac{s}{FB+Ts} \right) \quad (4)$$

$$T(s) = \frac{2K}{s} \left[-E(s,T) \cdot M(s,T) \left(\frac{s}{1+Ts} + \frac{s}{FB+Ts} \right) \right] \quad (5)$$

The Eq.(5) describes the model parameter T for constant adaptation of the reference model. The first step of real time application contains a simulation of process and reference model at the analog computer. Fig. 2.

There was used a very simple process model but still being able to produce a usable approximation of real process behavior. The factor K is together with the error function $f[e(t,T)]$ determinative for the speed of the identification process. Unstabilities can be avoided by matching to maximal possible parameter variations, disturbances and set point changes. One disadvantage is the fact that the function value $e(t,T)$ is sometimes too small during little disturbances. A "falling asleep of the algorithm" would follow. Therefore a gradient normalization is necessary for the disturbance dependent correction of the factor K in the software realization. There is assigned a maximum value limit at the same time.

The process data applied were found out in operating measurements in a ball mill 170 t/h [10].

SELF-OPTIMIZING ADAPTIVE CONTROL

The process parameter T obtained through identification is led on the one hand to the process reference model and on the other hand to a second process model. A second complete cascade control loop is realized with this process model as shown in Fig. 1. It is a fictitious control loop existing only in the software of the digital controller DR. Limited through its smallest possible cycling time, the computing time of the digital controller is about 100 times smaller than the real time. The parameters of the fictitious cascade controller are as long varied in each sampling interval till a minimum of the ITAE quality criterion is found.

$$Q = \sum_{0}^{t_1} |X_d| \cdot \Delta t \stackrel{!}{=} \text{Min} \quad (6)$$

Searching strategy is the gradient method. [1]

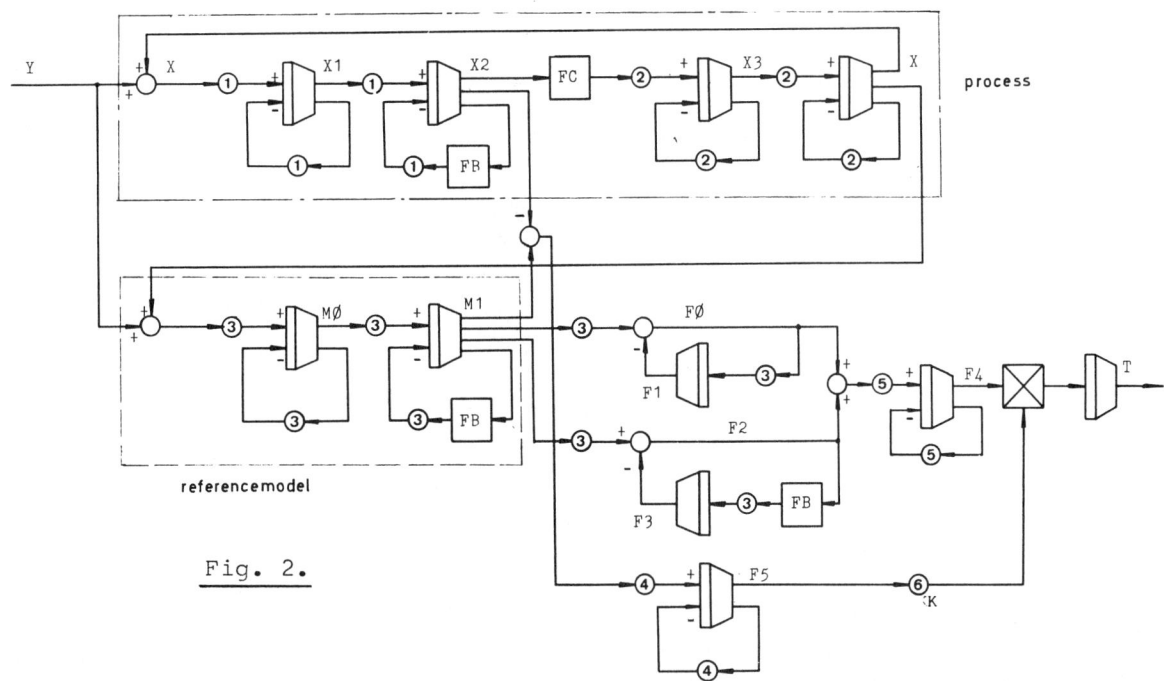

Fig. 2.

The following differential equation has been programmed for the slave controller (only P-controller):

$$KPH_{n+1} = KPH_n - B \cdot \frac{\Delta Q(KPH)}{\Delta KPH} \quad (7)$$

In a similar way the control parameters KI and KP from the master controller (PI) are cyclicly varied one after the other to the minimum of the quality function Q(KP, KI). The B-factor in Eq. ( ) is a variable value which is to calculate dependent on the particularly identified process parameter T. The calculated maximum value is reduced to a plausible limit value. Thereby it is achieved a quick and safe optimizing of the controller parameters for each identified process parameter T Eq. (5).

$$B = 200 \cdot \frac{1}{T} - 200 \quad \text{slave contr.}$$

$$BKI = 10^{-3} (\frac{1}{T})^3 \cdot 6,7 \quad \text{master contr.}$$

$$BKP = 10^{-3} (\frac{1}{T})^3 \quad (8)$$

The optimization of the controller parameter KPH is stopped when the changes between the last three KPH-values are smaller than 3%. These changes are calculated of the absolute value of the difference between the momentary and the preceding KPH-value. In connection with the slave controller optimization follows the master controller optimization. The last four quality values are always needed for the calculation of the KI- and KP-values. The optimization is finished when the same condition is fulfilled which has been described for the slave controller. Of course both parameters - KP and KI - must fulfill this condition now. In that case they are offered to the real - time controller (after appropriate calculation from computing time to real time). Afterwards the slave controller can again be optimized for the new identified process parameter T. The offered controller parameters are only accepted by the real-time controller when the amount of the difference to the just used controller parameter exceeds a selectable minimum value. Thus it is avoided to trigger control processes in the complete grinding circuit for no reason at all. The initial values of all controller parameters are chosen in a way that a strongly damped control process is guaranteed. The calculated maximum values of the controller parameters are reduced to plausible limit values. That goes also for the manipulated variable - as floating set point for the positioners of the weight-belt feed equipment in the fresh feed material flow.

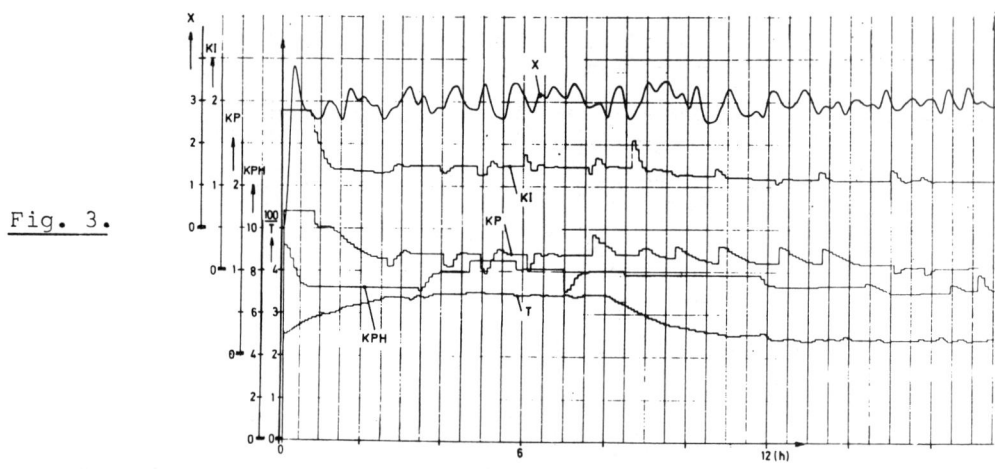

Fig. 3.

RESULTS

The control results of the cascade control are shown in Fig. 3. The adaptation of the slave controller parameter KPH was made at first. Start value KPH = 1o was too high and was diminished within 41 min to 6,6. Afterwards the master controller parameters were optimized. At first KI and then KP were varied by turns. The optimum value KI = 1,4 was identified after 1,33 h and the optimum value KP = 1,3 after 2,42 h. The corrections of both parameters which followed immediately, had no visible influence to the control process.
It is true that an identification of the process parameter T is finished not before 2,55 h but at this moment there are only slight parameter changes of the master controller. The optimizing of KPH is repeated not before 3,5 h of the grinding operation. Changes of T have caused that.

There do follow shorter optimizing cycles of all controller parameters, but the parameter variations are hardly accepted by the real-time controller because of insignificance.

A steady and quiet operation of the ball mill is also preserved if in the 8th operating hour a change of the grindability of the fresh feed material flow is showing in the process parameter T. The stability of the control loop is completely sufficient in the following 1o operating hours without remarkable controller parameter variations.

A control process is shown in Fig. 4. which was registered with the same start parameters of the real time controller like before but with a 25% worse (bigger) delay time constant. The new process behavior is in fact identified after 2 h (process parameter T = 2) but all con-

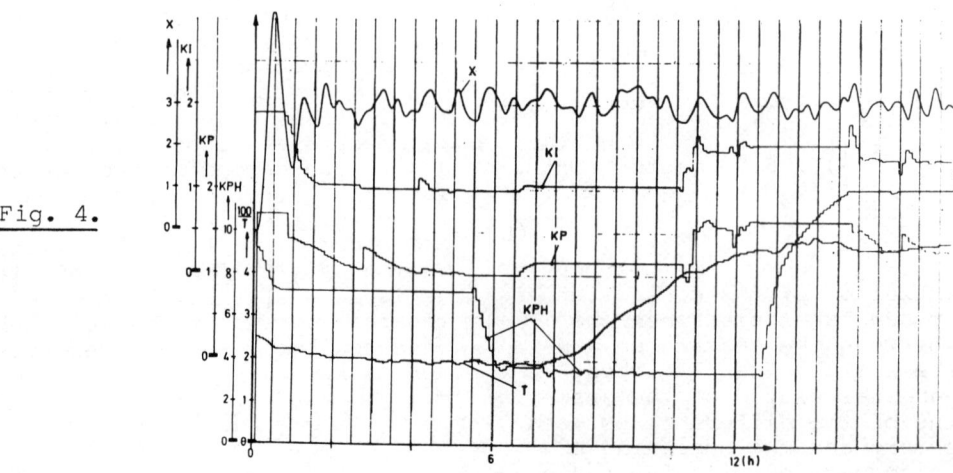

Fig. 4.

troller parameters are optimized not before 6,5 h. Variations in the grinding circuit led after 7 operating hours to a reduction of the process delay time constant of 4o% of the preceding value. The so changed process behavior is clearly recognized by the process parameter T. The identification is finished after 6 h, the oftimization 45 min later.

CONCLUSION

The on-line adaptation described here renders possibe an optimization of the controller parameters during the normal operation of a ball mill. Additional test signals are not necessary. The adaptation of the reference model to the process behavior does not effect an additional disturbance of the process. The same effect is found concerning the optimization of a fictitious control loop. Optimal controller parameters are identified within the control time of realtime control processes by means of time transformation in this control loop.

The aim: high operation security and good degree-of-fill of a ball mill without working with negative grinding progress only temporary, seems to be achieved here.

REFERENCES

1. Rake, H. (1967). Self-optimizing systems with the gradient method. Regelungstechnik, 5, 211-217

2. Bélanger, Kennedy (1972). Control of a closed circuit cement mill. Report of Dept. of Electrical Eng. McGill University.

3. Unbehauen, H. (1973). On-line identification methods. KfK-PDV 14, Karlsruhe GFR.

4. Schulze, H. (1973). Application of adaptive methods for the digital control of ball mills. German-French Colloquium: Industrial Application of adaptive Systems, University Freiburg, GFR

5. Schulz, R. (1973). Adaptive control of a raw meal grinding plant. German-French Colloquium: Indust. Application of Adaptive Systems, Freiburg, GFR. (University).

6. Keviczky, L., Udvardi, I, and Banyasz, C. (1976). Optimization of a cement mill by correlation technique. 8th AICA Congress Delft, Comp. 76, 737-745

7. Schulz, R. (1977). Possibilities for output control of ball mills. Cement Congress Düsseldorf, GFR.

8. Schulz, R. (1977). Controlling the power output of a ball mill. Zement-Kalk-Gips, 2.

9. Keviczky, Vajk, and Hetthésy, (1979). A self-tuning extremal controller for the generalized Hammerstein model. 5th IFAC Symp., TH Darmstadt, GFR.

10. Busija, R. (1981). Adaptive control of a non-linear process. Technical Report, Fachhochschule Bielefeld, GFR.

A MICROPROCESSOR-BASED ADAPTIVE COMPOSITION CONTROL SYSTEM

J. Hetthèssy*, I. Vajk*, R. Haber*, M. Hilger** and L. Keviczky***

*Department of Automation, Technical University of Budapest, Hungary
**Institute of Research and Planning in Silicate Industry, Hungary
***Computer and Automation Institute, Hungarian Academy of Sciences, Budapest, Hungary

Abstract. A general multivariable model of the mill-silo system is given. A general multivariable adaptive control system is described. Using a special linearizing and decoupling possibility single variable adaptive control loops are then introduced. Finally the applied µP system and practical results are presented.

Keywords. Self-tuning, composition control, cement plant.

INTRODUCTION

In the cement industry the computer control is mostly applied for the blending control. The appropriate control of the chemical composition of the ground mix of the raw materials considering its homogenization, before feeding it into the kiln, is a very essential problem in cement manufacture.

The simplified technological scheme of the composition control can be seen in Fig. 1. In the feeder tanks there are raw materials of different composition. The weigh feeders are controlled by computer. A conveyer belt feeds the rubble into the raw mill. Afterwards the ground mix of raw materials (rawmeal) gets into the silo. Before the silo the meal is sampled and the sample is analysed by an X-Ray Fluorescence Analyser (RFA) which provides the oxide compositions for the computer. The computer controls this automatic sampling, conveying, preparation and analysing process and computes the new set points (scale factors) for the weigh feeders using the so-called module values as references.

The plant is multivariable and a coupled one because the feeder tanks do not contain chemically homogeneous raw materials. If the components of the raw materials and their physical characteristics (e.g. grindability) change considerably, it is necessary to introduce such computer control which readjusts the regulator parameters.

A multivariable self-tuning adaptive regulator was proposed by Keviczky and coworkers (1978) to the control of cement raw material blending.

A single loop self-tuning regulator was applied by Westerlund und coworkers (1980).

APPLIED MODELS

The model of a mill-silo system is shown in Fig. 2, where $r(t)$ and $v(t)$ are the input feed and mill outlet (silo inlet), respectively. The vectors $\underline{ox}$ and $\underline{M}$ denote the oxides and moduli values, respectively. Furthermore $\underline{w}$ stands for the scale factors of the weighs, and $\underline{C}$ is the composition matrix. Linear noninteracting (diagonal) dynamics

$$\underline{G}_m(z^{-1}) = <\ldots, \frac{c_i}{1+d_i z^{-1}}, \ldots > z^{-d} \quad (1)$$

is assumed for the mill. This first order approximation with delay is proved to be good generally in the practice, because of the relatively large sampling interval (20-30 minutes) necessary to the RFA. (The faster dynamics of the mill are much more sophisticated, of course.) The input and output variables $\underline{r}'_{inp}$ and $\underline{r}'_{out}$ of this noninteracting multivariable linear part are material flows (lime, clay, pyrit) and the model is finally coupled for the oxides ($\underline{ox}_{inp}$ and $\underline{ox}_{out}$) because $\underline{C}$ is not diagonal.

The dynamics of a batch silo can be easily derived and is given by a time varying first order lag

$$\underline{\underline{G}}_s(z^{-1}) = \frac{b_o}{1+a_1 z^{-1}} \underline{\underline{I}} \qquad (2)$$

where $\underline{\underline{I}}$ is the unit matrix and

$$b_o = b_o(t) = \frac{v(t)}{q(t)} \qquad (3)$$

$$a_1 = a_1(t) = -\frac{q(t)-v(t)}{q(t)} \qquad (4)$$

Here t is the discrete time and

$$q(t) = \sum_{i=1}^{t} v(t) + q(0) \qquad (5)$$

is the instantaneous silo content. Fig. 3 shows the variation of the b_o and a_1 parameters with time. N is the assumed silo filling time, $\underline{\underline{G}}_s$ is sometimes called "silo-integration".

The blocks denoted by NL mean the non-linear computation of the moduli values, e.g.:

Lime standard:

$$ML = \frac{100 \, C}{2.85+1.1A+0.8 \, F} \qquad (6)$$

Aluminium modulus:

$$MA = \frac{A}{F} \qquad (7)$$

Silica modulus:

$$MS = \frac{S}{A+F} \qquad (8)$$

which are computed from the four most important oxides:

C - CaO
S - SiO_2
A - Al_2O_3
F - Fe_2O_3

in percantage.

CONTROL OBJECTIVES AND A GENERAL SOLUTION

The main goal of the blending control is to decrease the variance of the chemical composition of the raw meal rejecting the disturbances caused by the changes in the raw material compositions. Furthermore every full silo content is homogenized to reduce the composition variations around the average values applying batch-operaton silos. These double control objectives can be formulated using a general loss function

$$V = E\{\|\underline{y}(t+d)-\underline{y}_r\|^2 + \gamma\|\underline{y}_a(N)-\underline{y}_r\|^2 | t\} = \min \qquad (10)$$

where $\underline{y}(t+d)$, $\underline{y}_r$ and $\underline{y}_a(N)$ are the measured composition vector, the reference vector and the average composition vector at the end of a silo filling (t=N), respectively. Here N is the length of the silo filling and γ is a positive penalizing coefficient.

It can be proved that the minimization of V is equivalent to the minimization of

$$Q = \|\underline{y}(t+d|t)-\underline{y}_r\|^2 + \gamma\|\underline{y}_a(N|t)-\underline{y}_r\|^2 = \min \qquad (11)$$

where $\underline{y}(t+d|t)$ and $\underline{y}_a(N|t)$ are d and (N-t) steps ahead minimum variance estimates, respectively. The minimization of Q results in the following causal control law

$$\underline{y}(t+d|t) = \underline{y}_{rv}(t+d) = \underline{y}_r + K(\gamma,F,t) \cdot [\underline{y}_r - \underline{y}_a(t+d|t)] \qquad (12)$$

where

$$\underline{y}_a(t+d|t) = \frac{\underline{y}_a(t)q(t)+\sum_{i=t+1}^{t+d-1} v(i)\underline{y}_{rv}(i)}{q(t+d-1)} \qquad (13)$$

$$K(\gamma,F,t) = \frac{\gamma[\sum_{i=t+1}^{t+d-1} v(i)][F-q(t+d-1)]}{F^2+\gamma[F-q(t+d-1)]^2} \qquad (14)$$

and F is a full silo capacity. A good approximation of $\underline{y}_a$ can be obtained by the $\underline{\underline{G}}_s(z^{-1})$ "silo-integration". Using this simplification, Eq. (12) corresponds to a cascade closed-loop system, shown in Fig. 4, where a MIMO minimum variance (MV) regulator is in the inner loop and a time-varying proportional regulator, with gain $K(\gamma,F,t)$, is applied in the outer loop. The MV regulator can be implemented as a self-tuning (ST) regulator and the outer loop means a time-varying reference vector for it. This strategy introduced by Keviczky and coworkers (1978) is called MIMO-ST-MV-RAFT ensuring a required average for finite time.

The used

$$\underline{y}(t) = \underline{M}_{out}(t)$$
$$\underline{y}_a(t) = \underline{M}_{silo}(t) \qquad (15)$$

as output variables ensured to control the chemical compositions through the moduli values. The drawback of choosing these output variables is the strong nonlinearities introduced by the oxides/moduli computations.

A SPECIAL DECOUPLING POSSIBILITY

In this point a special decoupling possibility is introduced, whose applicability arised at the cement factory in Hejőcsaba (Hungary), this approach, however, is suggested for each other case, where the average chemical composition matrix has similar structure.

In our case $\underline{\underline{C}}$ has the following special structure with a good approximation:

$$\underline{\underline{C}} \cong \begin{vmatrix} C_1 & 0 & 0 \\ 0 & S_2 & 0 \\ 0 & A_2 & 0 \\ 0 & 0 & F_3 \end{vmatrix} \quad (16)$$

This structure ensures, that the most important oxides are

$$\tilde{C} \cong C_1 w_1; \quad \tilde{S} \cong S_2 w_2$$
$$\tilde{A} \cong A_2 w_2; \quad \tilde{F} \cong F_3 w_3 \quad (17)$$

which gives a very simple computation for the moduli values: Lime standard:

$$ML \cong \frac{100 \; C_1}{2.8 S_2 + 1.1 A_2} \left(\frac{w_1}{w_2}\right) \quad (18)$$

Aluminium modulus

$$MA \cong \frac{A_2}{F_3} \cdot \frac{w_2}{w_3} \rightarrow \frac{1}{MA} \cong \frac{F_3}{A_2} \left(\frac{w_3}{w_2}\right) \quad (19)$$

Silica modulus:

$$MS = \frac{S_2 w_2}{A_2 w_2 + F_3 w_3} \rightarrow \frac{1}{MS} \cong \frac{A_2}{S_2} + \frac{F_3}{S_2}\left(\frac{w_3}{w_2}\right) \quad (20)$$

The above formulae involve that introducing new input variables

$$u_1 = \frac{w_1}{w_2}$$
$$u_2 = \frac{w_3}{w_2} \quad (21)$$

the system is linearized and decoupled at the same time. (Do not forget that only two moduli values can be controlled simultaneously using these two variables!) It is reasonable to choose the reciprocal values 1/MA or 1/MS as output variables instead of the original ones. In our case the following quantities were chosen as output variables:

$$y_1(t) = \frac{ML(t) - ML_r}{ML_r} \quad (22)$$

$$y_2(t) = \frac{\frac{1}{MA(t)} - \frac{1}{MA_r}}{\frac{1}{MA_r}} \quad (22')$$

Here index r refers to the reference values. (Similar formula for MS can also be introduced as $y_2(t)$.)

THE FINAL CONTROL SCHEME

On the basis of the linearizing, decoupling possibility introduced in the previous section two separate SISO cascade control loops were applied to control $y_1(t)$ and $y_2(t)$. The block scheme of this control system is shown in Fig. 5. Here a single input single output minimum variance self-tuning (SISO-MV-ST) regulator is in the inner control loop. The outer control loop consists of a PI regulator with time varying parameters, based on a simple pole-cancellation. The pole to be cancelled is computed by the "silo integration" (4). The integral gain of the remaining system after cancellation can be chosen on the basis of the formula, ensuring 5% overshoot, suggested by Bányász and Keviczky (1982). The following parameters are obtained

$$p_o = \frac{1}{b_o(2d-1)} \doteq \frac{q(t)}{v(t)(2d-1)} \quad (23)$$

and

$$p_1 = p_o a_1 = -\frac{q(t) - v(t)}{v(t)(2d-1)} \quad (24)$$

The two ST regulators update 2x4=8 parameters (m=0, n=1, d=2 by channels). The general MIMO solution would have had 20 parameters. (Note, that either the time-varying P regulator with gain (14) or a simple constant PI regulator can be applied in the outer control loop.)

THE APPLIED µP SYSTEM

The adaptive composition control system was implemented on an INTEL SBC 80/20 system with 32 kbytes. The real-time operating system and language PCL'80 was applied which has special problem-oriented vector arithmetic modules for adaptive control. The application of a three-byte software floating point arithmetic proved to be acceptable.

The parameter and covariance matrix updates are performed by a UD filter and a continuous "burst supervision" is also applied. Fig. 6 shows the filling of two silos using the implemented control system. The applied reference values are $ML_r = 102$ and $MA_r = 0.9$.

REFERENCES

Keviczky, L., J. Hetthéssy, M. Hilger, J. Kolostori (1978): Self-tuning adaptive control of cement raw material blending. <u>IFAC Journal Automatica,</u> Vol. 14, pp. 525-532.

Westerlund, T., H. Toivonen, K.E. Nyman (1980): Stochastic modelling and self-tuning control of a continuous cement raw material mixing system. <u>Modelling, Identification and Control,</u> Vol. 1., pp. 17-37. 17-37.

Bányász. Cs., L. Keviczky (1982): Direct methods for self-tuning PID regulators. <u>6th IFAC Symp. on System Identification and Process Parameter Estimation,</u> Washington.

Hetthéssy, J., L. Keviczky, J. Pál, L. Varga (1980): Deterministic self-tuning regulation by PCL'80, <u>MIMI'80,</u> Budapest.

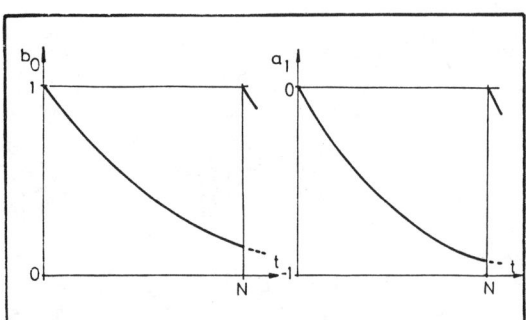

Fig. 3 Silo parameters versus time

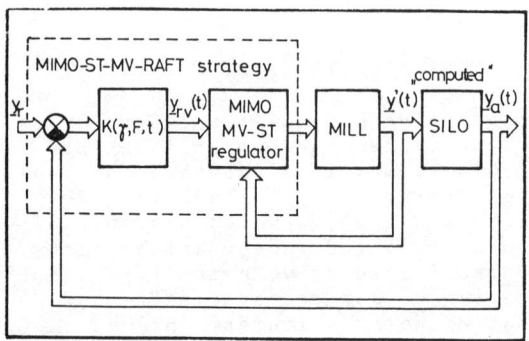

Fig. 4 Scheme of an optimal multivariable solution

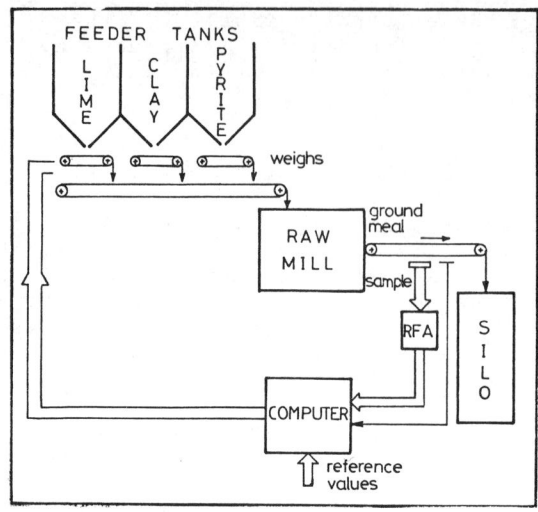

Fig. 1 Simplified technological scheme of composition control

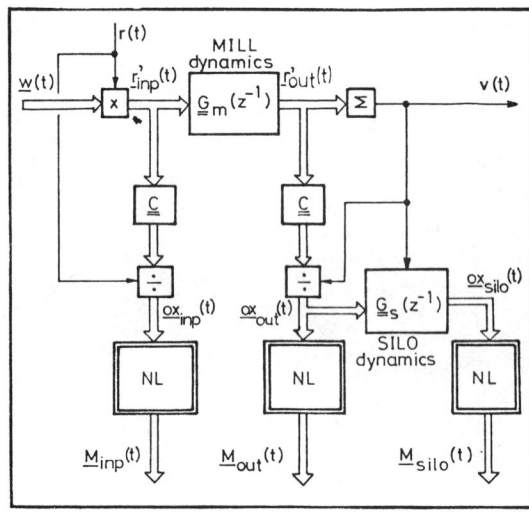

Fig. 2 Model of the mill-silo system

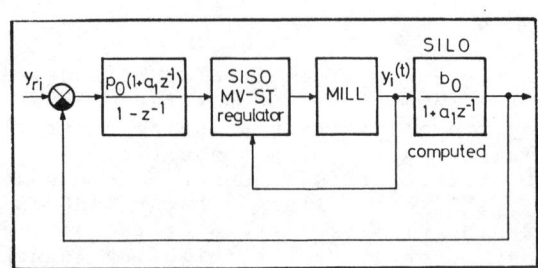

Fig. 5 The applied control scheme

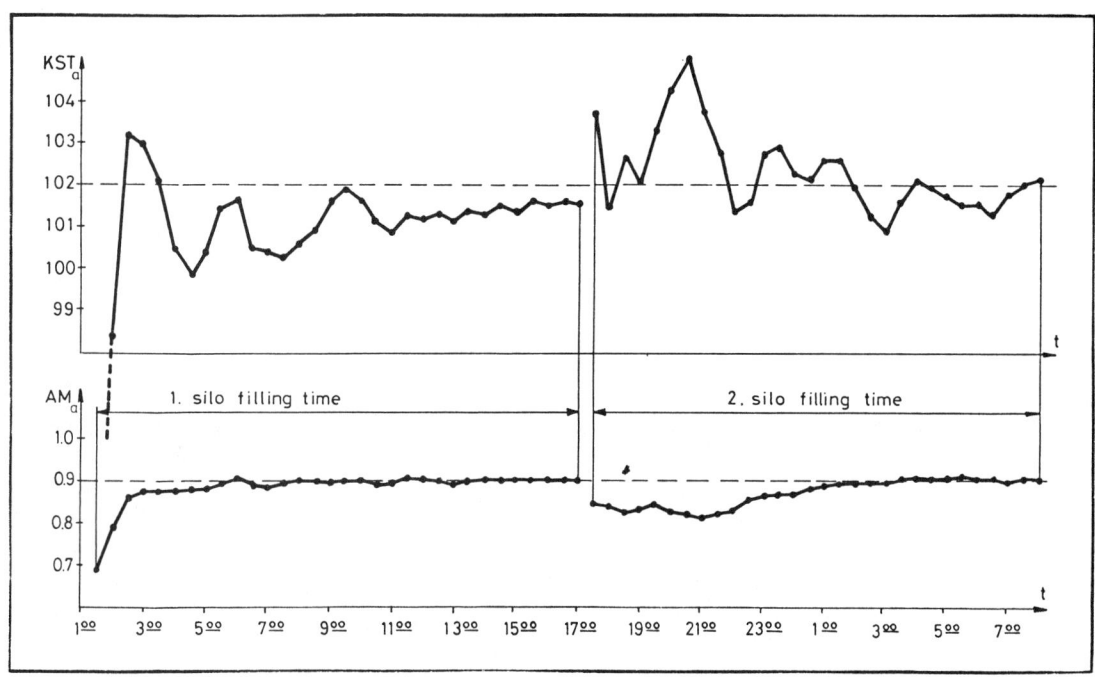

Fig. 6 Adaptive controlled average moduli values

EXPERIENCES FROM A DIGITAL QUALITY CONTROL SYSTEM FOR CEMENT KILNS

T. Westerlund

Department of Chemical Engineering, Abo Akademi, Biskopsgatan 8, 20500 Abo 50, Finland

Abstract. High energy costs during the last decade has in most cement factories resulted in a change from the wet mehod of cement making. Also the use of coal as fuel instead of oil has become ever more popular. These changes do not only have had a positive effect on the process although the fuel economy has been improved . The dry method implies homogenization of a dry raw material, which technically most often result in a more inhomogenious raw meal than the slurry used in the wet method. Also the change of fuel from oil to coal has lead to a more fluctuating energy flow to the cement kiln, resulting in undesirable kiln behaviour and deteriorated clinker quality. These aspects have motivated the development of efficient control strategies for cement kilns, in order to get the maximum profit of the process changes. In this paper a digital quality control systemwhich has succesfully been in on-line operation on two dry process cement kilns at the cement factory of the Lohja corporation in Virkkala, Finland, is presented. Each kiln have the annual capacity of 300.000tons of clinker and the control system has been in on-line operation on the first kiln since 1980 and on the other one since april 1982.

Keywords. Process control; quality control; optimal control;cement industry; industrial control.

Introduction

Cement kilns and the whole cement manufacturing process have been the objects for many digital control systems during the last decade (Higham, 1971;Kaiser, 1970;Otomo,Naka gawa, Akaike, 1972; Westerlund, 1981) The reasons can be several but the main reason surely is the fact that plant economics depends very tight on the production rate, fuel economy and product quality.

Although the available computer hardware today is almost at the same technical level for all control systems, most of the digital control systems differs a lot in the software design. This is typical for a large process with a lot of possible control variables and measurements but few direct measures of main qualities.

In this paper a digital quality control system is described in some detail.

The cement manufacturing process

In rough outline the cement manufacturing process can be classified in three main parts 1) Quarring and raw material mixing 2) The cement kiln process and 3) the clinker grinding process.

Today most of the cement is produced by the dry method. Dry raw material is homogenized as a dry raw meal or/and prehomogenized in big blending beds before grinding. The variations of the chemical composition of the raw meal is very much dependent of the mixing system and thus affecting the kiln in different ways.

The dry homogenized raw meal is preheated ususlly in two to four preheaters before the material eneters the cement kiln. The temperature of the raw meal entering the upper back end of the kiln is about 800 C. The raw meal then passes the rotary kiln in countercurrent to exhaust gas. The final material temperature is about 1450 C. The high temperature is needed in order to start the endo- and exothermic chemical reactions taking place in the raw meal. The product is called clinker and is a complex chemical compound of mainly dicalciumsilicate, tricalciumsilicate ,tricalciumaluminate, tetracalcium- aluminoferrite and a little rest of free lime. Before the clinker leaves the kiln process it is cooled in a clinker cooler in countercurrent to secondary air in order to improve the energy economy of the kiln. The clinker from the cement kiln is finally ground in closed circuit mills together with a small amount of gypsum before it is stored in cement silos before shipment in bulk form or bags. The strenght, setting time, heat of hydration, expansion et.c. of the cement is a function of the main components of the clinker, it thus is essential to have a proper raw meal combined with an efficient kiln control system in order to obtain the desired properties of the cement (Kaiser, 1970).

Control objectives and -structure

Ideal kiln operation will minimize fuel consumption while optimizing the production of quality clinker with respect to the minimum ratio fuel consumption/clinker production or maximum troughput. In order to achieve the optimal production it is implicitely understood that it is desirable to obtain stable thermal conditions in order to reduce the wear and tear on the lining and to cut the maintenance costs.

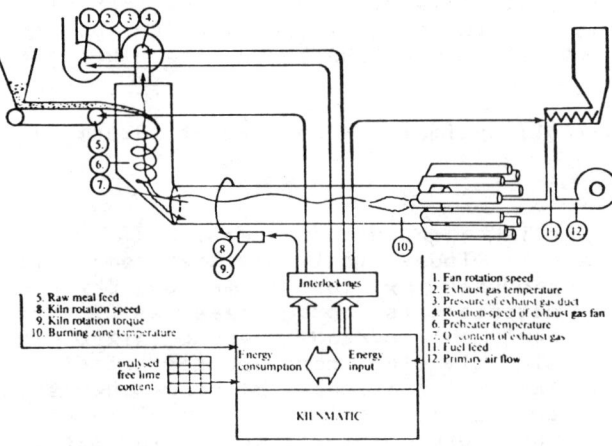

Figure 1.   Kiln   instrumentation.

When formulating the control objectives in a more quantitative form it is necessary to observe some of the special characteristics of the process.

Roughly speaking the cement kiln can be divided into two different dynamical systems. A fast and a slow dynamical part of the process. The fast dynamic part correspond to the exhaust gas passing the rotary kiln in only a few seconds. The slow dynamic part corresponding to the material flow passing the kiln in approximately half an hour.

It seems reasonable, when formulating the control objectives for a cement kiln, to build up a control hierarcy operating in different time scales, because of the special characteristics of the process-dynamics. The actual control system is structurally based on three levels In the lowest level (partial control level) we try to stabilize the thermal conditions of the cement kiln by using both analog and digital SISO PI-control loops (sampling time 1 sec) controlling variables corresponding to the fast dynamic part of the process. These control loops are for example:
* O2 content of exhaust gas controlled by the fuel feed
* Presure of exhaust gas duct controlled by the second exhaust gas fan speed
* Primary air flow controlled by the primary air fan speed
each of which usually are in the standard setup of cement kiln control loops.

The second level (overall control level) is the main control level for the slow dynamic part of the process (sampling time 5 minutes).

Two of the primary manipulated variables of the kiln is the fuel rate and the total exhaust gas rate, each of which has an important effect on the heat transfer and indirectly the reaction characteristics. In the lowest control level the oxygen consentration in the exhaust gas is controlled by the fuel rate. The manipulated variable thus is the exhaust gas fan speed, which then corresponds to the energy flow into the kiln. The conventional manipulated variables in rotary kiln operation also include the rotation speed, which controls the material recidence time and the feed rate of the raw material to the kiln, which controls the production rate. In the actual control system the kiln speed is synchronized with the raw material feed rate in order to make it possible to manipulate the feed rate of the raw meal without significant changes in the internal loading. Thus the feed of raw meal which corresponds to the energy consumption was taken as the second manipulated variable at this second control level. In order to stabilize the thermal conditions and to avoid problems in the alcali cycle it is necessary to keep the temperature of the combustion gas constant. It there fore seemed natural to incorporate the control of the exhaust gas temperature in the second control level. Furthermore, one of the variables to be controlled should include information about the clinker quality. As the clinker quality specified by the free lime content in the clinker can only be measured (off line) a while after the material has passed the kiln and is for practical reasons only analyzed every second every third hour it is difficult to use this quality measure directly for control.

It has however, been found (Gale, Stapelton, 1972; Westerlund, 1981) that the kiln rotation torque correlates to the burning zone temperature and the clinker quality. Physically the relationship has an explanation in the fact that increasing/decreasing energy flow into the kiln (de-/increasing free lime content in the clinker) leads to an in-/decreasing kiln rotation torque because the material in the burning zone tends to climb up the walls.

The second variable to be controlled at the second control level thus was taken as the kiln rotation torque which corresponds to the clinker quality.

The interaction between the manipulated variables and the variables to be controlled made it necessary to use a multivariable control strategy which is described in the next section.

Finally the third upper control level is an optimization level. Ideal kiln operation will minimize fuel consumption while achieving optimal production of quality clinker in respect to minimum ratio fuel consumption/clinker production or maximum troughput.

The quality of clinker is specified by an upper and a lower quality limit for the free lime content of the clinker. The energy consumption at the upper quality limit is however lower than at the lower quality limit (see figure 3). Fuel consumption can thus be minimized by maximizing the average free lime content in the clinker subject to the quality limit restrictions. This has been done by estimating the standarddeviation of the variations of the free lime content in real time and thereafter computing the target of the free lime content a factor multiplied by the estimated standarddeviation lower than the upper quality limit. The target for the kiln rotation torque is thereafter calcualted from the functional-realationship between the free lime content and the rotation torque. The adaption of free lime content target is based on (Westerlund, 1981),

$$S_t^2 = W S_{t-5min}^2 + (1-W)(LGFL-T)^2 \quad (1)$$

$$T = UL - F S_t \quad (2)$$

where W is an exponential weighting factor ($0 < W < 1$) which makes it possible to estimate a time varying standarddeviation. T is the target, LGFL the measured free lime content, UL is the upper quality limit and F a factor wich makes it possible to change the percent of free lime content measurements to be lower than the upper quality limit.

In order to achieve the optimum production rate a simple automatic experimental search procedure which examines the dependent variable with respect to the process constraints and adjust the material feed rate until the ratio fuel consumption/clinker production is minimized or kiln troughput maximized. In figure 1 the kiln instrumentation is shown.

Multivariable control model

As was mentioned in the previous section the second control level uses two manipulated variables the exhaust gas fan speed, U1 which corresponds to the energy flow into the kiln and the raw material feed rate, U2 which corresponds to the energy consumption of the kiln. Furthermore the variables to be controlled where the exhaust gas temperature, Y1 and the kiln rotation torque, Y2 which correlates to the clinker quality(see figure 2). As the variables do affect all each other it was natural to use a multi-variable process model which takes the interactions into account when designing a control strategy for the second control level.

It was found (Westerlund, 1981) that a so called controlled auto-regressive mowing average (CARMA) model could very adequatly describe the relationship between the variables. The CARMA model had the structure,

$$Y(k+1)=AY(k)+BU(k)+Ce(k)+e(k+1)+d \quad (3)$$

where Y is a vector containing Y1 and Y2, U is a vector containing U1 and U2 and e(k) is a sequence of independent random vectors with zero mean value and covariance,

$$E\, e(k)e(k)^T = R \quad (4)$$

A, B, C and R are 2x2 matrices. d is a vector with 2 elements.

The parameters were found by an identification experiment and by using the maximum likelihood method for parameter estimation (Ljung,1978) combined with Akaike's information theoretic criterion (Akaike,1974) for obtaining the number of parameters to be used in the model (Westerlund, 1981).

As the model do include information about the statistical properties of the disturbances the control objective was natural to formulate so as to minimize the sum of the variances of the control errors of Y1 and Y2 subject to constraints on the variances of the control signals.

$$J = \min (s_{y1}^2 + s_{y2}^2) \qquad (5)$$

subject to

$$s_{u1}^2 < g_1 \text{ and } s_{u2}^2 < g_2$$

The solution to the control problem formulation can be found by constrained optimization (Mäkilä, 1979; Westerlund, 1981). The Lagrangian of the problem is,

$$L = s_{y1}^2 + s_{y2}^2 + l_1 s_{u1}^2 + l_2 s_{u2}^2 \qquad (6)$$

where l_1 and l_2 are Lagrangian multipliers. Thus the lagrangian of the problem is equivalent to the linear quadratic gaussian control problem formulation which leads to the solution: a linear feedback from optimally estimated states and a Kalman filter producing the optimally estimated states (Aström,1970). Or if the transferfunction representation of the controller is used we obtain for the given model the controller (Westerlund, 1981),

$$U(k) = GU(k-1) + H_1 (Y(k) - Y_{sp}) +$$
$$H_2 (Y(k-1) - Y_{sp}) \qquad (7)$$

where G and H are 2x2 matrices with controller gains and Ysp a vector with setpoint values for Y1 and Y2.

Relation - kiln torque and free lime

The free lime content and the kiln torque was found to follow a functional relationship very adequatly for data with a time lag on approx. 20 minutes for the free lime content (see figure 2). Thus when specifying the target for the free lime the corresponding target for the kiln rotation torque can be calculated from the functional relationship. On the other hand estimates of the free lime content based on the measured kiln rotation torque can also be calculated from the function. As the free lime content is measured every second hour it was found natural to read the measurements of the free lime content into the digital control system and to update the parameters of the relationship in real time (by using the recursive least-squares method with exponential forgetting of past data). This adaption has also been found relevant in some cases where for example a so called ring of material is formed in the kiln. In such situations the kiln torque will arise without changes in the product quality. In such situations the control system could not make the correct decision without parameteradaption.

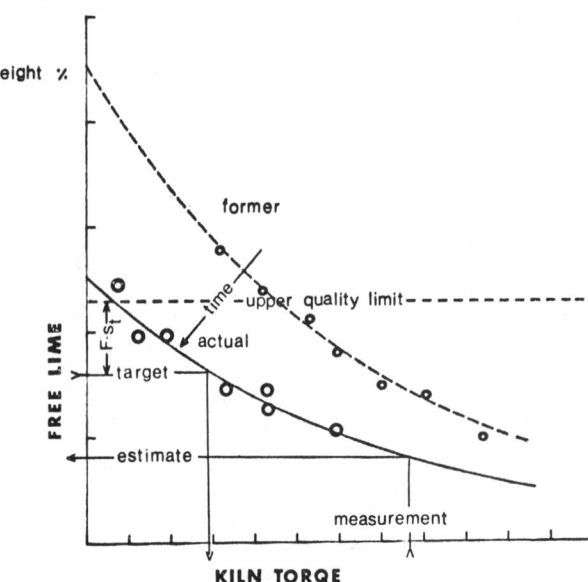

Figure 2. Relationship between the free lime content and kiln rotation torque.

Experiences from the control system

The control system has been in on-line operation since Oktober 1980 on kiln number seven at Oy Lohja Ab:s cement factory in Virkkala, Finland. The kiln is a dry process kiln with the annual capacity of 300.000 tons of clinker. It is 105m long and 5m in diameter. The kiln has two preheaters and a planetary cooler. The fuel is pulverized coal.

During the first year of on-line operation the improvements in reduced energy consumption has been about 2 % less energy per ton clinker. The clinker production has been increased by an average of 2.5 per cent. The savings in maintenenance costs have been estimated to be as high as 10 per cent. The volume of high quality clinker has been increased from approximately 48 per cent to 71 per cent and the volume of hard burnt clinker reduced from approximately 28 per cent to 13 per cent as is shown in figure 3.

As the variations in the product quality have been smaller the average free lime content of the clinker has been increased according to the target adaption, while the volume of second rate or rejected clinker has been decreased (se figure 3).

Quality improvements naturally have an effect on cement grinding capacity and energy consumption. As the amount of hard-burnt clinker (low content of free lime in the clinker) decreases, the energy requirements for clinker grinding are also reduced.

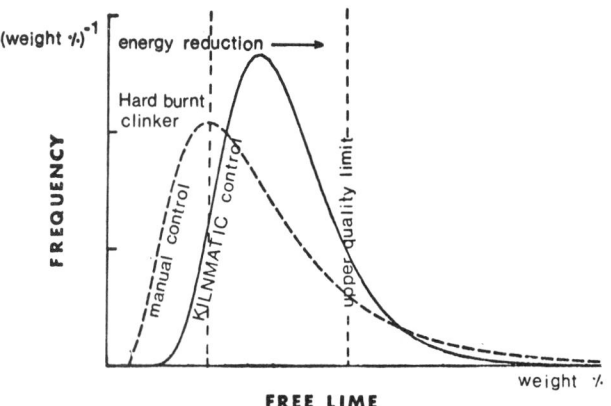

Figure 3. Frequency functions for the free lime content during manual and automatic control.

The improvements in the kiln profitability were surprisingly big during the first year of operation. It was therefore, motivated to implement the control system on Oy Lohja Ab:s other kilns. Since april 1982 the control system also has been in on-line operation on kiln number six at the cement factory in Virkkala.

References

Akaike H. (1974). A new look at Statistical Model Identification. IEEE Trans. Autom. Control, AC-19, 716-722.

Aström K.J. (1970). Introduction to Stochastic Control Theory. Academic Press, New York.

Gale W.M. and Stapelton C.A.(1972). Experimental verification of the correlation between the calcining rate and other kiln variables. 5th IFAC World Congress, Paris.

Higham J. D. (1971). Dynamic Computer Regulation of a Dry-Process Cement Kiln. Proc. IEE Vol 118, 609-619.

Kaiser V. A. (1970). Computer Control in the Cement Industry. Proc. IEEE, vol 58, 70-77.

Ljung L. (1978). Convergence Analysis of Parametric Identification Methods. IEEE Trans. Autom Control, AC-23, 770-783.

Mäkilä P. (1979). Optimal Structure Constrained Controllers Rep. 79-11, Abo Akademi.

Otomo T., Nakagawa T., Akaike H. (1972) Statistical Approach to Computer Control of Cement Rotary Kilns. Automatica, Vol 8, 35-48.

Westerlund T. (1980). A Digital Quality Control System for an Industrial Dry Process Rotary Cement Kiln. IEEE Trans. Automat. Control, AC-26, 885-890.

COMPUTER CONTROL OF A CEMENT PLANT

V. M. Dozortsev, E. L. Itskovich, I. V. Nikiforov and I. I. Perel'man

Institute of Control Sciences, Moscow, USSR

Abstract. The paper discusses the basic functions of a cement plant control by distributed computers. The attention is focused on optimization of two main processes, raw blending and klinker burning.

Keywords. Computer control; optimal control; simulation model; minimising of the production costs; autoregressive and moving average models; failure detection.

INTRODUCTION

The cement manufacture involves chemical technological processing of raw components in three successive stages. At the first stage, raw milling, lumps of lime, clay, chalk, and iron cinder are crushed, milled, blended, and fed to homogenising silos. The next stage is clinker burning. The raw blend is fed to a rotary kiln where it undergoes numerous physical and chemical transformations. Following the burning the resultant clinker is milled with various additives to obtain a final product. The grade and kind of the cement depends on the properties of the additives and the milling fineness.

Process automation in the cement industry has along history. Development of first process stabilisation systems was started over thirty years ago. Since then the automation has undergone many stages both in terms of the functions and of the control hardware. This paper will describe today's control principle employed in Soviet cement plants.

THE CONTROL SYSTEM CONFIGURATION

Because the technological flowchart is straightforward and the semi-fabricated products do not diverge, the control efforts concentrate in individual installation of lines rather than involve the entire plant which consists of several lines. A line for the purposes of this paper is made up of one rotary kuln and several raw and clinker mills. The number of the latterdepends on the output of the kiln. A plant can consist of several such lines. Field studies have revealed that automatic stabilisation of processes enables the personnel to interfere with the process rather rarely (several interferences per hour for all installations the line).Consequently, a centralised control of the entire plant is feasible.

The control systems take over:

- centralised monitoring of processes (about two hundred test points per line);
- interlocking of mechanisms and lines (in operations such as start, up stoppage or change of condition);
- stabilisation of operating modes in units (thirty to fifty control loops);
- optimisation of key assemblies in terms of specified agreed criteria (raw blending and klinker burning).

Cost analysis suggests that the most cost-effective, reliable, and simple way to perform these functions is at present a distributed computer network. This is a centralised two-level system. At the lower level the data gathering, interlocking, and process stabilisation algorithms are performed by microcomputers which are in the immediate vicinity of the equipment. These microcomputers which total around ten communicate with the central computer in the line control panel. The central computer is a mini-computer with disk and tape external memory and numerous terminals for the operator and plant managers. The central computer acquires measurement data from the micro-computers, controls their operation, and performs the algorithms of centralised monitoring, computation of techno-economic indices and production accounting, stores the entire accounting information, keeps the consumer up-to-date and responds to his queries, and, most important of all, performs algorithms of optimal control. The operator's panel consists largely of CRT's which act as a mimic panel, plot indices versus time, display current and accounting information, and supply verious data on demand. TV cameras enable the operator to see the state of the burning zone in the kiln and the most important pieces of equipment.

OPTIMAL RAW BLENDING CONTROL SUBSYSTEM

A simplified layout of raw blending is shown in Fig. 1. Lumps of raw materials are continuously fed in desired proportions into the raw mill where they are blended and milled. At the output of the mill the X-ray quantometer sampler determines the chemical composition of the blend. The blend is fed into the homogenising silo. There it is homogenised and fed to the burning area. The process is optimised if the blend at the silo output deflects from the specified technological standards by a minimal value.

The current chemical composition of components cannot be observed. Each is characterised in terms of the contents of four oxides, SiO_2, CaO, Al_2O_3, and Fe_2O_3, which will be denoted for brevity as S, C, A, and F. In real time only their concentrations in the blend can be measured. Consequently, what we have is feedback control with a considerable delay which is a function of the rate of raw movement in conveyors and mills of sample delivery to the quantometer, and the duration of the sample analysis. In addition to current values of oxide concentrations in the blend, only the average values of oxide concentration in each i-th component ($\bar{S}_i$, $\bar{C}_i$, $\bar{A}_i$, and $\bar{F}_i$) are available in advance. In many plants, however, the current values of oxide concentrations, (S_i, C_i, A_i, and F_i) are significantly different from their averages. Furthermore, as some rocks of the quarry are exhausted the averages can change but no information on this is usually forthcoming.

Another specific feature of this problem is the nonlinear relation between technological modules of the blend to be stabilised and oxide concentrations in the blend as expressed in the following indices:

the saturation coefficient
$$KS = \frac{100 C}{2,8 S + 1,1 A + 0,7 F}$$
the silicate module
$$SM = \frac{S}{A+F} \quad (1)$$
the alumina module
$$TM = \frac{A}{F}$$

The performance criterion of the raw area is quadratic

$$F = M\left\{\gamma_1\left(\frac{KS(t)-KS_N}{KS_N}\right)^2 + \gamma_2\left(\frac{SM(t)-SM_N}{SM_N}\right)^2 + \gamma_3\left(\frac{TM(t)-TM_N}{TM_N}\right)^2\right\} \quad (2)$$

where M is the mean value; γ_1, γ_2 and γ_3 are weighting factors which dictate the significance of the indices; KS_N, SM_N and TM_N are associated specified values of the modules as given by the line technology.

Still another feature of the problem is a another silo downstream of the raw mill and, consequently, of the control loop. Once the blend goes through this, the variations of the chemical blend composition change in a certain way. Most widespread silos operate "in closed mode". They are first filled with the blend, where it is stirres and then let into the burning area. At the latter two stages the blend flow from the mill is directed into another, empty, silo. The averaging in this silo cannot be described as continuous dynamic and so the linear control theory can hardly explain its properties. Still, the output of raw blending in the mill for feeding into the silo should be forecast in some way, which makes the control system take up still another activity.

The overall algorithmical structure of the problem is subdivided into several successive operations (see Fig. 2).

<u>Modules regulation</u>. Early investiga-

tors (Hammer H., 1975; Itskovich E. and co-worker, 1977) thought that nature of the process compels one to divide the overall control tasks into two; linear control of the modules and nonlinear programming which translates dummy module control signals into real signals that are fed into blend component feedres. This approach is made also possible by the fact that the number of the modules (KS , SM , and TM) exceeds that of independent control signals n - 1 (where n is the number of components) and n is usually equal to three.

Simulation model studies of various module control equations in the presence of specific disturbances, or variations in the composition of raw components have led to advice on what a reasonable control equation should be and how its parameters can be adjusted.

In design of controllers the "closed mode" homogeniser is especially difficult to simulate. Hammer H. (1975) and Lukanov V. and co-worker (1978) found that the simplest way to do this is to introduce an integral complement into the control algorithm. Let us consider a control channel for one of the modules

$$u_t = \sum_{i=1}^{p} \alpha_i u_{t-i} + \sum_{i=0}^{q} \beta_i \varepsilon_{t-i} + \tilde{u}_t , \quad (3)$$

where u_t is control at the t-th step;

α_i, β_i are controller gains;

p, q are the numbers of terms in sums;

ε_t is the misalignment at the t-th step, $\varepsilon_t = y_t - y^*$;

y_t is the measured value of the module at the t-th step;

y^* is the specified value of the module;

$\tilde{u}_t$ is the integral complement which is found from the formula

$$\tilde{u}_t = -C \cdot (t-1) \cdot (\bar{y}_t - y^*) + y^* \quad (4)$$

$\bar{y}_t$ is the module value computed from formula (1) from average concentrations of the blend oxides $\bar{S}(t), \bar{C}(t), \bar{A}(t)$ and $\bar{F}(t)$ for the first t steps of silo filling;

C is the factor of the integral complement.

Application of the algorithm (3), (4) may, however, result in dramatic changes in the final blend composition at the times of changing from one silo to another because of the integral complement (4) and imperfection in the blending.

Therefore one can use another way to simulate the silo whereby a priori information on the time of its filling and the averaging factor is used

$$K = \sqrt{\frac{D_y}{D_x}} ,$$

where D_x is the variance of the blend composition variations at the silo input;

D_y is same at the silo output.

Deflection of the current chemical composition from the setpoint value at the silo output consists of two components: the systematic error in blend averaging in each silo and random oscillations about this sample average

$$M(\varepsilon_t^2) = D_x \cdot K + M(\varepsilon_{\bar{y}}^2) , \quad (5)$$

where $\varepsilon_{\bar{y}} = \bar{y} - y^*$
and $\bar{y}$ is the average value of the module during silo filling.

In this approach the gains of the controller (3) are computed so as to obtain at the mill output a blend with a desired spectrum of composition variations. This spectrum should be such that a minimal total squared deflection ($M(\varepsilon_t^2)$) of the module from the setpoint is obtained. This approach has been expounded in (Le Min Tuan and co-workers (1981)). Simulation of such algorithms in specific examples suggests that they more than half the mean square deflection of the module from the setpoint in comparison with the conventional control algorithm.

The Nonlinear Programming Operation enables replacement of dummy controls u_t in formula (3) by actual control signals which are ratios of the raw components q_i. In doing so optimisation by the criterion (2) is performed with an allowance for the matrix A of the average component composition. The programming problem is formulated as follows

the criterion is

$$\min_{\bar{q}} \left\{ \gamma_1 \left(\frac{u_{KS} - KS(A,\bar{q})}{u_{KS}} \right)^2 + \gamma_2 \left(\frac{u_{SM} - SM(A,\bar{q})}{u_{SM}} \right)^2 + \gamma_3 \left(\frac{u_{TM} - TM(A,\bar{q})}{u_{TM}} \right)^2 \right\} \quad (6)$$

and the constraints are

$$\sum_{i=1}^{n} q_i = 1, \quad (7)$$

$$q_{i\,min} \leq q_i \leq q_{i\,max},$$

where

$$KS = KS(A,\bar{q}), SM = SM(A,\bar{q}), TM = TM(A,\bar{q})$$

the modules are represented as functions of the average composition of elements in the matrix A and the desired vector of control signals $\bar{q}$; u_{KS}, u_{SM} and u_{TM} are dummy controls computed from the control algorithms (3) in the channels KS, SM and TM; $q_{i\,min}$ and $q_{i\,max}$ are constraints on possible changes in the feed of the i-th component.

The problem (6), (7) is of low dimension ($\dim(\bar{q}) = 3 - 5$) and is easily solved by the method of penalty functions or by the Rosen method.

<u>The adaptation operation.</u> Because the raw component composition can vary randomly and no information on this may be forthcoming, the matrix of the average chemical composition in the criterion (6) becomes untrue at some unknown time. This results in systematic deflection of the blend modules from the desired values and a substandard blend. The adaptation prevents these undesirable phenomena. It is subdivided into two tasks: detection of the time of a systematic deflection and abolition of this deflection by changing the matrix of the average chemical composition of the components. Nikiforov (1980) found that the first task is performed by "detection of the desorder in the time series". An algorithm is developed which detects the time when the average multidimensional time series (of the modules to be controlled) is applied. This series has been found to be described by an autoregressive moving average equation of a low order. The disorder detection algorithm is a simple recurrent relation which is sensitive to a change in the value of any module following several measurements of the blend once the matrix A of the average chemical composition of the components has changed.

The second task is not performed unless a disorder misalignment signal arrives. It is tackled either by passive adaptation of the matrix A in the feedback loop or by an active experiment whereby special control signals for flows of indivudual components result in a new matrix A.

On the whole the blending control algorithm is supplemented with an instruction on adjustment of its paramenters depending on the process characteristics and disturbances. This enables easy use of the algorithms in various conditions (in different cement plants) and leads to a considerable technological and economic payoff.

OPTIMAL CLINKER BURNING CONTROL SUBSYSTEM

Clinker burning in a rotary kiln is controlled by two subsystems of which one employs a conventional stabilisation algorithm to maintain the desired temperature of the output gases by regulating the draught inside the kiln; the other, employs a new dynamic optimisation algorithm (Perelman I.I., 1978, 1982; Perel'man and co-worker, 1979) to control the fuel consumption for minimising the variable component of the clinker production costs (in other words, the costs of product unit burning are minimised). The algorithm embodies mathematical models which define the costs $S(t)$ at time t as a function of the temperature $T^0(t)$ of the material in the burning zone and fuel consumption $u(t)$ and the temperature $T^0(t)$ as a function of the current and past values of u and W. Here W is the raw flow at the kiln input.

It is assumed that in the stationary condition

$$S(t) = S_0 + f(t) + \alpha u(t), \quad (8)$$

where S_0 is the constant part of the costs; f is losse due to manufacture of rejects; α is a cost factor; αu is the fuel cost. Since the clinker grade is not determined until after a long delay, the current value of f is determined indirectly from the temperature in the burning zone. In the operating range of f a certain value $T^0 = a$ has been found below which the probability of obtaining a substandard product dramatically grows. On the other hand, with $T_o > a$ no further temperature increase is useful because the grade cannot be improved. Therefore in the vicinity of the nominal mode $f(t)$ is defined as

$$f(t) = \begin{cases} k(a-T^o(t))^2 & \text{with } T^o \le a \\ 0 & \text{with } T^o > a \end{cases} \quad (9)$$

where k is a constant factor.

In its turn, the dependence of $T^o(t)$ on the control signals applied to and the disturbances in, the kiln is estimated in a mathematical model in the form

$$\delta \tilde{T}(t) = y^{(u)}(t) + y^{(w)}(t) + \varphi(t). \quad (10)$$

Here $\delta \tilde{T}(t)$ is the estimate of the variation of the current value of $T^o(t)$ about a fixed nominal value $\hat{T}$; $\varphi(t)$ is a component of $\delta \tilde{T}(t)$ which is dictated by variations of unobserved factors such as changes in the physical and chemical properties of the raw and changes the kiln lining; and $y^{(u)}$ and $y^{(w)}$ are components due to variations δu and δw in the flows of the fuel and raw, respectively. These are given by models designed as linear sampled data filters

$$y^{(u)}(t) = \sum_{i=1}^{J_u} h_i^{(u)} \delta u(t-i\Delta);$$
$$y^{(w)}(t) = \sum_{i=1}^{J_w} h_i^{(w)} \delta w(t-\tau-i\Delta), \quad (11)$$

where Δ is the chosen period of time sampling $J_u \Delta$ and $J_w \Delta$ characterise the dynamic filter memory; τ is the transportation delay between the raw entering the kiln and the burning zone; $h_i^{(u)}$ and $h_i^{(w)}$ are experimental weighting coefficients of the model.

The flowchart of the dynamic optimisation subsystem is shown in Fig. 3. The input variables are u, w and T^o measured in the kiln. The subsystem operates with a sampling rate Δ.

At current time $t = j\Delta$ ($j=1,2,...$) the model M_1 which reproduces the dependence (10) and (11) and is connected in parallel with the kiln yields a current estimate $\hat{\varphi}(j\Delta)$ of the unobservable variable φ. The available data on the values of $W(j\Delta-i\Delta)$ and $\hat{\varphi}(j\Delta-i\Delta)$ ($i=0,1,...$) are processed by statistical extrapolation to obtain forecasts of these variables

$$\tilde{W}[(j+\lambda)\Delta|j\Delta] \text{ and } \tilde{\varphi}[(j+\lambda)\Delta|j\Delta], \lambda=1,2,...,L,$$

where L is the specified forecasting duration. The forecast and the sequence

$$U_{j+1} = \{\tilde{u}_{j+1}, \tilde{u}_{j+2}, ..., \tilde{u}_{j+L}\} \quad (12)$$

of the assumed future values of fuel consumption at times $(j+\lambda)\Delta$ are fed into a model M_2 which is similar with M_1 and the output of which is added to the forecast $\tilde{\varphi}$. As a result a forecast of temperature variations, consistent with a given U_{j+1}, is obtained and then (8) and (9) lead to the expected value of costs over the time period $[(j+1)\Delta, (j+L)\Delta]$:

$$Q(U_{j+1}) = \sum_{\lambda=1}^{L} \tilde{S}\{(j+\lambda)\Delta|j\Delta, U_{j+1}\}. \quad (13)$$

Numerical methods lead to the optimal value of the sequence (12)

$$U_{j+1}^* = \{u_{j+1,j}^*; ...; u_{j+L,j}^*\} = \\ = \arg \min_{U_{j+1} \in \Omega} Q(U_{j+1}), \quad (14)$$

where Ω is the specified value of admissible control values. The value $U_{j+1} = u_{j+1,j}^*$ obtained from (14) is introduced at time $t = j\Delta$ as a fuel flow setpoint over the period $[j\Delta, (j+1)\Delta]$. Then at time $t = (j+1)\Delta$ the procedure is reiterated.

It results in a new fuel flow setpoint $u_{j+2} = u_{j+2, j+1}^*$ over the period $[(j+1)\Delta, (j+2)\Delta]$, etc.

The forecast duration L in the criterion (13) is taken commensurable with the delay time of transient processes in the kiln. In our algorithm the range of Ω can be specified either as constraints on the maximal and minimal fuel flow and as a constraint on the magnitude of the flow change from one setpoint to another.

The subsystem is now operational and plays its way by reducing the fuel flow for maintaining the desired clinker grade.

CONCLUSIONS

The above interactive systems enables cost effective centralised control of industrial processes. For complete automation it should be supplemented in actual plants with subsystems for control of raw quarrycus and transportation to the plant and for marketing of the product and these subsystems built around minicomputers are now either under development or in operation.

REFERENCES

Hammer, H. (1975). Prozessrechner erleichtern die Kohmelaufbereitung in der Zementindustrie. Regelungtechnik, Heft 3, s.78-83.

Itskovich, E., Nikiforov, I. (1977). Typical Algorithms of Raw Blending Control in the Cement Industry. Proceeding of 5th IFAC/IFIP Conference on Digital Computer Applications to Process Control, Hague, Netherlands, pp.395-402.

Луканов,В., Маринов, Д., Дудников,Е. Ицкович, Э., Никифоров, И.(1978). Разработка типовой системы управления участком производства (на примере цементного завода). Москва, МЦНТИ, 72 с.

Леминь, Туан, Никифоров, И.В., Тихонов, И.Д. (1981). Выбор рационального алгоритма управления смешиванием сырьевых компонентов. Цемент, № 6, с. 18-21.

Nikiforov, I.V. (1980). Modification and analysis of the cumulative sum procedure. Automatica i Telemekhanika. Vol.41, No.9, pp.74-80.

Perel'man I.I. (1978). Dynamic optimization in automated Control systems of industrial processes using algorithms of conditional prediction. Automatica i Telemekhanika. Vol.39, No.9, pp.146-160.

Перельман,И.И. (1982). Оперативная идентификация объектов управления. Москва.: Энергоиздат,1982.

Perel'man, I.I., Dozortsev, V.M. (1979). Quasioptimal process control with production cost minimization.- Preprints of the 2 nd IFAC/IFORS Symposium of Optimization Methods.- Varna (Bulgaria), pp.271-278.

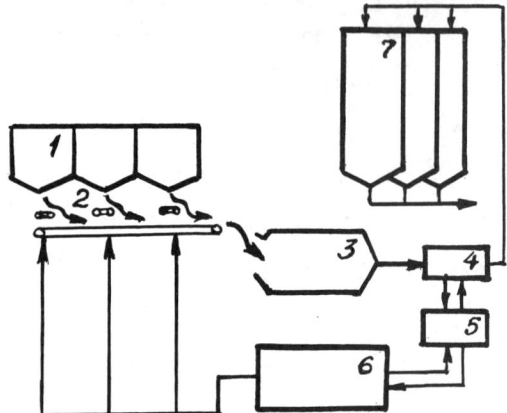

Fig.1. 1. raw bunkers; 2. batchers; 3. raw mill; 4. sampler; 5. quantometer; 6. computer; 7. closed silo

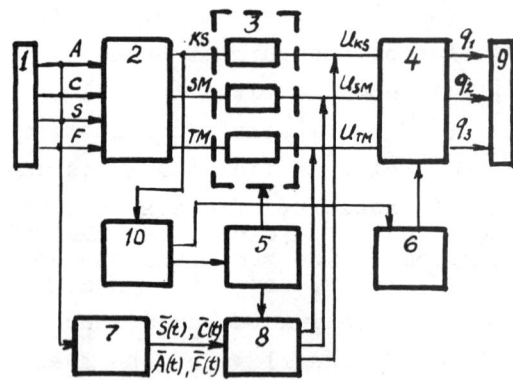

Fig.2. 1. quantometer; 2. computation of modules; 3. linear control of models; 4. nonlinear programming; 5. modules setpoint; 6. matrix A; 7. silo model; 8. computation of modules; 9. batchers; 10. adaptation

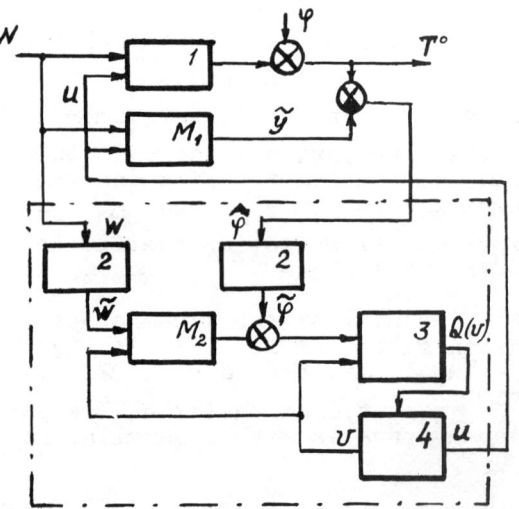

Fig. 3. 1.rotary kiln; 2.extrapolator; M_1 and M_2 - kiln models; 3. criterion computation unit; 4. optimization algorithm

THE DIFFFUSION OF INDUSTRIAL ROBOTS IN SWEDEN

J. Carlsson and H. Selg

Computers and Electronics Commission, Ministry of Industry, Stockholm, Sweden

Abstract. The paper will present a study of the diffusion of industrial robots in the Swedish industry during the 70's and a forecast for the 80's. Comparisons with the diffusion of NC-machines and CAD-systems are also shown. The role of the leading-edge companies for the diffusion process will be illustrated. Seven companies account for about 40% of the total robot population in Sweden.

The paper also presents the results of a study regarding the flexibility of industrial robots compared to NC-machines. The flexibility is measured with respect to production volume, batch size, number of part variations, operation cycle and reprogramming frequency. Based on this information conclutions are drawn regarding what type of production, companies and industry sectors that most likely can benefit from industrial robots.

The use of industrial robots as well as other automation equipment are still concentrated to a few leading-edge companies. With this background the barriers for achieving a wide diffusion will be discussed as well as policies for promoting the wide diffusion.

Keywords. Robots; computer-aided design; numerical control; machine tools; manufacturing process; forecasts; diffusion; versatility.

INTRODUCTION

The shift from belt-driven to electric-motor-driven machinery is an example of a major technological change in manufacturing that has brought forward spectacular improvements in productivity. For about a decade now another important shift, perhaps the most significant since the beginning of the industrial revolution, has been evolving at an ever-accelerating pace - the shift from rigid mechanized to flexible automated manufacturing. The major force driving the new manufacturing technology is the rapid advancement of computer and microelectronic technology.

Despite these rapid advances of computer technology, however, the process of developing the new computer-aided manufacturing technology is evolutionary rather than revolutionary. During the 60's and 70's the automation effort was directed towards various stand-alone machines. Industrial robots (abbreviated IRb), machine tools (NC)[1], material-handling equipment etc, began to be equipped with programmable control systems with a rapidly increasing degree of sophistication. Now that the wide diffusion of these machines has started to pick-up, the main target for the automation effort is to integrate various machines into larger and larger systems. This is a very complicated process which requires large investments in both capital and new management and technical skills. At the same time the greatest opportunities for productivity improvement lie in optimally interconnecting various processes into computer-integrated manufacturing systems.

---

[1] The abbreviation NC will be used for both numerical control and computerized numerical control.

In the engineering industries, and especially those subjected to strong international competition (automotive industry, computers and telecommunication, consumer electronics, household appliances etc), systems integration is regarded as the key to survival in the 80's.

Given the economic problems that most industrialized countries now face, productivity improvement is a main target of industrial policies. For this reason both Sweden and other countries have given high priority for promoting the development and the diffusion of IRb and computer-aided design/manufacturing (CAD/CAM) technology.

NEW MANUFACTURING ORGANIZATIONS

Background

New manufacturing technology such as the IRb is, however, not just a matter of highly efficient computer-aided machines. In fact, it is more a matter of developing new principles for organizing the manufacturing process. The systems-integration approach in manufacturing has also made it more urgent to develop more formal knowledge about the manufacturing process.

The objectives behind the new manufacturing principles are twofold:

a) New manufacturing organizations can increase productivity without any new investments in hardware. Through shorter inventory cycles, lead times, set-up times etc there are great potentials for improving both capital productivity and cash flow.

b) New manufacturing principles are often required in order to maximize the potential benefits from CAD/CAM.

Mixed batch manufacturing is often described as "an activity of organized chaos": constantly changing mixes of products, components, material, batch sizes, available machines and operators, urgency of order, inventory etc. The essence of manufacturing management is therefore to solve one "crisis" after another, using the solution most expedient at the time. In this context the computer has become the most important tool (but not the solution) to bring order out of this chaos.

Manufacturing management, such as process planning and reqirement planning (MRP), is particularly suitable for computerization because it deals mainly with collecting, processing and distributing information. For this reason manufacturing management and overall company management, including office administration, soon became prime targets for computerization. However, many systems for computer-aided process planning (CAPP) and material requirement planning (MRP) failed to generate the results that management had been led to expect. Even if there were technical deficiencies in the systems, the main reason for the failures were that too much emphasis was put on the tool (that is, the computer system) itself and not on how the tool should be used. Translations from manual systems into software did not work for the simple reason that the formal manual system did not work in the first place. Instead companies used informal ad-hoc systems where the rules could be changed rapidly to handle the many unpredictable day-to-day events on the factory floor. These flexible but unstructured methods of management were never picked up in the computer model. The companies ended up with a rigid computerized system which did not reflect the true manufacturing process.

Another reason for the failures was that the required input and output data often were neither timely nor accurate.

The experiences from computerized manufacturing management and factory automation are clear. The main barriers and bottlenecks do not stem from hardware and computer software but rather from the organizational structure of manufacturing. Much effort is therefore now being directed towards learning more about the manufacturing process in order to develop more realistic simulation models.

Product-oriented manufacturing

During the 70's inventory reduction has become more and more a prime target for industrial productivity programs. This is by no way surprising if we consider that the value of the inventory (supplies + raw materials + work in process + finished goods) in the Swedish engineering industry is twice the value of all machines and buildings. By adopting new production methods, including computerized inventory control systems, it has been possible to reduce inventory by 25% or more. The increased importance of a faster inventory turnover is also reflected in the new methods of organizing batch production.

For a long time the functional layout in batch production, that is, all machines of the same kind are gathered in groups, has been as natural as the transfer line in mass production. Through the functional layout, machine utilization can be kept high, but at the expense of complex routing of parts through the shop and large buffers and inventories.

Lead time is defined as the total time needed for material to be processed into a finished product. The lead time is thus the sum of net processing time and waiting time. In batch production with functional layout the ratio of net processing time to waiting time is usually one to some hundreds, or sometimes one to some thousands. Thus, it is much easier and far more efficient to try to reduce waiting time, which is a process that does not add any value to the product, than to increase the processing speed by more sophisticated manufacturing techniques.

In the new manufacturing methods the main principle is to organize the factory according to product oriented layouts. All machines needed to produce one product or one set of products are grouped together in a "subfactory", sometimes with its own administration. Each worker in product-oriented layouts attends several machines. In the functional layout we can with some simplification say that the materials wait for the machines while the machines in the product-oriented layout wait for the materials. The lead time can thereby be reduced dramatically. For instance in the middle of the 70's one sector in the Swedish electrical company ASEA switched to product oriented layout. This resulted in a reduction of lead time from 6-8 weeks to one hour.

The recent increase in interest rates has tended to raise even further the relative importance of a more rapid inventory turnover.

THE IMPORTANCE OF LEADING-EDGE COMPANIES IN THE DIFFUSION OF NEW MANUFACTURING TECHNOLOGY

The diffusion of new manufacturing processes among potential user firms is generally a slow process. Lags of 20-25 years between first and last adopters are common. For the purpose of analysis it is practical to divide the diffusion process into two steps; first diffusion and wide diffusion. The reason for this is that the first adopters - the technically leading-edge users - exercise a strategic role in determining the speed and direction for the subsequent diffusion. The results and experiences from first adopters may serve as important information to later adopters, thus reducing their risk.

The leading-edge companies are watched by a second group of fairly technically advanced and financially strong companies. If the leaders successfully apply the new technology it will soon be diffused to the companies in the second group. From this group the technology will then successively be diffused to industry as a whole via a third, fourth.. etc group of companies. The process of development and diffusion of a new manufacturing system is modelled in figure 1.

The aim of increased productivity in industry is realized primarily through the wide diffusion. The leading-edge companies are important because they start the diffusion process. But it is the speed of the wide diffusion that determines what productivity gains can be achieved on an aggregate level.

In the Swedish engineering industries six main leading-edge companies can be identified with respect to the use of IRb and CAD/CAM. These companies are: Ericsson, ASEA, Volvo, Sandvik, Saab-Scania and Electrolux. In 1979 these companies accounted for 25% of the total NC-stock, 40% of the total IRb-stock and for nearly 50% of the total number of installed CAD-systems.

The reason why the NC-share of these companies is considerably lower than the corresponding share for IRb och CAD is that NC-technology can now be regarded as mature and thus has already started to be diffused in industry as a whole. IRb and CAD on the other hand have just begun the phase of wide diffusion.

THE DIFFUSION OF INDUSTRIAL ROBOTS IN THE SWEDISH INDUSTRY

Swedish industry started to invest in IRb at the end of the 1960's. By 1970 the number of IRb installed amounted to 50. At this time a very rapid diffusion of IRb was predicted. In 1972 experts claimed that by the end of the decade there would be 25 000 IRb installed in Sweden. The actual outcome

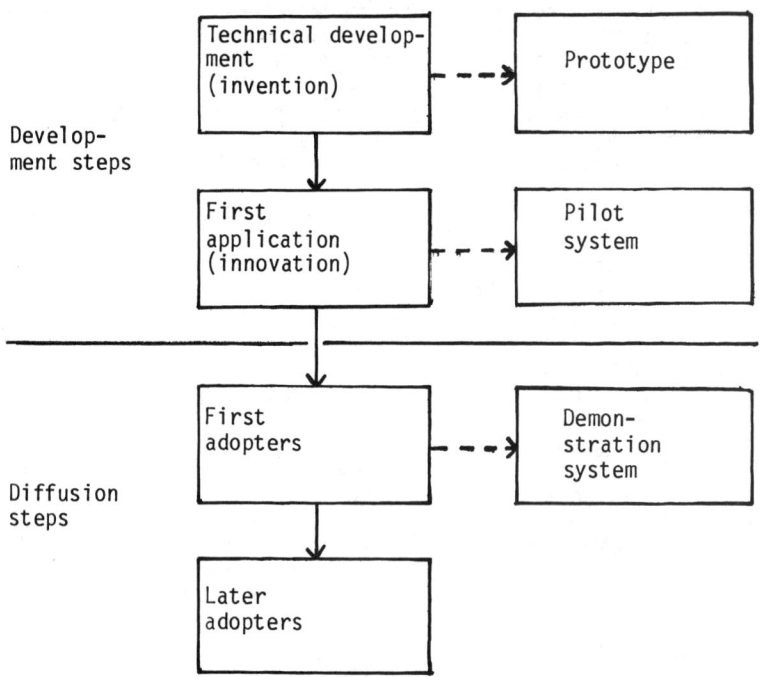

Fig. 1. The steps in the development and diffusion of a new manufacturing system.

was a growth rate of about 40% per year on average and an installed IRb-stock of nearly 1 000 units (1979), see table 1. We mention this forecast because it is a good example of how the barriers to the wide diffusion of a new technology are often underestimated. In any case, a growth rate of 40% per year and a stock of 1 000 IRb is impressive. Few other investment goods can show a growth rate of similar magnitude. Besides, considering the size of the economy Sweden probably has the highest IRb density in in the world.

TABLE 1 Number of Installed IRb in Sweden 1970-1979 and Forecast for the 80's

Company groups	1970	1973	1977	1979	1984	beginning of the 90's
The 7 largest IRb-users	30 (55%)	85 (63%)	310 (63%)	420 (45%)	800 (35%)	
The rest of industry	25 (45%)	50 (37%)	180 (37%)	520 (55%)	1 500 (65%)	
Total	55	135	490	940	2 300	6 000-9 000
Average yearly growth, %		35	38	39	20	17-26

Source: Computers and Electronics Commission

According to a more recent study the IRb-stock in september 1981 was estimated to 1 250.

Table 1 also shows that the use of IRb so far has been concentrated to a few large companies. The 7 largest IRb-users accounted in 1979 for nearly half the total IRb-stock.

During the 80's the IRb-stock will continue to expand rapidly. The net growth rate, however, will not continue to be 40% per year but rather in the order of 20%. There are several reasons that justify a lower growth rate in the 80's compared to the 70's:

a) Growth is calculated from a much higher level of stock.
b) The "easiest" installations will already have been done.
c) There are other automation alternatives besides general purpose IRb. No doubt there are great potentials to automate, for instance, assembly and inspection operations with sensor-equipped IRb. However, computer-aided special purpose assembly machines, redesign of products and components or new materials may in many situations be the most advantageous solution to improve the efficiency of assembly operations.

THE DIFFUSION OF NC-MACHINES AND CAD-SYSTEMS

NC-machines

NC-technique was introduced in Swedish industry around 1960. In 1970 the total NC-stock amounted to about 500 units, see table 2.

During the 70's the NC-stock increased on average by 25% per year and in 1979 the number of units was about 3 600.

As can be seen from table 2, the use of NC-machines during the 70's was largely confined to a few large companies. In 1970, 15 companies accounted for 70% of the total NC-stock. In 1979 the share for these companies had decreased to about 40%. Now that the phase of "wide diffusion" has commenced, it is mainly the small and medium sized companies that account for the increase in the stock.

During the 80's the growth rate of the NC-stock will decrease. This is natural for a technique that has matured and become "a conventional" technique. There are however two other strong reasons for the decreasing growth rate:

a) An increasing share of the NC-investments consists of machining centers,

TABLE 2 Number of Installed NC-machines in Sweden 1970 - 1979 and Forecast for the 80's

Company groups	1970	1973	1976	1979	1984	beginning of the 90's
The 15 largest NC-users	340 (71%)	670 (63%)	1 170 (56%)	1 490 (41%)	2 030 (34%)	
The rest of industry	140 (29%)	390 (37%)	930 (44%)	2 160 (59%)	3 980 (66%)	
Total	480	1 060	2 100	3 650	6 010	8 000-13 000
Average yearly growth, %		30	26	20	11	7-12

Source: Computers and Electronics Commission

each having a capacity of at least three conventional NC-machines.

b) NC-machines are successively being integrated into manufacturing systems which are operated in two or three shifts, which considerably increases the degree of utilization.

For these reasons, growth rates expressed in terms of the number of units strongly underestimate the growth rate of the total production volume that stems from NC-machines.

CAD-systems

During the first 70 years of the 20th century, design work was not affected by any significant technical change. The drawing-board, slide-rule, ruler and hand-books were the main tools. While the productivity of the manufacturing process, due to new technologies, increased in the order of 1 000%, the productivity increase in the design process has been estimated at only about 20%.

With the computer-aided design system a technical breakthrough has occurred which promises large potentials for productivity improvements. Studies that we and others have made of Swedish industry show that the highest productivity improvements are realized in connection with new product and component design and not as a result of investments in new efficient machines.

CAD results not only in higher productivity in the design process but also, and especially through the integration of CAM, in higher over-all factory productivity.

CAD-systems were introduced in Swedish industry in the middle of the 70's. So far it is mainly big corporations, especially in the electronics industry, which have invested in CAD. In 1979 the number of CAD-system (mini- and general purpose computer systems, not desk-top models) amounted to 60, representing a total investment of about 130 million Skr, see table 3.

About 200 CAD-systems are expected to have been installed in the Swedish industry by 1984.

THE ACTUAL VERSATILITY OF INDUSTRIAL ROBOTS IN COMPARISON TO NC-MACHINES

In 1979 a survey of installations of NC-machines and robots was undertaken by the Computers and Electronics Commission in order to examine the nature of the manufacturing process related to these installations. Data were collected regading

A. the goods manufactured
 - kind of goods
 - nature of the market
 - production volume (number of units/year)
 - productfamily size

TABLE 3 The Number and Value of CAD-systems Installed by 1979 and Forecast for 1984

Company groups	1979		1984	
	Number	Million Skr	Number	Million Skr
The 22 largest NC- and IRb-users	40	100	100	250
The rest of industry	20	30	100	175
Total	60	130	200	425

Source: Computers and Electronics Commission

B. the part passing the machine
- kind of part and its function
- operation complexity, measured in min/unit
- number of variants
- number of machine reprogramming per week
- batch size

The study showed a considerable difference in application between NC-machines and robots. The role of NC-machines in flexible manufacturing was far more important than that of robots. Table 4 presents a summary of the results in aggregated terms.

Thus the production volume related to the average NC-machine was less than 10 000 units per year. Average operation cycle exeeded 5 minutes, and the part manufactured appeared in a relatively big - more than 10 - number of variants. The machine was reprogrammed 2-10 times a week, and consequently the average batch size was low, less than 100 units.

In this sample the average robot was used where production volume exceeded 10 000 units per year. Operation cycle was relatively short, less than 5 minutes. The part appeared in few variants, or no variants at all. Reprogramming occured once a week or less. In many cases the robot was not reprogrammed at all. The average batch size amounted to more than 1 000 units.

The study thus showed that the versatility of industrial robots has in most cases been exploited only to a minor degree, or not made use of at all.

The robot has merely been serving as a fix automation equipment but with that difference that the robot can easily be resued in other applications.

The full use of the potential versatility of IRb will be one of the most challenging tasks during the 80's for companies that develop sofisticated flexible manufacturing systems.

WHERE ARE NC-MACHINES AND ROBOTS USED, AND WHY

In table 5 the engineering industry is broken down into major sectors according to the ISIC-nomenclature, and combined with the following four product categories:

o consumption goods (high production volumes)
o components (medium-high production volumes)
o investment goods (low-medium production volumes)
o consumer durables (high production volumes)

This classification has been used in order to find out if there exists a relation between nature of output and the implementation of certain production equipment.

The relative size in terms of value added of each sector, as well as shares of NC-machines and robots are added. Out of this information and the results of the NC and robot study presented above, the usefulness of NC and robot applications in various kinds of manufacture will be discussed.

In table 5, a very strong penetration of robots in the industry for fabricated metal products (sector 381) is observed. Consumption goods and components, both high volume categories, account for high shares while finished goods amount to relatively low shares. Remembering that robots at present are mostly used in high volume manufacturing the outcome thus seem quite reasonable.

TABLE 4 Main Tendencies regarding Applications of NC-Machines and Robots

	NC	Robots
Production volume (units/year)	< 10 000	> 10 000
Operation cycle (min/unit)	> 5	< 5
Part variation	> 10	1-5
Reprogramming frequency (times/week)	2-10	≤ 1
Batch size	< 100	> 1 000

Source: Computers and Electronics Commission

TABLE 5 Swedish Engineering Industry broken down

	Sector code*)				
	381	382	383	384**)	38
Share (%) of output:					
- Consumption goods	17	7	2	2	6
- Components	34	12	31	2	17
- Investment goods	39	72	54	63	60
- Consumer's durables	10	9	13	33	17
Share (%) of					
- Value added	20	29	19	23	100
- NC-machines	22	42	19	13	100
- Robots	51	15	9	22	100

*) 381: Manufacture of fabricated metal products, except machinery and equipment
382: Manufacture of machinery except electrical
383: Manufacture of electrical machinery, apparatus, appliances and supplies
384: Manufacture of transport equipment
Group 385 (Manufacture of professional and scientific, measuring and controlling equipment not elsewhere classified, and of photografic and optical goods) is of minor importance in terms of value added and employment and will not be discussed
**) Shipyards not included

In the industry for transportation equipment (sector 384), the robot share equals the share of value added. While this sector includes the automotive industry, traditionally on the technological frontline, one might have expected a higher share of robots. One reason is that this sector manufactures not only high volume products such as cars but also low volume investment goods such as trains and airplanes. Another explanation could be that many components, assembled in the sector 384, are manufactured (sometimes by employing robots) in other sectors, especially 381.

With investment goods as dominating category in the machine industry (sector 382), mostly owing to non-electrical machinery in the range of low-to-medium volumes, the low share of robots is not surprising.

Also in the industry for electrical machinery (sector 383), a much higher share of robots might have been expected. Here, massproduction is prevailing to an important extent. Bearing in mind that the manufacturing of electrical appliances, printed circuits, telecommunication equipment as well as consumer's durables is characterized by advanced automation, one may conclude that this is achieved by using other types of equipment than robots. In the electrical industries, transfer lines, sometimes controlled by computers, are widely used. As to consumer's durables (except TV- HiFi), the main manufacturing activity is assembly of parts and components often manufactured in establishments classified in sector 381.

Commenting upon the determinants of the diffusion of the NC-machines, will be the reverse of the robot discussion. Thus the investment goods-intensive sector 382 is a heavy user of NC-machines, where the flexibility of this equipment fully can be exploited in small batch manufacturing of special purpose components.

Also in sector 381, with metal working as predominant activity, NC-machines are used to an important degree. Accordingly, the assembly-intensive production of transport equipment in sector 384 accounts for a rather low share, despite the heavy use of NC-machines in manufacture of car engines, gear-boxes and transmission.

Again sector 383 produces a somewhat suprising result, this time by presenting a higher share of NC-machines than might have been expected. This is probably due to the existance of numerically controlled special purpose machines, above all in telecommunications, electronics and TV-Hi-Fi production.

THE SWEDISH INDUSTRIAL ROBOT INDUSTRY

Taking a broad definition of industrial robots we can distinguish three kinds of manufacturers:

o <u>Manufacturers of general purpose IRb</u>. In this group the main manufactures are ASEA (including the former Industrial System Division of Electrolux which was acquired by ASEA in 1981), Atlas-Copco and Kaufeldt. The production volume for 1982 can be estimated to about 1 100 units.

o <u>Manufacturers of special purpose IRb</u>, i.e. automatic loading/unloading equipment attached to specific machines. Volvo's "Doppin" for loading/unloading press machines is an example of special purpose IRb. Several hundred units of this "robot" have been sold to autobile manufacturers all over the world.

o <u>Manufacturers of programable material handling equipment</u>, i.e. auto carrier systems, computer controlled crane and wharehousing systems. This is an industry where Sweden is regarded to have a strong international position. The leading companies in this group are Volvo ACS, BT Lifters, Tellus, Digitron, ASEA and Moving. The "robots" produced by these manufacturers are essential components in Flexible Manufacturing Systems.

Among the general purpose IRb manufacturers ASEA is the incomparably largest and has the most sophisticated robot program. The production target for 1982 is over 1 000 units (including the MHU-models from Electrolux) which probably makes ASEA one of the three largest IRb manufacturers in the world.

With an export ratio of about 90 %, the international markets are absolutely essential to ASEA and for that matter the other robot manufactures as well[2]. For this reason ASEA has set up production facilities in both USA and Spain.

---

[2] Very high export ratios is a characteristic for Swedish industry. More than half of the industrial production is exported (the export is 1/3 of the GNP.

The domestic market is however important for the Swedish IRb-manufacturers for the following reasons:

a) The Swedish IRb-market materialized earlier than in most other countries (with the exception of Japan and USA).
b) Test market for new robot systems.

POLICIES FOR PROMOTING THE DIFFUSION OF INDUSTRIAL ROBOTS

IRb and other computer aided manufacturing equipment are important means for increasing productivity in industry. For this reason most countries have set up programs in order to speed up the diffusion. Japan and some other countries have or are planning Government programs that include investment grants to IRb-manufacturers as well as IRb-users, subventionized leasing arrangements, special tax deduction rates etc.

These types of actions have also been considered in Sweden but have been rejected for the following two reasons:

1. In general it is not the financing of IRb investments that is the obstacle. If the investment is calculated to give normal ROI there should be no problem in raising capital (from owner capital, commercial banks, the Government Investment Bank, regional development funds etc.).

2. The competetiveness against other solutions than IRb gets distorted. The IRb is far from being the only solution in order to improve the productivity or the working conditions in the manufacturing process.

The barriers for the diffusion of new manufacturing technologies, of which IRb is one, are according to our findings:

o The general technical and engineering knowledge in companies. Without this knowledge many firms, especially small firms, will not commence the "learning" process which starts with the following steps: a) awareness of new technologies, b) ask the question "is the robot something for me?" c) feasibility studies and investment calculations etc.

o Knowledge in assessing the prerequisites for a profitable IRb-investment. Critical factors are not only the cost of the IRb and calculated income but also present and future production volumes, the costs of tools and other peripherals, the integragration of the IRb with other machines and the time it takes to make the whole system operational.

In comparison to conventional machine investments, there is a lack of experience of how to calculate reliable investment functions. Present standard methods are not satisfactory. Often income as well as costs are underestimated.

o Lack of skilled labour (can be a bottleneck as well as an incentive to invest in IRb).

In short we therefore think that the barriers are rather "knowledge-based" than "capital-based".

For this reason the Swedish Government has recently presented a bill, based on recommendations from the Computers and Electronics Commission, with the following propositions:

1. The formation of three Engineering Development Centres (EDC) located at technical universities. The centres, which will be financed by the Government as well as the Association of Mechanical and Electrical Industries, will act as "a bridge" between universities and industry. They will undertake development projects either on their own or in cooperation with the universities or industry. They will also be able to carry out specific promotion programs directed towards the regional small and medium sized firms. The EDC:s will be based on an existing research organization. In the future the number of EDC:s will probably increase to five or six.

2. The formation of three CAD/CAM-centres as a joint-venture between universities and EDC:s.

3. Increased funds for education and R&D at universities and trade schools with respect to computer technology and new manufacturing technologies.

However, this new program does not indicate a sudden interest in robotics and CAD/CAM from the Government. It merely reflects a still higher priority given to these fields. The National Technical Board (which is an agency under the Ministry of Industry) has in recent years sharply increased its funding for R&D in robotics and CAD/CAM. Several large long term research programs are now in progress at universities and industry.

Finally, in order to stimulate leading-edge companies to invest in advanced and high risk pilot projects, thereby initiating new diffusion processes, the Government provides special risk-sharing capital through the Industrial Fund. However, as the Industrial Fund can be used for all kinds of advanced industrial projects IRb and CAD/CAM have to "compete" with projects within other technologies.

Copyright © IFAC Real Time Digital Control Applications
Guadalajara, Mexico 1983

A HIERARCHICAL DISTRIBUTED INFORMATION PROCESSING SYSTEM FOR FOREST MANIPULATION

P. Kärkkäinen and M. Manninen

Technical Research Centre of Finland, Electronics Laboratory, Oulu, Finland

Abstract. The paper describes a distributed microcomputer network for hierarchical control of manipulators. A sensory-interactive manipulator is a complex system where the information processing is executed at different levels. The control system is partitioned into three functional levels. The developed multiprocessor architecture is flexible to accomodate the data processing in different levels and the integration of intelligent sensors. The multiprocessor system has been implemented on a high power crane manipulator.

Keywords. Crane manipulator; microprocessors; multiprocessing systems; robots; sensors.

INTRODUCTION

A manipulator is a complex system consisting of several interacting subsystems. This paper concentrates on describing the organization of the manipulator control system and the distribution of the information processing based on a multiprocessor control architecture.

The early robot and manipulator systems relied on a large, high-speed computer using complex software such as real-time multitasking operating systems. However, the development of integrated circuits and microprocessors has offered the possibility to distribute and increase the computational capabilities of the manipulator at a moderate cost.

Also the greater use of sensory systems in manipulators has lately increased the demand on distributed processing. It is necessary that the processing of the sensory data must be carried out sufficiently quickly to be compatible with the real-time requirements of the application environment.

The hierarchical control structures of robots and manipulators have been investigated by some authors. Barbera and athors (1979) have partitioned the control system on five levels of hierarchy and used microprocessors on each level to execute the algorithms of that level. Shin and Malin (1980) have described a system having one microcomputer per degree of freedom and a coordinating microcomputer. A similar approach also has been used by Bisioni and Cassinis (1976). Luh and Lin (1979 have developed a scheduling algorithm for executing subtasks in parallel in multiprocessor controller. Saridis (1976) has applied hierarchical control theory on controlling of prosthetic arm, but has used a minicomputer to realize control.

However, the above solutions ignore how to integrate different kinds of sensors to the manipulator system. Thus, a different approach which allows for a flexible way to modify the basic system according to external requirements is needed. The system presented in this paper is based on the distributed microcomputer network where the sensor modules are connected to the control system simply through a standard serial communication line. The sensor interface module contains computational capability for preprocessing and on-line operation for matching the sensor to the total manipulator control system. The design results have been applied in a high-power supervisory controlled manipulator aimed for handling of timber.

This paper consists of three main sections. The first section describes the control hierarchy, the second the organization of the control system and the third contains the description of the experimental system.

CONTROL HIERARCHY

Processing levels

The generalized interaction diagram for the manipulator control levels is shown in Fig. 1. The control system has been separated into three hierarchical levels: sensor information processing level, coordinating level and action level.

In order to accomplish a complex manipulative task the control hierarchy must realize the following phases:
(a) Manipulator communication,
(b) Interaction with the environment,
(c) Coordinated motion control.

In supervisory controlled manipulator either analog joysticks or symbolic commands can be used for man-machine communication. The main function in the communication are the manual control and the teaching of the manipulator.

Interaction with the environment means that the manipulator has the ability to integrate the feedback signals from the sensor subsystem into the total system and update its strategies or sequences of control actions to accomplish the task.

The coordinated motion control allows the manipulator autonomously perform the given subtask without the assistance of the operator and then submit the control for manual operations.

The need to monitor the environment and adapt the manipulator operations based on the detected conditions imposes several requirements on the manipulator control system. In the information processing level the sensor data must be processed and analyzed and the feedback information passed into the coordination level in real time. In higher coordination level the results are interpreted and as a result either the control actions are modified or the new ones are determined to take next. At the lowest level the mechanism of the manipulator system is commanded to move according to the resulted sequence of operations.

Generation of the control actions

The control actions are generated towards solving the positioning and orientation problems of the manipulator. Next a heuristic approach has been described for formulating the motion control (see Fig. 2). A given task (e.g. manual postioning of the manipulator end-point or the repeating of the teached trajectory) is assigned to the manipulator by an operator. The input commands are recognised and the respecting function mode is invoked (manual, auto or teach mode). The input command results in some desired position and orientation coordinates of the manipulator. The state of the manipulator and the data from the external sensors affect on the calculations made on the coordination level. The homogenous coordinate transformations are then used to derive the motion in joint coordinates. These values are used in the action level together with the data of the internal state for joints servo control to move the manipulator to the desired target.

THE ORGANIZATION OF THE CONTROL SYSTEM

The multiprocessor network

In order to be able to meet the changing demands of the working environment as well as to cope up with possible future sophisticated demands, a multiprocessor control system with a modular hardware and software structure is suggested. The multiprocessor system for manipulator control is shown in Fig. 3.

The architecture of the control system is based on the hierarchical partitioning of the system as described in the previous section. The three main levels each have a microcomputer processing and executing the algorithms. The processor units are capable to execute trigonometric and floating point operations. The individual computers process in parallel and communicate with each other through common variables. The variables related to the sensor subsystem are obtained via servo control module which updates the variable table located in the common memory at regular intervals. The sensor data processing module then analyzes data and the coordination module takes care of that the manipulator system responds both to the internal and external sensory information so that the desired goal can be archieved. Besides of the execution of the servo algorithm, the servo control module has to communicate with the sensor subsystem and the man-machine interface.

The described control architecture provides a real time control behaviour in complicated and time critical processes. Modular systems with distributed intelligence have advantages with regard to
- system diagnosis
- controlled access of system parameters
- prioritized execution of the application tasks.

The sensor subsystem

The sensor subsystem is connected to the control system via serial line which circulates through every sensor interface module. By using RS-232 standard communication interface, a feasible way is achieved when connecting different kinds of sensors to manipulator system. Each sensor is connected to the manipulator system via a sensor interface module which is based on a single chip microcomputer. The sensor interface module provides possibilities for controlling and timing the transducer interface logic, preprocessing the sensed signal and for communication with the master unit. The microcomputer locating near the transducer eliminates the interface wiring and releaves the higher level computer from simple data manipulation operations. It is also possible to implement self-diagnostics to the sensor modules to increase the system reliability and maintenance.

The sensor interface module consists of three subunits: line interface, microcomputer and the sensor interface unit.

THE REALIZATION OF THE MULTI-PROCESSOR CONTROLS IN A FOREST MANIPULATOR

The manipulator overview

The forest manipulator has been constructed from a log-loader and it has been designed to be mounted on a forwarder for handling and processing of trees. The manipulator has six degrees of freedom and it is controlled by proportional type hydraulic valves. The experimental laboratory system used for the study of interactive control is shown in Fig. 4.

Control principle

The human operator, computer system and the mechanical boom system form a semi-autonomous manipulator system. The role of the operator is to supervise the manipulator system, to solve the problems caused by changes in the unknown environment and to relay the function plan to the control system. The computer system controls the mechanical manipulator according to the commands given by the operator via the control console and informs the operator of the state of the manipulator.

The man-machine interface consists of a terminal and a dedicated control console by which the operator can manually control manipulator motions, select computer assistance functions and observe control status. The control console includes a set of function buttons and a pair of spring-loaded joysticks. The joysticks are used for manual control in joints direct or coordinated control mode. The function keys are used for teaching and control mode selection (Manual/Automatic control). Using the function keys and joysticks the operator can take the manipulator through a sequence of tasks, selecting control modes and manual operations that are most appropriate for each subtask.

Implementation of the control system

The manipulator control system is based upon 16 bit microcomputer modules and associated peripheral units. In sensor subsystem single-chip 8-bit microcomputers are used. The modules used are VME bus (Motorola) compatible double Eurocards. The system architecture (Fig. 5) is analogous to the system layout shown in Fig 2. While the control system can contain all the modules mentioned in section 3.1, only the modules to fulfill the basic coordinated motion control was incorporated into the control unit. The M68VECPU100 acts as a controller in this VME bus system. The M68VERAM100 module provides storage capacity for system parameters.

The man-machine interface as well as the valve drive control has been implemented by the programmable I/O lines provided by the M68VECPU100. The valve drive control logic contains two cards with one having 8-bit DA conversion logic and the other current amplifiers for supplying the control current to the proportional valves. The panel interface provides all the necessary I/O-lines to interface pushbuttons, selectors and indicator lights mounted on the panel. It drives also an α-numerical display which outputs messages and information to the operator during the work phases.

In sensor interface modules single chip microcomputer 8051, (INTEL) was selected as a controller. The module can have on-board memory 12 kbytes of program memory (ROM) and 2 kbytes + 128 bytes 8051 internal data memory (RAM). In addition to address and I/O decoding logic the module has the standard baud rate generator. For manipulator internal state sensing incremental type transducers are used. The microcomputer keeps count of the pulses given by the transducer and converts the count to respecting 16 bit integer numbers containing the position data in radians as well as in sinus and cosinus values. Also the joint's velocity is detected by the incremetal transducer. By taking a mean of the succeeding pulse intervals an average velocity value is obtained. The master can read the sensor data by sending a frame containing the 4-bit sensor address and the 4-bit function code which expresses the type of the data wished to be transmitted by the sensor. The sensor module which has been addressed transmits back its address and function code and the preprocessed sensory data in two bytes. For four joints the data updating interval is obtained less than 20 ms at 19200 baud rate which is quite satisfactory for this kind of a large boom system.

CONCLUSION

This paper has introduced a consept for the modular construction of the manipulator control system. Further, an example to demonstrate the organization of the information processing based on the distributed microcomputer network in a forest manipulator was described. The presented three level computer hierarchy supports the real-time processing requirements and promises solution to problems concerning to the sensor interfacing and data processing.

The developed multiprocessor control system has proved to be suitable for the control of a large mechanical boom system. The basic computer modules are based on commercial products. The realized system has for the present only one VME bus compatible controller so further software modules have still to be implemented for the application comprising the modules required in three level computer hierarchy.

REFERENCES

Barbera, A.J., J.S. Albus, and M.L. Fitzgerald (1979). Hierarchical control of robots using microcomputers. Proc. of 9th Int. Symp. on Industrial Robots, Washington D.C., pp. 405-422.

Bisiani, R., and R. Cassinis (1976). The development of a multi-microprocessor system to be used in the control of an industrial robot. Mi-Mi '76, Zürich, pp. 58-63.

Luh, J.Y.S., and C.S. Lin (1979). Multiprocessor controllers for mechanical manipulators. Proc. of IEEE 3rd Int. Computer Software Applications Conference, Chicago, pp. 458-463.

Saridis, G.N., and H.E: Stephanov (1976). Hierarchical intelligent control of a prosthetic arm. Technical report, no TR-EE 76-21, Purdue University, West Lafayette.

Shin, K.G., and S. Malin (1980). A hierarchically distributed robot control system. Proc. of IEEE 4th Int. Computer Software Applications Conference, Piscotaway, pp. 814-820.

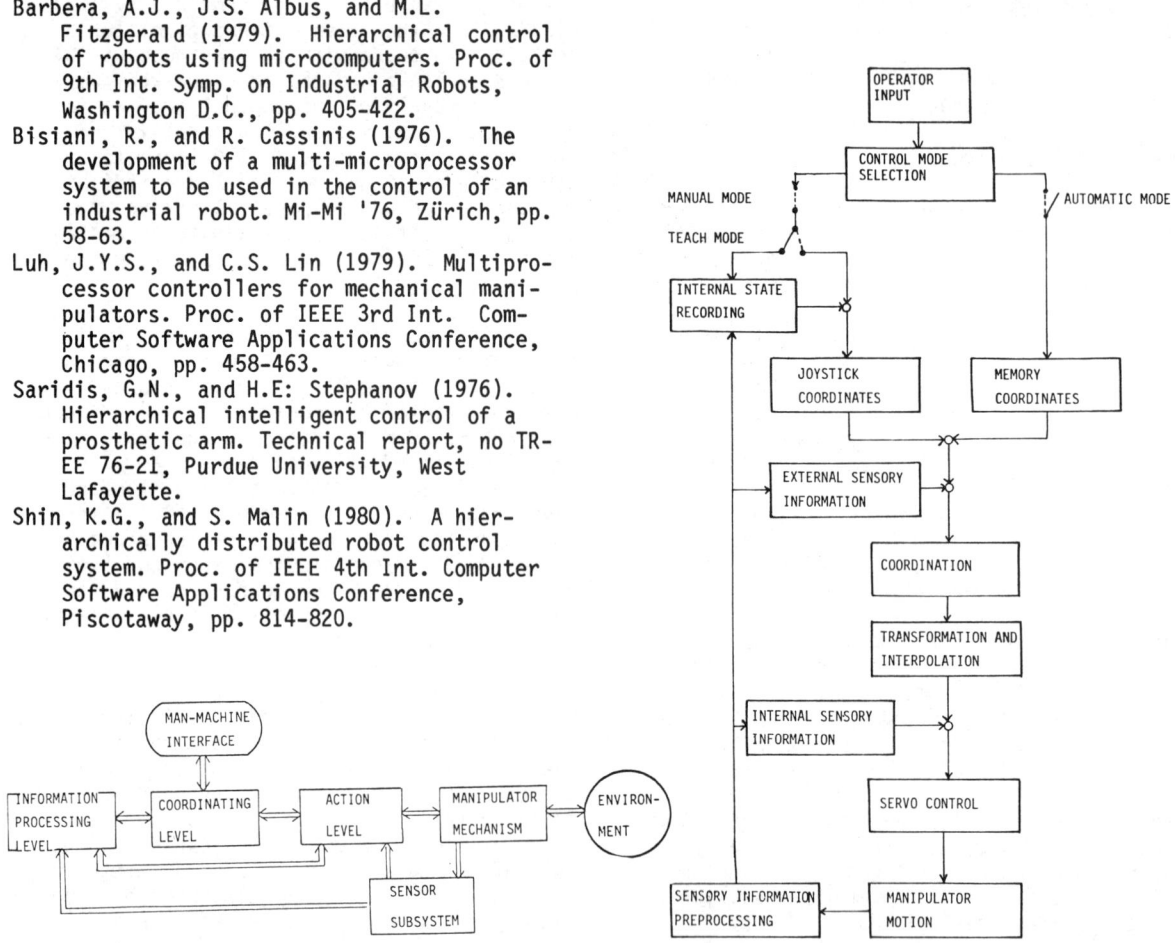

Fig. 1. The control levels in a manipulator system.

Fig. 2. The motion control algorithm.

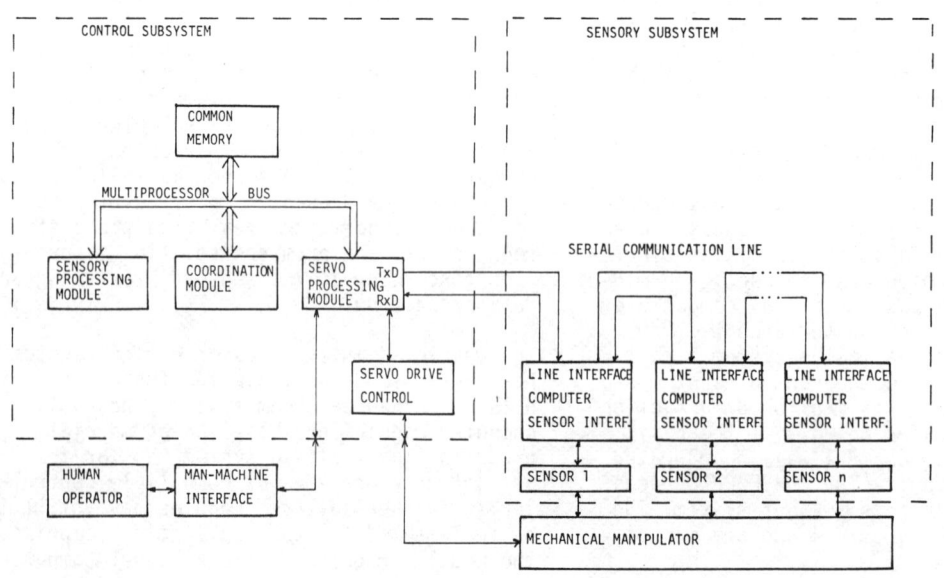

Fig. 3. The multiprocessor system for manipulator control.

Information Processing System for Forest Manipulation 241

Fig. 4. The experimental manipulator system.

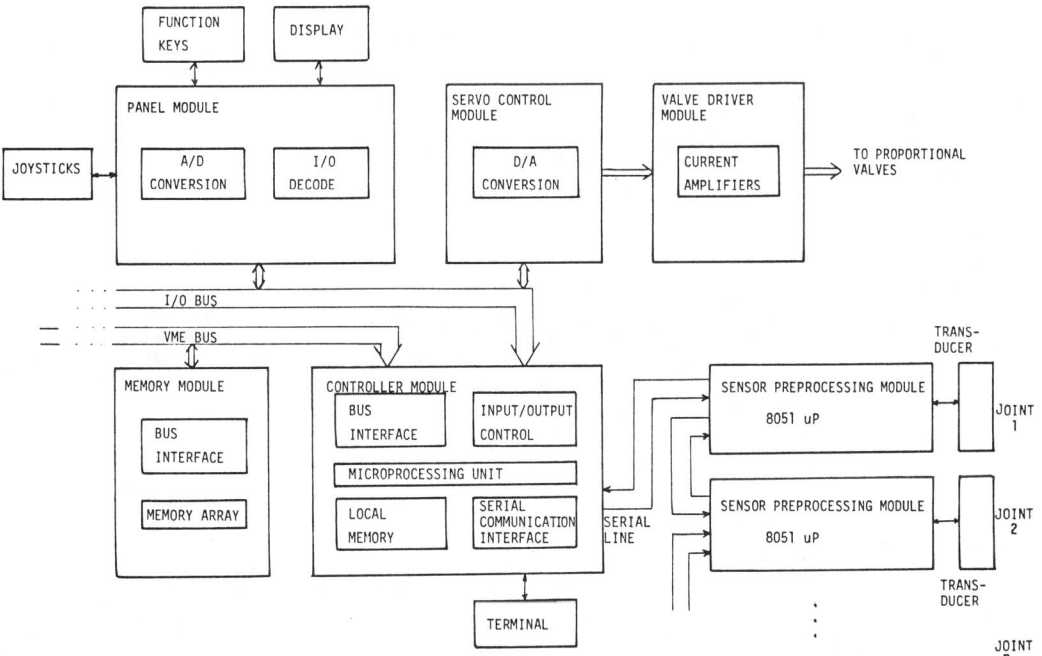

Fig. 5. The architecture of the control system in a forest manipulator.

Copyright © IFAC Real Time Digital Control Applications
Guadalajara, Mexico 1983

DIRECT DIGITAL ROBOT CONTROL USING A FORCE-TORQUE SENSOR

G. Hirzinger

*Deutsche Forschungs- und Versuchsanstalt für Luft- und Raumfahrt e.V. (DFVLR),
Institut für Dynamik der Flugsysteme, D-8031 Wessling, Federal Republic of Germany*

Abstract. The paper treats the analysis and design of external sensory feedback loops in robots. Main emphasis is laid on force-torque sensors but the techniques described are more general. It is shown that by using very simplified dynamical robot models the main effects occuring in a robot's reaction with its environment are amenable to a systematic treatment. Timing schemes based on the interactions between the robot's internal microprocessor and a supervisory control microprocessor are outlined and related to sampled data theory. Digital control techniques using z-transform and state space methods are applied to typical problems as are "moving to and pressing onto a surface with prescribed force", "motion speed control in a grinding process", and "robot teaching via force-torque-sensors".

Keywords. Force-feedback control; robot dynamics; robot-teach-in; direct digital control

I. INTRODUCTION

Control problems in robotics may occur on different levels. The lowest one is concerned with servo control of the joints, a very difficult task from the control theoretical point of view especially for fast motions due to nonlinear coupling and inertia variations. Yet in this paper we divert to the next higher level where we assume a given robot with a given (more or less satisfactory) joint control system and a sensor indicating the robot's reaction with its environment. So we arrive at an outer control loop that - with microprocessors involved - may be seen as a classical digital control system; nevertheless it is often very difficult to specify a control design goal properly. In a very advanced robot system these design goals may stem from decisions on higher levels of hierarchy not in discussion here; yet whatever these goals are and whereever they come from, a necessary prerequisite for them to be achievable is the availability of basic digital control modules. A few of such control modules are derived and analyzed in more detail here.

The sensor used here to monitor the robot's reaction with its environment is a three-axis-force-torque sensor. From many applications as occur e.g. in assembly a force-torque sensor is a powerful tool, partly even more powerful than a much more complicated visual system as we know from human experience. Yet as a short-come it provides information on the system's state (i.e. the robot's position and velocity) less complete than we are used to from normal sampled-data systems. As an example a force-torque sensor may get relevant information on the robot's position only when it is already "too late", that is the robot has hit some obstacle. Other sensors like proximity sensors avoid this drawback. As a common feature of force-torque and proximity sensors however, they provide only relative positional information between robot and object. Yet in most cases this is just what we are interested in and so it need not be seen as a great problem.

The force-torque sensor as developped in our institute is shown in fig. 1.

It is based on mechanical deformations that are measured by strain-gauges stuck on the four horizontal spokes and the four vertical stays. Via matrix transformations in a digital signal processing unit three mutually perpendicular forces and three torques are issued. For the examples given below it is of minor importance whether these calculations are performed in a separate processor (as is meanwhile our standard version) or are performed in a supervisory microcomputer that is charged with coordinate transformation and control algorithm, too (as was realized here).

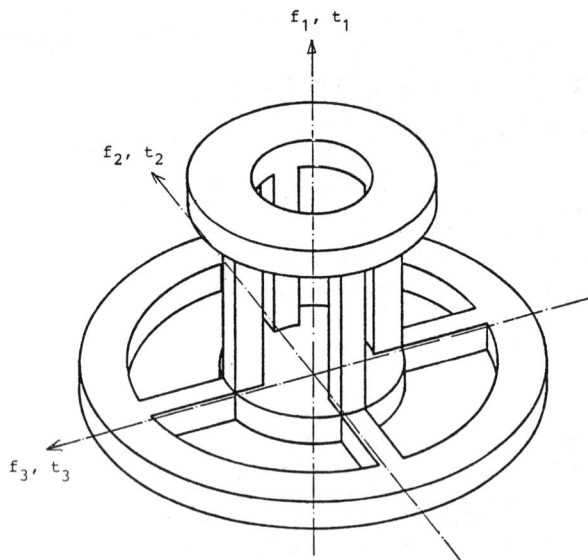

fig. 1 Sensor arrangement

II. THE ROBOT AS A SAMPLED-DATA SYSTEM.

In fig. 2 the structure of the closed loop is shown in a block diagram. We will discuss different applications, but in each of them the basic task is to make the robot (our dynamical system) to respond to sensor information (of the force-torque-kind) in an outer control loop.

A supervisory microcomputer (formerly an Intel 8085 with arithmetic unit, now an Intel 8086/87 combination) has fast, parallel access to the robot's bus via an interface board we designed in our lab. Timing is shown in fig. 3.

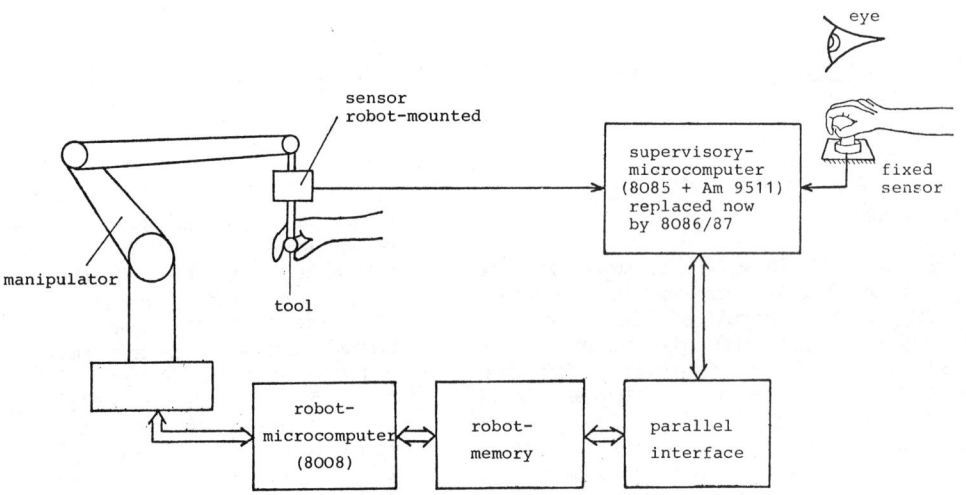

fig. 2 Closed-loop with force-torque sensor

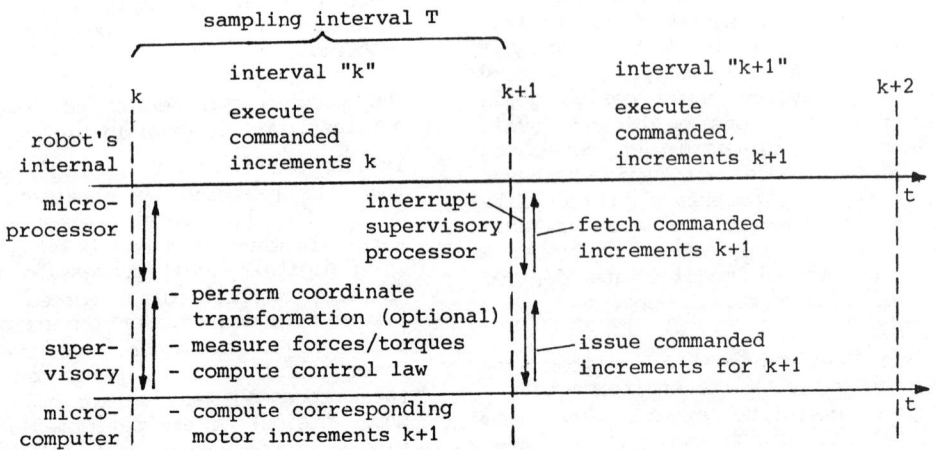

fig. 3 Timing diagram for processor actions

The robot's internal microprocessor acts as a slave who has just to execute the joint increment commands it receives from the supervisory control computer, which has information on the sensor signals and is able to work in cartesian space. The sampling interval T depends on the mode of operation. If no coordinate transformation is required we use a sampling interval of 20 msec. If coordinate transformation is involved we refer to a sampling interval of 90 msec in this paper (meanwhile by using the 16 bit-microprocessor it has been reduced to 30 msec, but this is of minor importance here).

Thus the robot may be seen as a normal sampled data system where every T seconds sensor signals are measured, a control law is computed and actuator commands are issued to the joints. The robot representing the "plant" responds in a certain way to be discussed in the following section.

III. ROBOT DYNAMICS

In fig. 3 the commanded set of joint increments corresponds (via a static transformation) to a commanded step or motion increment

$$\underline{u} = \begin{bmatrix} \delta \underline{x}_c \\ \delta \underline{\phi}_c \end{bmatrix}$$

in cartesian space ($\delta \underline{x}_c$ translational, $\delta \underline{\phi}_c$ rotational). A dynamical robot model relates commanded and executed motion. Setting up a complete six axis robot model that shows up flexibility and nonlinear effects in the joints is a tedious and even questionable job. Instead our considerations were as follows:

There are different main effects influencing the robot's behaviour. One is the joint motor dynamics, which may be assumed first order. Another one is the finite stiffness in the joints acting as kind of spring, so that in case of free motion acceleration of masses or moments of inerta and damping should result in second order systems. Yet, these systems would occur in each joint, so again we would arrive at fairly high order models. Instead of elaborating such a model we tried to measure the robot's dynamical behaviour. But to clarify the comments made, fig. 4 first is to indicate the different types of robot position in question. For simplicity let the scalar positional variable x stand for the six-dimensional vector comprising translational and rotational positions. In fig. 4 the meanings of the different robot (end-effector) positions are as follows:

x_c = commanded position

x_a = computational position due to actual joint positions

x = real robot position (e.g. forced by contact with a fixed surface)

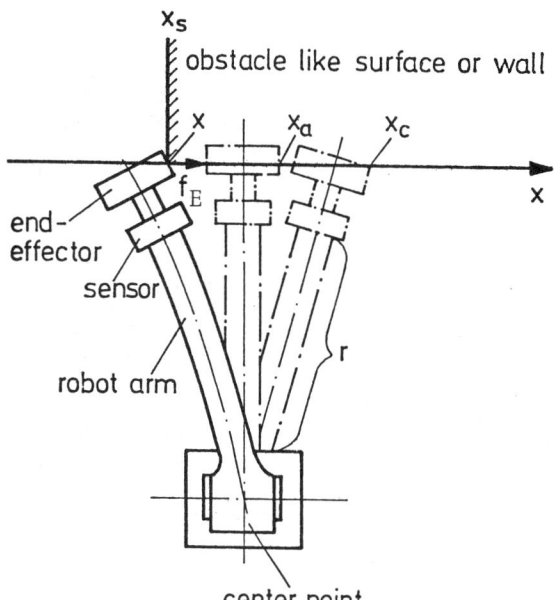

fig. 4 Simplified robot mechanics

In fig. 5 typical responses to step commands x_c are given for an ASEA robot, where a) and b) show free motion cases, the real robot positon x being measured optically by a TV-tracker, while c) was measured via the force-torque sensor, the robot being always in contact with a fixed surface. Assuming an overall stiffness coefficient E, the dynamics of x_a show up in the sensed force f_E for x = constant via the equation $f_E = E \cdot (x_a - x)$

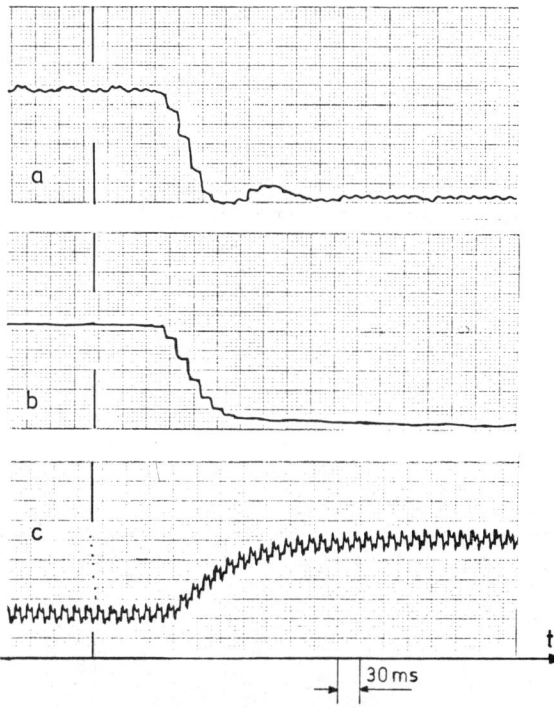

fig. 5 Typical reactions to commanded motion steps
a) free motion horizontal
b) free motion vertical
c) contact force response

Our arguments then were as follows. From fig. 5 c) we conclude that the overall joint motor dynamics may be approximated by a dead time and a first order system with time constant τ. This model is valid in case with a fixed surface. From fig. 5 a) and b) this same model seems sufficient in case of free motion (i.e. we assume $x = x_a$ then) especially for the applications mentioned later where at least in the final stage of motion there is always contact to some environment. Thus the overshoot effect in fig. 5 a) seems of minor importance then.

In case of reactions with non-fixed objects (pulling the robot's hand with a human hand or grinding a burr that acts like some tough fluid) the stiffness force f_E is modelled as to overcome the prevailing acceleration and damping forces, thus arriving at a relation between x_a and x.

But let us come back to the $x_c - x_a$ robot dynamics. A block diagram of this extremely simplified model is given in fig. 6.

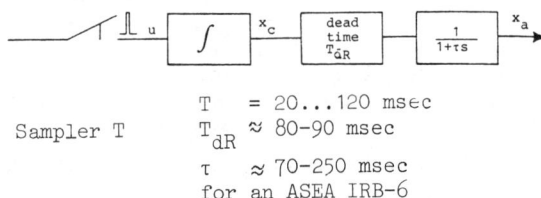

$T = 20...120$ msec
Sampler T $T_{dR} \approx 80-90$ msec
$\tau \approx 70-250$ msec
for an ASEA IRB-6

fig. 6 Simplified model relating commanded and actual positions as computed from actual joint positons.

This model is to indicate that at the beginning of a sampling interval a motion increment u in cartesian space is issued to the robot as a δ-impulse (realized as a set of joint motor increments) yielding a piece-wise constant commanded position (or/and orientation) x_c. This in turn generates the actual positon (or/and orientation) x_a via dead time T_{dR} and first order dynamics with representative time constant τ.

As already indicated we treat u, x_a and x_c as scalar variables though in general they represent six-dimensional vectors consisting of 3 translational and 3 rotational components

$$\left(i.e.\ u \triangleq \begin{bmatrix} \delta x \\ \delta \phi \end{bmatrix},\ \mathbf{x}_c \triangleq \begin{bmatrix} x \\ \phi \end{bmatrix}_c,\ x_a \triangleq \begin{bmatrix} x \\ \phi \end{bmatrix}_a \right)$$

yet the assumption confirmed by measurements, that the principle dynamics are very similar for all possible motions leads to this simplification. Thus we arrive a a robot transfer function $G_R(s)$ that is thought to be representative for all six components.

$$\frac{x_a(s)}{u(s)} = G_R(s) = \underbrace{e^{-T_{dR}\cdot s}}_{\text{dead time}} \cdot \underbrace{\frac{1}{s(1+\tau s)}}_{G_{RO}(s)\ =\ \text{robot dynamics}} \quad (1)$$

with a sampled input u.

A state space description (without dead time) is given by

$$\begin{bmatrix} \dot{x}_c \\ \dot{x}_a \end{bmatrix} = \underbrace{\begin{bmatrix} 0 & 0 \\ \frac{1}{\tau} & -\frac{1}{\tau} \end{bmatrix}}_{F} \cdot \underline{x}^* + \underbrace{\begin{bmatrix} 1 \\ 0 \end{bmatrix}}_{g} \cdot u \quad (2)$$

and the discrete version

$$\underline{x}^*_{k+1} = \underline{A}\ \underline{x}^*_k + \underline{b}\ u_k \quad (3)$$

where $\underline{A} = \begin{bmatrix} 1 & 0 \\ 1-e^{-\frac{T}{\tau}} & e^{-\frac{T}{\tau}} \end{bmatrix};\ \underline{b} = \int_0^T e^{\underline{F}(T-t)} \delta(t)\underline{g}$

$$= \begin{bmatrix} 1 \\ 1-e^{-\frac{T}{\tau}} \end{bmatrix}$$

In the sequel we describe different applications with force-torque-sensor feedback, the various extensions of the robot models implied and the digital control design aspects.

IV. MOVE TO AND PRESS ON A SURFACE WITH PRESCRIBED FORCE

Consider the following practically important task. The robot moves in some spatial direction along a straight line and expects some surface of a priori unknown position x_s where he is to press on with a certain force f_o. The sensor is mounted between robot and hand or end-effector. Of course after contact the settling time should be short and - if possible - the robot should be able to perform better if he has to repeat this task. For in a second or further approaches he would have information where the surface actually is.

The control scheme we realized for this problem makes use of the fact that reactions of the robot are not required in cartesian space here but only along an approximately straight line. This line may be generated by a fixed vector of joint increments which is multiplied by a factor due to the controller's output in each sampling interval. Thus no coordinate transformation is necessary here but the model fig. 6 is still valid with x_a and x_c being real scalar measures now along some spatial direction. A sampling interval of 20 msec according to the clock of ASEA's "point-to-point-linear"-instruction was chosen and a robot time constant $\tau = 125$ msec was assumed.

The basic idea was to develop a control scheme that makes use of the fact that though force is to be contolled finally, most of the time no force measurements are available, and when they occur, it may be too late to stop the robot in time. Thus a position control with force-based corrections was developed.

The controller structure including a state space model of the robot's motion is shown in fig. 7.

Due to an internal model estimates (marked by ^) of commanded and actual position as well as surface positions x_s are available. A predictive part allows to model and compensate the dead time $T_{dR} = n_d \cdot T$. This means that after the dead time T_{dR} has passed - in case there are no disturbances - the system behaves as if there were no dead time at all. T_{dR} is assumed to be an integer multiple n_d of the sampling interval $T (n_d = 4$ as we have $T_{dR} \approx 80$ msec and $T = 20$ msec here).

By iterating equ. (3) n_d times we get the predicted state $\hat{\underline{x}}^*_{k+n_d}$:

$$\hat{\underline{x}}^*_{k+n_d} = \underline{A}^{n_d} \tilde{\underline{x}}^*_k + [\underline{b}, \underline{A}\,\underline{b}, \ldots, \underline{A}^{n_d-1}\underline{b}] \begin{bmatrix} u_{k+n_d-1} \\ \vdots \\ u_k \end{bmatrix}$$

(4)

This predicted state is compared with the nominal steady state values $\hat{x}_s + f_o/E$, which indicate that to arrive at a prescribed force f_o, the robot must be commanded to a position behind that of the surface (x_s) so that via the stiffness E the proper force is generated. As we assume that in case of free motion $x = x_a$ (fig.4), we have to take in account the stiffness force only when the robot is in contact with the surface, i.e. $x_a - x_s \geq 0$ for motions in positive direction. This stiffness force is then fully transduced to the sensor if we neglect any small remaining motions. The position x_s of the surface (modelled as $\dot{x}_s = 0$) is estimated by a very simple

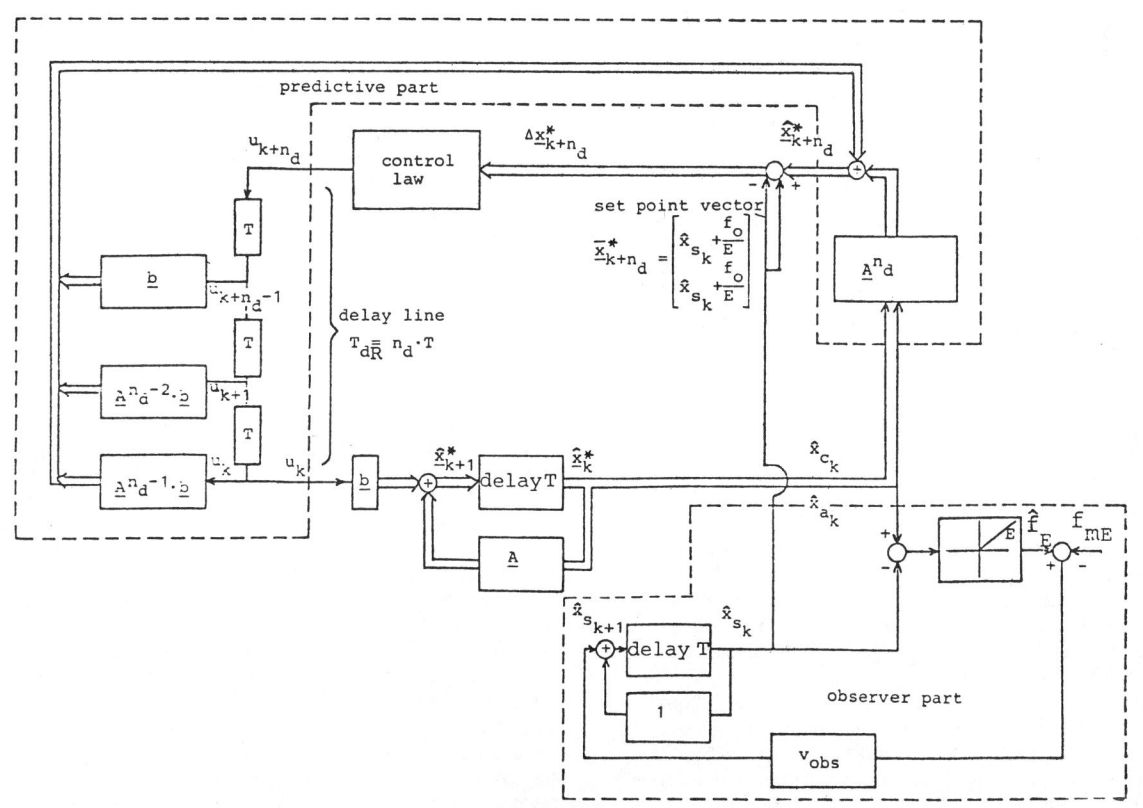

fig. 7 Control scheme

observer comparing measured stiffness force f_{mE} and estimated force $\hat{f}_E$. The difference is multiplied by an observer gain V_{OBS} and fed back to correct the estimate of the surface position. The corresponding equation is

$$\hat{x}_{s_{k+1}} = \hat{x}_{s_k} + \underbrace{(E(\hat{x}_{a_k} - \hat{x}_{s_k}) - f_{mEk})}_{\hat{f}_{Ek}} \cdot V_{OBS} \quad (5)$$

Concerning $\hat{x}_c$ and $\hat{x}_a$ no observer corrections are provided as $\hat{x}_c$ is exactly known and errors in $\hat{x}_a$ mainly stem from wrong modelling.

The equation (5) is active only if either f_E or f_{mE} exceed some positive threshold. The force-torque measurements are made at the very beginning in each interval (see fig. 3). Thus the assumption made in the observer equation (5), that measurements are made at time k (or at the beginning of interval k) to produce an estimate for k+1 is fulfilled. With this estimate a new control variable u_{k+1} is computed during interval k and can be commanded to the robot at time k+1.

For the observer gain V_{OBS} a value V_{OBS} = 0,5/E was chosen as a compromise between speed of observation and sensitivity to measurement noise without a detailed analysis. Different control laws were investigated, especially time-optimal control laws with saturation constraints. Yet due to the uncertainties in the robot model a less fast linear controller of the form

$$u_k = V_1 \cdot \Delta \hat{x}_{c_k} + V_2 \Delta \hat{x}_{a_k} \quad (6)$$

was realized. Gains V_1 and V_2 were computed thus that the closed-loop-system has two poles at z=0,5 in the z-domain.

The initial estimate of the surface position may be arbitrarily wrong. For a completely unknown surface position one may start with $\hat{x}_s = 0$ (corresponding to the robots position); then by the observation effect the robot moves "cautiously" along the given direction, until he senses the surface and slowly ends at the desired force. If he has to repeat this action, he starts now with the sensed surface position as estimate and moves much faster then. We verified this phenomenon (fig. 8 a)), in practical experiments with our ASEA-robot. Fig. 8 b) shows a force-recording for the contrary case, too. Here the robot has been given a wrong initial estimate $\hat{x}_s > x_s$, so the robot moves fast and suddenly hits the surface with overshoot in the force. But already in the second trial transients are soft due to the actualized estimate of surface position (fig. 8 c)).

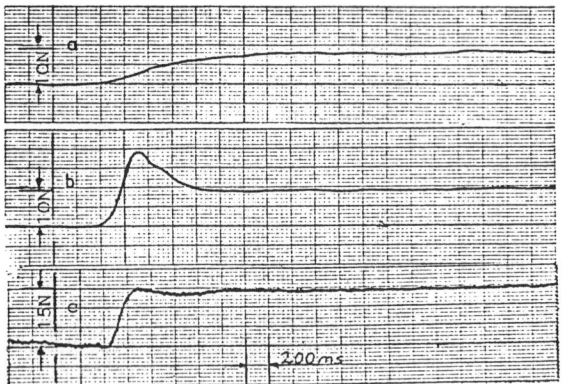

fig. 8 Force transients

a) surface is far behind expected, $x_s > \hat{x}_s$ (f_o=10N)
b) robot hits an unexpected surface $\hat{x}_s > x_s$ (f_o=10N)
c) robot makes a second trial with improved initial estimate $\hat{x}_s$ (f_o=15N)

An application of the techniques proposed here is that fast approaches to objects - including contact forces - are possible without crashing them. Of course the force transients are dependent on the difference between assumed and actual temporal behaviour of the robot with no force overshoot when there is good coincidence. Meanwhile we have installed a learning system so that the robot is able to adapt its model parameter τ by the sensory experience it makes, especially the force overshoot.

V. FOREWARD SPEED CONTROL IN A GRINDING PROCESS

Robots are ideally fit to release man from inhuman work like deburring cast iron motor blocks. In a German cast iron manufactury an ASEA-IRB 60 (the great version) has been deburring for two years now using our force-torque-sensor for sensing the grindstone wear and correcting this path correspondingly. As a second step we introduced burr-dependent forward speed control along a given (may be corrected) path. So like in the past section we may resign coordinate transformation, work with a sampling interval of 20 msec and scale the given "point-to-point linear" joint increments according to the requested path velocity.

But what is the design goal here? It has been pointed out in [8,9] that a reasonable design goal assuring good grinding performance is a constant burr volume decrease, i.e. the higher or broader a burr is, the slower should be the grinding machine's path velocity. Furthermore it has been pointed put in [8,9] that an appropriate measure for the burr volume decrease may be seen in the damping force $f_d = D(x) \cdot \dot{x}$, like a tough fluid onto the robot who moves with velocity $\dot{x}$. Roughly speaking the burr's shape is contained in the burr damping parameter $D(x)$, so that for greater burrs $\dot{x}$ must be decreased to keep the forces constant. Yet we must be aware here that the sensor which is mounted between the robot and the heavy grinding machine (fig. 4) measures not only the burr damping forces but also the acceleration forces of the grinding machine. The measurable force f_m (sensor output denoted by f_{ms}) is given by

$$f_m = f_d + m_g \ddot{x} \qquad (7)$$

where m_g is the grinding machine's mass. Strictly speaking f_m should be the force component in path direction but it has been shown in [8,9], that using the force vector's magnitude makes sense, too. Note that we have restricted again here to a one-dimensional representation justified by motions along a given path. For a more general treatment see [9], where it has been outlined, too, that it makes sense to use $f_m s$ instead of f_d as the control variable to be compared with some set point gained by experiments. The control loop is depicted in fig. 9. From fig. 9 we see that now the stiffness force f_E in contrary to the fixed surface case has influence on the real motion of the robot's hand with tool. The stiffness force is counterbalanced by the already mentioned burr damping force (exceeding by far the robot's own damping) and the acceleration term $\frac{\Theta(x)}{r^2(x)} \cdot \ddot{x}$, where Θ is the robot's moment of inertia around the center point and r the corresponding length (fig. 4).

Note from fig. 9 that for simplicity we have replaced the robot's $x_c - x_a$ -dynamics here by a single time constant $\tau^* = 300$ msec. which is to represent the combined effects of dead time and time constant as indicated in fig. 6. We could afford this simplification here because the robot does not work against a fixed wall as in the preceding chapter where uncorrect modelling is much more crucial.

The positional observer indicated in fig. 9 has to take care that the commanded position does not exceed the final position of the present point-to-point-linear instruction as given by the programmer. Thus it corrects the last position increment if necessary.

Another feature of the scheme fig. 9 is a nonlinear characteristic including saturation to take in account the maximum allowable joint angular increments.

Furthermore the depicted characteristics tries to avoid that the robot comes to rest anyhow, because then the material is heated excessively. Instead the robot runs

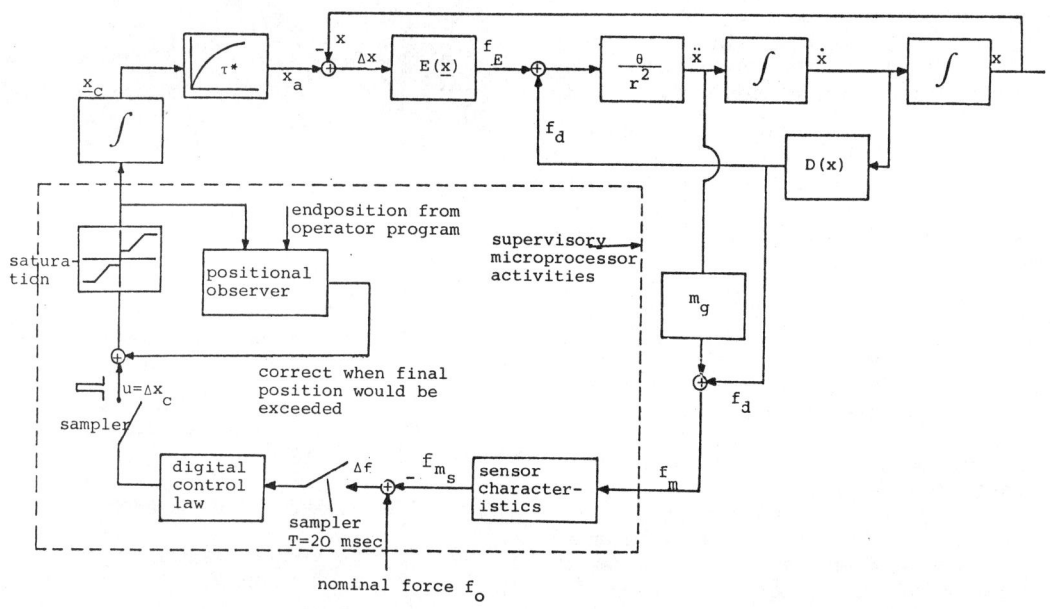

fig. 9. Structure of deburring control loop

into oscillating motions if it meets an unexpected voluminous burr similar to the reactions of a human worker.

For designing a control law several simplifications were made. All physical parameters like stiffnes or moment of inertia were assumed constant. Nonlinearities were neglected. The dynamical order of the system thus arrived at four as can be easily verified by hand of fig. 9. A discrete state space model was set up, and, via the Leverrier algorithm the z-transform of the open loop was derived [11].

It is well known from control theory that for an n-th order system a (n-1)-th order controller with 2n-1 free parameters allows to place all closed-loop-poles arbitrarily. Thus a third order controller was chosen and by placing all poles of the closed-loop-system at z = 0,6 we arrived at the following control algorithm (u as output in $\lfloor m \rfloor$, Δf as input in $\lfloor N \rfloor$)

$$u_k = 1,6\, u_{k-1} - 1,11\, u_{k-2} + 0,48\, u_{k-3}$$
$$- 7,7 \cdot 10^{-4}\, \Delta f_k + 9,32 \cdot 10^{-5}\, \Delta f_{k-1} \quad (8)$$
$$+ 1,16 \cdot 10^{-3}\, \Delta f_{k-2} + 6,67 \cdot 10^{-4}\, \Delta f_{k-3}$$

based on plant parameters

$E=10^4$ N/m; $D=4000$ kg/sec; $\Theta/r^2=200$ kg;

$m_g = 30$ kg, $\tau=0,3$ sec, $T=0,02$ sec.

The controller's transfer function shows up a pole near z=1, thus implying near-integral behaviour for avoiding steady-state errors. Equ. (8) implies -in contrast to the observer scheme as given in the preceeding section- force measurements Δf_k immediately at the beginning of interval k (see fig. 3). Indeed due to computation times this is impossible, so we make the force measurements a few milliseconds before time k without loss in control performance.

The practical implementation of this controller in the cast iron manufactory showed that indeed a much more efficient deburring performance resulted. The main phenomena to be observed are:

a) saving of time by accelerated motion when burrs are small

b) improvement of deburring tolerances by slowing down the robot motion in case of voluminous burrs.

VI. ROBOT-TEACHING VIA FORCE-TORQU-SENSORS

VI 1. General remarks

Teaching a robot's motion is a field where improvements are still necessary in the future. Former keyboards where each key represented a joint are replaced now by keyboards where a key represents translation or rotation around cartesian axis with fixed speed. But there remains awkwardness if the robot is to be taught some oblique or curved motion. Teaching via visual displays - if they are not threedimensional - in our opinion is not of great help here due to the spatial relationships involved. For us using a three-dimensional force-torque-sensor is a better means to tell the robot where to go or how to rotate with using only one hand.

By gripping the force-torque-sensor shaped as a hand-fitting knob (fig. 10) the forces exerted are intended to command the robot's translational motion while the torques are to command the rotational motion. By processing either only forces or torques we may separate the motions, by processing both a superposed motion will occur. Any point and orientation attained hereby or the whole path might be stored in the robot's memory.

fig. 10 Sensor knob for teaching purposes

Let us split up cartesian motion of the robot's hand into position vector $\underline{x}$ and orientation vector $\underline{\phi}$. The force-torque vectors $\underline{f}_{cart}$, $\underline{t}_{cart}$ in a spatially fixed cartesian system are then to cause velocities $\underline{\dot{x}}$ and $\underline{\dot{\phi}}$ respectively (or $\delta\underline{x}$ and $\delta\underline{\phi}$ in discrete version).

We investigated two approaches:

a) A sensor-knob is fixed on ground or on a teachboard. The forces/torques $\underline{f}_{sens}$, $\underline{t}_{sens}$ are either interpreted in the space fixed cartesian system or in a tool-based system. Only this latter case requires multiplication by the orientation matrix $\underline{\Theta}$ to arrive at $\underline{f}_{cart}$, $\underline{t}_{cart}$.

b) A sensor is mounted at the robot's end-effector or hand and allows to pull and turn the robot as wanted. The sensor-based coordinate system is varying and the forces/torques $\underline{f}_{sens}$, $\underline{t}_{sens}$ in the sensor based coordinate system have first to be multiplied by the orientation matrix $\underline{\Theta}$ to arrive at $\underline{f}_{cart}$, $\underline{t}_{cart}$.

In either case dynamics of the human arm are involved and appropriate models had to be taken in account.

Due to the need for cartesian transformation our timing diagram looks as depicted in fig. 11. Sampling time when using our 8-bit supervisory microprocessor was chosen 90 msec.

VI. 2. Model of the human arm

The model of human arm control for fast voluntary movements as applied here is based on investigations of LINN/FOSSIUS [6,7]. For free motion it looks as indicated in fig. 12.

This model is based on different assumptions verified by Linn's work:

a) In fast voluntary movements every 100 msec a pulsed input (= neuron firing) to the muscle fibres occurs, and a corresponding force/torque is generated with certain dynamics $G_m(s)$

$$G_m(s) = \frac{V_m}{(s + \frac{1}{\tau_m})^2} \qquad (9)$$

where τ_m is assumed to be 60 msec.

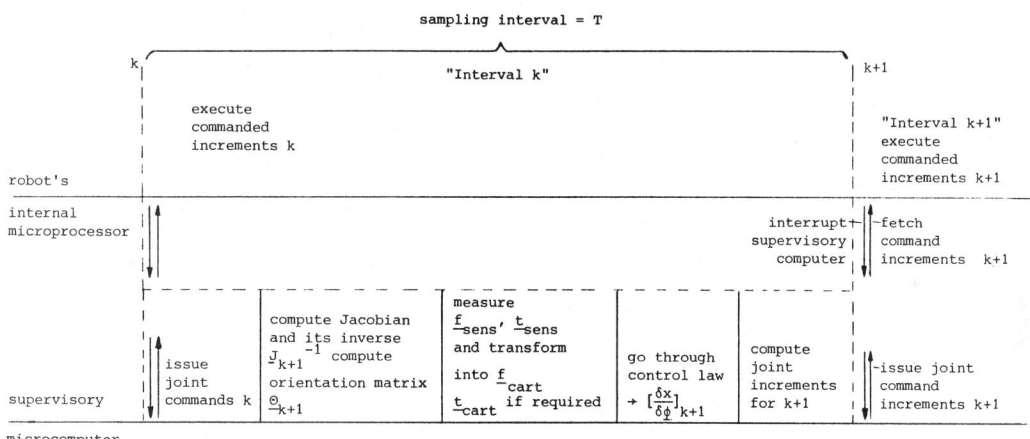

fig. 11 Timing diagram for processor actions

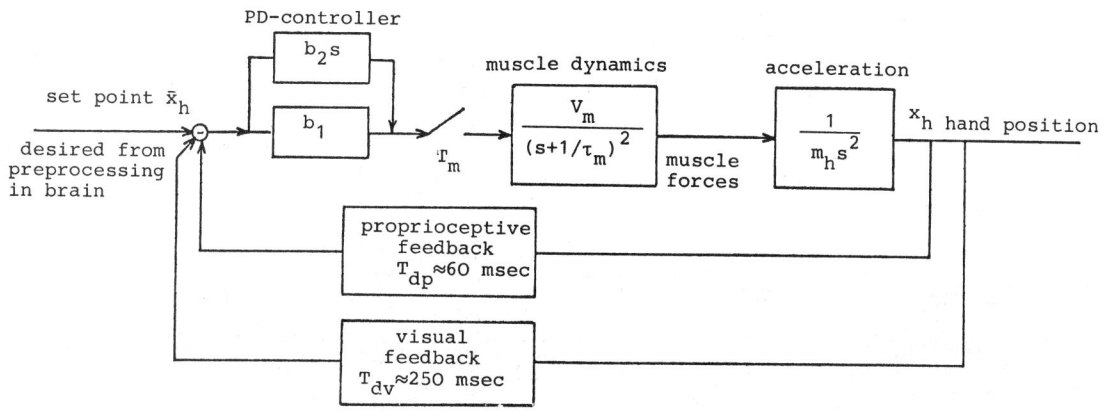

fig. 12 Motion control of human arm [6,7]

We sense the pulsed action of the fast fibres as the so-called "tremor" preventing us from keeping our hand completely quiet.

b) arm position and (variable) set point position are compared and the difference is processed in a PD-controller. The set point is generated by preprocessing hierarchies, i.e., may result from predictions of an object position that has to be tracked. In LINN's work mainly the proprioceptive channel is emphasized and experimentally verified. Proprioceptive sensors (e.g. muscle spindles) measuring position and velocities are distributed in the human arm. The deadtime of $\approx$ 60 msec as assumed in the proprioceptive feedback path is considerably smaller than that in the visual feedback path ($\approx$ 240 msec). Apparently if the arm can be freely moved, the proprioceptive channel is dominant due to its faster reactions, however in case of the fixed sensor where we have to observe the robot motion visually, clearly the optical path is the relevant one. We presumed the same controller there as in the proprioceptive case. As a general saying we found in addition to LINN's stability boundaries, that the arm loop in fig. 12 is fairly insensitive to variations in the parameters b_1, b_2, V_m (-which are adaptable by the brain-) though it is only weakly stable for the whole stabilizing parameter range ($0,1 \leq b_1 \leq 0,6$; $0,2 \leq b_2/\tau_m \leq 4$; $7,2 \leq V_m \cdot \tau_m^2 \leq 210$ $\lfloor N/m \rfloor$).

For our investigations we assumed the stable values $b_1 = 0,15$; $b_2/\tau_m = 0,6$; $V_m \cdot \tau_m^2 = 20 \lfloor N/m \rfloor$.

VI.3. Teaching the robot's path with a space-fixed sensor

Only visual feedback by observing the robot's motion is assumed. As a difference to fig. 12 we don't have free motion of the human hand. Instead the human hand is gripping the sensor and the forces/torques are transduced to the sensor by small arm motions via finite stiffness E_h thought to be concentrated in the hand's inner surface. Of course there is considerable damping D_h in the hand's skin; we assumed critical damping $D_h = 2\sqrt{m_h \cdot E_h}$ but we found that stability of the closed loop as given in fig. 13 is not sensitive to these parameters.

For performing a stability analysis of this loop in the z-domain (as is usual with sampled data systems) the following assumptions were made:

- The robot sampling interval T = 90 msec was used for muscle control too, i.e. T_m = 90 msec. Synchronous sampling was assumed.

- The robot's dead time was set $T_{dR} = T$. The force measuring delay (see fig. 11) was aggregated with the visual perception dead time to yield a representative dead time $T_{dV} = 3T = 270$ msec.

As is wellknown, a dead time $T_d = n_T \cdot T$ in a sampled data system corresponds to the z-transform $F(z) = z^{-n_T}$ and thus can be easily handled. In fig. 13 the z-transforms of the continuous transfer function products $G_m(s) \cdot G_h(s)$ and $G_{RO}(s) \cdot G_{PD}(s)$ have to be found in order to arrive at a complete z-domain description of the loop.

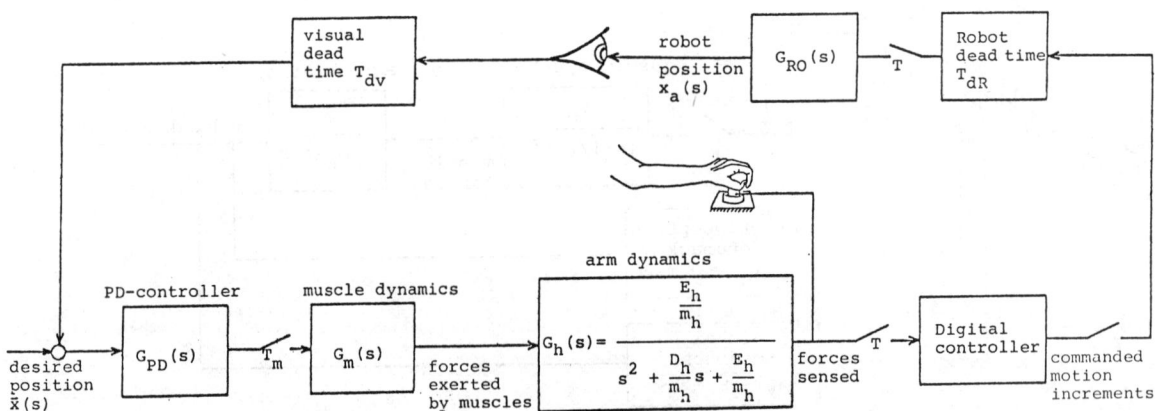

fig. 13 Closed loop when teaching via space-fixed sensor. This is a representative scheme for the force loops, by appropriate normalization the torque loops are similarly treatable.

Pure proportional feedback of the form

$$\begin{bmatrix} \delta \underline{x} \\ \delta \underline{\phi} \end{bmatrix}_K = \begin{bmatrix} \underline{V}_{diag} \end{bmatrix} \cdot \begin{bmatrix} \underline{f}_{cart} \\ \underline{t}_{cart} \end{bmatrix}_k \quad (10)$$

was assumed, so that root locus techniques were applicable. Let V be the scalar gain for each of the spatial directions concerning force-feedback. Then with parameters $E_h = 0,5 \cdot 10^4$ $[N/m]$, $m_h = 2$ kg, and the assumed robot's time constant $\tau = 125$ msec we arrive at the open-loop z-transfer-function G(z).

$$G(z) = 31,47 \cdot V \cdot \frac{(z+73,7)(z-0,13 \pm j0,12)^2(z-0,73)}{z^3(z-0,01)^2(z-0,22)^2(z-1)(z-0,49)} \quad (11)$$

A root-locus plot for the closed-loop is given in fig. 14 indicating that stability is guaranteed up to gains $V \approx 2 \cdot 10^{-4}$ $[m/N]$. This is in good agreement with our laboratory experiments indicating that a gain of $0,2 \cdot 10^{-3}$ $[m/N]$ assures stability and allows fast elegant motions for approaching objects to be gripped. Concerning the torques and corresponding rotational commands the experimentally found optimal gains were about $2^0/Nm$.

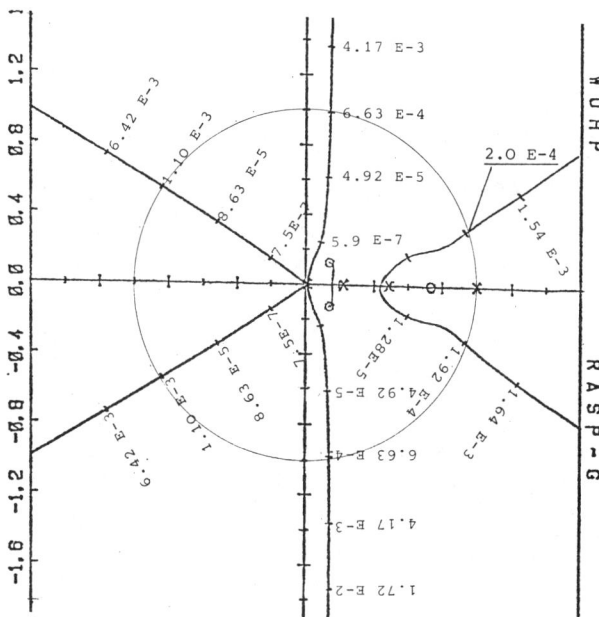

fig. 14 Root locus plot for the closed loop fig. 13

VI.4. TEACHING THE ROBOT'S PATH WITH A ROBOT MOUNTED SENSOR

This case is a bit more complicated than the static sensor case, not so much because we now have a steadily changing sensor-based-coordinate system, but more because we now have to deal with complicated mutual reactions between human arm and robot arm. We assume that the sensor is either mounted at the robot's hand in form of a sensorknob again or that it is integrated in the robot's hand so that by pulling or torquing the robot's hand corresponding motions are executed by the robot. Again we consider only one representative coordinate for stability considerations without loosing generality, as we assume similar behaviour in all spatial directions. Fig. 15 shows up the relevant contact behaviour between human hand and robot, fig. 16 gives the block diagram for the corresponding dynamics.

Fig. 15 is to be seen in context with fig. 11. By the muscle dynamics as described in section VI.2., a force f_w due to human will is generated based on the difference between desired position $\bar{x}$ and actual hand-position x_{ah} according to the human arm's joint positions. The hand's contact surface position x_h is different from x_{ah} due to the finite stiffness E_h assumed to be concentrated in the hand's inner surface. The hereby generated force $f_h = E_h \cdot (x_{ah} - x_h)$ counteracts the arm driving force f_w. On the other side it moves the robot from a position x_{aR} as given by the actual joint positions to the surface constrained position x_R against the spring force $f_R = E_R \cdot (x_R - x_{aR})$ and the inertia and friction of the robot's arm.

The reaction force f_h between human hand and robot is measured by the force-torque-sensor and - via some control law - commands the robot's joints. Their dynamics is represented by the transfer function $G_{RO}(s)$ given by equ. 1).

Stability of this coupled system fig. 16 was investigated for a purely proportional feedback law using state space techniques. For this purpose we neglected the set point input $\bar{x}$, and exchanged the positions of $G_{PD}(s)$ and the proprioceptive dead time, so that the PD-controller does not occur as a dynamical system but only as an additional output vector $\underline{c}_{PD}^T = \lfloor -b_1, -b_2 \rfloor$ that feeds back the two state variables (position and velocity) of the human arm model.

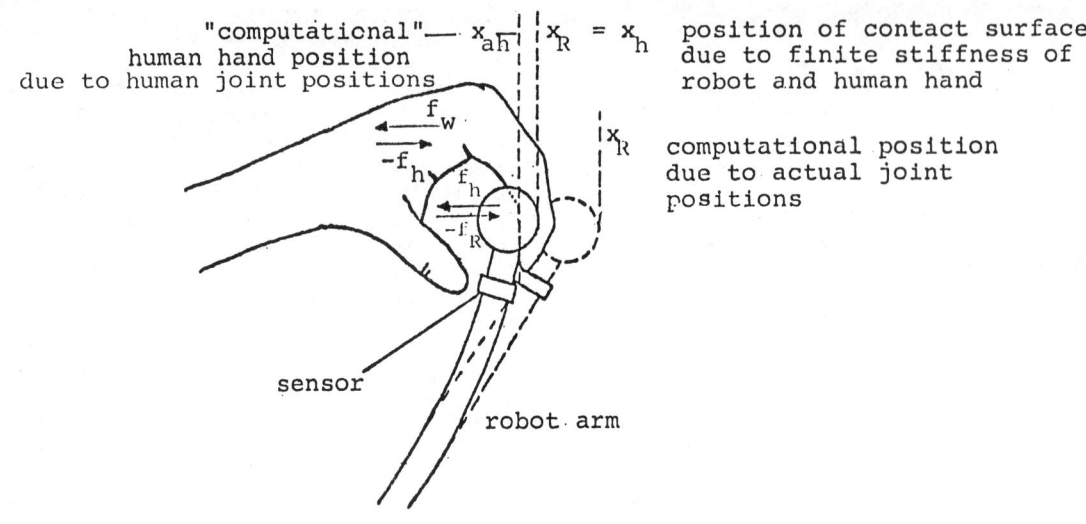

fig. 15 The effect of finite stiffness in robot arm and human hand

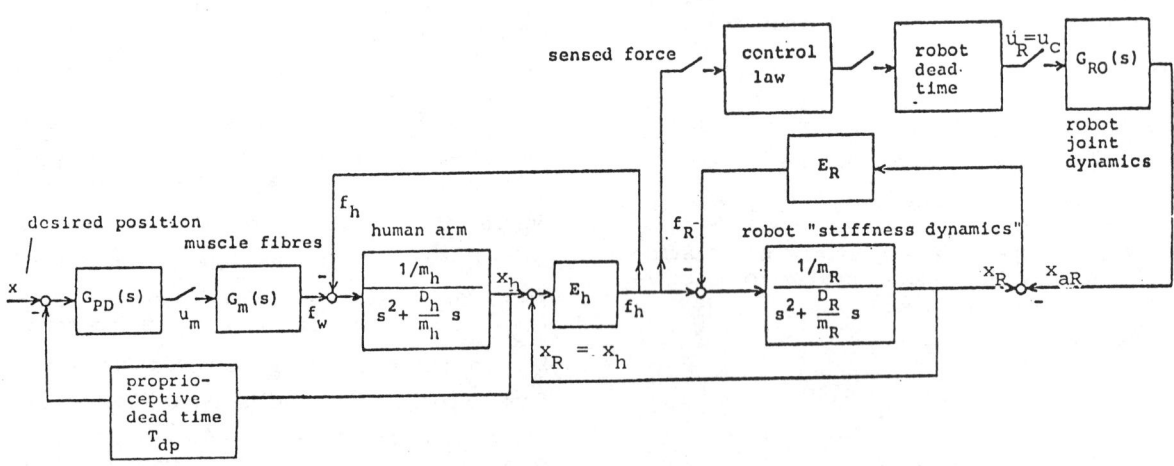

fig. 16 Dynamical interactions in the robot-mounted-sensor case

An overall dynamical model is constructed in two steps:

An overall dynamical model is constructed in two steps (for details see |10|)

a) Second order models are set up for the continous-time systems "muscle fibres", "human arm", "robot stiffness" and "robot joints" as shown in fig.16. Mutual couplings are expressed in an eigth order model with two pulsed inputs u_R and u_m; this leads to a pulsed input description similar to that given in equ. 3.

b) we now close the discrete-loops via two additional state variables representing the robot deadtime and the proprioceptive deadtime and a digital controller of arbitrary order. Thus we arrive at a (10+n)-th order system dynamics where n is the controller's order.

It should be once more emphasized that the loop depicted in fig. 17 is thought to be representative for each of the six force-torque components, though only force-feedback is treated here in more detail.

A systematic controller design would have to minimize the hand's reaction force f_h for a class of typical movements, thus generating the feeling that the robot follows the human hand in a fast and smooth way. In this early state of our investigations we only performed a stability analysis for pure proportional feedback gains V relating motion increments to sensed forces Stability was attained up to gains $V \approx \cdot 10^{-3} \lfloor m/N \rfloor$ (see fig. 17) for the estimated parameter set
$E_h = 0.5 \cdot 10^4 \lfloor N/m \rfloor$, $D_h = 200 \lfloor Nsec/m \rfloor$,
$E_R = 10^4 \lfloor N/m \rfloor$, $D_R = 750 \lfloor Nsec/m \rfloor$, $m_h = 2$ kg,
$m_R = 50$ kg; again the stability margin is not very sensitive to these parameters.

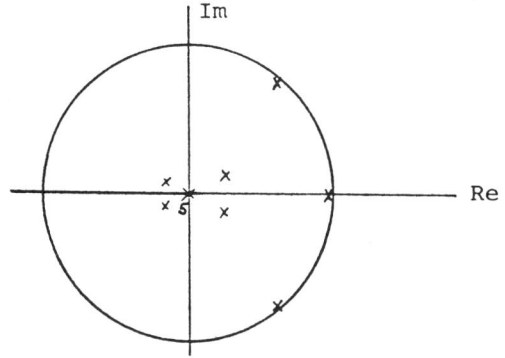

fig. 17 Pole distribution for feedback gain $V = 10^{-3} \lfloor m/N \rfloor$

In our laboratory experiments with the ASEA robot we found gains of $\approx 0.5 \cdot 10^{-3}$ $\lfloor m/N \rfloor$ to be optimal for translational motions, thus confirming the theoretical results despite the considerable, position-dependent variations in robot stiffness and time-constants. As an interesting effect, if the hand's stiffness is set to infinity in the theoretical model, the system is not stabilizable with pure proportional feedback. Indeed we found experimentally that, if one tries to grip the sensor knob mounted at the robot's gripper very tightly and rigidly (thus increasing the human hand's stiffness considerably) human hand with robot tend to start unstable oscillations.

VII. CONCLUSION

Sensory feedback is one of the most challenging fields in robotics today. The main aim of this paper is to show that complex signal processing (as pattern recognition) or hierarchical information structures and robot languages are not sufficient to assure fast reliable robot reactions to sensor information. It is of crucial importance here to treat the robot as a dynamical system that cannot react instantaneously to a command. Considering the robot including its internal joint control system as a dynamical plant that responds to cartesian commands of a supervisory computer we arrive at a normal sampled data system. It is the purpose of this paper to emphasize that in many applications the problem is not to set up a complicated, mathematically derived high-order robot model but to take in account robot dynamics at all with even very simplified low order models that are derived from measurements and reflect the most important phenomena only. It was shown that with these very simplified models feedback designs were accomplished that make the robot's behaviour come near to the desired one and furthermore yield sufficient coincidence between theoretical results and experimental experience.

VIII. ACKNOWLEDGEMENT

The author is greatfully indebted to Mr. Plank, Mr. Heindl and Mr. Brunet who contributed to this paper essentially. Mr. Plank was involved in designing and realizing the digital control laws in section IV and V, while Mr. Heindl and Mr. Brunet contributed in analyzing and implementing the "force-torque-teach" method.

LITERATURE

[1] Craig, J.J., Raibert, M.H.
A Systematic Method of Hybrid Position/Force Control of a Manipulator.
IEEE Computer Software Applications Conference, Chicago, Nov. 1979.

[2] Schmieder, L., Mettin, F., Vilgertshofer, G.
Kraft-Drehmoment-Fühler.
Patent E01L1/22 applied 6.6.79.

[3] Schott, J.
Tactile Sensor with Decentralized Signal Conditioning, 9th IMEKO World Congress, Berlin, May 1982.

[4] Hirzinger, G.
Force feedback problems in robotics. Second IASTED Symposium on Modelling, Identification and Control, Davos, March 82.

[5] Hirzinger, G.
Robots with Force-torque-sensing.
Process Automation, Vol. 4, 1982.

[6] Linn, K.O.
Untersuchung des Kraftverlaufs bei Willkürbewegungen.
Dissertation, Universität Karlsruhe, 1974.

[7] Linn, K.O., Vossius, G.
The relevance of the so-called tremor for the control of voluntary movements. 6th IFAC Congress, Boston 1975. Session 40.4.

[8] Plank, G.
Geschwindigkeitsregelung eines Industrieroboters beim Schleifvorgang.
Diplomarbeit Technical University Munich.

[9] Plank, G., Hirzinger, G.
Controlling a Robot's Motion Speed by a Force-Torque-Sensor for Deburring Problems.
4th IFAC-IFIP Symposium on Information Control Problems in Manufacturing Technology.
11.-14. Oktober 1982, Gaithersburg, Maryland, USA

[10] Hirzinger, G.
Robot-Teaching via Force-torque-Sensors, Sixth European Meeting on Cybernetics and Systems Research, EMCSR'32, Vienna, April 13-16,1982.

[11] Ackermann, J.
Abtastregelung, Berlin, Heidelberg, New York: Springer Verlag (1972).

TRACKING CONTROL SYSTEM FOR COMPLEX SHAPE OF WELDING GROOVE USING IMAGE SENSOR

M. Kawahara

Technical Research Institute, Hitachi Zosen Corporation, Osaka, Japan

Abstract. This paper discusses an automatic weld-line tracking control system, in which a light sectioning technique by laser and image sensor is used for imaging the sectional pattern of joint groove and a microcomputer is used for image processing and producing the control output. The image processing method used here is applicable to various shapes of joint grooves, and even effective for the presence of previous weld passes. Another feature of this system is that the picture pickup optical system is intentionally put out of focus; by processing this out-of-focus image, the detection resolution better than that determined from the number of image sensor elements has been successfully achieved. An automatic tracking controller based on this system has been applied to the large-diameter steel pipe production line of a steel mill, proving that it is practical and useful, and contributes to improvement of the weld quality and work efficiency.

Keywords. Computer control; Microprocessors; Pattern recognition; Position control; Robots; Sensors; Tracking systems; Welding; Image processing.

1. INTRODUCTION

Before assembling into a steel structure by welding, its individual materials, such as steel plates, are gas-cut or machined to form a joint groove for welding. However, these materials are often of insufficient accuracies in terms of the straightness, flatness and/or bending curvature, resulting in errors in weld line formation. There are also other errors that add to these preprocessing errors, including setting errors of the rails of a welding machine, temporary assembly errors by tack-welding, and errors arising from thermal distortion during the welding and deflection of the materials due to their own weight. In view of these, introduction of an automatic tracking controller that controls the welding torch position according to the weld line is desirable for uniform weld quality, relieving the welding operative from burden, and for increasing the work efficiency.

The strong field demand under these circumstances has led to development of various sensors of non-contact type for tracking control. In recent years, more emphasis has been placed on development of optical non-contact sensors. (see for example Arata, 1977; Hyosha, 1977). The author (Kawahara, 1981) also developed a practical system consisting of an optical sensor and a tracking controller, which exhibited the most acceptable performance for standard V-groove joints.

However, the actual groove shapes are various; besides the standard V-groove, narrow square groove joints of 0° groove angle and similar joints of small groove angles are now widely used. And moreover, it has become necessary to track and weld prebeaded joints in the production line where products are made by several welding processes. In the prebeaded joint, the previous weld bead is not uniform in shape, and sometimes the groove is almost filled up by previous bead. In other words, the groove of a prebeaded joint takes a series of variously changing sectional shapes, which none of the then existing sensors can cope with.

Discussed here is a new tracking control system, that expands the applicability of the light-sectioning type weld-line sensing to cover a wide range of groove shapes, including not only square and similar acute-angle grooves but also complex shape grooves of prebeaded joints. The system is unique in using the out-of-focus imaging intentionally and achieving a higher resolution than that determined from the number of image sensor elements by processing the out-of-focus image.

An automatic tracking controller based on this system has been applied to the large-diameter steel pipe production line of a steel mill, proving that it is practical and useful, and contributes to improvement of the weld quality and work efficiency.

The discussion is made in four sections and in the following order: general description, image processing, experimental results, and application to steel pipe production.

2. GENERAL DESCRIPTION OF TRACKING CONTROL SYSTEM

Figure 1 shows the principle of tracking control by groove shape detection. The light sectioning technique for groove shape detection consists in the use of a laser and an image sensor. That is, a fanning out flat beam from the laser irradiates the groove at a suitable angle of incidence and at right angles to the weld line, so that a bright line describes the contour of the groove as viewed from top. The laser used is a He-Ne laser of 5 mW output, and the image sensor is a two-dimensional matrix sensor (32 × 32 elements) of RETICON make. The image sensor receives the image of this bright contour line and converts it into video signal. This video signal is suitably preprocessed by the preprocessor and fed to the image processor.

(a) In-focus image

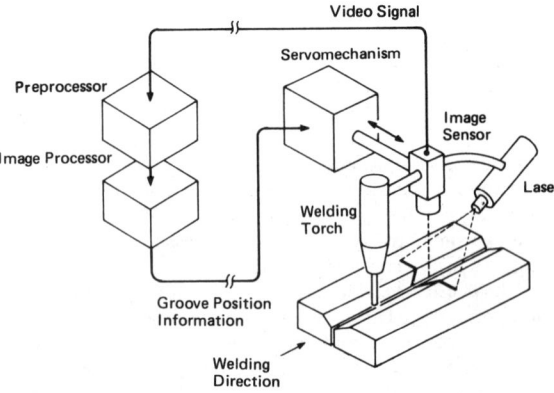

Fig. 1. Sketch to show principle of groove shape detection and tracking control.

The preprocessor processes the video signal from the image sensor for the image processor, and contains circuits for: binary digitization of video signal, automatic threshold setting for binary digitization, and image superposition for noise removal.

The image processor consisting mainly of a microcomputer (INTEL 8080A), processes the preprocessed video signal and accurately computes the deviation of the groove center position from the center of the sensor imaging area according to the software, and produces control output with a gain appropriate for the servomechanism. The servomechanism holds an integrated assembly of the laser, image sensor and welding torch and drives this assembly until the above-mentioned deviation becomes zero. In this type of closed loop, the center line of image is always in agreement with the groove center, i.e., the welding torch can always track the center of the groove whatever its variations in position and width may be.

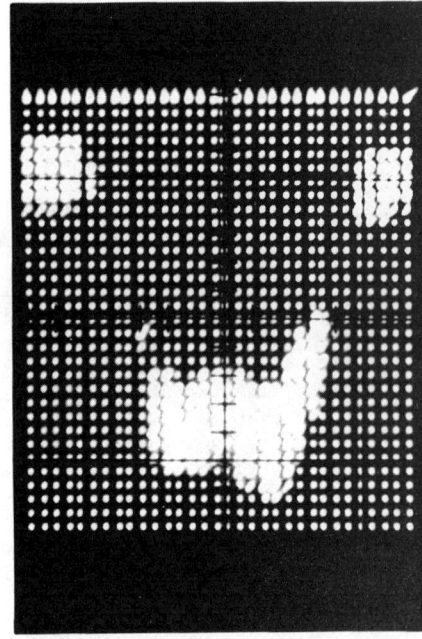

(b) Out-of-focus image

Fig. 2. Binary images of a beaded joint groove with low bead height.

In the image processing, emphasis is placed on two points. The first point is to correctly determine the groove center even for a beaded joint by distinguishing its slightly remaining right and left edges from the bead. The second point is to increase the overall resolution of position measurement higher than the resolution as determined from the number of image sensor elements. Let the image sensor element matrix be N × N and the joint area picked up by the image sensor be A × A (mm^2), then the resolution determined from the number of image sensor elements is A/2N (mm). With N = 32 for the image sensor used and taking A = 32 mm, for example, we have a resolution of 0.5 mm. Increasing the number N for higher resolution means a longer image processing time and an increased equipment cost. In our image processing method, improvement of the resolution for high-accuracy groove center position computation is achieved, instead of increasing the number N, by increasing the number of picture elements (pixels) forming each groove edge in the image and computing the edge position as the average of many data. While keeping the laser beam thickness as thin as possible, the optical system for imaging is intentionally put out of focus to increase the number of groove edge pixels. The out-of-focus image obtained in this way takes a spread-out form of the original sharp shape, with reduced brightness. Figure 2 compares two binary images of the same groove obtained by preprocessing the picture taken: (a) in-focus condition and (b) out-of-focus condition, indicating that the latter provides a much greater number of data pixels for groove center position computation. Figure 3 also makes a similar comparison for a beaded joint groove with bead height as high as base metal surfaces. This figure shows that binary image (b), though out of focus, retains essentially the same information as in-focus image (a) and does on the delicate change in the edge profiles.

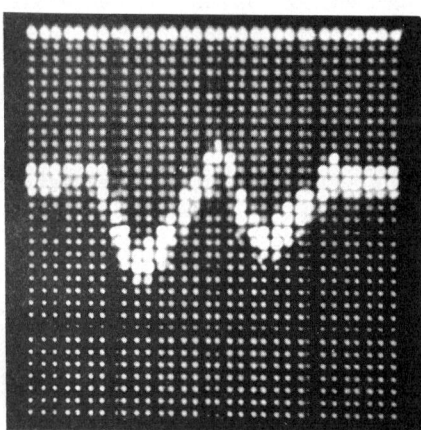

(a) In-focus image

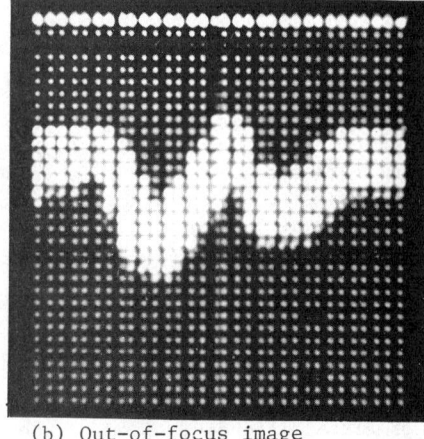

(b) Out-of-focus image

Fig. 3. Binary images of a beaded joint groove with high bead height.

3. IMAGE PROCESSING METHOD

3.1 Pattern Screening and Compression

The pattern screening and compression are processed to remove non-essential data from the image data matrix and to reduce it to an essential set of data for simplification of later processing, i.e., groove center position computation. Figure 4 shows an example of image data matrix. The picture coordinate system is as follows: the upper left-hand corner of the picture area is the origin (0), from which X-axis extends horizontally to the right and Y-axis extends vertically downward. For each row of the image data matrix, the number, n_d, of data pixels (image matrix elements that are in Logic One state, receiving the laser light reflected from the joint) is counted, starting from the first row (Y = 1) and on. As understood from Fig. 4, the surface of steel plates can be recognized from several consecutive rows of n_d sufficiently greater than zero, and the other rows are non-essential. Therefore, as n_d counting proceeds from the first row, second row and so on, the row that first shows the inequality $n_d \geq \delta$ (δ = judgement parameter) is judged to be the start of steel surface portion data. In Fig. 4, the row of $Y = Y_1$ corresponds to this. The rows from this on are picked up as significant data until a row that does not satisfy the inequality $n_d \geq \delta$ is found, i.e., rows of $Y = Y_1, Y_2, Y_3 \ldots Y_k$ ($n_d < \sigma$ for $Y = Y_{k+1}$) are picked up, or until k reaches k_{max}.

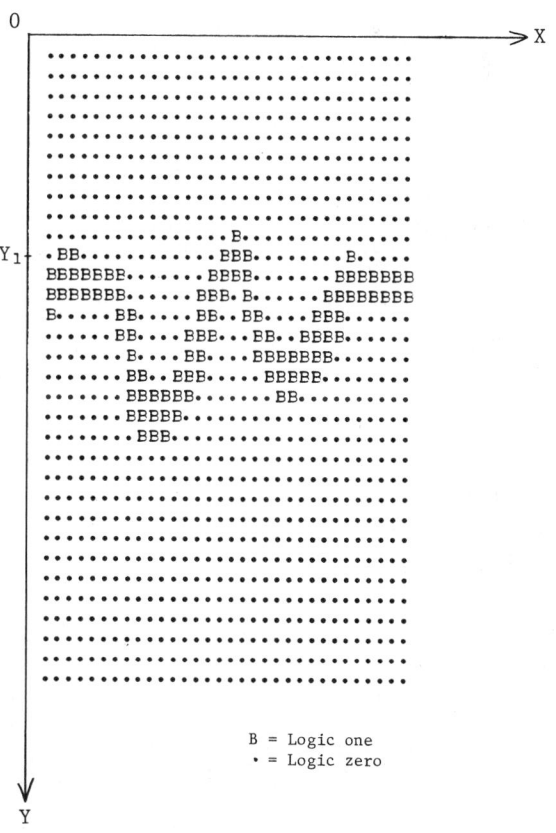

Fig. 4. Image data matrix.

The k rows of steel surface data as screened above are then OR-operated on each column and compressed into a single row of data.

3.2 Pattern Classification

Considering the varying factors, such as the groove shape, presence of bead and base metal surface conditions, the image data compressed into a single row is assumed to take one of eight basic patterns:

(i) D_1, B_2, D_2
(ii) B_1, D_1, B_2, D_2
(iii) D_1, B_2, D_2, B_3
(iv) B_1, D_1, B_2, D_2, B_3
(v) D_1, B_2, D_2, B_3, D_3
(vi) $B_1, D_1, B_2, D_2, B_3, D_3$
(vii) $D_1, B_2, D_2, B_3, D_3, B_4$
(viii) $B_1, D_1, B_2, D_2, B_3, D_3, B_4$

where: B_1 = number of consecutive Blank (Logic Zero) pixels as counted from the most left of the row;
D_1 = number of consecutive Data (Logic One) pixels as counted from the most left of the row, or which follow B_1;
B_2 = number of consecutive Blank pixels following D_1;
D_2 = number of consecutive Data pixels following B_2;
B_3 = number of consecutive Blank pixels following D_2;
D_3 = number of consecutive Data pixels following B_3;
B_4 = number of consecutive Blank pixels following D_3.

Among these basic patterns, pattern (i) represents basically the narrow groove joint with D_1 and D_2 being the base metal surfaces and B_2 being the groove. Patterns (ii) through (viii) are variations of pattern (i) with loss of data pixels at the right or left end of the row, bead height in the groove, and others. If the sensed image pattern does not agree with any of these basic patterns, then the data will be ignored as abnormal and not processed.

3.3 Computation of Groove Center Position

The left and right groove edge positions are determined in terms of the number of pixels as counted from the most left of the picture (X-axis), then the median value between them is computed as the groove center position.

First, the left and right edge positions, X_L and X_R, are determined from the compressed image data.

For pattern (i) or (iii)   : $X_L = D_1$
                             $X_R = D_1+B_2+1$
For pattern (ii) or (iv)   : $X_L = B_1+D_1$
                             $X_R = B_1+D_1+B_2+1$
For pattern (v) or (vii)   : $X_L = D_1$
                             $X_R = D_1+B_2+D_2+B_3+1$
For pattern (vi) or (viii) : $X_L = B_1+D_1$
                             $X_R = B_1+D_1+B_2+D_2+B_3+1$

$$\ldots (1)$$

Next, for each of the k rows of joint surface data screened as mentioned in 3.1, the left and right edge positions, X_{Li} and X_{Ri}, are determined according to:

$$X_{Li} = \max \{ X \mid V(X, Y_i) = 1, X_L - d \leq X \leq X_L \}$$
$$X_{Ri} = \min \{ X \mid V(X, Y_i) = 1, X_R \leq X \leq X_R + d \}$$

$$\ldots (2)$$

$$i = 1, 2, \ldots, k.$$

where $V(X, Y)$ represents the logic level (Logic Zero or One) of an element at position (X, Y) of the binary image, and d is the parameter to define the permissible range of spread of X_{Li}, X_{Ri}.

From the edge position data $\{X_{Li}\}$ and $\{X_{Ri}\}$ obtained using Eqs. (2), the groove center position is computed according to:

$$X_0 = \frac{1}{2k_L}\sum_{i=1}^{k_L} X_{Li} + \frac{1}{2k_R}\sum_{i=1}^{k_R} X_{Ri} \quad \ldots (3)$$

$$k_L, k_R \leq k$$

Equation (3) means to compute the arithmetic mean $\bar{X}_L$ and $\bar{X}_R$ of $\{X_{Li}\}$ and $\{X_{Ri}\}$, respectively, then to compute the median value between $\bar{X}_L$ and $\bar{X}_R$. The number of data operated on, k_L for the left edge and k_R for the right edge, may not agree with the number of rows, k, picked up in the screening process, but may be smaller than k with some data rejected as being out of range in the equations (2) computation process.

The X_0 value as obtained in the various conventional image processing methods is equivalent to computing it as $(X_L + X_R)/2$ from X_L and X_R calculated using Eqs (1). Comparison of X_0 values according to Eq. (3) with conventional X_0 values is discussed in Section 4.

4. EXPERIMENTAL RESULTS

4.1 Test of Groove Center Detection Accuracy

The method used to determine the accuracy of the groove center detection is as follows. A joint groove test piece is secured on a precision slide table, so that the groove position can be changed in directions perpendicular to the weld line. For values of the groove position displacement x (mm) from the assumed center position (x = 0 mm), the image sensing and processing equipment computes the groove center position X_0. There should hold a linear relationship between x and X_0, but no ideal linear relationship will be obtained due to quantization and image processing errors. Therefore, for a group of data (x, X_0), a straight regression line is fitted, and deviations of the individual X_0 values from this regression line are calculated to obtain a distribution of these deviations. Then the largest value among these deviations is defined as the detection error.

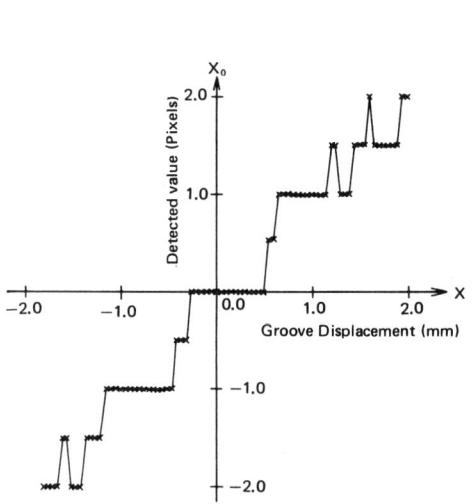

(a) In-focus imaging.

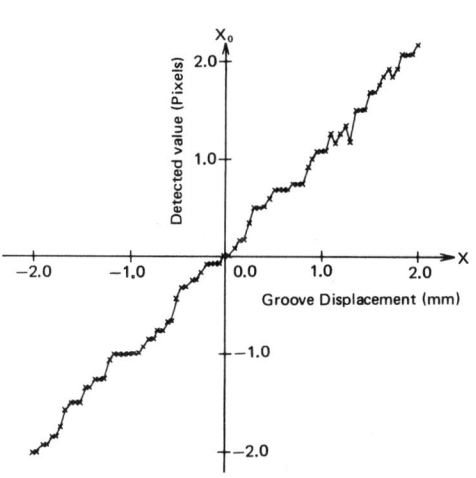

(b) Out-of-focus imaging.

Fig. 5. Groove center detection results.

Figure 5 shows the groove center detection results for two cases of imaging: (a) in-focus condition and (b) out-of-focus condition. In case of (a), errors due to quantization of the picture to pixels by the image sensor are clearly seen. Whereas, in case of (b), nearly a straight-line relationship is seen, indicating that the out-of-focus imaging reduces the quantization errors and achieves a high accuracy in groove center detection.

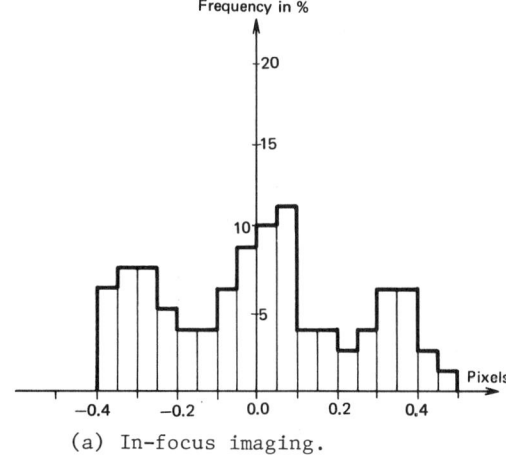

(a) In-focus imaging.

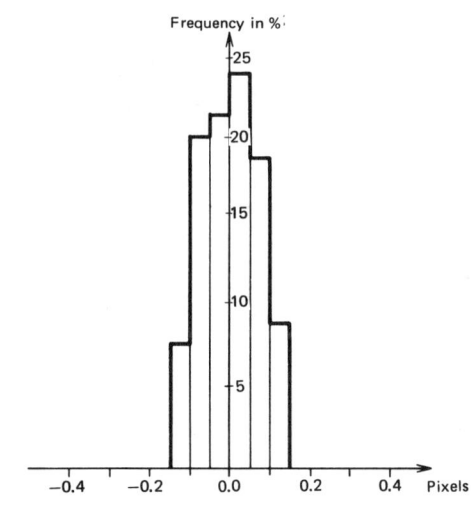

(b) Out-of-focus imaging.

Fig. 6. Distribution of groove center detection error.

Figure 6 shows a frequency distribution of X_O deviations in pixels from the regression line, for Fig. 5 cases (a) and (b). The detection error is 0.50 pixel for case (a) in-focus imaging, and 0.15 pixel for case (b) out-of-focus imaging. From these results, it is clear that the out-of-focus imaging and processing much improves the detection accuracy. The time required for imaging and preprocessing is 23 ms and the time required for image processing by the microcomputer is 135 ms, a total of 158 ms.

4.2 Test of Tracking Control Accuracy

The method used to test the accuracy of the tracking control is as follows. A joint groove test piece is secured on a precision slide table as mentioned in 4.1. Next, while the tracking control is in progress, the groove position displacement x (mm) is changed and the resultant amount of travel

x_t (mm) of the sensor head is measured. The tracking control accuracy is quantitated as follows. The groove position displacement x and the resultant sensor head travel x_t should ideally be equal in value (mm), but, in actuality, they are not because there are detection errors as discussed in 4.1 and servomechanism errors. In view of this, a linear regression of x_t on x is fitted to a group of data (x, x_t), and deviations of the individual x_t values from this regression line are determined, of which the largest deviation is defined as the tracking control accuracy. The test specimen used is the same as that used in the detection accuracy test (4.1).

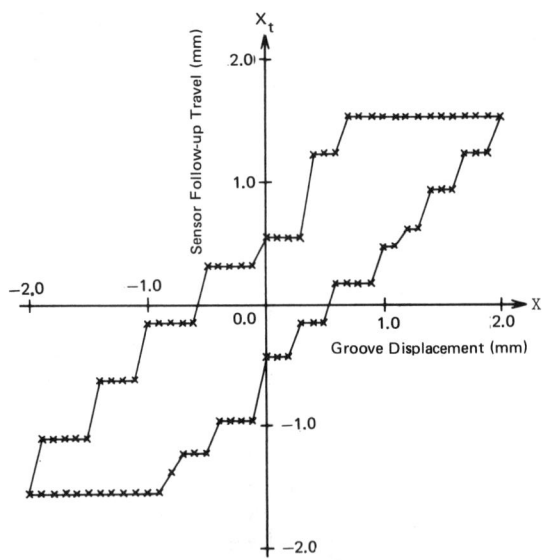

(a) In-focus imaging.

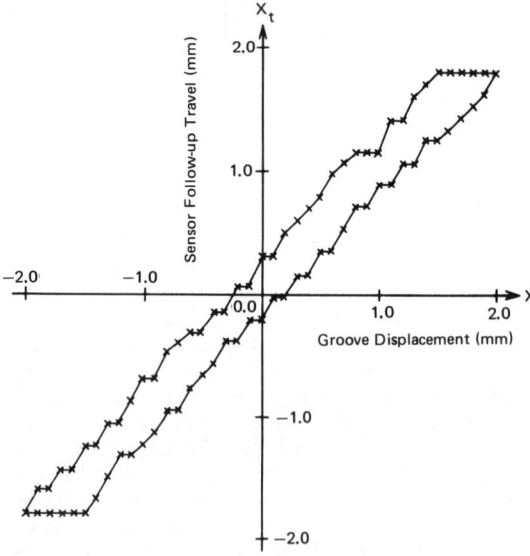

(b) Out-of-focus imaging.

Fig. 7. Tracking control results.

Figure 7 shows the travel of the sensor head following variations in the groove position for two cases of (a) in-focus imaging and (b) out-of-focus imaging. Although both diagrams exhibit a hysteresis characteristic, case (b) is much smaller in hysteresis width than case (a). The principal cause of hysteresis is a dead flat portion around the origin (o, o) of the groove center detection characteristic, as seen in Fig. 5 (a) and (b). During the tracking control, the operation point is always found near this origin and it will never go far away from the origin. Therefore, the size of this flat portion, or the dead zone, can affect greatly the tracking control accuracy. From this reasoning, the hysteresis of Fig. 7 (b) seems to be a little greater than expected. The reason for this is found in the servomechanism characteristic, the second cause of hysteresis. That is, in order to prevent the hunting, the servomechanism has a properly adjusted gain as well as a suitable dead zone around its operation point. The hysteresis seen in Fig. 7 (b) is due to this dead zone introduced into the servomechanism characteristic. In Fig. 7, case (b) shows more of a straight-line relationship between input and output than case (a), which is due to less quantization errors in the image processing method.

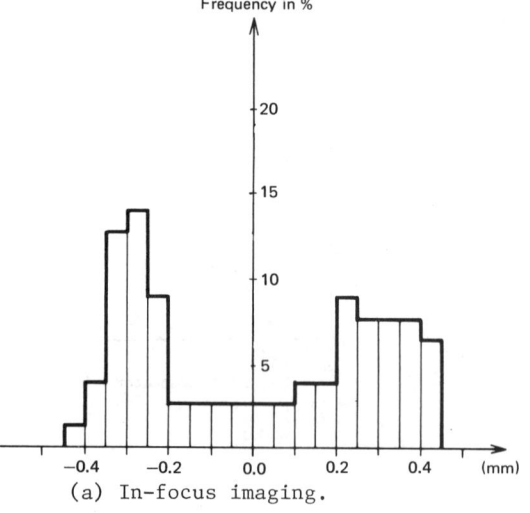

(a) In-focus imaging.

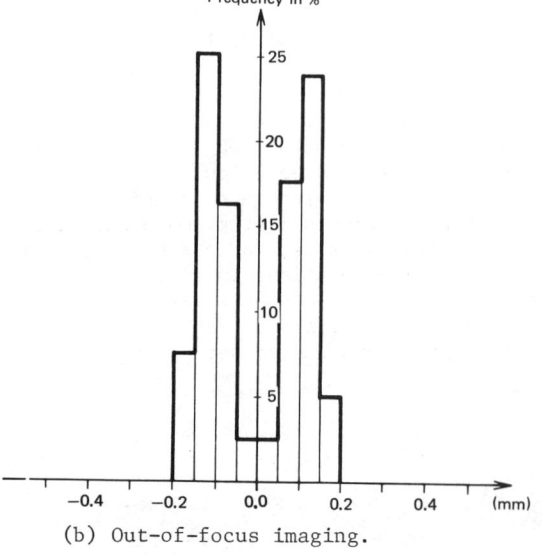

(b) Out-of-focus imaging.

Fig. 8. Tracking error distribution.

Figure 8 shows a distribution of tracking errors as obtained for cases (a) and (b) of Fig. 7. The error distribution in case (b) is much narrower than in case (a). The tracking control accuracy is ±0.45 mm for case (a) in-focus imaging, and ±0.20 mm for case (b) out-of-focus imaging, indicating that the out-of-focus imaging is quite effective in improving the tracking control accuracy.

5. APPLICATION EXAMPLE TO LARGE-DIAMETER STEEL PIPE PRODUCTION

The groove shape sensing and tracking control system discussed here has been applied to large-diameter steel pipe production lines. In the production line, thick steel plate is bent into a cylindrical shape; and the seam is welded from the inside first. Next, the same seam is welded from the outside. The steel pipe is further adjusted for final shape, checked and inspected, surface-treated and finished, then leaves the line as product. This tracking control system is applied to the outside welding process of the above steel pipe production line. Figure 9 shows a sketch of this outside welding process, to which pipe with its seam pre-welded from the inside is delivered on a running cart, so that the pipe runs immediately below the outside welding machine. The machine is a three-electrode submerged-arc welder. The steel pipe on the running cart is roughly position-adjusted so that the weld line will be about in 12 o'clock position, but the weld line itself is not straight due to some manufacturing errors, such as twist, nor has a simple, uniform groove shape as it is pre-welded from the inside.

In this process, because of the complex groove shape, various approaches for tracking control were made, but none of them could satisfy the required control accuracy. The tracking control system introduced this time has achieved the target control accuracy of 0.5 mm in automatic tracking outside seam welding, improving both weld quality and work efficiency.

6. CONCLUSIONS

The automatic weld-line tracking control system as developed and discussed in this paper uses a combination of laser and image sensor for groove profile detection and a microcomputer for image processing. The system is applicable to a variety of groove shapes, even to prebeaded joints, whose groove profile is complex and considered difficult to track. Also, by making the picture pickup optical system out of focus and processing the out-of-focus groove image, the error of the groove center position detection is reduced to 1/3 of that in in-focus imaging.

The system has been successfully applied to large-diameter steel pipe production, improving both weld quality and work efficiency, and demonstrating its usefulness.

We will continue efforts toward size reduction of the groove sensor head.

REFERENCES

Arata, Y., and K. Inoue (1977). Application of digital computer to pattern measurement and processing in automatic control system of welding. J. Jap. Welding Soc., 46, 129-134.

Hyosha, K. (1977). Welding control system using ITV. R&D Kobe Steel Engineering Reports, 27, 81-84.

Kawahara, M., and H. Matsui (1981). Tracking control system for arc welding using image sensor. Proceedings of the 8th Triennial World Cogress of IFAC, Kyoto, Japan, Vol. 4a, Pergamon Press, Oxford.

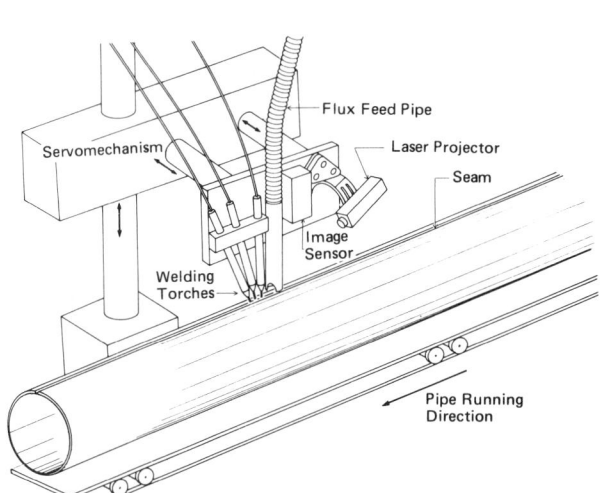

Fig. 9. Process of automatic seam welding.

A MULTI-MICROCOMPUTER-BASED ROBOT CONTROL SYSTEM

K. W. Plessmann

Department of Process-Computer Application, Aachen Technical University, Aachen, Federal Republic of Germany

Abstract. When designing a multi-computer system it is necessary to keep in mind that the system should be expandable from a single computer up to full scale just by having to plug in additional components or boards and not by having to change the layout of the system or the basis of the application. Usually this is known as modularity. If the computer is dedicated to a special field of application, an additional design aspect is of great importance. This is the portability of the software, which means that the software should run without major modifications under almost all circumstances. This paper deals with the design and layout of a multi-computer system which is dedicated to robot control. It uses microprocessors as central units and is based on the design aspects mentioned above.

Keywords. computer architecture, computer interfaces, control equipment, microprocessor, robots, multiprocessing system, parallel processing.

INTRODUCTION

The development of robot control systems from point-to-point - PTP - programmed with a teach-in keyboard to threedimensional attitude control systems - continous path - which are programmed by using a special high level language or a playback device, is a substantial step in the aspect of computer performance. Not only the underlying system design of the robot, together with the measurement equipment and the actuators, is of great importance for the industrial application of the total system; it is also important to guarantee the necessary performance of the driving computer.

Looking ahead to robots, which are sensor guided or use video input together with pattern recognition systems, the required performance of the computer has to grow by a magnitude of 10 or more. In the last four years the capability of microprocessors has not only grown in data width and directly addressable memory space, but has also increased in data throughput. The peripheral devices and family components are mainly programmable - i.e. they have own intelligence. In this context arithmetic processors are of particular interest, which in some systems have been introduced as co-processors. Keeping in mind that the increase in performance demanded for robot control will keep growing in the future it makes sense to design a multi-computer system using these high integrated components. Such a system will be able to cope with the increase in performance demand, which will come along with more sophisticated robot control systems, especially those sensor- and videoguided.

The use of more than one computer for the control of a robot system has already been realized some years ago /1/. In this case a hierarchical structure was chosen due to the different levels of application. The hardware of this type was very specialized and an upgrading was not a simple task. This was implied by the dedication of the computer layout to a special application, disregarding the fact that hierarchy of levels actually is a question of programming. This solution is similar to other applications, where mainly two computer systems were used /2,3/. Here also the design was determined by the system to be controlled and not by the idea of adaptability and expandability.

Microcomputers mainly of the 8 bit type are used in all areas of multi-computer application for robot systems /1,3-6/. Considering the performance required the step to a 16 bit system is a proper means to obtain higher accuracy and better performance. Also more sophisticated components with some measure of own intelligence can be utilized. With respect to resolution even 16 bit usually are not enough, particularly if discretization and rounding errors in the arithmetic are considered. Up to now only very few of these systems are available, more are announced.

DESIGN OBJECTIVES

When designing a computer for robot control application the following points should be considered:

i) Designing and realizing the computer as modular as possible will provide the following advantages:
 - The performance of the control system required for a specific application can be tailored to meet the demands. A non-modular design will offer a certain performance, which in most cases will either be too small or too large but will seldom be exactly what is needed.
 - The computer system will consist of several subsystems, each of which accomplishes a specific task in different robot control applications.

ii) In conjunction with a multi-computer system this concept of modularity should provide the possibility to include additional processing units - even after the control system of the application has been designed.

iii) Software, designed for a single computer system, should be portable or at least easily transferable - i.e. some parts of the software must be adapted - to support the step from one to more computers. The modern modularization of software, mainly written in high level procedural languages, constitutes a means to subdivide the total package into independent entities.

Along with these general objectives for a multi-computer system, which cannot be fulfilled at the same time with today's computer structures, some important problems arise, for instance:

i) the synchronization between the computers in the system
ii) the interfacing to other systems already in use, e.g. an 8 bit computer for PTP on the basis of a sequential control algorithm.

In the design stage the consideration of these points is of great importance for the software production as well as for the assignement of a task to one of the computers in the system. Especially the synchronization should be done on a hardware basis, so that the user does not have to care about this in the development of the software. This is usually a part of the operating system or at least of some system routines. Last but not least the design of a multi-computer system should be based on components readily available. But this is the easiest part of all.

THE MULTI-COMPUTER SYSTEM

Taking into account the objectives mentioned above a good realization of such a multi-computer system is one that

i) is self-contained in each of its single computers and

ii) is tied together by a special bus to which the I/O sub-systems are connected.

While i) supports the modularity, ii) will fulfill most of the design objectives mentioned above. Considering the different areas of application of a computer in robot control - which cover techniques from PTP to video guided systems - a multi-computer for this purpose must be expandable. This means for instance that for PTP only one computer is utilized and for a video-guided robot up to four units might be required. A subdivision, concerning the computational activities for a robot, is given in table 1. Besides this the table furthermore gives information about the sampling period for 8 and 16 bit data width on the basis of microprocessor-components.

A four computer system obviously fits into the spectrum of requirements defined by the robot and its environment. This takes into account the fact that video-guided systems have a higher need for computer performance. But until now this is the domain of research and not yet of direct application. Preprocessing could be a solution to cope with the performance required by this tool until multi-computer systems have come into general use.

THE SYSTEM

Figure 1 shows the overall system where PS_i is a self-contained processing element. The system consists of a maximum of four - PS_0,... ..., PS_3 - as well as of the I/O sub-systems, which are connected to a data and signal transfer unit called external bus. In the upper part of this figure the input-output modules are shown, here called general process modules, since this computer system may be used for general process automation as well. On the right hand side a connection to a different computer system is shown, e.g. the multi computer is connected to a system already in use. This multi-computer system might be linked to the existing systems as an additional element for coordinate transformation or a special preprocessing unit. All components of the multi-computer system communicate by a bus used for data and signal exchange, which has been designed as simple as possible to ease the bus arbitration for the modules attached to it.

Due to the problem of synchronization between the different single computers and between these and the I/O system the following strategies were used:

1. The external bus is subdivided into two different parts, one for the communication between the single computers - as shown in figure 2 - and the other for the data exchange between each of the single computers and any of the modules of the I/O system. In the upper part of figure 2 these are represented by the basic modules for a robot control application.

2. Data transfer between the single computers is performed in a block mode manner, where a block consists of the destination and the source address, a word count, N datawords with a width of 16 bit, a checksum and a trailing character - e.g. the EOT character in the ASCII set.

3. The data transfer between a single computer and an I/O module is performed in the conventional way, i.e. by a register to register transfer controlled by interrupts.

4. The bus access sequence is controlled by the elements for the data block transfer, so that only that computer tries to get an appropriate number of buscycles, which has a block of data prepared for output to another single computer of the system or wants to exchange data with a module of the I/O system.

The easiest way to implement communication between the single computers is to use a FIFO. Elements of this type

- collect the incoming and outgoing data blocks
- have signals - when using components which are now on the market - to do the synchronization.

While figure 3 shows the connection of the PS_is, figure 4 gives an insight into the bus adaption of a selected subsystem concerning the data transfer between a single computer and an I/O module. The main concept of this multi computer is illustrated by discussing the bus and its signals. It consists of 69 lines, not considering the power supply lines. The bus is subdivided into

- 2 address busses to access a single computer from another single computer and the I/O modules
- 2 data busses of 16 bit width each for the data transfer
- coordination and hand-shake signals
- signals for exceptions, e.g. fail
- 7 interrupt request lines.

The connection of all modules is based on the following strategy:

1. When sending data to another module it is collected in a FIFO or a special register.
2. When the FIFO has a data set or the register contains valid data for I/O module, the bus arbitration logic tries to access the bus for data transfer.
3. Access-rights to the bus will be given in a daisy chain mode, in order to avoid conflict situations if more than one single computer tries to get bus cycles.

This last point is of great importance. Hereby it is possible to expand a system and enhance its performance by plugging in single computers into the bus connectors without modifying the structure or even changing one of the signals. In the case of a system with only one processing unit this unit always has access rights to the bus. For a system with more than one module the access to the bus is governed by the daisy chain signals. These signals are shown in figure 3 and 4 and are named RBPI and RBPO - computer bus priority in and out - respectively EBPI and EBPO - bus priority in and out.

The structure of this system based on this synchronization is shown in figure 5.

For incoming data transmitted by the other elements - computer and I/O - a similar technique is utilized. Due to the structure of the data transfer block generated, the bus arbitration logic is capable to extract addresses - source and destination - from the block and to switch the signal logic to the accept or pass-by mode. While the second is not interesting for consideration the first activates the appropriate FIFO or the bidirectional register of figure 6 for I/O. This register works like a FIFO with the difference that only a depth of one is accessible. The basis of this difference to the processor - processor interface is caused by the usage of the elements of the overall system. While processors are exchanging messages to communicate what will control the program flow, I/O components are sending or recieving data. These data will be used in algorithms or are produced by them. The only thing common to both data and messages is the structure of the objects by which they are spread over the system.

In the system interrupts for synchronization between computer performance demanded and computer activity is organized on two levels. On the first level all of the elements can send interrupt requests - with a maximum of seven interrupt groups, which are assigned by the user via jumpers. If they are honoured by the addressed processor an entry to an interrupt vector will be transmitted on the second level via the databus. This allows a maximum in flexibility and a minimum in expenditure. Furthermore, if there are no conflicts in accessing the bus the processors in the system can work absolutely in parallel, even when they are exchanging messages or driven by interrupts or sending and transmitting data, while the I/O system is capable to generate interrupts to support their own needs.

THE SUBSYSTEM

Any of a maximum of four subsystems is self-contained with respect to processing data and all the elements that constitute a computer. As shown in figure 7, they consist of a processor - the M68k is shown in figure 7, but the system architecture does not depend on this microprocessor - and memory together with a floating-point unit and some special I/O components. They are all collected in a local bus that is designed to minimize the number of lines and signals but which is of no importance in this context.

The I/O interface is utilized in order to avoid the connection to the external bus for small systems, so that a data link to key-boards and other user-oriented peripherals may be connected directly. The parallel I/O lines can be used to adapt special interfaces. In this sense a subsystem is similar to a single-board computer having an interface and data-transfer components to a special bus.

An additional feature is the expansion module, which is used to enlarge the memory, when this is necessary. Figure 8 shows a block sketch of this module, which will be connected to a subsystem via a connector by a flexible cable. The module itself resides in the same backplane as the subsystem, which physically holds the external bus to link it to power.

Another feature accessable by the same transfer mechanisms is of some importance in the context of data exchange between processors and I/O modules and vice versa. This is the connection to other computers. Very often, especially in robot control application, a control system starts with a simple 8 bit micro-processor, e.g. in a PTP environment. When performance is to be extended, it is easier for the user to do this with the already existing system in order to hold its functions by connecting it to a more powerful parallel system. Usually this is achieved by parallel or serial data links. A different method is shown in figure 9.

This module, running in the environment of the external bus, consists of a microprocessor, some I/O elements for data transfer in both directions and another system bus. This interface is not a general one and must be redesigned whenever a link of this manner is taken into account.

PROGRAMMING

The implementation of the user system must be seen under two aspects, which depend on each other:

i) the application-oriented part, representing the robot and its tasks to be controlled by the compiler system
ii) the system part, responsible for the interaction, which is the message and data transfer.

Looking at the high data throughput of todays microprocessors and the fact that more powerful components are announced, it does not make sense to write **programms** on a machine-oriented level - MOL. High level languages - HLL - are easier to use and are oriented closer to the problem, even if they are only procedural.

In conjunction with this system programming is done with FORTRAN and/or PASCAL by using a conventional development station as it is used for all other microprocessor applications. It is not necessary to subdivide a problem as a function of the number of processors in a real system. The only system dependences the programmer has to keep in mind are:

- only processes can be implemented in parallel,
- interprocess communication is restricted to a special procedure,
- input/output has to use a set of procedures when applying the external bus and the I/O modules; the modules of the subsystem's I/O may be programmed in a conventional manner.

The process of implementing the user-oriented software package under these restrictions will be done by invoking different libraries holding modules with the same procedure identifier but with different bodies. This is shown in figure 10, where the complete process of mapping a program from source to absolute object is derived in close detail.

In a similar manner
- timer
- floating-point processor or floating-point software

may be used. The predefinition of procedure identifier for special purposes is an appropriate means to programm an embedded computer. Thus the user has to test the performance given by a computer system in the real world and compare it with the demand of the system to be controlled. If the performance is sufficient, there is no need for modification, if not, an additional subsystem has to be installed. In this case the complete user package has to be recompiled and linked with a different library - if at all - to support the hardware structure on the logical level. The development of software for robot control purpose must not only be seen from the point of view of mapping the algorithms onto the absolute object and to verify the corrections - even when it is not necessary to decompose the total system into parts for a parallel system, as in this context. With respect to the following points portability is of steadily growing importance:

- Using modules in future application or under changing constraints
- Prefabricating modules out of which the package may be configured
- Simplifying the design of hierarchical systems and supporting system integrity
- Cutting costs and reducing development time.

While portability is an objective which is far away from realization, adaptability - i.e. transferring a program from one computer to another with some modification due to system specifications - may be achieved in the real world of controlling a robot or in other industry-oriented areas. By writing a program in a HLL instead of a MOL the first step towards this aim is taken. Using no language specific feature of the I/O - as in FORTRAN - would be the second step. Since the hardware discussed above supports adaptability even in those parts normally found in operating systems in connection with parallel processors, this could be defined as a higher level of adaptability.

Portability of user packages may be possible on the basis of roboter languages /9/. But comparing this with what is already known about procedural HLL - especially what is called a dialect of a language - portability is pretty low.

APPLICATION

Looking back to the system and its architecture, the multi-computer system may be used not only for robot control systems but also in general control configurations. Only the expandability to four processors may be a constraint, but this side comes from the subdivision of areas of application of robots.

Usually, as may be seen from the development over the last 5-6 years, the introduction of robots for automation purposes in the industry started with a PTP system, in many instances programmed with a teach-in keyboard. This type of system can be controlled with a single processor configuration. Only when high resolution and/or high speed is necessary a 16 bit component comes into use. For more complicated systems, especially when coordinate transformation - cylindric or spheric - has to be done in the system, a two-processor environment has to be considered. While PTP, multi-point or continuous path program run on the first subsystem together with all the activities to drive the robot, the transformation can take place on the second subsystem in full parallelity. Depending on the reaction time governed by the dynamics of the robot the floating-point operations may be programmed or a processor has to be installed. It is possible to achieve a sample time of 6-7 ms for the appropriate coordinate transformations of a 5-axis robot with an attitude control system on the basis of a M68k equipped with an arithmetic processor. This is of some importance due to the high resolution that comes along with this performance of the system. The product of velocity and resolution, which could be used as a measure for the robot "activity", is close to the magnitude of one higher than the systems which are now in use.

The third subsystem in the party will come into action as soon as sensors for robot guidance come into use. Once again a subdivision not only into different computers in the controlling equipment, but also in splitting the algorithms into independent modules is of some interest for the user. This is due to the hierarchy of processes acting to produce control signals from sensors in such a way that the command variable is fed into the process where the control algorithm is implemented. Thus the step from continuous path or attitude control is only a short one. While the first takes the command variable from the sensor, the last holds this value in the memory.

Under the aspect of pattern recognition or for other reasons, when more performance or independency to the run-time situation is necessary - e.g. on-line interactive programming or robot HLL with direct interaction to the running system - the fourth subsystem may be attached. Finally it should be mentioned that the I/O structure of the subsystem is not only capable to drive user-oriented subsystems, but also data-peripherals or special equipment. This could be a content-addressable memory under the aspect of pattern recognition problems.

CONCLUSION

In this paper a multi-computer system dedicated to robot control with limited expandability up to four processors was discussed. Programming and application and the problems related to theses fields were mentioned or described in order to give an understanding of the interior interactions and dependencies.

In this context the point of view of programming is quite important. The subsystems can be seen under the aspect of object-oriented methods. This has to do with hiding a progam and/or data structures - generally spoken: objects - in a system, to give access only to those components of the total system which have access rights. Methods like these increase the security and data consistency, supporting parallel processing. Some of the necessary structures were presented in this paper.

REFERENCES

/1/ Barbera,A.J. et al:
"Hierarchical Control of Robots using Microprocessors"
Proc. 9th Int. Symp. on Indusrial Robots, 1979

/2/ Veda,M. et al:
"A Real-Time Shape Recognition of Moving Objects"
Proc. 11th Int. Symp. on Industrial Robots, Tokyo 1981

/3/ Hoifoldt,J.R., Hakonson, H.:
"SIMAGE - Si's Image Recognition System"
Proc. 11th Int. Symp. on Industrial Robots, Tokyo 1981

/4/ Sasaki,N., Matsushima,K.:
"Industrial Robots Controlled with High Speed Micro-Computer"
Proc. 11th Int. Symp. on Industrial Robots, Tokyo 1981

/5/ Iigima,J. et al:
" Elementary Functions of a Self-Contained Robot"
Proc. 11th Int. Symp. on Industrial Robots, Tokyo 1981

/6/ Fugii,S. et al:
" Computer Control of a Locomotive Robot with Visual Feedback"
Proc. 11th Int. Symp. on Industrial

Robots, Tokyo 1981

/7/ Plessmann, K.W. et al:
"Grenzen der Modularisierung"
to be published in "Industrieanzeiger"
Girardet-Verlag, FRG 1982

/8/ Courtney,J., Herlin,G.:
"A Portable Robot Software System"
Proc. 11th Int. Symp. on Industrial
Robots, Tokyo 1981

/9/ Bison,P. et al:
"The Formal Definiton of YML and a
Proposed Portable Implementation"
Proc. 11th Int. Symp. on Industrial
Robots, Tokyo 1981

/10/ Malms,M.:
"Ein inhaltsadressierbares Speicher-
system zur Unterstützung zeitkriti-
scher Prozesse der Informationswie-
dergewinnung in Datenbanksystemen"
Doctorial Thesis at the Aachen Technical University, Aachen, FRG 1982

area of application	robot co-ordinates	number of computing elements	sampling period [msec]
PTP	cartesian (c)	1	8 bit = 30; 16 bit = 6
		1 or 2	8 bit = 30; 16 bit = 6
multi-point	c	1 + 1 per axis	8 bit = 4-5; 16 bit = 4-5
	spheric (s)	1 or 2	8 bit = 4-5; 16 bit = 4-5
attitude control	c	1 + 1 per axis	8 bit = 60; 16 bit = 40
	s	2 + 1 per axis	8 bit = 60; 16 bit = 40

Table 1

A Multi-microcomputer-based Robot Control System

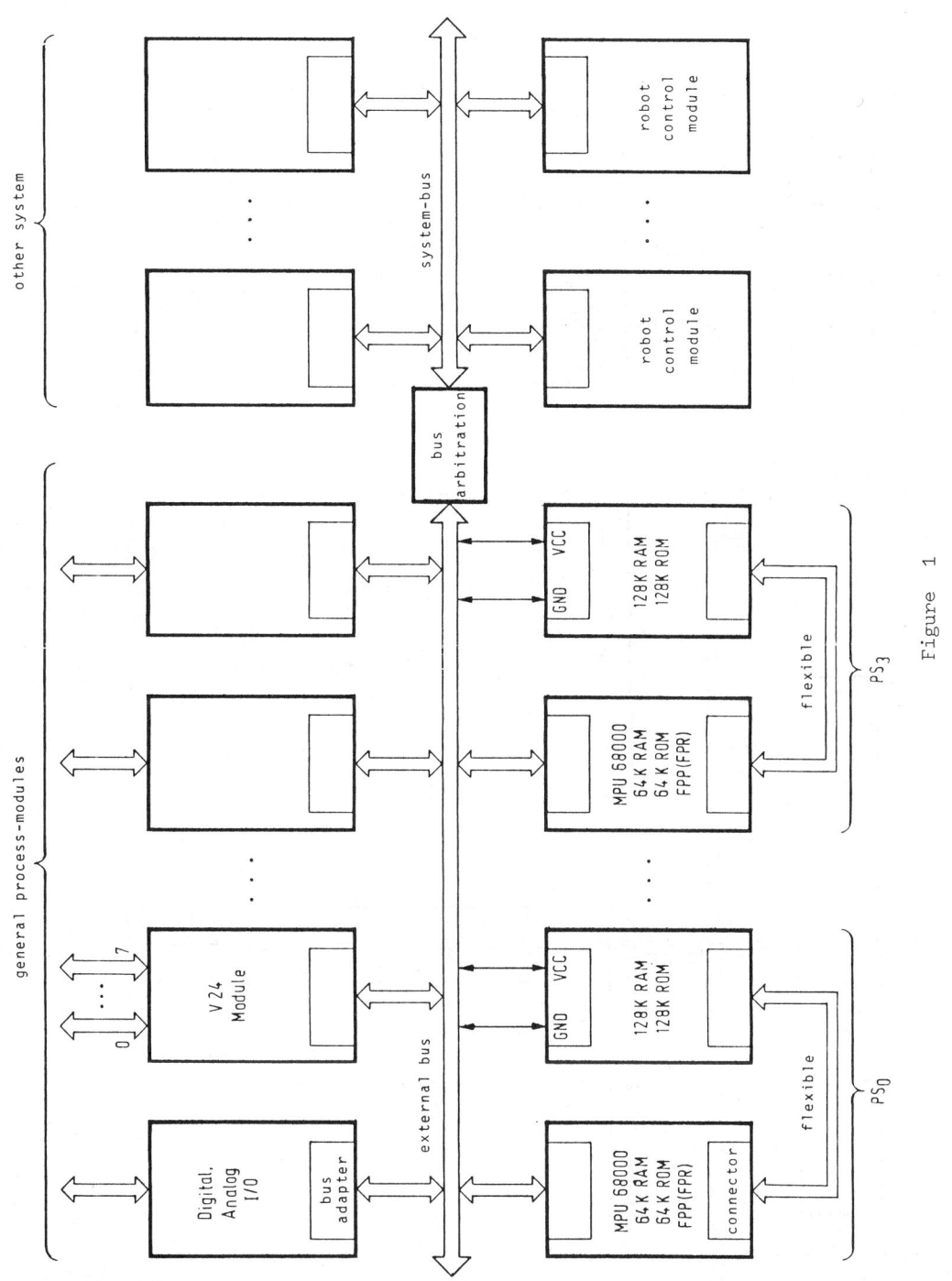

Figure 1

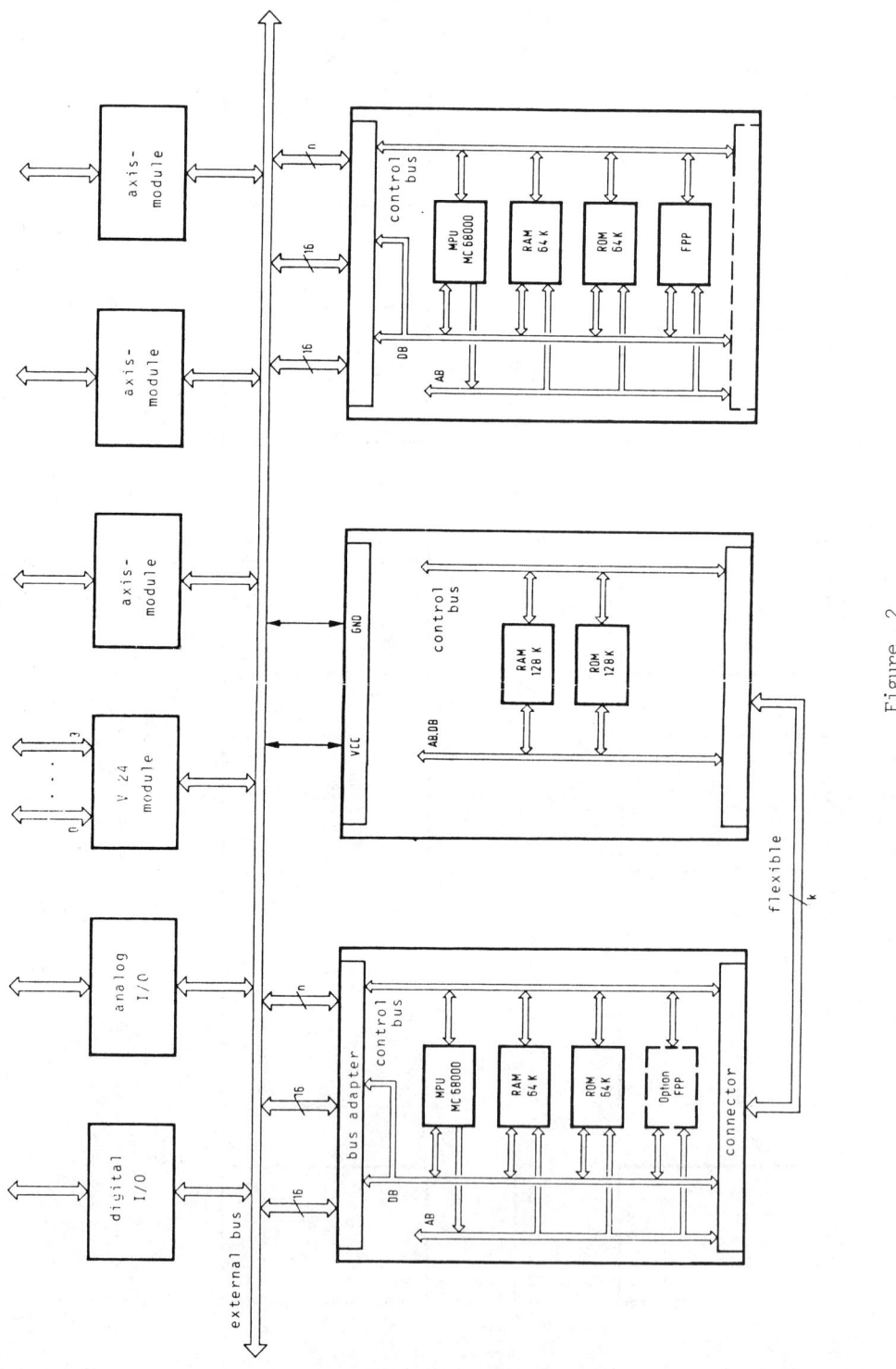

Figure 2

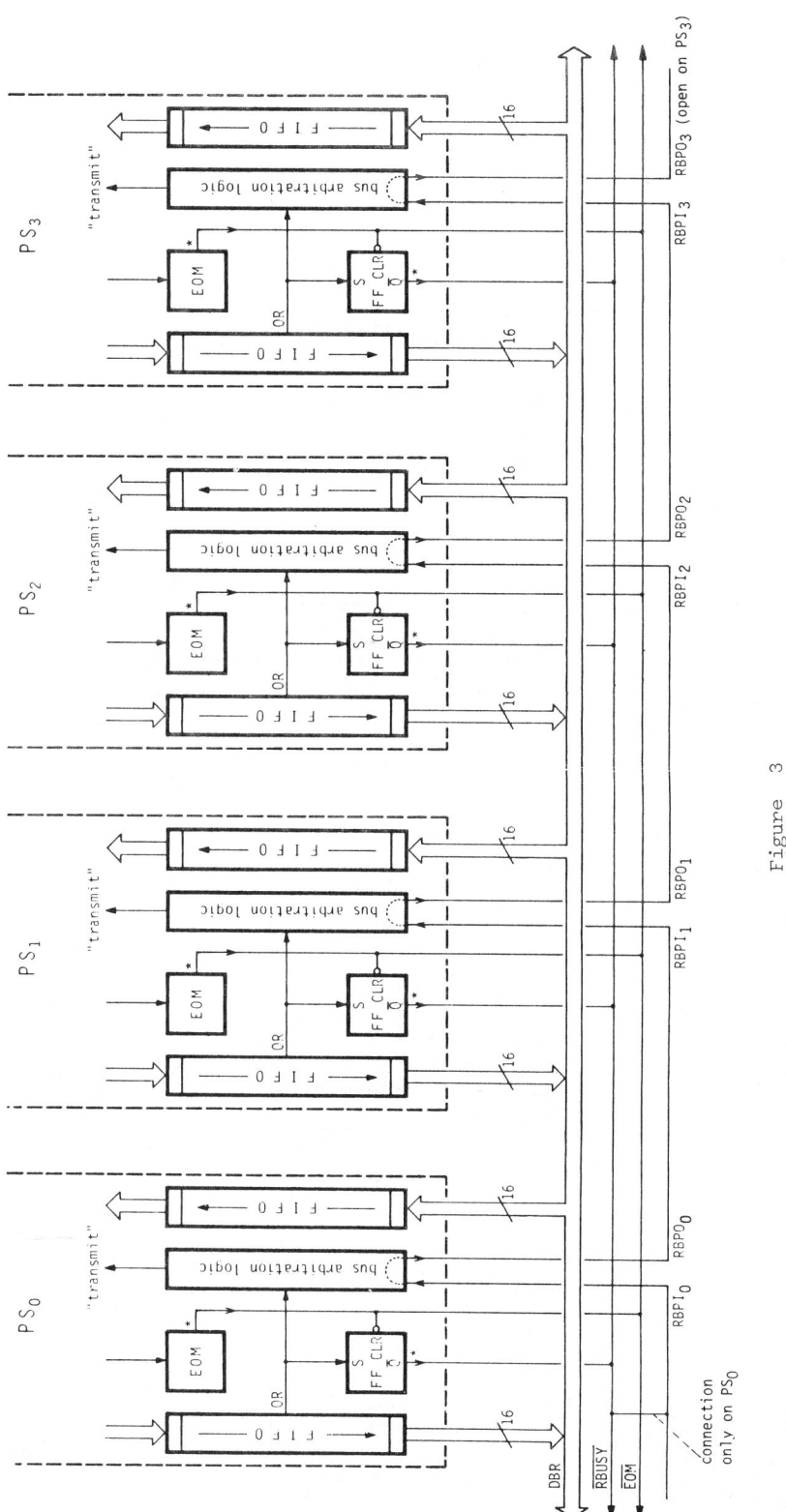

Figure 3

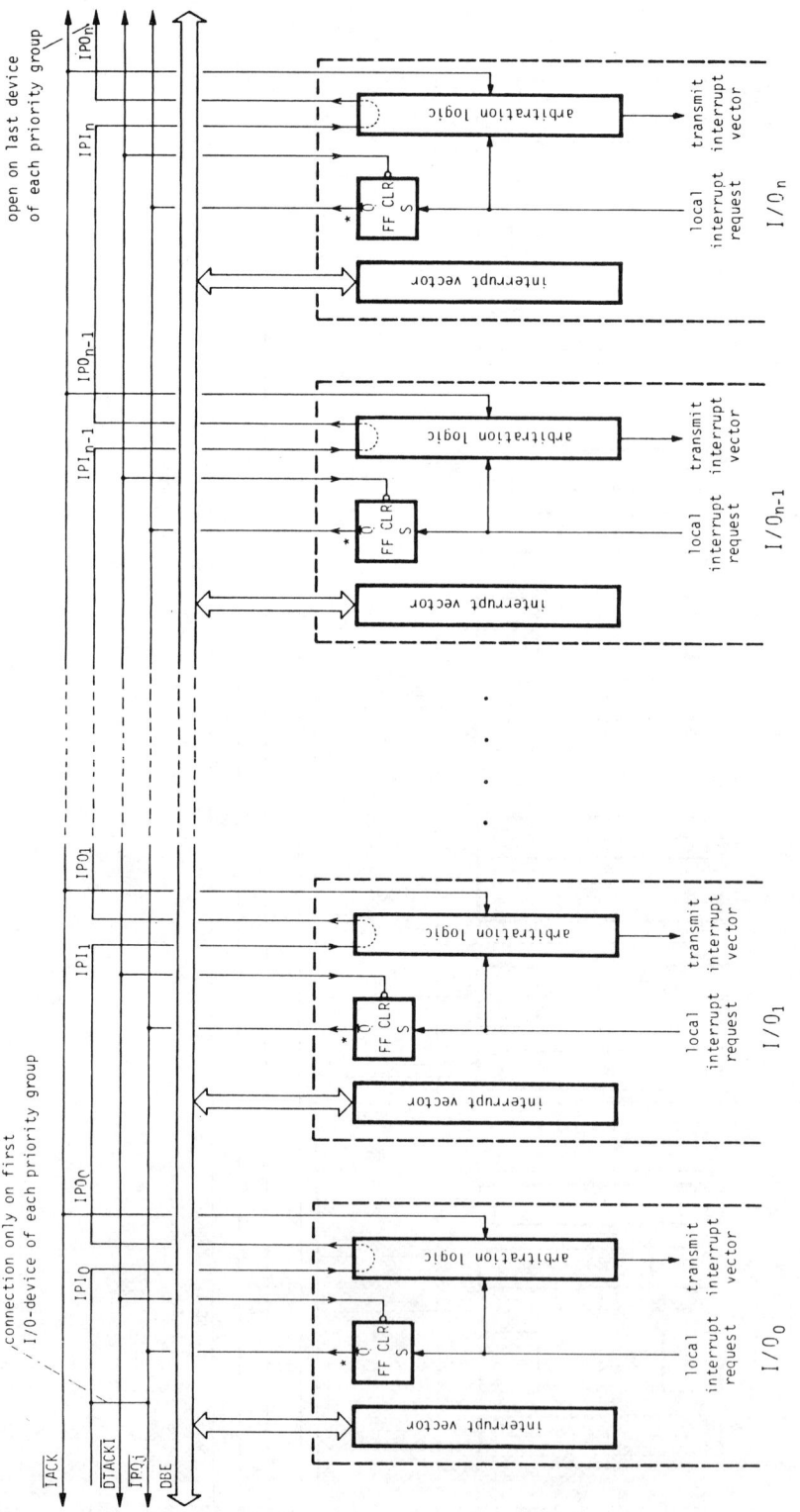

Figure 4

A Multi-microcomputer-based Robot Control System

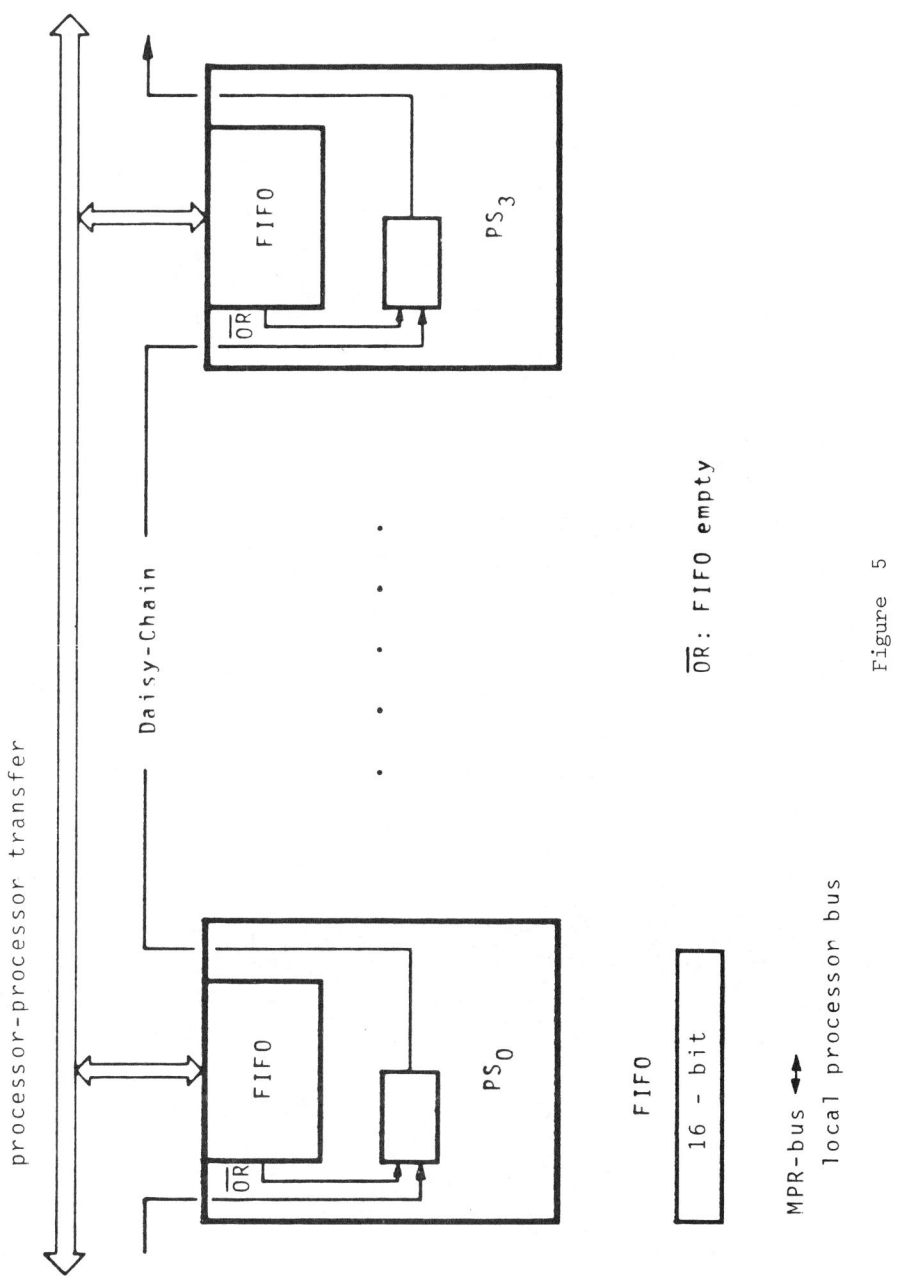

Figure 5

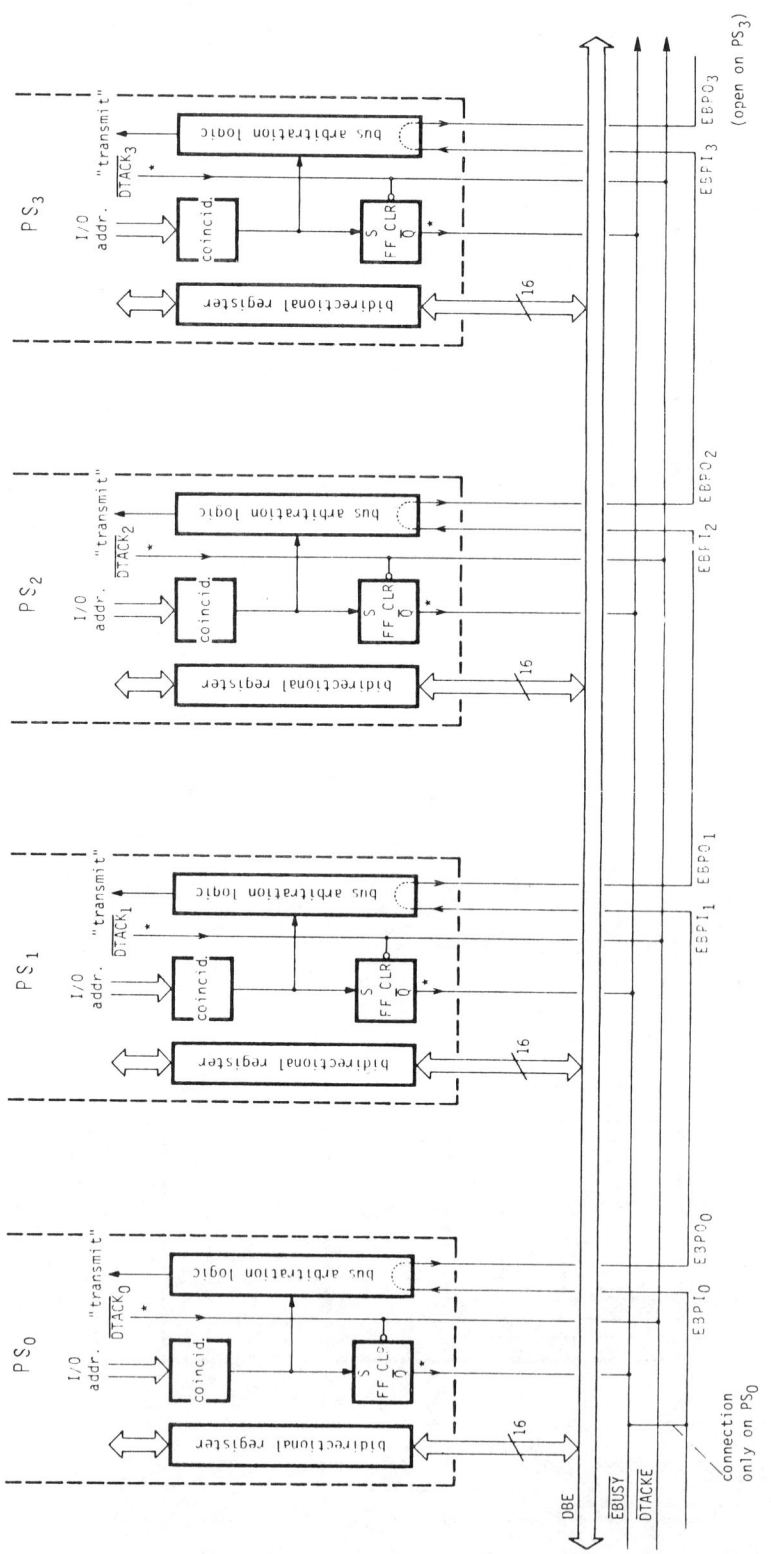

Figure 6

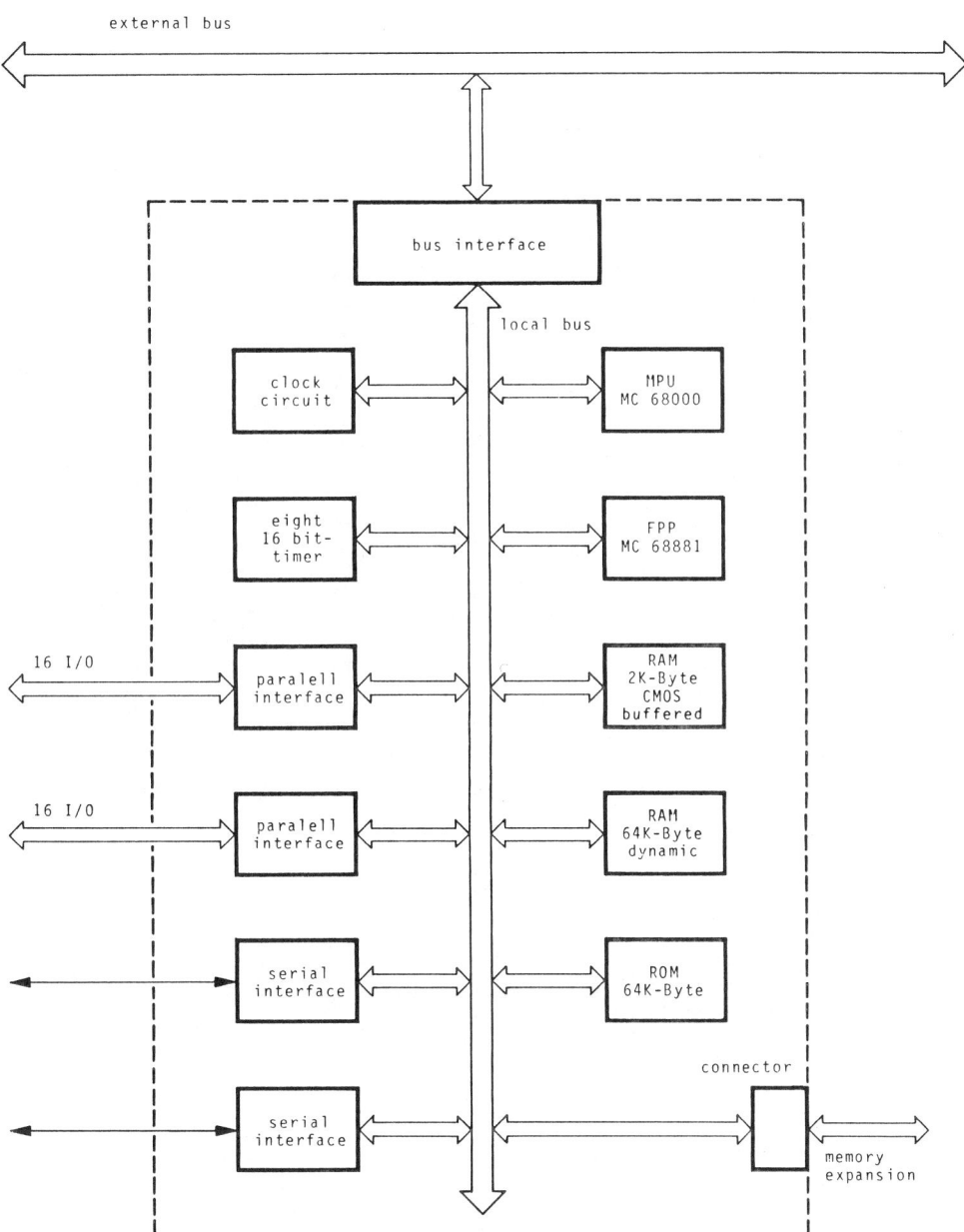

Figure 7

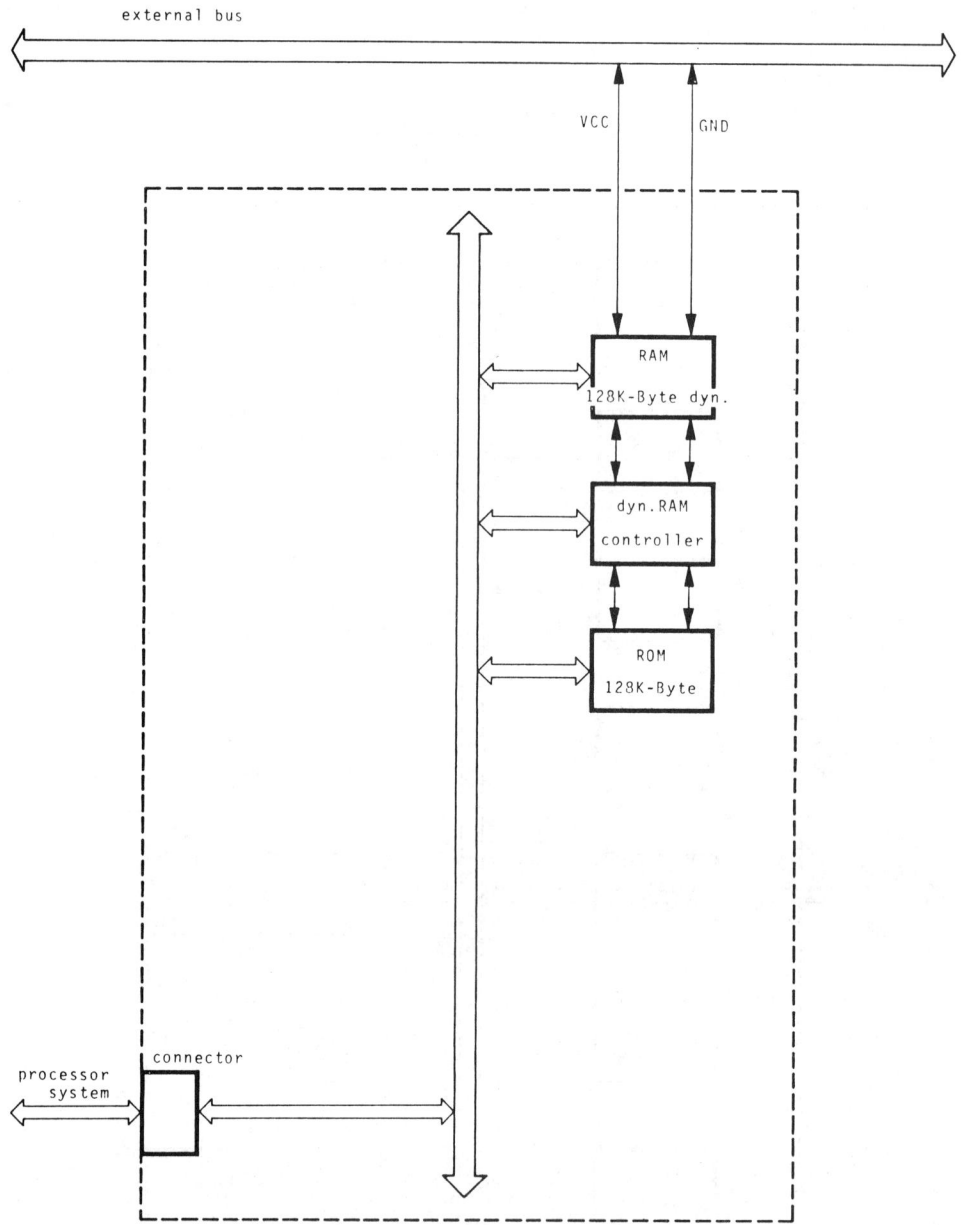

Figure 8

A Multi-microcomputer-based Robot Control System

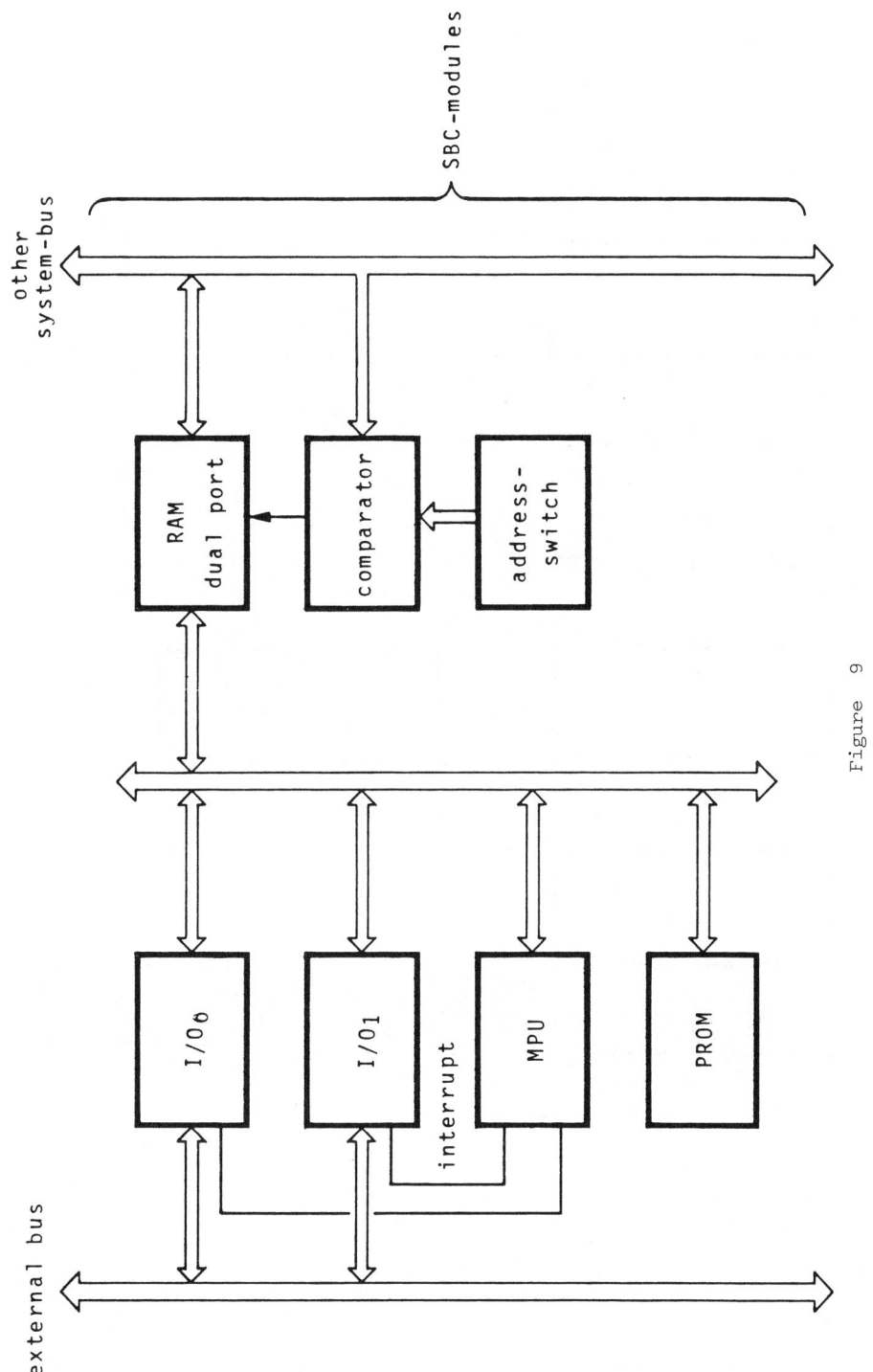

Figure 9

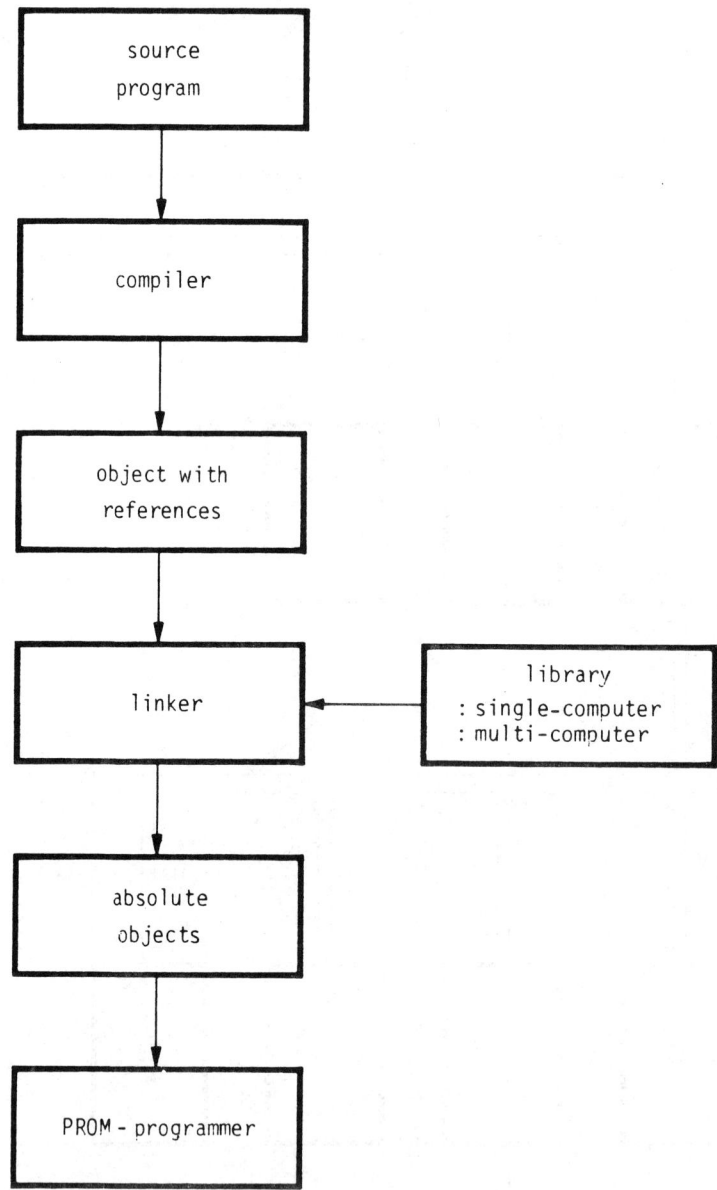

Figure 10

DISCUSSION

SESSION TA2: ROBOTICS

Paper: TRACKING CONTROL SYSTEM FOR COMPLEX SHAPE OF WELDING GROOVE USING IMAGE SENSOR

Authors: *M. Kawahara* (Technical Research Institute, Hitachi Zosen Corporation. Osaka, Japan)

Discusser: *KW Plessmann*
c/o A Achen University
Templer Graben St.
5100 Aachen
Federal Republic of Germany

Questions or Comments:

1) Sampling-time?

2) Function between accuracy and point of sensing?

Author's reply:

1) The time required for imaging is 23 ms and the time required for image processing by microcomputer is 135 ms; so sampling time of this control system is 158 ms.

2) The distance between the detecting point and the welding point is about 10 cm in this application. The error due to this distance is less than 0.1 mm so it is a negligible factor. But in the case that it is not negligible, it still is no problem because, by arraging software, the control output for every detecting point can be delayed until the welding torch reaches to the point. And moreover, driving a welding torch and sensor as one body, sensor imaging area does not go far from groove even if a welding line is strongly curved. Then this sensor can track groove in any case.

Paper: A MULTIMICROCOMPUTER BASED ROBOT CONTROL SYSTEM

Authors: *KW Plessmann* (A Achen University, Templer Graben St, 5100 Aachen Federal Republic of Germany

Discusser: *G. Guilles*
Université de Lyon I
Laboratoire d'Automatique,
Bat 721, 43 Bonlevard du 11 novembre 1918, 69622 Villeurbanne
Cedex, France

Questions or Comments:

1) Has you system been used for kinematic control, or also for dynamic control, involving a dynamic model of the robot?

Author's reply:

1) The system was used and is in use for dynamic control. But we known that a dynamic model of the robot would be of greal value to support the different control functions. Therefore we have started some activities in this direction, where one of the computers will simulate the robot, manily the kinematics. The bases of this will be done with CSMP, because the compiler we have is producing FORTRAN output that makes the implementation much easier.

Discusser: *JP Thomese*
ENSEM BP 850
54011 Nancy Cedex
France

Questions or Comments:

1) How is the software designed?

2) How are the different tasks distributed?

Author's reply:

1) The software is task-oriented. When two or more tasks are running on one computer the message interchange will be done by use of a procedure, mentioned in the context of my paper. The programmer has to use a special procedure. When implementing the same package on two or more computers, a re-link - with a different library - is necessary, while the message interchange procedure is now supporting the multicomputer environment, e.g. the interface to the external bus to transfer the message.

2) Task-distribution is application dependent, due to the number of computers in the system.

ON-LINE SCHEDULING FOR TRANSPORTATION OF RAW MATERIALS

K. Azumi*, Y. Yamamoto**, S. Ishikawa**, Y. Maeda** and Y. Ienaga***

*Sumikin System Development Ltd., Japan
**Wakayama Steel Works, Sumitomo Metal Industries Ltd., Japan
***Central Research Laboratories, Sumitomo Metal Industries Ltd., Japan

Abstract. In the process of raw materials handling, the smooth charging of raw materials to the next process (blast furnace, sintering plant or coke oven) is very important. For this reason, the scheduling of raw materials transportation needs to be consistant and effective for regular and efficient production in the next process. This problem is a network flow problem which must simultaneously determine the following correlated items:

(1) raw material bins or brands requiring transport;
(2) order and routes of transport;
(3) start time and duration of each transport.

As sub-systems of the raw materials computer control system, two kinds of scheduling systems have been developed and successfully applied in our steel works. One uses the heuristic logic similar to "Branch and Bound Methods" aiming at maximizing transportation efficiency, and the other uses decision tree analysis for direct control of transportation equipment.

Keywords. On-line scheduling; transportation control; network flow problem; digital computer applications; raw materials transportation.

INTRODUCTION

In the raw material yards of the steel works, numerous kinds of brands are stocked. Each of them must be delivered at the proper time to the bin of the next process by a conveyor network. This movement involves various kinds of work, is broad in range, and involves the frequent movement of equipment. Therefore, even skilled operators can not do it satisfactorily. Moreover, the reduction of production cost and stabilization of quality have been increasingly necessary in recent years. For this, it has become necessary to realize a reliable and efficient transportation scheduling system utilizing a process computer for the following reasons.

(1) In raw material bins, the amount in store must always be kept over a certain level, generally 50% of maximum stock capacity, in order to achieve stabilization of grain size of the raw materials. This relation is shown in Figure 1.

(2) The number of transportation routes is sometimes less than the number of required brands for the next process; therefore, the competition for the use of transportation equipment occurs.

(3) For changes of conditions such as equipment failures and change of mixing ratios, rapid adaptation is necessary.

We have developed an on-line scheduling system for providing information to operators and direct control of transportation equipment. Two kinds of scheduling algorithms are introduced, one is the heuristic similar to "Branch and Bound Methods", and the other uses decision tree analysis.

SOFTWARE APPLICATION

Figure 2 gives a layout of the raw materials yards at our steel works. We have developed a raw materials computer control system for this whole process. This system includes the following sub-systems of raw material traffic control.

(1) On-line scheduling for raw material transportation

(2) Direct control of transportation equipment

Each sub-system considers the scheduling problem with respect to a different object. Figure 3 shows an outline of the above sub-system and relative functions.

ON-LINE SCHEDULING FOR RAW MATERIALS TRANSPORTATION

Description of the Problem

The objects of this schedule are as follows:

(1) to always keep the storage level of bins over a certain control level;

(2) to minimize the loss of time by loader moving, route switching and competition for the equipment for transportation.

This problem is a kind of network flow problem which simultaneously determines the following correlated items:

(1) raw materials bins or brands requiring transport;

(2) order and routes of transport;

(3) start time and duration of each transport.

Formulation of the problem by Mixed & Integer Programing is possible, but it is difficult to attain the solution fast enough for use in an on-line system. In one scheduling system , the two-stage algorithum is used, in which the items (1), (2) are determined by heuristic logic, in the first stage, and them start time and duration/quantity of each transportation by linear programming are determined in the second stage.

For the following reasons, we have developed a unique heuristic logic which simultaneously determines the above three items, and also uses a basic rule which is to determine the maximum quantity of material transportable for each transportation.

(1) To transport the greatest quantity useable each time is the most efficient method. In this case the solution by linear programming will give us approximately the same results.

(2) The determination of order and route is more important than that of duration when the order is determined for the efficiency of the transportation schedule.

(3) The order and duration of the job are closely related when seeking the optimum point, therefore, they must be treated simultaneously.

Algorithm

Figure 4 shows the calculation flow of this algorithm.

Preparative routine

In this routine, the real time data such as stock level of bins, state of transportation, and state of equipment, and on-line input data such as mixing plans are used.

(1) Selection of raw material bins requiring transport

For the ith raw material bin, the retention time T_L^i is calculated as follows:

$$T_L^i = \sum_j t_j^i$$
$$\sum(C_j^i \times t_j^i) = A^i - L^i \quad (1)$$

t_j^i ; interval of jth consumption rate C_j^i
C_j^i ; jth consumption rate
A^i ; current stock level of ith bin
L^i ; control level of ith bin

If the transportation equipment is not useable at the time equal to T_L^i because of equipment maintenance, the T_L^i is moved to the point T^* as shown in Figure 5.

The bins with $T_L^i \leq$ Tmax. are then chosen as the ones requiring transport, where Tmax. is the threshold value set for 24 hours in our system. The transportation unit decided by route and bin/brand is here called a "job". From T_L^i and their available routes, whole jobs are selected and written on the job table. At the same time, the unuseable time of equipment and routes are registered on the route check table.

(2) The minimum transportation duration and the available start time

If the transportation duration is too short, the efficiency of the transportation is lowered, because of an increase in the ratio of loss time. Because of this consideration, we introduce the conception of the minimum transportation duration. The available start time T_S^i of the job is calculated from formula (2) (see Figure 6).

$$T_S^i = (Tmin. \times (S^i - C^i) + A^i - U^i)/C^i \quad (2)$$

Tmin.; the minimum transportation duration
A^i ; current stock level of ith bin
C^i ; consumption of ith bin
S^i ; transportation capacity for ith bin
U^i ; upper limit of ith bin

This job must start during the time from T_S^i to T_L^i, and finish during the time from $(T_S^i$ + Tmin.) to T_E^i. This time domain is called here the available transporation time domain $\widetilde{T}^i$.

(3) Description of job combination on Gantt chart

As shown in Figure 7, all jobs are described on the job scheduling table which is like a Gantt chart. Of course, jobs currently operating are also entered in this table.

If any required bin/brand can be transported by M routes, and N bins/brands are required, the number of permutations and combinations of the jobs are as follows:

$$\bar{O}(N) = N! \times M^N \qquad (3)$$

In formula (3), M^N is the number of combination of the jobs, and $N!$ is the number of permutations of one combination. Of course formula (3) is the maximum number considered theoretically. We must calculate the objective function for $N! \times M^N$ cases to obtain the real optimum solution. However, it is very difficult to calculate all cases fast enough to use in an on-line system as shown in the following example.

"Example" (using a likely number)
- if $N=8$, $M=3$
- number of cases $= 8! \times 3^8$
  $\doteq 2.6 \times 10^8$
- approximate processing time by usual process computer >30 hours

Division of the whole problem into sub-problems

The reduction of N in $N! \times M^N$ is very effective in order to reduce the processing time. There are two methods for reduction of N, one is a division of the time domain, and the other is selection of preferable brands from all brands. These methods are determined by the characteristics of the problem as follows:

(1) having the minimum transportation duration

(2) the requirement of transportation from each bin is discrete upon the time axis.

Consequently the available transportation time domain $\widetilde{T_i}$ is distributed discretely. It is not always necessary to treat all combinations simultaneously. In the case that lots of $\widetilde{T_i}$ overlap, the feasible region becomes very narrow, and it is possible to reduce the number of combinations.

This algorithm eliminates impossible combinations.

Optimal routine of sub-problems

This routine produces the optimal solution of the sub-problems. It is not always necessary to calculate all combinations included in the sub-problem because, the relation which indicates competition or preference between each two jobs shows the unnecessary combinations. Thus, we introduce the job relation function as follows:

$$\left. \begin{array}{l} Re\ (Job\ A_i,\ Job\ B_i) = \bigcirc \\ Re\ (Job\ A_i,\ Job\ B_i) = \times \\ Re\ (Job\ A_i,\ Job\ B_i) = \rightarrow \\ Re\ (Job\ A_i,\ Job\ B_i) = \leftarrow \\ Re\ (Job\ A_i,\ Job\ B_i) = \emptyset \end{array} \right\} \qquad (4)$$

$\bigcirc$; non-competition for routes of Job A_i and Job B_i, the efficiency does not depend on order

$\times$; $T_S^{A_i} + T_{min} > T_L^{B_i}$ and $T_S^{B_i} + T_{min} > T_L^{A_i}$, infeasible combination

$\rightarrow$; $T_S^{A_i} + T_{min} < T_L^{B_i}$ and $T_S^{B_i} + T_{min} > T_L^{A_i}$, order is fixed as Job A_i before Job B_i

$\leftarrow$; $T_S^{A_i} + T_{min} > T_L^{B_i}$ and $T_S^{B_i} + T_{min} < T_L^{A_i}$, order is fixed as Job A_i after Job B_i

$\emptyset$; $T_S^{A_i} + T_{min} < T_L^{B_i}$ and $T_S^{B_i} + T_{min} < T_L^{A_i}$, the efficiency depends on order

In these relations, "$\bigcirc$", "$\rightarrow$" and "$\leftarrow$" reduce $N!$, and "$\times$" reduces M^N. To study all relations, we make a job relation matrix. It is a symmetrical matrix, and the number of factors/relations are as follows:

$$\bar{O}(N) = \sum_{i=1}^{N-1} (i \times M^2) \qquad (5)$$

For the remainder of combinations, this routine attempts to maximize the efficiency function.

$$\sum_i T^i / T_E \rightarrow MAX. \qquad (6)$$

T^i; transportation duration of ith job
T_E; operation interval of this schedule

The optimal solution is registered on the job scheduling table, and after the prediction of the bin inventory has been performed, the algorithm returns to the first routine. In our problem, the processing time for 8 hours scheduling is about 30 seconds by an on-line process computer.

DIRECT CONTROL OF TRANSPORTATION EQUIPMENT

The equipment for charging the raw materials to the bins is used jointly between transportation routes. When manually performing this process, the mixing between different brands and the interruption of the charging operation occurred sometimes, because of complex competition for equipment.

Figure 8 shows the outline of this system applied to the coal delivery process. In Figure 8, the horizontal conveyor (row $1\sim4$) and the vertical conveyor (column $1\sim6$) actually cross each other, so it is possible to charge any brand on any horizontal conveyer into any bin. In this process, transportation routes are switched by movement of the scraper up and down.

Usually, a brand is assigned to several bins, it is necessary to switch the vertical conveyer continuously without interruption until each bin reaches its desired level. The object of the direct control system of transportation equipment is as follows:

(1) maximization of the quantity handled in one charge;

(2) prevention of the mixing of different brands.

Therefore, the scheduling system is needed to control dynamically the competition for equipment.

Charing Schedule

This problem is a network flow problem too. Therefore, we can apply basically an algorithm similar to the previous problem. However, this schedule must be processed fast enough to apply for real time operation. Consequently, this algorithm decides successively the charging order of the bin for the transported brand along the decision tree, and we can obtain a schedule for 3 hours future within 2 seconds.

Algorithm

The time points required to determine the charging bin are as follows:

(1) arrival of the head of raw material's load to the bin;
(2) arrival of the end of raw material's load to the bin;
(3) when stock level of the bin goes over the upper limit;
(4) when stock level of the bin falls below the control level.

These points are shown on a Gantt chart as shown in Figure 9.

Step 1... Regarding the earliest end point on the chart, the decision tree is made as shown in Figure 10. A node is defined by the brand and the equipment/conveyer it is using the time of the earliest end point. For the node of the brand which has the earliest end point (which we call the "root"), which means the bin/conveyor needs to changed, we determine the nodes of the conveyor which must be used by the above brand at subsequent stages and connect the nodes sequentially. Doing this, forms the branches of the decision tree which is made by repeating the process until all nodes are determined as shown in Figure 10.

Step 2... Calculation of the estimation value for each pass from root to the Nth stage

$$E_i = \sum_{\ell=1}^{N} (Z_S/Z_M)_\ell^i \times C + ED^i \quad (7)$$

Z_S; storage level of the bin before changing at the earliest end point
Z_M; storage level of the bin after changing at the earliest end point
C; constant value
ED^i; penalty value for a dead lock pass
i; ith pass
ℓ; ℓth stage

In Figure 10, pass ① is called the dead lock pass, and it is very difficult to change the bin without mixing with another brand.

Step 3... Registration of the minimum pass ⓘ to the scheduling table

If pass ③ is selected, Figure 11 replaces Figure 9.

Step 4... Repetition of Step 1 to Step 3 to obtain a 3-hour schedule.

If competition for equipment will occur, according to a temporary chart, where some transportation must be stopped at the half-way point, it is necessary to consider various roots. Where there are two nodes at Stage 1, the algorithm evaluates the possibility of the use of alternative decision trees. The charging schedule is begun according to the following timing.

(1) arrival of the raw material's head or end
(2) start or end of equipment failure
(3) changing of auto/manual mode
(4) constant interval

Direct Operation of the Equipment

This function directly operates the charging equipment called scrapers following a schedule.

Usually, as shown in Figure 11, for example, at one changing point EP_1, plural brands and equipment must be operated on the sequence decided by the current conditions. We can easily obtain this sequence by the calculation of the matrix and vector which show the position and state of equipment/scraper.

For example, in the case of Figure 11 (the sequence at EP_1):

$$(M_1) = \begin{bmatrix} 000100 \\ 001000 \\ 100000 \\ 010000 \\ 000000 \\ 000000 \\ 000000 \\ 010000 \\ 101000 \\ 000100 \\ 000000 \end{bmatrix} \quad (M_2) = \begin{bmatrix} 010000 \\ 000100 \\ 001000 \\ 000001 \\ 010000 \\ 000000 \\ 001000 \\ 000100 \\ 000000 \\ 000001 \\ 000000 \end{bmatrix} \quad (8)$$

(M_1); status of scrapers before changing
(M_2); status of scrapers after changing

"1" indicates the scrapers are down and "0" indicates the scrapers are up.

$$(M_3) = (M_2) - (M_1) = \begin{bmatrix} 0 & 1 & 0 & -1 & 0 & 0 \\ 0 & 0 & -1 & 1 & 0 & 0 \\ -1 & 0 & 1 & 0 & 0 & 0 \\ 0 & -1 & 0 & 0 & 0 & 1 \\ 0 & 1 & 0 & 0 & 0 & 0 \\ 0 & 0 & 0 & 0 & 0 & 0 \\ 0 & 0 & 0 & 0 & 0 & 0 \\ 0 & -1 & 0 & 1 & 0 & 0 \\ -1 & 0 & -1 & 0 & 0 & 0 \\ 0 & 0 & 0 & -1 & 0 & 1 \\ 0 & 0 & 0 & 0 & 0 & 0 \end{bmatrix} \quad (9)$$

(M_3); movement and sequence of scrapers for changing at EP_1

"-1" shows lifting the scraper, "1" shows dropping the scraper and "0" shows no change. In addition to the previous functions, this system includes the following functions.

(1) stopping of the conveyor when trouble (equipment failure, over filling of the bin etc.) occurs.
(2) out put of the command to stop the transportation.
(3) man-computer communication by way of a CRT.

RESULTS

As shown in Figure 12, the utilization of on-line scheduling for transportation and direct control of transportation equipment has reached nearly 100%. From this we have obtained the following results:

(1) decrease to nearly zero in the amount of time which the stock level falls below the desired control level (see Figure 13);
(2) improvement in transportation efficiency by 20% reduction in distance of loader travel;
(3) reduction of man power requirements (from 4 to 0);
(4) reduction in the amount of mixed raw materials by 30%.

CONCLUSIONS

In the raw materials preparation process of the steel works, two kinds of applications of network flow problem are described in this paper.

One is the on-line scheduling for raw material transportation aiming at maximizing transportation efficiency. The other is the real time scheduling for direct control of transportation equipment aiming at minimizing competition of equipment.

Both schedules produce the feasible (optimal) solution with sufficient speed to be useful by using an algorithm.

When we design a scheduling algorithm for a practical problem, it is important to reduce the processing time especially, because it is usually necessary to recompute the schedule within the actual period of use due to unpredictable trouble.

Usually, as shown in this paper, the processing time for scheduling can be decreased by looking as closely as possible at all of the individual constraints.

REFERENCES

Tokuyama, H., and Watanabe, H., (1973). On-line scheduling for the transportation of raw material in the yards of steel works.

Society of Instrument and Control Engineers

Tokuyama H., and Ienaga, Y., (1978). On-line scheduling for the transportation of raw material in the yard of iron works.

IFORS, Montreal. P.336-352

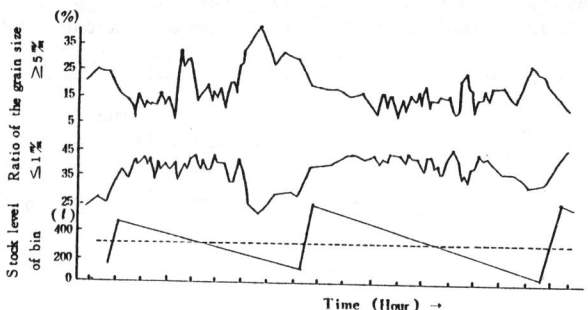

Fig. 1 The relation between stock level of bin and grain size distribution of raw materials

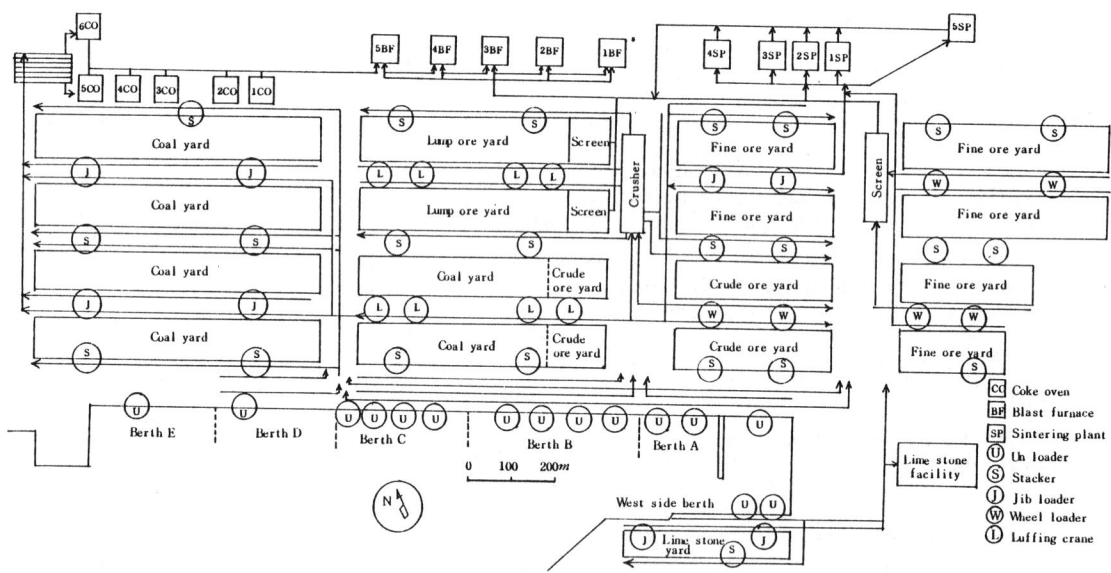

Fig. 2 Layout of raw materials yard at Wakayama Steel Works

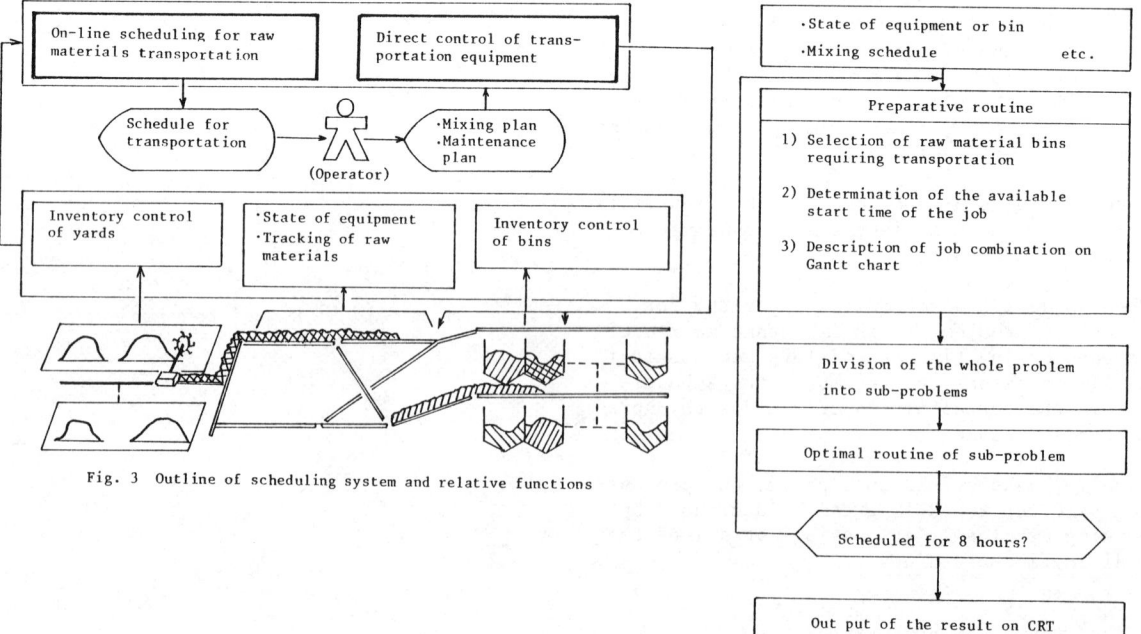

Fig. 3 Outline of scheduling system and relative functions

Fig. 4 Calculation flow of the algorithm

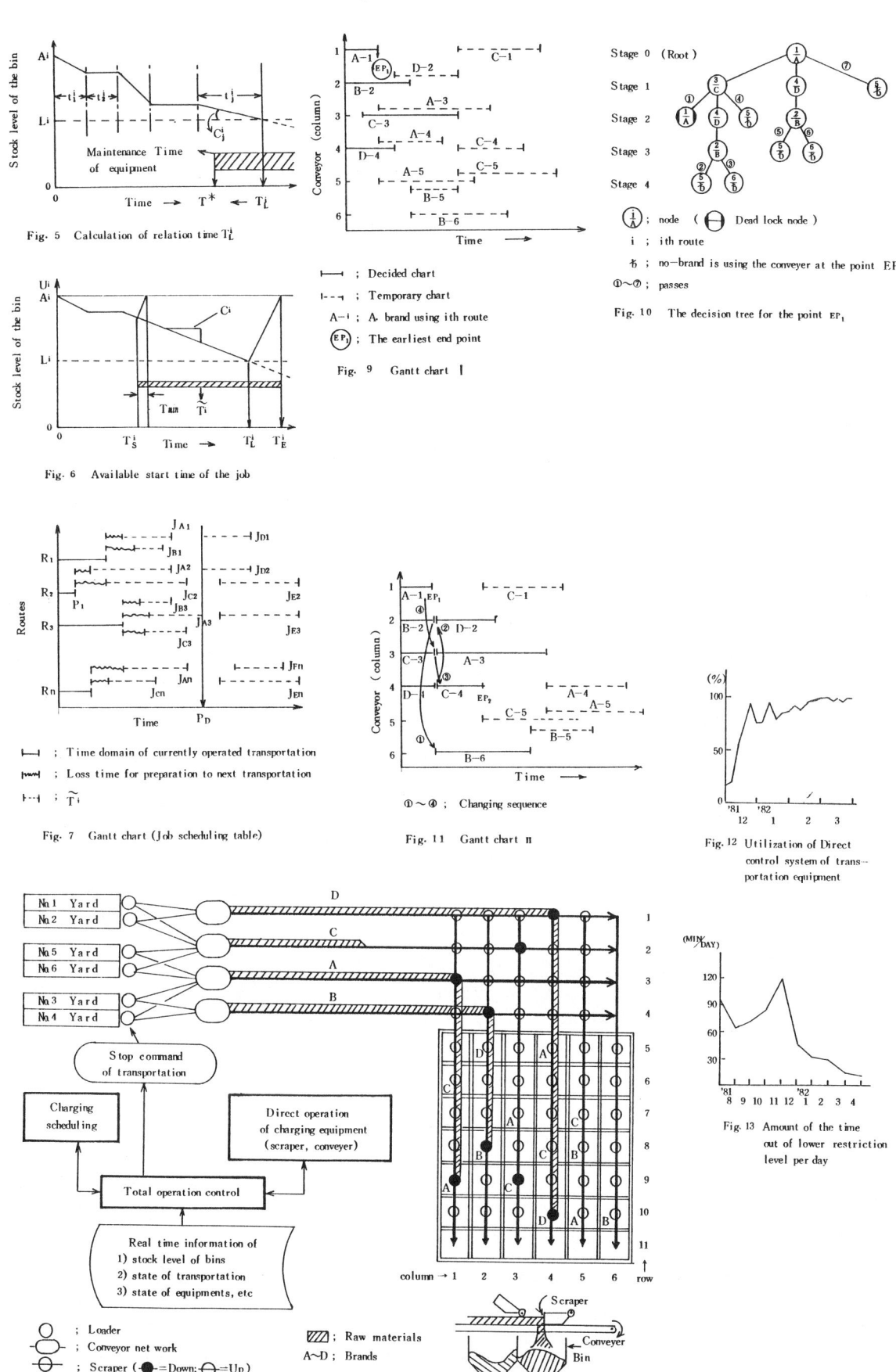

OPTIMIZATION CONTROL FOR COMBUSTION AIR IN REFUSE INCINERATORS

M. Kawahara* and K. Uosaki**

*Technical Research Institute, Hitachi Zosen Corporation, Osaka, Japan
**Faculty of Engineering, Osaka University, Osaka, Japan

Abstract. The direct optimization control method has many advantages in applications to real processes, though most of which are limited to be stationary.
This report aims at a practical application of the direct optimization control to nonlinear non-stationary systems. A refuse incinerator plant has been taken as the object of control and a direct optimization control system has been constructed. The system consists of two stages: the optimization stage under constraints; and the correction stage of constrains themselves depending on the state of process.
It has been clarified that the proposed control method is expedient for application of the direct optimization control method to the control of non-stationary systems.

Keywords. Control engineering computer applications; Hierarchical systems; Industrial control; Optimization; Process control; Time-varing systems.

1. INTRODUCTION

Direct optimization control has been used generally for steady state process (see for example Bernard, 1959; Weiss, 1961).
It has the following advantages:
(1) Process model is not required to apply this method.
(2) There is no need to taking nonlinearity into consideration.

In spite of these very attractive advantages, application of the direct optimization control is strictly limited due to the following disadvantages:
(1) The time interval of control should be determined so that the response settles to the stationary state.
(2) Number of necessary observations for optimization increases rapidly in the case of multi-variate processes.
(3) There is no guarantee of convergence to the true optimum point in the case of multimodal objective function.
(4) There is a possibility to make erroneous optimizing operations in the case of non-stationary processes.

This report aims at a practical application of the direct optimization control to non-stationary systems while keeping its advantages. A refuse incinerator has been taken as the object of the control. Based on the result of preliminary experiments, a new control system has been designed, and its appropriateness has been confirmed from the result of control experiments in the refuse incinerator.

2. OUTLINE OF REFUSE INCINERATOR

The outline of an incinerator is shown in Fig. 1. The grate system consists of a predrying grate, a main grate and a complete combustion grate. The refuse supply rate can be changed by adjusting the shaking speed of each grate. Supplied air for predrying and combustion is pre-heated and fed from under the grate. Among various kinds of supplied air, both the amounts of combustion air A and B are particularly effective to control the refuse combustion.
The reason why the automatic control is difficult lies in inhomogeneity of shape, calorific value and moisture content of refuse. These factors cannot be measured in advance. Since there is no a priori information about the properties of refuse, operators are compelled to take the trial and error method to keep the state of combustion well. Generally, they use the following criteria to operate incinerators:
(1) State of the refuse at the rear end of the predrying grate: It should already start burning.
(2) Burning state of the refuse on the main grate: It should be uniform. The center of combustion should not be on the front side nor on the rear side of the main grate.
(3) State of the flame: It should be powerful. At the same time, it should be adjusted so that light refuse is not blown up.
(4) State of the refuse at the rear end of the complete combustion grate: Combustion should be terminated.

The temperature of an incinerator declines due to excessively high or excessively low amount of combustion air. There may exist the optimal amount of air for the refuse of different properties. Thus, it can be said that the relationship between the amount of combustion air and the state of combustion is nonlinear and non-stationary. It is required to optimize the amounts of combustion air A and B in order to maintain stabilized state of combustion.

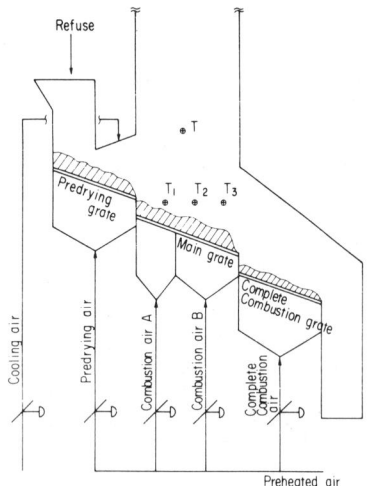

Fig. 1. Outline of a refuse incinerator.

3. DIRECT OPTIMIZATION CONTROL SYSTEM FOR REFUSE INCINERATORS

3.1 Construction of the Control System

Figure 2 shows the direct optimization control system. This system consists of two stages: the evolutionary optimization under constraints, and the correction of constraints themselves depending on the state of the process. In the ordinary operation, the hill-climbing type optimization control under constraints is carried out. When the state of process changes too sudden for optimizing operation to follow, the constraints are changed and the feasible search region for optimum point is forcibly shifted.

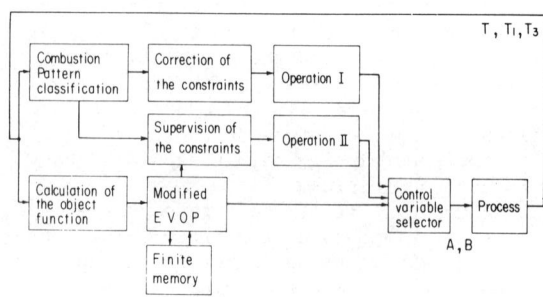

Fig. 2. Combustion air optimization control system.

This control system has the functions to classify combustion patterns by using temperatures of an incinerator and to establish the feasible region in the search area depending on the combustion pattern. When the feasible region is changed and the seeking point is deviated from the region, the seeking point is brought into the feasible region by Operation I. If the seeking point is likely to go out of the feasible region, the seeking point is controlled to be kept in the region by Operation II.

3.2 Objective Function

According to skilled operators' knowledge and preliminary experiments, following objective function is established.

$$P(t) = \min \{T(t), T_c\} - \alpha[\max \{\overline{T_1}(t) - T_0, 0\}], \quad (1)$$

where $T(t)$: temperature at the top of incinerator, T_c: desirable temperature for T, $\overline{T_1}(t)$: average of the temperature T_1 on main grate at $(t-\Delta t, t)$ (average value is taken since T_1 has larger variation than T), T_0: reference temperature for T_1, α: weighting coefficient, and Δt: control interval. The measured points of temperature are shown in Fig. 1. The first term of Eq. (1) is to measure quantitatively whether the combustion is powerful; $T(t)$ is to be kept as close as possible to the desirable temperature T_c. The meaning of the second term is as follows: The center of combustion can be estimated by the temperature pattern determined by T_1, T_2 and T_3. Judging from the temperature pattern, the amount of air is controlled to locate the center of combustion on the efficient position of the grate. Since T_1, T_2 and T_3 are not independent, it can be expected that T_2 and T_3 are controlled by maintaining T_1 at a certain reference temperature T_0 and that combustion is carried out with suitable center of combustion for various refuse. At the same time, sudden decline of temperature T_1 does not affect badly to the optimization process, even a large amount of refuse falls from the predrying grate.

3.3 Optimization Algorithm

Several optimization methods have already been proposed. For direct optimization control of an actual process, the extremum seeking must be carried out safely in a stationary process. Therefore, the random methods cannot be applied because they in some cases may select a seeking point that will drastically decrease the value of objective function.
Here, Simplex EVOP (Evolutionary Operation, see for example Matsubara, 1970) has been adopted as a basic scheme of extremum seeking. A polyhedron having n+1 vertexes in n-dimensional Euclidean space is defind as a Simplex. For instance, a triangle is a Simplex in 2-dimension. The simplex EVOP operation is carried out by transferring the vertex (X_w) which gives the worst value of objective function to the position of a mirror image ($\hat{X}_w$) with respect to the center of gravity (X_g) of the remaining vertexes:

$$\hat{X}_w = 2X_g - X_w. \quad (2)$$

Thus the Simplex can be moved to the optimum
point in the n-dimensional space. As stated
in Section 2, it is difficult by directly
applying this EVOP to optimize such processes
with a strong non-stationary trend.
The author (Kawahara, 1976) proposed the
following modified EVOP algorithm which
is applicable to the optimization of
non-stationary system.
The objective function $P(t)$ is calculated
after waiting one control interval Δt, then
the oldest data among the finite memory is
discarded and the latest data is stored newly.
When three values of objective function at
the corresponding vertex of the Simplex are
obtained, the optimizing operation is carried
out and the Simplex is shifted to the direction of the optimum point. After optimizing
operation, the process is controlled at
first under the condition corresponding to
the new vertex of the Simplex. If data corresponding to the remaining vertexes of the
Simplex are available in the memory, these
data will be used in the next optimizing
operation instead of controlling the process
again under these conditions. The capacity
of the memory depends on the non-stationarity
of the process. In the case of refuse incinerator the memory is designed so that extremum
seeking data are memorized during three control intervals.

3.4 Operation I and Operation II

Either Operation I or Operation II is carried
out according to the state of combustion,
particularly combustion at the rear side of
the main grate, while combustion at the front
side of the main grate reflects on the objective function.
Figure 3 shows the method to classify the state
of combustion into the combustion on the
front side (State I) or the combustion on the
rear side (State II).
In the following two cases, the feasible region
of EVOP is shifted toward the decreasing
direction of the combustion air B, namely the
amount of air B is decreased by ΔB through
Operation I.

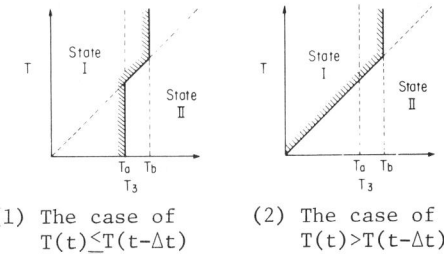

(1) The case of $T(t) \leq T(t-\Delta t)$
(2) The case of $T(t) > T(t-\Delta t)$

Fig. 3. Classification of the state of combustion.

(1) When the combustion has changed from State II to State I.
(2) When the temperature T has decreased continuously for a period of $2\Delta t$ in combustion State I.

In the opposite case, when the combustion is
changed from State I to State II, the amount
of air B is increased by ΔB as Operation I.

Operation II is executed in the following
two cases:
(1) When $T(t) > T(t-\Delta t)$ in combustion State II
or when $T(t) > T(t-\Delta t)$ and $T_3(t) > T_3(t-\Delta t)$
in combustion State I, the shift of
seeking point which decreases the amount
of air B is prohibited.
(2) When $T(t) < T(t-\Delta t)$ in combustion State I
or when $T(t) < T(t-\Delta t)$ and $T_3(t) < T_3(t-\Delta t)$
in combustion State II, the shift of
seeking point which increases the amount
of air B is prohibited.

4. RESULTS OF CONTROL EXPERIMENT

The experimental result without Operations I
and II is presented in Fig. 4. It is shown
that the state of combustion has gradually
changed toward the state of combustion on
the front side (predrying side). Along with
this phenomenon, air A increased while air B
decreased. This is well acceptable as the
result of control on the state of combustion
on the front side. However, there is no
brake to stop the front sided combustion
tendency and this shows a trend of further
increase of air A. As the next step, almost
all of the refuse was burnt out and the
temperature of incinerator decreased due to
excessive amount of air. After this stage,
when the refuse newly arrived at the main
grate, combustion became powerful because of
sufficient amount of air supply, and the

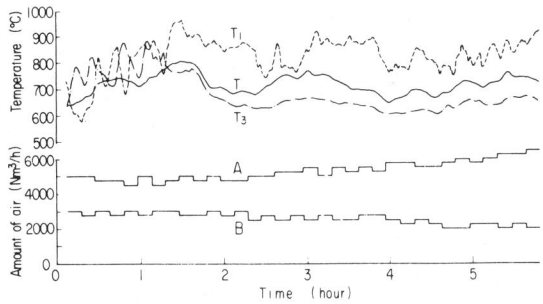

Fig. 4. Experimental result without Operations I and II.

temperature of incinerator increased again,
and the amount of air A further increased.
As a result of the repetition of these phenomena, the temperature of incinerator T
varied largely. Therefore, it is obvious
that combustion state is not stable if
Operation I and Operation II are not applied.
Figure 5 shows a typical example of experimental results using the proposed direct
optimization control system. This result
indicates that the control system is sufficiently responding to the sudden change in
the state of combustion, and is maintaining
the temperature of incinerator in a stabilized way.

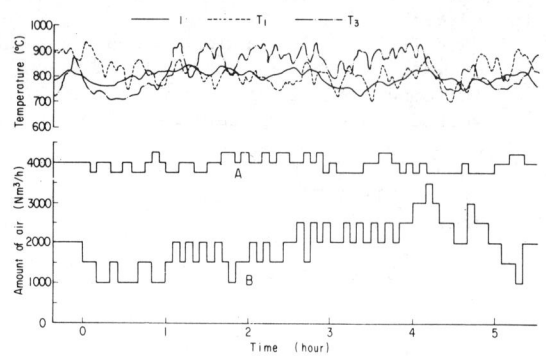

Fig. 5. Experimental result using the proposed method.

During the control experiment over twenty hours, 87% of operation were done by optimization treatment and 13% were done by Operations I and II.
Table 1 shows the quality of control of three experiments using the proposed method, compared with the quality of manual control. Thus, it is confirmed that the quality of control is nearly same for both cases.

5. CONCLUSIONS

The direct optimization control by extremum seeking has so far been effective only for the stationary system. In order to apply the direct optimization control method to a refuse incinerator which is a strongly non-stationary system, a new control method has been proposed. This control system consists of two parts. One is the modified evolutionary operation method with constraints and the other is the method of adjusting the constrains themselves using combustion patterns. The feasible search region is set in extremum seeking area. When the state of process changes suddenly, the feasible region is forcibly shifted. In order to detect the sudden change in the state of process, the state has been classified into several patterns and a method has been introduced to judge whether there is a change in the pattern.

According to experimental results, it has been confirmed that the quality of control is nearly same as that of the case of operation by a skilled operator, judging from the degree of variation of the temperature in the incinerator.

It has also been clarified that the method proposed in this paper is suitable for application of the direct optimization method to the control of non-stationary system. And this method is applicable not only for the refuse incinerator but also for any other process control, provided that the state of process is classified into several patterns and the expedient range of control variables for each pattern is available.

REFERENCES

Bernard, J.W., and F.J. Soderquist (1959). Control Engineering, 6, 124.
Kawahara, M. (1976). Optimization Control for Non-Stationary System. Preprints of 20th JAACE Convension, pp. 49-50 (in Japanese).
Matsubara, M. (1970). Process System Engineering. Asakura Book Co., Tokyo. pp. 291-294 (in Japanese).
Weiss, E.A., D.H. Archer, and D.A. Burt (1961). Hydrocarbon Processing & Petroleum Refiner, 40, 169.

Table 1 The Quality of the Optimization Control

	Experiment I	Experiment II	Experiment III	Manual control by skilled operator
Control duration (hours)	5.5	8.75	6.75	24
Mean value of the temperature T (°C)	780	771	765	795
Standard deviation of the temperature T (°C)	36	40	39	38

CONTROLLING A DISTRIBUTION CONVEYOR BY A DEDICATED MICROPROCESSOR

L. E. M. Boullart

Laboratory for Automatic Control, State University of Ghent, Grotesteenweg-Noord, 2, B9710 Zwijnaarde, Belgium

Abstract. The paper describes an industrial application to setup a microcomputer system to perform a real time control of the last part of a production line of wash-leathers. The line consists of a measuring unit linked to a conveyor to perform an automatic distribution into packaging units of the leathers depending on measuring values. The small 8-bit microcomputer controls the whole section : i.e. measuring unit, conveyor transport, unloading mechanism and furthermore the account of the number of stored units.

Keywords. Conveyor, Transportation control, Microprocessors, Microprogramming, Computer control, On-line operation.

INTRODUCTION.

In the production of natural wash-leathers the ultimate form and size of the leathers can vary enormously due to different parameters (first of all the size of the sheeps itself). After production the leathers may be sold as they are, or they can be cut previously according to some standard forms. In any case, the quality of the leather, and thereby its commercial value is directly determined by its surface. Selling from factory is done by packaging units, where each unit contains leathers within certain surface bounds. Thereby every type of unit ("class") stands for leathers of a certain quality. Each packaging unit contains a fixed number of leathers, so productions figures are calcutaled following the number of packages of each quality.

PROBLEM DEFINITION.

From the introduction it stands that the last step in the production unit of wash leathers will be a surface measuring unit, followed by a distribution into the packaging units according to the results of the measurements, which should be as accurate as possible. The goal of the project was to automate completely this measuring/ distribution station. In order to achieve this, the plant management engaged an engineering bureau, which in turn consulted the Laboratory for Automatic Control of the University of Ghent. The engineering bureau designed the measuring unit and the distribution conveyor, while the Laboratory for Automatic Control designed the microprocessor control unit and associated software.

The whole measuring/ distribution system should operate as follows :

- an operator will place manually each wash-leather sequentially on the measuring unit;
- the leather passes under a row of sensors, and the microprocessor catches the reading, which is equal to the surface of the leather;
- the leather is picked up by a pneumatic pick-and-place unit, which puts it on a peg, hanging on a conveyor chain;
- the leather is transported over the conveyor, while continuously monitored by the microprocessor-unit;
- when arriving at the packaging station corresponding to its quality (class), the microprocessor steers the taking off (pneumatic) mechanism;
- while controlling, the microprocessor should account fo the number of leathers, deposit in each packaging station, and give an audible signal to the personnel when shipping units are reached.

THE MEASURING UNIT.

The measuring unit is schematically depicted in fig. 1. As can be seen, it consists of a table with an endless rolling carpet of metal gauze, to which the wash-leather is sucked by a vacuum pump. The leather passes with a (known) constant speed trough the measuring lath, where one side contains a lamp and the other side a series of photodiodes. The surface is

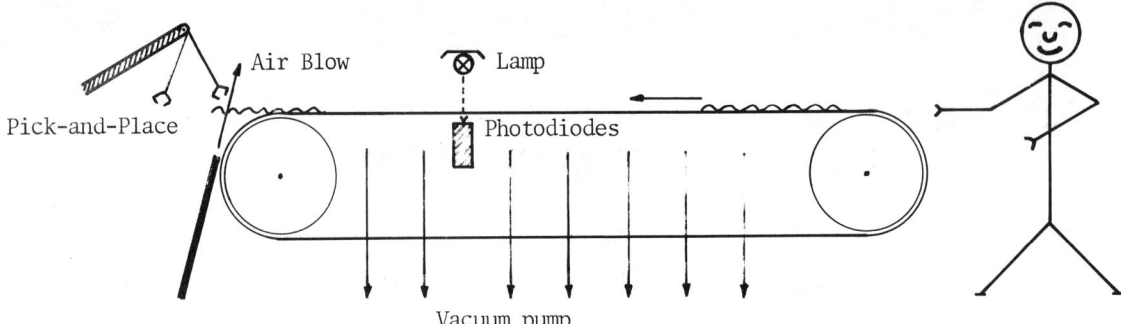

Fig. 1. : The measuring Unit.

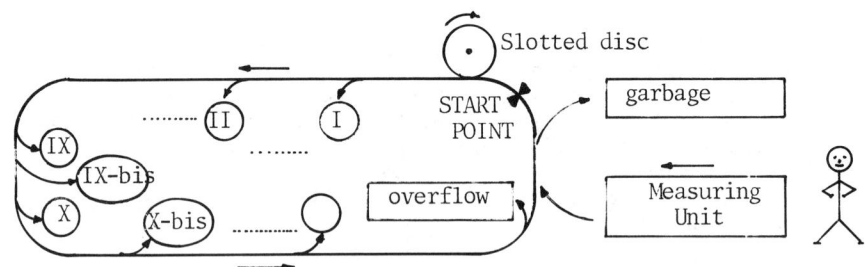

Fig. 2. : The conveyor unit.

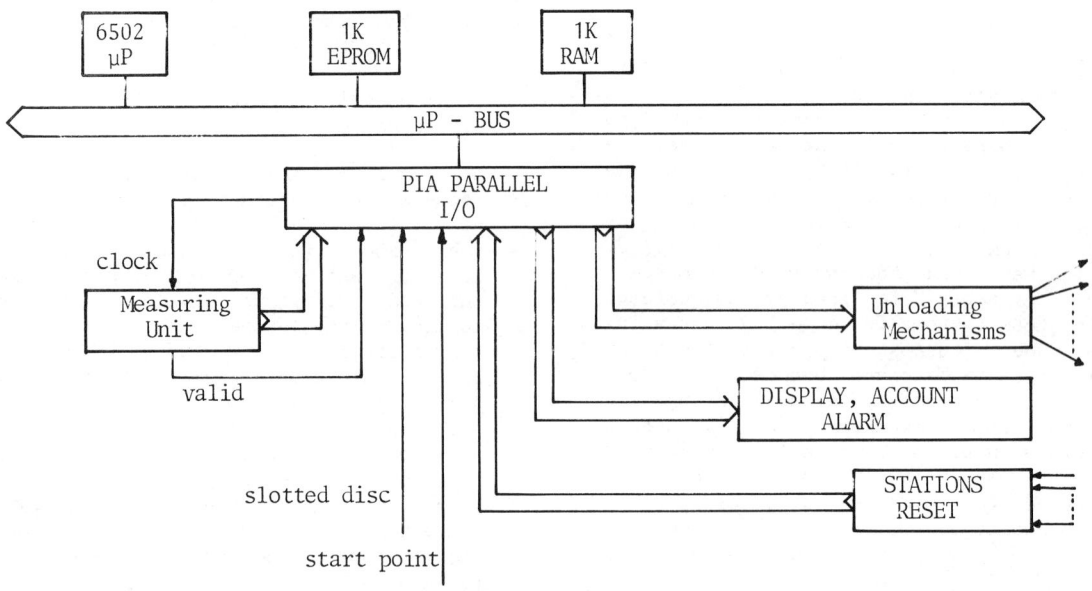

Fig. 3. : The microprocessor configuration.

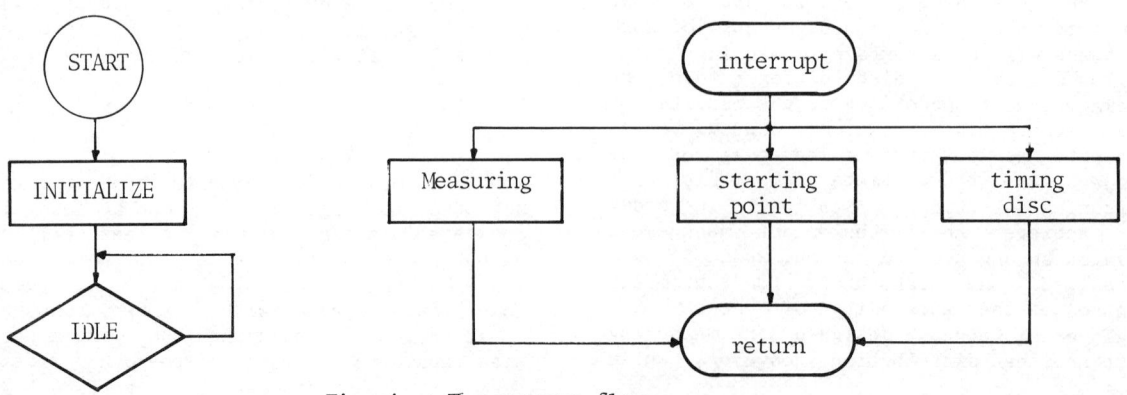

Fig. 4. : The program flow.

measured by integrating the number of dark diodes with a fixed time-slice. This is accomplished by the microprocessor unit as follows:

- the microprocessor program is triggered when the number of dark diodes is above a critical level;
- from that time, at regular interval the number of dark diodes is serially shifted out of the measuring unit into the microprocessor, up to the end of the leather.

At the end of the rolling carpet, the leather is blown off by heavy air flow; at that point the pick-and-place unit intercepts the leather and hangs it onto the conveyor.

After the measuring, the microprocessor program knows the exact surface of the wash leather, and immediately attributes it a quality label by performing a look-up into a memory table, which contains the classe's limits. This label will be deterministic for the packaging station where it will be delivered further on.

THE CONVEYOR.

The transport of the leathers is performed by a conveyor which is schematically depicted in fig. 2. It consists of an endless chain, provided with pegs to hold the leathers during transport. At regular points in the chain, packaging stations are situated, each equipped with a pneumatic unloading mechanism.

Immediately after the measuring unit a special "garbage" station is situated, to accept all leathers of minor quality : i.e. leathers with a surface smaller than a predetermined value. Also, for security reasons, at the end of the chain, a station is situated to accept all leathers, which by certain reasons (e.g. overflow : see further) could not have been unloaded.

The microprocessor has to follow every leather very precisely during the whole transport, for the command to the unloading mechanism has to be given at the exact position. Therefore a high degree of synchronisation is absolutely necessary. This is accomplished in two ways :

- a very precise start point at the chain actively alerts the microprocessor when passing by;
- in order to be independent of speed variations of the conveyor chain, a "slotted" disc delivers timing pulses to the microprocessing unit; the frequency of those pulses thereby becomes proportional to the conveyor speed.

Inside the microprocessor unit, every packaging station is located at a number of timing pulses from the start point. After measuring, the microprocessor gets such a number for every quality label, and during processing, it accounts continuously for the total number of timing pulses. When this total matches the predetermined value, the unloading sequence is performed. The exact distance in timing pulses was roughly measured at the beginning, and fine tuned in situ during the set-up phase.

THE MICROPROCESSOR UNIT.

The microprocessor unit was configured with an 8-bit 6502-microprocessor (Synertek e.o.) running at 1 MHz. It consists of a modular Eurocard system, designed and constructed in the laboratory itself. The system configuration is given in fig. 3.

Besides the usual boxes (CPU, RAM and EPROM), the most important element in this application is the parallel input/output unit (PIA : "peripheral input/ output adapter"). It connects the system with the various measuring and steering points :

- information is received from the measuring carpet by outputting a series of clock marks to it, and shifting in the replying photodiode pulses; a new measurement is announced by a "valid-reading" line, connected to the interrupt mechanism of the processor;
- the unloading mechanism is activated by a timed bit on a selected output port;
- via displays, the operators are warned when one of the packaging stations is full; after emptying the station, the operator signals this to the microprocessor by pushing the "reset" switch of the cleaned station;
- every leather deposit at a station, an electro-mechanical counter (for each quality) is activated; those counters provide accurate production figures
- in order to keep a perfect synchronisation, both synchronisation signals (slotted disc, and starting point) are connected directly to the interrupt lines.

The whole configuration is constructed with only 3 Eurocards.

THE MICROPROCESSOR PROGRAM.

The microprocessor program, written in 6502-assembly language, runs completely interrupt driven. This means that after the initialisation phase, the mainstream of the program remains idle, and only responds to external interrupts. There are three possible interrupt routines : a new measuring value is announced; a leather passes the starting point and the slotted disc sends a timing pulse (fig. 4.).

The program controls the conveyor as follows. When a new measurement has been performed, the value, after conversion to the corresponding quality label is put aside in a waiting queue. After reception of the starting point (i.e. when the leather passes by), the distance of the expected delivery address is added to the queue element. From then on, at every interrupt from the timing disc, each waiting leather in the queue is checked to verify if unloading is necessary. If so, the unloading sequence is iniated, and the leather is taken off the queue.

Besides some small possible alarm checks, there is one important exception handling in the program. The question indeed arises, what to decide when a leather has to be unloaded, but the station is fully packaged, and thereby blocked until manual intervention. This may occur rather frequently, because most wash-leathers turn to have medium sizes, located in only two or three stations. To handle this, such "medium" stations were doubled, (i.e. 2 consecutive stations with the same quality); so, when one station was blocked, leathers had to be diverted alternatively to the next station in line. For the rare occasion, leathers were blocked at other stations, they were simply diverted the overflow area, from where they can be fed again into the measuring unit. This whole exception handling turned out to be rather complicated to incorporate into the program.

THE DEVELOPMENT PHASE.

The microprocessor control program was developed in a combined mini/micro environment at the laboratory site. Once written in 6502 assembly language, it was entered into the PDP 11/34 minicomputer. After assembling with a general macro/cross-assembler, it was further down-loaded into a 6502-configuration via a serial line. This target configuration contained besides the necessary CPU and memory a special "monitor" card. In this way a conversational debugging via a terminal was possible.

The process to be controlled itself had to be simulated. For the fast measuring unit control, this was done on patchable general purpose logic present at an analog computer in the laboratory. For the (slower) transportation part, this was done by a PDP-11/34 Fortran program and a hardware connection with a parallel interface connection between the mini- and microcomputer.

After initial debugging, the final tuning was conversationally done in situ by a VDU-terminal.

CONCLUSIONS.

After initial tests and fine tuning of the system, a comparative evaluation with the non automated operation could be done. This manual system operated as follows:

- there was one operator at both sides of the measuring unit; the latter had a readout display on top of it;
- the operator which accepted the leathers, selectively deposed the leathers on different heaps (qualities);
- for every class, a primitive abacus mechanism was used for accounting;
- when packaging was full, the measurements were interrupted for cleaning the heaps.

Due to the automation, production figures were increased:

- the feeding operator could speed up the rate of measurements, for he didn't have to wait for his colleague (at least 30% increase);
- only one operator is continuously present, while intervention of another (for emptying the stations) is only partially required;
- there is no need for interruptions whatsoever.

The application shows that, although at a first glance there are a lot of tasks to be fulfilled, this can be done with a small program (< 1 K bytes!) in an 8-bit microcomputer. When the software is well-structured and when the right development hardware/software is present, a straightforward design and implementation in a laboratory environment is possible. Furthermore an in situ fine tuning can easily be done, which demonstrates the flexibility of a programmed implementation above a pure hardware design. Last but not least, the elegant handling of the exception situations (overflows) shows its full power.

Concerning future extensions of the systems, a new production unit will be planned. The microprogram will be extended with another task to perform the complete steering of the pick-and-place unit between measuring and conveyor section. It will be quite a challenge for the program to still make it without loosing synchronisation. If this should fail, there is an escape possible, by doubling the operating speed of the CPU-unit from 1 to 2 MHz.

ACKNOWLEDGEMENT.

The author explicitly wishes to thank ir. M. Moortgat, from the Moortgat Engineering Bureau, for his enthousiastic cooperation. He indeed offered us the chances for an interesting project, to which he himself devoted an unlimited amount of his time.

DISTRIBUTED TRAFFIC CONTROL SYSTEM

M. Nakai and M. Kasahara

Technical Research Laboratory, Kinki Nippon Railway Co. Ltd., 10-1, Amagatsuji-kitamachi, Nara-city, Japan

Abstract. The newly developed distributed train traffic control system adopting micro-computers is introduced, in contrast to the conventional system, which integrates signal control distributed at stations into the operation dispatching center using remote-control apparatus. In the new system, the micro-computers installed at each station control signals following to the train timetable, and the central computer at the operation dispatching center manages the train schedule through the whole line and transmits the train timetable data of operation plan or operation changes, in place of signal control data, to the micro-computers at stations. The authors start the test operation of the prototype equipment for one station.

Keywords. Computer control; railways; traffic control; train control; transportation control.

INTRODUCTION

The urban or suburban type railway is required to operate the trains frequently and rapidly for both near-distance and far-distance passengers. To satisfy these requirement from passengers, by means of setting up the train classification, e.g. normal or express, and installing the interlocking machines at the intermediate stations for passing of trains, the transportation service is hightened the efficiency and installations are utilized effectively.

In the railway management the transportation plan i.e. the train schedule should be settled in advance, and the train operation should be rectified according to the daily demands. Moreover, in case of the disordered schedule caused by a train accident and other reasons, the train schedule should be revised forcasting every possibility in advance, for a train cannot pass another train at arbitrary places as trains are restricted on the railway.

The operation dispatching center is the place to supervise and estimate the traffic of trains, and the signal boxes at stations are the places to control the signals according to the dispatching. The signal boxes, however, are installed here and there along the railway line, and the dispatching is informed by telephone if signal men are in the boxes, or the signals are remote-controlled from the center if signal men are not in the boxes. The system composed of the supervision of traffic and the control of signals in this way is generally called Centralized Traffic Control (CTC) system. The configuration of the conventional CTC system has the form in which the signal boxes are integrated into the operation dispatching center, and becomes a large scale remote-control system.

In contrast to the conventional CTC system, the authors propose the distributed traffic control system using the up-to-date technology of micro-computer and optical transmission. In this system, the central subsystem executes the train schedule management and the local subsystem for a station executes signal control. As hardware structure, it consists of the functionally distributed system using micro-computers. For the communication network between stations, the data-way transmission method using the optical fiber cable is adopted.

Prior to the construction of the whole system, one prototype of the system is developed and installed at Higashi-ikoma station, a typical standard station of Kinki Nippon Railway in Japan, and started its test operation.

In this paper, the authors mention the new train traffic control system of the distributed control method and the test system.

NEW TRAIN TRAFFIC CONTROL SYSTEM

Realization of Train Operation

Fig. 1 shows the flow of the train operation

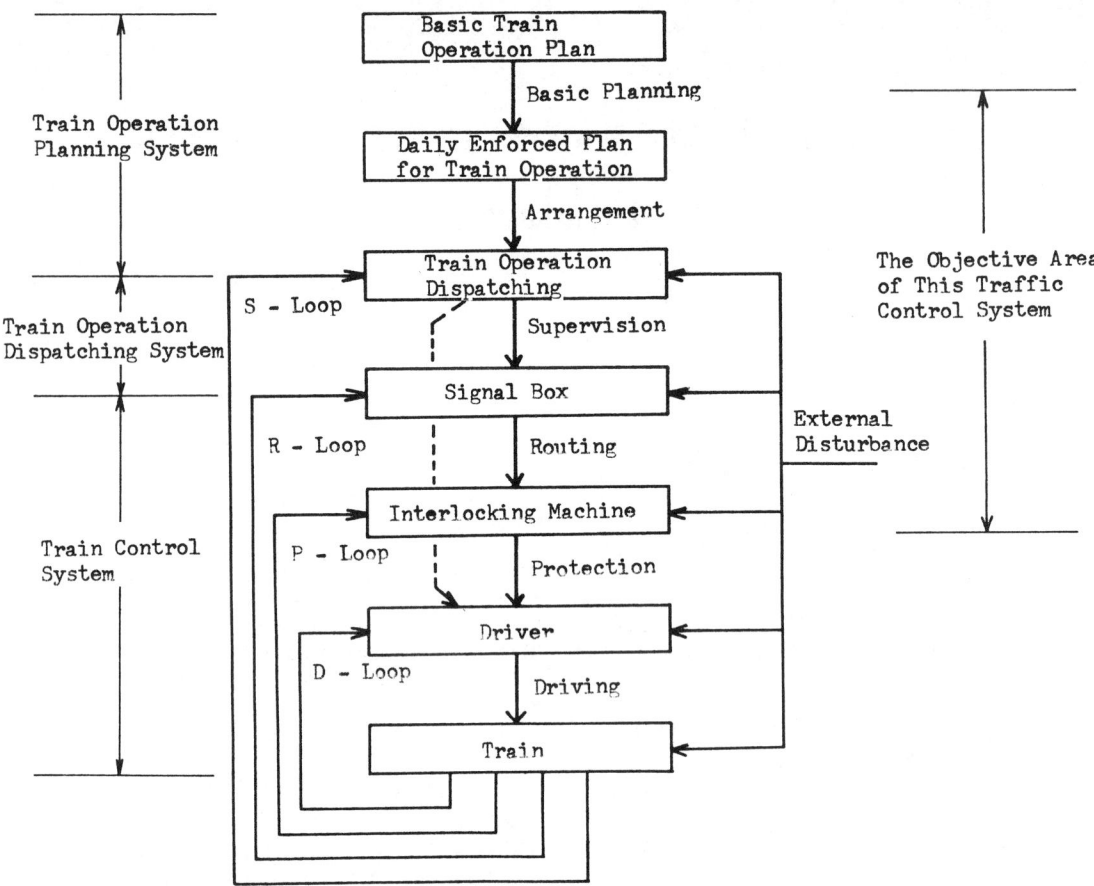

Fig. 1 Train Operation Flow

in which the train schedule is made according to transportation demands and then trains are actually operated. The information or dispatching flows downward as shown in Fig. 1, and the feedback information, which shows the result caused by external disturbance, effects the decision of the next action.

Preparation of train operating planning. The basic train operation plan is prepared founded on the characteristics and data of the railway equipments, the station equipments, and the railcars, and then the utility plan of the railway resources is provided. These plans are expressed as the diagram of train schedule.

The daily enforced plan i.e. daily train schedule is decided according to the passengers' demands occured seasonally or provisionally such as extra train, chartered train or trial running in addition to the basic plan.

Operation dispatching. The characteristic of the railway is that a train is not permitted to drive freely because of the restriction of one dimensional railway. So, in order to maintain the transportation efficiency, it is indispensable to operate trains following to the plan prepared in advance. Moreover, in case of the disorder caused by the external disturbance, it is required to dispose of the disorder promptly and properly, which is main function of the operation dispatching.

The operation dispatching is to supervise the train traffic according to the train schedule, and also in case of the traffic disorder by unexpected accident or external disturbance, it is required to detect precisely the traffic conditions at the right moment and to estimate and determine the suspension of scheduled trains or the operation of extra trains. Thus, the train operation is conducted by informing these results to drivers through the dispatching radio and to signal men through the dispatching phones.

Train control. The train control is divided into two groupes, the signal control and the train drive. A train cannot control the course itself, but the course is determined by the direction of the switches, and this movement is controlled by signal control. The schedule of the train operation is announced to passengers by timetables of stations, at the same time, the schedule is delivered to drivers by staffs. Trains are driven by the operation of drivers following to the staffs.

The "ATO" (Automatic Train Operation) is the system to automate the driving and stopping of the train by drivers' operation. The "CTC" (Centralized Traffic Control) is the one to integrate the operation of signal controlling conducted by signal men. The object of this traffic control system includes, as shown in Fig. 1, the preparation of daily enforced plan, operation dispatching, the signal control and the related operations.

Function and Structure of the System

The fundamental functions of the train traffic control system systemizing the previously described duties are as follows.

(1) Preparation of the enforced schedule for control revised daily.
(2) Maintenance of current train timetable.
(3) Supervision of train traffic.
(4) Signal control.
(5) Transmission of information.
(6) Recording of train operation.

The conventional CTC systems, as shown in Fig. 2, transact all functions above at the center, and execute so-called remote-control; the information of the control and the status data at the signal equipment are transmitted between the center and stations.

In contrast to conventional systems, the authors propose the distributed traffic control system, as shown in Fig. 3, which consists of micro-computers at each station and has train timetable data of each station. This system distributes functions; the central computer manages the train schedule, and local micro-computer at each station controls signals; the functions of the station side are (3) and a part of (4) above mentioned. In this system, the information transmitted between the center and the station is the train timetable data and the supervision data. As the high-speed transmission channel adopts the optical fiber cable and configures loop-type, the data transmission between stations are capable.

The distinctive feature of the distributed system compared with the centralized system follows:

(1) Train operation is available to run on in case of the trouble at the central equipment as the local micro-computer has its train timetable data.
(2) Duty of the central equipment and transmission line is reduced.
(3) Each local micro-computer copes with different demands at each station of various structure and operation.
(4) Besides the information of signal control, various information transmission such as offering transmission of command and information of train traffic to each station is capable.
(5) The expansion of the system is flexible.

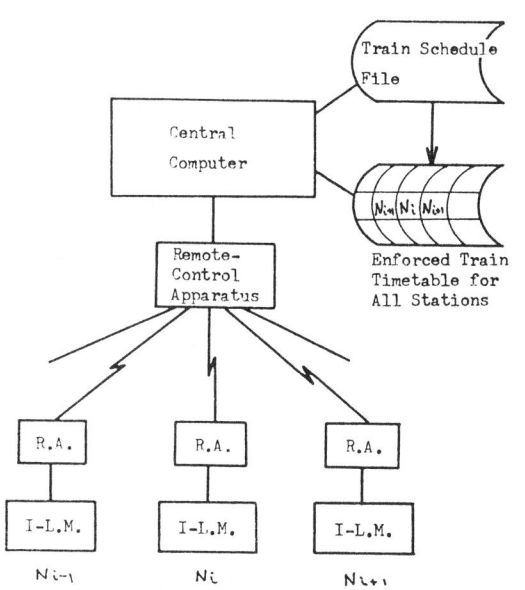

R.A.: Remote-Control Apparatus for Station
I-L.M.: Interlocking Machine
N_i: Signal Station

Fig. 2 Centralized Traffic Control System

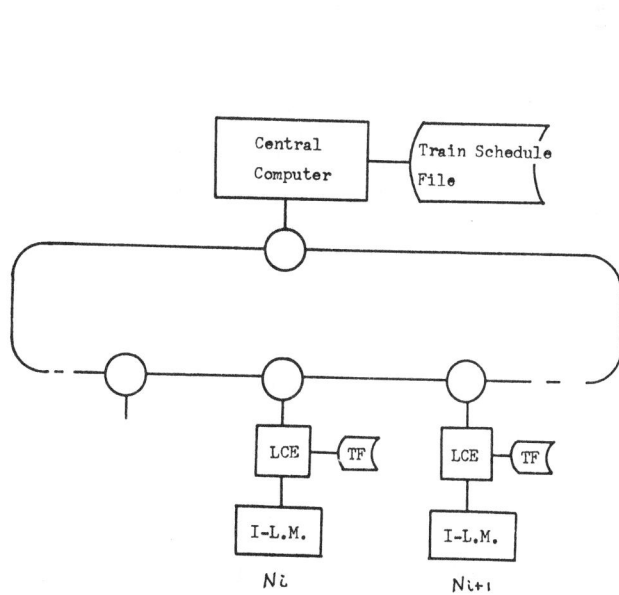

LCE: Local Control Equipment
TF: Enforced Train Timetable for Station
I-L.M.: Interlocking Machine
N_i: Signal Station

Fig. 3 Distributed Traffic Control System

TEST SYSTEM

As the structures of stations differ from another because of transportation demands, so it is important to standardize the local equipments for the sake of design and maintenance because the system controls many stations. Therefore, prior to the construction of the new whole system above, authors start to put it into practical use installing the test system at one station.

Scale of Test Station

Station structure. Higashi-ikoma station of Kinki Nippon Railway has the typical struc-

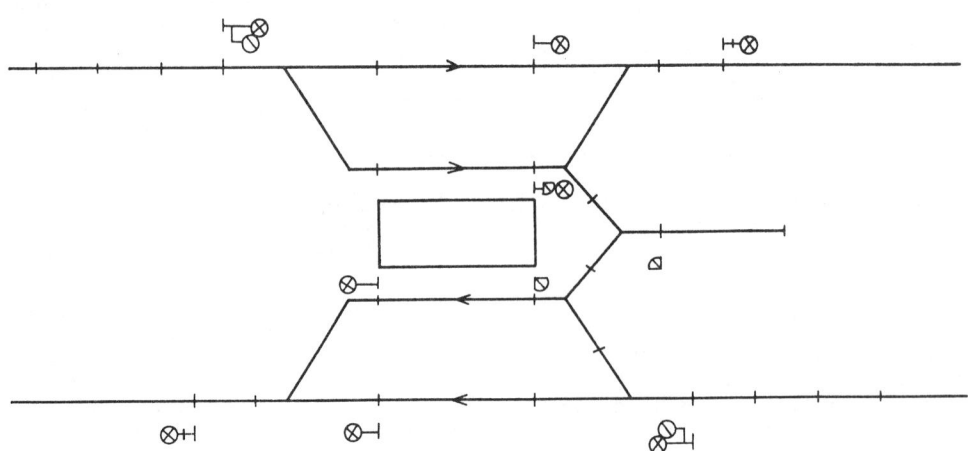

⊗⊣ Entering or Starting Signal
◁ Shunting Signal

Fig. 4 Route Structure of Higashi-ikoma Station

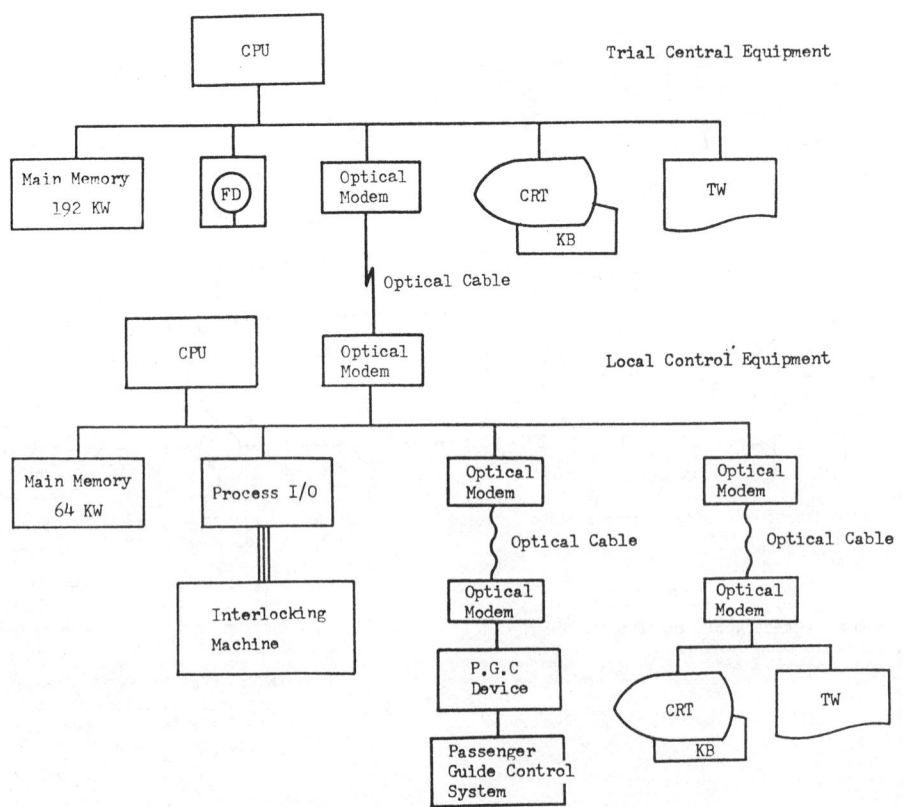

Fig. 5 Hardware Structure of Test System

ture as the double tracked signal station i.e. one overtaking track for each up and down bound track and one lead track as shown in Fig. 4. This station has 12 routes (signals) and 5 switches.

Traffic situation. The train schedule consists of three kinds of schedule, weekday, Saturday and Sunday. The number of trains is about 550 per day and about 40 per hour in rush hour though it differs corresponding to the kind of train schedules. Some trains pass the station, some stop and start at the station and others shuttle at the station.

Structure of the Hardware

The local control equipment and the trial central equipment in stead of the proper central equipment are installed as shown in Fig. 5. The local control equipment is a 16 bits microcomputer (MELCOM 350-2010) with 64 KW main memory, and is connected to the interlocking machine through a process I/O. Besides, through the transmission channel, the local control equipment is connected to a intelligent terminal which is a man-machine interface at the station, the passenger guide control system, and the trial central equipment. The intelligent terminal is an ordinary personal computer with a 12 inches color CRT display and a printer (See photo 1). The passenger guide control device controls the guide announcement equipment and the passenger guide display equipment, changing the train data and timing information from the local control equipment.

The trial central equipment consists of a mini-computer (MELCOM 350-2100) with 192 KW main memory and two flexible discs, a 20 inches color CRT display and a printer. The CRT display is used for traffic display for supervision, alarm display for trouble occurrence, and operation display (See photo 2). The trial central equipment and the local control equipment are connected by the communication line.

The optical fiber cables are used for the communication channels among the local control equipment, the trial central equipment, intelligent terminal and also the passenger guide control device.

Functions of Local Control Equipment

Fig. 6 shows the software structure conceptionally. The local control equipment has following functions.

Signal control.

(1) Train is tracked by the change of track circuits and the delayed time of the train is calculated by its results.
(2) Signal is controlled by the train timetable data stored in the local control equipment.
(3) Interlocking machine is supervised by this equipment in irregular actions.

Management of the train timetable of the station. Local control equipment stores enforced train schedule transmitted from the center, manages the schedule according to modified requirement from the center or station intelligent terminal.

Passenger guide control. The local control equipment sends to the passenger guide control device the train information data of train timetable and timing data of the track circuit change.

Information transmission. The information of traffic condition, interlocking machine trouble, and train timetable data are transmitted between the central and local control equipment.

Man-machine interface. Display of traffic condition and alarms, and revised operation of train timetable are shown by the CRT display of the intelligent terminal.

Photo 1 Intelligent Terminal

Photo 2 CRT Display of Trial Center

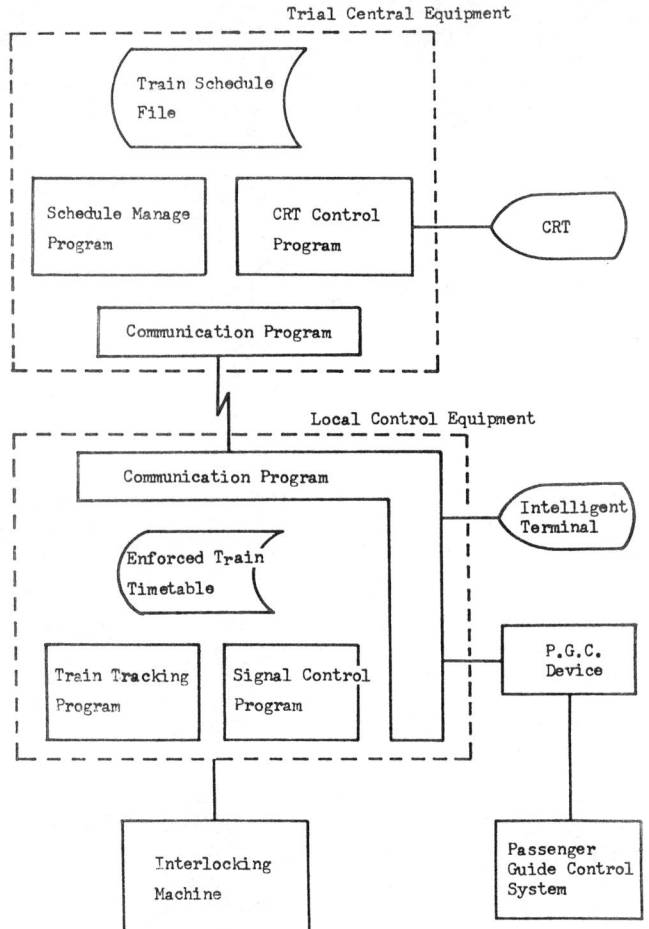

Fig. 6 Software Structure of Test System

Functions of Trial Central Equipment

The trial central equipment supervises the traffic of one station and manages the train schedule in place of the proper central equipment.

Train schedule management.

(1) Train schedule file includes basic schedule, weekly schedule, seasonal schedule and daily schedule for seven days. The enforced train schedule is composed of these schedule files.
(2) The enforced train timetable is prepared, as mentioned above, and transmitted to the local control equipment before the first train of the day.
(3) In case of changing the schedule after starting the train operation, the train timetable is modified at the central equipment and also transmitted to the local control equipment.

Man-machine interface. Operator can watch the train traffic condition and conduct the train change by the CRT display.

CONCLUSION

The distributed train traffic control system is introduced, and the test operation of the prototype equipment is started. The new system, which manages the train schedule at the center and controls signals by means of the micro-computers at each signal station, is identical to the conventional railway structure.

ACKNOWLEDGEMENTS

The authors acknowledge the members of the train traffic control system development project promoted by Kinki Nippon Railway Co., Ltd., Japan.

REFERENCES

Yasuhara, M. and T. Nakahara (1975). *Traffic Control*. Corona, Tokyo.

DISCUSSION

SESSION TA3: INDUSTRIAL APPLICATIONS I

Paper: OPTIMIZATION CONTROL FOR CUMBUSTION AIR IN REFUSE INCINERATORS

Authors: M Kawahara (Technical Research Institute, Hitachi Zosen Corporation, Osaka, Japan)

K Uosaki (Faculty of Engineering, Osaka, University, Osaka, Japan)

Discusser: Jaime Alvarez G
Apdo 14-740
CIEA-IPN
07000 México, DF

Questions or Comments:

1) Does the method you have used, assure convergence in all cases?

Authors' reply:

1) Mathematically speaking the method does not assure convergence. But practically speaking, it is the only one method that can be applied to actual processes as a direct optimization method and the seeking point can converge in many cases.

Discusser: D. Tabak
Ben Gurion U, EE
Beer Sheva 84105
Israel

Questions or Comments:

1) If the manual and the optimization approaches gave the same standard deviation, then what is the benefit of the optimization procedure? Why is it worth while?

Authors' reply:

1) Generally a skilled operator is the best controller. And the amount of combustion air in refuse incinerators has not been optimized automatically so far. Our system can optimize it automatically as well as a skilled operator does; therefore it is worth while.

TRENDS IN DIGITAL CONTROL APPLICATIONS IN PULP AND PAPER INDUSTRY

P. Uronen

Department of Process Engineering, University of Oulu, Oulu, Finland

Abstract. The first applications of real-time digital control to pulp and paper industry stem from early 60's. Then a centralized hardware was applied to quite large and poorly defined problems. These early projects were not very successful; one of the main reasons being the unreliable and limited hardware. During 70's the so-called packaged systems entered into the market. These were built by using the existing minicomputers and those systems were dedicated to smaller and well-defined problem areas; for example dry end controls of paper machine, batch-digesting, Kamyr digester, bleach plant etc. Today we are ahead a third generation of these systems in the Pulp and Paper Industry. The microprocessor technology has created the powerful microcomputers and the distributed digital instrumentation systems. This has lead to the concept: decentralized control and centralized reporting and supervision. This means that the lowest level of the control hierarchy will be handled by a digital instrumentation system and with a local datanetwork (data highway, bus) a hierarchical structure including area control and coordination, production planning and scheduling and management information system at the higher levels a total mill-wide hierarchy will be built. There is also clearly a trend towards more integration between process control functions and other EDP activities inside the organization. These existing systems and future needs and trends are discussed in the paper.

Keywords. Digital control, pulp industry, paper industry automation, computers, hierarchical systems.

INTRODUCTION

During the last 20 years the most characteristic trend in Pulp and Paper Mill Control has been the ever increasing tendency to use digital computers and modern control methods. In connection with this, many special instruments for the needs of this industry have also been developed. Here we can state that the attempts to use computerized control have stimulated the evolution of new instruments and vice versa: many computer systems have become practical and profitable only after the development of some key instruments.

The very first process computer installation in a paper mill was the paper machine control system installed in 1961. On the pulp mill side the first installation dates from the year 1962 (Kamyr control). The first bleach plant applications date from the year 1966.

The total number of process computer applications in the pulp and paper industry is quite difficult to estimate very accurately but based on published reports (Keyes, 1975; Gee and Chamberlain, 1977; Uronen, 1981) and discussions, we can say that 2600 is a relatively close number today. Annually about 300 new systems will be purchased.

Figure 1 represents the rate of development of these systems from the beginning until today (Uronen and Williams, 1978).

A very interesting feature concerning the type of existing applications, the method of implementation, and the tasks of the systems is the large number of dedicated, mainly mini-computer-based, packaged systems delivered mainly by a quite limited number of specialized vendors. According to Keyes (1975) 69.5 % of the process computer systems in the pulp and paper industry of the United States are dedicated packaged systems and 9.5 % are modular systems, i.e., packaged but user expandable or modifiable. The situation in other countries is quite similar. However, in Scandinavia the percentage of packaged systems is somewhat smaller (Eriksson, 1977). These systems have generally been successful: the average return of investment of these installations in the USA after taxes has been 59.14 % (Keyes, 1975).

The development of dedicated packaged process computer systems has some historical background. In the beginning of the early 1960's the computer equipment was relatively large with difficult and complex programming systems demanding highly skilled personnel. In principle, these machines were capable of controlling and optimizing several subprocesses or units on a time-sharing or multi-programming basis. That also was tried in the first projects in this area. However, most of these projects then were not successful because:

1) the reliability of systems was not high enough;

2) the lack of models and instrumentation;

3) the lack of competent specially trained staff;

4) the underestimation of difficulties and resources needed;

5) the difficult and unefficient programming systems;

6) the top management was not directly involved.

After these first attempts (and some disappointments), the evolution in process computer hardware and software has been concentrated around the use of minis and micros. In the middle 1960's, the problem of computerization in pulp and paper mills was attacked from another direction, i.e., one defined smaller, well-limited potential targets from among the different subprocesses in the pulp and paper mill. One also put emphasis on the required instrumentation and the necessary process models, economical benefits, etc. This development and the trend towards "standardized" systems which could be used in most pulp and paper mills without major modifications have lead to the dedicated packaged process computer systems which are common today controlling well-defined subprocesses, for example, batch digesting or the wet end of a paper machine. They can be purchased on a turn-key or fixed price basis. Thus, no special personnel at the mill are needed. This has also made it possible for smaller companies to use these systems because the maintenance is taken care of by the vendor in most cases. These package systems have been successful because:

1) they were developed through cooperation between computer specialists and user engineers to solve well-defined problems;

2) they had the necessary instrumentation and process models as background;

3) the computers used were already reliable enough and capable enough at that time to handle these smaller problems;

4) the economic results could be clearly verified and in a reasonable period of time;

5) the problems involved with the man-machine interface were noted and solved using the technology then available.

The benefits achieved by these systems can be divided into the following classes:

1) enforced control
   a) better tuning of conventional controllers,
   b) recording of alarms and deviations,
   c) recording of operator changes,
   d) control by exception.

2) improved control
   a) cascade, ratio, bias, blending and batching control,
   b) indirect variable control,
   c) feedforward control,
   d) multivariable control,

3) advanced control
   a) non-linear control,
   b) programmed adaptive control,
   c) self-tuning control,
   d) time optimal control,
   e) sampled data control,
   f) direct optimization.

4) supervisory control
   a) start-up and shut-down situations,
   b) emergency situations (back-up),
   c) logical sequencing,
   d) programmed sequencing,
   e) batch cycle programmed control,
   f) performance evaluation,
   g) automation of certain standard procedures (spoolings, circulations, etc.).

5) data and control communication
   a) data display and entry,
   b) operating reports,
   c) special reports,
   d) communication with higher level systems.

This classification is based on computer functions compared mainly with the possibilities of conventional analog control systems. If we on the other hand take the view of the mill manager he will have another list, for example as follows:

1) Savings in raw materials, chemicals and energy
   a) higher yield
   b) smaller variations
   c) operation closer to limits
   d) more accurate dosaging
   e) also small disturbances and changes will be compensated for
   f) no running according to "easiest way"
   g) disturbance compensation
   h) smooth and stable operation
   i) close control of rate and grade changes
   j) balance and effectiveness calculations
   k) optimal load allocation between parallel units and machines.

2) Higher production capacity
   a) smooth and stable operation
   b) elimination of disturbances
   c) optimal loading
   d) throughput maximation programs
   e) effective use of buffer storages
   f) more accurate grade and rate changes
   g) coordinated use of production equipment.

3) Decrease of losses
   a) close monitoring of effluents and emissions
   b) early detection and alarming
   c) disturbance elimination
   d) preventing of overflows etc. by coordinated operation
   e) minimization of human factors

4) Others
   a) decrease in mill personnel
   b) safety
   c) longer lifetime of equipment
   d) improved process knowledge
   e) diagnostics (process equipment and instruments).

These two lists, of course, include both the same basic factors; the former is a control engineering approach and the latter projects these factors into operational results. As already stated these systems have proven to be profitable investments in pulp and paper mills all over the world and they are therefore widely accepted and used and we can expect the continuous and rapid increase of these systems also in the future.

Concurrently with the evolution of process control systems discussed above there also has been a big change in the "business" or "administrative" computers. They have become very powerful with much improved operating systems capable of simultaneous running multiple programs and effective use of the methods of operations research and scientific and mathematical analysis relating to the management and decision making in a large industrial complex. Their performance/cost ratio has thus vastly improved.

These computers are also widely used in the pulp and paper industry today. The tasks of these systems include payroll, inventory, bookkeeping, accounting, techno-economical calculations etc. The people taking care and using these systems include programmers, mathematicians, system analysis and hardware specialists. They do not in most cases include process and production people or those with financial and managerial expertise. They form a data processing department totally isolated from process control systems which are considered to be production tools. Figure 2 (Uronen and Williams, 1978) represents this situation having several computer systems in pulp mill, in paper mill and in administration. All these computers, however, in most cases work separately with none or with minimal real time communication or data exchange between them.

THE PRESENT SITUATION

We can say today that the control engineering and process automation in all process industries, not only in pulp and paper, are under rapid development stages: there is a remarkable change from analog techniques into the distributed digital control with a motto: "distributed control and centralized supervision". This means that the field control stations are distributed and installed near to the sensors and actuators making thus the response of the individual loops very fast. Also the central control room in its earlier form has disappeared and the new control room technique utilizes multicolour videos and other modern display and communication techniques. The main functions in the central control room include higher level controls i.e. plant wide controls, optimization and reporting. This new era started in 1975 when the first digital instrumentation system (Honeywell TDC 2000) was introduced. Since that there are more than 100 different digital instrumentation systems in the market. It is to believe that heavy competition and the demand on full range services by the users will in the future mean that the number of these vendors will be remarkably smaller than today.

An other remarkable feature in the present trend in process automation is the discretizing of the signals and previously continuous functions. For example the conventional PID-controller is now a digital control algorithm in a microprocessor chip. The digital signal transmission has many important advantages like: accuracy, reliability, error-checking and correction etc. Figure 3 shows a typical arrangement where digital process control is applied to an analog process environment: the border between the digital and analog parts of the whole system is changing further and further in favour of the digital part: more and more of the equipment and functions will be digital in the future. During this decade a big part of transmitters and actuators will be digital.

The idea of distributed control can be realized in many different ways and using different architectures. Figures 4, 5 and 6 represent the typical analog centralized control and two different distributed control architectures, where the distribution is on one or two levels. The big advantages of the digital distributed control are: speed, accuracy, reliability, self-diagnostics, flexible control algorithms and savings in cabling costs. The vital part of these systems is the high speed data transmission bus (Data highway): this is realized with a coaxial cable and different transmission protocols are used. The international standardization in this area would be useful and important. The further integration of the control activities in different departments and at different levels of organization having computer systems of different age and from different manufacturers make a local data network by using this kind of transmission buses a necessity. I believe that the building of mill-wide local data networks and integrating the process control activities horizontally and vertically into the management and ADP functions of the mill will be one of the key development areas and directions. However, the real implementation of these new technologies varies very much in different countries and in different industries: the chemical industry being in the leading group world-wide. It can also be stated that in connection with big new projects it is natural in most cases today to select the digital instrumentation system; in old plants this can in most cases happen only in connection with a major renewing project.

As always in introducing a new technology there has also been missbelieve and reluctant attitudes towards the distributed control: one example has been the motor and other high voltage operation switches and push-buttons: it has been difficult to implement these also with the microprocessor techniques. However, now the first experiences from this kind of installation are available and they are very positive, so the trend will continue to integrate also these functions into the digital instrumentation.

CONSTRAINTS FOR THE AUTOMATION

The control theory has always been far ahead the real applications; during the whole history of industrial automation there has been discussion about the gap between theory and practice. When the computer technology has progressed as it is, we can say that there are not any more so much constraints in applying more and more the so-called modern or advanced control methods (multivariable, optimal, stochastic etc. control). This means that the computing, data processing and storage are not any more limiting factors, but there are still others: economic potential and educated and skilled people to apply, update and maintain these algorithms and strategies. Today we can also see a remarkable interest to apply for example self-tuning and adaptive controllers to certain industrial applications.

Further there are a lot of CAD systems for the design of control systems but these are so far mainly used in universities and research institutes. Just during last few years there has been raising interest towards these systems also in the industry. The real constraint in applying these control strategies is the economic potential: there must be something to be gained with them the life time installed costs or the control system recognized, i.e. the reliability, operability and maintainability of the control system are the central criteria.

The software continue s to be a more and more central constraint in the development of automation, much more than the HW or interface. It has been stated that in many installations the SW costs are more than 2/3 from total investment and even 70 % from the SW costs have been occured after the installation i.e. the tuning, start-up and development costs can be a real trap.

Human engineering will become more and more critical; here we have an other big gap of automation: the gap between human factors theory and automation system design. It has been said that the job of a human operator in industrial control consists 90 % of boredom and 10 % of terror. The big question is: how to eliminate or decrease the probability of errors in emergency situations, or can we really rely at all on human operators in these situations? Is this i.e. the emergency control and start-up and shut-down situation control an other big automation possibility? Further: how could we utilize most efficiently the experiences and skills of the operators?

In programming languages there has been certain development but even the new control language ADA is not the final and ideal solution, but it seems to be better and more efficient than anything else in this area so far.

I have already stated before that the education and training of specialists in all levels in this field seems to continuously be one important limitation. The measurements have always been and remain one central constraint for the automation and control: "you can not control better than you can measure". The most part of process measurements will continue to use the well-developed conventional sensors, but it is to believe that the further development of microelectronics and applications of solid state physics will create new measuring principles and sensors especially in the area of continuous on-line measurement of the material characteristics (analysis). Further these transmitters will include more data processing capabilities and also several variables and variables calculated or derived from these can be transmitted to the control system: the sensors will include more and more intelligence. Automatic calibrations and self-diagnostics are also typical features in these transmitters. More and more integration in some applications can also bee seen: for example in flow measurements the whole control loop can be integrated in the measuring element or in the actuator.

FUTURE TRENDS AND CONCLUSIONS

In the preceeding discussion it was already stated that classical process control will be more and more integrated both horizontally to other operative functions (material handling, maintenance, marketing) and vertically to management and ADP functions (Uronen, 1981). Thus we will have a hierarchical plant or corporate wide control system as depicted in Figure 7. An interesting new development at the top of the hierarchy is DSS (Decision Support Systems) to help top management in unstructured decision making situations. This also necessitates the local data networks as stated already. The use of computers in industrial automation will continue to increase; one statistics shows that in 1980 there were about 22500 minicomputers in these applications world wide and 680000 different micros respectively. The annual rate of increase has been ca. 20 %.

Other typical trends briefly:

- The instrumentation will be digital.
- More and more digital transmitters and actuators will be developed and used.
- The use of fiber optics will increase especially in sensors.
- More intelligent sensors will be developed.
- Multivariable transmitters will be developed.
- Application of speech recognition in man-machine communication will be a reality.
- One loop microprocessor controller will be common.
- Modular application oriented HW solutions to specific problems will be developed.
- Modern control methods will be applied on broader basis.
- Plant and corporate wide hierarchical control and information systems will become more and more important.

The progress will, however, be evolutionary in most cases due to the constraints mentioned already before.

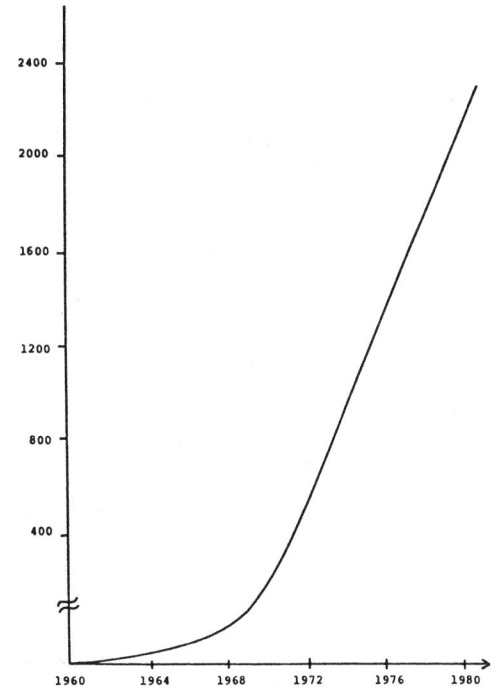

Fig. 1. Development of computer control installations in pulp and paper industry

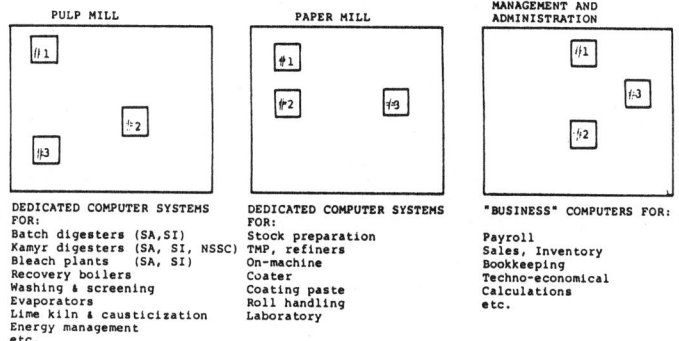

Fig. 2. Computer systems in the pulp and paper mill today.

REFERENCES

Eriksson, L. (1977). Integrerade datorsystem i våra fabriker - nuläge och utveckling. Svensk Papperstidning, Nr. 18, 565-570, 602 (in Swedish).

Gee, J.W., and Chamberlain, R.E. (1977). Digital Computer Applications in the Pulp and Paper Industry. IFAC/IFIP International Conference on the Digital Computer Applications in the Process Industry. Hague.

Keyes, M.A. (1975). Computer Control Census, Tappi, Vol. 58, No 6, 71-74.

Uronen, P. (1981). Integrated Computer Systems in Pulp and Paper Industry. International Institute for Applied Systems Analysis (IIASA), WP-81-68, May.

Uronen, P., and Williams, T.J. (1978). Hierarchical Computer Control in Pulp and Paper Industry. Purdue Laboratory for Applied Industrial Control, Purdue University, West Lafayette, Indiana, USA, Report No. 111.

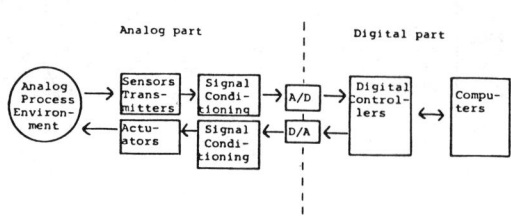

Fig. 3. Digital and analog parts of automation.

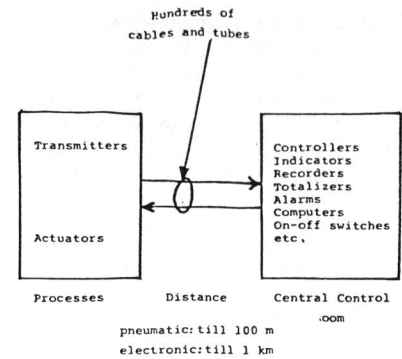

Fig. 4. Centralized Analog Control.

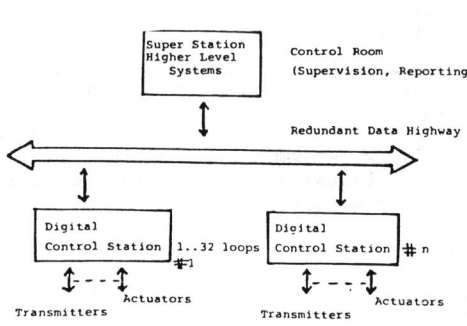

Fig. 5. Distributed Control System (1-level distribution).

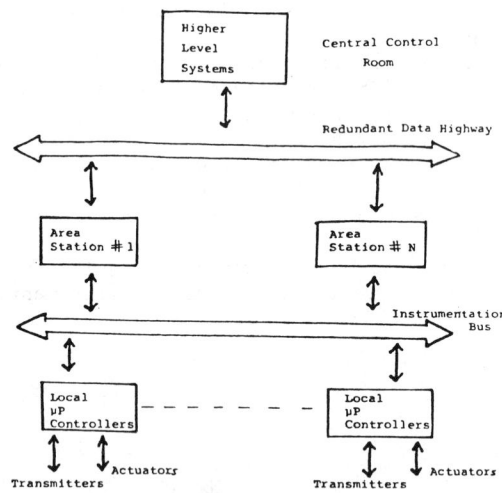

Fig. 6. Distributed Control System (2-level distribution).

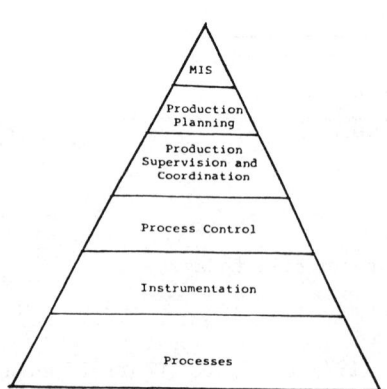

Fig. 7. Total Mill Control Hierarchy.

EVOLUTION OF DIGITAL CONTROL IN ENERGY CONTROL CENTERS

J. L. Carpentier

Electricité de France, 2 rue Louis Murat, 75008, Paris, France

Abstract. The evolution of digital control in energy control centers of electrical power systems is reviewed. Electrical power systems being large scale systems, decision processes are decomposed in space and time. In this decomposition, energy control center digital control essentially concerns on-line and one hour ahead decision processes for the system considered as a whole. The various historic steps of digital control in energy control centers are described : early stages, computer aided human decision off-line then on-line, computer decision making, computer automatic decision application. The present state of the art of digital control applications in closed loops is presented, including the present standard implementation and two advanced applications, one taking account of transmission security, the other one of load change anticipation constraints. Then, an analysis of the present needs, new concepts and developments under study are presented.

Keywords. Computer control ; digital control ; frequency control ; load dispatching ; on-line operation ; power system control ; voltage control ; automatic generation control.

INTRODUCTION

Electric power system operation is a typical example of an integrated large scale process where control cannot be performed as a whole but must be decomposed in time and space [1]. Natural cycles and physical response times allow to decompose the control problems vs time into problems more or less loosely coupled. A typical decomposition vs time of the main control problems to solve for the system considered as a whole may be e.g. as follows :

(a) Every year (or more), find a strategy for water and nuclear fuel management, for power plant and transmission network maintenance.
(b) Every week, and more accurately every day, solve unit commitment, (thermal and hydro) power scheduling and network topology problems for the next period.
(c) One hour or less ahead, schedule powers, voltage magnitudes, network topology in order to meet the forecasted load for the next period within security and economy.
(d) On a real time basis, but with reponse times from a few seconds to a few minutes, adjust powers, frequency, voltage magnitudes, network topology in order to meet the actual load within security and economy.
(e) Avoid bad consequences of electromechanical transients in the system, spanning over seconds.

(f) Avoid bad consequences of electromagnetic transients along electric lines, involving microseconds or at most a few milliseconds to die after a surge.

Decomposition may also be applied in space : subsystems as power plants and substations may be controlled on a nearly local basis, only a few control variables being common with those of the system considered as a whole.

In this paper, we shall limit out scope to the control actions possible to perform in an energy control center ; they concern the system considered as a whole, and not control in local subsystems [2] [3] ; from time scale point of view, due to data transmission delays, actions are limited to those having response times larger than a few seconds ; in particular, corrective actions concerning electromechanical transients need response times too small for a control center : they belong to "emergency control techniques" [4] , where actions are performed locally although on a global theoretical basis. Among the control actions left (a to d above), only the real time actions d may involve an automatic closed loop and apparently belong to our topic ; as a matter of fact, d is the executive stage of the slower cycles and, in particular cycle c (one hour and less ahead) may be considered as the preparation of cycle d (real time) : in these conditions, a complete view of digital real time control needs the study not only of real time actions but also of one hour and less ahead actions.

Within this scope, 4 main kinds of decision processes may be distinguished :
1 - Human decision (without any computer help),
2 - Computer aided human decision,
3 - Computer automatic decision making,
4 - Computer automatic decision application.
Only the process 4 is a closed loop automatic control, but the other decision processes still have a large place in control centers and use digital techniques, so that we have to consider them as control loops, even if human beings are included in these loops. Evolution goes towards the process 4, but very slowly. So, we shall, in turn, present the following sections, according to the historical evolution :
- Control variables and objectives.
- Early stages of control in energy control centers.
- Computer aided human decision.
- Computer decision making.
- Present on-line standard digital automatic control.
- A present application of advanced automatic control, with transmission security.
- A present improvement of automatic control, taking an account of load change anticipation .
- Needs for future.
- New concepts and developments under study.

REAL TIME CONTROL VARIABLES AND OBJECTIVES FOR AN ENERGY CONTROL CENTER

In an energy control center, for real time and one hour ahead control, the usual control variables are :
- the real powers of thermal units and of hydro units provided with equivalent costs if any,
- the generator voltage magnitudes, the status of reactive sources as capacitors and reactances,
- variable transformer ratios and shifts,
- network topology.

The usual objective consists of minimizing the operation cost while meeting the real and reactive loads and security constraints. "Security constraints" means that some variables as line currents, voltage magnitudes, must remain within given limits in the intact system and under contingency, i.e. after the sudden trip of one (or sometimes several) element of the system. An important numerical problem appears here : the variables to keep within limits - specially under contingency - are not straightforward to compute vs. the control variables, so that computer aid will be indispensable to prepare and perform security control or monitoring.

EARLY STAGES OF CONTROL IN ENERGY CONTROL CENTERS

First Energy Control Centers

First energy control centers mainly gathered analog measurements from the system as frequency, generated powers, voltage magnides, line flows, and some breaker status around a mimic board giving an overview of the system. The main active control components were a phone network and the dispatcher himself, who tried to adjust power to load, to find satisfactory network topology and voltage levels.

Local control loops existed for generator frequency and voltage magnitude and applied the same basic theory as now in the so-called "primary control". Frequency is kept near its nominal value using a proportional regulator acting on turbine valve ; in the same way, voltage magnitude is kept near a given value using a regulator acting on rotor current. Response times of primary controls are around a few seconds for frequency and a fraction of a second for voltage magnitudes. In order to get stable real and reactive power balance between units, a proportional control is used, leaving a slight error for the controlled variable. In these first applications, regulator set points were set manually, following general rules or dispatcher orders.

Secondary Frequency and Power Exchange Control (LFC)

The local primary frequency control presents two main drawbacks : a substantial error may be left for frequency and, when a load change occurs, powers allocation between the various units is neither accurate nor well known. In order to remove these drawbacks, a global integral frequency control was imagined ; various alternatives exist, a typical one being as follows : a signal proportional to the integral vs. time of the frequency deviation is built up at the control center and sent to each plant on control ; a "participation factor" is allocated to each unit ; in the plants, for each unit on control the integral signal is multiplied by the participation factor and acts on the set point of the primary frequency regulator. The result of this 2 - loop control is an integro-proportional control with a response time between 1 and 2 minutes, giving a theoretical zero frequency deviation - in practice a better quality control - and mainly an allocation of power between units accurate and well known.

This principle is valid for an insulated power system ; in practice areas (or companies, or countries) are interconnected with other power systems, and the previous integral control may be extended to control at a time frequency in every area and power exchanged between areas ; for this purpose, it is sufficient, in each area, to add a signal proportional to the exchanged power deviation to the frequency deviation. Each regulator can operate only from measurements from inside the area, which gives a complete decentralization of the control system.

The implementation of this frequency and power exchange secondary control, so-called "Load Frequency Control" or "LFC", was performed with an analog technology in the fifties [5] [6] [7] . It was the first global control loop in power systems. Various improvements came with years [8] , and even appear now, as area control performance measurement and corrective financial control [9] .

First Generation Economic Dispatch and Automatic Generation Control.

Necessity of an economic dispatch function. LFC gives a means to allocate powers to various units through the values of the participation factors. More accurately, starting from a state where the generated powers have steady state powers P^o, the participation factors allocate the changes in generated powers corresponding to the load changes. Load changes may be decomposed into two components, a random component (with pseudo periods around a few minutes) and a trend. The load random component induces stresses in the generating units, and it is interesting to share these stresses among the greatest number of units possible ; on the other hand, the trend component corresponds to steady state fuel expenses and it is interesting to balance it by only changing the powers of the most economical units : from this analysis, it may be concluded that not one but two control actions appear necessary, one concerning the load random component and easily achieved by LFC, and a new one concerning the load trend component, where decisions depend on economy : this new control function is called "economic dispatch" (ED or ELD).

Notice that in order to be able to perform LFC in good conditions, often a "regulating bandwidth" B_j is allocated to each unit j on control, equal to the product of the participation factor by the maximum expected amplitude of the load random component : in these conditions, the maximum steady state power of unit j is not its physical maximum P_j^M but $P_j^M - B_j/2$ and in the same way its minimum steady state power is $P_j^m + B_j/2$ instead of P_j^m ; it may be perceived that regulating bandwidths are costly, for they restrict the possible steady state powers of units.

Economic dispatch solution. From the previous analysis, economic dispatch consists of finding the steady state powers P^o within allowed bounds for P^o (taking the bandwidths into account if defined) which minimize the operation cost. If there were neither losses nor security constraints, this would be very simple, since the sum of the generated powers would be equal to the load. In the first ED methods applied, no security constraints were taken into account ; losses were taken into account thanks to approximations allowing to express the losses as quadratic functions of the generated powers, with the famous "B coefficient" formulas [10] . So, the ED problem was reduced to minimize a function of variables subject to one quadratic equation ; it could easily be solved by Lagrange's method and gave the very simple following result :

$$\frac{\partial F/\partial P_j}{1 - \partial L/\partial P_j} = \lambda \quad (1)$$

with :
- $\partial F/\partial P_j$: incremental production cost of unit j
- $\partial L/\partial P_j$: "incremental loss" of unit j, i.e. partial derivative of the losses vs. P_j, which is a linear function of P (through B coefficients).
- λ is a constant value independant of j.

First analog automatic generation control. At the end of the fifties and in the sixties, ingenious analog control systems were implemented, performing LFC and ED at a time, building up the so-called "Automatic Generation Control" (A G C) [10] [11] : a genuine analog computer, able to represent the incremental cost curves and the B coefficient formulas, solved the equations (1) and sent orders to the units meeting LFC and ED functions.

These control systems represent the top achievement of analog techniques. Such systems are still working. They have the merit to provide a continuous control, but are neither flexible nor accurate.

COMPUTER AIDED HUMAN DECISION

In addition to lack of accuracy, analog techniques are not able to take an account of transmission security contraints nor of voltage magnitude and reactive power problems. While analog control systems were implemented, in the sixties, network analysis digital techniques developped considerably, first for off-line purposes, but brought tools usable later on-line.

Development of Digital Techniques for Network Analysis

Load flow computations are the corner stone of network analysis ; from the system loads and control variables, they allow to compute the voltage phase angles θ and magnitudes V at every bus, so-called system "state variables", and from which any variable, as e.g. a line current, may easily be deduced. As soon as 1963 [12] [13], the solution of the corresponding set of non-linear equations was solved by Newton Raphson method, using matrix sparsity properties [14]. Static security analysis consists of computing the state of the network after the sudden trip of anyone element of the system (or sometimes of couples of elements belonging to a given list). Many methods exist, often using the Woodbury formula, profiting by the knowledge of the state of the intact network before tripping to shorten the computation duration [15] [16]. Transient security analysis evaluates if a danger exists due to the electromechanical transients after an incident. Simulation is used for this purpose, but computations are very long. Other techniques, approximate, look more suitable for on-line application: Liapunov method, or better, the use of a Liapunov-like transient security index [17] easy to compute vs the system state before incident. Short-circuits are now very fast to compute using sparse inverse matrix technique [18].

System Monitoring with Computer Aided Human Decision

The previous off-line network analysis techniques, plus some specific tools, allowed to build up a monitoring system where the dispatcher receives data with improved quality, takes decision himself but with the aid of a computer able to simulate the consequences of a decision, especially on security. Such a process typically includes the main following functions [19] :

a) Supervisory control and data acquisition ("SCADA") : through a data transmission network, telemeasurements and telesignals are cyclically gathered at the control center, compared to thresholds, then visualized.

b) Network modeling and state estimation must provide the dispatcher with reliable and accurate data ; network modeling must update the network model (topology, measurements) ; then state estimation must provide reliable and accurate estimates of bus voltage phase angles and magnitudes. This is achieved by overmeasuring the system with more measurements than necessary to solve a load flow problem ; this redundancy enables a kind of cross-checking among measurements to eliminate any gross measurement or system configuration error. Various methods exist [15] ; weighted least square method associated with sparse inverse technique [20] [21] seems the most often applied today.

c) Bus load forecast provides bus load values for sometime ahead, using simple methods.

d) Real time load flow applies load flow techniques to find the system state corresponding to forecasted loads.

e) Security analysis : static security analysis is applied to the system with actual or forecasted loads resulting in a warning to the dispatcher in case of constraint violation. In practice, nothing similar was implemented for transient security analysis, although the use of a transient security index could be fruitful.

f) Short-circuits computations may be used to check if, after change in network topology, the short circuit powers remain inside their allowed limits.

This monitoring process represents, from the point of view of transmission security, a stage that a lot of operation people consider as modern and satisfactory. Of course the quality of data is better and security is evaluated, but the dispatcher is still alone to find the decision to take when something is wrong, and often to apply it. The response time of the control loop including a human being remains long (e.g. 10 minutes or more) and the solution has no reason to be economical nor even well defined : despite its popularity, perhaps due to the fact that human habits were not too much perturbed, from process control point of view, this monitoring process is very far from being satisfactory.

COMPUTER AUTOMATIC DECISION MAKING

Development of Digital Techniques for Decision Making.

Optimal power flows. From 1962 [22] a new kind of decision making techniques, so-called "optimal power flows", was developped. Given real and reactive loads, and network topology, they are able to find the values of the system control variables, such as real generated powers, reactive source voltage magnitudes, variable transformer ratios and shifts, which minimize the operation cost while meeting the loads and the security constraints : they basically solve the system control problem for all control variables except network topology. The problem statement may be complete (real and reactive control variables) or partial (real only or reactive only control variables). The mathematical problem to solve is a non-linear mathematical program with an enormous lot of constraints (100,000 for security in a large scale system), only a very little number of which happily play an active part. Various kinds of solutions are used [22 to 27]; they may be filed into two different method families, "non-compact" and "compact methods"[28]. Non-compact methods apply mathematical programming algorithms to the whole set of control and state variables and practically cannot handle security constraints under contingency ; compact methods proceed in two steps : first they build a "reduced model" where all the useful (i.e. active and nearly active) constraints are expressed vs the system control variables, second they apply mathematical programming to the reduced model. These two steps, reduced model building, reduced problem optimization may be either performed only once in simplified linearized models (giving a linear program to solve), or, better performed several times iteratively till the reduced model is exactly equivalent to the whole problem ; compact methods can handle security constraints under contingency very easily. Both families of methods have been used for years for off-line studies ; now, compact methods appear very suitable for on-line purposes, as they easily handle security constraints and also because the reduced model, specially when non-linear, is valid on a large region and may be used alone for on-line purposes between complete computations.

It must be understood that a compact optimal power flow program, in its reduced model building step, includes more than security analysis ; indeed, this step includes :
1 - A load flow program, giving the state variables vs the latest values of the control variables,
2 - An automatic security analysis selecting the useful constraints to include into the reduced model,
3 - Sensitivity computations giving Taylor series developments of the constraints and of losses vs the control variables ; in particular, non-linear methods as [27] include the computation of the equivalent of exact B coefficients for losses.

The reduced model optimization step changes the control variables in order to minimize the operation cost while meeting the real power balance equation and the useful security constraints : compact optimal power flow programs represent a degree of integration never reached before and are first class tools for control, more especially as computation durations now allow to run these programs on minicomputers witch acceptable delays.

Corrective switching. When a security constraint is violated, optimum power flows allow to perform corrective rescheduling on powers and voltages in order meet the constraint again. Corrective switching tries to do the same, acting on network topology, which may be more economic. The problem is much more difficult because highly combinatorial ; solutions are not yet ready to use, but major progresses seem near to appear in this field [29].

System Monitoring with Automatic Computer Decision Making

This is a straightforward application of optimal power flow techniques : having performed network modeling and state estimation, the actual system state is known ; then bus load forecast for a very near future is carried out and an optimal power flow is performed from the corresponding data, giving generated powers, voltage magnitudes and other miscellaneous control variables economically optimal and meeting security, which helps the dispatcher to monitor the power system. A few applications of this kind have just been implemented in control centers, e.g. in France with a linear program for real powers only and in Ivory Cost, with a reactive only model using the Differential Injections method [27], which is a typical non-linear compact method. In both cases, programs are run on 16 bit word minicomputers. In the latter case, which gives the greater bulk of computation, run times are around one minute.

To our knowledge, corrective switching has not yet been implemented in a control center ; it is a question of method efficiency which should be solved within a few years, the most attractive way of use being including corrective switching as a subroutine of optimal power flows.

Even with only optimal power flow techniques, this kind of monitoring represents an important step towards secure and economic control. Computerized decision making improves the decision quality and lessens the response time. But it is still monitoring, the dispatcher having to apply the computer decisions himself ; a last step must be done to get an economic and secure automatic control : to close the loop.

PRESENT STANDARD ON-LINE DIGITAL CONTROL

Let us close the loop and look at present standard on-line digital control, leaving two advanced applications for the next sections. 3 kinds of implementations are now currently on service : digital load frequency control (without economic dispatch), digital automatic generation control without transmission security and secondary (regional) voltage control.

Digital Load Frequency Control

It is nothing but the transposition of analog LFC defined above to digital technology : instead of continuous control loop, sampled numerical data are used, with a numerical integration vs time. Sampling period are usually between 2 and 10 seconds. The only noticeable problem met with sampling came from the fact that the sampling instants mays be different for the various tie-lines, which produces a noise term in their sum. This difficulty was solved by filtering.

Standard Digital Automatic Generation Control.

Principle. The main feature of the present standard digital AGC [30, 31, 32] is a hierarchy in the control levels : at the top level, the local primary frequency control ("unit control") is activated without any delay ; then LFC is activated e.g. every 5th second ; at last, ED, with the lowest level, is activated e.g. every 5th minute. The origin of this hierarchical structure is certainly the idea that economy need not so small response times as power balance and perhaps also the lack of speed of economic dispatch models. With this principle, AGC controllers, located in the energy control centers are built up according to the following pattern :

1 - A LFC input signal is elaborated, as usual proportional to the frequency deviation and to the sum of tie line powers : it is the so-called "Area Control Error " ACE = $K \Delta f + \Delta P_e$, which enters the AGC internal logic block every 5th second.

2 - Every 5th minute, the ED block receives the sum of the actual generated powers, i.e. the system load ; the ED block computes the most economic unit power or "basic" power P_j^o for each unit on control ; in the standard implementation, no care is taken of security constraints, but losses are represented, e.g. by B coefficient formulas, and a digital solution of the equations (1) is found. Each 5th minute, the ED basic power P^o are entered as input data into the AGC internal logic block.

3 - With the two input signals ACE and P^o, the AGC internal logic block builds up the "unit set points" P_j^s, control signals sent to each unit every 5th second (either as power set point P_j^s or more often as increment ΔP_j^s). The P_j^s are function of ACE and P^o which may be defined according to a lot of alternatives, e.g. :

$$P_j^s = \begin{cases} P_j^o + E_j (\sum_j P_j - \sum_j P_j^o) \\ + R_j F(ACE) + R'_j F'(ACE) \end{cases}$$
(2)

where, for the unit j, P_j^s is the unit set point, P_j^o the last ED solution, P_j the actual power, E_j the economic participation factor, F (ACE) the "filtered" ACE for normal LFC, F'(ACE) the "filtered" ACE for emergency conditions, R_j and R'_j the LFC participation factors for normal and emergency conditions. The function F (ACE) may include multiplication by a factor, time constant filtering, integration, derivation,etc. F '(ACE) becomes different of zero only when |ACE| overpasses an emergency threshold. It may be noticed that in formula (2) the unit set point P_j^s is the sum of 3 elementary requirements relative to ED, LFC under normal conditions, LFC under emergency. In other kinds of AGC controllers, P_j^s may be a logical combination of several requirements ; in particular, AGC control is called "mandatory" if the ED requirement is always applied, as in formula (2) ; it is called "permissive" if the ED requirement is applied only if the ED and LFC requirements have the same sign. In a typical implementation, AGC control is mandatory if |ACE| remains lower than a given value and permissive on the contrary.

Practical properties. The function giving the unit set point P_j^s vs ACE and P^o is particularly intricate (and could be still more in the case of a pool, we did not describe in order to keep simple !). Two main reasons may explain this complexity : the first one is the need to adapt AGC to the disturbance size, i.e. to strengthen the control when the disturbance increases; this need is in Nature and is met satisfactorily by adopting measures as the third term of aquation (2) ; the second reason is due to the hierarchical structure of this AGC system : indeed, with respective periods of 5 seconds for LFC and 5 minutes for ED, there may be inconsistency between both requirements. Formula (2) and analog ones are built up to try to avoid discontinuities in the control set points when a new ED solution is computed and to avoid contradictory requirements between ED and LFC, inducing unit power oscillations. These contradictory requirements appear as the result of a conflict between physical and economic needs ; in particular, usually the LFC participation factors R_j and R'_j are determined by unit response times, whereas E_j depends on unit cost curves : fast response units may have an increase then a subsequent decrease in their powers. The use of logic combinations may help to get a better control quality in some particular cases, e.g. permissive control allow to get an efficient control when ACE becomes large, but it may introduce new discontinuities. As a whole, despite the complexity of the control system, due to the ED discontinuity, oscillations of unit powers cannot be avoided, which is not desirable for unit wear, maintenance and even operation cost.

To sum up the properties of this present standard AGC, it may be stated that :
- This system control have the great merit to exist, showing a closed loop control including economy is feasible, even if not perfect : algorithms are working, plants follow control and operation people did not foment a revolution when the control system was implemented.
- On the negative side, 3 main insufficiencies may be mentionned, leaving 3 problems to solve :
1 - "Economic dispatch discontinuity", covering the drawbacks just described, the too large period for ED resulting in unit wear.
2 - The absence of security constraints : security constraints were not taken into account, so that uncontrollable network overloads may occur.

3 - The "pin-point" aspect of economic dispatch. A reproach was recently formulated against economic dispatch and optimal power flow problem statements : they only consider the actual load and minimize the operation cost for the present time only : it is a "pin-point" optimization without any care for load variations one hour or more ahead ; due to limits on unit power ramps, in some particular cases, a sequence of instanneous optimizations does not result in a global cost minimization or even would lead to costly changes in the on-off unit state. Although rather scarce in many companies, this kind of constraints may appear and, in a way or another, would need a "dynamic dispatch" spreading on 1 hour or more.

Hereunder, we shall present an implemented solution of problem 1, an experimented solution of problem 2 and new concepts having for objective to solve the 3 problems at a time.

Secondary Regional Automatic Voltage Control

3 possible levels may be imagined for voltage control : "primary control", at unit level, controlling the stator voltage magnitude ; "secondary control", controlling voltage magnitudes and reactive powers for a few power stations ; "tertiary control", providing secure and economic voltage magnitudes and reactive powers in an area or a country. To our knowledge, tertiary control has not yet been implemented. At least one implementation of secondary voltage control is on service [33] according to the following principle : In a "zone" including 2 to 10 power stations, all unit reactive powers are rated with the same percent of their possibilities ; in these conditions, in a zone, only one degree of freedom is left : the rate of reactive generation, so-called "zone signal". The value of this zone signal is built up in a voltage secondary controller, in order to control the voltage magnitude of the most important node of the zone, so-called "pilot-node". In practice 3 control loops exist : 1) The usual primary control loop, controlling the unit voltage magnitudes (reponse time 0.3 sec.) ; 2) an intermediate "reactive power controller" receives the zone signal, computes the corresponding desired reactive output for each unit, taking various constraints into account, and acts on the primary control set point till the desired reactive power is reached ; this intermediate controller is located in the plants for thermal and nuclear units, in district control centers for hydro plants ; 3) the voltage secondary controller itself, located in the area control center, receives the value of the pilot node voltage magnitude, builds up the zone signal and sends it to the intermediate reactive power controllers (reponse time 150 sec.). This secondary control is digital, based on microprocessors in its present implementation.

Such a secondary voltage control was built up having in mind that reactive power have more or less a local character. It shares reactive power reserves equally between the units of a zone, which is favourable to security. It simplifies operation by reducing the number of variables to control ; in particular, in future , it will simplify the technological implementation of a tertiary voltage control. In operation, this control gives satisfactory results provided that zones and pilot nodes are well chosen vs network topology. On the other hand, the equal rating rule for generated reactive powers introduces artificial constraints which may be (scarcely but sometimes) harmful : all these problems left should be solved by a convenient secure and economic tertiary voltage control.

ADVANCED AGC WITH TRANSMISSION SECURITY

To introduce transmission network security into AGC, it is sufficient to replace the usual ED function without security by an ED function including an optimal power flow. A few implementations are on service, the most complete to our knowledge being that of New York Power Pool [34] . In this application, the ED function, run every 5th minute, includes :
1) A security analysis, with 1rst and even a few 2nd contingency analysis, giving a list of the 20 most critical security constraints and the (linear) corresponding sensitivities (computation time between 3 sec. and 4 min.) which are used as input data to the optimal powers flow itself
2) An optimal power flow using a sequence of linear programs ; penalty factors $1 - \partial L/\partial P_j$ appearing in the power balance are computed from B coefficients and kept constant along the computation ; a sequence of linear programs (and not a single linear program) is used to take an account of the quadratic character of the cost function, which is locally linearized for each linear program (computation time around 20 sec. for 80 units).
The results of this ED function are basic generated powers P^o meeting economy and security, which are entered into the internal logic block of a typical AGC controller such as described previously (even a little more intricate as it is a pool). Two special features may be noticed : very powerful computers are used (with a speed as large as IBM 370-168), and no state estimation is performed, which does not appear necessary to the operation people.

Even if the optimal power flow process is somewhat approximate and could be improved, this implementation, just as some others going in the same direction, represents a considerable progress in comparison with the standard AGC solution, making the synthesis between automatic control and security : it is an example of what should be generalized in future.

ADVANCED AGC WITH LOAD CHANGE ANTICIPATION

Another type of AGC was imagined [35] and temporarily experimented in Wisconsin Electric Power Compagny. Its main objective was to solve the problem mentionned aboved as the "pin-point" aspect of ED, i.e. to take an account of unit power ramp limits in accordance with load change forecasts for 2 hours ahead. For that purpose, a new technique was imagined, "Dynamic Dispatch", which is run every 5th minute, instead of the usual ED function : first load forecasts are performed for 2 hours ahead, with 5 min. steps (around 10 sec. computation time) ; then, from these forecasted loads, using Dynamic Programming with relaxation, the Dynamic Dispatch algorithm finds the unit basic powers minimizing the total generation cost over the 2 hours and taking the unit power ramps into account (around 45 sec. computation time). Then the basic powers P^o for the first 5 min. point considered are entered into the internal logic block of a typical AGC controller. One cancels the "pin-point" aspect of ED, but is must be noticed that transmission security was not taken into account, which seems impossible even in future, due to the nature of the method used. This experiment also included applications of optimal control, in order to improve the control system dynamics. Results do not seem clear about the practical interest of such applications.

NEEDS FOR FUTURE

In the standard AGC implementation, 3 problems are left to be solved to be satisfactory : ED continuity, transmission security and anticipation. Even if solutions exist for each one of the two last problems separately, the main need for future would be an AGC including the solutions of these 3 problems at a time, with complete consistency. As extensions of AGC, automatic load shedding may be useful in emergency cases and would be easy to implement, corrective switching would be of first interest but still need method improvements. For voltage and reactive powers, an automatic tertiary voltage control is needed, finding secure and economic voltage magnitudes consistent with AGC solutions and applying them automatically : due to the speed needed, this is still a challenge : as a whole, for real and reactive power problems, the main need is more consistent and faster methods.

NEW CONCEPTS UNDER STUDY

New concepts for a modern AGC are now under study at Electricité de France ; their main objective is to build up an AGC system including the solution of the 3 problems mentionned above at a time. Two main methodological results have been obtained and are presented in the two next sections.

Simultaneous solutions of ED Continuity and Transmission Security in AGC : "Fast On-line Optimal Power Flows."

In order to get transmission security, compact optimal power flow techniques will be used [27], providing every 5th minute not only the optimal and secure basic generator real powers corresponding to the bus loads, but also a reduced model of the system valuable during the 5 following minutes ; this reduced model, which includes security constraints, and the validity of which was extended when loads and frequency change [36], is used to compute a secure ED solution every 5th or 10th second. In order to compute that solution, a new method, so-called "Fast On-line Optimal Power Flow " ("FOLOPF") was imagined ; it uses a new parametric quadratic programming algorithm built up at this occasion [37]. Experiments were carried out with the EDF present large scale network on scientific computers and gave very fast computation times (less than 0.1 second, i.e. at least 100 times shorter than usual optimal power flows) ; the corresponding duration times on a fast on-line minicomputer and for the EDF 1995 network should not reach 1 second : this speed allows to perform a secure ED and LFC at a time every 5th or 10 th second, which solves both ED continuity and security problems .

Simultaneous Solutions of Anticipation, ED Continuity and Transmission Security in AGC.

The "FOLOPF" algorithm has "pin-point" properties : like usual optimal power flows, it only concerns one instant and not a 2 hour period, so that anticipation appears a priori impossible to introduce. A very recent result, which will be published in details later (probably at the 1984 PSCC) shows it is possible to separate "pin-point" and "dynamic" dispatch problems in a consistent manner. Indeed, it was shown that, provided that operation costs are at least quadratic, solving a dynamic dispatch problem or a sequence of usual "pin-point" ED problems gives the same solution if the unit cost of the pin-point problems are modified, adding "anticipation overcosts" which are straightforward to compute from the values of the dual variables of the dynamic dispatch problem. So, the following organization is foreseen : every hour, a "secure dynamic dispatch" is performed, computing anticipation overcosts two hours ahead for each unit and each 5 minute time interval ;

then these overcosts are used in the pin-point secure ED every 5th minute during the next hour. Mathematical programming with partitionning will be used to solve secure dynamic dispatch. It could be shown theoretically and checked experimentally that anticipation overcosts are very stable vs load changes and keep valuable even if load prediction was not perfect : anticipation costs are the link between dynamic dispatch and pin-point models and will allow to build a consitent AGC with anticipation, transmission security and ED continuity.

Sequence of Main Models to be used in the New Consistent AGC

In the new consistent AGC, the following sequence of computations is foreseen :
- Every 5th or 10 th second : AGC executive routine, including decomposition of ACE and critical line current deviations into oscillatory and trend components ; application of FOLOPF algorithm to the trend components, within unit possibilities ; application of LFC algorithm to the oscillatory components within the left unit possibilities ; sending the sum of both results to the plants.
- Every 5th minute : real only optimal power flow, using anticipation overcosts.
- Every 30th minute : real reactive optimal power flow.
- Every hour : dynamic dispatch and real only optimal power flow with optimal allocation of regulating bandwidths.

These new concepts will certainly allow to build up a solution for AGC which is satisfactory and very consistent, even with reactive problems, reduces LFC necessary bandwidths (which are costly) and leaves the field for improvements open : indeed, optimal power flow and FOLOPF algorithms may be extended to include automatic load shedding, preventive transient control [35][39] and even corrective switching. At last, FOLOPF techniques could be extended to reactive powers [40], but this has not yet been studied, a considerable bulk of detailed studies being still left to be done.

CONCLUSION

Slowly but surely, real time digital control in energy control centers goes towards consistency. The present implementations include exemplary advanced solutions but are still far from being completely satisfactory ; the main challenge is a consistent AGC taking security and load change anticipation into account, with a continuous ED allowing fast response and avoiding unit unuseful oscillations. In this way towards progress, on the technical point of view, computation methods remain the greatest abstacle : they need be faster and more consistent ; attractive new solutions seen to take form, but a lot of work is still necessary. And in the way towards progress, the main obstacles are perhaps not technical : the main problem left to be solved is certainly that of human being, who must remain the power system real time operation manager ; but this would be the subject of another paper !

REFERENCES

1. G.Quazza, "Large Scale control problems in electric power systems", state of the art lecture, IFAC symposium on large scale systems theory and applications, Udine, 1976, pp 1-18.
2. Y.Sekine, "Digital control applied to network protection", IFAC symposium on real time digital control applications, Guadalajara, January 1983.
3. B.Bouzin,"Digital control in nuclear power plants", IFAC symposium on real time digital control applications, Guadalajara, January 1983.
4. J. Zaborsky, "Emergency control during stability crises by tracking the observation decoupled reference", IFAC symposium on real time digital control applications, Guadalajara, January 1983.
5. Concordia and Kirchmayer, "Tie-line power and frequency control of electric power systems", IEEE Trans., June 1953.
6. F. Cahen and A.Chevallier, "Le réglage puissance-phase ; nouvelle méthode pour le réglage automatique de la fréquence d'un réseau comportant de multiples usines génératrice", Bulletin de la Société Française des Electriciens, oct. 1953.
7. F. Cahen, A. Chevallier, R.Robert, B.Favez and J.Carpentier, "Les problèmes de réglage automatique de la fréquence et des échanges de puissance dans les grands réseaux. CIGRE 1956.
8. N.Cohn, "Control of generation and power flow in interconnected systems", Wiley, New-york 1966.
9. N.Cohn, "Area control performance measurement and corrective control in interconnected systems" IFAC symposium on real time digital control applications, Guadalajara, January 1983.
10. L. K. Kirchmayer, "Economic operation of power system", Wiley, New-York 1958.
11. L. K. Kirchmayer, "Economic operation of interconned Systems", Wiley, New-York 1959.
12. J.Carpentier and M. Canal, "Ordered eliminations", proc. PSCC 1, London 1963.
13. J.Carpentier, "Application of Newton's method to load flow computations", Proc. PSCC 1, London 1963.
14. B. Stott, "Review of load flow calculation methods" Proc. IEEE, July 1974.
15. A.S. Debs and A.R. Benson, "Security assessment of power systems", Engineering foundation conference on systems engineering for power, status and prospects, Henniker, N.H. 1975.
16. J.Carpentier, "System security in the Differential Injections method for optimal power flows", Proc. PSCC 5, 1975.
17. M. Ribbens Pavella,P.G. Murthy, J.L Horward and J.L.Carpentier,

"Transient stability index for on-line security assessment and contingency evaluation", Internatinal Journal of Electric Power and Energy Systems, Vol. 4 N° 2, April 1982.

18. K. Takahashi, J.Fagan, M. Chen, "Formulation of a sparse bus impedance matrix and its application to short-circuit study", Proc. PICA 1973, pp 63-69.

19. T.E. Dy Liacco, "Security functions in Power System Control Centers", 2^d international symposium on control conters for electric power systems, Caracas, Nov. 1979

20. F. Broussolle, "State estimation in power systems : detecting bad data through the sparse inverse matrix method, IEEE Trans. PAS, Vol. 97, pp 678-682, May 1978.

21. F. Broussolle, A. Le Roy, "State estimation and unobservable networks", PSCC 7, pp 982-988, July 1981.

22. J. Carpentier, "Contribution à l'étude du dispatching économique", Bulletin de Société Française des Electriciens, Ser. 8, Vol. 3, August 1962.

23. D. W. Wells, "Method for economic secure loading of power system", proc. IEE, Vol. 115, N° 8 (August 1968), pp 1190-1194.

24. H. W. Dommel and W.F. Tinney, "Optimal power flow solution", IEEE trans. PAS Vol. 87, 1968, pp. 1866-1876.

25. A.M. Sasson, "Non-linear programming applications to power systems", Proc. symposium Helors - IFORS, Athens 1968.

26. J. Carpentier, C. Cassapoglou, C.Hensgen, "Injections Differentielles, une méthode de résolution générale des problèmes de dispatching économique sans variables entières utilisant le procédé du gradient réduit généralisé, Proc. Symposium Helors IFORS, Athens 1968.

27. J. Carpentier, "Differential Injections method, a general method for secure and optimal load flows", Proc. PICA (1973), pp 255-262.

28. J. Carpentier, "Optimal power flows" (survey), International Journal of Electrical Power and Energy Systems, Vol. 1, N° 1, April 1973, pp 3-15.

29. H.J. Koglin, H.Muller, "Overload reduction through corrective switching actions", International conference on power system monitoring and control, IEE, London, June 1980, pp 159-164.

30. H. Glavitsch and J.Stoffel, "Automatic Generation Control, a survey", International Journal of Electrical Power and energy Systems, Vol. 2, N° 1, January 1980, pp 21-28.

31. F.P. de Mello, R. J. Mills and W. F. B'Rells, "Automatic Generation Control, parts I and II, IEEE Trans. PAS 92, 1973, pp 710-724.

32. G. Schellstede and H. Wagner, "Design aspects of a software package for Automatic Generation Control with instantaneous economic dispatch and load forecasting functions", IFAC symposium on automatic control in power generation distribution and protection, Pretoria 1980, pp 61-69

33. G. Simonnet, "The secondary voltage control of EDF network", IFAC symposium on automatic control in power generation, distribution and protection, Pretoria 1980, pp 87-92.

34. A.J. Elacqua and S.L. Corey, "Security constrained dispatch at the New-York Power Pool, IEEE paper N° 82 WM 084-2 (1982 Winter power meeting).

35. R. Raithel, S. Virmani, S. Kim, D. Ross, "Improved allocation of generation through dynamic economic dispatch", Proc. PSCC 7, pp 273-279.

36. J.L. Carpentier, "Principle of a secure and economic automatic generation control", IFAC symposium on automatic control in power generation, distribution and protection, Pretoria 1980, pp 463-471.

37. J.L. Carpentier, G. Cotto, P.L. Niederlander, "New concepts for automatic generation control in electric power systems using parametric programming", IFAC symposium on real time digital control applications, Guadalajara, January 1983.

38. J.Carpentier, "Prospects for security control in electric power systems", IFAC 8th world congress, Kyoto, August 1981.

39. M.Ribbens Pavella, P.G. Murphy, J.L. Horward and J.L. Carpentier, "On-line transient stability assessment and contingency analysis", 1982 CIGRE report N°32-19.

40. - J. Carpentier, "Optimal voltage scheduling and control in large scale power systems", IFAC Symposium on computer applications in large scale power systems, New Delhi, August 1979.

AREA CONTROL PERFORMANCE MEASUREMENT AND CORRECTIVE CONTROL IN INTERCONNECTED SYSTEMS

N. Cohn

1457 Noble Road, Jenkintown, PA 19046, USA

Abstract. In an interconnected electric power system, the parameters system time deviation and area inadvertent interchange can be decomposed into components respectively caused by regulating deficiencies in each of the individual control areas. The component of an area's inadvertent interchange caused by the area itself is designated its "primary" inadvertent. The remaining components of that area's inadvertent interchange caused respectively by regulating deficiencies in each of the other areas are designated "secondary" components. The primary and secondary components can serve as the basis for an equitable debit/credit dollar payment technique for unscheduled transfers, to replace the present practice of "repayment in kind". In addition, either the area component of system time deviation or the area primary component of inadvertent interchange can be used unilaterally for single-step corrective control to simultaneously reduce to zero the area component of system time deviation, its primary inadvertent, and all secondary inadvertents it previously caused in remote areas.

Keywords. Large scale systems, power system control, distributed parameter systems, control system analysis, frequency control, power control, energy control, load dispatching.

INTRODUCTION

Control practice on US-Canada interconnected systems is for each control area to utilize distributed frequency biased net interchange control for the regulation of inter-area bulk power transfers, and to allocate total area generation to area participating sources to achieve area economic, security and environmental objectives. The evolution of real-time power system control over the years, identifying the contributions of many individuals and groups and leading to the presently prevailing practice is summarized in Cohn (1983), presented earlier at this symposium.

An inherent element of this type of control is that if all areas fulfill their inter-area responsibilities, then scheduled system frequency and individual scheduled area net interchange exchanges will be maintained. In the event that one or more areas fail to do so, then system frequency and area net interchange transfers will be off schedule, and system time deviation and unscheduled flows will develop and accumulate. Because of the interactive nature of system operation, unscheduled -- or inadvertent -- exchanges will involve both the area or areas that were at fault, as well as the areas that provided assistance to them. There will then follow two resultant needs: (1) The need to provide for correction of the system time deviation, and (2) The need to provide for compensation for or correction of the unscheduled transfers.

Techniques have been developed to achieve these needs (NERC, 1982), and have been satisfactorily used for many years. In Cohn (1982) some new thoughts and suggestions have been developed, which it is believed would provide improvements over presently utilized corrective techniques.

The proposed new methods are based on decomposition of system time deviation and area inadvertent interchange into their respective area-caused components. These components are then utilized to identify quantitatively the effect of poor performance in individual areas, to provide a basis for an equitable debit/credit, penalty/reward compensation and payback technique for unscheduled bulk energy transfers, and to provide a unilateral single-step corrective control method that would eliminate present time deviation and unscheduled energy transfer corrective techniques.

The new component concepts are summarized and additionally analyzed in this paper. Mathematical derivations are separated from the paper's text, and are tabulated in an attached appendix.

Basic Control Relationships

The basic relationship for frequency bias net interchange control is shown in (1) of the appendix. In order to be comprehensive and universally applicable, it includes, in addition to the net interchange, frequency and frequency bias parameters, provision for any applicable measuring or schedule setting errors, and any corrective control offsets to either the net interchange or frequency schedules. In (2), any measuring or schedule setting errors for net interchange or frequency are respectively combined. There then results (3) which is a practical operating equation, with the net interchange and frequency parameters as actually measured or set, complete with their respective errors, and with only any corrective control schedule offsets that may apply shown as additional parameters. It will be understood that (1), (2) and (3) are equivalents.

System time deviation accumulations are defined, for a 60 Hz system, by (4). (For a 50 Hz system, the units conversion factor would be 72 instead of 60.) Area inadvertent interchange accumulations are defined by (5).

Assumptions

In the decomposition derivations, the following conditions for the designated time period that is involved are assumed to be applicable:

1. All control areas have remained in synchronism, i.e. they all have a common steady state frequency, common frequency schedule and common system time deviation.

2. The algebraic summation of assigned net interchange schedules for all areas is zero.

3. The algebraic summation of true net interchange of all areas is zero.

4. The algebraic summation of inadvertent interchange for all areas is zero.

5. The bias settings for all areas are known.

Definition of Area Regulating Deficiencies

For purposes of this paper, the term "area regulating deficiencies" is defined as the summation in MWh for the designated time period that is involved, of all area control errors, measurement errors, schedule setting errors, and schedule offsets for corrective control. The latter factors are included since they are intended to create area errors to compensate for previously created control, measurement or schedule setting errors, and hence are indeed, though self-created, regulating deficiencies.

It should at this point be noted that the bracketed terms on the right-hand side of (8), beyond the summation sign, as well as all the terms on the left-hand side of (10) are mathematical representations, in MWh, of area regulating deficiencies accumulated in time t in hours.

Decomposition of System Time Deviation

It will be clear from the development of (8), (11) and (12) that in an interconnected system of N areas, system time deviation can be decomposed into N components, to whose algebraic sum it is equal. Each component is caused by, and related to, the summation of regulating deficiencies of a specific area, as shown in (8). This particular relationship, though of interest, is lacking in practical use since many of its factors are unknown and non-measurable.

The development of (14), however, all of whose parameters are known or measurable, produces a useful quantitative relationship for each area's component of system time deviation.

Decomposition of Area Inadvertent Interchange

From (18) it will be clear that in an interconnected system of N areas, each area's inadvertent interchange can be decomposed into N components, to whose algebraic sum it is equal. One component, designated the primary component, is caused by the regulating deficiencies of the area itself. There are, for a given area, (N-1) additional components, each designated a secondary component, and each caused respectively by the regulating deficiencies of each of the other areas.

This particular relationship, as was true with (8), though of interest, is lacking in practical value since many of its factors are unknown and unmeasurable.

It will be desirable here to assign symbology for the components of area inadvertent interchange.

A system of identifying double subscripts will be used. The first subscript will identify the area that has the inadvertent component. The second subscript will identify the area that caused it. The appendix identifies primary and secondary inadvertent components more explicitly for a local and a remote area.

Equations for the Primary Component

From the development of (18), (19), and (20), there is obtained (21), an equation for primary inadvertent, all of whose terms are known or measurable. This provides a means of determining the quantitative measure of the effect, in megawatt hours, of primary inadvertent energy transfers made by the area in question caused by its own regulating deficiencies.

For the proposed new technique of payment for inadvertent in dollars, it will be necessary to know, not only the primary inadvertent of an area, but the secondary inadvertent it has caused in all of the remaining

areas. This is determined by decomposing
the primary inadvertent of an area into
(N-1) related secondary inadvertents, each
of course of opposite sign, and to the sum
of which the primary is equal in magnitude.

Equation for the Secondary Component

Following the development in the appendix
from (18) to (22) we have an equation for
secondary inadvertent that, though of interest, contains -- like (8) and (18) -- unknown and unmeasurable parameters and therefore, of itself, does not have practical
use. However, from (22) and (23) there is
obtained (24) which does indeed contain only
known or measurable parameters, and therefore provides the means of making a quantitative determination of the secondary inadvertent in one area caused by the regulating
deficiencies in another area. The same
information is provided by (26) but in terms
of the primary inadvertent in the remote
area. Similar information is provided by
(27) and (28) respectively but with the area
identities reversed.

Relationship of New Equations to NERC "Control Error"

It will be noted that the new components
equations (14), (21) and (27), applying
respectively to area n component of system
time deviation, area n primary inadvertent,
and the secondary components of inadvertent
caused by area n in each remote area i, all
contain the term $(I_n - B_n \varepsilon/6)$. This is the
same term used in NERC (1982) to calculate
"control error" in area surveys.

I think it is of importance to note that
this term, per (23), is a measure of the
summation of area regulating deficiencies in
the time period involved. It must then be
multiplied by suitable factors, as in (14),
(21) and (27), to obtain measures of the
effects of area regulating deficiencies.
This step yields area component of system
time deviation, in seconds for (14), primary
inadvertent in MWh for (21) and secondary
inadvertent in each remote area in MWh for
(27) respectively.

It will be recognized that for many areas,
the bias ratio Y_n is small, but for some
areas it may be as large as several percent.
But whether large or small, it is important
to use, so that, for example, in the proposed debit/credit technique now to be discussed, dollar value amounts to suppliers
and receivers will have zero sum.

DEBIT/CREDIT PAYBACK TECHNIQUE

The availability of area-caused primary and
secondary inadvertent components makes possible a new technique for payment of unscheduled energy transfers. Some operating
people have expressed dissatisfaction with
the limitations and possible inequities of
the present practice, and it is believed
that the proposed new technique can provide
a fully equitable payback arrangement. The
present technique and the proposed new one
will both be briefly discussed.

Present Payback Practice

Each day is divided into "on-peak" and "off-peak" periods. Each area is expected to
track its inadvertent accumulations separately for each of these two general
periods, and pay back inadvertent accumulations "in kind", i.e. at an on-peak or
off-peak period consistent with the on-peak
or off-peak period respectively during which
the inadvertent was accumulated.

Potential difficulties with this technique
are: (1) Inaccuracies in allocating accumulated inadvertent interchange to on-peak
or off-peak accounts. (2) During a 24-hour
daily span, there are significant differences between incremental energy costs
during just two designated on-peak and
off-peak periods, each of which must be
several hours in duration. Thus, payback
during a designated on-peak or off-peak
period may not really constitute "in kind"
payback, with corresponding inequities for
the exchange. (3) Payback may be delayed
over relatively long periods of time, and
conceivably be achieved by poor regulation
in one direction matched later by equally
poor regulation in the opposite direction,
resulting in no net inadvertent that would
require payback, but significant assisting
burdens having meanwhile been placed on
other areas of the interconnected system.

Proposed New Payback Technique

With this proposed new technique, it is
suggested that there no longer be on-peak or
off-peak periods as such. Instead, each day
would be divided into appropriate segments
of time, say 24 1-hour segments. Inadvertent accumulations would then be monitored
for each of these time segments and be recorded at some selected central clearing
agency, such as a selected area, or a
financial institution, or a selected middle
man. Dollar values would then be assigned
to each hour's transactions. The dollar
values could be flexibly different for each
pair of companies, if so desired. Further,
unscheduled energy taken by an area, representing for that area primary import and for
other areas a related secondary export,
would be priced high, preferably higher than
a scheduled purchase. Unscheduled energy
into the system, representing primary export
for that area and related secondary import
for other areas, would be assigned a low
value, preferably lower than the price for
scheduled sales.

Probably a two-tier pricing system, uniform
each hour for all companies, would be a
reasonable approach.

At the end of each hour, each area would earn credits for exports and debits for imports, at high or low values respectively depending on who caused the unscheduled transfer.

Four specific examples involving high value local primary import/remote secondary export, and low value local primary export/remote secondary import are included in the appendix to demonstrate the nature and equitability of the technique.

A very important point to note is that the initial debit/credit, penalty/reward transaction for a given unscheduled energy transfer, does not, despite the dollar payment, fully complete the transaction. Sooner or later, at the choice of the area at fault, a second step is required to provide the requisite correction of time deviation and inadvertent accumulation. Such correction can only be initiated and completed by the area that caused the unscheduled transfer in the first place. When such second corrective step is undertaken there will again be primary inadvertent and related secondary inadvertents created, which for balance will have to be equal in magnitude and opposite in sign to the initially created inadvertent. For this corrective payback, there will again be a dollar payment, high for primary import, low for primary export, at the rates applicable to the hour in which the correction is made.

In summary, the area at fault would be paying the net difference between the primary import value and the primary export value. The areas providing bias assistance would be receiving the difference between the applicable secondary export value and the applicable secondary import value.

Another significant point to be made is that a remote area that does not provide its programmed bias assistance during a period of another area's need, does not receive full compensation even though it is assigned a secondary inadvertent credit per (27) or (28). What happens is that the area's failure to provide programmed bias assistance results in that area developing a primary inadvertent of its own per (25), which cancels part or all of its dollar compensation depending on whether it provides some or no assistance.

CORRECTIVE CONTROL

Techniques for achieving corrective control for system time deviation and area inadvertent interchange have been well developed and applied over the years. It is felt that the new components concept which provides means for identifying and measuring the magnitude of individual area causes of these accumulations can provide improved methods of unilateral correction, wherein the corrective effort is limited to the specific area or areas that caused the accumulation.

Present Corrective Techniques

Present corrective control techniques are described in NERC (1982).

Time deviation correction. Time correction is achieved by uniform offset of frequency schedule in all areas. The offset, in magnitude and direction, is introduced manually on orders from the central timekeeping area on the US-Canada Eastern and Texas systems. It is done automatically on the Western system. In each instance, participating areas all share in the corrective burden, even though all may not have shared in causing the error.

It can readily be shown that when all areas participate in frequency schedule offset and regulate effectively, there is no change in net interchange flows as system frequency is changed in order to correct system time. This has, quite properly, been the supporting basis for the present universal participation by all areas in time correction.

The new components concept makes it clear, however, that though total area inadvertent is not changed by this present technique, the relative magnitude of an area's primary and secondary components is indeed altered. An area with a primary component of the same algebraic sign as system time deviation, will have its primary component reduced by time correction. An area with no primary, or primary of the opposite algebraic sign, will have its primary increased by time correction. Subsequent additional corrective action will be needed to reduce the new increment in primary.

Area Inadvertent Correction

A current technique for inadvertent correction is for two areas, which have inadvertent in opposite directions, to simultaneously offset net interchange schedules to mutually reduce respective inadvertents.

A possible limitation to this method is that corrections are based on total area inadvertent, including both primary and secondary components, whereas an area can only itself correct for its own primary. Its secondary components must be corrected by the remote causing area.

Proposed New Control Techniques

It will be noted from the relationships of (33) and (34) that for a given area, its component of system time deviation, its primary component of inadvertent interchange and the secondary component of inadvertent interchange it causes in other areas are linearly related, and when one is zero, all are zero.

This suggests that corrective control action in an area to reduce either its area component of system time deviation or its primary inadvertent to zero will, in the process of this single-step unilateral corrective control being carried out, reduce all of the components caused by that area to zero.

Relationships for appropriate offsets of area frequency schedule or area net interchange schedule are developed in (35), (36) and (37) and in (39) and (40) respectively. Operating equations including corrective control factors are (38) and (41) respectively.

It will be clear from (43) and (44) that the two operating equations, (38) and (41), are equivalent. Either may be used. Both have had introduced into them a factor equal to the accumulated area regulating deficiencies whose effect is to be duplicated with opposite algebraic sign, divided by the time within which it is desired to achieve the correction.

Equation (45) shows (38) and (41) in an alternative form, wherein it will be noted that the additional corrective factors are the time integrals of true net interchange deviation from schedule and true frequency deviation from schedule, each divided by the time during which the correction is to be achieved.

CONCLUSION

This paper has described the following:

1. In an interconnected system of N areas, techniques for computing from known or measurable parameters, (a) the N components of system time deviation, each caused by the regulating deficiencies of its respective area, and (b) the N components of each area's inadvertent interchange, including a primary component caused by the regulating deficiencies of that area, and (N-1) components, each respectively caused by the regulating deficiencies of each of the other areas. These components are quantitative measures of the control performance of each area,

2. A debit/credit, penalty/reward dollar payment technique for unscheduled energy transfers, and

3. A unilateral single-step corrective control technique which will simultaneously correct for an area's component of system time deviation, its primary inadvertent, and the secondary inadvertents it has caused in all other areas, and will do so regardless of any corrective control that may or may not be in operation simultaneously in other areas.

APPENDIX

Tabulation of the basic control relationships for each area of an interconnected system of N areas, utilizing, as in the US-Canada interconnections, distributed frequency biased net interchange control in all areas, derivation of equations for decomposing system time deviation and area inadvertent interchange into constituent components, relationships for a proposed debit-credit dollar payment technique, and derivations of equations for offsetting area frequency and net interchange schedules to achieve single-step time-related corrective control are outlined in this appendix. Symbology used is that of Cohn (1982).

Basic Control Relationships

The basic control equation for each area of US-Canada interconnected systems is:

$$E_n = (T_n + \tau_{1n} - T_{on} - \tau_{on} - \hat{\tau}_n)$$
$$-10 B_n(F + \phi_{1n} - F_o - \phi_{on} - \hat{\phi}_n) \quad (1)$$

where,

E_n is the area control error in MW,
T_n is the actual area net interchange, in MW, power export is positive,
T_{on} is the scheduled area net interchange, in MW,
B_n is the area frequency bias setting, in MW/0.1 Hz, and has a negative sign,
F is the system frequency, in Hz, and
F_o is the scheduled system frequency in Hz,
τ_{1n} is any error in measurements of T_n,
τ_{on} is any error in setting T_{on},
$\hat{\tau}_n$ is any offset in T_{on} for corrective control,
ϕ_{1n} is any error in the measurement of F at area n,
ϕ_{on} is any error in the setting of F_o at area n,
$\hat{\phi}_n$ is any offset in F_o at area n for corrective control

The operating objective in each area is to adjust area generation to reduce E_n to zero.

Let τ_n represent the difference between any net interchange schedule setting errors and any net interchange measurement errors at area n, so:

$$\tau_n = \tau_{on} - \tau_{1n},$$

Also, let ϕ_n represent the difference between any frequency schedule setting errors and any frequency measurement errors at area n, so:

$$\phi_n = \phi_{on} - \phi_{1n}$$

Then (1) may be written:

$$E_n = (T_n - T_{on} - \tau_n - \hat{\tau}_n)$$
$$-10 B_n(F - F_o - \phi_n - \hat{\phi}_n) \quad (2)$$

As a practical operating equation, (2) may be written as:

$$E_n = (T_n' - T_{on}' - \hat{\tau}_n) - 10 B_n(F_n' - F_{on}' - \hat{\phi}_n) \quad (3)$$

where,

T'_n is net interchange as actually measured at area n, and is equal to $T_n + \tau_{1n}$,

T'_{on} is net interchange schedule as actually set at area n, exclusive of any offset for corrective control, and is equal to $T_{on} + \tau_{on}$,

F'_n is system frequency as actually measured at area n, and is equal to $F + \phi_{1n}$,

F'_{on} is system frequency schedule as actually set at area n, exclusive of any offset for corrective control, and is equal to $F_o + \phi_{on}$.

System time deviation ε, in seconds, accumulated in time span t, in hours, is given for a 60 Hz system by:

$$\varepsilon = 60 \int_0^t (F - 60)\, dt \qquad (4)$$

Area inadvertent interchange, I_n, in MWh accumulated in time span t, in hours, is given by:

$$I_n = \int_0^t (T_n - T_{on})\, dt \qquad (5)$$

Decomposition of System Time Deviation

Summing (2) for all N areas yields:

$$\sum_{n=1}^N E_n = \sum_{n=1}^N T_n - \sum_{n=1}^N T_{on} - \sum_{n=1}^N (\tau_n + \hat{\tau}_n)$$

$$- \sum_{n=1}^N 10 B_n (F-F_o) + \sum_{n=1}^N 10 B_n (\phi_n + \hat{\phi}_n) \qquad (6)$$

By definition, $\sum_{n=1}^N T_n = 0$, $\sum_{n=1}^N T_{on} = 0$, and F and F_o are common in all areas. Then, let

$B_s = \sum_{n=1}^N B_n$, and assuming fixed bias settings in each of the control areas, (6) becomes:

$$F - F_o = -(1/10 B_s) \sum_{n=1}^N [E_n + \tau_n + \hat{\tau}_n$$

$$-10 B_n (\phi_n + \hat{\phi}_n)] \qquad (7)$$

For purposes of this paper, "regulating deficiencies" of an area during a designated time period is defined as the net summation of any control errors (non-zero E_n), any errors in setting its net interchange schedule or measuring its net interchange (non-zero τ_n), any errors in setting its frequency schedule or measuring its frequency (non-zero ϕ_n), any corrective control offsets to its net interchange schedule (non-zero $\hat{\tau}_n$) and any corrective control offsets to its frequency schedule (non-zero $\hat{\phi}_n$).

Equation (7) shows the cumulative effect on system frequency of regulating deficiencies, i.e. non-zero E_n, τ_n, $\hat{\tau}_n$, ϕ_n or $\hat{\phi}_n$ in any of the areas.

Taking the time integral of (7) in accordance with (4), yields:

$$\varepsilon = -(6/B_s) \sum_{n=1}^N [\int_0^t E_n dt + \int_0^t (\tau_n + \hat{\tau}_n)\, dt$$

$$-10 B_n \int_0^t (\phi_n + \hat{\phi}_n) dt] \qquad (8)$$

It will be clear from (8) that system time deviation consists of components, each related to the aggregate of the regulating deficiencies in each of the areas. The equation, however, includes many unknown and unmeasurable parameters, and therefore of itself has no practical use.

Now, take the time integral of (2), thusly:

$$\int_0^t E_n dt = \int_0^t (T_n - T_{on}) dt - \int_0^t (\tau_n + \hat{\tau}_n) dt$$

$$-10 B_n \int_0^t (F-F_o) dt + 10 B_n \int_0^t (\phi_n + \hat{\phi}_n) dt \qquad (9)$$

Substituting I_n and ε from (4) and (5) respectively for the net interchange and frequency integrals in (9) and rearranging, yields for a 60 Hz system:

$$\int_0^t E_n dt + \int_0^t (\tau_n + \hat{\tau}_n) dt - 10 B_n \int_0^t (\phi_n + \hat{\phi}_n) dt$$

$$= I_n - B_n \varepsilon / 6 \qquad (10)$$

Equation for area component of system time deviation. As noted from (8), ε is made up of components, ε_n, in number equal to the number of control areas, and the following may be written:

$$\varepsilon = \sum_{n=1}^N \varepsilon_n \qquad (11)$$

From the viewpoint of each individual area n, considering all other areas as areas i each with a time deviation component of ε_i, the following applies:

$$\varepsilon = \varepsilon_n + \sum_{\substack{i=1 \\ i \neq n}}^N \varepsilon_i \qquad (12)$$

Also, from (8) and (12), for individual area n:

$$\varepsilon_n = -(6/B_s) [\int_0^t E_n dt + \int_0^t (\tau_n + \hat{\tau}_n) dt$$

$$-10 B_n \int_0^t (\phi_n + \hat{\phi}_n) dt] \qquad (13)$$

Now, substituting (10) in (13), there is obtained, for every area n of a 60 Hz system, the following quantitive, measurable relationship for its component of system time deviation, caused by the aggregate of its own regulating deficiencies, i.e. non-zero E_n, τ_n, $\hat{\tau}_n$, ϕ_n and $\hat{\phi}_n$, during the designated time span:

$$\varepsilon_n = -(6/B_s)(I_n - B_n \varepsilon/6) \quad (14)$$

Decomposition of Area Inadvertent Interchange

Now, substitute (7) in (2), yielding;

$$E_n = (T_n - T_{on} - \tau_n - \hat{\tau}_n)$$

$$+ (B_n/B_s) \sum_{n=1}^{N} [E_n + \tau_n + \hat{\tau}_n - 10B_n(\phi_n + \hat{\phi}_n)]$$

$$+ 10B_n(\phi_n + \hat{\phi}_n) \quad (15)$$

Let area bias ratio, Y_n, = B_n/B_s, then (15), rearranged, becomes:

$$T_n - T_{on} = [E_n + \tau_n + \hat{\tau}_n - 10B_n(\phi_n + \hat{\phi}_n)]$$

$$- Y_n \sum_{n=1}^{N} [E_n + \tau_n + \hat{\tau}_n - 10B_n(\phi_n + \hat{\phi}_n)] \quad (16)$$

Let all areas other than area n become areas i. Equation (16) then becomes:

$$T_n - T_{on} = (1 - Y_n)[E_n + \tau_n + \hat{\tau}_n - 10B_n(\phi_n + \hat{\phi}_n)]$$

$$- Y_n \sum_{\substack{i=1 \\ i \neq n}}^{N} [E_i + \tau_i + \hat{\tau}_i - 10B_i(\phi_i + \hat{\phi}_i)] \quad (17)$$

Equation (17) shows that area unscheduled power flow is caused by the algebraic summation of local and remote effects.

Taking the time integral of (17) in accordance with (5) yields:

$$I_n = (1 - Y_n)[\int_0^t E_n dt + \int_0^t (\tau_n + \hat{\tau}_n) dt - 10B_n \int_0^t (\phi_n + \hat{\phi}_n) dt]$$

$$- Y_n \sum_{\substack{i=1 \\ i \neq n}}^{N} [\int_0^t E_i dt + \int_0^t (\tau_i + \hat{\tau}_i) dt$$

$$- 10B_i \int_0^t (\phi_i + \hat{\phi}_i) dt] \quad (18)$$

Equation (18) shows that area inadvertent interchange has local and remote causes. But like (8) it contains unknown and unmeasurable parameters, and hence of itself has no practical value.

Symbology of area inadvertent components. Components of area inadvertent interchange will be designated I with double subscripts. The first subscript will identify the area that has the component. The second subscript will identify the area that caused it. This convention will apply to both primary and secondary components. Thus:

- I_{nn} is the primary component of area n inadvertent interchange.
- I_{ii} is the primary component of area i inadvertent interchange.
- I_{ni} is a secondary component of area n inadvertent interchange caused by the primary inadvertent of area i. In an interconnected system of N areas, area n has (N-1) such secondary components.
- I_{in} is a secondary component of area i caused by the primary inadvertent of area n. There are (N-1) such components in each area i.

Equation for primary component of area inadvertent interchange. Equation (18) for area n inadvertent interchange can be written:

$$I_n = I_{nn} + \sum_{\substack{i=1 \\ i \neq n}}^{N} I_{ni} \quad (19)$$

The equation for area n primary inadvertent, from (18) and (19) is:

$$I_{nn} = (1 - Y_n)[\int_0^t E_n dt + \int_0^t (\tau_n + \hat{\tau}_n) dt$$

$$- 10B_n \int_0^t (\phi_n + \hat{\phi}_n) dt] \quad (20)$$

Substituting (10) in (20) there is obtained, for every area n of a 60 Hz system, the following measurable relationship for its component of primary inadvertent interchange, caused by the aggregate of its own regulating deficiencies, i.e. non-zero E_n, τ_n, $\hat{\tau}_n$, ϕ_n, and $\hat{\phi}_n$, during a designated time span:

$$I_{nn} = (1 - Y_n)(I_n - B_n \varepsilon/6) \quad (21)$$

Equation for secondary components of inadvertent interchange. For area n, each of the (N-1) secondary components of its inadvertent interchange identified in (19) is defined in (18) for each remote area i, thusly:

$$I_{ni} = -Y_n[\int_0^t E_i dt + \int_0^t (\tau_i + \hat{\tau}_i) dt$$

$$- 10B_i \int_0^t (\phi_i + \hat{\phi}_i) dt] \quad (22)$$

Equation (10), applicable to area n, may be rewritten comparably for each area i, thusly:

$$\int_0^t E_i dt + \int_0^t (\tau_i + \hat{\tau}_i) dt - 10B_i \int_0^t (\phi_i + \hat{\phi}_i) dt$$

$$= I_i - B_i \varepsilon/6 \quad (23)$$

Substituting (23) in (22) there is obtained for every area n remote to area i the following relationship for a secondary component of its inadvertent interchange caused by regulating deficiencies in area i during a designated time span:

$$I_{ni} = -Y_n (I_i - B_i \varepsilon/6) \qquad (24)$$

For area i, its primary component of inadvertent interchange in the manner of (21) is:

$$I_{ii} = (1-Y_i)(I_i - B_i \varepsilon/6) \qquad (25)$$

Thus (24) may be written:

$$I_{ni} = -[Y_n/(1-Y_i)]I_{ii} \qquad (26)$$

Similar to (24), the secondary inadvertent at area i caused by area n is given by:

$$I_{in} = -Y_i(I_n - B_n\varepsilon/6) \qquad (27)$$

Or, in the manner of (26):

$$I_{in} = -[Y_i/(1-Y_n)]I_{nn} \qquad (28)$$

Debit/Credit Technique

Assume that for a designated time span, such as one hour, a two-tier dollar value payment is to apply to unscheduled energy transfers, thusly:

λ_H is the dollar per MWh value for primary import into an area, and related secondary export from all other areas. This would be of a relatively high value.

λ_L is the dollar per MWh for primary export from an area, and related secondary import in all other areas. This would be of a relatively low value.

Consider dollar debits and credits for four cases, namely, area n has primary import, area n has primary export, area i has primary import, and area i has primary export.

For ready representation of the dollar value of such transactions, let V_{xxx} represent the transaction value. The first subscript will indicate whether it is of the high or low value. The second subscript will identify the area that receives the credit. The third subscript identifies the area that receives the debit.

Area n has primary import. All other areas, areas i, have related secondary export. Such unscheduled transfers have high value, i.e. high penalty for area n, and high reward for each area i, thusly:

$$V_{Hin} = I_{in} \lambda_H \qquad (29)$$

This transaction provides high credit to each area i, high debit to area n.

Area n has primary export. All other areas, areas i, have related secondary import. Such unscheduled transfers have low value, i.e. low reward for area n and low penalty for each area i, thusly:

$$V_{Lni} = -I_{in} \lambda_L \qquad (30)$$

This transaction provides low credit to area n, low debit to each area i.

Area i has primary import. All other areas, areas n, have related secondary export. Such unscheduled transfers have high value, i.e. high penalty for area i and high reward for each area n, thusly:

$$V_{Hni} = I_{ni} \lambda_H \qquad (31)$$

This transaction provides high credit to each area n, and high debit to area i.

Area i has primary export. All other areas, areas n, have related secondary import. Such unscheduled transfers have low value, i.e. low reward for area i, and low penalty for each area n, thusly:

$$V_{Lin} = -I_{ni} \lambda_L \qquad (32)$$

This transaction provides low credit to area i, and low debit to each area n.

Corrective Control

This section will derive equations for the schedule offsetting factors of (3), $\hat{\tau}_n$ and $\hat{\phi}_n$, either of which will then provide, in a selected time period, unilateral single-step simultaneous correction of previously accumulated area component of system time deviation, area primary inadvertent, and secondary inadvertent caused by area n in all remote areas i.

Relations between components. From (14) and (21) the following relationship between ε_n and I_{nn} applies:

$$I_{nn} = -(1-Y_n)(B_s \varepsilon_n/6) \qquad (33)$$

Similarly, from (14) and (27) the following relationship between ε_n and I_{in} applies:

$$I_{in} = (B_i/6)\varepsilon_n \qquad (34)$$

From (33) and (34) it will be noted that ε_n, I_{nn} and I_{in} are all linearly related, and when one is zero all are zero.

Corrective control by offset of area frequency schedule. From (8), it will be clear that for a 60 Hz system a frequency schedule offset in area n of $\hat{\phi}_n$ effective for H_n hours will provide a related corrective time deviation component of $\hat{\varepsilon}_n$ in accordance with:

$$\hat{\varepsilon}_n = -(6/B_s)(-10B_n)(\hat{\phi}_n H_n) \qquad (35)$$

Or,

$$\hat{\varepsilon}_n = 60Y_n \hat{\phi}_n H_n \qquad (36)$$

To reduce a prevailing ε_n to zero, $\hat{\varepsilon}_n$ should be equal to $-\varepsilon_n$. Thus, from this relationship and from (36) the proper corrective values for frequency schedule offset is:

$$\hat{\phi}_n = -\varepsilon_n/(60\ Y_n H_n) \quad (37)$$

With such a frequency schedule offset, and with no need for a net interchange schedule offset (i.e. $\hat{\tau}_n=0$), operating equation (3) becomes:

$$E_n = (T'_n - T'_{on}) - 10B_n[F'_n - F'_{on} + \varepsilon_n/(60Y_n H_n)] \quad (38)$$

Corrective control by offset of area net interchange schedule. From (18) it will be clear that a net interchange schedule offset in area n of $\hat{\tau}_n$ effective for H_n hours will provide a related corrective primary inadvertent component of $\hat{I}_{nn}$ in accordance with:

$$\hat{I}_{nn} = (1-Y_n)\hat{\tau}_n H_n \quad (39)$$

To reduce a prevailing I_{nn} to zero, $\hat{I}_{nn}$ should be equal to $-I_{nn}$. Thus, from this relationship and from (39) the proper corrective value for net interchange schedule offset is:

$$\hat{\tau}_n = -I_{nn}/(1-Y_n)H_n \quad (40)$$

With such a net interchange schedule offset, and with no need for a frequency schedule offset (i.e. $\hat{\phi}_n=0$), operating equation (3) becomes:

$$E_n = [T'_n - T'_{on} + I_{nn}/(1-Y_n)H_n] - 10B_n(F'_n - F'_{on}) \quad (41)$$

Either (38) or (41) will operate in area n to provide simultaneous corrective control of previously accumulated ε_n, I_{nn}, and I_{in} in all areas i, while maintaining generation control response to its own load changes and providing frequency bias assistance to areas in need.

Equivalence of both corrective offsets. To demonstrate the equivalence of (38) and (41) it is necessary to show that the net interchange schedule offset of the latter is the same as the frequency schedule offset of the former multiplied by $(-10B_n)$, i.e. that $\hat{\tau}_n$ of (40) is equal to $(-10B_n)$ times $\hat{\phi}_n$ of (37). Thus, the identity that must be proved is:

$$-I_{nn}/(1-Y_n)H_n = -10B_n(-\varepsilon_n)/(60Y_n H_n) \quad (43)$$

Substituting on the left side for I_{nn} from (21) and on the right side for ε_n from (14), and clearing, there results,

$$(I_n - B_n\varepsilon/6)/H_n = (I_n - B_n\varepsilon/6)/H_n \quad (44)$$

The equivalence of (38) and (41) is thus established.

Equations (38) and (41) may also each be written:

$$E_n = (T'_n - T'_{on}) + \int_0^t (T_n - T_{on})dt/H_n$$
$$- 10B_n[(F'_n - F'_{on}) + \int_0^t (F - F_o)dt/H] \quad (45)$$

REFERENCES

Cohn, Nathan (1982). Decomposition of time deviation and inadvertent interchange on interconnected systems. Part I: Identification, separation and measurement of components. Part II: Utilization of components for performance evaluation and corrective control. IEEE Trans. on Power Apparatus and Systems, PAS-101, Nos. 5 and 8, 1144-1169.

Cohn, Nathan (1983). The evolution of real time control applications to power systems. IFAC Symposium on Real Time Digital Control Applications. Guadalajara, Mexico, Jan. 15-21.

NERC (1982). Operating manual, North American Electric Reliability Council, Princeton, New Jersey.

EMERGENCY CONTROL DURING STABILITY CRISES BY TRACKING THE OBSERVATION DECOUPLED REFERENCE

J. Zaborszky

Department of Systems Science and Mathematics, Washington University, St. Louis, Missouri, USA

Abstract. In the last 10-12 years techniques were developed at Washington University for the control of the power system during Stability Crises in order to prevent system breakup. These techniques derive from the concept of observation decoupled reference (or local equilibrium reference). When the system state coincides with this reference it is at its stable equilibrium. While the reference has no physical identity or interpretation otherwise it can nevertheless serve to guide the system state to its stable equilibrium. This result can be acheived either by using a full set of simple control tools —one at each unit-but strictly local feedback or by using a sparse set of more sophisticated control tools along with communication between the units and the control center and central processing. The letter results, which are new and as yet unpublished, are especially appropriate for using an HV-DC system to stabilize an associated AC system. In this paper this entire area of work is summed up, its essential core exposed and some of its newest results introduced.

Keywords. Observation decoupled; stability; emergency control; braking resistors; HV-DC; centralized control; decentralized control.

1. INTRODUCTION

Emergency is a condition of the power system which makes continued normal system operation impossible without extraordinary remedial measures because of severe violations of allowable limits of frequency, voltage, power flow, current, fuel supply, temperature etc. affecting all or part of the components of the system. Most emergencies are attributable to three classes of phenomena:
1.) trapped energy 2.) unbalance of power and 3.) disintegration and thus they can be grouped into three types of crisis conditions.

1. Stability Crisis. An ongoing dynamic (short or long range 2-15 sec) condition involving electromechanical energy swings of sufficient intensity to break up the system.

2. Viability Crisis. An existing or imminent unbalance between generation, load and transmission which causes violations of allowable limits (of frequency, voltage, power, current, temperature, torque angle, boiler output etc.) so severe that it can be tolerated at most for some specified time interval.

3. Integrity Crisis or In Extremis Condition. An actual disintegration of the system involving loss of load at a substantial level and/or islanding and/or blackout.

These classes of crises were previously defined in the framework of "Operating the Power system by Decision and Control (Zaborszky, 1981b)" in association with the Normal and the less severe Defect conditions, six System conditions in all. In the basic spirit of digital computer based operation these six conditions can be assigned to six basic regimes of control (Zaborszky, 1981b).

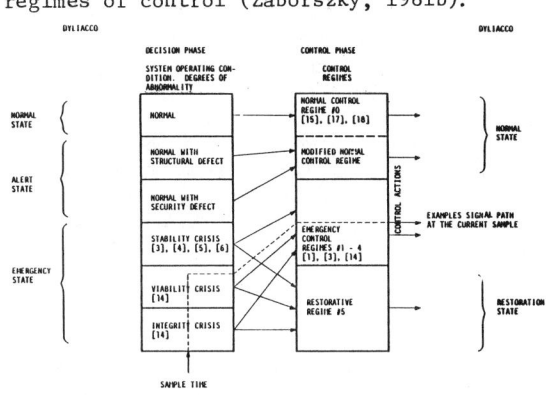

Fig. 1. The basic decision and control operation of the power system.

Extensive research is in progress at Washington University on most aspects of this overall operating scheme. Results of the last dozen years were reported in about 30 publications. Control during Stability Crises (Regimes 1 and 2) is one major component of this work which has undergone a gradual development of its concepts and techniques during the work of the last decade.

This paper summarizes and discusses this development of digital control approaches to Stability Crises and Stabilization in its matured form and context, including some of the newest previously unpublished results on the use of sparse control tools with central data processing.

2. THE STABILITY CRISIS AND THE APPROPRIATE SYSTEM MODEL

Stability Crisis is a dynamic condition which is violent enough to disrupt the system unless control measures are taken. Such a condition usually arises in the aftermath of heavy disturbances; short circuits aggravated by malfunction of the protection equipment, loss of load or generation. During such events the load distribution on generating machinery is abruptly altered with the result that certain machine rotors acquire and others lose kinetic energy relative to the basic value at 60 Hz. The accompanying speed difference results in increasing angles across transmission lines. When these angles exceed the angle of peek power transmission for the lines of a critical cutset cummulative separation may result. The principal dynamic elements in a stability crisis, at least for the first one or two seconds are the machine inertias and the angle dependent power flows on the transmission lines in the role of nonlinear springs. Other dynamic elements like the magnetic fields in the generators, the exciters, the governors gradually come into effect in one or two seconds. The network acts as instantaneous coupling. The loads have their own dynamics but are often approximated by constants or by impedances.

If synchronous machines are represented by their "voltage behind the transient reactance", which is considered constant during the short (2 sec) period of the control action, if the loads are moved to this internal point of the machine and if the buses are eliminated by network transformation, then an approximate model, some times called the machine angle model, results as shown in Fig. 2. It was shown (Zaborszky, 1979, 1981a) that the results are also applicable to detailed dynamic models.

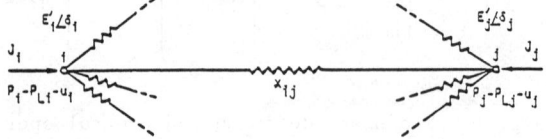

Fig. 2. A line i-j in the machine angle model of the interconnected power system.

For this simplified situation the three components that is units, lines and loads, collapse into a set of coupled differential equations.

$$J_i \ddot{\delta}_i = P_i - P_{Li} - \sum_{j=1}^{N} C_{ij} \frac{E_i E_j}{x_{ij}} \sin(\delta_i - \delta_j) - u_i,$$

$$i \varepsilon [1, N-1] \quad (1)$$

where E_i is the internal generator voltage at bus i and is considered constant

δ_i is the phase angle of node i referred to $\delta_N = 0$

P_i is the mechanical power input of machine i which remains constant during the control action

P_{Li} is the load at bus i which may be considered constant or may vary with its own dynamics

x_{ij} is the reactance of the transmission line i-j

u_i is the control power input at bus i such as may be produced by connecting a special purpose damping resistor or by interrupting the load for a short time (fraction of a second) or by initiating fast valving which withdraws a pulse of the mechanical power input in about 0.5 seconds or by a thyristor device such as an HV-DC converter or a Static Var Control device.

$$C_{ij} = \begin{cases} 1 \text{ if there is a transmission line from bus i to bus j} \\ 0 \text{ otherwise} \end{cases}$$

The <u>state vector</u> z for this model of the power system is

where $\quad z^T = \{\delta^T, \dot{\delta}^T\}$, T for transpose $\quad (2)$

$$\delta = \{\delta_i | \{\delta_i\} \varepsilon \Omega', i \varepsilon [1, N-1]\} \varepsilon R^{N-1}$$

with $\quad \dot{\delta} = \{\dot{\delta}_i | i \varepsilon [1, N-1]\} \varepsilon R^{N-1} \quad (3)$

$$\Omega' = \{\delta \varepsilon R^{N-1} | |\delta_i - \delta_j| < \pi,$$

all i, j such that $C_{ij} = 1\} \varepsilon R^{N-1}$ (4)

The <u>stable equilibrium point</u> $\delta_0 = \{\delta_{0i} | i \varepsilon [1, N-1]\}$ of the power system is then defined by substituting $\ddot{\delta}_i = 0$ and $u_i = 0$, all i in (1)

$$P_i - P_{Li} - \sum_{j=1}^{N} C_{ij} \frac{E_i E_j}{x_{ij}} \sin(\delta_{0i} - \delta_{0j}) = 0,$$

$$i \varepsilon [1, N-1] \quad (5)$$

These of course are N-1 <u>coupled</u> nonlinear algebraic equations. Computing the solution is nontrivial even for this highly simplified model. For a more realistic model the solution-<u>a Load Flow solution</u>-is a major computational problem which has a large literature. The point is that it is not feasible to compute the position of the stable equilibrium point on line during an emergency. Such a stable equilibrium point will, however, exist uniquely in Ω' for the power system, at least in its undisturbed state, because of the way it is operated.

3. THE EQUILIBRIUM REFERENCE STATE SPACE

The <u>Observation Decoupled Reference</u> (Zaborszky, 1981a, 1981c, 1981d) or <u>Local Equilibrium Reference</u> (Zaborszky, 1979), δ_{ei}, will now be defined for bus i by the ith equation of the following set of <u>decoupled</u> equations

$$G_i = P_i - P_{Li} - \sum_{j=1}^{N} C_{ij} \frac{E_i E_j}{x_{ij}} \sin(\delta_{ei} - \delta_j) = 0,$$

$$i \in [1, N-1] \quad (6)$$

$$|\delta_{ei} - \delta_i| < \pi/2 - \varepsilon, \quad \varepsilon > 0 \quad (7)$$

The set of decoupled equations (6) will then define the vector

$$\delta_e = \{\delta_{ei} | i \in [1, N-1]\} \quad (8)$$

called the Observation Decoupled Reference vector or Local Equilibrium Reference vector.

The Observation Decoupled Reference vector has some very important and useful characteristics:

<u>Property I</u>: $\{\delta_{ei}, \dot{\delta}_{ei}\}$ can be determined for machine i in small scale computation at bus i. From routine voltage and current measurements performed at bus i and from the transmission line equation in case of δ_j or δ_{ei}. δ_{ei} can be estimated by local dynamic filtering based on the limited frequency range of the electromechanical oscillations.

<u>Property II</u>: If a solution δ_{ei} exists at bus i for given E_j and δ_j, $\{j \in [1,N-1] | C_{ij}=1\}$, then such a solution is unique. Furthermore $\dot{\delta}_{ei}$ also uniquely exists if $\dot{\delta} \in R^{N-1}$ and δ_e do.

<u>Proof</u>: Let

$$I_{ib} = |\sum_{j=1}^{N} C_{ij} \frac{\hat{E}'_j}{\hat{x}_{ij}}| \quad (9)$$

and

$$\delta_{ib} = \arg(\sum_{j=1}^{N} C_{ij} \frac{\hat{E}'_j}{\hat{x}_{ij}}) \quad (10)$$

where $\hat{E}'_j = E'_j \varepsilon^{i\delta_j}$, $\hat{x}_{ij} = x_{ij} \varepsilon^{i\pi/2}$. Then (6) can be written as

$$P_i - P_{Li} - E'_i I_{ib} \sin(\delta_{ei} - \delta_{ib}) = 0 \quad (11)$$

$$|\delta_{ei} - \delta_{ib}| < \pi/2 - \varepsilon, \quad \varepsilon > 0 \quad (12)$$

Equation (10) has a unique solution whenever $E_i I_{ib} \cos \varepsilon > P_i - P_{Li}$.

Differentiating (6) with respect to time

$$\frac{dG}{dt} = \frac{\partial}{\partial \delta} G(\delta, \delta_e) \dot{\delta} + \frac{\partial}{\partial \delta_e} G(\delta, \delta_e) \dot{\delta}_e \quad (13)$$

$\dot{\delta}_e$ can then be written as

$$\dot{\delta}_e = -[\frac{\partial}{\partial \delta_e}G(\delta, \delta_e)]^{-1}[\frac{\partial}{\partial \delta}G(\delta, \delta_e)]\dot{\delta} \quad (14)$$

Since $\partial G/\partial \delta_e$ is a diagonal matrix, it is easy to see that it is also nonsingular given condition (7). Hence a unique $\dot{\delta}_e$ will exist if δ_e does.

<u>Property III</u>: δ_e is an instantaneous bounded function of t, continuously differentiable in t except at discrete points where parameters (C_{ij}, P_i, P_{Li}) in (6) may change abruptly in the course of the disturbance.

<u>Proof</u>: Immediate form preceding.

<u>Property IV</u>: $\{\delta_e, \dot{\delta}_e\} \subset \Omega_\delta \times R^{N-1}$ is a statespace for the dynamic system (1)-(4). It will be called the <u>Equilibrium Reference State Space</u>.

Specifically the transformation

$$\{\delta, \dot{\delta}\} \rightarrow \{\delta_e, \dot{\delta}_e\} \quad (15)$$

on the domain

$$\Omega_\delta = \{\delta \in \Omega' | \text{there exists } \delta_e \in R^{N-1} \text{ such that}$$
$$G(\delta, \delta_e) = 0\} \quad (16)$$

is a diffeomorphism and hence $\{\delta_e, \dot{\delta}_e\}$ carries the same dynamic information as set $\{\delta, \dot{\delta}\}$. Consequently $\{\delta_e, \dot{\delta}_e\}$ is a valid statespace equivalent to $\{\delta, \dot{\delta}\}$.

<u>Proof</u>: It was previously proven in (Zaborszky, 1981d) that transformation $\{\delta, \dot{\delta}\} \leftrightarrow \{\bar{\delta}, \dot{\bar{\delta}}\}$ where $\bar{\delta}=\delta-\delta_{ei}$, $\dot{\bar{\delta}}=\dot{\delta}-\dot{\delta}_{ei}$, is a diffeomorphism on Ω_δ. This differs from the formulation $\{\delta, \dot{\delta}\} \rightarrow \{\delta_e, \dot{\delta}_e\}$ by addition of the unit transformation which itself is of course a homeomorphism.

It is important to observe that δ_e does not identify with any physically meaningful quantity except when $\delta=\delta_e$ that is when the system state is lined up with the Observation Decoupled Reference state everywhere. When this happens the system is clearly in a equilibrium point by (5) and the following property applies:

<u>Property V</u>: If the system is in an equilibrium state δ_0 such that $\delta_{0i}=\delta_{ei}, i\in[1,N-1]$ and δ_{ei} exists, all $i\in[1,N-1]$ then $\delta_0 \in \Omega_\delta$ and it is stable.

<u>Proof</u>: Previously published in (Zaborszky, 1981d).

4. RECOGNIZING THE PRESENCE OF STABILITY CRISIS USING THE OBSERVATION DECOUPLED REFERENCE

The undisturbed power system is operating at its stable equilibrium that is $\delta=\delta_e=\delta_0$ and $\dot{\delta}=\dot{\delta}_e=0$ or $\bar{\delta}=\delta-\delta_e=0$ and $\dot{\bar{\delta}}=\dot{\delta}-\dot{\delta}_e=0$ on the undisturbed system. Furthermore δ_e and $\dot{\delta}_e$ are discontinuous at the time of structural changes of the system. Thus monitoring appropriate norms $||\bar{\delta}||$ and $||\dot{\bar{\delta}}||$ on the system will give immediate indication of stability crisis on the system by observing nonzero norms. Likewise nonzero components $\bar{\delta}_i$, $\dot{\bar{\delta}}_i$ at bus i will give local indications. These can be used to trigger control measures.

5. STABILIZING THE SYSTEM BY TRACKING THE OBSERVATION DECOUPLED REFERENCE

As was established in Properties I-V the Observation Decoupled Reference Vector, $\{\delta_e, \dot{\delta}_e\}$, has no specific physical meaning until and unless it overlaps exactly with the system state in all of its components. Then and only then it identifies the stable equilibrium point of the system. At all times $\{\delta_e, \dot{\delta}_e\}$ is mathematically well defined by a diffeomorphism on the conventional state space $\{\delta, \dot{\delta}\}$. Furthermore $\{\delta_e, \dot{\delta}_e\}$ is smoothly time varying (except at instants of structural change of the system when it can be discontinuous). While it has no physical identity it is quantitatively known. Its individual coordinate components can be computed at the individual buses from measurements made at these buses. It then readily follows that if it were possible to monotonically drive the system state $\{\delta_i, \dot{\delta}_i\}$ toward the observation decoupled reference $\{\delta_e, \dot{\delta}_e\}$ without leaving the domain Ω_δ. Then eventually $\delta=\delta_e$, $\dot{\delta}=\dot{\delta}_e$ that is the stable equilibrium would be reached. During the chase, of course, δ_e remains a moving target which does not identify the equilibrium point until $\delta=\delta_e$ and $\dot{\delta}=\dot{\delta}_e$ is reached. Conversely by this the stable equilibrium point is found and reached without any prior knowledge of its location. This means that there is no need to execute a load flow during the control. This would be hopeless computationally any way on the available time scale. Only the existence of a stable equilibrium is required for the process to work. Note that $\delta=\delta_e$, $\dot{\delta}=\dot{\delta}_e$ also means that the conventional state space and state vector $\{\delta, \dot{\delta}\}$ is brought into coincidence with the Equilibrium Reference State Space and state vector $\{\delta_e, \dot{\delta}_e\}$. Then the homeomorphism connecting the two becomes the identity transformation.

6. THE BASIC CONTROL LAW

Now the Euclidean distance between $\{\delta, \dot{\delta}\}$ and $\{\delta_e, \dot{\delta}_e\}$ is

$$V = \frac{1}{2} \sum_{i=1}^{N} [(\delta_i - \delta_{ei})^2 + (\dot{\delta}_i - \dot{\delta}_{ei})^2] \quad (17a)$$

Clearly

$$V=0 \text{ iff } \delta=\delta_e=\delta_0, \ \dot{\delta}=\dot{\delta}_e=\dot{\delta}_0=0 \quad (18)$$

that is V=0 identifies the stable equilibrium point in domain Ω_δ.

Furthermore, δ_0, the equilibrium angles, are unknown until $\delta=\delta_e$, however $\dot{\delta}_0=0$ is known all along as the equilibrium velocity (deviation from synchronous). Thus it is possible to incorporate this into (17a) directly as

$$V = \frac{1}{2} \sum_{i=1}^{N-1} [(\delta_i - \delta_{ei})^2 + \dot{\delta}_i^2] \quad (17b)$$

It was shown in (Zaborszky, 1981a, 1981c) that (17a) <u>Alternative a</u> and (17b) <u>Alternative b</u> are equivalent conditions.

Consequently the following applies.

<u>BASIC CONTROL LAW</u>: Find control vector $u(\delta, \dot{\delta}, \delta_e, \dot{\delta}_e; t) \in R^r$, $r \le N$ 1.) such that either <u>Alternative a</u> (using (17a))

$$\dot{V} = \sum_{i=1}^{N} [(\dot{\delta}_i - \dot{\delta}_{ei})((\dot{\delta}_i - \dot{\delta}_{ei}) + (\ddot{\delta}_i - \ddot{\delta}_{ei})] \le 0 \quad (19a)$$

or <u>Alternative b</u> (using (17b))

$$\dot{V} = \sum_{i=1}^{N} [(\delta_i - \delta_{ei})(\dot{\delta}_i - \dot{\delta}_{ei}) + \dot{\delta}_i \ddot{\delta}_i] \le 0 \quad (19b)$$

is satisfied along the system trajectory $\{\delta(t), \dot{\delta}(t)\}$, all $t \ge 0$ and 2.) such that $\{\delta(t), \dot{\delta}(t)\} \in \Omega_\delta \times R^{N-1}$ all $t \ge 0$ (t=0 is the present time).

Control satisfying both conditions (that is $\dot{V} \le 0, \{\delta, \dot{\delta}\} \in \Omega_\delta R^{N-1}$ all t) will clearly lead the system to its stable equilibrium in Ω_δ. No proof is needed. It will have to be proven however that any proposed implementation of this law will satisfy the two conditions.

7. IMPLEMENTATION OF THE CONTROL LAW

Now it is necessary to explore if and how such a control law may be implemented. Since the control law is sufficient various gradations of sufficient implementation can be arbitrarily selected.

For instance the <u>Alternative a</u> condition

$$V_i = \frac{1}{2}[(\delta_i - \delta_{ei})^2 + (\dot{\delta}_i - \dot{\delta}_{ei})^2] \quad (20a)$$

$$\dot{V}_i = (\dot{\delta}_i - \dot{\delta}_{ei})[((\dot{\delta}_i - \dot{\delta}_{ei}) + (\ddot{\delta}_i - \ddot{\delta}_{ei}))] \le 0$$

all $i \in 1, 2, \ldots, N$ all $t \ge 0$ \quad (21a)

or the <u>Alternative b</u> condition

$$V_i = \frac{1}{2}[\delta_i - \delta_{ei}]^2 + \dot{\delta}_i^2 \quad (20b)$$

$$\dot{V}_i = (\delta_i - \delta_{ei})(\dot{\delta}_i - \dot{\delta}_{ei}) + \dot{\delta}_i \ddot{\delta}_i \le 0$$

all $i = 1, 2, \ldots, N$ \quad (21b)

are clearly sufficient to assure conditions (19a) or (19b) although more restrictive. (19a) or (19b) mearly require that the Eucledian distance V be monotonically reduced along the trajectory of the systems while (21a) or (21b) require each component associated with a subsystem to be so reduced. While this may seem too sufficient it will be shown that effective control is practical along these lines. Three grades of implementation will now be distinguished.

7.1 Grade 1 Implementation

Implement the Basic Control Law through condition (21a) or (21b) using strictly local means of measurement, computations and control at each unit i.

This approach has clear advantages. No communication between units or between units and the control center are required no central computations. The required local computations are simple so it is feasible to operate the control on the split second time scale required for stabilization. Also the control commands are of the "larger than" variety so crude devices like damping resistors and load skipping are applicable. Disadvantages are the requirement that control tools must be available at every generating station. The optimal size of resistor seems to be 1/4-1/3 the unit capacity. While this requirement may seem excessively expensive the cost could be more than offset by a truly major saving in transmission system expansion. Tenfold increases of critical clearing time were consistently found in simulation (Zaborszky, 1981a) thus loading of existing systems could be substantially increased without strengthening the system.

7.2 Grade 2 Implementation

Implement the Basic Control Law through conditions (21a) or (21b) combining local measurements and computation with central computation of the control commands for sparsely placed control tools.

The principal advantage of this solution over Grade 1 is that control tools are not needed at every generating station. In fact it will be shown that as few as 4 controls are sufficient to stabilize a 118 bus system. However there are major tradeoffs. A fast communication system is needed both ways between the generator buses and the control center which remains reliable in emergencies. There is need for central computation and in fact the overall computation load is much larger than in Grade 1 although apparently still quite feasible. The control tools must be capable of continuously varying control power. This calls for thyristor type devices which are certainly much more expensive than damping resistors although they are needed in smaller numbers. On the other hand if the system already contains a (multiterminal) HV-DC component this type of control is a natural way to utilize it for stabilizing the AC part of the system. In fact this approach seems to represent the first real solution for utilyzing the HV-DC system for stabilization.

7.3 Grade 3 Implementation

Implement the Basic Control Law through conditions (19a) or (19b) using a combination of local and central data processing and sparsely placed control tools.

This would be a direct implementation of the Basic Law without the additional requirement that each component of V monotomically decrease. The obvious advantage would be the more efficient use of the controls for stabilization, thus a larger class of stability cirses could be successfully handled. The disadvantage is the requirement of a much more complex central computation of control commands then in Grade 2 control. It is doubtful that such a computation could be fitted into available computer capacity on the short time scale (a very few seconds) required in stability crises. No control law for Grade 3 control was worked out or proposed so far.

Some details of the three Grades of control will now be surveyed then compared on the basis of simulation.

8. GRADE 1 CONTROL

Stabilization by strictly local feedback (Zaborszky, 1981a, 1981c, 1981d). The effectiveness of either of conditions (21a) or (21b) using strictly local feedback was previously demonstrated (Zaborszky, 1979) by simulation experiment. The second, or b, alternative, (21b), nevertheless seems better suited for the local feedback approach because it does not contain $\ddot{\delta}_{ei}$. This quantity is relatively hard to estimate accurately. Furthermore controls, u_i, at different generators are directly and instantaneously coupled in acceleration by (1) consequently less satisfactory coordination with the other buses should be expected in strictly local control when such control is based on $\ddot{\delta}_{ei}$. On the other hand coupling through δ_{ei} and $\dot{\delta}_{ei}$ as in case of (21b) depend dynamically rather than instantaneously on u_i. Consequently a u_i control applied at bus i will have gradually developing effects in δ_k, $\dot{\delta}_k$, δ_{ek}, $\dot{\delta}_{ek}$ at other buses k and will be detected there by local measurements.

Using condition (21a) i.e. alternative a, then $\ddot{\delta}_i$ can be eliminated from (21b) by (1) as follows:

$$V_i(\delta,\dot{\delta}) = [(\delta_i - \delta_{ei})(\dot{\delta}_i - \dot{\delta}_{ei}) + \dot{\delta}_i \frac{1}{J_i}(P_i - P_{Li} + \sum_{j=1}^{N} C_{ij}\frac{E_i E_j}{x_{ij}}\sin(\delta_j - \delta_{ei}) - u_i)] \quad (22)$$

This form then contains only a local cluster of state variables δ_i, $\dot{\delta}_i$, δ_j, δ_{ei}, $\dot{\delta}_{ei}$ and local imputs, P_i, P_{Li}, u_i all known essentially instantaneously at the local bus i. The remaining difficulty is that δ_i, δ_j, δ_{ei} depend on the entire system state and on the entire system dynamics, at least the former of which is unknown at locally at bus i. Yet predicting the trajectory is necessary to assure that the trajection will stay in Ω_δ. Fortunately the power system also has a network structure which assures that distant events or variables cause little local effect. This allows the establishment of a reasonably low bound value for the dynamic influence of locally unknown states. Let

$$\delta_{ei} = g_i(\delta), \quad \dot{\delta}_{ei} = \frac{\partial g}{\partial \delta}\dot{\delta} = \sum_{j=1}^{N} C_{ij}\frac{\partial \delta_{ei}}{\partial \delta_j}\dot{\delta}_i \quad (23)$$

since g_i as defined implicitly in (6) by $G_i=0$, is a diffeomorphism by Property III. But by (6) and (14)

$$\frac{\partial \delta_{ei}}{\partial \delta_j} = -\left[\frac{\partial G_i}{\partial \delta_{ei}}\right]^{-1}\frac{\partial G_i}{\partial \delta_j} = \begin{cases} 1 & j=i \\ \dfrac{\dfrac{E_i E_j}{x_{ij}}\cos(\delta_{ie}-\delta_j)}{\sum_{\substack{k=1 \\ k\neq i}}^{N}\dfrac{E_i E_k}{x_{ik}}\cos(\delta_{ei}-\delta_k)} & j\neq i \end{cases}$$

(24)

Thus

$$\frac{\partial g_i(\delta)}{\partial \delta_i} = 1 \quad (25)$$

$$\left|\sum_{j=1}^{N} C_{ij}\delta_{ej}\frac{\partial g_j(\delta)}{\partial \delta_i}\right| \leq ||\delta_{ej}||\cdot\left|\sum_{j=1}^{N} C_{ij}\frac{\partial g_j(\delta)}{\partial \delta_i}\right|$$

$$\leq \frac{\pi}{2}\left|\sum_{j=1}^{N} C_{ij}\right| \triangleq K_i \quad (26)$$

Note that $\sum_{j=1}^{N} C_{ij}$ is the number of neighboring buses directly connected to the local bus i. This number is about 4 on the average. Equation (22) then simplifies to

$$J_i \dot{V}_i(\delta,\dot{\delta}) \leq \dot{\delta}_i(v_i - u_i) \quad (27)$$

where

$$v_i = (P_i - P_{Li}) + J_i(\dot{\delta}_i - \dot{\delta}_{ei}) +$$

$$\sum_{j=1}^{N} C_{ij}\frac{E_i E_j}{x_{ij}}\sin(\delta_j - \delta_{ei}) - K_i \operatorname{sgn}\dot{\delta}_i \quad (28)$$

Which leads to proposing the following

CONTROL LAW FOR STRICTLY LOCAL CONTROL (GRADE 1 CONTROL)

$$u_i > v_i \text{ if } \dot{\delta}_i > 0$$
$$u_i < v_i \text{ if } \dot{\delta}_i < 0$$

(29)

---

where u_i is an otherwise arbitrary, continuously differentiable function of $\{\delta,\dot{\delta}\}$ in $\Omega'_\delta \times R^{-1}$.

By the preceding control law (21b) will assure $\dot{V}\leq 0$ at all $\{\delta,\dot{\delta}\}$ where $\{\delta_{ei},\dot{\delta}_{ei}\}$ exists. It is not apparent however that it will be possible at all future instances to assure $\dot{V}<0$ because it is not clear that under control law (29) the trajectory will not leave the domain of existence for $\{\delta_e,\dot{\delta}_e\}$. This latter fact requires proof and such proof was presented earlier in (Zaborszky, 1981d).

Since the control laws represent sufficient conditions, various versions beside the one shown in (29) can be proposed as has been done in (Zaborszky, 1981a, 1981c, 1981d).

Control law (29) cannot be carried out unles there is an appropriately strong control, u_i, available at all i=1,2,...,N. If (29) is not satisfied for some i, stabilization would not be guaranteed (although it is probably feasible to pinpoint buses by off line simulation experiments where the control is relatively inactive and thus disposable).

9. GRADE 2 CONTROL

Stabilization by feedback using centrally additionally coordinated sparse controls.

Coupling between controls at the various units is direct and instantaneous in $(\ddot{\delta}_i - \ddot{\delta}_{ei})$ and hence effective central coordination of the controls could be accomplished easier using alternative a, or condition (21a).

Here $\delta,\dot{\delta}$ are known from ultrafast state estimation (Zaborszky, 1980). $\ddot{\delta}$ is defined by the dynamic equation (1). It is then possible between (1) and (9)-(12) to obtain a solution in the form

$$\ddot{\delta}_i - \ddot{\delta}_{ei} = f_i(\delta,\dot{\delta}) + \sum_{j\neq i} F_{ij}(\delta,\dot{\delta})(\frac{1}{J_j}u_j - \frac{1}{J_i}u_i)$$

(30)

Furthermore while f_i and F_{ij} as stated, are functions of the entire state, $\{\delta,\dot{\delta}\}$, it turns out that, because of the structure of the system, their instantaneous values $f_i(t)$, $F_{ij}(t)$ at time t can be computed from the knowledge of a limited local cluster of states, that is

$$f_i(t) = f(\delta_i(t),\delta_j(t),\dot{\delta}_i(t),\dot{\delta}_j(t)), \quad (31)$$

$$F_{ij}(t) = F_{ij}(\delta_i(t),\delta_j(t),\dot{\delta}_i(t),\dot{\delta}_j(t)) \quad (32)$$

where j are all buses connected to bus i, and k are all the buses connected to buses j. In other words knowledge of the state components for two tiers of buses around i make it possible to compute $\ddot{\delta}_i - \ddot{\delta}_{ei}$ at bus i. Furthermore explicit equations for f_i and F_{ij} can be obtained (Zaborszky, 1981b). Since the state $\{\delta,\dot{\delta}\}$ would be known from ultrafast estimation. (Zaborszky, 1980), $(\ddot{\delta}-\ddot{\delta}_{ei})$ can be computed unit by unit from state values on the two first tiers around the unit. Hence the instantaneous value of $\dot{V}_i$ can also be computed from (21a). Then controls can be selected to keep $\dot{V}_i\leq 0$.

Since the controls work on the acceleration level their effect is instantaneous. This process can be condensed (Zaborszky, 1982a) into the computation of the coefficients A and b for the following:

$$\boxed{\begin{array}{c}\text{CONTROL LAW FOR SPARSE}\\ \text{CONTROL TOOLS (GRADE 2 CONTROL)}\\[4pt] Au \leq b \ , \ A = [A_{ij}] \ , \ b = b_i]\\[4pt] A \in R^{m(N-1)} \ , \ u \in R^{N-1} \ , \ b \in R^m\end{array}}$$

(33)

where u has sparse nonzero values in R^{N-1}. m represents the combined number of inequality and equality constraints.

The inequality constraints are of the form

$$U_i^m \leq u_i \leq U_i^M \tag{34}$$

which usually are applicable to the controls. Equalities like

$$\sum_{i \in D} u_i = 0 \tag{35}$$

which would apply for instance to the converter currents of an HV-DC system that may represent a group of controls for this approach.

The type of equation represented by Grade 2 Control Law (33) is not necessarily easy to solve but fortunately an algorithm was proposed recently which provides a way for solution on the order of a computation time of 10 microseconds for systems of a few hundred buses (Fan, 1956; Hiebert, 1980).

Following the control law (33) then will assure $\dot{V}<0$ at all instants of time but it does not per se assure that future segments of the trajectory will not escape Ω_δ where δ_{ei} exists. This latter fact must be proven independently as is done in (Zaborszky, 1982a).

10. COMPARISON OF PERFORMANCE UNDER STRICTLY LOCAL (GRADE 1) AND CENTRALLY COORDINATED SPARSE (GRADE 2) CONTROLS

Two grades of control are introduced in the preceding: Grade 1 by strictly local feedback which depends on simple control tools (like resistors) at each generating unit and Grade 2 which may use only a few control tools of a more sophisticated nature (thyristor devices such as HV-DC converters) along with some central processing and communication between units and control center. The effectiveness of each of these approaches was previously demonstrated by simulation (Zaborszky, 1979, 1980, 1981a, 1981c, 1981d, 1982a). A direct comparison of the two follows. The results shown for Grade 2 control here were not published previously.

The IEEE 118 bus system is shown in Fig. 3. (Data listed in (Zaborszky, 1979) among other sources). Since this system in its original form is exceptionally stable it was necessary to modify it for studies of stability crisis. This was accomplished by increasing the impedances on respectively lines 15-33, 15-34, 23-24 and/or lines 70-74, 69-74, 69-77, 69-80 by a factor of ten. This then results in a system consisting of two or three very loosely coupled areas which actually too little basic static stability to be operated in the context of state of art operation and protection practices. The simulation runs to be presented illustrate that such weak systems could be operated if emergency control, as proposed here,

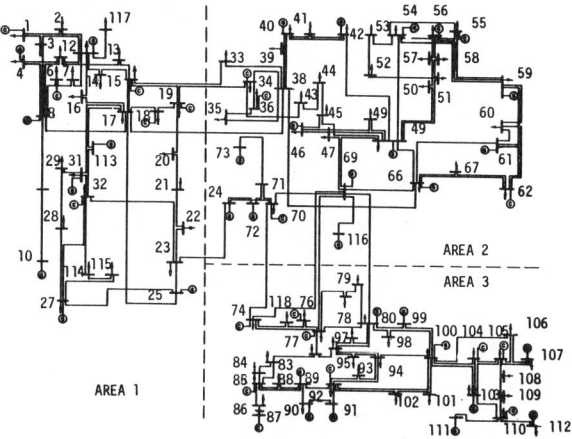

Fig. 3. The modified IEEE test system.

is used to handle stability crises. Another modification on the 118 bus system is the elimination (as an approximation) of very short branches such as single transformers between buses since these tend to bias undesirably the meaning of the observation decoupled reference, the basis of the control.

A complete dynamic simulation of the bus angle model (Zaborszky, 1979) of the 118 bus system was used with the Philadelphia Electric Co. Stability Program to obtain the swing curves of Figs. 4-12. The control was computed using the equations presented in this paper for the, approximate, machine angle model. Subroutines written for this latter purpose were inserted between integration steps of the Philadelphia program.

Figures 4-12 will now be presented in sequence. Comparison and discussion is provided by the captions.

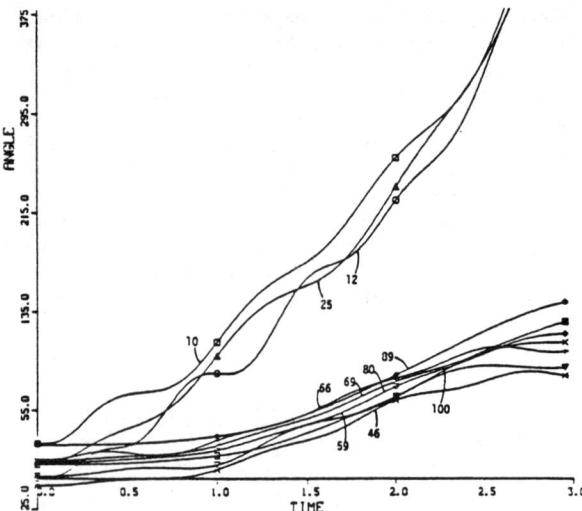

Fig. 4. 12 cycle 3 phase fault at bus 17. Line 17-38 and 300 MW load at bus 17 lost. System not controlled. Rotor angle history for ten machines from the system identified by their bus numbers. Area 1 separating from the rest of the system after about 1 second.

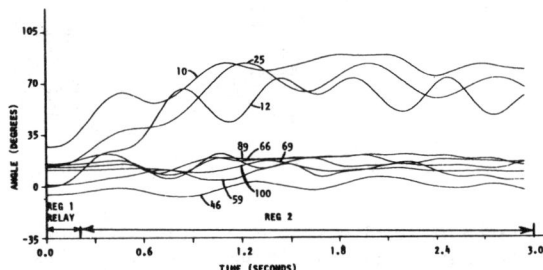

Fig. 5. Same scenario and rotor angle history for same 10 machines as in Fig. 4 but with Grade 1 control utilyzing damping resistors of about 1/3 of unit power rating at each unit and load skipping at each unit. System stability is preserved but system is not viable - area 1 is running about 70° ahead of the rest of system because of loss of 300 MW load. This gives insufficient static stability. The system needs to be made viable.

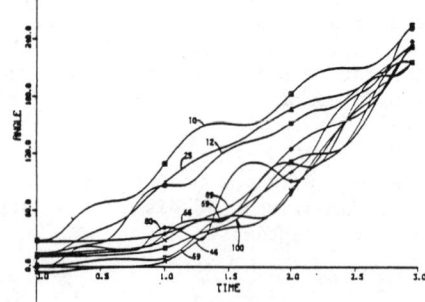

Fig. 6. Same scenario and rotor angle history for the same 10 machines as in Fig. 4 but with Grade 2 control consisting of four HV-DC converters as indicated below. System stabilized but accelerating because of 300 MW load loss on Area 1. The HV-DC system does not inject net power so it is unable to control net acceleration unlike resistors in Fig. 5 under Grade 1 control. On the other hand 70° torque angle is eliminated between Area 1 and the rest of the system through redistribution of power in HV-DC. Control locations 12, 25, 66, 80 with respective p.u. load limits of (-20,+20), (0,20) (-20, +20).

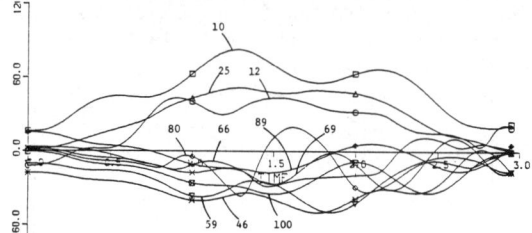

Fig. 7. Same as Fig. 5 but center of angle subtrackted. Angular history of relative machine angles under Grade 2 control by four HV-DC terminals. Stabilizing effect is equal to Grade 1 control in Fig. 5.

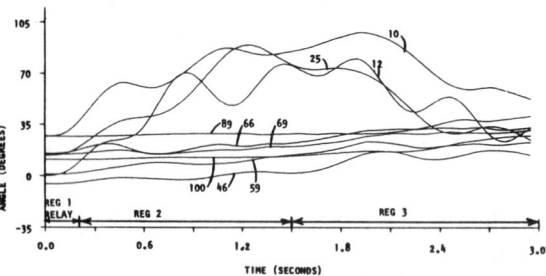

Fig. 8. Same as Fig. 5 but 300 MW generation runback between 1.75 and 2 sec. for viabilization. 70° torque angle for Area 1 eliminated.

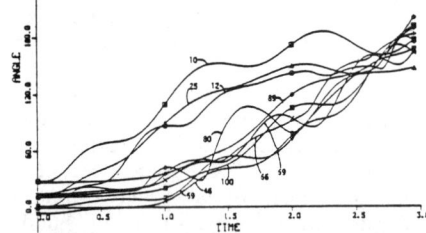

Fig. 9. Same as Fig. 6 but 300 MW generation runback between 1 and 2 seconds. Acceleration is eliminated but average angle still increasing linearly because of off synchronous speed. This average angle relative to synchronous reference has no physical significance whatever. Frequency increment would soon be eliminated by Automatic Generation control.

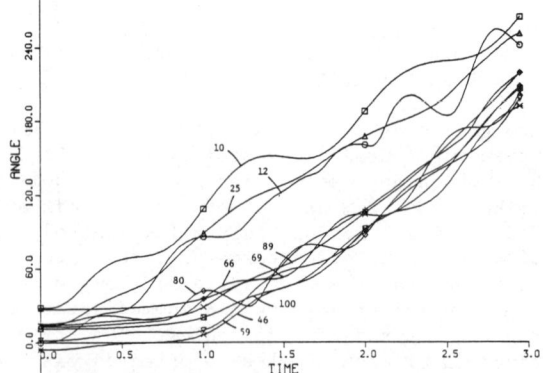

Fig. 10. Same as Fig. 5 but loading limit on HV-DC converters reduced to 10 pu. Stabilization effect is becoming marginal.

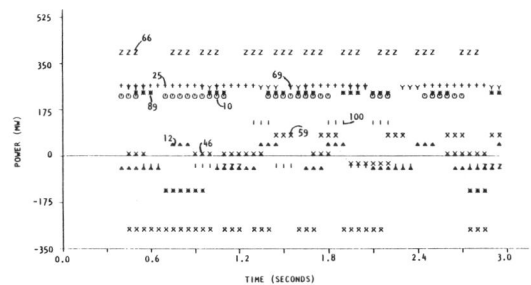

Fig. 11. History of control actions for Grade 1 control at each machine. Same scenario as in Fig. 5, control history given for the 10 machines in Fig. 5.

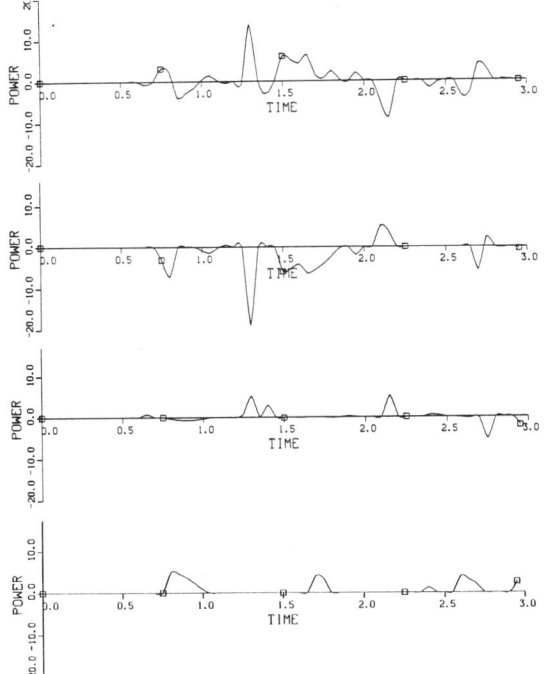

Fig. 12. History of the four controls in Grade 2 control. Same scenario as in Fig. 6. Control better coordinated and smoother than in Fig. 11, as expected because of central coordination and continuously variable controls.

11. CONCLUSION

It is shown in this paper that tracking the Observation Decoupled Reference, or aligning the conventional State space with the Equilibrium Reference Statespace is an effective means for stabilizing a power system in a severe stability crisis, whether control is used at every machine under strictly local feedback (Grade 1) or just a few controls (4 in the example) are used under central control (Grade 2). In fact the simulation experiments indicate essentially equivalent results for either Grade 1 or Grade 2 control. The tradeoff is that Grade 1 needs a control tool at every generating unit but no communication to the control center and no central processing. Furthermore the control tools used in Grade 1 control can be simple and relatively inexpensive such as damping resistors. Grade 2 requires sophisticated, continuously variable controls as exemplified by thyristor devices but in small numbers. It also requires fast communications from all units to the center which will remain reliable even during emergencies.

The choice between the two approaches would seem to depend on the circumstances of the particular application but if an HV-DC system (one line or multiterminal) is available then Grade 2 control could be very effectively used as illustrated by simulation here. In fact it is believed that this represents the first real solution for utilyzing HV-DC for stabilizing the power system.

Thus the Observation Decoupled Reference has some valuable characteristics which can be readily exploited for stabilizing the power system. It is probable that other large nonlinear systems which have a structure consisting of individual dynamic subsystems with network type coupling could be similarly treated.

Fan, K. (1956). On systems of linear inequalities. In H. W. Kuhn, and A. W. Tucker (Ed.), Annals of Math Studies, #38 (1956), Linear Inequalities and Related Systems. pp. 96-156.

Hiebert, K. L. (1980). Solving systems of linear equations and inequalities. SIAM J. of Numerical Analysis, 17, 447-464.

Zaborszky, J., G. Huang, and L. J. Chiang (1982a). On line stabilization of the large HV-AC-DC power system by using observation decoupled reference as a system wide target. Submitted to 1982 IEEE PES Winter Meeting, also SSM Report No. 8203.

Zaborszky, J., K. V. Prasad, and K. W. Whang (1981a). Stabilizing control in emergencies, part II: control by local feedback. IEEE Trans. on PAS, May 1981, 2381-2389.

Zaborszky, J., K. V. Prasad, K. W. Whang, M. Ilic-Spong, G. Huang, and L. J. Chiang (1982b). towards a comprehensive analysis and operating practice of the large compound HV-AC-DC systems. DOE Final Report, Feb. 1982, Contract No. ED-78-D-01-3090.

Zaborszky, J., and K. W. Whang (1981b). Future role of the control center in the emergency state control of the power system. IFAC 8th Triennial World Congress, Vol. XX, pp. 41-48, Aug. 24-28, 1981, Kyoto, Japan.

Zaborszky, J., K. W. Whang, and K. V. Prasad (1979). Monitoring, evaluation and control of power system emergencies. Systems Engineering for Power: Emergency Operating State Control, Davos, Switzerland, Sep. 30, 1979 to Oct. 5, 1979.

Zaborszky, J., K. W. Whang, and K. V. Prasad (1980). Ultra fast state estimation for the large electric power system. IEEE Trans. Autom. Control, 25, 839-841.

Zaborszky, J., K. W. Whang, and K. V. Prasad (1981c). Stabilizing control in emergencies, part I: equilibrium point and state determination. IEEE Trans. on PAS, May 1981, 2374-2380.

Zaborszky, J., K. W. Whang, K. V. Prasad, and I. N. Katz (1981d). Local feedback stabilization of the large interconnected power system in emergencies. Automatica, 17, No. 5, 1981, 673-686.

DIGITAL CONTROL APPLIED TO POWER SYSTEM PROTECTION

Y. Sekine* and T. Matsushima**

The University of Tokyo, Tokyo, Japan
**Toshiba Corporation, Tokyo, Japan*

Abstract. Recently, microprocessor-based protective relay (digital relay) has reached the level of practical use. The digital relay is expected to be a major weapon against various kinds of problems in the power systems. This paper summarizes the trends in digital relay technology and several applications are described to protections of transmission lines, power system stability, and frequency drop as well as to fault location.

Keywords. Digital relay; Microprocessor; Power system protection; Transmission line; Power swing; Self-monitoring.

INTRODUCTION

Recent advancement in the microcomputer field has had a great impact on every phase of industry, including electric power system protection.

Increased demand for electric power has brought changes in the power supply structure, with the introduction of large remote power stations and long-distance multi-terminal transmission lines. The difficulty of protecting this power system consequently has increased, a problem compounded by the need for more reliable, high-performance protective facilities. For a power system to meet these requirements, digital relays using microprocessors will be needed and many studies have been done worldwide on this during the last 10 years.[1] From a viewpoint of practical use, however, most of the studies were not sufficiently advanced.

In Japan, the need for digital relays is great because of the severe power system environment. Several studies of digital relays are now being conducted, in Japan, and some digital relays are already used practically, making good use of the following general features:

i) Higher performance with the intelligent functions of microprocessors.
ii) Automatic self-monitoring, bringing higher reliability by detecting any equipment trouble.
iii) Reducing size of facilities by replacing wired logic with software.
iv) Cutting costs with standardized hardware and software.

This paper describes the present technical state and some application examples of digital protective equipment.[2]

TRENDS IN DIGITAL RELAY TECHNOLOGY

The functions of digital relays are mainly realized through software, while standardized hardware can be applied widely to various kinds of relays. In this section, current and future trends in digital relay technology are described from a hardware viewpoint.

Digital relay design

Figure 1 shows a block diagram of a general digital relay. Design features are briefly described below:

i) Analogue filter. In general, higher harmonics contained in sampled data cannot be distinguished from the lower frequency component. For example, a (720-60) Hz component sampled at 720 Hz cannot be distinguished from a 60 Hz component. The analogue filter must eliminate the higher harmonics from input signals to avoid this type of aliasing error.

ii) Sampling. The 30 electrical degree sampling interval is widely used because it is well suited to relay characteristic calculations. However, there are some cases where the sampling interval of 90 or another electrical degree is chosen. It is easy to get any phase angle data with software using the two data whose phase difference is 90 electrical degrees. Each voltage and current datum in the same equipment should be sampled synchronously. Synchronously sampled data are suitable for relay calculation.

iii) A/D converter. In many cases,

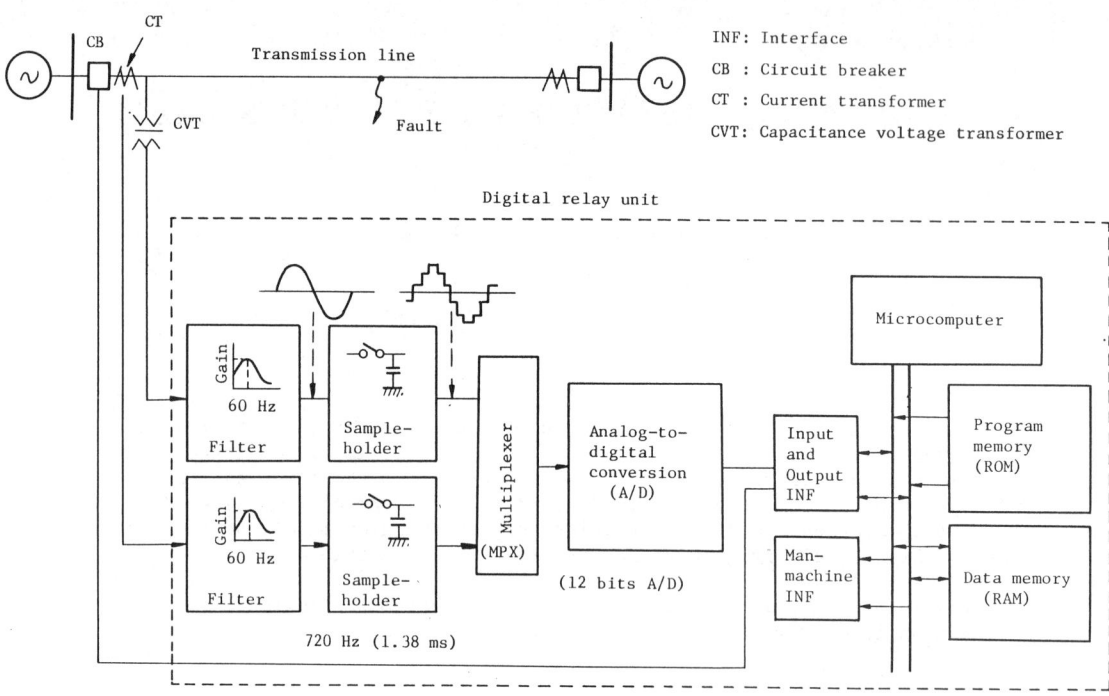

Fig. 1 Basic block diagram of digital relay unit

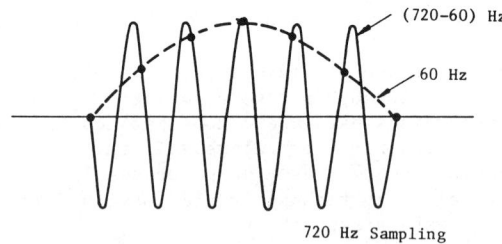

Fig. 2 Aliasing error

Table I Microcomputer performance

Items	Performance
Process technology of LSI microprocessor	Shottky TTL
Control structure	Microprogram control
Word length	16 bits
Number of basic instructions	53
Basic instruction cycle	0.25 µs
Add execution time between registers	0.5 µs
Multiply execution time between registers	4.95 µs

analogue data are converted into 12-bit (sign + 11 bits) bytes. The least significant digit of a byte is about 0.05% of the full scale. Considering the error level of analogue circuits, 12-bit data length is appropriate.

iv) Microcomputer. An example of specifications for a microprocessor and memories is shown in Table 1.

Calculating speed and function

Digital relays are real time systems. A microprocessor executes a program using the newest several data samples on input voltage and/or current to calculate relaying characteristics. In the case that 60 Hz input voltage and current are sampled at 30 electrical degree intervals, one sampling interval is 1.39 ms. From the viewpoint of operating speed, it is desirable to complete the relaying program in one sampling interval. Thus the calculation speed of a microprocessor directly affects operating time and the number of elements calculated in a digital relay, so the fastest type of microprocessor should be used.

Automatic self-monitoring

Automatic self-monitoring plays a very important role in a digital relay. Relaying equipment does not produce a tripping signal except in rare cases where faults occur in the power system. So, it is very significant to monitor themselves automatically in normal operating state. Even conventional analogue type relaying equipment manufactured today is equipped with effective self-monitoring systems. But, the digital relay has more sophisticated self-monitoring functions for the following reasons:

i) Higher and more complex functions are performed by the microprocessor.

ii) Input data are recognized precisely up to the least significant digit level.

Table II shows self-monitoring procedures. All troubles in a digital relay can be found easily through these.

Table II Self-monitoring functions

Hardware block	Diagnosis	Check contents
Analog input	Zero sequence detection	Summation of three phases, Va+Vd+Vc=3Vo
Microcomputer	Diagnosis program	Function confirmation of all instructions
Memory	ROM check	Summation of all ROM content
	RAM check	Read-in and read-out of constants
Manmachine interface	Memory comparison	Comparison of duplicated memory contents
	Range and rationality check	Range check of set values and big-small check between set values
Tripping circuit	Circuit connection check	Confirmation of closed circuit by flowing small current
Total system	Continuous monitoring of relay output	Confirmation if relay output is initiated over constant time
	Automatic inspection	Confirmation if relays operate normally by using artificial input

Heat problems and microprocessor selection

The first requirement a microprocessor for the digital relay should satisfy is high calculating speed. Today the fastest available microprocessor is a bipolar type. A bipolar microprocessor, though fast, consumes a relatively high power and is not suitable for higher integration. Heat caused by high power consumption influences microprocessor life and reliability.

Several kinds of cooling systems are now used in digital relays. The heat pipe method takes heat from the IC (integrated circuit) surface to a radiating fin on the outside of the equipment through heat pipes. The heat pipe is filled with a heat medium and has high heat conductivity. The heat sink method disperses heat directly to printed circuit boards or equipment frames. But neither method is a perfect solution to heat problems.

Recent advances in integrated circuit technology are great, and the same is true regarding the calculating speed of MOS (metal oxide semiconductor) microprocessors. While today's MOS microprocessor has relatively slow calculation speed, it consumes less power and is suitable for higher integration. Fig. 3 shows the development of MOS microprocessor calculation speed. The figure predicts that MOS calculation speed will equal that of today's bipolar unit in a few years. By replacing bipolar microprocessors with MOS types, heat problems will be greatly reduced.

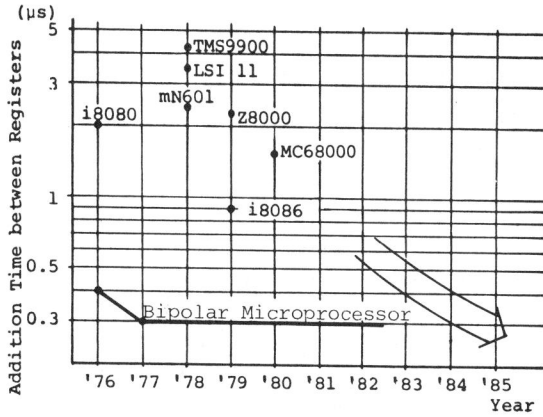

Fig. 3 Yearly change of the addition time between registers of microprocessors

Surge withstand capability

Digital relays are installed in a very high voltage and current environment, while they use very low-level (5 V) and high-speed (about 10 MHz) signals. Thus these fundamental countermeasures against surges are important:

i) Using surge-absorbing capacitors for CT and CVT circuits.
ii) Insulating trip or other input/output signals with auxiliary relays or photocouplers.
iii) Separating clean signal circuits from power lines.
iv) Using shielded or twisted pair wires for fine signals.

Field experience shows that there is no significant difference between conventional static and digital relays with these fundamental countermeasures.

DIGITAL RELAY APPLICATIONS

Today's electric power systems suffer various kinds of problems and newly developed digital relays are expected to be a major weapon against them. Many universities, utilities and manufacturers have studied how to thus utilize the digital relay and some of the digital relays have already been put to practical use.

Current differential protection for EHV transmission lines

Recent changes in the power system have a tendency to make protection of transmission lines difficult. For example, on a multi-terminal line, fault current can flow in an external circuit from one of the terminals even in an internal fault case. When this happens, conventional protection methods such as directional comparison, phase comparison method are useless. Current differential protection is almost perfect for multi-terminal transmission lines. But, in the past, it was not

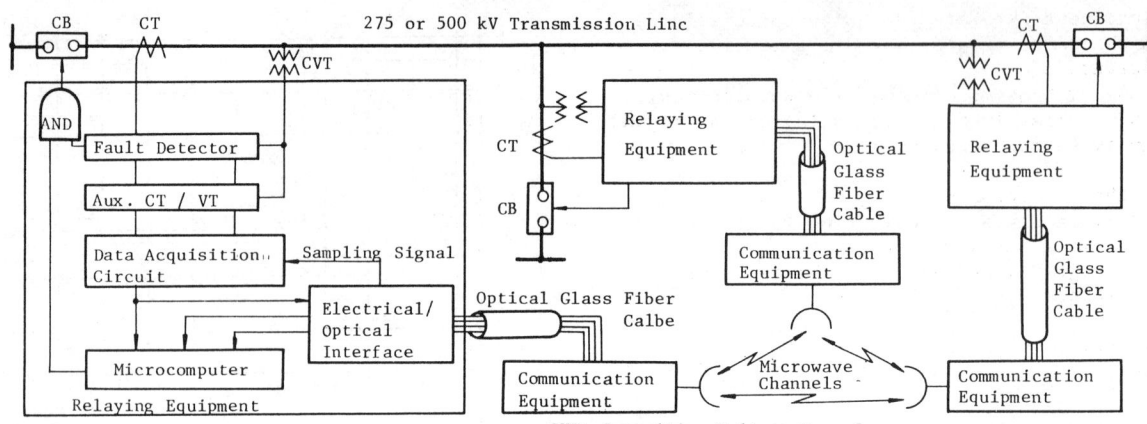

Fig. 4 System configuration of the current differential relaying system (CDR).

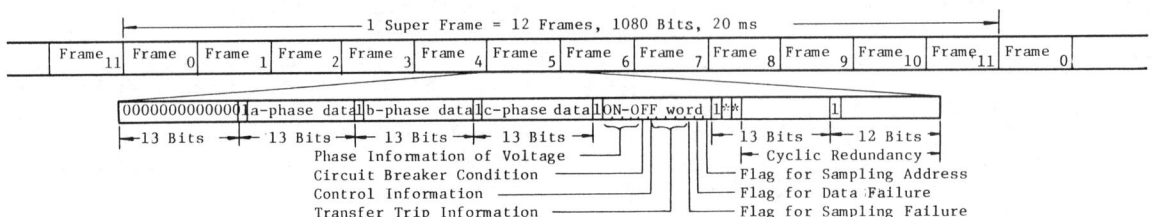

Fig. 5 Data transmission format for CDR.

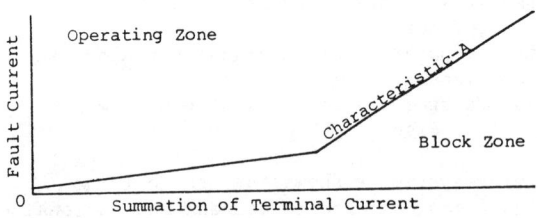

Fig. 6 Ratio differential characteristics of the CDR.

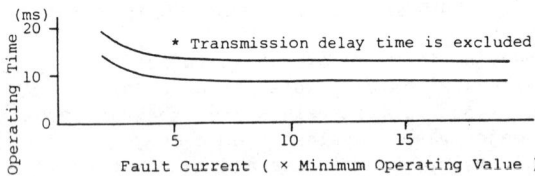

Fig. 7 Operating time characteristic of the CDR (50 Hz).

practical for long-distance lines because of difficulties in current wave data transmission.

Fig. 4 shows the current differential protection system for EHV transmission lines that was completed in 1980 for practical use. The system employs PCM (pulse code modulation) method for current wave data transmission and a microprocessor for executing differential relay programs. Principal features of the system are:

i) Current waveform data of each terminal are sampled synchronously to facilitate current vector summation. Errors of sampling timing among terminals are kept within 10 μs.

ii) The data transmission format is shown in Fig. 5. Redundant checking bits ensure high-quality data transmission.

iii) Transmitted data are treated as digital data by a microprocessor. Thus the system exhibits no more errors and needs no extra conversion time.

iv) Any system trouble can be detected by the automatic self-monitoring function that is realized by the high performance microprocessor.

This current differential relay (CDR) has a ratio characteristic calculated from equation (1), where the vector sum of each terminal current functions as an operating quantity and the scalar sum functions as a restraining quantity. Equation (1) is calculated by a microprocessor using sampled data of each terminal current.

$$|\dot{I}_A + \dot{I}_B + \dot{I}_C| - K_1(|\dot{I}_A| + |\dot{I}_B| + |\dot{I}_C|) - K_0 > 0 \quad (1)$$

where $\dot{I}_A, \dot{I}_B, \dot{I}_C$: Current of each terminal
K_1, K_0 : Constant

Fig. 6 shows the characteristics. In a large current zone, the ratio is set larger so that CT errors may increase. Operating time with appropriate fault current is within 20 ms including 5 ms transmission time lag. With this current differential relay, multi-terminal transmission lines are protected reliably.

Distance relay

Distance protection, which detects a fault point by the impedance of a transmission line from a relay point, is one of the most fundamental protections.

In the future, when digital relays are widely used, distance protection will still serve an important function as a backup protection using local voltage and current information.

Digital distance relaying equipment is expected to have the following advantages:

i) Improved performance on distance measuring and operating time, etc.
ii) Reduced equipment size
iii) Automatic self-monitoring to maintain high reliability and reduce maintenance time and maintenance cost.

Two examples of digital distance relays are described below.

Example 1. Relaying elements for 10 transmission lines are calculated in one microprocessor. The equipment is applied to a 77 KV resistance grounded system for practical use. Relaying characteristics are simple as shown in Fig. 8.

Example 2. Minute simulation of a certain power system indicated that the characteristics shown in Fig. 9 are proper for the system. The characteristics were easily realized with a microprocessor and the equipment was applied to a 275 KV direct grounded system for a field test. [6]

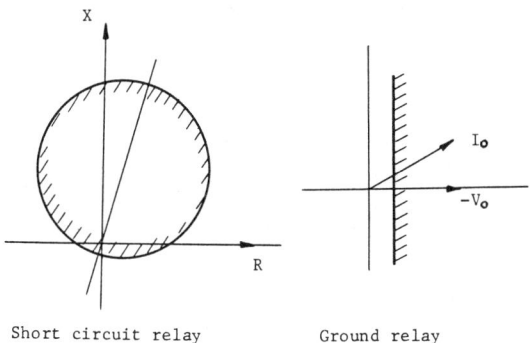

Fig. 8 Distance relay characteristics (I)

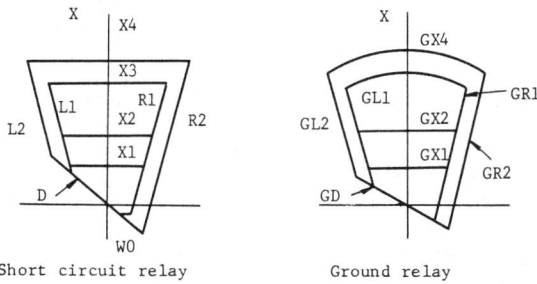

Fig. 9 Distance relay characteristics (II)

Fault locator

Fault locators, which measure the distance to a fault point on a transmission line, conventionally used the following methods. One is to detect the surge voltage caused by a fault at both ends of a transmission line and the time difference of the surge detected between the two terminals determines the fault point. Another method is to generate pulse signals and transmit them when a fault occurs and the pulse return time from the fault point determines the fault point. However, these are not economical systems because they require data transmission facilities or blocking coil equipment. Recently, a simple fault locating method using one terminal's voltage and current data has been studied and the prototype equipment using a microprocessor was put into a field test. [7,8,9]

The fundamental equation is shown in Eq. (2).

$$x = \frac{\text{Im}(V_s \cdot I_s''^*)}{\text{Im}(Z\, I_s \cdot I_s''^*)} \qquad (2)$$

where
- x : distance to fault point
- V_s : voltage at terminal
- I_s : current at terminal
- I_s'': difference between pre-fault and post-fault currents (fault component current)
- Z : transmission line impedance per unit length
- $*$: conjugate component
- Im : imaginary component

Equation (2) is given by solving the wave equation. The following assumptions are made to solve the wave equation:

i) The transmission line is sufficiently short (namely $\tanh \gamma x \doteqdot \gamma x$, where γ is a propagation constant)
ii) The angles of fault point voltage V_F and fault component current I_s'' of the locating terminal are equal.

In Eq. (2), I_s'' eliminates an error component caused by fault resistance and load flow and microprocessor memory functions make it possible to calculate the difference current I_s''.

The microprocessor in this equipment calculates compensations for several error components.

i) Effect of other phase currents. Each phase transmission line has self and mutual impedances. Both self and mutual impedances, whether they are balanced or not, are used to compensate the effect of other unfaulted phases currents.
ii) Approximation error. Error caused by the approximation $\tanh \gamma x \doteqdot \gamma x$ cannot be neglected for a transmission line longer than 100 km. Error is compensated automatically using a simplified formula.
iii) Effect of fault current in an overhead

ground wire. some ground fault current returns through the earth and the rest through the overhead ground wire. The ratio changes according to a fault point, and this change causes the distance measuring error, which is compensated using an off-line simulation result.

This fault locator is showing excellent results in the field test.

Power swing protection

Interconnection of electric power systems is highly beneficial not only because of the economic merits due to peak demand leveling but also the reliability merits due to emergency power supply. However, interconnection may sometimes cause fault cascading bringing the whole system into blockout. A small power swing in one part of the power system may grow and propagate to neighboring power systems finally causing the collapse of the whole system. To prevent such phenomena, power swing protection equipment has been developed using a microprocessor.

The principle of the protection is to recognize peaks and bottoms of power swings by observing power flow at the tie line. When the pattern of the power swing satisfies certain conditions stated below, the equipment deliver a signal to trip off the tie line. The tripping condition is that the amplitude of the power swing (from bottom to peak and peak to bottom) exceeds a predetermined value and continues to increase for three swing cycles successively. This condition may be changed depending upon given power system constants.

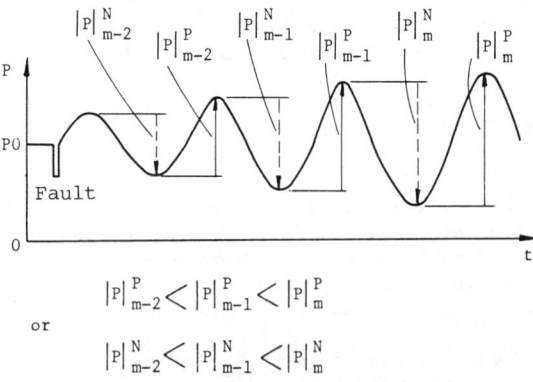

Fig. 10 Principle of power swing protection

Out-of-step prevention system

Fig. 11 shows a model power system, where groups A and B are large generator blocks, and groups C and D are load blocks with pumping-up power stations and small scale power plants, respectively.

If a CB (circuit breaker) failure occurs at the substation E after a transmission line fault and the line fault cannot be cleared for a long period, out-of-step between group A and the other groups is anticipated. In this case some generators in group A should be shed. This out-of-step prevention system determines the appropriate shedding procedure for CB failures.

Fig. 12 shows that a shedding value is determined according to pre-fault power flow and the kind of fault. Generators to be shed are determined in advance and when a fault with CB failure occurs these are tripped immediately.

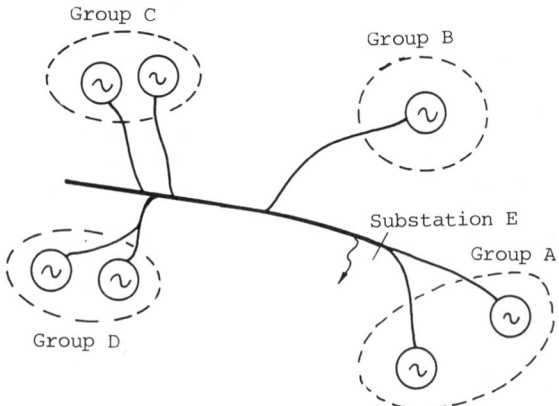

Fig. 11 Model network

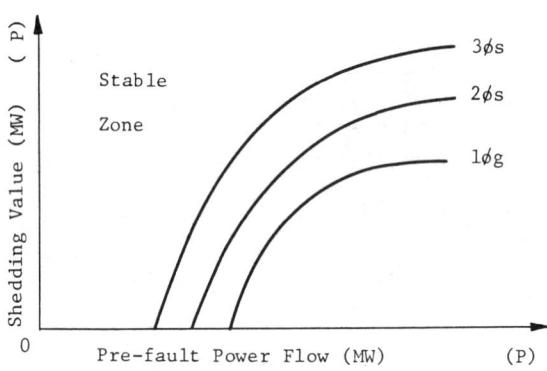

Fig. 12 Power shedding pattern

Frequency relay

Advanced industrial plants and nuclear power stations require high-quality frequency control in the power system from both demand and supply sides. In case a generator group is shed by some chance causing unbalance between supply and demand, a certain load should be shed preceding additional generator shedding by detecting the frequency drop accurately. With these backgrounds, high-performance frequency relays have been developed using a single-chip microcomputer and are now commercially used.

Fig. 13 shows a block diagram of the system. The principle is to measure the zero-crossing interval of input voltage. Detecting accuracy is ±0.05 Hz using a 1 MHz clock. Four cycles of confirmation time prevents false operation due to temporary power system fault or reclosing.

Single-chip microcomputers with built-in program memories and random access memories are compact, but their applications are now rather limited only to simple cases because of limitations of memory size and calculation speed.

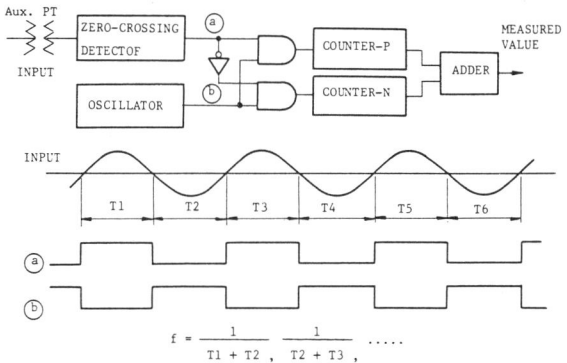

Fig. 13 Block diagram of the digital under frequency relay

CONCLUSIONS

Current technology and applications of digital protective equipment have been described. Many engineers are studying how to make further use of microcomputers to solve various problems in power system protection. Microprocessors continue to improve in regard to calculation speed, size, power consumption and other functions. Considering this, we should note that digital protective equipment is still in the beginning stage. From now on, their application should be expanded with the development of other related equipment such as current transformer, voltage transformer and fiber optics.

REFERENCES

1. D. I. Rummer, M. Kezunovic, "A Survey and Classification of the Digital Computer Relaying Literature". IEEE PES Summer Meeting, Paper No. A79 417-7, Vancouver, B.C., July, 1979.

2. Y. Sekine, T. Tamura, et. al. "Microprocessors in Electric Power Systems" International Journal of Electric Power and Energy Systems, Jan., 1981.

3. T. Takagi, M. Yamaura, "Digital Differential Relaying System for Transmission Line Primary Protection Using Travelling Wave Theorem" IEEE. PES Winter Meeting, Paper No. A79 096-9, New York, N.Y., Feb., 1979.

4. Y. Akimoto, et. al "Microprocessor Based Digital Relays Application in TEPCO" IEEE PES Winter Meeting, 81 WM 119-7, July, 1980.

5. Y. Akimoto, et. al. "Digital Current Differential Carrier Relaying System for EHV Transmission Line" IFAC '81 Kyoto Congress CS-2.3.1.

6. M. Kamiya et. al. "Development of Digital Back up Protective Relaying Equipment for EHV Power Systems" IFAC '81 Kyoto Congress CS-2.3.2.

7. T. Takagi, et. al. "A New Algorithm of An Accurate Fault Location for EHV/UHV Transmission Lines: Part I - Fourier Transformation Method" IEEE PES Summer Meeting, 80 SM 648-6

8. T. Takagi, et. al. "A New Algorithm of An Accurate Fault Location for EHV/UHV Transmission Lines: Part II - Leplace Transform Method" IEEE PES Summer Meeting, 81 SM 411-8

9. T. Takagi, et. al. "Development of a New Type Fault Locator using the One-terminal Voltage and Current Data" IEEE PES Winter Meeting, 82 WM 088-3

DIGITAL CONTROL IN NUCLEAR POWER PLANTS

B. Bouzon

Electricité de France, 2 avenue du Général de Gaulle, 92141, Clamart, France

Abstracts. This document presents the latest automatic control structures used in the programmable control systems of 13.00 MW nuclear power plants constructed by Electricité de France. The impact of this technological innovation goes beyond a straightforward design modification ; in addition to the new range of processes made possible, it permits far-reaching changes in the working methods employed at the design office and in the field.

Keywords. Power station control ; nuclear plants, numerical controls ; on line operation ; computer aided design.

I General architecture of nuclear power plant controls and scope of programmed logics.

1.1. General description.

The controls of a 13.00 MW nuclear power plant are designed to allow a 2 to 3 - man team to control ;

- 1400 actuators (pumps, fans, motor-drive valves...) with a total power of 96 MW from a control room incorporating :

- 760 on-off switches and 50 analog controls using the data given by :

- 1.000 status lights, 240 indicators, 80 recorders and one central data system which processes 5.000 logic signals and 800 anolog signals.

These controls are entirely automatized for all requisite actions during normal operation (above 15 % full power) and for all protection actions arising from accidents or incidents. The object of the large quantities of instrumentation and controls in the control room is to allow the operator to check that automatic actions are normal and that the relevant measures are taken in the event of failure.

1.2. Command control architecture

Logics are processed in two main units :

- Logics for normal operation are processed in a group of programmable logic bay units which use the "controbloc" equipment designed by CGEE Alsthom on the basis of an EDF specification.

- The protection controls are divided into a programmable logic unit and a cabled static unit called SPIN, which was specially designed for this purpose by Merlin Gerin.

- Bailey analog modules are used in the first 8 1.300 MW units. Digital modules of the new Bailey range, called "micro z", will be used in the next 14 units.

- Centralized data processing is provided by two computers (one SINTRA front acquisition computer CLX and one SEMS data processing computer SOLAR).

Note that outside these main units specific functions such as : control rod control and position indication, periodic core neutron flux mapping... etc are processed using digital controls.

II Non-reactor protection controls : controbloc

2.1 Controbloc

Controbloc is a programmable sequence controller which was developed in response to nuclear power plant control requirements :

- from the availability and reliability standpoint,

- from the point of view of exchange capacity between programmable controller units to ensurethe control of a unit of this kind.

It has been made fail-safe by use of the following features :

- dual control structure : controbloc features two full identical sets of data processing and exchange structures. Both operate synchronously. A command can only be issued if the output receives the same order from both structures. Any discrepancy will inhibit an output until the fault has been found.

- continuous data checking : the status of each value from one structure is continuously compared with the value from the other structure during the various stages of data collection, transfer, processing and restitution. Any discrepancy is rapidly investigated and displayed. When discrepancies are detected, a predetermined logic status is applied at the input to bring the corresponding process device to the state affording greater safety. If no logic status has been preset for the variable, controbloc will maintain the value of its last normal status.

- two cycle cross checking Every status change of input or output variables is enabled only when the changed status has been maintained through two processing cycles. These systematic checks eliminate transient interferences such as : external interferences sizeable enough to propagate through input interfaces, contact transient signals, controller internal interferences.

- galvanic insulation from the process : input and output interfaces are fitted with isolation amplifiers providing a galvanic insulation from the process.

- total self-checking coverage : since it is designed as a programmed cycle controller, controbloc has a large number of checks inserted through its cycles to verify at any moment that electronic data processing and exchange components are operating or available.

Controbloc is made up of bays including :

- one electronic block containing two identical and redundant structures

   - a power supply

   - a dialog keypad

   - four interface cubicles for wired connections (256 input and output leads per bay)

Wired and multiplexed connections are provided on the rear panel.

Automatic control unit logic is programmed into "reprom" memories to permit the tailoring of each bay to a specific application.

Controbloc is made up of L.S.I components.

The dual structure of controbloc permits on-line maintenance of automatic control functions once a first fault has been detected (exceptional operation on one bus). It is possible to merely change the PCB indicated by the bay monitor screen to clear the fault without interrupting operation.

The scope for linkage by multiplex connections to other controbloc units or to other digital units such as "reactor protection" or "centralized data processing" make it possible to satisfy control room architecture requirements.

2.2 System architecture

Automatic controls are processed in two set of controbloc bays. One set is for train A and the other train B

The train A and train B bays are installed in separate rooms.

Each unit can be divided into :

   - bays for automatic controls of different systems

   - bays for generating and processing "shared data". These are used to transmit data concerning several systems to all the automatic control bays.

   - bays used to collect alarms from all automatic controls and to distribute them to the different control room display screens (alarm polling units) (UGA).

Some controbloc bays for decentralized systems may be installed locally.

2.3 The benefits of controbloc operation.

Because of their large data transfer capability, controbloc units allow use of sophisticated logic and consequently improve the quality of information conveyed to the operator (sophisticated alarm blocking logic).

The use of digital technology and data transfer by multiplex connections permits more detailed data display on screens. It has proved possible, for example, to replace the 1.200 900 MW PWR control room alarm windows by 8 screens.

This use of multiplex connections involves a simplification of the wiring layout, which makes it easier to respect the regulations governing separation between redundant trains. Separation of inputs and outputs also reduces the risks created by proximity of cables (e.g in the control room). It is interesting to note that controbloc has had a great impact on work in design offices and methods used in the field.

The programmable automatic control typically works in a sucession of three operations :

   1°) Drawing of an automatic logic control diagram corresponding to the functionnal requirements expressed by a functional diagram. This stage is the same as that for electromagnetic relays.

   2°) Conversion of this logic diagram into "program" equations using the language of the programmable controller in question.

   3°) Use of a keyboard to feed these equations into the automatic control memories.

From the moment the automatic logic diagram is drawn, equation writing and typing of the equations on the keyboard are straightforward logical operations which can be carried out by computerized means.

Since the main difficulty in using computerized tools is the feeding of input data, it was considered necessary to try to start directly from the initial form of the data, in other words from the logic diagram before entering the computerized system.

This approach led to the formulation of the following process :

   - direct acquisition of the logic diagram by a graphical screen acquisition system (APPLICON system).

The diagram is traced on screen by means of prepared symbols (macro instructions), keys and a pointer.

   - Based on this acquisition, the computer system transforms the logic diagram into equations and transfers the equations into a reprom memory.

   - The reprom memory is then introduced into "controbloc" bay which is interfaced with the process to be automatized to obtain the desired automation.

This method does away with tedious manual operations such as the writing and typing of equations. It also has the advantage of creating systematic continuity between the diagram and the content of the "reprom". From the quality assurance standpoint, this continuity is valuable since it is impossible to check the logic contained in a "reprom" by inspecting the wires, as was the case for electro-magnetic relay circuitry.

In fact it is primarily the large scope for additional processing permitted by the computing of the data which gives this method its high profitability and productivity.

We have selected two examples among the many additional processing operations permitted by this system. These examples relate to two site operations :"checking of the wiring of the automatic controls to the process" and site startup tests. They show that the impact of the methods employed goes beyond the context of design offices and even mofifies startup operations.

As regards wiring verification, it was possible to elaborate "test" software based on the equation software resulting on the initial acquisition of the diagram. The "test" software will identify the outputs of a given automatic control in a controbloc bay and automatically create the software of a test reprom, which, when fed into this bay at the startup stage energises these outputs in sequence. It therefore allows a check on the correct running of the cables as far as the load.

For startups, since 1.300 MW plant automatic controls are provided by about 80 "controbloc" bays, it is clear that these 80 modules (which control several thousand actuators) are not started up simultaneously. When part of the logic is actuated, therefore, it is necessary to know which data coming from the other bays or even from other "reproms" of the same bay are not available at the time

The listing of these elements and the creation of a reprom for simulating the missing elements (equivalent to the "straps" of electromagnetic relay cir - cuitry) are performed automatically by the appropriate software

From a different point of view, the availability of a scale 1 simulator in the design office introduced a number of interesting possibilities. The organisation of the controbloc equipment in bays permit the reconstitution in the design office of a configuration identical to that which will exist later in the power plant. Only the programming of small reprom memories introduced in the printed circuit boards indivi - dualise the bay by introducing the automatic logic of the circuit in question. It is therefore possible to construct a simulator in the design office, for checking all the logics before their dispatch to the site. Availability of a few bays is sufficient, into which the "reproms" for all the bays of the plant can succesively be fed.

The configuration arranged by the design office includes three controbloc bays for all the systems of the plant including any connections to other systems contained in other controbloc bays and connections with the alarm polling bay and with the "general control" bays, which circulate overall data to all controbloc bays.

It incorporates a console equipped with control buttons and lights identical to those in the control room as well as an alarm screen and a relay rack of the same type as those controlling the equipment supplying power to the plant actuators. This configuration is therefore complete and allows full scale testing of the automatic controls.

This possibility has enormous advantages. Firstly, it allows the design engineer to check his logic much earlier. Instead of waiting for the relay panels to be constructed (internal wiring) and field-connected for static testing of the automatic controls (as in electromagnetic designs), this check can be carried out as soon as programmed in the design office. The reprom is immediately re-written which avoids the need for modification of on-site wiring during the system startup period, i.e during the critical part of the process.
It also permits complete freedom from the phase of on-site verification of correct automatic control operation during initiation of the process since the verification has already been made in the design office.

During the initial startups, there was some reluctance to accept the complete validity of checks carried out in the design office.

The experience currently being acquired in large numbers of bays proves that this arrangement is fully satisfactory.

III The integrated digital protection system (SPIN)

- Since reactor protection signals are generally processed from analog signals, they are generated by a different programmable control unit called the integrated digital protection system (SPIN).

The SPIN, which is based on 4-fold redundancy, is a programmed digital system for processing all NSSS protection logic. It is equipped with a computerized automatic test system.

The SPIN consists of :

- four redundant data acquisition and processing units for the protection system (UATP)

- two independant engineered safety features logic units (ULS)

- four sensors are utilized for protection and control. Signals generated by the four sensors are routed directly to the control system.

The sensors are connected to the UATP by means of specific components which enable signals to be suitably edited prior to acquisition and processing. These components are used for routing signals to the control system.

Reactor trip signals are processed by the UATPs which each generate a trip signal which is routed to one set of two breakers.

Two safeguards logic units are used for automatic actuation of engineered safeguard features (since there are two engineered safeguard features logic trains). Signals for automatic actuation of engineered safeguard features (safety injection, etc.) are processed by the data acquisition and processing units which each generate an actuation signal which is routed to the safeguard logic units. Signals from the data acquisition and processing units are routed through 2/4 logic gates at the input to each safeguards logic unit. All signals processed by the safeguards logic units are processed by means of cabled logic circuitry consisting of fail-safe electro-mechanical modules. The manual control for the system is located as far downstream of the safeguards logic units as possible. Consequently, each train may be actuated independantly of the other.

3.1 Protection system data acquisition and processing units : UATP

The function of each unit is as follows :

- acquisition of analog and digital signals,

- analog to digital conversion,

- multiplexing of signals,

- computation and comparison with setpoints,

- logic processing which takes account of interlocks,

- logic processing of blocks on protection system data acquisition and processing unit functions,

- signal exchange with other protection system acquisition and processing units and with other stations (switchgear, control room) by means of data transfer units.

A tester-simulator may be connected to a data acquisition and processing unit for the periodic functional tests.

A protection system data acquisition and processing unit consists of various components ensuring specific functions (logic processing, data transfer, etc.), each unit is controlled by a MOTOROLA 6800 type microprocessor.

3.2 Safeguards logic units

The safeguards logic unit consists of two identical logic units to which logic signals from the data acquisition and processing units are routed. The safeguards logic units and data acquisition and processing units are connected in a configuration whereby each safeguards logic unit is connected to all four data acquisition and processing units. This configuration increases the safety of the safeguard function (possibility of actuating the two safeguard trains despite simultaneous unavailability of two data acquisition and processing units).

The tester-simulator is utilized for periodic functional testing of the safeguards logic units.

3.3 Improvements permitted by the SPIN

The main improvements are as follows :

- possibility of processing analog inputs by complex functions which are more representative of real physical phenomena, with the accompanying possibility of recovering the power margins needed to cover the flaws in previous processing methods,

- excellent separation between the 4 redundant trains by the use of optical fibers,

- automatic tests by the tester.

IV Computer system

4.1 Functions

4.11 System functions

The centralized data processing system (KIT) acquires, processes and displays plant operating data to the operator in a format which is either more complete or more sophisticated than the format used for direct display in the control room, or than display which is decentralized by means of switchgear (alarm windows, screens or annunciators).

The main functions of the computer system are as follows :

- aid to direct (or indirect) control of the plant unit,

- aid to analysis of plant operation after the event on a short or long term basis (less than 24 hours or more than 24 hours) subsequent to either normal operation or an incident,

- storage of data on memory, including possible local processing of neutron flux data required by the in-core instrumentation (INCORE or RIC system) and development of an information carrying medium permitting detailed utilization of these data in a computing centre.

The computer system is not a safety system. It is not indispensable and breakdown does not adversely affect plant operation.

4.12 Real time functions

Real time functions consist of the following :

a) acquisition and storage on memory (limited capacity) of digital data status changes detected or generated by the switchgear (controbloc), recording of status changes (logbooks and status display panels).

b) acquisition, processing, storage on memory (limited to 24 hours) and monitoring of analog data including that which is converted to digital data elsewhere. These data may be subjected to surveillance providing alarm actuation if a setpoint is exceeded or if there is a discrepancy between certain parameters, and such data may result in the initiation of "internal" measurements.

c) processing functions : aid to unit operation, diagnosis of accident conditions, processing of in-core data, daily logging, miscellaneous computations, etc.

d) display of data, mainly in the control room, on screens or by means of printers, using various formats (printouts, cards and mimic panels, etc.). Display semi-automatic or automatic (monitoring mode - messages) or subsequent to operator dialogue (monitoring mode or history recording, status panels or a specific request).

e) archival storage, on a data carrier which can be utilized as input to a system other than the KIT system for deferred analysis, of all data acquired over a 24-hour period.

4.2 System layout

The KIT system consists of two interconnected levels each built around a small computer : an "acquisition" level known as level 1, and a processing level, known as level 2, to which all data acquired or developed by level 1 are routed together with data from the in-core instrumentation system.

a) Level 1

Level 1 includes the data acquisition peripherals

- 80 asynchronous-series connections by means of which a maximum of 700 data status changes may be acquired.

- 5 asynchronous-series connections carrying a total of approximately 600 digital parameters (SPIN and in-core thermocouples)

- 1092 individually-wired analog inputs (average scanning rate of approximately 70 points/sec)

- 32 individually-wired digital inputs.

These data are processed prior to operational use at this level and are displayed on two semi-graphical screens and/or 2 alphanumeric printers for aid to direct plant control, depending upon the status of the dialogue desk with function selector keys (and alphanumeric keyboard). Virtually all data processed or acquired at this level is transferred to level 2.

b) Level 2

Level 2 receives all data routed from level 1 and in addition to providing conventional monitoring (logbooks, trend, etc.), provides the following :

- display of 24-hours history recordings or monitored variables and the last 10.000 changes in status of digital data. The level 2 is accordingly equipped with an internal floppy disk storage capacity of 50 M octets.

- graphical display of plant variables (monitoring mode or history recording).

- display of data on status and/or plant variables using cards and mimic display updated to real time (on screen).

- display of various boards.

The following display is available for level 2 : 3 colour semi-graphical screens (identical to those of the N1), electrostatic and alphanumeric printers, two dialogue panels with function selector keys and track ball (plus alphanumeric keyboard).

V Conclusion.

Other digital systems are used in the instrumentation and control of nuclear power plants for specific functions : control rod control and position indication, periodic core flux mapping, control of handling facilities... All these functions were previously carried out with cabled static relay circuitry or even electromagnetic relay circuitry. The progress made in this new generation of power plants, of which the first (Paluel 1) has passed the startup test stage without difficulty, is a revolutionary step from the technological standpoint and from the point of view of design office methods and new processing possibilities.

The next stage is already being studied from two angles :

- the use of rapid connections to improve data transmission between the programmable control modules,

- the use of interacting screens for a more compact presentation of control room devices permitting the creation of a "seated" control room.

DISCUSSION

SESSION TP1: POWER SYSTEMS I (INVITED SESSION)

Paper: EVOLUTION OF DIGITAL CONTROL IN ENERGY CONTROL CENTERS

Authors: J Carpentier (Electricité de France, 2 rue Luois Murat, 75008, Paris, France)

Discusser: R Nieva
Instituto de Investigaciones
Eléctricas, Cuernavaca, México
Apdo Postal 475
México

Questions or Comments:

1) Fuel consumption modeling for fossil-fueled generating units is of obvious relevance to economic dispatch. At present the models used in economic dispatch calculations (input-output, incemental heat-rate curves) are valid for steady state operation of generating units.

What is your opinion on the validity of these models for economic dispatch? Also, do you foresee new developments aimed at merging models that represent more accurately the fuel consumption process of fossil-fueled generating units, operating in load following mode, with economic dispatch algorithms?

Author's reply:

1) In France, the specialists of thermal plant operation do not seem able to give consumption data in other conditions than steady state operation; as a consequence, no special development is foreseen now to introduce the relevant costs in economic dispatch models. However, it seems necessary to update the static results used in economic dispatch periodically to take account of the full measurements performed on the last operations periods.

Discusser: José Calderón
Loma Perpetua 50
México, DF 1620
México

Questions or Comments:

1) From my point of view it is difficult to run EDC at the same frequency of AGC (eg 2-10 sec). CFE's approach uses constrained participation factors as well as sensitivity factor base in the load forecast to extrapolate the base generating points and tie lines schedules. The EDC calculation uses the solution of the state estimation, with runs every 15 minutes.

Author's reply:

1) In the new ED proposed, it is not necessary to perform state estimation to get the relevant data, but the ED is associated to a closed loop control where only ACE and current near their allowed limits are necesarry as accurate and frequent information, the philosophy being to operate the system with the least necessary data. In these conditions state estimation is necessary only very 5th minute or less frequently, although an on-line ED is performed every 10 th second

Discusser: Enrique Morfin
Santa Ana 135, Col. Las Fuentes
45070 Guadalajara, Jalisco
México

Questions or Comments:

1) My question is about sampling times of "standard" AGC and ED. Is there a way to find optimal time?. Also for the optimal power flow. I think it is difficult to run ED every 10 seconds.

Author's reply:

1) The sampling period of LFC depends on the time lag of this particular control, which is about one minute (ie, much larger than the time lag of primary frequency control, which is a few seconds). In these conditions, the sampling period of LFC must be much smaller than 1 minute, e.g. between 2 and 10 seconds. Then, to avoid oscillations, ED and LFC must have the same sampling rate, which drives to perform both LFC and ED at the same sampling rate; eg, 10 seconds. To run a secure ED every 10 th second, new computational techniques, so-called "Fast on-line Optimal Power Flows", had to be refined. They give an optimal power flow solution for a 500 bus network in less than 1 second.

Paper: AREA CONTROL PERFORMANCE MEASUREMENT AND CORRECTIVE CONTROL IN THE INTERCONNECTED SYSTEMS

Authors: N Cohn (Consultant, 1457 Nobel Road, Jenkintown, PA 19046 USA)

Discusser: Miguel García Rubio
Depto de Ingeniería Eléctrica
CIEA-IPN
Apdo Postal 14-740
07000 México 14, DF, México

Questions or Comments:

1) In summary, what does "area N" have to know so that the proposed control scheme can work?

Author's reply:

1) To utilize the proposed decomposition technique, area n needs to know, for the time space that is to be covered: the bias setting, B_n, for area n; the total of the bias settings, B_s, for all areas; the inadvertent interchange, I_n, for area n; and the time deviation, ε, for the system. With this data, the primary component of inadvertent interchange for area n and the time deviation component for area n can be computed. To determine the magnitude of secondary inadvertent caused in any remote area i by primary inadvertent in area n, the additional information needed by area n is the bias setting of area i, B_i.

Discusser: Dr. John Zaborszky
Dept of Systems Science
and Mathematics, Box 1040
Washington University
St. Luois, MO 63130

Questions or Comments:

1) I wish to make some observations of a philosophical nature. It appears that N. Cohn's two most important contributions, that is the Area Control Principle and the Decoupling Principle for inadvertent tie line exchange, presented in this paper, have ramifications far beyond their original context as the foundations of the state of art Load Frequency Control.

If one rethinks the Automatic Generation Control from scratch with implementation using current digital control technology {1} the conclusion emerges that the Area Control Principle is still the logical foundation of LFC and will remain so as long as the structure of the American power industry remains unchanged. The U.S. power system consists of independent companies which are widely interconnected for improved economy and mutual help, but which operate and control individually withour central direction. The Area Control principle simply requieres that each area covers its own load at scheduled frequency and scheduled net tie line load. Consequently there is an equilibrium at scheduled frequency and scheduled tie line loads.

Paraphrased one can say that each Area operates and controls independently but conforms to a protocol. The result is very satisfactory overall system operation. In his recent keynote address at the IEEE Decision and Control Conference in December 1982, Tibor Vámos advocated this approach to control, Cooperative Control, as the foundation for the control of very complex large scale systems. It certainly has major advantages over the hierarchical approach in many situations and will probably play a major role in large scale system control in the future. The Area Control Principle is then the first manifestation of this type of control.

{1} "A Reevaluation of the Normal Operating State Control of the Power System Using Computer Control and System Theory", J Zaborszky and associates.

Part I: Power Industry Computer Applications Conference, Cleveland, Ohio, May 15-18, 1978

Part II: IEEE Trans. PAS, January 1981

Part III: IEEE/PES Winter Meeting, New York, Jan/Feb 1983

Author's reply:

1) Prof. Zaborszky remarks are appreciated

Paper: EMERGENCY CONTROL

Authors: J Zaborszky (Department of Systems and Mathematics, Washington University, St. Luois, Missouri

Discusser: J. Carpentier
2 Rue Louis Marat
75008 Paris, France

Questions or Comments:

1) For a large scale system, with eg. 100 plants, how many control devices would be necessary and where would they be located to apply the emergency control you presented?

Author's reply:

1) The author apreciates the opportunity provided by the question to give additional information on the method described in the paper. The answer depends on which of the two methods described is used:

When the local feedbacks method is used control tools are required at every generating station, thus 100 in the example. On the other hand a small microprocessor is the only other equipment required at the local generating station. The control tools would preferably be located near or at the generating station. No comunication links outside the generating station are needed and of course there is no central computation. Since the stabilization is very effective (critical closing times can increase as much as ten times) such stabilizing control can take the place of strengthening the system by building additional transmission lines. This many provide economic justification for the investment in control tools. Furthermore off-line simulation studies can be used to establish which of the controls are inactive in a practical sense during important contingencies. Such control tools could then be actually ommitted.

2) When the sparse control techniques is used, then a limited number of controls is only needed, probably not more than 10 (the system used for illustration in the paper has about 40 synchronous machines and is stabilised by as few as 4 controls). But there control tools must be more elaborate, on the order of thysristor devices such as HV-DC terminals. Also there is need for fast and reliable communication between the generating stations and the control center and there is a small amount of central computing in addition to a small microprocessor at each generating station.

COMPUTER CONTROL OF SIMPLE VARIABLE FLOW PROCESSES

C. P. Jeffreson

Department of Chemical Engineering, University of Adelaide, Adelaide, South Australia, Australia

Abstract. The effect of variable production rate on the feedback control of simple constant holdup "injective" flow processes is considered. As process stream flowrate varies through such systems it is well known that both the gain and the process dynamics vary and this may lead to instability particularly under low flow conditions. The conditions under which a very simple <u>self adjusting</u> feedback control law is appropriate and necessary are derived for a number of heat and mass transfer processes including two simple distributed parameter systems. It is shown that "perfect mixer", or first order lag processes do not require adaption of controller parameters for flowrate although high order systems generally do. The theory is illustrated by the experimental computer control of an air heater system and some implications for specification of digital controllers discussed.

Keywords. Self-adjusting systems; flow processes; computer control; distributed parameter systems; concentration control systems; temperature control.

INTRODUCTION

A considerable number of processes installed in the chemical and allied industries involve the continuous injection of heat or matter into a flowing stream. Four examples of such continuous flow, "injective" processes are listed below.

1. Neutralization of liquid wastes and other products with acid or alkali.
2. Injection of spray cooling water into waste gases (for example, into Basic Oxygen Furnace off-gases or water into superheated steam).
3. "Injection" of thermal energy into a continuously flowing gas or liquid stream (for example continuous water heating).
4. Addition of surfactant to a continuous ball mill effluent.

Each of these processes may be represented by the generalised intrumentation diagram of Fig. 1 which shows a simple feedback controller manipulating the flow of an "additive" (containing a high concentration of a particular molecular species, or thermal energy at a higher temperature than the incoming stream) to achieve a desired additive concentration or temperature at the vessel outlet. The transfer of heat and/or matter to and from solid surfaces may occur within the vessel or the vessel may be a simple mixer. We assume that chemical reaction does not occur in the vessel and assume negligible heat loss. An essential feature of "injective" processes as defined here is that the flux of heat of matter entering the vessel of Fig. 1 is independent of the temperature or concentration in the vessel and may be directly manipulated.

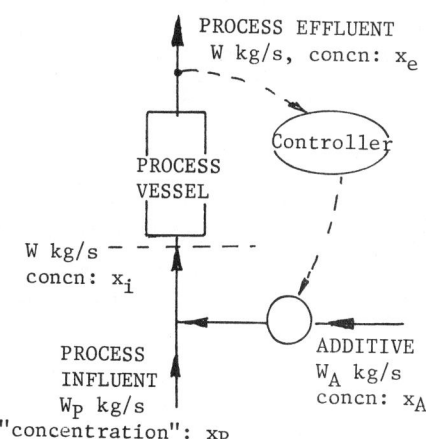

Fig. 1. General "Injective" Process Under Feedback Control

When the demand for product varies widely, it may not be possible to ensure stable operation over all flowrates unless controller settings are re-adjusted to allow for this change in flowrate. During startup and shutdown, it is usually necessary to manipulate the additive manually and this may result in a prolonged startup or shutdown procedure. Furthermore, the "turndown" or rangeablility of the process may depend on the control characteristics of the process (as well as on the resolution of actuators and measurement devices).

Niemi (1981) has suggested a method by which the measured process stream flowrate (W kg/s) could be used to adjust the parameters of a conventional three term controller. Unfortunately, his paper neglected to allow for the variable _gain_ of the process whilst correctly providing for the changing dynamics of the process. In this paper we correct this error and attempt to explore in greater detail the conditions under which the feedback controller of Fig. 1 should allow for variable flow conditions. We also extend this analysis to simple heat transfer processes. Marsland (1975, 1981) has described the correct application of a self adjusting controller to a steam desuperheater.

We shall also show that such flow-variable controller settings are not necessary for certain first order lag systems and outline some implications for specification of process computer systems.

CONCENTRATION CONTROL SYSTEMS
Material Balances

Assuming no loss of matter from the process vessel of fig. 1, an overall, unsteady state materials balance may be applied to yield:

$$W_A + W_P = W + dM/dt \qquad (1)$$

where:
 M is the total mass, kg, of fluid held in the vessel at time t,
 W_A is the flowrate of an "additive" stream kg/s,
 W_P is the influent process stream flowrate kg/s,
 and
 W is the effluent stream flowrate kg/s.

An identical equation may be written for an injective heat transfer process.

An unsteady state balance carried out on the additive molecular species of concentrations x_A, x_P and x_e kg/kg in the additive, process influent and process effluent streams respectively becomes:

$$W_P \cdot x_P + W_A \cdot x_A = W \cdot x_e + d(M\hat{x})/dt \qquad (2)$$

where:
 $\hat{x}$ is the _volumetrically averaged_ concentration of the additive species in the vessel at time t, in kg/kg.

Analogous equations may be written for the corresponding thermal process where the x's are _enthalpies_ (or specific heat - temperature products) and the rate of accumulation term of equation (2) must include the heat capacitance of solid surfaces to which heat transfer may occur. In this case, we ignore heat losses from the vessel. For certain thermal processes, such as that used in the experimental work for this paper, $W_A x_A$ becomes rate of heat input in watts which may be manipulated directly.

Finally, we assume that the holdup, M kg of material in the vessel is constant. The consequences of relaxing this assumption will be examined in a later paper.

NOTE

When additive flowrate W_A is manipulated, outflow W will vary. We hence provide a _perfect_ flow controller to keep total product flow W constant by manipulating process inflow W_P to compensate for changes in W_A.

Considering the concentration control process, equations (1) and (2) become:

$$W_P \cdot x_P + W_A \cdot x_A = W \cdot x_e + M \cdot d\hat{x}/dt \qquad (3)$$

where:

$$W(t) = W_A(t) + W_P(t) \qquad (4)$$

at all times t.

Preliminary Problem Definition

A number of disturbances may affect the controlled system. Disturbances in product stream _flowrate_ will be deferred to the next main _section_ of the paper; here, we consider only:

 changes in influent concentration x_P (or enthalpy),
or
 changes in additive concentration x_A (or enthalpy).

Denoting "design" or reference conditions by superscripted bars over the variables (for example, $\bar{W}$), we wish to specify the feedback controller transfer function required at some other steady flowrate W, where k is the "turn down ratio":

$$\bar{W}/W = k \qquad (5)$$

At each flowrate considered, we suppose the system to be operating steadily under feedback control. The same disturbance, $x_P(t)$, t ≥ 0, is applied in each case. We may also be interested in the response of the system to changes in setpoint.

Process Transfer Functions and Gains

Before considering this problem in detail, it will be found useful clearly to separate _steady state_ gains of the process from the unity gain process transfer function $G_P(s)$. Considering the average concentration x_i of additive entering the vessel just after the injection point, we write, for any time t:

$$W_P \cdot x_P + W_A \cdot x_A = W \cdot x_i = (W_P + W_A) x_i \qquad (6)$$

assuming a negligible rate of accumulation in the entrance section shown in fig. 1 between mixing point and vessel entrance.

Hence equation (3) may also be written:

$$W \cdot x_i(t) = W \cdot x_e(t) + M d\hat{x}/dt \qquad (7)$$

It is obvious that the effluent concentration

will equal influent concentration x_i at steady state in the absence of heat loss or chemical reaction. We define the following transfer functions relating changes in W_A, x_p and x_A to resultant changes in x_e:

Process transfer function:
$$\frac{x_e(s)}{W_A(s)} = \frac{x_i(s)}{W_A(s)} \cdot \frac{x_e(s)}{x_i(s)} = K_p \cdot G_p(s) \qquad (8)$$

First concentration disturbance transfer function:
$$\frac{x_e(s)}{x_p(s)} = \frac{x_i(s)}{x_p(s)} \cdot \frac{x_e(s)}{x_i(s)} = K_{D_1} \cdot G_p \qquad (9)$$

Second concentration disturbance transfer function:
$$\frac{x_e(s)}{x_A(s)} = \frac{x_i(s)}{x_A(s)} \cdot \frac{x_e(s)}{x_i(s)} = K_{D_2} \cdot G_p(s) \qquad (10)$$

where K_p, K_{D_1} and K_{D_2} are real numbers or gains, and $x_e(s)$, $x_i(s)$, $W_p(s)$ etc are Laplace transformed perturbations from the steady state.
For this class of injective process, the (unity gain) dynamic parts, $G_p(s)$, of system and disturbance transfer functions are all identical assuming negligible actuator and entrance section dynamics.

The advantage of this approach is that steady state (linearised) gains are clearly separated from the (unity gain) vessel transfer function $G_p(s)$. Note that $G_p(s)$ could be determined at constant flow by injection of a test tracer concentration or energy function $x_i(t)$ into the vessel operating with a steady throughput of W kg/s. Injection of a delta function would yield, in the majority of cases, the residence time distribution function for a mixing proeess (Wen and Fan 1975).

The linearised gains K_p, K_{D_1} and K_{D_2} may be obtained, for small changes about each steady state operating condition by partial differentiation of the material balance, equation (6) ie:

$$K_p = \frac{\partial x_i}{\partial W_A} = \frac{W_p}{W^2} \cdot (x_A - x_p), \qquad (11)$$

$$K_{D_1} = \frac{\partial x_i}{\partial x_p} = \frac{W_p}{W}, \qquad (12)$$

and $\quad K_{D_2} = \frac{\partial x_i}{\partial x_A} = \frac{W_A}{W} \qquad (13)$

Note the following important, well known control characteristics of these gains:

Linearised Process Gain K_p is inversely proportional to total demand flowrate W since W_p is directly proportional to W at constant effluent concentration,

The First Disturbance Gain, K_{D_1}, is independent of flowrate under steady state conditions

and

The Second disturbance gain, K_{D_2}, is also independent of flowrate.

Variable process gain is the most important result from a control viewpoint since it directly, through loop gain, affects stability.

OPEN LOOP SYSTEM DYNAMIC MODELING
"Time-scale" Transformation

Consideration of the phenomological equations describing the transfer of heat or matter to and from process streams within the vessel of fig. 1 generally involves the writing of a heat and/or material balance over an element, δ Mkg of process fluid or solid at any point in the vessel through which fluid is flowing. Provided such rates of change (which express the relation between the nett flux of material or energy into and out of each such element) may be shown to be directly proportional to flowrate, then a change in flow will result in a change in the "time scale" of the ultimate response of concentration or enthalpy to any particular disturbance. In other words, the dynamics of the process speed up or slow down with flow rate. One of the effects of a change in flowrate may thus be regarded as analogous to the simultaneous changing of all capacitors in an electronic analogue computer simulation. This concept was found quite useful in considering the control of thermal regenerator or blast furnace stove systems (Jefferson 1979 a,b)

Example 1: Axial Dispersion Model of Pipeline Mixers

As an example, consider a materials balance over an additive species entering and leaving the differential element δz of a pipeline mixer illustrated by fig. 2.

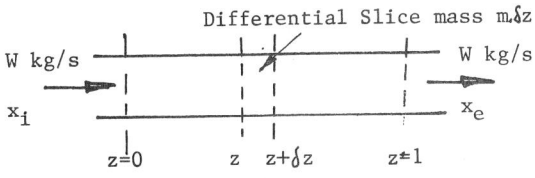

Fig. 2 Pipeline Mixer

Then defining an effective radially averaged axial dispersion coefficient $\mathcal{D}$ m²/s (Wen and Fan), the average rate of accumulation of additive within the element equals the net inflow due to convective transfer plus the net inflow due to dispersion where A_c is the cross sectional area of the pipe:

$$A_c \delta z \, m \frac{dx}{dt'} = Wx_z - Wx_{z+\delta z} + \mathcal{D} \rho A_c \left[\frac{\partial x}{\partial z}\bigg|_{z+\delta z} - \frac{\partial x}{\partial z}\bigg|_z \right] \qquad (14)$$

This results in the following partial differential equation for the concentration of additive at a normalised distance of z from the injection point at (real) time t's:

$$\frac{\partial x}{\partial t'} = \frac{W(t)}{m}\left[\frac{1}{P}\frac{\partial^2 x}{\partial z^2} - \frac{\partial x}{\partial z}\right] \quad (15)$$

where we have defined normalised distance z such that $0 \leq z \leq 1$. The axial dispersion "Peclet number" is defined by:

$$P = \frac{LW}{A_c \rho \hat{\Delta}} = \frac{L\hat{v}}{\hat{\Delta}} \quad (16)$$

where $\hat{v}$ is the average velocity of fluid across the section. Provided that the dispersion coefficient is directly proportional to flowrate, P is independent of flow and the rate of change of concentration at every point z in the mixer is <u>directly proportional</u> to flow.

When P approaches infinity, the above equations reduce to the "plug flow" advection equation with delay time equal to the mean holdup time in the pipeline.

Example 2: <u>Heat Transfer in Packed beds and Thermal Regenerators</u>

The following equations have been widely used (Schmidt and Willmott, 1981, Kays and London, 1965) to describe the transfer of heat to and from a flow section packed with a porous medium such as the parallel-plate system illustrated by fig. 3. Similar equations may be used to describe one side of a conventional "heat exchanger" or recuperator.

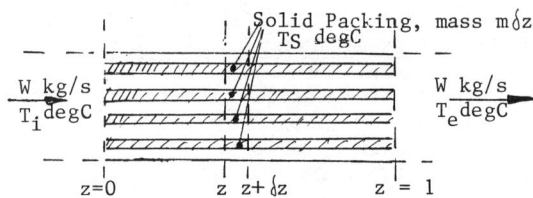

Fig. 3 Parallel Plate Packed Regenerator.

In this case, <u>energy</u> balances may be written expressing the rates of change of enthalpy (or temperature at constant specific heat) in both the solid and the fluid elements contained within the differential slice illustrated at any normalised distance z, $0 \leq z \leq 1$, from the heat injection point. For the fluid, assuming a total mass m in the section:

$$\frac{\partial T}{\partial t'} = \frac{W(t)}{m}\left(-\frac{\partial T}{\partial z} + \Lambda(T_s - T)\right) \quad (17)$$

Whilst for the solid:

$$\frac{\partial T_s}{\partial t'} = \frac{W \cdot c_f}{M \cdot c_s} \cdot \Lambda \cdot (T - T_s) \quad (18)$$

In equations (17) and (18), fluid and solid temperatures are $T(z,t')$ and $T_s(z,t')$ respectively at any real time t' s, c_f and c_s are fluid and solid specific heats $J/^\circ C$-kg respectively, M and m are the masses of packing and fluid in the section and Λ, the "reduced length" or "number of transfer units" is related to the mean heat transfer coefficient h, $w/^\circ C$-m^2 between fluid and packing and to the total surface area for heat transfer by:

$$\Lambda = h \cdot A / W \cdot c_f \quad (19)$$

Equations (17) and (18) show that once again, the rates of change of fluid and solid temperature are both directly proportional to fluid flowrate provided that the heat transfer coefficient is also directly proportional to flowrate in the bed ie, Λ is constant. These equations are used to describe the experimental system of this paper and the validity of the above assumption will be tested later.

Example 3: <u>"Perfect Mixer" Systems</u>.
In this case the volume element taken may extend over the entire contents of the vessel since the concentration (or fluid enthalpy) is assumed uniform throughout the vessel. A similar balance leads to:

$$\frac{dx_e}{dt} = \frac{W}{m} \cdot (x_i - x_e) \quad (20)$$

so that once again rates of change throughout the vessel are all proportional to flowrate. Note that the addition of first order chemical reaction (eg removal of the additive species at a rate proportional to concentration) will not usually satisfy the time scaling transformation.

Example 4 <u>Mixed Models</u>.
Other, more complex mixing systems have been modelled by subsystems of plug flow and/or perfect mixing regions interconnected by branching and recombining convective recycle and parallel streams. As pointed out by Niemi, provided all such convective and diffusional transfer rates remain strictly proportional to overall vessel throughput, then obviously all rates of change of concentration and/or enthalpy will be proportional to flowrate.

CLOSED LOOP CONTROL SYSTEM MODELING
Transfer Function Analysis

Consider a <u>closed loop</u> linearized representation of the system of fig. 1. As we have shown above, systems which suffer a time scaling directly proportional to flowrate have an open loop, unity gain, dynamic, "mixer", transfer function $G_p(s)$ at flowrate W which is related to the transfer function $\bar{G}_p(s)$ at the reference flowrate $\bar{W}$ by the following equation:

$$G_p(s) = \bar{G}_p(ks), \quad (21)$$

where $k = \bar{W}/W$.

We have assumed in fig. 4 that the controller transfer function has the form $KG_c(s)/s$ where

K is real and

$$\lim_{s \to 0} G_c(s) = 1$$

In other words we have assumed some form of integral action.

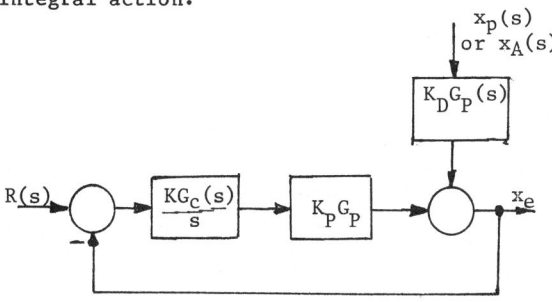

Fig. 4 Simplified Block Diagram System of Fig. 1.

Actuator and measurement dynamics have been assumed negligible and their steady state gains incorporated into the "controller gain" K.
Then considering the regulator response of the system to a disturbance in either x_A or x_P, where the system is initially operating at steady state, conventional block diagram reduction yields:

$$x_e(t) = \mathcal{L}^{-1} x_p(s) \cdot \frac{K_D \cdot G_p(s)}{1 + K_L \cdot G_L(s)/s} \quad (22)$$

where:

$$K_L \cdot G_L(s) = K \cdot G_c(s) \cdot K_p G_p(s), \quad (23)$$

and

$$G_p(s) = \bar{G}_p(ks)$$

Then from equations (11) and (12) or (13):

$$K_p = k \bar{K}_p \text{ and } K_D = \bar{K}_D \quad (24)$$

for either disturbance.

Hence the zero initial condition response $x_e(t)$ at flowrate W may be related to the corresponding response at flowrate $\bar{W}$ to the same disturbance by:

$$\boxed{x_e(t) = \bar{x}_e(t/k)} \quad (25)$$

provided that:

$$x_p(s) = k \cdot \bar{x}_p(ks), \quad (26)$$

$$K_L = \bar{K}_L, \quad (27)$$

and

$$G_c(s) = \bar{G}_c(ks). \quad (28)$$

Equation (26) is valid for step changes and (25) follows from

$$x_e(t) = k \mathcal{L}^{-1} x(ks)$$

the Laplace transform time scale theorem.

From a practical viewpoint, equation (28) implies that all controller parameters associated with the Laplace transform variable s should be multiplied by the turndown ratio $k = \bar{W}/W$. Once this time scaling has been done, equation (27) indicates that the overall controller gain should be divided by k or that the overall controller gain should be directly proportional to flowrate to ensure constant loop gain.

Thus, a conventional two term (PI) controller transfer function at flowrate W should become:

$$K_c \left[1 + \frac{1}{T_R s} \right] = \frac{\bar{K}_c}{k} \left\{ 1 + \frac{1}{k \bar{T}_R s} \right\}. \quad (29)$$

This may be contrasted with Niemi's recommendation implied by his equation (26):

$$K + K_I/s = \bar{K} + \bar{K}_I/ks \quad (30)$$

Fig. 5 shows the simulated response of a perfect mixer system to a step disturbance under PI control where:

$$\bar{G}_p(s) = 1/(\bar{T}s + 1), \quad (31)$$

and $\bar{T} = m/\bar{W}$.

The two disturbance were introduced at flowrates of $\bar{W}$ and $\bar{W}/2$ where the controller was tuned to a damping ratio of 0.2 at the reference flowrate. As would be expected, the compensated system has the same damping ratio as the reference system, however the speed of response has been slowed by a factor of two.

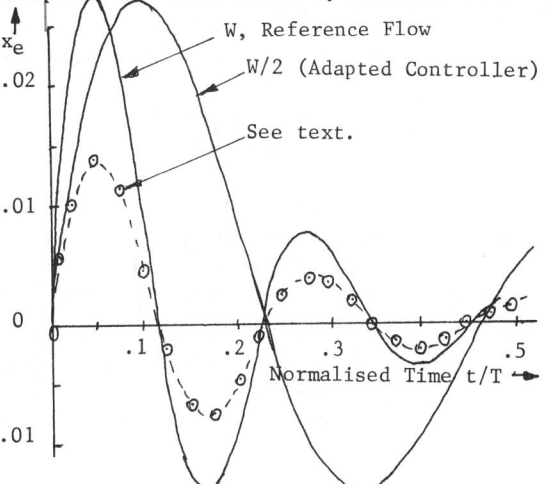

Fig. 5 Normalised Response $x_e/\delta d \cdot K_D$ Perfect Mixer Under Closed Loop PI Control at Flows $\bar{W}$ and $\bar{W}/2$ with Controller Adaption

Since disturbance gains are unaffected by flowrate for this particular system, the magnitude of the response is unchanged at the reduced flowrate at corresponding normalised times, t. Responses marked thus: ⊙ will be discussed later.

Integral Error Performance Criteria

Since the use of the correct flow compensated controller must result in a simple stretching or compression of the time scale of the closed loop system then the Integral absolute and integral square errors IAE and ISE respectively at any flowrate W must be simply related to those occuring at the reference flowrate $\bar{W}$ by:

$$IAE = k.\overline{IAE}$$
and
$$ISE = k.\overline{ISE}. \quad (31)$$

The same (step) disturbance has been assumed at each flowrate. According to these criteria of system performance therefore, the performance of a flow-compensated system must be degraded as flowrate is reduced although the same relative stability will be maintained. Such a degradation in performance will, in general, be preferable to instability which, as the following examples show, will occur for high order systems tuned to achieve maximum performance at the reference flowrate.

CLOSED LOOP PERFORMANCE EXAMPLES

Example 1: Pipeline Mixer System.

Laplace transformation of equation (15) using the time scale $t = t'/(m/W)$ and the following simplified boundary conditions:

$$\lim_{z \to \infty} x(z,t) \text{ is finite}$$

and

$$\lim_{z \to 0} x(z,t) = x_i(t),$$

yields the following transfer function at any flow W:

$$G(z,s) = \exp\left(\frac{Pz}{2}\left[1 - \sqrt{1 + \frac{4sk}{P}}\right]\right) \quad (32)$$

$$= \bar{G}(z,sk)$$

Jeffreson (1976) has derived the frequency response, proportional action critical frequency and maximum allowable loop gain for this system.

In "tuning" the system using the Zeigler Nichols tuning rules, PI loop gain K_L and reset time T_R become, under reference flow conditions:

$$\bar{T}_R = \bar{P}_u/1.2 = 2\pi/1.2\,\bar{\omega}_c$$

where:

$$\bar{\omega}_c = \pi(1 + 4\pi^2/P^2)^{1/2}, \quad (33)$$

$$\bar{K}_{L_{max}} = 0.45\, K_{L_{max}},$$

with

$$\bar{K}_{L_{max}} = e^{-P/2}.\exp\left(\frac{P}{2\sqrt{2}}\left[\sqrt{(1+x^2)} + 1\right]^{1/2}\right) \quad (34)$$

and

$$x = 4\omega_c/P.$$

Assuming a constant Peclet number of 30, a simple calculation shows that the closed loop system would become unstable under PI control without controller adaption when the flow is reduced to about 54% of the reference value. Controller adaption for flowrate is hence essential for this system.

Example 2: "Packed Bed" System

Figure 6 illustrates a temperature control loop similar to that used in the experimental part of this paper. It consists of a temperature control loop closed around a packed bed such as the parallel plate system of fig. 3. The controller manipulates a source of "additive" heat input.

For feedback control of such a system, the transfer function relating inlet temperature changes $T_i\,^\circ C$ to resultant fluid temperature changes at the outlet becomes:

$$G_p(s^*) = e^{-s^*} \cdot \exp\left[-\frac{\Lambda\, s^*\, V_H}{\Lambda + s^*\, V_H}\right],$$

with $\quad V_H = Mc_s/m_f C_f$

where time t^* has been normalised relative to the mean fluid holdup time m/W in the system. For most gas/solid systems of practical significance, the ratio V_H is very large so that the initial dead time may be ignored.

The transfer function then becomes:

$$\bar{G}_p(s) = \exp\left[\frac{-\Lambda s}{\Lambda + s}\right] \quad (35)$$

where time has now been normalised relative to the thermal residence time of the packing, $Mc_s/\bar{W}c_f$ at reference flow $\bar{W}$.

Provided Λ is independent of flowrate, the transfer function at flow W is related to that at a reference flowrate of $\bar{W}$ by equation (21).

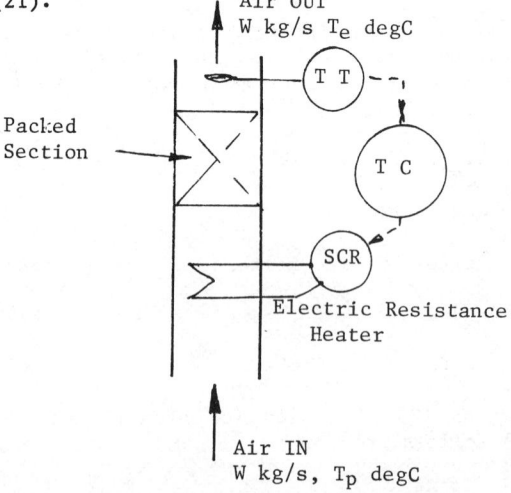

Fig. 6 Experimental Packed Bed Temperature Control System.

Proportional action loop gain and critical frequency for this system have also been derived:

$$\omega_c = \frac{\Lambda^2}{2\pi}\left[1 - \sqrt{1 - \frac{4\pi^2}{\Lambda^2}}\right], \quad (36)$$

$$K_{L_{max}} = \exp\left(\frac{\omega_c^2/\Lambda}{1 + \omega_c^2/\Lambda^2}\right) \quad (37)$$

provided $\Lambda > 2\pi$.

Similar calculations to before show that if controller adaption is not applied, a system with Λ of about 20 under reference conditions will become unstable when flowrate has been reduced to about 66% of the reference flowrate assuming Zeigler Nichols' tuning rules were applied at the reference flowrate.

"Perfect Mixer" Systems

When the process transfer function may be regarded as that of a perfect mixer or first order lag, theoretical analysis and simulation shows that uncompensated performance may actually improve as flowrate is reduced, ie controller adaption is not necessary. This is an important example of the relation between process and control system design.

Thus for a first order lag (where $x=x_e$ in equ (2)) under PI control, the usual equations for damping ratio, ζ and natural frequency result:

$$\zeta = \frac{1 + K_L}{2\sqrt{K_L}}\sqrt{\frac{T_R}{T}}$$

and $\quad \omega_n = \sqrt{\frac{K_L}{T_R \cdot T}}$

If no controller adaption is used, it is easy to show that:

$$\zeta = \frac{1 + k \cdot K_L}{2k\sqrt{K_L}}\sqrt{\frac{T_R}{T}}$$

and

$$\omega_n = \overline{\omega}_n$$

ie, natural frequency is unchanged and damping ratio little changed as flow decreases especially for $\overline{K}_L$ large.

However, in the absence of flow adaption, increased loop gain results in a reduced peak height so that (as indicated by the responses denoted by $\odot$ in fig. 5) the integral absolute error (IAE) and integral square error (ISE) decrease with increasing k or decreasing flowrate. This may be more readily seen from the formulae for the reglator IAE and ISE resulting from unit step changes in either disturbance:

$$IAE = \frac{K_D}{K_L} \cdot T_R \cdot \frac{1 + \beta}{1 - \beta}, \quad (38)$$

with $\quad \beta = \exp\left(\frac{-\zeta\pi}{\sqrt{(1-\zeta^2)}}\right)$,

and $\quad ISE = \left(\frac{K_D}{K_L}\right)^2 \cdot T_R^2 \cdot \frac{\omega_n}{4\zeta}. \quad (39)$

Note that for a flow adapted controller, all parameters in eqs (38) and (39) remain unchanged with the exception of $T_R = kT_R$ and $\omega_n = \overline{\omega}_n/k$.
Hence
$$IAE = k \cdot \overline{IAE},$$
and
$$ISE = k \cdot \overline{ISE}$$
as before.

For the non-adapted system, substitution into equations (38) and (39) shows that performance improves because of increased loop gain despite small decreases in damping ratio.

Hence it appears to be good design practice (from the viewpoint of control at least) to ensure that mixers approximate as closely as possible to first order lags!

FLOW DISTURBANCES
USE OF FEEDFORWARD ACTION

Up to date we have considered the response of the steadily operating system to a disturbance in process concentration x_p at fixed flowrate. Since controller adaption requires measurement of process flowrate it is obvious that addition of a feedforward signal to the controller which is proportional to W_p will eliminate the effects of this particular disturbance since, according to equation (6), the change in additive concentration x_i entering the plant may be made zero for an injective system. Note that for an injective system with negligible actuator dynamics, dynamic compensation will be unnecessary in this case.

EXPERIMENTAL VERIFICATION
Apparatus

The packed bed system shown in Fig 6 was used to verify the general adaptive feedforward control strategy proposed for all the injective systems described in the Introduction (except for perfect mixing systems). Complete details of the system have been described by Possingham (1981) except for subsequent modifications to the computer system itself.

The main temperature control loop consisted of a Nichrome wire heater of total resistance 50 Ω strung across the 141 mm diameter test section and separated from the 1 mm dia platinum resistance probe by 27 layers of 16 x 16 x 0.020" stainless steel wire mesh packing. The Nichrome wire heater consisted of bare wires arranged in two layers and strung in zig-zag fashion across the section. The temperature probe extended to the centre of the section with flow normal to the probe. Heater power was manipulated by a 0 to 5 v signal from one of the PDP11/03 computer's 12 bit D/A (AAV11) converters through a phase angle controlled SCR regulator. A linear resistance to emf temperature transmitter with input range 18 to 54°C was used. All measurement signals

ranged from 0 to 5 v corresponding to the
range of the 12 bit ADV11 converters. The
sampling interval chosen, (80 ms) was much
faster than necessary and resulted in effectively continuous control.

The flow control loop consisted of a BS 1042
standard Venturi meter and Fischer and Porter
electronic DP cell on the suction of a variable speed blower which forced air upwards
through the test section. An armature
current controller manipulated blower speed
through a DC motor. This loop was designed
to be much faster than the main temperature
loop, allowing "step" changes in flow to be
applied. A sampling interval of 160 ms was
found to be adequate nonetheless.

In order to eliminate the effects of inlet
temperature drift and to allow step changes
in inlet temperature to be applied to the
main temperature loop, another temperature
control loop was installed between the blower
and the main loop. This "environmental" loop
was identical to the main loop except that
only one layer of wire gauge were installed
between the SCR heater and measurement
bulb. The much faster speed of this loop
ensured that on the time scale of the main
temperature loop, inlet temperature changes
could also appear to be steps when desired.

Computer Software

The Central Electricity Generating Board's
"DDACS" operating system described by Marsland (1981) was used for all control and
logging functions. Appropriate DDACS
"SCHEMES" were written to provide bumpless,
balanceless auto/manual transfers for each
loop, the bumpless balanceless introduction
of feedforward action and digital ramping of
set points or manual regulator outputs via 16
line parallel digital I/O hardware. Digital
inputs were also used to start and stop DDACS
integrators and to apply or remove feedforward action. DDACS linear interpolation
TABLES were used to linearise the relationships between each temperature controller
output and power. An input TABLE and a
square root block linearised the DP cell
output/flow relationship. Since the PID
controller software blocks accept controller
gain, reset time, and derivative time from
the SCHEME, it was a simple matter to allow
the flowrate variable to set controller parameters adaptively when desired.

Experimental Results

Plots of open loop gain against the reciprocal of flowrate were almost perfectly
linear as would be expected from equation
(11). Open loop gain and dynamic measurements were made at the same time by applying
a step in heater power to the main temperature loop with the controller on manual.

Use of a DDACS integrator to evaluate the
hatched area between the reaction curve and

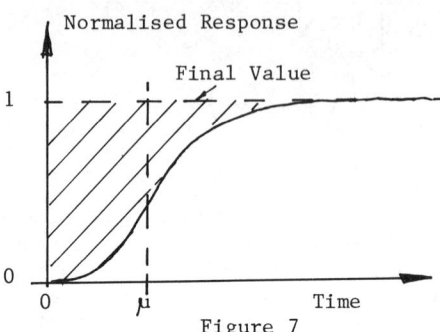

Figure 7
"Reaction Curve" Integration

the final value shown on fig. 7 provides the
sum of measurement and heater time constants
and dead times and the first moment of the
impulse response of the bed transfer function. This was done by storing the reaction
curve or step response in a DDACS table and
running a real time SCHEME later to normalise
the response to a zero to one scale and evaluate the integral.

Jeffreson (1970) provides the generalised
theory of this method of dynamic testing.
For this system, assuming the response curve
to be normalised to 0 to 1, the integral
should provide the value:

$$\mu = T + \frac{Mc_s}{Wc_f} + D \qquad (40)$$

where T is the sum of first order lags in
       measurement and/or actuator,
       Mc_s/Wc_f is the mean thermal residence
       time in the packed bed, or first mom-
       ent of the bed impulse response
and   D is the sum of any pure dead times
       in the system (expected to be zero).

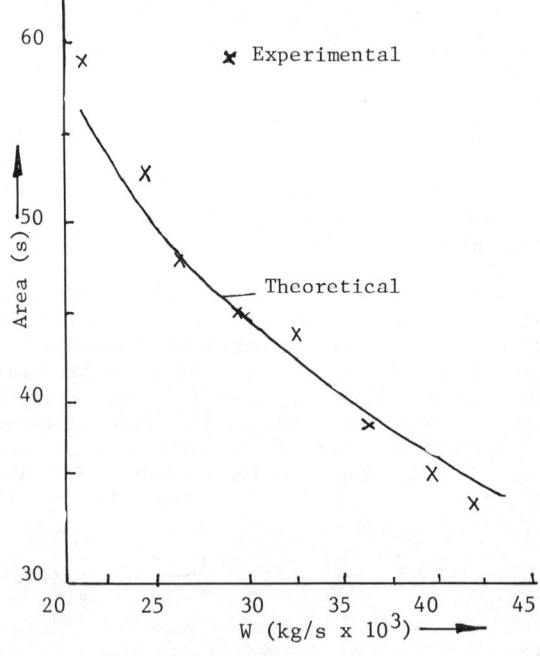

Figure 8
Experimental and Theoretical
First Moments from Reaction
Curve Integration

independent measurements of measurement bulb first order time constants were made at a number of flowrates by removing the bulb, heating it and immersing it in the constant temperature, constant flow air stream. The result of adding these experimental time constants to the theoretical value of Mc_s/Wc_f and plotting against experimental values is shown in fig. 8. Excellent agreement confirmed the expected wire specific heat of 502 J/°C-kg and the absence of significant actuator lags.

A series of closed loop "ultimate sensitivity" tests under proportional control resulted in the correspondence between theoretical and experimental values tabulated below:

TABLE 1 Maximum Loop Gain and Ultimate Period vs. Flow

FLOW kg/s $\times 10^{-3}$	MAX. LOOP GAIN Theory	Exp	ULTIMATE PERIOD (s) Theory	Exp.
19.5	2.92	2.73	96.7	89.1
24.4	3.18	3.16	80.5	73.9
29.3	3.43	3.39	69.2	64.1
34.1	3.67	3.55	62.2	56.6
38.6	3.85	3.85	56.1	51.5
43.5	4.05	4.01	52.3	47.4
48.4	4.22	4.29	44.7	43.6

"Theory" used the experimental measurement time constants and the phase shift and magnitude ratios from the transfer function of equation (35). The values of Λ used were calculated from the heat transfer versus Reynolds number correlations summarised by Kays and London in their fig. 7-8. Good agreement resulted for the maximum loop gains although experimental ultimate periods were about 6% below theory.

Finally, the smoothed values of $K_{L_{max}}$ and ω_c at a reference air flowrate of 4.35×10^{-2} kg/s were used to calculate reference flow Ziegler Nichols' PI controller settings. Without controller adaption, instability was observed at a flowrate of 75% of the reference flow.

The simple adaption of equation (29) was then implemented and a DDACS integrator and absolute value block used to evaluate integral absolute errors following step changes in inlet air temperature over flowrates ranging from 4.8×10^{-2} kg/s down to 1.9×10^{-2} kg/s. The values of IAE were divided by the inlet temperature change and by the turndown ratio k for comparison with the prediction of equation (31). Remarkable agreement was observed, with a maximum range over the entire flowrate of 28.4 s to 31.3 s. Hence flow-adaptive adjustment of controller gain and reset time resulted in regulator "performance" which was (to experimental accuracy) proportional to flowrate for a turndown range of 2.5 to 1. Limitations imposed by the (square root dependent) flow measurement, the resolution of the A/D converters and the minimum controllable armature current prevented exploration of higher turndowns.

As would be expected for such a simple system, feedforward action completely cancelled out the effects of flow changes. Further details are provided by Possingham. In later work the feedforward gain was set automatically by allowing DDACS to compute values of the actuator-process gain product with feedforward turned out.

More Recent Experiments

Although the simple adaptive adjustment of controller settings according to equ. (29) results in approximately constant relative stability over the turn down ratio tested, examination of Table 1 shows that this is in fact, largely fortuitous. Strictly, simple controller adaption of this type assumes that maximum loop gain is constant over all flowrates and that critical frequency varies directly with flow. For this system, the square root dependence of heat transfer coefficients (both measurement and process) on flowrate means that maximum loop gain decreases as flow decreases as the system becomes more like a large dead time. Hence controller gain adaption should not only allow for process gain variations predicted by equ. (11) but also for this decrease in allowable loop gain at low flowrates. For this particular system, the relationship between measurement and process time constants ensures that an inappropriately high controller gain is offset by a lower reset rate, $1/T_R$ at low flowrates.

Further experiments are in progress using tighter controller tuning and the use of simple empirical relations between flowrate and maximum controller gain and critical frequency. Preliminary results are quite encouraging and suggest that for many industrial applications, a pair of Ultimate Sensitivity" tests at the maximum and minimum expected flowrates should be sufficient to provide a more precise relation between the controller settings and flowrate than suggested by equ. (29).

CONCLUSIONS, FURTHER WORK

For the constant holdup, high order "injective" processes considered in this paper we have shown that simple flow adaptive controller adjustment is necessary if the control system is to cope with a wide range of production rates. It is therefore surprising to note that a number of commercial single loop digital controllers do not allow continuous adaptive adjustment of controller settings.

Very simple relationships between controller settings and flowrate such as that of equ. (29) are strictly only possible when heat and mass transfer fluxes within the process vessel are directly proportional to flowrate; the most important factor for high order processes appears to be the inverse relation between flowrate and process gain.

For low order processes the analysis in the paper suggests that controller adaption may not be necessary at all to ensure good performance over all flowrates.

For equipment such as counter flow liquid liquid heat exchangers, where the flux of "additive" heat depends on process temperatures and cannot be non-interactively manipulated, adaptive controller adjustment becomes very complex unless suitable control strategies, under investigation, are adopted. The effects of flow variable holdup mass are also of interest.

REFERENCES

Bartlett, L.A., C.R. Marsland and R.D. Watkins (1975). Commissioning...of Secondary Superheater ..on Thorpe Marsh No.2 Unit CEGB Report SSD/NE/R305.

Jeffreson, C.P. (1970). Dynamic Testing, a Unification. Chem.Eng.Sci.,25, 1319-1329.

Jeffreson, C.P. (1976). Controllability of Process Systems. Ind.Eng.Chem.Fund., 15,171-179.

Jeffreson, C.P. (1979a). Feedforward control of Blast Furnace Stoves. Automatica,15, 149-159.

Jeffreson, C.P. (1979b). A Computer Control System for Blast Furnace Stoves. Proc. 7th Aust. Conf. on Chem. Eng'g 107-110.

Kays, W.M. and A.L. London (1964). Compact Heat Exchangers, Second Edition. McGraw-Hill. New York.

Marsland, C.R. (1981) Commissioning of Secondary Superheater System. Chapter 15 in Proc. Digital Process Control Adelaide, University of Adelaide.

Niemi, A.J. (1981). Invariant Control of Variable Flow Processes. Proc. IFAC 8th World Congress, Kyoto, Japan. Pergamon New York.

Possingham, G.D. (1981). Feedforward Feedback Adaptive Control of a Variable Flow System. Honours Report, Chem Eng Dept., University of Adelaide.

Schmidt,F.W. and A.J. Willmott (1981). Thermal Energy Storage and Regeneration. Hemisphere Press. Washington.

Wen, C.Y. and L.T. Fan (1975). Models for Flow Systems and Chemical Reactors. Marcel Dekker. New York.

REAL TIME DIGITAL MULTIVARIABLE CONTROL FOR A FERMENTATION SYSTEM

J. Carrillo, J. Alvarez and J. A. Gallegos

Department of Electrical Engineering, Advanced Studies and Research Center, Mexico City, Mexico

Abstract. In this paper, the application of a real time optimal control algorithm to a continuous culture fermentation process is described. This algorithm can be applied to a certain class of discrete multivariable nonlinear systems. The process parameter variations problem is discussed and a control scheme using a real time identification algorithm is proposed. This algorithm allows the control always uses the real parameter values leading to a very robust control structure.

Keywords. Nonlinear systems; optimal control; identification; fermentation process; computer control.

INTRODUCTION

There are very few papers dealing with real time control applications for fermentation processes. Those described in Alvarez and Alvarez (1982); Ribot (1976); Sevely and co-workers (1981) are some attempts to control this kind of processes. The main problems still unsolved are originated by the high nonlinearity of the plant model and the typical parameter variations presented when changes in the environment ocurr.

The classical PID algorithm has not given very performant results when applied to fermentation processes; then the use of control algorithms for nonlinear systems is necessary. It is desirable these algorithms be able to be implemented on a digital computer because of the facility for programing there any control algorithm. In Alvarez and Alvarez (1982) an algorithm with such features is presented.

The work presented here tries to improve the results obtained by Alvarez and Alvarez (1982). Moreover, the problems for implementing the algorithm on real time are discussed.

CONTROL ALGORITHM

The control law used here is described in Alvarez and Alvarez (1982). A brief summary of the results is given here. The proposed control algorithm has been designed based on a discrete nonlinear model and leads to a very robust global scheme under the presence of internal (parametric) or external disturbances.

The class of nonlinear systems considered are those described by the following equations:

$$\underline{x}_{k+1} = \underline{a}(\underline{x}_k) + B(\underline{x}_k)\underline{u}_k \quad (1)$$

$$\underline{y}_k = C\underline{x}_k \quad (2)$$

where $\underline{x}_k \in R^n$, $\underline{u}_k \in R^r$ and $\underline{y}_k \in R^m$ are the state, control and output vectors, respectively; $\underline{a}(\underline{x}_k)$ and $B(\underline{x}_k)$ are vector and matrix functions of class C^∞. C is a constant matrix of proper dimensions. Moreover, in this paper the following hypothesis are proposed:

H1) The state $\underline{x}$ is measurable. This condition is necessary in order to obtain an optimal controller.

H2) m = r, that is, the system has an equal number of inputs and outputs. This is a frequently used hypothesis in the design of control algorithms for multivariable systems.

The control law is designed using optimal control theory and in consequence a performance index must be given. The following performance index has been proposed in Alvarez and Alvarez (1982)

$$J = \sum_{k=0}^{N-1} \{\underline{e}_k^T Q \underline{e}_k + (\underline{e}_{k+1} - \underline{e}_k)^T(\underline{e}_{k+1} - \underline{e}_k)\} \quad (3)$$

where Q is a positive definite matrix, N is the observation horizon and $\underline{e}_k = \underline{z}_k - \underline{y}_k$ is the error between the reference vector $\underline{z}$, and the output vector $\underline{y}$, at time k.

The obtained control law, when (3) is minimized, will allow to drive the output vector to a given one (reference). Moreover, it is possible to maintain the excursions of the control vector below a certain level from one sampling time to the next.

In Alvarez and Alvarez (1982) the control scheme depicted in Fig. 1 is proposed. Such a scheme permits to eliminate the effect of constant external disturbances. It is also possible to set the transient response of the overall system.

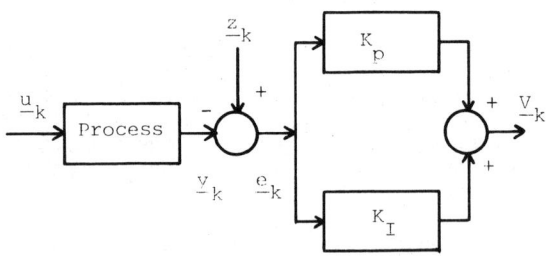

Fig. 1 Control scheme with gains K_p and integrators of gain K_I.

It can be shown in Alvarez and Alvarez (1982) that, if the observation horizon is infinite, the control law is given by:

$$\underline{u}_k = \frac{(CB(\underline{x}_k))^{-1}}{K_p + K_I} \left\{ K_p \underline{y}_k - (K_p + K_I) Ca(\underline{x}_k) - (I + \tilde{M})(\hat{\underline{z}} - \underline{v}_k) + (K_p + K_I) \underline{z}_{k+1} - K_p \underline{z}_k \right\} \quad (4)$$

where $\hat{\underline{z}}_k$ is the reference vector for $\underline{v}_k$ that must be constant. $\tilde{M}$ is a constant matrix that affects the closed loop system dynamics. $\tilde{M}$ is given by the following expression:

$$\tilde{M} = \text{diag}(\tfrac{1}{2}\sqrt{q_i(q_i + 4)} - \tfrac{1}{2} q_i - 1)_{m \times m}$$

where q_i is the weighting factor corresponding to the ith error component ($Q = \text{diag}(q_i)$).

In order to apply this control law to a given system it is necessary to have the model of the process in the form given by (1) and (2) and then experimentally to adjust the values of Q, K_I and K_p.

FERMENTATION PROCESS MODEL

The process model, which has been simulated in an analog computer, is described by the following equations:

$$\dot{x} = (\mu - D) x \quad (5)$$

$$\dot{s} = D(S_a - s) - \frac{\mu x}{R} \quad (6)$$

$$\mu = \mu_m \frac{s}{K + s} \quad (7)$$

where x and s are the biomass and substrate concentrations (g/ℓ) respectively, D is the dilution rate (h^{-1}), S_a is the feed substrate concentration and μ_m, R, and K are characteristic parameters of the process. A detailed description of the process is given by Alvarez (1978).

The system described by expressions (5), (6) and (7) may be seen like a process whose inputs are D and S_a, the states are x and s and the model parameters are μ_m, R and K. The model is nonlinear not only in the states but also in the parameters. From a physical point of view the selection of D and S_a as the control variables is adequate because of the facility to handle them.

Discrete Model

The forward Euler algorithm to discretize the analog process model has been applied. In spite of the existance of better choices, it has been found this algorithm well adapted to this problem. Then we have obtained the following equations:

$$\underline{x}_{k+1} = a(\underline{x}_k) + B(\underline{x}_k) \underline{u}_k$$

$$\underline{y}_k = C \underline{x}_k$$

where $\underline{x}_k = (x_k \; s_k)^T$; $\underline{u}_k = (D_k \; D_k \cdot S_{a_k})^T$

$$a(\underline{x}_k) = \begin{pmatrix} x_k + H\mu_k x_k \\ -\frac{H}{R}\mu_k x_k + s_k \end{pmatrix} \; ; \; B(\underline{x}_k) = H \begin{pmatrix} -x_k & 0 \\ -s_k & 1 \end{pmatrix};$$

$$C = I \; ; \; \mu_k = \mu_m \frac{s_k}{K + s_k}$$

H is the sampling period.

CONTROL LAW

The control law obtained (4) has been applied to a simulated continuous culture fermentation process. The case when regulation is desired will be shown, then the output reference is a constant vector: $\underline{z}_k = (x^*, s^*)^T$. Also, it is assumed that $\hat{\underline{z}} = \underline{z}_k$.

Application of equation (4) to the process model gives the following espressions for D_k^* and $S_{a_k}^* = \frac{(D_k \cdot S_{a_k})^*}{D_k^*}$

$$D^*_k = \frac{1}{K_p + K_I} \left\{ \frac{1 - K_I}{H x_k} x^* + \frac{K_I}{H} + (K_p + K_I)\mu_{ck} - \frac{1}{Hx_k} v_{xk} + \frac{\tilde{K}_x}{Hx_k}(x^* - V_{xk}) \right\} \quad (8)$$

$$S^*_{ak} = s_k - \frac{1}{(K_p + K_I) HD^*_k} \left\{ (1 - K_I) s^* + K_I s_k - \frac{(K_p + K_I)H}{R_c} \mu_{ck} x_k - v_{sk} + \tilde{K}_s(s^* - v_{sk}) \right\} \quad (9)$$

where $\mu_{ck} = \mu_{mc} \frac{s_k}{K + s_k}$. The subindex c is associated to the controller parameters. v_{xk} and v_{sk} are the added state components when the PI block is included. $\tilde{K}_x$ and $\tilde{K}_s$ are given by expressions of the form

$$\tilde{K}_i = \frac{1}{2}\sqrt{q_i(q_i + 4)} - \frac{1}{2} q_i - 1 \quad (11)$$

where q_i is the ith component of the diagonal of the Q matrix.

REAL TIME SIMULATION RESULTS

A 8-bit microcomputer (CROMEMCO D-II) to program the control algorithm and an analog computer (EAI-1000) to simulate the process have been used.

Several problems that are present in a real time control application must be mentioned. Some of these problems are numerical; they are due to the finite word length of the A/D and D/A converters used (8 bits).

On the other hand, analog simulation of some nonlinearities of the process (multipliers and dividers) is not very accurate.

These two error sources make impossible to simulate the process exactly. In consequence, the model employed by the controller is not exactly the same that the simulated one.

The control laws given by (8) to (10) has been applied to the process represented by equations (5) to (7), simulated in the analog computer. The nominal parameter values has been chosen as $\mu_m = 0.2$, $R = 0.3$, $K = 0.1$. The reference vector has been fixed to $z = (x^*, s^*)^T = (3, 1)^T$. Moreover, $Q = I$, $K_I = 1$ and $K_p = 10$.

Fig. 2, shows the system evolution when the reference vector is changed from $(3, 1)^T$ to $(2.4, 1)^T$. It is clear that the process output is driven to the correct values in about 6 h.

In Fig. 3 it is shown the time response of the inputs and outputs when the value of μ_m is changed from 0.2 to 0.24 (20 percent) and the controller does not detect this change. The outputs are affected, but after a certain time the process returns to its nominal operating point.

Fig. 4 shows outputs and control inputs when R is changed from 0.3 to 0.36 (20 percent). It can seen that only the substrate concentration (s) is affected. It returns to its nominal value in about 9 hours.

Parametric Variations

The proposed control structure is robust for small process parameter variations. However, when these variations are very big and they are not detected by the controller, the global structure may become unstable.

In this case it is possible to propose a better control structure. This structure has an on-line identification loop which updates the controller parameters, as it is shown in Fig. 5.

However, the main problem is the nonlinearities present in the process model. This fact may become very difficult the on-line identification.

Identification Algorithm For The Fermentation Process

We have used a very simple identification algorithm for the particular model given by (5) to (7). This algorithm only identifies μ_m and R. It is obtained as follows:

From (5) and (7):

$$\dot{x} + Dx = \mu_m \frac{sx}{K + s}$$

If $w_1 = \dot{x} + Dx$, then

$$w_1 = \mu_m \frac{sx}{K + s} \quad (12)$$

From (12):

$$\mu_m = \frac{w_1(K + s)}{sx} = \frac{w_1}{\emptyset} \quad (13)$$

where $\emptyset = \frac{sx}{K + s}$

Considering (6) and (7) we arrive to:

$$R = \frac{w_1}{w_2} \quad (14)$$

where $w_2 = D(S_a - s) - \dot{s}$

We can obtain the discrete version of these equations by means of the forward Euler algorithm:

$$w_{1k} = \frac{x_k - x_{k-1}}{H} + D_k x_k \quad (15)$$

$$w_{2k} = D_k(S_{ak} - s_k) - \frac{(s_k - s_{k-1})}{H} \quad (16)$$

$$\emptyset_k = \frac{s_k x_k}{K + s_k} \quad (17)$$

μ_m and R are obtained from equations (13) and (14).

This simple identification algorithm has been tested with a digitally simulated process. - The results obtained have been very good. - Later on, the global control scheme, with the identification algorithm included, has also been tested on digital simulation and the results obtained have also been very good. - Fig. 6.a shows the system evolution when a change in μ_m from 0.2 to 0.24 ocurrs and the identification algorithm has not been included in the control scheme. Fig. 6.b shows the system evolution for the same experiment when the identification algorithm has been included. As it is shown, the system remains at its operating point when the parameter identification is made.

The identification algorithm uses the derivative of x ($\dot{x}$). Then, if a real time simulation of the overall system is made, some problems will arise concerning physical realization of the algorithm.

CONCLUSIONS

In this paper the problem of controlling a certain kind of fermentation processes is discussed. These processes are highly nonlinear, so the use of conventional controllers is not adequate. Then, the control design is made considering the process model is nonlinear.

The work presented here has tried to improve the results obtained by Alvarez and Alvarez (1982) by introducing a on-line identification scheme which updates the controller parameter values. The identification algorithm used is very simple; however, the preliminary results obtained has been satisfactory; then it could be expected better results when the identification algorithms used be more suitable to the particular system to be controlled.

REFERENCES

Alvarez, J. (1978). Identificación de procesos de fermentación y optimización estática en cultivo continuo. Ph. D. Dissertation. CIEA-IPN. México.

Alvarez, J., J. A. Gallegos (1982). Optimal control of a class of discrete multivariable nonlinear systems. Application to a fermentation process. To be published in ASME Journal of Dynamic Systems, Measurement, and Control.

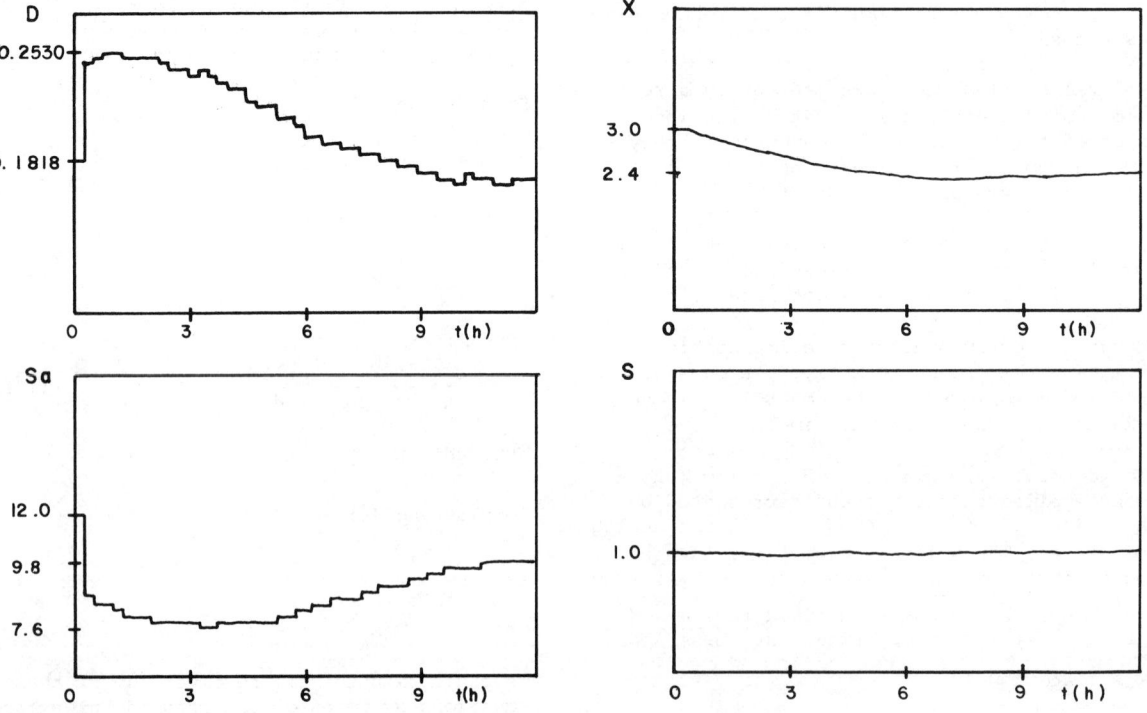

Fig. 2 System evolution for a change in the reference vector.

Ribot, D. (1976). Commande numérique et optimisation d'une unité pilote de fermentation continue. Ph. D. Dissertation. Paul Sabatier University. Toulouse, - France.

Sévely, Y., J. B. Pourciel, G. Rauzy, J. P. Bovee. (1981). Modelling, identification and control of alcohol fermentation process in a cascade reactor. <u>Proc. VIII IFAC World Congress</u>. Kyoto, Japan.

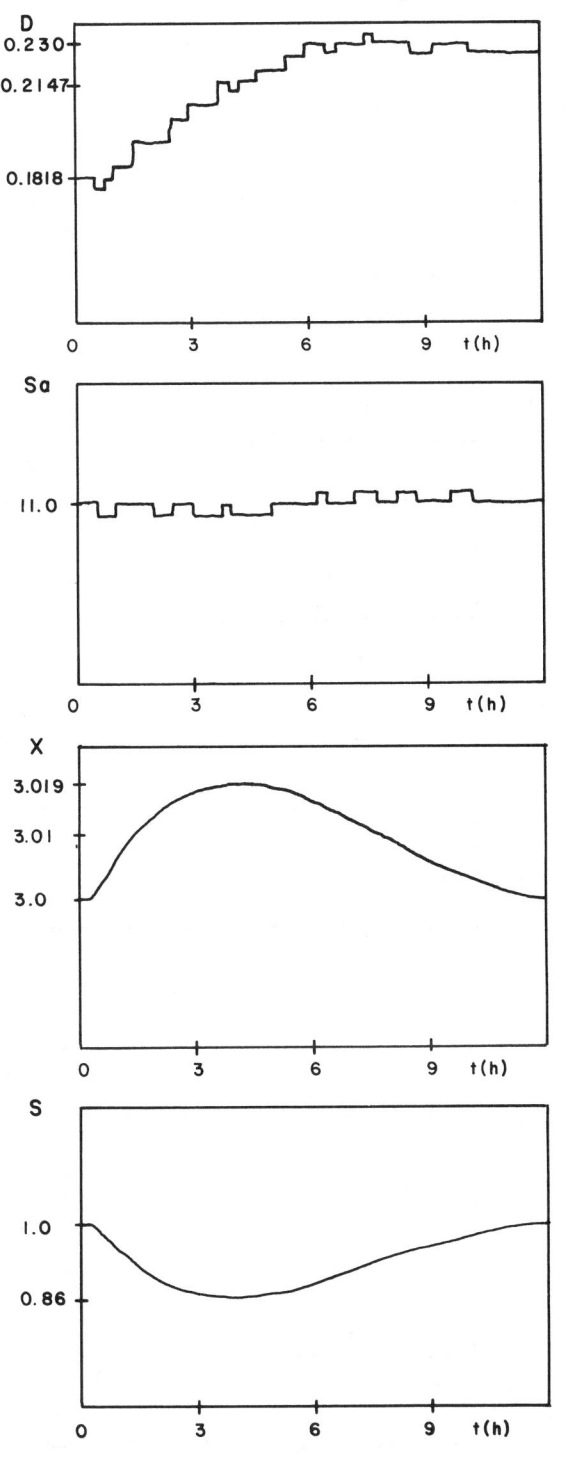

Fig. 3 System response for a change of 20% in μ_m.

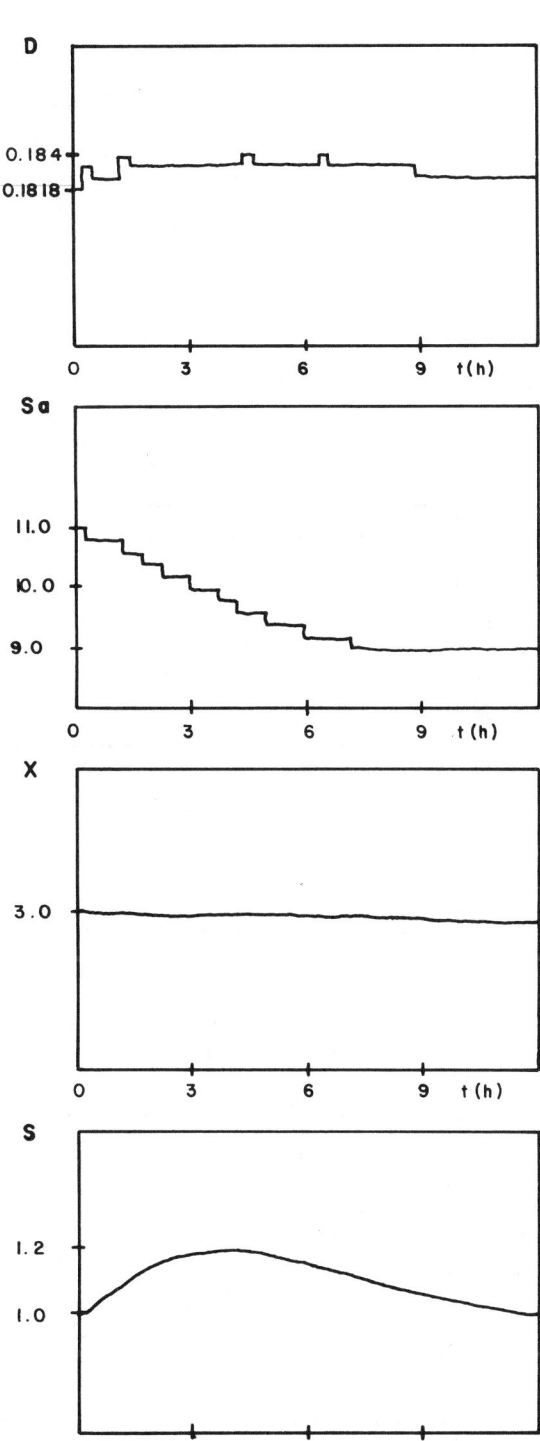

Fig. 4 System response for a change of 20% in R.

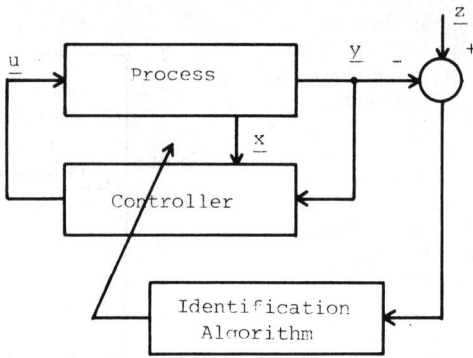

Fig. 5 Identification and control structure.

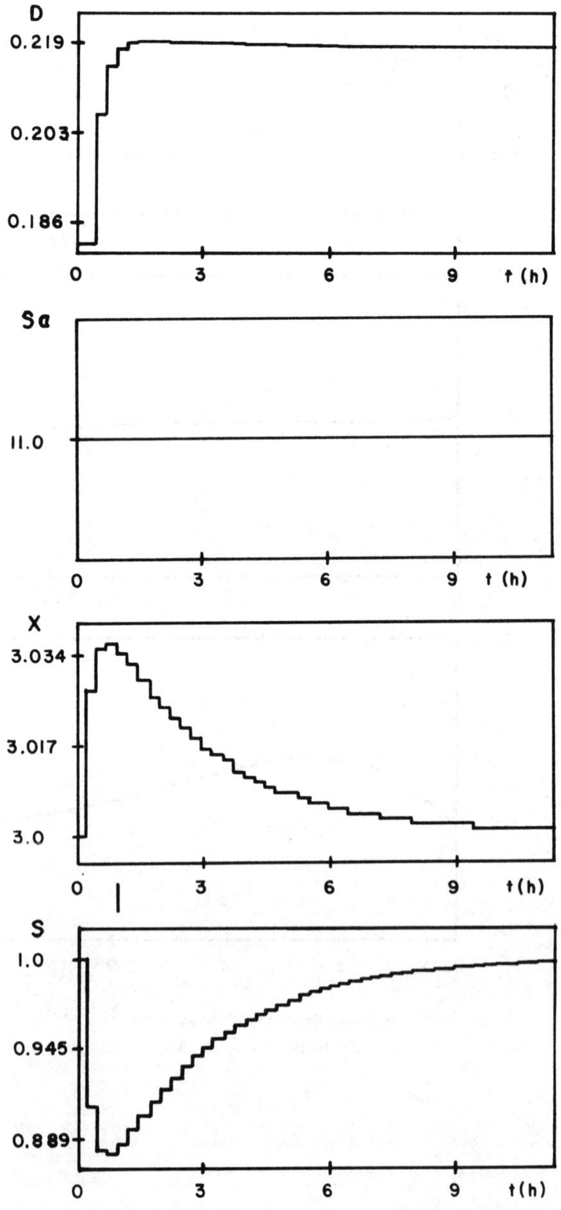

Fig. 6.a System evolution for a change of 20% in μ_m <u>without</u> identification.

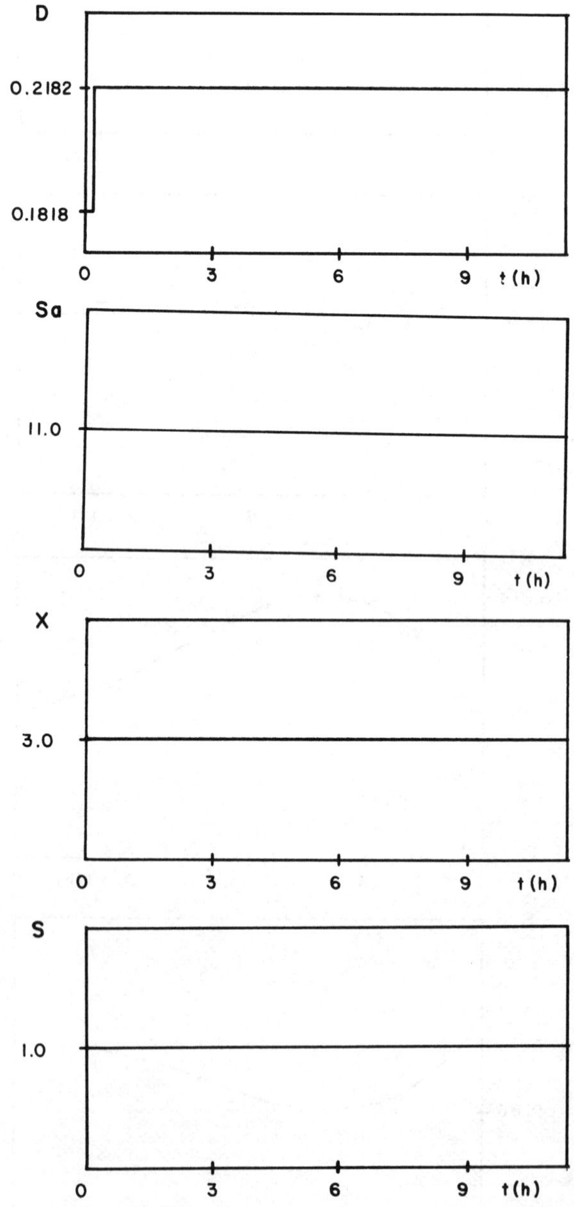

Fig. 6.b System evolution for a change of 20% in μ_m <u>with</u> identification.

A DIGITAL APPROACH TO MONITORING AND CONTROLLING FIBERFILL PLANTS

C. Gressel and A. Cohen

Department of Electrical and Computer Engineering, Ben Gurion University of the Negev, Beer Sheva, Israel

Abstract. Fiberfill factories are typical non-woven textile plants. They generally are old and poorly controlled carded fabric producers. A new digital approach is discussed for solving the "weight" and "hand" enigmas and to improve product quality. This overview analyzes computerized algorithms and methods for controlling weight, conveyor speed, polymerization temperature, and resin flows. For each parameter a digital transform was engineered to respond quickly and to adapt to mechanical aberrations, changing line frequencies and noisy environment. The I/O interface to all monitoring and controlling functions is through a Systems Timing Controller, compatible to any eight or 16 bit computer bus. In tangible benefits, the estimated savings per card are $ 40,000 per annum.

Keywords. Textile industry; process control; computer control; data acquisition; weight control; temperature control.

INTRODUCTION

Non-woven textiles are a very important branch of the textile fabric industry. Fiberfill, a non-woven textile product, is the synthetic replacement for goose down used for filling blankets, dolls, and jackets. Other growing users of fiberfill are automotive insulators, asphalt road bed reinforcers and airfilter manufacturers. This is a large volume, low-percentage-profit industry which clearly needs a simply integrated control system.

This overview is of a complete system for monitoring and controlling the functions and parameters in a typical fiberfill plant using an all digital approach designed around a single component, a Systems Timing Controller (STC). (AMD, 1980).

STC's are a sophisticated derivative second generation programmable counter-timer computer interface. These multifunction units can be software configured to serve as frequency counters and generators, auromatic timers and one shots, real time clocks and with almost endless computer interface options.

The card (see Fig. 1) is the basic tool in the non-woven and conventional fabric industry for converting raw fiber into a thin, clean, finely combed, lightly interlocked fiber web. It consists of rough and finely toothed rolls and drums revolving at varied speeds. The combing operation of the rolls and drums is designed to give the fibers a

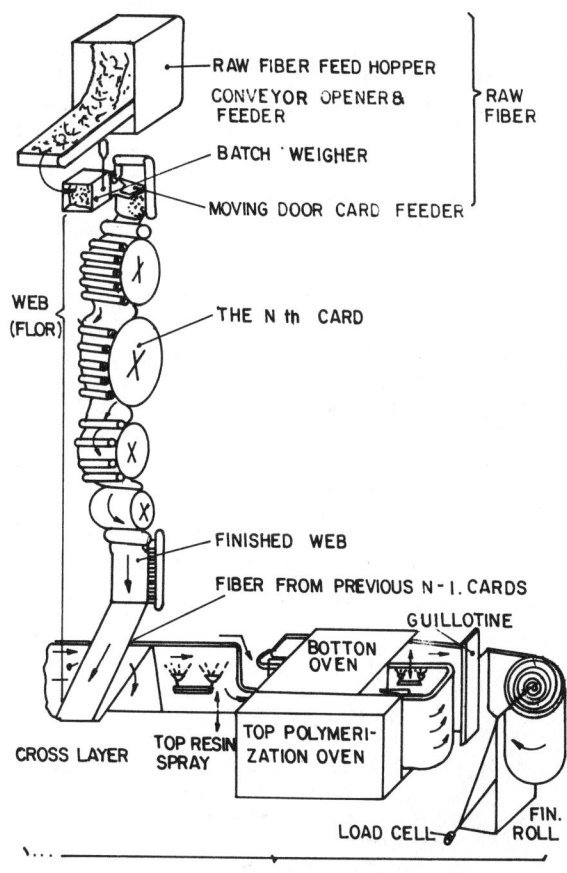

Fig. 1. A fiberfill plant

gentle monodirectional orientation with individual fibers interlocking at the sharp angles (with crimped fibers) or screwed together (with helix shaped fibers) (Schlese, 1968).

Typically, bales of raw fiber are mixed in a separate section of the plant (not in picture) then blown into the raw fiber hopper at the entrance to the card apparatus. From this hopper the fiber is fed into a moving pusher which goes forward every half cycle to receive a new batch of fiber from the weighing hopper.

The output of the card, a thin fragile web, is then cross layed onto a perpendicularly aligned conveyor which can simultaneously be fed by several cards. The crosslaying apparatus continuously goes back and forth the desired width of the final product laying this thin web. The final wadding is composed of many layers of webbed material.

Generally, the wadding is not stable and must be sprayed with a plastic binder. When properly applied the final user will not have a blanket or jacket where the fill is either lumped together in bumps, or where static charges shoot the individual fibers through the textile cover to form a stand of ugly hairlike fibers. The lumps would appear typically in a cheap blanket and the hairs are often seen in poor quality jackets. To assure a launderable properly cured polymer, this prayed wadding must be dried, then held at a polymerization temperature for a minimal curing time. Undercuring means that there are too many unlinked polymers. Overcuring in all cases is a waste of energy and with some proprietary binders can cause an irreversible process of polymer link breakdown.

The wadding is sprayed first on the top side, cured in the top oven, flipped over, sprayed again with resin binder, once more to be cured and dried in the bottom oven. The finished product is now ideally of proper "weight" (an industry misnomer for weight per square meter) and "hand" (a term for the feel of the product, denoting hardness of the wadding caused by the binder).

At the process output, the finished material should be cut to a measured length, weighed, and dispatched.

The bulkiness of the fill precludes large stocks of finished goods and forces constant changing of weight and width of output. Changeover on the ordinary uncontrolled system generally entails production stoppage, a trial and error period of readjustment of the conveyor drive train, the spray system, and the curing oven.

The motivation to keep close tolerance on weight, flow and temperature parameters on such a plant is overwhelming. Material economies can easily save 5% of the gross operating costs. The intangible savings from improved quality and production methods may amount to even more.

This solution is totally dependent on a host computer with extensive decision making capability. However, by using STC's, the computer time is kept to a minimum.

DESIGN CONSIDERATIONS

The fiberfill plant environment is generally unpleasant and electrically noisy with possible high voltage interference through electronic power supplies and atmosphere. The conservative retrofitter will opt for a bootstrapping digital computer using all existing monitoring and control devices. To make the expanded control system inherently fail safe the computer will override the manual setting making relatively small changes, and then at the first stage of development, only after receiving operator approval. Should the computer be inoperative, the plant must continue operating as well as or better than in the pre-computer era.

Knowing that this process is slow moving (almost all the time constants are much larger than 10 seconds) it is easily seen that conventional fast sampled A/D conversions are inferior to methods measuring elapsed time between pulses or, when convenient, simply counting frequency over reasonably gated periods. Simple, cheap, unshielded twisted pair copper lines are literally unaffected by noisy environments when they transmit these high SNR signals. Using a frequency or elapsed time base for calculations, signals can be digitally averaged over long periods of time (seconds as opposed to microseconds with sample and hold devices on common A/D's).

To simplify interfacing, all retrofitting instrumentation is designed to be compatible with STC I/O counting devices, software configured for each application.

CONTROLLED PARAMETERS

A direct solution to the weight problem becomes self evident when the operator uses the weight equation for control instead of the accepted trial and error methods. Adjusting weight has a direct effect on polymerization of the binding resins and consequently, on the oven controls.

The weight of the product in a plant with N similar cards as a function of the observable parameters is,

$$W_m = \frac{\tau_d}{\ell_d \ell_w} \{ \eta_f \sum_{r=1}^{N} \frac{\overline{g_r}}{\tau_r} + \eta_s \Omega_s \} \qquad (1)$$

where -

W_m = weight per area of finished product (g/m²)

ℓ_d = distance between two imaginary cogs on the final conveyor (m)

ℓ_w = width of wadding (m)

τ_d = interval between passage of adjacent imaginary cogs on the final conveyor (m)

η_f = fiber yield = $\dfrac{\text{fiber in final product}}{\text{raw fiber input}}$ (p.n.)

$\overline{\tau_r}$ = average period of batch cycle of r'th card (s)

$\overline{g_r}$ = average batch weight of fiber dump in the r'th card (g)

η_s = spray yield = $\dfrac{\text{dry weight of applied binder}}{\text{weight of aqueous solution}}$ (p.n.)

Q_s = total aqueous binder monomer solution flow per total system (g/s)

N = number of cards feeding system (p.n.)

Controlled weight is crucial to the economic feasibility of a plant. Obviously, fine tuning can be achieved simply by controlling one parameter, τ_d. Controlled quality demands that all variables be kept under vigilant surveillance.

Weight

The plant. Conventionally, the raw fiber is mechanically weighed and pulse fed to the card. The mechanical batch weigher is fed by the conveyor feeder at a random rate as chunks and chains of raw fiber fall into the batch hopper. When the static weight setting is reached the pendulum (scale beam) slowly moves up to the microswitch which stops the conveyor motor. This scale pendulum acts more sluggishly as the weight of the batch increases (see Fig. 2). This simultaneously increasing delay of response with increasing weight is caused both by the increasing length of the pendulum and more drastically by the nonlinear damping effects of the increased weight on the mechanical parts of the scale. The card serves as a temporary repository for passing fibers which cover the rolls and drums and can be reasonably treated as a single pole filter. Knowing the above mentioned problems of mechanical variance of the hopper weight and the single pole filter assumption for the card, it is easily understood why small batches (small τ_r and small g_r) will promise more uniform web with less ripple than infrequently dumped very large batches (Gressel, 1982).

At the output, promptly weighing a precisely measured and cut length of wadding for every finished roll is the most sensible safeguard for monitoring the weight equation.

The controller (operator or computer) can fine tune the system each time a roll is weighed by adjusting the speed (τ_d/ℓ_d in the weight equation) of the final conveyor.

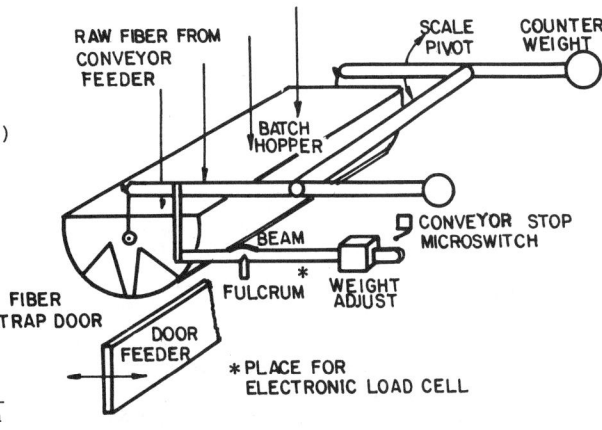

Fig. 2. Raw fiber batch scale.

The digitizers. The same type switching amplifier can adequately monitor the electronic load cell at the apex of the roll scale or at the fulcrum of the batch scale. To assure a well working plant with an inoperative computer, a novel roll weight digitizer was developed (see Fig. 3). Simultaneously, the digitizer drives a local display and at each weighing cycle sends a burst of pulses corresponding in number to the measured weight, adapted from Motorola (1979).

This method assures that the operator will read the same number both on the display and on the printed weight slip.

The electronic weigher drives the STC directly. Here the amplified signal is converted to frequency using a popular proprietary voltage controlled oscillator, Raytheon (1981). In case of computer failure the system reverts to the mechanical system previously described.

The STC interface. Both signals can be handled ideally by the STC. At each roll weighing cycle a burst of pulses is collected by a counter. After each burst the counter number is transferred to a holding register waiting for a computer check. See Mode N AMD (1980).

For batch weighing the STC counter is configured as a frequency meter, using the same mode as above. Here the counter is software configured to include a 16 bit comparator and an alarm register. If on any 20 msec sample the frequency count exceeds the alarm register a signal is dispatched to stop the fiber loading conveyor. Once initialized this operation is automatic and does not occupy the computer. However, the last weight is always available for computer surveillance.

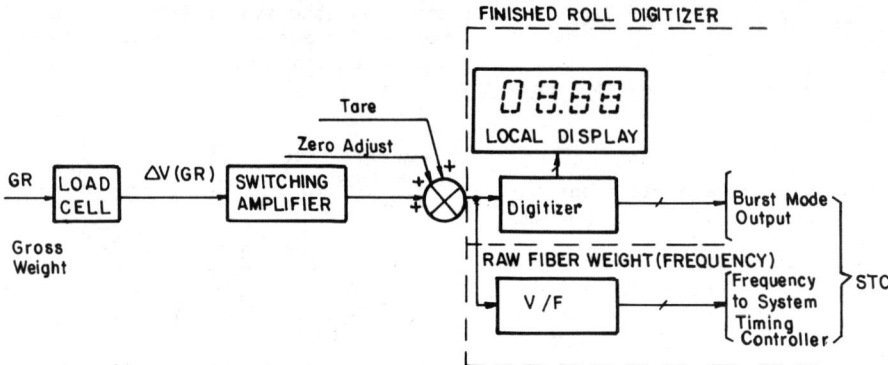

Fig. 3. Load cell amplifier schematic

Weighing procedure. The first improvement to conventional input weighing procedure entails quickly stopping the feeding conveyor when the static predetermined weight setting is reached. Further refinement implies accurately weighing the portion of fiber which has actually fallen into the hopper. This portion will invariably be larger than the preset weight as fiber falls into the hopper after the stopping signal. For up to 100 Hz sampling rates the frequency output from the pulse amplifier provides an ideal STC output. In the electronic version, the static weight setting is a computer derived digital comparator preset. Whenever fiber batch weight frequency exceeds the alarm preset number, a stopping signal is sent to the conveyor. Ideally, this preset comparator number could be changed for every batch to conform with actual previosuly delivered portions of fiber, to alleviate web weight ripple.

For final rollweight determination the STC counts each burst from the local digitizer. The computer receives a signal from the roll lift piston. Once weight signals are stable, several samples are averaged, registered in the billing computer and a proper weight slip is printed.

Measuring elapsed time. For slow moving processes, measuring time elapsed between two pulses with an accurate clock can provide a quicker precise rate indication than long term averaging of pulse counts. In the weight equation τ_d must constantly be monitored and $\bar{\tau}_r$ should be known to estimate W_m. The batch dumping period is dependent on the speed of an AC motor, and is therefore a function of line frequency and an unchangeable gear ratio. Conversely, τ_d is a controllable parameter indicative of conveyor speed. In the weight equation, fiber weight is the bulk of the total product weight and for quick adjustments of conveyor speed W_m can be approximated by Eq. (2), which neglects resin weight.

$$W_m = \tau_d \{k_1 \sum_r^N \frac{\bar{k}_r}{\tau_r}\} \qquad (2)$$

where k_1 and $\bar{k}_r$ are machine constants and τ_d and τ_r are inverse functions of line frequency. τ_r is uncontrollable and in some cases τ_d is measured and controlled using an absolute clock, i.e. the new proprietary DC motor control mechanisms. In such cases W_m, the material weight, becomes an almost linear function of line frequency. The simplest "decoupling" solution is to use line frequency dependent motors exclusively and to measure τ_d with a line frequency synchronized clock.

In many industrial areas line frequency aberrations are frequent and vary up to 10% from maximum to minimum. On a local installation with such a DC drive, product variations from day to night often averaged more than 5%.

Measuring elapsed time to determine conveyor speeds and batch rates on cards with a programmable counter in the STC is straightforward. The general configuration would be as in the weight examples. The gate would be software or hardware configured to be edge activated. The counting source is software routed to a multiplied line frequency clock.

Monitoring and controlling τ_d, τ_r, and ℓ. On most installations conveyor speeds are controlled by a motorized bolt adjusting mechanism. Typically, this is a 3-phase motor where two (accelerate, decelerate) separate relays must be operated. Driving these relays with an accurately timed one shot and a good control algorithm can bring such a system to close tolerance speed (weight) in one to three rapid iterations.

It has been found that on one popular make of speed controller, the proper length of the one shot signal is a function of present speed and the previous adjustment. An acceleration pulse of interval t following another acceleration pulse is far more effective than an acceleration pulse of identical t following a deceleration pulse. This can easily be explained mechanically, but complicates computer algorithms. Obviously, accomodating these nonlinear quirks is only possible with an easily programmed computer.

In the STC, the bolt controlling software

configuration is tantamount to two programmable one shots. In each case a counter is software configured to toggle a flip flop for a measured count of clock pulses. As the counter decrements to zero, a pulse toggles off the output flip flop.

Aside from the problems of changing line frequency, the path to absolute speed and length measurement is strewn with obstacles, as generally, tachometry cannot be easily performed directly on fluffy unbound wadding. Assuming stationary mechanical ratios between the measured output of the speed drive mechanism and the actual conveyor screen which carries the wadding is presumptive. Drive chains have been known to stretch up to 105% of their original length, without operator's noticing. Mechanically, this means an overall speed amplification of 105% with aberrations each time a link settles into a cog. In practice the relative speed of a conveyor provides sufficient information for controlling the weight equation.

For accurate weight measure, the true cut length of material in the roll must be known. A pulse generating disc can be depressed on bound output wadding to supply the odometric (distance travelled) input for measuring the length of the roll. Here an STC counter decrements from a preset number to zero to control the guillotine cutoff to proper roll length. Except for initialization when the preset number is loaded, this process does not demand computer intervention. At any instant the counter contents can be read through the Hold Register into the CPU without affecting counter data.

The odometer readout can be used to calibrate ℓ_d, the real distance between imaginary cogs on the final conveyor, compensating for slowly changing mechanical constants, resulting from normal wear and tear.

Measuring Flow of Binder Resin Solution

Knowledge of flow is necessary to solve the weight and temperature equations and no less to assure acceptable hand and loft (proper product bulk after laundering). Improperly working spray guns invariably cause unacceptable hand and poor product. Analysing flow as a function of time using simple mathematical tools (Rake, 1979) can pinpoint faults in spray gun operation. Because of problems of keeping clean resin pipelines, the metering method chosen should be non-invasive with minimum chance for build-up of particles in the line.

For this application a special electromagnetic flow meter (EFM) was designed and built. EFM's are generally used for much larger flow rates only, because of inherent complexities resulting from bulky coils and very low SNR's (Hogrefe, 1976).

This EFM accurately detects dynamic flow changes at sampling rates of 25 Hz (see Fig. 4). Because of parasitic sinewave and galvanic and other non-symmetrical noise signals, accurate signal processing demands synchronized filtering techniques. These disadvantages are more than offset an EMF's naturally output a linear signal for all flow rates unaffected by temperature or viscosity variations for all "slightly conducting" solutions.

For proper dynamic detection, the unit transmits two signals to the STC. The Synch output gates the flowmeter counter to accept pulses from one complete sine cycle. This, in essence performs a numerical integration over the noise aberrations of one complete period.

The same frequency output can be used for dynamic rate analysis and total flow (volume) measure. For totalizing, several counters are software cocatenated. In this configuration, the first cascaded counter serves as a calibration prescaler, whilst the others serve as counters which the CPU would read to determine volume of binder consumed since STC initialization.

Polymerization

Binder curing. Polymerization is performed in a chamber with forced hot air circulation. Experiments have shown that for a large class of proprietary binders, the linking of the binder monomers to a launderable polymer progresses at a rate corresponding to the Ahrrenius equation (Kice, 1974; Allewelt,

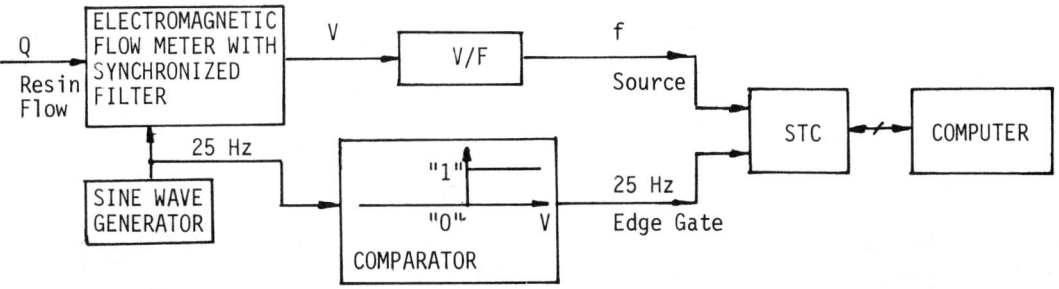

Fig. 4. Dynamic flow detection with EFM

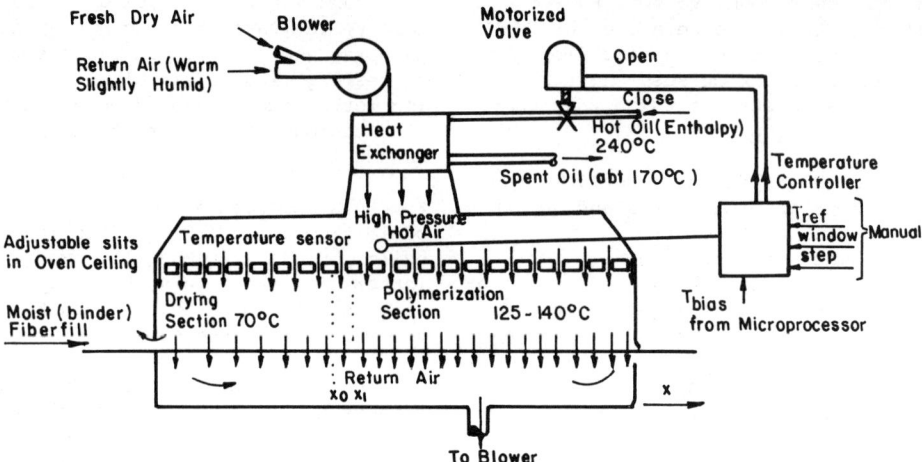

Fig. 5. Schematic - one chamber of polymerization oven

1964)

$$\text{Rate} = A \exp(-B/T) \quad (3)$$

where A and B are essentially constant for a given process and T is absolute temperature.

It can easily be shown (Gressel, 1982) that for this class of proprietary preparations in practical polymerization ovens, the polymer linking is dependent only on the time that the uncured product resides at final curing temperature. It is therefore essential that a multi-point temperature monitor "follows" the moving wadding through the drying then curing stage and determines accurately the temperature profile of the oven. The practical result of this profile measurement is to know the amount of time, t_{pol}, that the binder resides in the oven chamber at full temperature. Once known, the computer can send a temperature correcting signal to the hot oil enthalpy controller (Fig. 5). This signal, T_{bias}, offsets the manually set reference signal to the temperature controller and in a given oven is a function of t_{pol} and the reference temperature setting, T_{ref}.

Calculating T_{bias}. Eq. (4) is a linear piecewise approximation of the strict solution of Eq. (3) based on the extrapolation for this class of binders of the rule of thumb that for every 5°C increment, the time of residence, t_{pol}, is halved

$$T_{bias \text{ in the } i^{th} \text{ segment}} = T_{bias \text{ io}} - \frac{t_{pol}}{2^{i+1}} \quad (4)$$

where T_{bias} is the bias at the beginning of the ith segment. Adopting this solution allows a quick approximation of T_{bias} in crude machine language using rotate operations for division. In Fig. 6 one can see this linearization of the curve for the popular acrylic resin binders with a set temperature setting T_{ref} of 137.8°C compared to the exact solution in broken line.

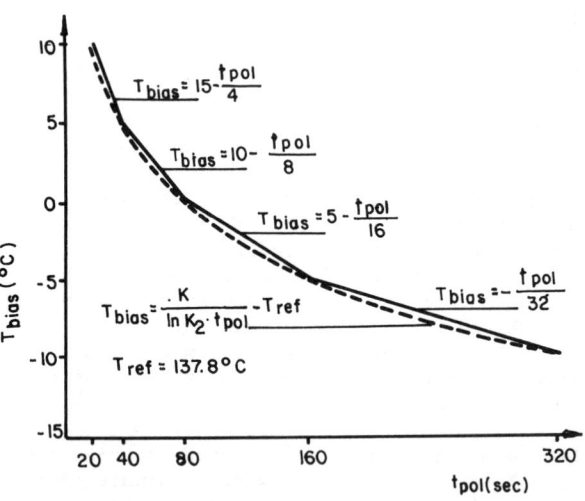

Fig. 6. Temperature bias as a function of residence time of resin

In a well kept oven a meticulous operator might be enticed into assuming that t_{pol} can be calculated as a function of oven length, the amount of resin flow and the speed of the conveyor. He would assume that his oven is stable in time. As stray resin powders build up in all of the oven orifices forced air passage is blocked, making the temperature profile a partial function of time and a parameter that must constantly be assessed.

Temperature measurement and control. An inexpensive multipoint time division multiplexed temperature sending device for determining oven profile was developed. Each successive signal is gated and marked (see Fig. 7) with a low frequency marker. Temperature signals are differentiated and evaluated at the STC. Each burst from the scanner transmits either a clear one of eight

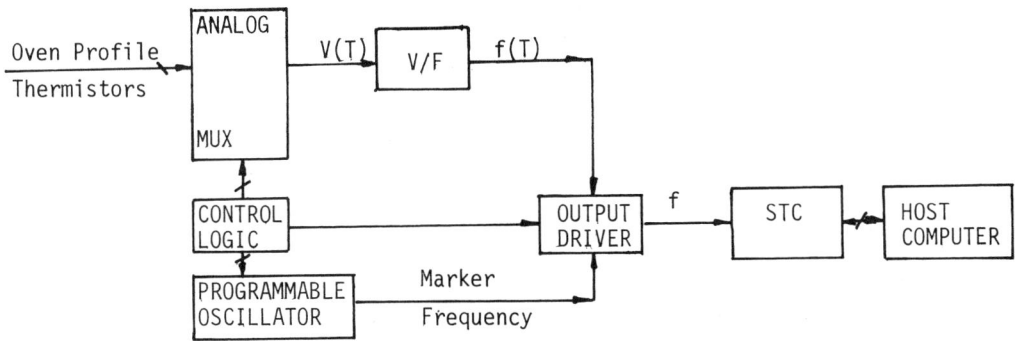

Fig. 7. Multipoint thermistoric temperature scanner for an STC.

device marker or the corresponding scanned temperature signal from an inexpensive thermistor. Algorithms for automatic calibration of thermistors and for making precision linear networks using the cheapest nonuniform off the self thermistors were developed based on Trolander (1971).

Temperature bias signals are easily transmitted from the STC to the hot oil controllers as simple frequencies. A control frequency to be converted at the controller to a voltage reference is generated at an STC counter. This counter is software configured to serve as a divide by N frequency divider. An STC down counter is loaded at every half cycle with N/2. At count zero an internal flip flop is toggled. The result is a clean square wave of a frequency corresponding to T_{bias}.

All commercial enthalpy controllers can accept these analog signals, either directly or by biasing the incoming sensor signal.

CONCLUSIONS

Several efficient tactics for controlling an enigmatic fiberfill plant to achieve uniform quality in a retrofitted old or completely new plant were described.

Emphasis was placed on several tachometric, thermal, and weight quirks in such plants.

An overview of the methods for solving these problems and the developed circuits was given.

A uniform strategy was developed using a systems timing controller which should ease developing a computer system in many similar slow moving processes.

It is almost impossible to execute a plan such as this without using a digital computer with capacity for decision making and the ability to change to many simultaneous parameter changes.

REFERENCES

Allewelt, A.L. (1964). Acrylic polymers for spray bonding of polyester fiberfill. Resin Review, Vol. XIV, No. 3.

AMD (1980). System Timing Controller Am9513, Advanced Micro Devices, Inc., Sunnyvale, Cal.

Gressel, C (1982). Controlling the Production of Fiberfill (Hebrew). M.Sc. Thesis, Electrical and Computer Engineering Dept., Ben Gurion University of the Negev, Beer Sheva, Israel.

Hogrefe, W. (1976). Magnetische-induktive Durchflussmesser, Regelunstechnische Praxis, Heft II, pp. 299-304.

Kice, J.L. and E.N. Marvel. (1974). Modern Principles of Organic Chemistry. MacMillan Publishing Co., New York.

Motorola. (1979). Linear Interface Integrated Circuits. Technical Information Center, Motorola, Phoenix, pp. 849-859.

Rake, H. (1979). Step response and frequency response methods. Tutorial presented at the 5th IFAC - Symposium on Identification and System Parameter Estimation, Darmstadt.

Raytheon. (1981). Voltage to Frequency Conversion-4153. Raytheon, Mountain View, California.

Schlese, G. (1968). Production of carded fiber webs for non-woven fabrics. Chemiefasern, November.

Trolander, W., D.A. Case and R.W. Harruff. (1971). Reproducibility, stability and linearization of resistance thermometers. Paper presented at the 5th Symposium on Temperature, Washington, D.C., June 21-24.

A PILOT-SCALE DISTILLATION FACILITY FOR DIGITAL COMPUTER CONTROL RESEARCH

J. L. Marchetti, A. Benallou, D. E. Seborg and D. A. Mellichamp

Department of Chemical and Nuclear Engineering, University of California, Santa Barbara, Santa Barbara, CA 93106, USA

Abstract. A new pilot-scale multicomponent distillation facility and associated computer control system are described. The control system design utilizes a hierarchical structure, with a microcomputer for direct digital control of secondary process variables and a general-purpose, real-time computer for high-level control of the primary variables. The facility is used for advanced process control research and real-time system instruction. Some typical results incorporating a new multivariable predictive control algorithm are presented.

Keywords. Computer control; direct digital control; distillation; distillation column control; hierarchical systems; multivariable control systems; process control; supervisory control

INTRODUCTION

A pilot-scale distillation column designed to fractionate a ternary mixture of alcohols has been installed in the Department of Chemical and Nuclear Engineering at the University of California, Santa Barbara. The experimental unit will serve four important functions: 1) provide a convenient demonstration unit for advanced control concepts of a general nature such as multivariable and adaptive control, 2) help to critically evaluate promising control strategies for multicomponent distillation problems, 3) serve to develop and evaluate multicomponent distillation models, both for control studies and for the advancement of modeling methodology for distillation columns and large-scale systems generally, 4) play an important role as a demonstration unit in applications of real-time computing, both for teaching and research.

In choosing a distillation pilot-plant unit for research studies, we have been guided by two main considerations: i) previous successful experience with a dual-effect evaporator unit that was used for advanced process control studies at the University of Alberta by Fisher and Seborg (1976), and ii) the importance of distillation processes, both as a fundamental chemical engineering process and because they represent a major energy user in any modern, petroleum-based society. As the price of refined products and the cost of energy have increased, the economic incentives for improving the control of distillation columns have become more compelling (Luyben, 1975; Fauth and Shinskey, 1975). Shinskey (1977), in particular, cites the relationships of distillation control to productivity and energy conservation.

The existence of a very large number of published articles dealing with various aspects of distillation column control also is ample evidence of the importance that industry and the academic control community have given to this subject. Top and bottom product composition control of even a binary separation has proven to be particularly difficult because of the interactions between control loops, cf. Rijnsdorp (1965), Rosenbrock (1962), and Davison (1970). Because of these interactions, many studies have been concerned only with the control of overhead product composition, cf. Wood and Pacey (1972), Shunta and Luyben (1971), and Merluzzi and Brosilow (1972). Multicomponent columns, such as the UCSB pilot unit, are much more typical of those found in industry; because stream analysis must be used to measure compositions of important streams, they also are much more difficult to control.

Implementation of advanced control strategies on most large-scale processes, distillation included, requires a computer control system. Characteristics of the control systems are closely related to both the process design and to the control objectives. Kramer (1980) cites three classes of computer control applications that conventionally have been used with distillation systems:
(i)   Direct digital control,
(ii) Supervisory control,
(iii) Optimal and other advanced (dynamic) control methods.

As will be discussed in more detail below, a typical distillation system lends itself to a combination of classes (i) and (ii) in industrial situations and (i) and (iii) in academic control research. The reasons are that most columns typically involve a number of control loops which merely regulate secondary variables in the system, e.g., liquid levels in the reboiler and distillate receiver. Set points for primary variables, such as energy input at the bottom of the column or product returned as reflux at the top, are adjusted periodically. With supervisory control systems, the period is quite long, typically once per day, and the adjustments are based on optimization of a high-level economic model that yields optimum steady-state operating conditions to use as set points for the primary variables. With optimal and other dynamic control methods, the period usually is set as short as the cycle time ("analysis time") of the product stream analyzer, often a sampled output device such as a gas chromatograph.

Digital control systems which require, at the same time, regulation of a number of routine secondary loops and high-level optimization or dynamic manipulation of primary loops lend themselves admirably to a hierarchical type of organization (Mellichamp, 1983). For this reason the UCSB distillation facility has been designed with a hierarchical configuration, one that easily permits the implementation and testing of advanced control algorithms in a high-level, real-time computer while handling routine control objectives in a low-level, microcomputer-based unit.

Figure 1 gives an overview of the computer-controlled pilot plant, clearly showing the high/low-level computer organization. The control system consists of a microcomputer (Moore Industries MI 1002) which performs the low-level (DDC) control functions, and a host computer (Data General Eclipse S/130) used for implementation of advanced control algorithms, mass and energy balance calculations, and the generation of periodic reports. The control computations performed by the Eclipse computer generate the set points for several of the low-level microcomputer control loops. Each of the two computers has been equipped with a number of peripheral devices including operator consoles, graphics terminal, line printer, etc. A Hewlett Packard 5840A gas chromatograph with automatic liquid sampling system is used to obtain periodic composition measurements of the two distillation product streams. Individual component concentrations are communicated directly to the high-level computer via a 600 baud ASCII link.

At the present time, the MI 1002 digital communication interface is still under development; hence communication between the Eclipse S/130 and the MI 1002 necessary for implementation of the hierarchical system implementation is performed through the respective computer analog interfaces. This communications alternative and the detailed design and functions of each of the computer systems are discussed subsequently after the distillation column is described.

THE PILOT-PLANT DISTILLATION UNIT

Preliminary criteria for the design of the multicomponent distillation system were not tied to a particular production rate or quality of the product mixtures. However, the desired system had to furnish the flexibility of operation needed for research equipment and to satisfy constraints due to limitations on space, available steam, cooling water consumption, safety considerations, and storage capacity for the chemicals, to mention only the most important ones.

From the viewpoint of process dynamics, it was desirable to have small residence times in both the condenser/reflux-drum system and the bottoms-accumulator/reboiler system so as to reduce the overall dynamic response of these units vis-a-vis those of the trays as much as possible.

The column was designed to fractionate a ternary system of alcohol isomers: n-butyl, s-butyl and t-butyl alcohols. After a careful screening of many alternatives, the above ternary mixture was selected based on the following key considerations:
a) Availability of equilibrium data for the ternary system for all three binary pairs.
b) absence of azeotropes.
c) suitable boiling point range (allows 20 psi steam to be used in the reboiler and cooling water in the condenser).
d) safe operation, i.e., low toxicity and relatively low flammability.
e) suitable relative volatilities (separation is neither too easy nor too difficult).
f) ease of on-line composition analysis (via gas chromatograph).

Column

The installed column consists of twelve sieve trays, six inches in diameter, separated by ten-inch dividers. Each divider contains i) a four-inch high stainless steel section with appropriate connections for sampling vapor and liquid streams, and ii) a standard six-inch glass pipe section to allow inspection of the internal process. The stainless steel section also has connections to locate thermocouples in both the vapor and the liquid phases. Moreover, a feed connection stub-in has been provided on each tray, increasing the flexibility of the unit for a variety of research studies. Sieve trays (perforated plates) with liquid crossflow were used in this unit. These trays are most commonly specified for new distillation column designs because they provide 10 to 20% more separation efficiency at optimal column loadings and cost 50 to 70% as much as bubble-cap trays. A circular overflow weir and tubular downcomer can be adjusted to accomodate different operating conditions, for instance to increase the tray liquid hold-up. The column and all heated auxiliary units are insulated with commercial fiberglass pipe insulation.

A preheater in the feed line, a condenser and receiver for the distillate vapor, and a natural recirculation reboiler have been installed in order to provide for flexible operation. Three micropump magnetic drive gear pumps are used in the pilot plant: i) the feed pump installed in the storage room approximately 150 ft from the column is dedicated exclusively to supplying the column feed line, ii) the bottoms pump, located in the column area, returns the bottoms product to the storage room, and iii) the distillate pump supplies pressure to return both the distillate product to the storage room and the reflux stream to the top of the column. A fourth pump in the storage room is used to transfer material from one storage tank to another or to recirculate the contents of the feed tanks. Figure 2 gives a schematic view of the column and its main auxiliaries. Marchetti (1982) gives complete details of the system.

Column Auxiliaries

The condenser installed in the distillation unit is located vertically with cooling water in the shell side and alcohol condensate in the tubes. Following current practice for vertical-tube condensers, the vapor and liquid flow cocurrently downward. Since pressure drop is not a limiting consideration, this configuration can result in higher heat transfer coefficients than shell-side condensation; it also has particular advantages for multicomponent condensation. A reflux drum (distillate receiver) with low capacity provides minimum hold-up of reflux/distillate liquid so as to minimize the associated time constant.

A thermosiphon or natural recirculation reboiler is connected to the lower part of the column through a 2-inch diameter pipe which feeds the reboiler with the liquid bottoms mixture and a 4-inch diameter pipe that returns vapor and recirculating liquid to the column. In addition to the heat transfer considerations, the reboiler was designed to meet the capabilities for recirculation as well as the reduced hold-up criteria noted above in connection with the reflux drum.

The feed preheater is used to heat the feed stream from room temperature to just below the mixture boiling point. This heat exchanger operates with steam in the shell side and the ternary feed mixture in the tube side. The feed preheater, condenser, and reboiler, were constructed of Type 316 stainless steel.

Since the column is designed for operation at atmospheric pressure, a vent line and knock-out condenser are provided to vent noncondensable vapors to the environment safely. Because both distillate and bottom products are quite hot, these streams are cooled to 25°C before returning them to the storage area.

A separate (isolated and explosion-proofed) feed and product storage area is provided for the pilot-plant unit. The large capacity of the feed and product tanks permits the column to be operated continuously between 8 and 12 hours, depending on characteristics of the operating point.

Process Instrumentation

Since this distillation pilot plant serves as an experimental unit for modeling studies and applications of advanced control techniques, it was heavily instrumented to meet the needs of future operations, including the requirements of any control tests. Every significant process variable is recorded, logged, or displayed to keep track of the operating conditions. Table 1 lists the key process variables. Figure 3 shows schematically the main facility and the installed instrumentation. Each of the stages except the top one has been fitted with a type J thermocouple to sense the liquid temperature. These signals are recorded using an analog multi-channel recorder to give a continuous indication of column temperature profile.

Temperature transmitters have been connected to the thermocouple installed on tray number eleven and to a resistance thermometer device (RTD) in the lower part of the column, respectively. These two temperature signals are extremely important from the control point of view because they can be used as controlled variables when direct composition analysis is not suitable for control purposes (as is discussed later in the experimental section).

The lower part of the column is fitted with two additional instruments which should be mentioned: i) a pressure transmitter connected below the first tray for measurement of column ΔP, ii) a level transmitter that serves the bottoms sump level controller.

Four primary process lines serve the needs of the distillation column: the feed line, the distillate product line, the bottoms product line, and the reflux line. Each of these lines has a flow transmitter connected to an orifice meter for measurement of flow rate and an automatic valve for control of flow.

Steam available at 100 psig is reduced to 40 psig through a pressure regulator in the main supply line before it splits into two branches, one supplying the feed preheater and the other providing steam to the reboiler. A normally-closed solenoid valve installed in the main line is manually activated from a remote switch panel. This valve has proved to be extremely useful, not only during start-up and shut-down operations, but particularly for emergency situations. Elements common to these two lines are a control valve, pressure indicator, temperature indicator and steam trap. The single significant difference is the pressure transmitter connected to the reboiler inlet for measurement of steam pressure in the reboiler.

Water is used as the cooling medium for three different heat exchangers: the column main condenser, the knock-out condenser, and the cooling tank for the product lines. The main condenser has an important effect on the column operating condition while the other two exchangers can be regarded as necessary auxiliary elements with no important effect on the process. Consequently, only the water line to the main condenser has been instrumented; a control valve, a rotameter and two dial thermometers satisfy the current needs.

Available compressed air at 160 psig is reduced to about 20 psig through a pressure regulator. A manifold feeds the current-to-pressure transducers that provide the interface between the controllers and the final control elements, the seven pneumatic control valves shown in Figure 3.

Provisions for automatic sampling of the bottoms and distillate lines are included in the instrumentation, as noted above. Since the reflux stream has the same composition as the distillate stream, and since the feed stream does not change (and can be analyzed manually at the beginning and at the end of each run), there is no need to automate the sampling of these two lines. Each sampling line has a flow regulator attached to a rotameter to maintain constant flow and to allow visual inspection of its working condition.

A Hewlett Packard 5840A Gas Chromatograph which includes an automatic sampling system is programmed to sample both product streams alternatively. Currently, the sampling period has been reduced to about 140 secs, the minimum time required for chromatographic analysis and reporting of a single stream result. The gas chromatograph can be operated manually from a terminal or automatically from the high-level computer.

Low-Level Control Loops

Seven low-level control loops are used during regular operation of the distillation column. Table 2 summarizes This multiloop configuration. All of these loops are serviced by the low-level microcomputer system, as is discussed below. A considerable amount of trouble initially was encountered in tuning the PI or PID controllers for several of these loops (Marchetti, 1982). Primary sources of difficulty were associated with:
i) Valve stiction combined with oversized valve trim resulting in limit cycle behavior,
ii) interactions between loops, particularly Loops 1 and 6 where the single distillate pump supplies upstream pressure for both the distillate and reflux control valves,
iii) process noise, associated with steam supply pressure in Loop 2,
iv) excessive thermal capacitance associated with the heat exchanger metal and cooling water in the distillate condenser.

Problems i) and ii) were eliminated by using smaller control valve trims. Problem iii) has been attributed to a defective steam supply pressure regulator. It will be replaced in a future modification. Problem iv) was substantially eliminated by redesigning the cooling water supply line to reduce the lag associated with that stream. However, of the seven loops, the distillate temperature controller (Loop 7) remains the most difficult to tune and the one most subject to disturbances. Eventually it may be necessary to replace the condenser with one of smaller capacity.

COMPUTER CONTROL SYSTEM

The Low-Level (DDC) Computer

The Moore Industries MI 1002 controller is a general-purpose process control microcomputer with a read/write bubble memory capable of handling up to 112 analog and/or digital inputs and outputs and up to 24 control loops. In addition, a number of algebraic manipulations can be performed using the 96 available function blocks. This microcomputer, which is located in the proximity of the distillation unit, is used for monitoring and controlling it in a dedicated fashion. As indicated in Figure 1, the MI 1002 has two peripheral devices for operator communications, a color graphics terminal and a line printer.

The color graphics terminal is used for routine MI 1002/operator communications. These communications are supported by full bar graph displays of the key process variables, alarm variables, and control loops. Several operator-selectable display modes can be used interchangeably depending on the level of information sought by the operator at a given time or depending on the state of the process. These display modes range from an overview of all the process and alarm variables to a bar graph display of a single control loop or process variable.

The line printer is used to obtain periodic reports of operator-selected process variables or controller outputs and to produce a hard copy of any alarm reports during process operation. This type of information is particularly useful for analysis of nonstandard process conditions.

In the present column control configuration, the MI 1002 microcomputer performs the following tasks:

Data acquisition. The key process variables are logged by the low-level microcomputer and are used for control computations, alarm checking, and generation of periodic reports. These variables are shown in Table 1; the data are made available to the MI 1002 through its analog interface (signal level: 1-5 volts, 16-bit converter, 0.0025% resolution). Some of the variables in Table 1 play an important role in the safe operation of the column. These variables are checked for alarm conditions at each sampling time.

Alarm checking. Table 3 shows the process variables which are regularly checked to insure safe operation of the distillation unit. During operation the column pressure is not allowed to exceed a certain limit and the reboiler and distillate drum levels are kept between pre-set low and high limits. In addition the feed and distillate flow rates are also regularly checked in order to determine instrument malfunctions resulting in aberrant flow measurements. When a variable enters an alarm state and remains there for an operator-prespecified time interval, the digital output of the alarm function is set. The system then automatically transmits this information to the operator through the graphics terminal, and an alarm report is generated via the line printer. Typically, an alarm report shows the date and time when a given variable enters its alarm state or leaves it. In addition, the output of the alarm function can be used to take a control action which would prevent prolonged process operation under abnormal conditions. An example of such a control action would be to close the solenoid valve supplying steam to the reboiler if the column pressure remains in its alarm state for an extended time period.

Signal conditioning and direct digital control. In addition to data logging and alarm checking, the DDC microcomputer is responsible for low-level control. As discussed above, seven loops are used to control the key process variables shown in Table 2. Before beginning control computations, some of the measurements are filtered using the signal conditioning functions available on the MI 1002. A typical direct digital control loop arrangement is shown in Figure 4. Note that the output from the MI 1002 is a sequence of pulses which is sent to the Direct Digital Interface (DDI). The DDIs are manual/automatic override stations that also can be thought of as the latching components for the final control elements. The DDI output is an analog signal in the 4-20 ma range which drives a current-to-pressure transducer connected to a pneumatic control valve.

All control options available on the low-level microcomputer are versions of the PID discrete algorithm including several practical features such as:

· velocity and position algorithms
· anti-reset windup
· anti-derivative kick
· error limiting capability.

In addition, the sampling period is varied internally to keep it as small as possible, i.e., the system does not remain idle between two control computations. When the number of control loops is large, a maximum sampling period of 5 seconds is imposed.

Operator assistance during start-up and shut-down. The DDC machine is also used to assist the operator during start-up and shut-down by supervising the operation and using the alarm functions discussed earlier. At the present time, the MI 1002 only reports process conditions and alarms in an "operator-assist" fashion; decisions are left to the operator. However, the control system design provides the necessary hardware for automated start-ups and shut-downs, i.e., for on/off control of all pump drive motors and operation of the steam-line solenoid valve. TTL-compatible relay switches allow the DDC machine to start or stop any of the feed, distillate, or bottoms pumps, as well as to open or close the steam-line supply valve. Note that these electrical connections are made in such a way as to allow operator manual override via the DDIs.

In the dual composition control scheme used by Benallou (1982), the set points of five of the DDC microcomputer control loops are established by the operator (Loops 3-7). These would only be changed to simulate a disturbance to the column, for example, a change in feed flow rate. Once the system start-up is complete, the set points of the primary loops (reflux flow rate and reboiler steam pressure) are driven by the high-level or host computer described in the next section.

The High-Level Computer

The host machine (Data General S/130 Eclipse) is located in the Real-Time Laboratory which is housed one floor above the distillation area. In addition to performing the high-level functions necessary to control the distillation unit, the host computer is responsible for a number of additional tasks. It is a general-purpose machine connected to a number of peripheral devices and processes through standardized interfaces. Figure 5 illustrates the present system. The system software consists of the Data General Advanced Operating System (AOS) supporting multiuser, multitask operations in a real-time environment.

The signals available to the high-level computer are in the 1-5 volt range. If these signals were sent directly to the machine, the noise to signal ratio would not be adequate. Consequently, differential amplifiers and low-pass filters were installed at the host computer ADC ports. The purpose of the differential amplifiers is to raise the signal level from 1-5 volts to -10 to +10 volts, permitting utilization of the full ADC span in the host machine.

A multitask program in the host computer contains several tasks dedicated to operation of the distillation unit:

Operation of stream analyser and logging of results. As shown in Figure 1, the sampling system consists of a terminal interface (HP 18833A digital interface) which links the host computer to the stream analyser (HP 5840A Gas Chromatograph) equipped with an automatic liquid sampling valve. The host computer both commands the system to take a sample at operator- or program-specified time intervals and collects the results of the gas chromatograph analysis. In the present configuration, two streams, the distillate and bottoms, are sampled on an alternate basis. The analysis and reporting time for a single stream is 140 seconds.

Data acquisition and storage. The key process variables for the distillation unit (see Table 1) are made available to the host computer through its analog interface. The system also collects the distillate and bottoms composition measurements via the gas chromatograph terminal interface. Data collected by the host machine are stored and used to perform a number of operations including advanced control calculations and the generation of the column status reports.

Mass and energy balance computation and status reports. At operator request, the host computer uses the most recently logged data to compute stream densities and enthalpies. Material and energy balances are then calculated and the results are transmitted to the operator through the CRT terminal in the form of a column status report. The status report includes the temperature, composition, density, enthalpy, and flow rate of each column stream, as well as the results of mass and energy balance computations. Condition reports are used by the operator to check the consistency of measured data and the approach to steady-state operation. The operator can save the status reports on disk as data files for later analysis.

Advanced control calculations. In addition to the auxiliary functions discussed above, the multitask program also performs the advanced control algorithms. The set points for the low-level microcomputer reflux flow rate and reboiler pressure control loops are generated by the host computer periodically and sent to the low-level microcomputer as analog signals (Figure 6). This type of communication between the high and low-level machines is a straightforward alternative to digital communication (currently not available since the MI 1002 digital communication interface is still under development by Moore Industries).

Operator-host communications. The multitask program also handles routine communications between the operator and the high-level machine. These communications include requests to:
- perform mass and energy balances,
- generate condition reports,
- save condition reports,
- open or close high-level control loops,
- change or display the value of a given parameter or variable.

CURRENT RESEARCH

At the present time the distillation unit is being used by one M.S. and five Ph.D. students to verify their theoretical modeling and control developments. Table 4 summarizes the areas of study and personnel involved.

EXPERIMENTAL CONTROL STUDIES

During the past five years, several research and development groups in both university and industry, have investigated predictive control techniques. This approach is characterized by the use of a multi-input/multi-output process model based on a convolution-type representation whose parameters can be obtained easily by step-input testing of the actual process. The resulting deterministic model typically requires from 20 to 40 parameters to represent each input/output pair; hence a computer is required to perform the matrix operations associated with prediction of the future process trajectory. Knowledge of the predicted trajectory, however, permits the manipulated inputs to be computed so as to optimize a particular transition from one operating state to another, to avoid hard constraints on the process outputs, etc. Unlike traditional multiloop and many modern multivariable control techniques, there is no requirement that the number of process inputs and outputs be equal; hence an additional advantage of such predictive methods is that many process inputs can be utilized for control, and a strict pairing of an input with an output is avoided.

References concerning output-predictive techniques are predominately those related to Model Algorithmic Control (Mehra and others, 1978), to Model Predictive Heuristic Control (Richalet and co-workers, 1978), and to Dynamic Matrix Control (Cutler and Ramaker, 1980). Marchetti presents a complete review of recent activities in this field and describes several modifications to the methodology. In an experimental evaluation of multiple prediction control (MPC) techniques, Marchetti compared an MPC algorithm with standard well-tuned, multiloop (proportional integral) control of the pilot-plant distillation column for the case where temperatures at both top and bottom of the column are controlled. In this case the temperatures are approximately related to the product stream compositions and can be measured continuously. Hence implementation of the control algorithms is not restricted to the measurement cycle of the gas chromatograph.

Figures 7 and 8 show a typical comparison of PI and MPC responses for a sampling period of 10 seconds. In Fig. 7 the bottoms temperature set point has been increased by 1.5°C in Fig. 8, the top temperature has been reduced 1.5°C. In both cases the faster response of the multivariable algorithm (MPC) is apparent. The better decoupling obtained with the MPC algorithm is particularly apparent in Fig. 8 where there is relatively little change in top temperature (none is desired) compared to the multi-loop PI control. These results illustrate the comparative advantages that advanced control methods can yield; in this case, however, a high-level digital computer is required to obtain them.

CONCLUSIONS

The investigation of advanced multivariable control techniques requires a sophisticated computer system. In the case of the UCSB distillation facility, a commercial microcomputer control unit has been interfaced directly to the process to provide direct digital control of the multicomponent distillation column, alarm checking for safe operation, data logging and start-up and shut-down operator assistance. In addition, a high-level computer is used for data storage, generation of mass and energy balances, and to calculate setpoints for two of the DDC control loops in a hierarchical control scheme. In the present system, any control strategy which uses reflux flow rate and reboiler steam pressure as manipulated variables, and distillate and bottoms compositions as controlled variables can be accomodated merely by changing the high-level control task. Other control configurations also can be implemented easily. Communication between the high and low-level machines is presently accomplished through the system analog interface; however, a digital communication interface will be installed in the near future.

The computer system configuration used in the UCSB distillation facility is an example of a hierarchical control system which simplifies the application of advanced multivariable control schemes by separating high- and low-level functions between two or more control computers.

REFERENCES

Benallou, A. (1982). Ph.D. Dissertation, University of California, Santa Barbara.

Cutler, C.R., and B.L. Ramaker (1980). Dynamic Matrix Control. A computer Algorithm, Joint Automatic Control Conference Preprints, paper WP5-B, San Francisco. (Also presented at 83rd National AIChE Meeting, Houston (1979).)

Davison, E.J. (1970). The Interaction of Control Systems in a Binary Distillation Column, Automatica, 6, 447-461.

Fauth, G.F., and F.G. Shinskey (1975). Advanced Control of Distillation Columns, Chem. Eng. Prog., 71, 49-54.

Fisher, D.G., and D.E. Seborg (1976). Multivariable Computer Control-A Case Study, North-Holland/American Elsevier, Amsterdam.

Kramer, A.R. (1980). Fractionating Column Control, In M.R. Skrokov (Ed.), Mini- and Microcomputer Control in Industrial Processes, Van Nostrand-Reinhold, New York, 81-127.

Luyben, W. L. (1975). Steady-State Energy Conservation Aspects of Distillation Column Control System Design, Ind. Eng. Chem. Fundam., 14, 321-325.

Marchetti, J.L. (1982). Ph.D. Dissertation, University of California, Santa Barbara.

Mehra, R.K., W.C. Kessel, A. Rault, J. Richalet, and J. Papon (1978). Model Algorithmic Control Using IDCOM for the F-100 Jet Engine Multivariable Control Design Problem. Alternatives for Linear Multivariable Control with Turbo Engine Theme Problem, (Ed. Said, Peczkowski and Melsa) National Engineering Consortium, Chicago.

Mellichamp, D.A., (Ed.) (1983). Real-Time Computing with Applications to Data Acquisition and Control, Van Nostrand Reinhold, New York.

Merluzzi, P., and C.B. Brosilow (1972). Nearly Optimal Control of a Pilot Plant Distillation Column, AIChE J., 18, 739-744.

Richalet, J., A. Rault, J.L. Testud, and J. Papon (1978). Model Predictive Heuristic Control: Application to Industrial Processes, Automatica, 14, 413.

Rijnsdorp, J.E. (1965), Interaction in Two-Variable Control System, I and II, Automatica, 3, 15-52.

Rosenbrock, H.H. (1962). The Control of Distillation Columns, Trans. Inst. Chem. Engrs. 40, 35-53.

Shinskey, F.G. (1977). Distillation Control, McGraw-Hill, New York.

Shunta, J.P., and W.L. Luyben (1971). Studies of Sampled-Data Control of Distillation Columns. Feedback Control of Bottoms Composition with Inverse Response Behavior, Ind. Eng. Chem. Fund., 10, 496-493.

Wood, R.K., and W.C. Pacey (1972). Experimental Evaluation of Feedback, Feedforward and Combined Feedforward-Feedback Binary Distillation Column Control, Can. J. Chem. Eng., 50, 376-384.

TABLE 1   Key Process Variables

Variable	Sensor Type	Transmitter (see Fig. 3)
Feed flow		FT1
Reflux flow	Orifice	FT2
Distillate flow	meters	FT3
Bottoms flow		FT4
Top temperature	Thermocouple	TT1
Bottoms temperature	Resistance	TT2
Feed temperature	Thermometer	TT3
Distillate temperature		TT4
Column pressure	Pressure	PT1
Reboiler steam pressure		PT2
Bottoms level	Differential	LT1
Distillate drum level	Pressure	LT2

TABLE 2   Low-Level Computer Control Loops

Control Loop	Controlled Variable	Manipulated Variable
1 (PI)	Reflux flow rate position	Reflux valve
2 (PI)	Reboiler steam pressure	Steam valve
3 (PI)	Feed flow rate position	Feed valve
4 (PID)	Feed temperature position	Steam valve
5 (PI)	Reboiler level	Bottoms flow rate
6 (PI)	Reflux drum level	Distillate flow rate
7 (PID)	Distillate temperature	Cooling water

TABLE 3   Variables Checked for Alarm Conditions

Process Variable	Type of Alarm
Steam pressure	High
Column pressure	High
Distillate drum level	High and Low
Bottoms level	High and Low
Feed flow rate	High and Low
Distillate flow rate	High and Low
Bottoms flow rate	High and Low
Distillate temperature	High

TABLE 4   Research Projects Using the Distillation Column Facility

Research Topic	Graduate Student
Modeling and control of bilinear systems	A. Benallou (PhD)
Multivariable predictive control	J. Marchetti (PhD) P. Maurath (PhD)
Self-tuning control	I.L. Chien (PhD)
Control strategies for nonlinear systems	S. Wong (PhD)
Controller design for large-scale systems	C. Naumchik (MS)

A Pilot-scale Distillation Facility   393

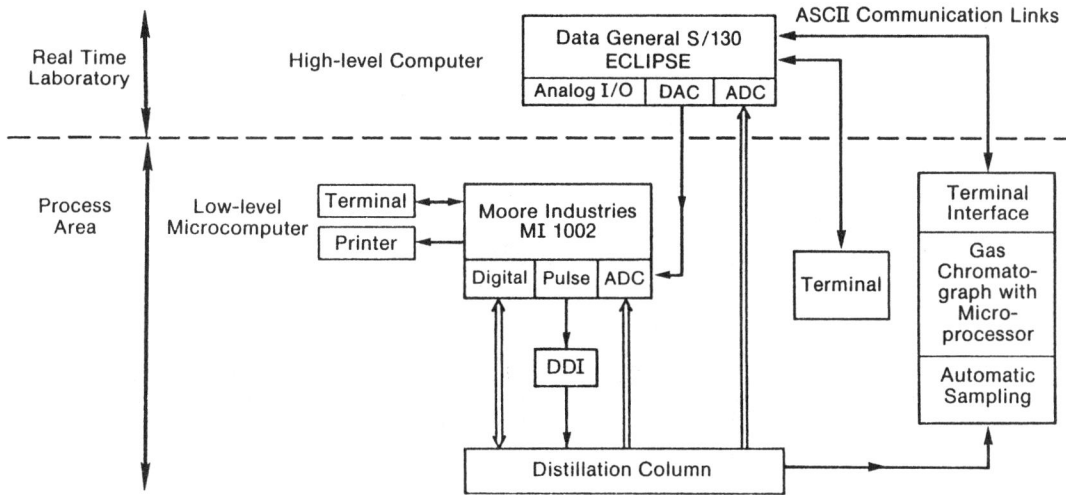

Fig. 1. Hierarchical control configuration for the UCSB distillation facility

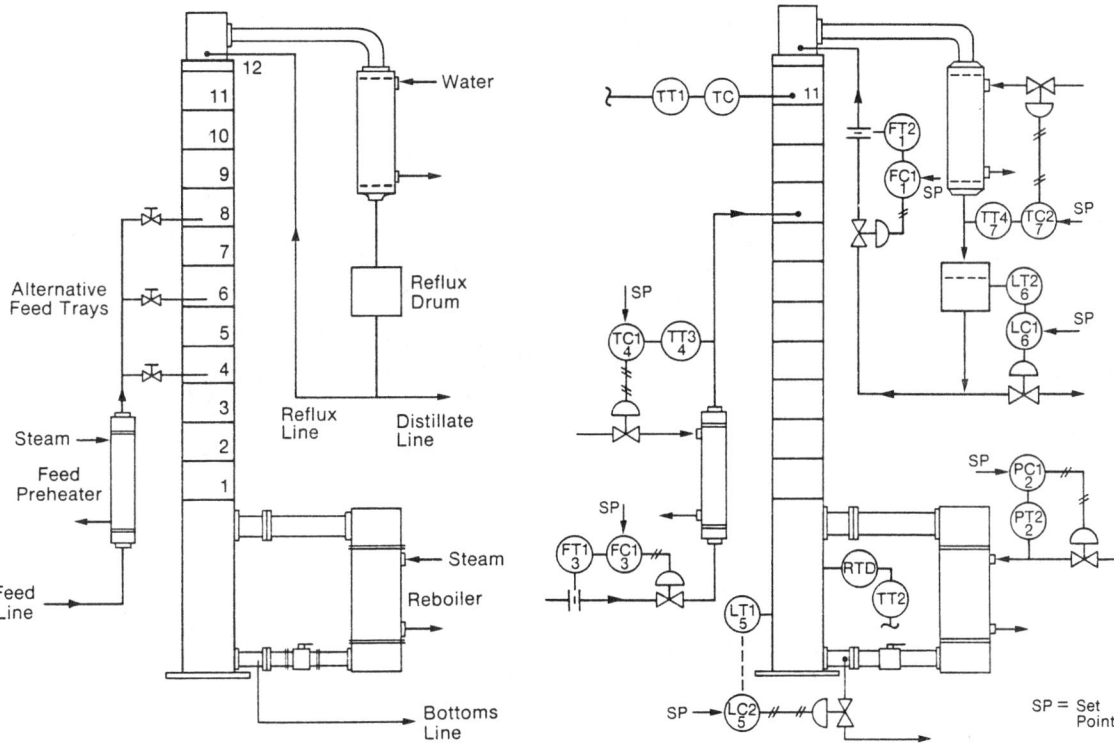

Fig. 2. Pilot-scale distillation unit

Fig. 3. Low-level control loops implemented via microcomputer.

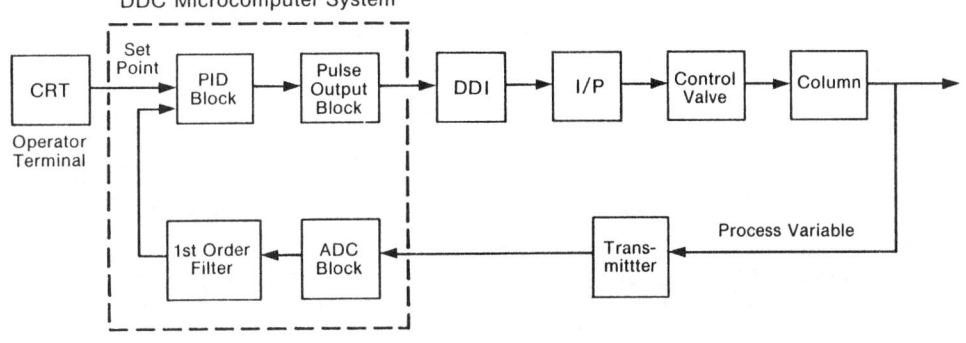

Fig. 4. Typical DDC control loop

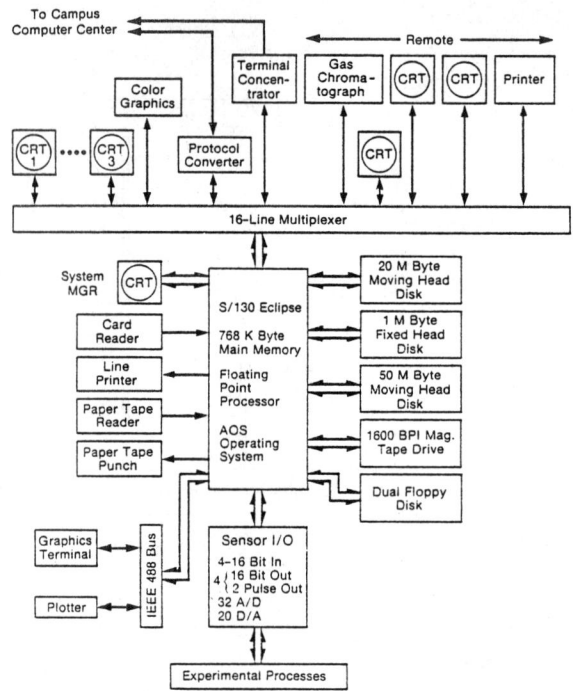

Fig. 5. High-level real-time computer

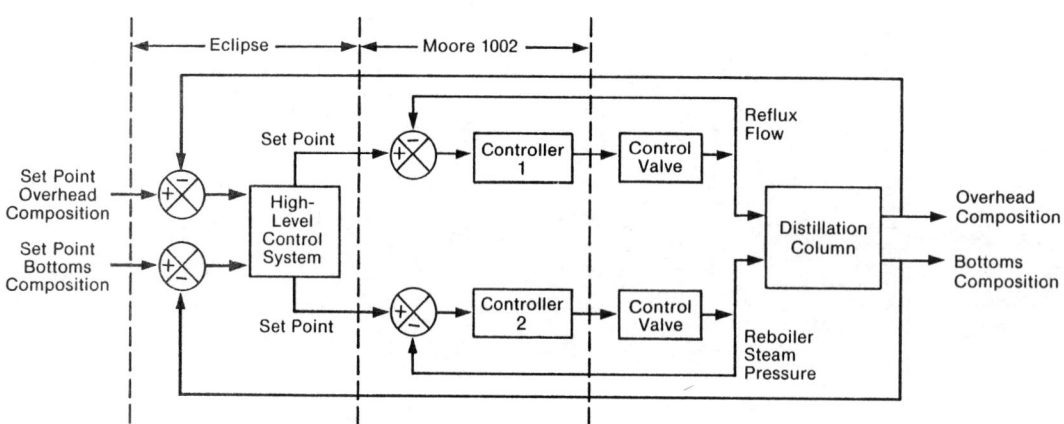

Fig. 6. Arrangement of distillation column primary loops

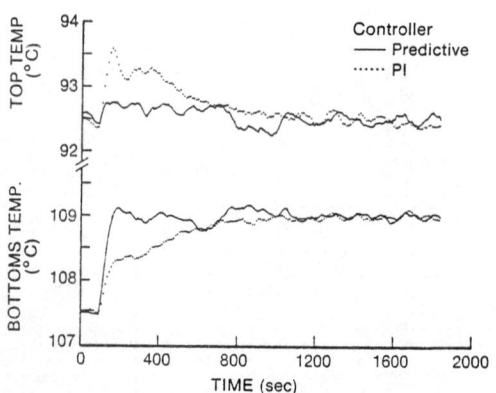

Fig. 7. Comparison of MPC and multi-loop PI control for a +1.5°C step change in bottoms temperature set point

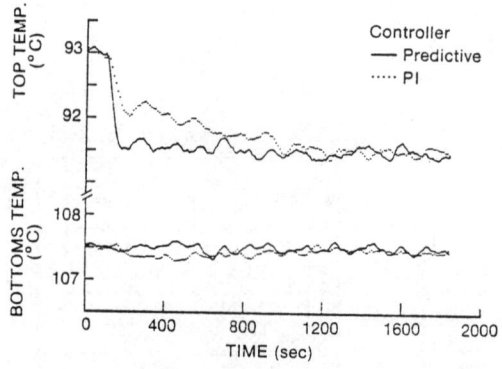

Fig. 8. Comparison of MPC and multi-loop PI control for a -1.5°C change in top temperature set point

AN ADAPTIVE FEEDFORWARD CONTROL ALGORITHM FOR COMPUTER CONTROL OF WASTEWATER NEUTRALIZATION

R. A. Balhoff* and A. B. Corripio**

*Shell Development Company, P.O. Box 1380, Houston, TX 77001, USA
**Department of Chemical Engineering, Louisiana State University, Baton Rouge, LA 70810, USA

Abstract. The purpose of this paper is to show the development and testing of an adaptive feedforward control of a wastewater neutralization process. The adaptive controller is compared to a nonlinear proportional-integral-derivative (NPID) controller developed by Shinskey and Myron (1970). The process and controllers were simulated on a digital computer. The adaptive controller utilizes two pH probes, a feedforward probe and a feedback probe, with the latter used in the adaptive calculations of the feedforward gain. The feedback measurement provides the adaptive controller with a form of reset action.

The process simulated for control combines a strong (hydrochloric) and weak (carbonic) acid neutralized by a strong base (sodium hydroxide). The adaptive controller was shown to give superior responses both for step changes in the strong acid and in the weak acid (buffer) concentration. One particular advantage of the adaptive controller is that it requires only two tuning parameters, while many controllers in use today require five or more tuning parameters.

The sensitivity of the adaptive control to changes in certain parameters (probe noise and lag, valve hysteresis and lag, and dead time) was studied, as was the effect of a step change in flow rate to the system. Noise in the feedforward pH probe and the dead time between the reagent addition and the feedback probe had the largest effect on the adaptive controller performance.

Keywords. Adaptive control; pH control; feedforward; nonlinear control systems; self-adjusting systems; PID control.

INTRODUCTION

Waste neutralization has long been an extremely difficult process to control. This is due mainly to the wide variations in process gain caused by changes in the pH and in the relative amounts of strong and weak acids and bases in the waste stream. Obviously, these characteristics place severe demands on the typical linear controller.

Most of the industry has attacked the pH problem by using large settling ponds with reagent addition between each pond to bring the pH within the acceptable range. As restrictions on the effluent pH continue to tighten and costs escalate, the necessity for a more efficient control scheme becomes more apparent.

In 1970, Shinskey applied a more sophisticated control than the standard linear PID controller to an idustrial wastewater treatment. The scheme employed three controllers: a feedforward, an adaptive nonlinear feedback, and a proportional feedback controller. Several parameters had to be adjusted, most of them manually, to accomodate the changes in the process parameters. In 1973, he further modified this technique by eliminating the trim controller and adding some additional parameters to the remaining scheme (crossover frequency, low-frequency gain and adaptive reset) to try to handle the nonlinearities and rapid changes in process conditions. In 1979, Shinskey simplified his design in his control book, but gave no indication of its control performance.

A number of approaches to solve the pH control problem have been presented in the recent literature by Niemi and Jutila (1977), Gupta and Coughanowr (1978), Bucholt and Kümmel (1979), Davalloo and Nowroozi (1979), Richter and co-workers (1979), and Gray (1980). Trevathan (1978, 1979) reviewed the literature on pH control, and provided general guidelines for improving the controllability of a pH process but referred to pH control as "one of the most difficult single dimension control problems in the process industries despite considerable research and publication efforts..." (1978).

The studies cited have either used too simple a system or have developed schemes which require manual adjustment of several control parameters as process parameters change. These systems are not very efficient in the complex and rapidly changing process conditions usually present in wastewater neutralization processes. In an attempt to solve some of the shortcomings of these previous control schemes, an adaptive feedforward controller has been developed using a digital simulation of a wastewater neutralization process.

After a brief definition of the problem we will present the development of the adaptive feedforward controller and the equations for the nonlinear proportional-integral-derivative (NPID) controller proposed by Shinskey and Myron (1970). This controller has been implemented on many industrial processes and is used to gauge the effectiveness and stability of the proposed controller. Then we will compare the performance of the adaptive controller to that of the NPID controller for various levels of the strong and weak acid concentrations, and will present the effect of these acid levels on the tuning of the adaptive controller.

DEFINITION OF THE PROBLEM

The problem of waste stream neutralization consists of mixing a waste stream which is too acid or too basic with a strong basic or acid stream in order to bring its pH within a range around neutrality (pH=7.0) which is acceptable for environmental protection purposes. The pH, a measurement of the hydrogen ion concentration, is given by the following function:

$$pH = -\log_{10} [H^+] \quad (1)$$

where the hydrogen ion concentration, $[H^+]$, is in gmoles per liter. This logarithmic function causes the pH measurement to be more and more sensitive to the hydrogen ion concentration as it approaches neutrality, as the change of one pH unit requires smaller changes in concentration. This increase in sensitivity would continue beyond the neutral point were it not for the buffering action of the water dissociation equilibrium, given by the reaction:

$$H_2O \rightleftarrows H^+ + OH^- \quad (2)$$

which imposes the following relationship between the concentrations of the hydrogen and hydroxyl ions (OH^-):

$$[H^+][OH^-] = K_W = 10^{-14} \quad (3)$$

Thus, on the basic side of neutrality, there is an excess of hydroxyl ions and the pH sensitivity decreases as their concentration increases.

In addition to the equilibrium of the water dissociation reaction, additional equilibrium relationships are imposed on the ion concentrations by the presence in the wastewater stream of weak acids or bases or of their ions. In our study we used carbonic acid as the weak acid because carbonates are commonly present in wastewater streams. The equilibrium reactions are:

$$H_2CO_3 \rightleftarrows HCO_3^- + H^+ \quad (4)$$

$$HCO_3^- \rightleftarrows CO_3^= + H^+ \quad (5)$$

The equilibrium relationships for these reactions are:

$$\frac{[H^+][HCO_3^-]}{[H_2CO_3]} = K_{b1} = 4.46 \times 10^{-7} \quad (6)$$

$$\frac{[H^+][CO_3^=]}{[HCO_3^-]} = K_{b2} = 5.61 \times 10^{-11} \quad (7)$$

For the strong acid in the waste stream we used hydrochloric acid (HCl), and for the strong base in the control stream we used sodium hydroxide (NaOH). Selected neutralization curves for this system, calculated from the simultaneous solution of the species continuity equations and Eqs. (1), (3), (6) and (7), are shown in Fig. 1. These curves match those published by Shinskey (1973). The slope of each line is proportional to the sensitivity of the pH to the addition of reagent at a given concentration of the weak acid. This is the process gain. It is evident that the gain varies by several orders of magnitude with pH and that, at a given pH, the gain varies significantly with the weak acid concentration. It is precisely this wide variation in process gain that makes neutralization such a difficult control problem.

PROCESS SIMULATION

The detailed equations used in the simulation of the neutralization process are given by Balhoff (1982). In the simulation all tanks are modeled as perfectly mixed, with a dead time of one second to simulate the effect of imperfect mixing. Flow in the lines from one point to another is assumed to be plug flow (no axial mixing); this means that the lines are simulated as pure dead time.

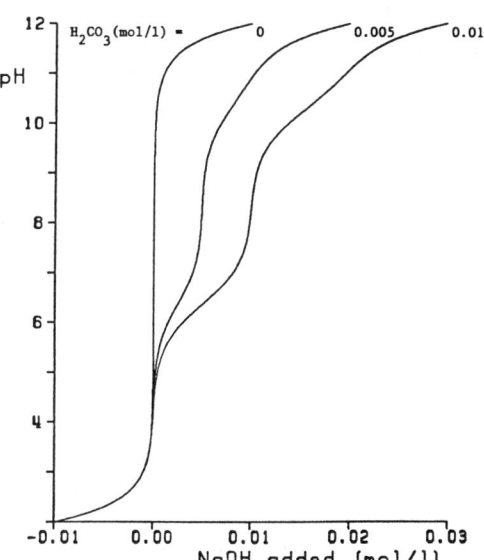

Fig. 1. Neutralization curves at various levels of weak acid concentration. NaOH added is relative to amount required to neutralize to pH 7.

The pH probes were simulated as linear first-order lags with a constant time constant as suggested by Marra (1979). However, one run was made with the time constant being different in each direction, as suggested by McAvoy (1979); this variation did not seem to significantly affect the performance of the control system in this case. A uniformly distributed random signal with an adjustable maximum amplitude and zero mean was added to the calculated value of the pH to simulate pH probe noise.

The control valve actuator was simulated as a first-order lag with a dead band on the input to simulate hysteresis or valve sticking. The magnitude of the hysteresis effect was varied by adjusting the amplitude of the dead band as a fraction of the full valve travel. Hysteresis is caused by dry friction between the packing and the valve stem.

The numerical solution of the set of ordinary first-order differential equations was carried out by the analytical solution of each first-order equation over a short integration time step. In this method the input variables and interaction variables between equations are assumed to be constant over the short integration step. This method gave good results in this case because there was very little coupling among the model equations.

CONTROL VALVE SEQUENCING

Since the amount of electrolytes present in a waste stream can vary over a very wide range, the control valve must also be able to accurately deliver reagent over a wide range. Shinskey (1973) shows that the best way to achieve the necessary rangeability is by sequencing (or coupling) two (or more) equal-percentage valves installed in parallel. Two equal-percentage valves having rangeabilities of 50:1 could be sequenced to achieve a rangeability of up to 2500:1. In practice, a reasonable overlap of the valves is desirable, reducing the rangeability to 1000:1. Figure 2 shows an example of this sequencing on a semi-log plot. This arrangement would require the larger valve to be twenty times the size of the smaller valve. Note that only one valve should be open at any given time. A pressure switch can be used with solenoid valves to select the proper valve depending on the controller output. For example, the large valve should open when the control signal rises to 57% (maximum flow for the small valve), while the small valve closes, whereas when the control signal drops to 43% (minimum flow for the large valve), the large valve should close as the small valve opens. This procedure eliminates the switching back and forth between the valves when the controller signal is around the switch point.

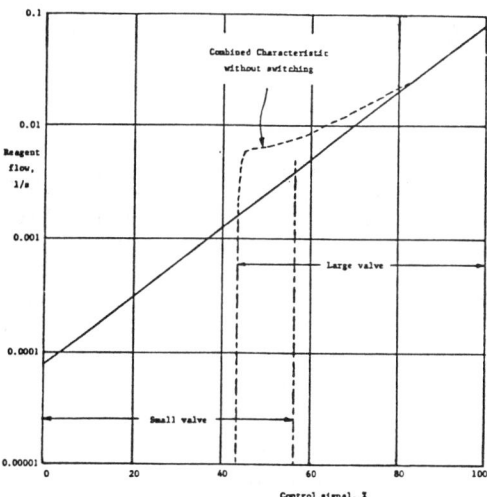

Fig. 2. Sequencing of two parallel equal-percentage valves (Shinskey, 1973).

ADAPTIVE FEEDFORWARD CONTROLLER

As has been pointed out, the gain of a pH process varies greatly depending on the pH and on the buffering of the solution. This means there is a need to adjust the controller to compensate for these changes in process gain. Ideally, the controller should adapt itself to the process it is controlling. In this case, the gain of the controller would have to change inversely with the changing process gain. The following is a development of a controller which does exactly that, adapts its gain to compensate for the changing process gain.

Simple Feedforward Model

If the logarithm of the reagent required for neutralization of a strong acid (no buffer) is plotted against the pH of the incoming stream, a straight line is obtained when the pH is outside of the neutral range. This is also approximately true for solutions which have the same relative amounts of strong and weak acids. These relationships can be seen in Fig. 3 for a pH set point of seven.

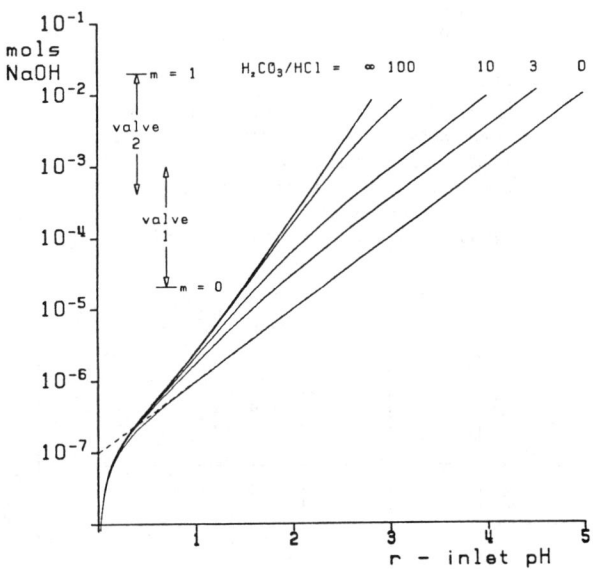

Fig. 3. Moles of NaOH required to neutralize one liter of the incoming solution to the set point of 7.

For an equal percentage valve, there is a linear relationship between the valve position, m, and the logarithmic scale of the y-axis shown in Fig. 3. (The two sequenced equal percentage valves drawn on this figure show the range of the valve position, zero to one.) This means that the linear relationship holds for the valve position versus incoming pH. The simple feedforward controller takes advantage of this linear relationship. Thus, the equation for the simple feedforward controller is as follows:

$$m = m_0 + K_{cf} \cdot (r - pH_0) \qquad (8)$$

where m = output of controller
m_0 = controller bias
K_{cf} = feedforward gain
r = pH set point
pH_0 = incoming pH.

The controller bias, m_0, should be set to the value of the valve position which corresponds to the flow rate intercept for the system being controlled (e.g., 10^{-7} mol NaOH in Fig. 3). The value for m_0 can be calculated from the system parameters for a given line in Fig. 3 by the following formula:

$$m_0 = \ln(M_0 F_{in} R/(F_{vmax} C_{NaOH})/\ln(R) \qquad (9)$$

where M_0 = moles of NaOH at the intercept in Fig. 3
F_{in} = inlet flow rate (l/s)
F_{vmax} = maximum reagent flow rate from the equal percentage valve (l/s)
C_{NaOH} = concentration of the reagent (mol/l)
R = rangeability of equal percentage valve (i.e., 1000).

For the strong acid-strong base system, M_0 has the value of 10^{-7} moles of NaOH.

The feedforward gain, K_{cf} should be set to the slope of valve position versus pH line (the change in "m" divided by the change in pH). The process gain, K_p, is the reciprocal of this slope and is given by

$$K_p = (r - pH_0)/(m - m_0). \qquad (10)$$

If the degree of buffering changes appreciably, the process gain, K_p, changes because of the change in the slope of the operating line. For a constant controller gain, K_{cf}, the reagent delivery will be grossly in error. Therefore, systems utilizing this feedforward scheme require frequent manual adjustment of the feedforward gain to obtain satisfactory control. What is needed is for the process gain to be accurately estimated and automatically updated.

Estimation of Process Gain

The process gain can be estimated from the generalization of Eq. (10) to yield

$$K_p = (pH_1 - pH_0)/(m - m_0) \qquad (11)$$

where pH_1 is the pH after the reagent addition. This estimation can be achieved by utilizing a scheme such as the one shown in Fig. 4. Good mixing between the stream and the added reagent is essential and must be obtained by the use of an in-stream mixer which gives a high degree of mixing with a minimum of lag and dead time. A mixing tank between the reagent addition and the feedback probe would be undesirable because changes in the valve position would not directly and immediately affect the feedback value.

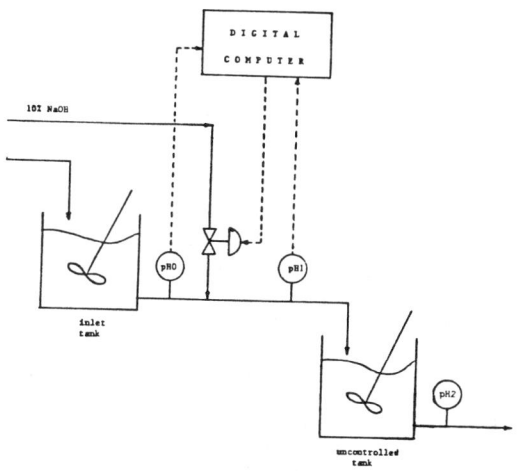

Fig. 4. Adaptive feedforward control scheme.

Updating of Controller Gain

The gain estimate obtained from this method is not exact. This is due to several factors such as probe noise and lag, dead time and lag between the probes, and valve dynamics.

Since each estimate of the process gain incorporates these inaccuracies, substituting this process gain directly into the control equation would result in highly erratic behavior. To smooth this behavior, a digital filter can be used to exponentially update the estimate of the process gain used by the controller, such that

$$K_p(n) = K_p(n-1) \cdot (1 - a_1) + K_p \cdot a_1, \quad (12)$$

where a_1 is an updating parameter which should be adjusted to obtain a rapid, stable response to load changes. Eq. (12) is the formula for a digital exponential filter for which "a_1" is equal to one minus the filter constant, that is

$$a_1 = 1 - \exp(-T/\tau_f), \quad (13)$$

where T = sampling interval (seconds)
τ_f = updating time constant (seconds)

It may be useful, at times, to look at the tuning in terms of this time constant rather than the updating parameter. If the updating parameter is less than 0.1, the updating time constant is essentially the ratio of the sample time to the updating parameter. This approximation provides a quick estimation of the time constant since most of the values of the updating constant are less than 0.1.

Summary of Adaptive Controller Algorithm

The equations for the adaptive algorithm were presented in the preceding section. The procedure for implementation of the adaptive controller is 1) estimate the process gain using Eq. (11), 2) average the process gain using Eq. (12), and 3) use the reciprocal of this gain in the feedforward equation [Eq. (8)] to obtain the controller output (valve position). This results in the following equation for the feedforward controller:

$$m = m_0 + (1/K_p(n)) \cdot (r - pH_0(n)). \quad (14)$$

These steps are executed at regular intervals of time T.

SHINSKEY'S NONLINEAR PID CONTROLLER

It is desirable, when developing a new method, to have a standard for comparison. This helps to identify the relative strong and weak points of the proposed method. Ideally, the standard should be fairly advanced yet well established and used in numerous industrial applications.

The control method chosen as a standard in this work was developed by Shinskey and Myron (1970) and meets the qualifications listed above. The method is a nonlinear proportional-integral-derivative controller utilizing a characterized pH. Shinskey reported the industrial application of this controller in this same paper (1970). This controller will be briefly discussed below. For a more detailed treatment consult Shinskey's book on pH and pIon control (1973).

Characterized pH

Since the neutralization curve for most systems is highly nonlinear, control with a linear PID controller is very unstable and/or sluggish. This is because the loop gain, which is the product of the controller and the process gain, can vary over several decades as the pH or the buffering changes. To try to alleviate this problem, Shinskey proposed a nonlinear function to compensate for the nonlinearity of the process. The function was a simple combination of three straight lines as shown in Fig. 5. The dead-band width and gain are adjustable to customize it to the neutralization function being controlled.

These constants must be continually updated manually if the amount of buffering is changing with time. The function can be defined mathematically as

$$f(e) = \text{sign}(e) \cdot \max(e-b, G_b e) \quad (15)$$

where e is the pH error (set point - pH), b is half the width of the dead band and G_b is the gain within the dead band.

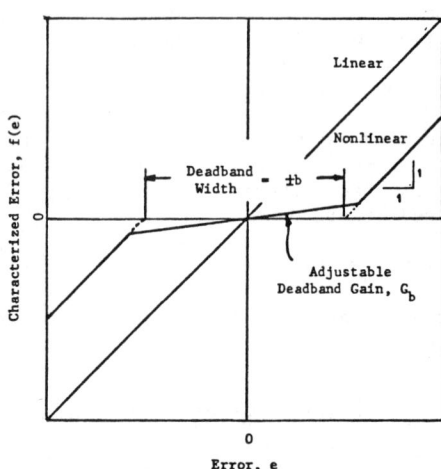

Fig. 5. Characterized pH function for the nonlinear PID controller.

This characterized pH is used as the controlled variable in the PID controller instead of the actual pH, having the effect of lowering the controller gain around neutral in order to maintain a more uniform loop gain.

At very low buffering, it is difficult to properly fit the highly nonlinear pH function with this simple function. Also, this control scheme requires the independent adjustment of five interrelated control parameters, a difficult task, at best.

Valve Characterization

The use of equal-percentage valves introduces a gain variation with load which is undesirable. It is, therefore, necessary to compensate for this variation. Shinskey suggests the use of an analog divider which approximates the equal-percentage characteristics. For the purposes of this simulation, however, the nonlinearity was exactly reversed digitally.

Tuning of the Controller

The tuning procedure outlined by Shinskey was followed for the most part in obtaining reasonable tuning parameters for each case. The procedure used is as follows:

1) set the derivative time to zero and the integral time to a very large value,

2) increase the controller gain until undamped oscillations are obtained,

3) set the dead-band width slightly less than the amplitude of the oscillations,

4) set the derivative time equal to the period of oscillation divided by four times π,

5) set the integral time equal to the time constant of the controlled tank,

6) adjust the controller gain and the dead-band gain until rapid, uniform damping is achieved.

The procedure used by Shinskey suggests a much shorter integral time than that in step 5. It, however, was difficult to obtain stable control at low buffering with such a short integral time. An integral time approximately equal to the time constant of the control tank was found to provide the best control for this sytem.

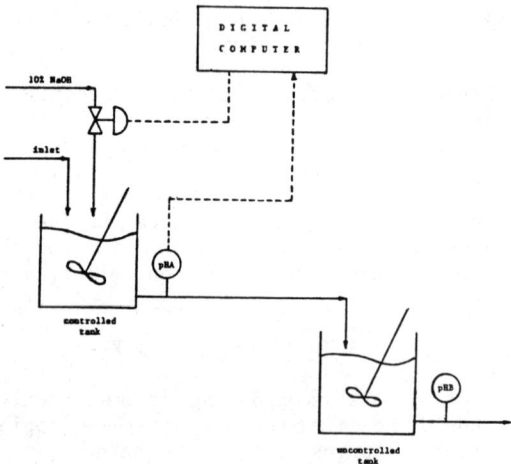

Fig. 6. Nonlinear PID control scheme (Shinskey, 1973).

COMPARISON OF NONLINEAR PID AND ADAPTIVE CONTROLLERS

Basis of Comparison

Figure 6 shows the configuration of the nonlinear PID controller. The nonlinear control loop is applied to the first tank.

The second tank, which is uncontrolled, serves the purpose of averaging any oscillations on the pH of the stream leaving the first tank. This is essentially the configuration implemented by Shinskey in one of his industrial applications (Shinskey, 1970). The time constants of the tanks and the associated dead times chosen for this study are two minutes and six seconds, respectively. These values are very close to those used by Shinskey in the above application, however, the tanks used are slightly smaller. This will have the effect of making it somewhat more difficult to control the pH.

In order to facilitate comparison between the controllers, the same flow configuration has been used for the adaptive control scheme. This control scheme is diagrammed in Fig. 4. The main differences between the two configurations are the location of the control valve and the algorithm determining the position (opening) of the valve. The adaptive control scheme also requires an additional pH probe.

TABLE 1 Parameter Values Used in Base Cases

Parameter	Value	Units
Sampling interval	1	s
Amplitude of noise in probes after mixed tanks	±0.01	pH units
Amplitude of noise in adaptive feedback probe, pH	±0.1	pH units
Probe lag	1	s
Valve hysteresis (fraction of total valve movement)	0.005	
Valve lag	1	s
Dead time between reagent addition and adaptive feedback probe, pH	1	s
Volume of tanks	120	l
Inlet flow rate	1	l/s
Dead time associated with tanks (percent of tank retention time)	5	%
Adaptive feedforward intercept, m_o	-0.84	

For the cases studied, the inlet pH will always be acidic, requiring only the addition of NaOH. For a system in which the incoming pH can be acidic or basic, an identical control algorithm can be used to control a valve which adds acid. The concepts developed are the same for both cases. The parameter values used in the comparison cases are summarized in Table 1.

Response to a step change in strong acid concentration

Figures 7 and 8 show the response of the well-tuned controllers with and without buffer (carbonate), respectively, for a step increase in the strong acid concentration from a pH of 3.3 to 3.0. Table 2 gives a brief description of the parameters listed in the legend. The solid lines and the broken line on these graphs represent the measurements from the adaptive controller. Dotted lines (light and heavy) represent values from the NPID controller. The important quantities for comparison of the controllers are the pH values leaving each system. These are pH2 (heavy solid line) for the adaptive controller and pHA (heavy dotted line) for the NPID controller. The other pH

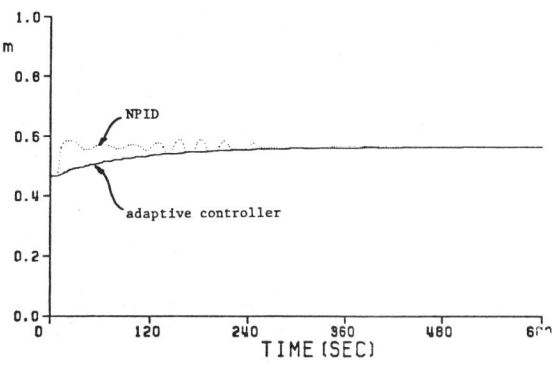

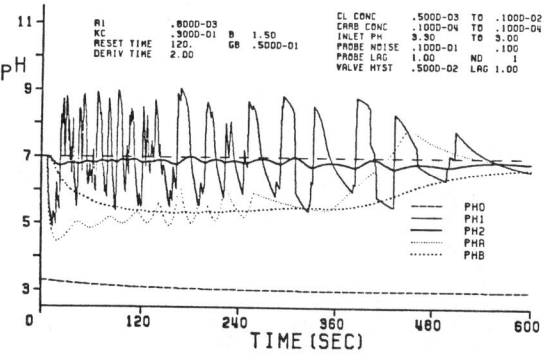

Fig. 7. Response to a step change in incoming pH from 3.3 to 3.0 for a carbonic acid concentration of 10^{-5} mol/liter.

TABLE 2 Description of Parameters in pH Versus Time Legend

A1	adaptive updating parameter
KC	nonlinear PID controller gain
RESET TIME	NPID controller integral time, s
DERIV TIME	NPID controller derivative time, s
B	NPID controller dead-band width, ±B pH units
GB	NPID controller dead-band gain, fraction of controller gain
CL CONC	concentration of chloride ion in incoming stream, mol/l, before and after step change at time = 0
CARB CONC	total concentration of carbonates (carbonic acid + bicarbonate ion + carbonate ion) in incoming stream, mol/l, before and after step change at time = 0
INLET PH	incoming pH before and after step change at time = 0
PROBE NOISE	maximum amplitude of probe noise, pH units, for probes after a tank and the probe after the adaptive reagent addition, respectively
PROBE LAG	probe time constant, s
ND	dead time between adaptive probes (pH0 and pH1), seconds
VALVE HYST	valve hysteresis, fraction of full movement
VALVE LAG	valve time constant, s

measurements, however, will also be presented to provide a better understanding of the behavior of each of the controllers.

From Figs. 7 and 8 it is evident that the adaptive controller (solid lines) gives tighter control than the NPID controller (dotted lines). Notice that, for this system, neither controller handles the case of zero buffering very well. The size of one or both tanks could be increased to improve this situation. Fortunately, a totally unbuffered system is rare in industry.

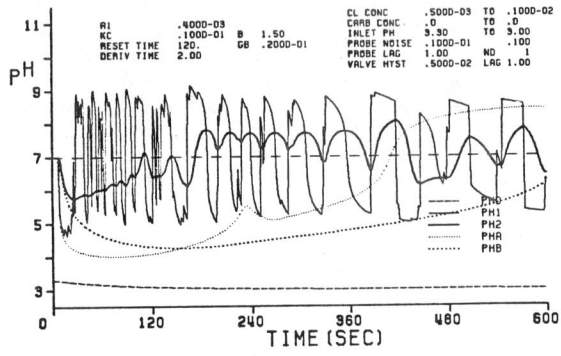

Fig. 8. Response to a step change in incoming pH from 3.3 to 3.0 for an unbuffered solution (no carbonic acid).

If one looks at Fig. 7 or 8, it first appears that the large changes in pH are caused by substantial changes in the valve position. However, this is not the case as is illustrated by the upper graph in Fig. 7 (valve position versus time). This figure shows that the valve position is very stable for the adaptive controller (more so than the NPID). The question is then: what causes these large, erratic oscillations? The answer lies in the fact that this is an in-line controller (no axial mixing between the reagent addition and the feedback measurement), therefore small changes in the valve position cause large variations in the resultant pH. These oscillations are not important, however, as they are damped out in the averaging tank (as shown by the heavy lines in Figs. 7 and 8).

A great deal about the two controllers can be learned by studying Fig. 7. The near-vertical changes in pH for the adaptive case are caused by small valve movements (the valve remains stationary for most of the time because of valve hysteresis). The more gradual decreases in pH are caused because the load coming into the adaptive controller is increasing with time due to the lag created by the first tank. Note that the slope of these changes decreases as the output of the first tank lines out (bottom dashed line in Figs. 7 and 8).

For the NPID controller, the series of oscillations in Fig. 7 between the times of 30 and 300 seconds are caused by the imperfect fit of the characteristic pH function to the actual titration curve. This high gain area is an indication of the nonconstant loop gain which could not be accurately characterized with the nonlinear function.

The valve movement for the NPID controller is slightly more oscillatory than that of the adaptive controller. Notice, also, that the NPID valve moves more rapidly to the new position. This is because the NPID controller must control the entire mixing tank, whereas the adaptive controller needs only to increase the valve position to handle the more gradual change in load experienced at the first pH probe.

Response to a Change in Buffer Concentration

At higher buffer concentrations than that shown in Fig. 7, both the NPID and the adaptive controllers could be tuned to produce good control of pH. However, this required that each controller be retuned for each buffer concentration. It is therefore of interest to compare the responses of the two controllers to a change in buffer concentration, as then we can see how they perform when they are not optimally tuned for the prevailing process conditions. Figure 9 shows the responses to a decrease in buffer concentration by a factor of ten.

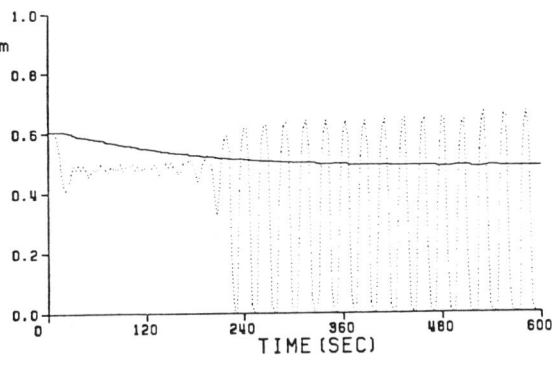

a) valve position versus time

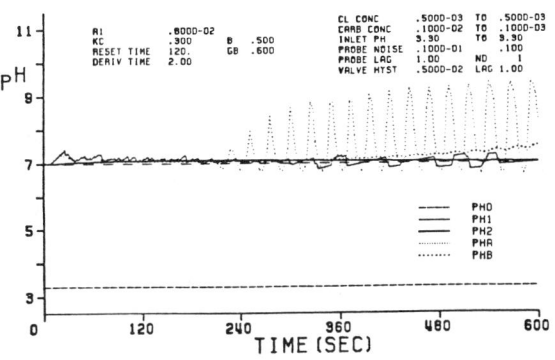

Fig. 9. Response to a step change of the carbonic acid concentration from 10^{-2} to 10^{-3} mol/liter.

On decreasing the buffer concentration the adaptive controller handles the step change with very little deviation from set point (0.1 pH units) and a very smooth valve movement. By contrast, the NPID becomes very oscillatory and unstable after approximately two minutes (one time constant of the tank). The valve cycling evident in Fig. 9 for the NPID (dotted line) is highly undesirable since it tends to wear out the valve. It is also interesting to note that appreciable valve cycling was occuring even when the control seemed perfectly stable. This is due to the damping by the control tank which can mask large valve movements in the NPID control scheme. Valve cycling is more likely to occur with the NPID controller because the outlet pH is relatively insensitive to valve movements as compared to the adaptive controller. Also, the NPID controller is more sensitive to buffering changes and it is, therefore, more likely to become either unstable or sluggish.

STABILITY LIMITS FOR THE ADAPTIVE CONTROL SCHEME

For a given system, if the updating parameter in the adaptive controller is too large, the response becomes so oscillatory that the pH of the outlet stream from the averaging tank drifts away from the set point and/or the valve begins to cycle. If, on the other hand, the updating parameter is too low, the response becomes sluggish and the controller cannot keep the final pH within the allowable range upon a load or gain change. Within these two extremes there is a range of values of the updating parameter for which the adaptive controller performance is satisfactory.

The acceptable range of the updating parameter, a_1, is plotted in Fig. 10 as a function of the ratio of the concentration of the carbonate (buffer) to that of the strong acid.

Figure 10 shows the effect which the strong acid (HCl) concentration has on the upper and lower limits of the updating parameter a_1. It is obvious that, for the most part, a lower updating parameter (slower gain updating) is required for lower buffer concentrations. However, notice that the lower limit curve passes through a maximum. This occurs when the carbonate concentration is approximately three times the HCl concentration, which is around the concentration ratio at which the slope of the pH versus "reagent required" line changes from strong acid dominant to weak acid dominant. In other words, when a strong acid step change effects a substantial change in the feedforward gain of the system, at the low updating parameter values, the controller gain is not corrected rapidly enough and

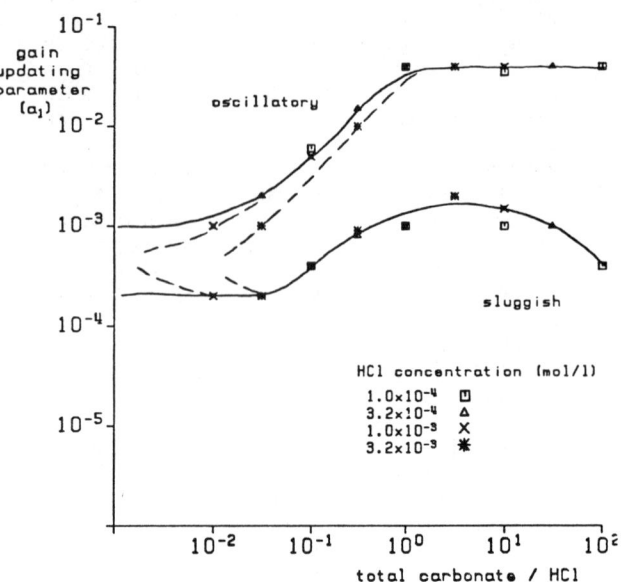

Fig. 10. Upper and lower limits of adaptive controller updating parameter versus the ratio of weak to strong acid concentrations.

the pH is forced out of the acceptable range. Because of this the lower limit of the updating parameter is higher around the concentration ratio of 3. The shift of the upper limit to a higher asymptote at high buffer concentrations reflects the greater stability of these systems.

The correlation of the updating parameter limits with the concentration ratio of weak to strong acid is, as shown in Fig. 10, better at higher levels of buffering. The dashed lines in Fig. 10 show that, in the low buffer region, the acceptable range of the updating parameter is narrower the higher the strong acid concentration. This is because, at low buffering, the gain around neutral is very high and the process is much more sensitive to errors in the valve position. At the higher HCl concentrations, the valve is opened considerably more, and the error caused by valve hysteresis creates larger errors in the flow. This, in turn, causes larger oscillations in pH and therefore the upper limit of the acceptable range is reduced considerably.

SENSITIVITY STUDIES

The sensitivity of the adaptive controller performance to several process variables was studied. The variables of interest were the amplitude of the noise on the pH measurements, the time constants of the pH probes and of the control valve, the dead time between the two probes, the volumes of the tanks, the amplitude of the hysteresis on the valve actuator, the flow rate of the waste stream, and the value of the feedforward intercept parameter m_0 [Eq. (8)]. In each case the values of these variables were increased and decreased from the base values listed on Table 1 and their effect on the updating parameter limits (Fig. 10) was determined.

The adaptive controller performance was found to be most sensitive to the amplitude of the noise on the feedforward pH measurement (pH0 in Fig. 4). Reduction of the amplitude of this noise resulted in a significant improvement of the control performance at low buffer concentrations. In contrast, the control performance was not affected by the amplitude of the noise on the feedback pH measurement (pH1 in Fig. 4), since the feedback action is highly filtered by the digital filter on the controller gain.

Large values of the time constants of the pH probes, the control valve, or of the dead time between the two probes resulted in a significant reduction in the upper limit of the gain updating parameter. However, the control performance was not affected by having a nonlinear probe lag with different time constants in each direction of pH change.

A small amount of valve hysteresis was found to be beneficial to the adaptive controller because it served as a nonlinear filter for the pH measurement noise. Acceptable pH control was obtained for changes in inlet flow rate. The adaptive controller also performed well when the mixing tank upstream of the feedforward probe was removed.

Except for very low buffer concentrations, the limits on the gain updating parameter are not sensitive to the value of the feedforward bias parameter m_0 as long as it is near the intercept of the neutralization curve (Fig. 2). This is a significant result because it means that the adaptive controller has only one critical tuning parameter, the gain updating parameter a_1.

We must point out that the absence of a mixing tank between the feedforward and feedback probes of the adaptive controller is essential to its success. This means that the neutralization reactions must necessarily be fast, with reactions times of just a few seconds. Thus the adaptive controller will not probably be applicable to systems containing solid reagents such as lime.

SUMMARY

A new algorithm for the control of pH in

wastewater neutralization systems has been presented and its performance compared to that of Shinskey's nonlinear PID controller. The new controller was found to perform as well as or better than the nonlinear PID controller. In addition, the adaptive controller requires the adjustment of a single tuning parameter as opposed to five parameters for the nonlinear PID controller. Furthermore, the adaptive controller does not require retuning over a wide range of buffer concentrations, while the PID controller must be retuned when the buffer concentration changes. The adaptive controller does not cause as much oscillation of the control valve as the PID controller.

REFERENCES

Balhoff, R.A. (1982). *Adaptive Feedforward Control of Wastewater Neutralization*. Ph.D. dissertation, Louisiana State University, Baton Rouge, Louisiana.

Bucholt, F. and M. Kümmel (1979). Self-tuning control of a pH-neutralization process. *Automatica*, 15, 665-671.

Davalloo, and Nowroozi (1979). Neutralizing industrial wastes with conventional controllers: A pilot plant study. *Instrum. Technol.*, 26, No. 9, 75-78.

Gray, M. (1980). Microprocessor characterizes pH ahead of controller for easy tuning. *Control Eng.*, 27, No. 1, 79-81.

Gupta, R. and D.R. Coughanowr (1978). On-line gain identification of flow processes with application to adaptive pH control. *AIChE J.*, 24, 654-664.

McAvoy, T.J. (1979). Dynamic modeling of pH in aqueous systems. *Proc. of the Workshop on Industrial Process Control*, Tampa, Florida, November 1974, AIChE, New York, 35-39.

McMillan, G.K., H.A. Crosby and R.C. Waggoner (1980). Process simulation of a waste neutralization process. *ISA Trans.*, 19, No. 4, 79-80.

Marra, P. (1979). Analysis and control of continuous neutralization: A case study. *Proc. of the Workshop on Industrial Process Control*, Tampa, Florida, November 1974, AIChE, New York, 40-47.

Mellichamp, D.A., D.R. Coughanowr and L.B. Koppel (1966). Characterization and gain identification of time-varying flow processes. *AIChE J.*, 12, 75-82.

Mellichamp, D.A., D.R. Coughanowr and L.B. Koppel (1966). Identification and adaptation in control loops with time-varying gain. *AIChE J.*, 12, 83-89.

Myron, J., and F.G. Shinskey (1979). Advanced control systems for waste pH processes. *Proc. of the Workshop on Industrial Process Control*, Tampa, Florida, November 1974, AIChE, New York, 29-34.

Niemi, J. and P.K. Jutila (1977). Process models and control of acidity. Computer applications in the analysis of chemical data and plants. *CHEMDATA*, 77, Finland.

Niemi, J. (1978). Process dynamic approach to pH control. ISA Annual Conf., Paper No. 838, Philadelphia, PA, 63-68.

Richter, J., C.D. Fournier, R.H. Ash and S. Marcikic (1979). Feedforward, adaptive pH control. *Proc. of the Workshop on Industrial Process Control*, Tampa, Florida, November 1974, AIChE, New York, 48-53.

Shinskey, F.G. and T.J. Myron (1970). Adaptive feedback applied to feedforward pH control. *Advances in Instrumentation*, Proc. ISA Conf., 25, Part 1, p. 565/1-7.

Shinskey, F.G. (1973). *pH and pIon Control*. Wiley, New York.

Shinskey, F.G. (1979). *Process Control Systems*, 2nd ed., McGraw-Hill, New York.

Trevathan, V.L. (1978). Advanced control of pH. ISA Annual Conf., Paper No. 839, Philadelphia, PA, 62-82.

Trevathan, V.L. (1979) Characteristics of pH control. *Proc. of the Workshop on Industrial Process Control*, Tampa, Florida, November 1974, AIChE, New York, 24-28.

INFERENTIAL CONTROL APPLIED TO INDUSTRIAL AUTOCLAVES

J. R. Parrish and C. B. Brosilow

Department of Chemical Engineering, Case Western Reserve University, Cleveland, Ohio, USA

Abstract. Inferential control systems will significantly outperform conventional feedback control with higher order and dead time processes. In order to realize the advantages of Inferential Control, several issues must be addressed regarding practical implementation of the control system. These are:

A. Elimination of control degradation due to saturation of the control effort.

B. Bumpless Manual-automatic switching

C. The design of cascade control systems.

Algorithms are presented for dealing with these issues. The methods have been applied to an industrial autoclave plant, with positive results.

Keywords. Cascade control; delays; models; industrial control; saturation

INTRODUCTION

Inferential Control (alias Internal Model control or Smith-Predictor) is potentially superior to conventional PID feedback for high order processes or those with significant process dead times. In order to achieve superior operation, however, the following practical problems concerning implementation must be overcome:

1. Prevention of control degradation due to controller saturation.

2. Bumpless auto-manual transfer.

3. Design of cascade control systems.

BACKGROUND

Inferential Control was originally proposed by Brosilow and Joseph (1978) as a means of controlling unmeasurable process outputs from secondary measurements. A block diagram of the Inferential Controller is shown in figure 1. The effects of disturbances on the primary outputs is estimated using an estimate of the effect of disturbances on secondary (measurable) outputs. The effect of the disturbances on the secondary outputs is estimated by subtraction of the control effort effect ($\hat{y}$) from the process measurement (y). The Inferential Controller (I.C.) of figure 2 is identical to that system of figure 1, except that the primary outputs are

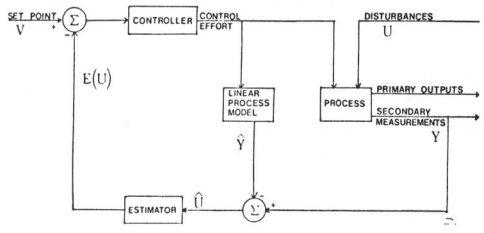

FIGURE 1. INFERENTIAL CONTROLLER
(using secondary outputs)

measured directly. The estimator is not needed when primary measurements are used and is replaced by a unity gain.

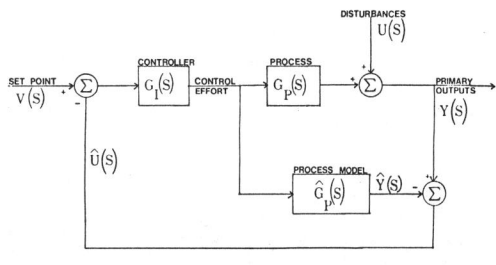

FIGURE 2. INFERENTIALL CONTROLLER
(using primary outputs)

The controller (G_I) of figure 2 is chosen as:

$$\hat{G}_I(s) = \hat{G}_p^{-1}(s) F(s)$$

$$y(s) = \frac{1-F(s)}{1+(G_p(s)\hat{G}_p^{-1}(s)-1)F(s)} U(s) + \frac{G_p(s)\hat{G}_p^{-1}(s)F(s)}{1+(G_p(s)\hat{G}_p^{-1}(s)-1)F(s)} V(s)$$

Several properties of the system are discernable from the above equation. These are;

1. If the model is both perfect and invertable then the filter is selected as a unity gain and

$$y=V \text{ for any } u$$

2. If the filter has a unity steady state gain, there will be no steady state error (y=V @ steady state) even if the model is not perfect. This result is significant because no integrator is present in the controller. It is the structure of the system which enforces this behavior, and not the accuracy of the model.

Brosilow (1979) suggests a two step method for design of the controller with imperfect models. The first step is the design of the controller assuming a perfect, invertable model. The second step is to design the filter to account for the non invertability of the model and to desensitize the system to modeling errors. For a simple linear process, the procedure may be carried out as follows:

$$\hat{G}_p = \frac{N(s)e^{-\hat{D}s}}{D(s)}$$

$$G_I = \hat{G}_p^{-1} F = \frac{D(s)e^{+\hat{D}(s)}}{N(s)} F(s)$$

In order to make the controller realizable, the filter is chosen as:

$$F = \frac{e^{-\hat{D}s}}{(\varepsilon s+1)^n}$$

where: n = degree of D(s) - degree of N(s)

and $n \geq 0$

The value of the filter time constant (ε) is chosen to provide a stable, well behaved system in spite of modeling errors.

The value of the filter time constant provides a direct measure of the model uncertainty in the design. The filter itself is approximately the closed loop response of the system for processes near the model $\hat{G}_p$. The actual response of the system is:

$$\frac{y}{V} = \frac{\hat{G}_p^{-1} G_p F}{1+(\hat{G}_p^{-1} G_p - 1)F} \approx F$$

THE TREATMENT OF SATURATION EFFECTS IN THE I.C.

A highly desirable feature exhibited by the I.C. system is the absence of a steady state error (offset). This is accomplished through the basic structure of the system (figure 2) and not through the presence of an integrator in the controller (cf. background). Although no integrator is present explicitly, offset is prevented by positive feedback from the process model. This positive feedback acts to force the process output to the set point in a manner similiar to an integrator.

In order for the control system to perform as designed, the input (from the controller) to the model and the plant must always be equal. During saturation, the process input is limited in magnitude by physical bounds. If the linear model is not subject to the same constraints, then the model will no longer follow the process input and the positive feedback will force the controller output to increase without bound. This situation is analogous to wind-up problems in controllers with integral action.

In the following, two methods are proposed for the elmination of controller windup.

Control Effort Limiter

Placement of a limiter in the control loop (figure 3) will prevent the controller output from increasing without bound during periods of saturation. This addition will correct the most serious problems associated with saturation, and the overall system response will be greatly improved. However, even with this scheme some undesirable properties of the system may be observed during the return from saturation. These properties are best examined by an example.

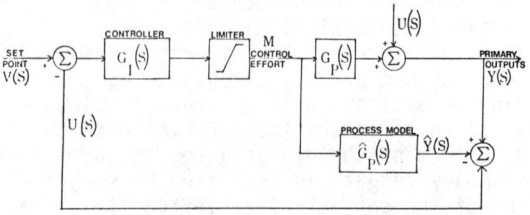

Figure 3 — Control Effort Limiter

Consider the case where the controller (G_I) is a simple lag as shown below:

$$G_I = \frac{1}{as+1}$$

where: (a>0)

Suppose the controller input (e) has been sufficiently high for a period of time so as to cause the controller output (r) to be above the saturation value (M_{max}) as illustrated in Figure 4. At this

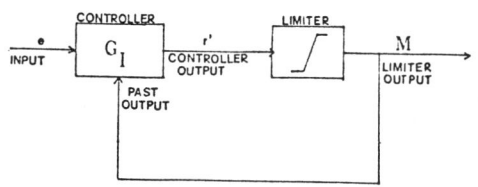

FIGURE 5 — LIMITER OUTPUT FEEDBACK

WHERE G_I = DISCRETE OPERATOR ON THE PAST OUTPUT VALUE AND THE PRESENT INPUT STATE

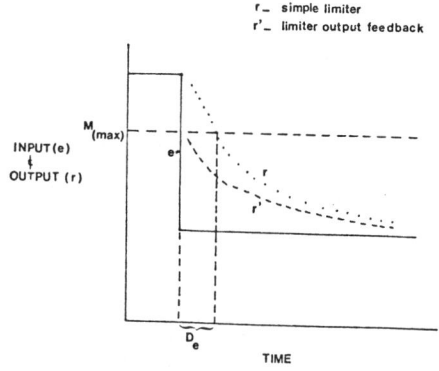

FIGURE 4
RESPONSE OF THE CONTROLLER ELEMENT TO A STEP INPUT

point, a step change in the controller input occurs. The step change forces the controller output (r) to return to the linear operating region.

The time required for the controller output to attain the saturation value (M_{max}) is an effective dead time (D_e) in the controller. The dead time is due to the lag action of the controller and wind-up of the system which is not completely eliminated by addition of the limiter. The apparent dead time degrades the system response by preventing rapid recovery from saturation.

CONTROLLER LIMITER OUTPUT FEEDBACK

It is possible to fully eliminate the effects of wind-up in the I.C. This is accomplished through the use of past value of the limiter output rather than the past value of the controller output in the controller algorithm. This type of controller is shown conceptually in figure 5. Using the output feedback approach, the controller responds as if the last input had caused the controller output to reach the saturation value. The response of the algorithm to a step change input is shown in Figure 4. This algorithm should be applied to controllers of the form:

$$G_I = K_I \frac{(\tau_2 s+1)}{(\tau_1 s+1)}$$

with: $\tau_1 > \tau_2$

The reason for his narrow application is as follows.

Controller induced dead time will occur only with controllers dominated by lag action. The lag action is due to the filter, which is designed based on the model uncertainty. The dominance of lag action is then due to a large degree of model uncertainty. In cases with a large degree of model uncertainty, the process model is normally chosen of the form:

$$\hat{G}_p = \frac{\hat{K} e^{-\hat{D}s}}{\hat{\tau}s+1}$$

corresponding to a controller of the form:

$$G_I = \frac{1}{\hat{K}} \frac{(\hat{\tau}s+1)}{(\varepsilon s+1)}$$

The algorithm suggested for the practical implementation of the output limiter feedback concept is given below (refer to figure 5):

$r' = (\hat{\tau}(e-e_o) + e(\Delta t) + \varepsilon M_o)/((\varepsilon + \Delta t)\hat{K})$

$M = r'$ if (low limit $\leq r' \leq$ high limit)

$M =$ low limit if $r' <$ low limit

$M =$ high limit if $r' >$ high limit

$M_o = M$; $e_o = e$ (Δt = sampling period)

It is recommended that the output feedback algorithm (i.e. the 2nd algorithm) be utilized when the ratio of the lag time constant to the lead time constant exceeds two. For ratios less than two, the use of output limiter feedback may cause undesirable overshoot in the controller action (figure 6).

A simple limit (the 1st algorithm) should be used in cases where the ratio of the lag to the lead time constant is less than two. In these situations there is normally very little or no apparent controller dead time, and no advantage would be gained through the use of the output feedback algorithm.

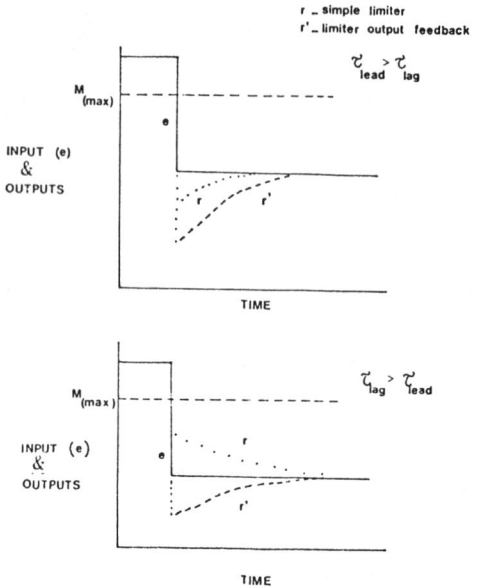

Figure 6
Control Response to a Step Change Input

BUMPLESS TRANSFER

Manual Operation

The proper operation of the I.C. requires a valid estimate of the disturbance $(\hat{u})$ during automatic operation. A valid estimate of the disturbance means that $(\hat{u})$ depends only on the disturbance itself and the process-model mismatch. Therefore, a valid estimate will require that the process and model receive the same control effort. If at any time the proces and model do not receive the same control inputs, undesirable transients will remain in the system due to the past discrepancies. These considerations lead to the recommended manual control configuation of the I.C. as shown in figure 7.

This type of operation allows the model to follow the process control effort at all times. With this configuration, a valid estimate of the disturbance $(\hat{u})$ is always available (after dissipation of the initial start-up transients). The control effort consistent with automatic operation (M_a) is then also available.

Transfer to Automatic

During resumption of automatic control from manual control, a bump to the system will occur if a difference between signals M_a and M_m exists. This disturbance may be undesirable and can be eliminated in a manner similiar to that used with three-mode controllers. The recommended procedure is illustrated in figure 8.

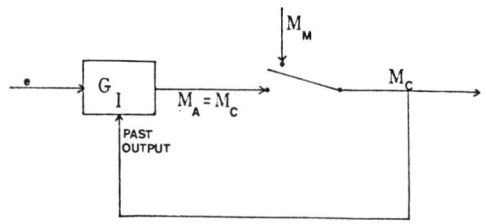

Figure 8 — Bumpless Transfer Operation

The method of transfer consists of maintaining the controller output (M_a) equal to the process input (M_c) at all times. In this way, manual to automatic switching will be bumpless. This procedure may be implemented using a controller algorithm similiar to that proposed to reduce saturation effects. An example of the algorithm applied to the first order system;

$$\frac{M_a}{e} = \frac{1}{\hat{K}} \left(\frac{\hat{\tau}s+1}{\epsilon s+1} \right)$$

is as follows;

M_m = manual input

$M_a = (\hat{\tau}(e-e_o) + e(\Delta t) + \epsilon M_o)/((\epsilon+\Delta t)\hat{K})$

$M_c = M_a$ if the control is automatic

$M_c = M_m$ if the control is manual

$M_o = M_c$

$e_o = e$ $\quad \Delta t$ = sampling interval

This method will reduce the bump of the auto-manual switching by limiting the rate of control effort change with the lag action of the controller. Figure 9 illustrates the effect of a bumpless transfer on the control effort during manual to automatic switching.

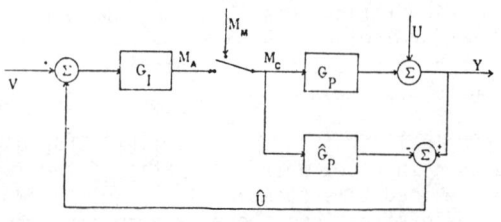

Figure 7 — Auto-manual Operation

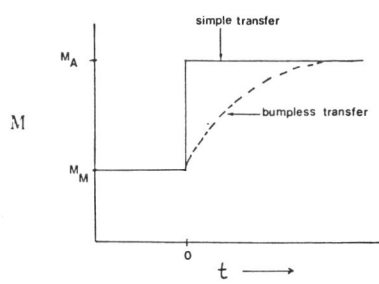

FIGURE 9 — CONTROL EFFORT CHANGE DURING MANUAL-AUTO SWITCHING

INFERENTIAL CASCADE CONTROL

The conventional cascade loop of figure 10 has found application throughout the process industry.

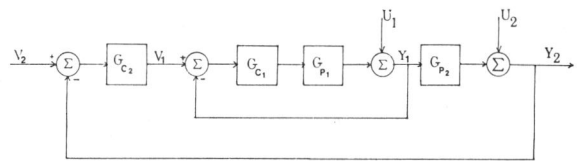

FIGURE 10 — CONVENTIONAL CASCADE CONTROL

The reason for the addition of the inner loop is to reduce the effect of the disturbance (U_1) on the final output (y_2). In order to accomplish this goal, the process inner loop should be faster acting than the outer loop process (G_{p_2}).

If the inner loop controller (G_{c_1}) and process (G_{p_1}) are sufficiently fast acting, the effect of U_1 on y_2 will be greatly reduced. However, if G_{p_1} is not sufficiently fast, there may be little or no advantage in using a cascade control scheme.
The speed with which the inner loop acts is often only restricted by saturation constraints. The existence of saturation will always slow down the response of the inner loop. Processes requiring frequent set point changes and/or which experience large disturbances are those most subject to saturation of the inner loop.

The main problem caused by the inner loop saturation is degradation of the outer loop control. In the conventional system (Fig. 10) inner loop saturation will cause the outer loop to windup. To reduce the windup (and subsequent overshoot) the speed of response of the outer loop controller must be reduced. Inner loop saturation tends to destabilize the configuration of figure 10.

In order to eliminate the windup effects of the conventional system, the outer loop controller is usually implemented as in figure 11.

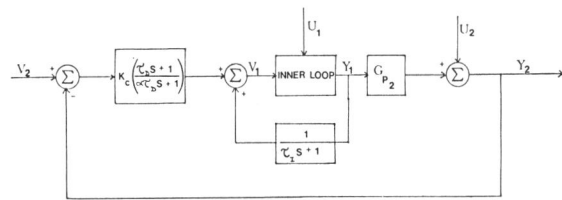

FIGURE 11 — THREE MODE CASCADE CONTROL

where: K_c = controller gain

τ_D = derivative time constant

τ_I = reset time
$\alpha \approx .05$

Using the intermediate process output (y_1) to implement integral action eliminates outer loop windup. The configuration of Figure 11 is stabilized (more or less) by inner loop saturation. Although the modified P.I.D. scheme has these desirable properties, it may be difficult to tune with the saturation effects. The problems with tuning the outer loop make it difficult to fully utilize the stabilizing influence of the inner loop saturation. The I.C. of figure 12 will deal effectively

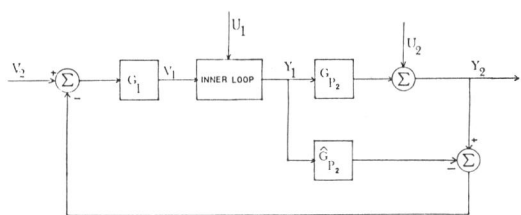

FIGURE 12 — INFERENTIAL CASCADE CONTROL

with the problems presented by the inner loop saturation. The configuration is inherently anti-windup because both the process and model receive the same input (y_1). The process and model receiving the same input is a sufficient condition for elimination of windup, as discussed previously in regards to control degradation due to saturation. In addition, the presence of saturation causes the inner loop to act as a filter.

The inner loop will filter out large amplitude, high frequency fluctuations in the setpoint (V_1). This property provides additional stabilization of the outer loop by the inner loop. For this reason, the filtering action within the controller may be reduced or eliminated.

The inferential cascade controller will have two advantages over the conventional cascade of figure 11:

1. The I.C. will outperform the conventional controller if the process has dead time.

2. The I.C. will be easier to design and tune than a conventional system when the inner loop is subject to saturation.

The second advantage is due to the basic I.C. design. The controller is specified as;

$$G_I = \hat{G}_p^{-1} F$$

where: F = filter function

The degree of stability of the system is selected through a single parameter, the filter time constant. The saturation of the inner loop will serve to reduce the filter time constant. The stability of the conventional system depends on all three controller parameters, which may be difficult to select. This difficulty is due to the non-linear behavior (saturation) of the inner loop.

APPLICATION OF I.C. TO INDUSTRIAL AUTOCLAVES

The inferential control enhancements described above have been applied to the control of 400 industrial autoclaves. A simple diagram of a single unit is given in figure 13.

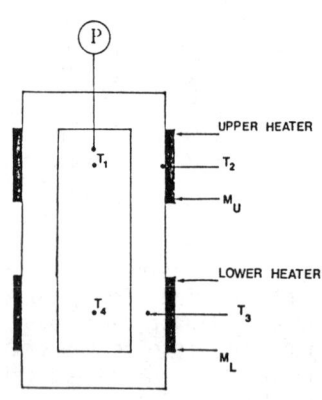

FIGURE 13 — INDUSTRIAL AUTOCLAVE

Two sets of resistance heaters (upper and lower) are used to control the vessel wall temperatures (T_2 and T_3). These temperatures will in turn affect the internal temperatures (T_1 and T_4). All four temperatures are measured. The wall temperatures have little effect on each other, while the internal temperatures are strongly coupled. The main source of disturbances to the process is changes in ambient conditions. These changes directly effect the wall temperatures.

Specific constraints on the process include:

1. Internal vessel pressure $< P_{max}$

2. Wall temperatures $< T_{max}$

In addition, persistent cycling of the wall heaters should be avoided.

The control objective of the process varies throughout the production cycle. During the initial phases, the skin temperatures (T_2 and T_3) are ramped at given rates to predetermined values. The wall temperatures are then maintained until a discrete change in the internal temperatures (T_1 and T_4) is observed. At that time, it is desired to begin tracking of pre-programmed trajectories for the internal temperatures.

Step tests of the process indicate nominal responses as shown below:

$$\begin{bmatrix} T_2 \\ T_3 \end{bmatrix} = \begin{bmatrix} \frac{30}{s+1} & 0 \\ 0 & \frac{30}{5s+1} \end{bmatrix} \begin{bmatrix} m_u \\ m_L \end{bmatrix}$$

$$\begin{bmatrix} T_1 \\ T_4 \end{bmatrix} = \begin{bmatrix} \frac{.2e^{-.1s}}{.4s+1} & e^{-.25s} \\ 0 & e^{-.25s} \end{bmatrix} \begin{bmatrix} T_2 \\ T_3 \end{bmatrix}$$

Under cascade control, the autoclave is to follow a desired trajectory for the temperature difference (ΔT) and T_4. The response is:

$$\begin{bmatrix} \Delta T \\ T_4 \end{bmatrix} = \begin{bmatrix} \frac{-.2e^{-.1s}}{.4s+1} & 0 \\ 0 & e^{-.25s} \end{bmatrix} \begin{bmatrix} T_2 \\ T_3 \end{bmatrix}$$

where: $\Delta T = T_4 - T_1$

The main source of disturbance to the process was noted as ambient air temperature and flow changes. This observation along with the process responses indicate an advantage in the use of cascade control. The scheme chosen is as shown in figure 12.

In the case of the T_4 control loop;

$V_2 = T_4$ setpoint

$\hat{G}p_2 = e^{-.25s}$

$G_I = \hat{G}p_2^{-1} F = 1/(\epsilon_1 s+1)$

$Y_1 = T_3$

$Y_2 = T_4$

In the case of the ΔT controller:

$V_2 = \Delta T$ setpoint

$\hat{G}p_2 = -.2e^{-.1s}/(2s+1)$

$G_I = \hat{G}p_2^{-1} F = (-5(.4s+1)/(\varepsilon_2 s+1))$

$\quad = -5(.4s+1)/(\varepsilon_2 s+1)$

$Y_1 = T_4$

$Y_2 = \Delta T$

The inner loop of the above schemes is chosen as a conventional (P.I.D.) control loop. This loop is subject to frequent saturation, particularly during setpoint changes. The outer loop is chosen as an Inferential controller, due to the dead time and for ease of tuning.

Frequent saturation of the inner loop tends to stabilize the outer loop of control scheme (as discussed previously). The stabilizing influence of the saturation allows selection of a smaller filter time constant than that allowable without saturation.

In the case of the T_3, T_4 loop, the model uncertainty is approximately

$$\hat{G}p_2 = \hat{K} e^{-\hat{D}s} \quad \begin{array}{l} .3 < \hat{K} < 2. \\ .1 < \hat{D} < .5 \end{array}$$

The nominal values of $\hat{K}=1$ and $\hat{D}=.25$ were the most frequently observed. In the absence of saturation, a filter time constant 1.0 of would be required to stabilize the system with the given model uncertainty (Appendix). However, due to the inner loop saturation a filter time constant of zero was found to work well (no stability problems). This result implies that the process may be forced as hard as possible, with the necessary stabilization accomplished via the inner loop.

The case of the (T_2, ΔT) loop yields a result similiar to that of the (T_4, T_3) loop in regards to controller filtering. However, in this case the (ΔT) signal is corrupted by high frequency noise. The noise is filtered in the feedback loop. The filtering was accomplished by a simple lag;

$$F_n = \frac{1}{\tau_F s+1}$$

The noise filter time constant (τ_F) was chosen as 0.1 hours.

The controller for the (T_2, ΔT) loop was chosen as:

$$G_I = \hat{G}p_2^{-1} F = (\frac{5(.4s+1)}{} e^{.1s})(\frac{e^{-.1s}}{\varepsilon s+1})$$

$$= 5. (\frac{.4s+1}{\varepsilon s+1})$$

The approximate uncertainties in the process model are:

$$\hat{G}p_2 = \frac{\hat{K}e^{-\hat{D}s}}{\hat{\tau}s+1} \quad ; \quad \begin{array}{l} -.4 < \hat{K} < -.05; \\ .2 < \hat{\tau} < 1.2; \\ .05 < \hat{D} < .5 \end{array}$$

Based on these uncertainties, a filter time constant of (2.2) is needed to stabilize the system without other stabilizing influences. With the noise filtering and inner loop saturation influences, the controller was found to require an arbitrarily small filter time constant. The filter time constant was chosen as 5% of the lead time constant:

$$G_I = 5 \left(\frac{.4s+1}{(.05).4s+1)}\right) = 5 \left(\frac{.4s+1}{(.02)s+1}\right)$$

As in the case of the (T_L, T_D) loop, the inner loop stabilizes the system.

Figure 14 illustrates a typical processing

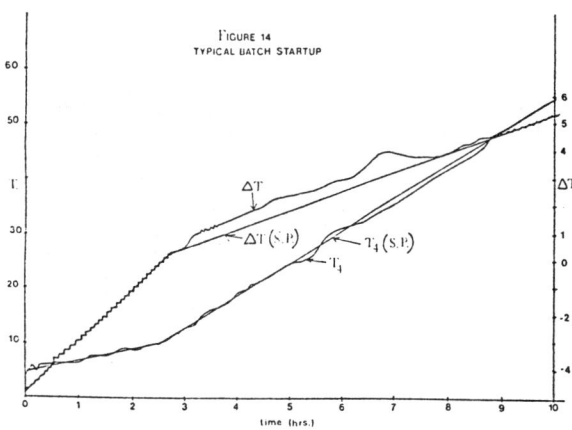

FIGURE 14
TYPICAL BATCH STARTUP

run. The output tracks the setpoint fairly well, with deviations due primarily to control effort saturation. Figure 15 makes a comparison of the I.C. and P.I.D. cascade control (circa, 1974) results, for constant control setpoints. The noise exhibited with the P.I.D. controller is likely due in part to the form of its implementation.

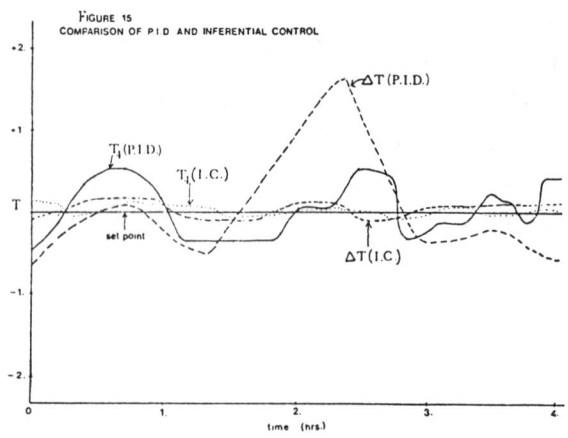

Figure 15. Comparison of P.I.D. and Inferential Control

The P.I.D. cascade was implemented conventionally (c.f. fig. 10), and was "detuned" to allow for inner loop saturation. For this reason, the P.I.D. cascade was not used during the start-up period when frequent setpoint changes are required. Therefore, no comparison is available for the I.C. response of figure 14.

CONCLUSIONS

Simple methods for implementation of Inferential Control schemes have been applied to an industrial autoclave process. These methods include considerations of control degradation due to saturation and auto-manual transfers. The recommended structure for Inferential cascade control is found to be easier to design and tune than a comparable three mode control system, and may give considerably better performance.

The design and implementation of I.C. systems can be somewhat more complex than the use of three mode controllers. The decision to use I.C. ultimately depends on the economic trade off between increased control system complexity and better control system performance. This paper gives the reader a qualitative "feel" for the complexities and potential benefits of the Inferential control schemes, which may aid in making the decision whether or not to employ Inferential Control.

BIBLIOGRAPHY

Parrish, J.R., 1982, "The Use of Model Uncertainty in Control System Design with Application to a Laboratory Heat Exchanger Process", M.S. Thesis - Case Western Reserve University.

Brosilow, C.B., 1979, "The Structure and Design of Smith Predictors from the Viewpoint of Inferential Control".

"Foxboro Fox-3 Software User's Guide", 1981, The Foxboro Company.

APPENDIX

The value given for the filter time constant is based on a maximum magnitude peaking criterion. In both cases the filter time constant was selected according to:

$$\max_{w,D,K,\tau} \left[G_L(jw) \right] = 1.2$$

where:
$$G_L(jw) = \frac{G_p(jw) \cdot \hat{G}_p^{-1}(jw) \cdot F(jw)}{1 + (G_p(jw) \cdot \hat{G}_p^{-1}(jw) - I)F(jw)}$$

$\hat{G}_p$ = the fixed model = $\dfrac{\hat{K} e^{-\hat{D}s}}{\hat{\tau}s + 1}$

G_p = the uncertain plant = $\dfrac{K e^{-Ds}}{\tau s + 1}$

Lower bound $\leq D \leq$ Upper bound

Lower bound $\leq K \leq$ Upper bound

Lower bound $\leq \tau \leq$ Upper bound

DISTRIBUTED MICROCOMPUTER CONTROL IN REAL TIME OF THE PROCESS OF FERMENTATION OF SUGAR CANE DERIVATIVES (BY-PRODUCT)

S. Teijero Páez* and J. Olivera Reyes**

Industrial Control and Instrumentation Enterprise, SIME, Cuba
**Department of Electronics, Faculty of Electronics, Cibernetic and Communications, ISPJAE, Cuba*

Abstract. The main features of a distributed microcomputer system applied to the fermentation process of the by-products of the Cuban sugar cane industry are analized in this paper. The main features of the analyzed technological process are exposed in it, giving emphasis in the main variables which characterizes it. The features of the intended parallel processing system are studied. They function under the control of a processor called micromonitor. The main real time functions intended as well as a study of the programs of the processor diagnostic which permits to find out the possibles errors passing to a control form less efficient during the time it lasts.

Keywords. Distributed microcomputer systems, real time systems, digital control in real time, data collection and processing of the information, programming in real time.

INTRODUCTION

In Cuba, tropical country having a great sugar cane industry, the production of the forage yeast from the final honeys results particularly interesting, taking into account the impossibility of having vegetable proteins of low cost caused by the climated characteristics. Besides, it is a way to become independent of the foreign market variations with regard to the import of protein materials adequated for the production of forage. These factors, as well as being it a located production, the mean technological complexity which diversifies the industrialization of the sugar cane and being, besides, a working source, make it recommendable for the developing countries.

The production of forage yeast from the final honeys of the sugar cane is characterized for being a continuous process comprising the following 4 basic units: raw material preparation, the fermentation, the recovering of the yeast, the concentration and the drying.

The purpose of this paper is to expose the main features of a control system of distributed microcomputers in real time applied to the first two basic units of the production of forage yeast giving emphasis by considering it the heart of the production process in the fermentation control.

In the general architecture of the designed system is emphasized the use of three processors and a monitor processor called micromonitor. The program of coordination of the different processors runs in the micromonitor being designated the particular memory of each program for the programs attending each specific task.

One of the aspects of great interest featuring the hardware of the system is the utilization of a common memory zone with its control unit which permits a multidirectional access to this memory from the micromonitor and the different processors. Another aspect of great interest featuring the hardware of the system is that one of the processors constitutes an intelligent terminal for the inlet-outlet operations related the operator of the process. The aspect of the hardware is another element ensuring the constant parallel processing of the system.

The main functions of the real time assigned to the system are discussed. They are distributed in data collection, primary processing of the information and alarm of the process, digital control and man-machine communication.

Inside the function of the digital control are discussed the features of the program system for the digital control of the processes in real time using the technique of programming in block diagram and determines its functions in a programming language of specific purpose which has been developed recently with the collaboration of the authors.

The functions of the micromonitor and how arranges it the tasks in the different processors are analyzed.

Finally, the main features of the diagnostic and test system of the different microprocessors comparing the structure are given. This system foresees that the error of one unit does not affect the functioning of the rest and, besides, that the functions of the affected unit are being assumed by the rest as the error is being solved.

SOME CONSIDERATIONS ABOUT THE TECHNOLOGICAL PROCESS

The production of forage yeast from the final honeys of the sugar cane characterizes by the continuous process in which the honey gives carbohydrates as an energy source and growing factors in the form of vitamins and minerals. The development of this production comprises 4 basic units of the process as follows: preparation of honey, fermentation, recovery of the yeast and concentration and drying

Figure 1 shows the diagram in blocks of the forage yeast production from the final honeys of the sugar cane.

Fig. 1

The molasses is the main substratum for the developing of the yeast. Because of its character, the honey contains a great number of impurities in colloidal and suspension form, as well as volatile organic acids and a high microbian flora that could affect the good conduction of the process. This is why, usually, a purification process has been used for which it has used multiple variants. One of the variants more used is the method of the hot acid, by which it is attained to eliminate a great part of the impurities in the honey as well as to reduce the calcium content. The experimental results in the application of this method report that in accordance with the pH value, the speed of the sedimentation varies, as well as the final bulk (volume) of the sediment. However, the temperature, although it has influence on the sedimentation speed, does not affect neither the final volume of the sediment, nor the calcium content of the floating. The sterilization process is the last step in the treatment of the honey. This operation is to eliminate the flora microorganisms of the honey, whose value is about 10^3 microorganisms by gram of physical honey in Cuba, permitting a better control of the process of fermentation, to increase the sediments and to raise the quality of the final product. It is common to take as design basis in the systems of continuous sterilization of honeys, temperatures of $120°C$, 4,5 pH and 4-5 minutes for the time of retention.

The fermentation is the main operation of the process, the quality of the product, as well as the efficiency in the assimilation of the substratum will depend on mainly the successful course of this operation. The process of yeast production is aerobic and exergonic, being necessary to supply great volumes of air, as well as to dispose of some exhaust system of the heat, to keep the adequate temperatures for the process.

The main propagation in the process of fermentation is performed in a fermentor equipment of great capacity, equipped with a stirring-aeration system, level control, pH, foam, water cooling system, dosage meters and others. The capacity of the fermentor equipment of producing yeast efficiently is featured by some indexes which have to be taken into consideration in its design and must be checked periodically. These indexes are the following:

Specific consumption of energy. It is determined as that energy in kilowatts per hour (kW.h) that is necessary to produce 1 kilogram of dry yeast. The fermentor equipment for the production of dry yeast from the final honeys swings in the value of 0,40 to 0,50 kW.h/kg of dry yeast.

Biochemical coefficient of the oxygen transference. It is the oxygen the one that could be used for the functions of the metabolism for each kW.h supplied in stirring-aeration. The normal value is from 2 to 2,5 kg oxygen/kW.h.

Specific consumption of air. It is the volume of measured air in temperature condition and standard pressure ($0°C$ and 760 millimeters of Hg) needed to produce 1 kg of dry yeast. For mechanically stirred systems is from 11 to 15 Nm^3/kg of dry yeast, corresponding to an utilization of oxygen supplied from 22 to 30 percent.

For only aerated systems, the index is from 30 to 40 Nm^3 corresponding to an utilization of 10 to 15 percent of the oxygen contained.

Ratio of oxygen transference. It is the maximum capacity of oxygen transference by unit of time and volume.

Volumetric productivity. It represents the increase of yeast by volume unit of fermentation and by unit of time. For the production from final honeys, the figure for the systems of high concentration are found between 2 to 2,8 kg of dry yeast/m^3.h and for low concentration between 1,5 to 2 kg of dry yeast/m^3.h.

From the above mentioned it could be deduced that constitute important variables for the two analyzed basic units the following:

Preparation of honey. Temperature, pH and time of retention.

Fermentation. Level, temperature, pH and the different fermentation indexes.

These variables constitute the center of the control system of the distributed microcomputers in real time developed, and, in some cases will be measured in a continuous form (temperature, level, pH, air consumption) and in others will be introduced to the system through the analysis coming from the laboratory of the chemical control.

ARCHITECTURE OF THE CONTROL SYSTEM IN REAL TIME

The basic principle of the application of the microprocessors to the control of the industrial process is the development of the centralized direction and distributed control systems which features in having the shared functions of control in different processors with communication between them. In a similar system could be ensured efficiently a parallel processing, attaining the organization and performance of the different functions in the different processors with one monitor of similar characteristics to the ones used in the centralized function systems. In this case the special function of this monitor is the coordination of the different processors. Immediately will be discussed the conception and realization of the system of distributed microcomputers for the control in real time of the fermentation process of the Cuban industry of the sugar cane by products.

Figure 2 shows the general architecture of the system where it is outstanding the use of three processors CPU_1, CPU_2, CPU_3 and of one processor monitor called micromonitor (Monitor CPU).

The coordination program of the different processors runs in the micromonitor and in the particular memory of each processor will appear reflected the programs for attending each particular task. The processors communicate through a common memory which could be accessed directly by a processor through its control unit and under the command of the micromonitor. The processor asks for attention of the micromonitor through the interruptions PIT and the micromonitor activates or stops the tasks in each one of the processors using the interruptions MIT.

Fig. 2

One of the aspects of great interest of the structure shown in figure 2 is the utilization of one zone of the common memory with its control unit. By this way, it is possible a multidirectional access to this memory from the micromonitor and the different processors.

The processors communicate through the common memory which appears for each processor as a prolongation of its own memory. The essence of this solution is as shown in figure 3.

Fig. 3

By this way, when a processor needs to work with the common memory requests to the controller of the buses. This request is made through the bit more significant of the bus of addresses. When this request exists, the control unit of the common memory stops momentarily the run in that processor and it enters in a wait for memory up to this one is free to be accessed for that processor. When the request is accepted the bus of the processor joins to the bus of the common memory up to finish the memory cycle. The memories used are from 200 to 300 nanoseconds of the memory cycle permitting in the cycle the instruction of a simple instruction of reference to memory of the processors used, to carry out more than 10 accesses to the common memory by other processors. It gives the possibility of carrying out the real parallel processing, although if the number of processors increase the probability of a longer wait will be higher

The requirement of the processors for using the common memory has the priority. This way may minimize the worst case of access time to memory, being it of little times of memory cycle depending on the number of processors.

For attaining all the organization of the accesses to memory, a unit control of memory is used with a quick logic with circuits TTL being neglectable the delays of time of the control unit compared with the speed of the microprocessors with MOS technology.

FUNCTIONS OF THE CONTROL IN REAL TIME ASSIGNED TO THE SYSTEM

The functions carrying out the designed system of distributed microcomputers are of the information-computational and of control types. These functions are:

1. Data collection and primary processing of the information.
2. Digital control.
3. Man-machine communication.

Data collection and primary processing of the information.

In the structure of the distributed microcomputers shown in figure 2, the function of data collection and primary processing of the information is carried out by the microcomputer called CPU_1. This microcomputer has an analog entry card which permits multiplex the different channels of entry under the control of the processor for its passing through the amplifier of programming gain, a sampled and hold amplifier, the analog-digital converter and from there to the inlet-outlet ports of the microcomputer.

This microcomputer has a zone of RAM, where stores the values of different inlet channels read in one zone of the ROM where appears the programs for the data collection, primary processing of the information and alarm of the process.

The programs for the data collection, primary processing of the information and alarm of the processes gives the possibility of carrying out the primary processing of the information and find out situations of alarm. Besides, these arranged data by variables permit the printing of periodical reports or to the request of the operator that is other of the functions of the system. In the structure of the system exists a data bank where the data information, the variables and its state are stored in the memory of CPU_1, and, besides, this reproduced data bank in the common memory, so that to be accessed by the rest of the processors.

The programs for the data collection, the primary processing of the information and alarm of the processes have among their functions the verification of the reading, the conversion to the engineering units, verification of the speed of change, the digital filtered the verification of the process limits and the calculus of some statiscal parameters.

The verification of the reading is carried out in order to know if the reading performed is over or down the limits of the instruments. If this happens is evident that some element of the chain of measurement is broken and the measurement should not be considered valid, so discarding the channel and inform the operator of the error. In the data bank appears the minimum limit value and the maximum limit of the instrument stored for each analog variable.

The verification of the speed of change consists in checking in each measurement the ratio of change of the variable, permitting to analize its performance. The ratio of the change is determined by the following equation:

$$\text{Change} = \left| (Y_n - YF_{n-1}) \right| \quad (1)$$

where

Y_n - is the read and processed value in that cycle.

YF_{n-1} - is the filtered value in the precedent cycle.

If the variable does not require the digital filtering, it is taken instead of YF_{n-1} to Y_{n-1}.

The user must specify for each variable which is the permissible maximum change. In case of exceeding and alarm state is declared. The alarm of this type could be interpreted of two different ways;

1. that the variable is affected by noise;
2. that the variable is associated to an instability state.

This alarm state does not stop the processing of the variable. The objective of the digital filtering is to eliminate the influence of the noises in the useful signal and the usual errors in the value of the controlled parameter. In this case a filter is used which is expressed mathematically by a discret differential equation of first order of the form;

$$Y_k = a(Y_{k-1} - X_k) + X_k \quad (2)$$

$$0 < a < 1 \quad (3)$$

$$a = \frac{1}{1 + \frac{T}{T_1}}$$

where

Y_k - filtered value

Y_{k-1} - filtered value in the precedent instant

X_k - read value without filtering

a - constant of the filter

T - period of sampling

$1/T_1$ - frequency of the cut

The verification of the process limits when finding out the alarm state does not declare up to does not pass N successives measurements in this state. The N value is fixed by the user among the variable data. Once declared, the alarm when the variable returns to the normal state this one stops. The maximum and minimum values are fixed by the user and appear in the cutting of each variable.

The statistical parameters which calculate the system are the average value and the typical drift which are used in the printing of periodical reports and in the calculus of some address indexes of the process. The average value is calculated by a recursive algorithm having the following general expression;

$$\overline{X}_k = \overline{X}_{k-1} + \frac{1}{N}(X_k - \overline{X}_{k-1}) \quad (4)$$

where

$\overline{X}_k$ - is the average value in K instant

X_k - is the measured value in K instant

$\overline{X}_{k-1}$ - is the average value in K-1 instant

N - number of measurements made up to the K instant

The typical drift is used for knowing how scatter are the measured values in the interval of time in relation to the average value. That is why in some occasion results an index of the quality control.

The calculus algorithm is;

$$S_k^2 = S_{k-1}^2 + \frac{1}{N-1}(X_k - \overline{X}_k)^2 \quad (5)$$

where the average value is thought to be calculated previously. The conversion to engineering units is performed taking into account the features of the sensors. The linear conversion is used mainly. It is solved through the equation;

$$Y = Ax + B \quad (6)$$

where

Y – converted value to engineering units
X – read and rectified value
A – pending
B – intercept

The pending and intercept values are calculated by classical methods and they are fixed values for each channel which are stored in the data bank.

Finally, storage of the read and processed value is carried out in a data table. This value stores in case of not finding out any alarm state in the form of a normalized integer or in engineering units depending on an intermediate value or a value that will be appeared in any periodical report. The CPU_1 processor asks for attention of the micromonitor through PIT_1 interruption if it is found out the alarm state in any variable, so that the micromonitor through the MIT interruptions could inform to CPU_2, as it would be seen lately carries out the function of digital control, it is a controlled variable in order to disconnect up to the alarm lasts and to CPU_3 in order that prints to the operator the alarmed variable and the hour in which this alarm occured. The CPU_1 processor once processed all the variable and stored in its data bank will ask to the controller of the access buses to the common memory through the bit more significant of the address bus. The control unit of the common memory will stop momentarily the running in this processor, passing it to the waiting state by memory up to the common memory stays free to be accessed by the processor. When this happens, the CPU_1 processor will dump the content of its memory, that is, the values of the analog channels read and processed in the RAM.

Digital control.

The function of the digital control in the distributed structure of the system is carried out by the microcomputer called CPU_2 which takes for its function from the common memory the stored data for CPU_1. The digital control system of processes in real time (SCDP) used, makes it possible to apply the programming techniques in diagrams of blocks and to determine the functions in the programming language of a specific purpose. This system was developed recently by the authors' collaboration of this paper and their main features appear explained in reference 2. The main advantages of the system are as follows;

1. To give the possibility that without stopping the running of the microcomputer to change constants and the connection of the blocks;
2. To change the form of the control algorithm;
3. To facilitate the manual and automatic operation to a minimum;
4. To reduce to a minimum the requirements of the memory;
5. To carry out a set of typical instructions (blocks) to many applications to reduce the time of starting the system.

This system uses a determine number of blocks which permits to carry out the most common functions in the application of digital control. To each one of these blocks is associated one instruction and the user with the instruction set should carry out his programs. The periodical task in this system is known as basic program and the main program of control (PPC) should order the execution of the basic programs. Since several periodical tasks are determined, a monitor of real time (MTR) has to be used, whose structure appears explained in reference 1. The basic instruction of the system has two main parts; the corresponding part to the definition of the inlets, outlets, number and code and the corresponding part to the parameters.

Block Number	Code	Inlet Block Number 1	Inlet Block Number 2	Parameters

The instruction data are stored separated. A data table is determined for each instruction (TDB) and the data table set is ordered according to the block number from 1 to 128, in which each block has its TDB located in a fixed position. The TDB has a pointer used to chain the instructions of the program, this facilitates the searching of the instructions by PPC. In this system are permitted five different basic programs, the maximum number of blocks to be used in each program is 128 and it is important to fulfill that the total number of blocks of the five basic programs been smaller than 128. The block number does not indicate execution order, but it is given by the chaining of blocks.

A program stacking in the memory the instruction data given by the user in the logic order or calculus is used. This program determines the chaining of the blocks.

The PPC for starting the execution of the blocks needs to know the number of the first block to be executed in a basic program. For this, it uses a table of transference and associates this table to the called ones of the MTR. When PCC is executed, it gets from the table of blocks, the data of the block in question and keep them in the inner registers of the microprocessor. The pointer indicates the next block to be executed and it is kept in the memory, if the pointer is zero indicates that the block taht was executed is the last one. The code of the block is used for determining which subprogram should be executed. It is carried out using the address table of subprograms in which for each ASCII in capital letter exists two locations storing the address of the first subprogram instruction.

The parameters of the blocks are stored in the parameter table and the start address of the block parameters are stored in the table of the parameter pointers. This address could be obtained easily with the block number.

The initial part of the subprogram of the blocks consists in obtaining the inlet values and locates the block parameters in the table. This part is general for all the blocks, although exist a few that does not have inlets and others having only one that its initial part is more simple.

The final part of this subprogram consists in storing the outlet value of the block in the corresponding place of the table of blocks. The system may have a maximum of 26 blocks with different characteristics. However, only 13 have been developed, remaining other 13 to the disposition of the user in order to program his own blocks.

Reference 2 shows the designed blocks

some data of interest about them, as well as an example of a digital control loop and its corresponding block diagram programmed with this system.

The first instruction in the basic program determines the data in relation to the periodical task; they consist in a number of basic program (1 to 5), priority of the task and the execution period. END is the last instruction.

The system permits to determine other blocks as for example, logic functions AND, OR, NOT and others, that render possible to create a sequential logic control or to point out states in a control panel.

This system gives the possibility to the user of changing the structure of control, block parameters and to create his own programs. The conversations will be carried out through the common memory. Using the CPU_3 the operator dispose of a zone in RAM of the common memory in order to deposit the changes and the new programs. The CPU_2 necessarily when adjusting the digital control will have to ask for access to the common memory in order to obtain this information. Besides, the CPU_2 once calculated the values of the digital control will deposit in the common memory the new values in order to be printed periodically or in case of being demanded to the process operator.

Man-machine communication.

Its function in the system is carried out by the microcomputer CPU_3. By the man-machine communication programs the operator may: change the form of control algorhithm; change the parameters in a same structural control; pass the system to the conventional automatic secondary system or to a hand operation; change the limits used for the verification of the reading; carry out adjustments in the changing rate of each variable; modify the number of successive measurements (N) which are desired to carry out once found out and alarm state; change the maximum and minimum values of the alarm of each variable; introduce to the system the results of the analysis of the laboratory; request a report of the instantaneous value of each variable.

Additionally the intelligent peripheric issues periodical reports in hard copy and inform about the alarm states with the real hour of occurrence of it.

FEATURES OF THE MICROMONITOR

The four processors used by the structure of the distributed microcomputers shown, carry out the function of the CPU monitor or micromonitor.

The basic functions of the micromonitor are: organization of the tasks in the different processus; execution of the micromonitor modules; handling of the interruptions.

Organization of the tasks. Figure 4 shows the different tasks and states in each processor.

Fig. 4

One task is passive when it has an inactive state; and it is active when it is ready for being running or executing. When a task is running and it is interrupted, it falls in a waiting state due to a time of delay of an event or a resource. The organizer part of the micromonitor that starts and stops the tasks is stored in the ROM of each processors.

Modules of the micromonitor. The calling to micromonitor could be performed by a task, executing it in a processor through the PIT interruptions. The micromonitor functions are executed through modules which offer the following possibilities: to change the state of the task; handling of the events; task syncronization; resource handling.

<u>Interruption handling</u>. The micromonitor handles outside interruptions as events thus, the tasks could be used in the same way as other events. By this way the user does not need to write the conventional program of the interruption handling.

DIAGNOSTIC SYSTEM AND COMPUTATIONAL SECONDARY SYSTEM

The functions of the diagnostic system are the following: to supply a test of the confidence level (connection tests of the feeding, routine tests of the processors when they are in the operations without occupation of charges and tests); to find out the errors to the level of the LSI chips (central processor, memories and others).

The system of distributed microcomputers designed has a computational secondary method which consists in double the program of the common memory. By this way, if by the verification of the autodiagnostic is found out the error of any processor, it is possible that another assumes the functions from that one is out of order, permitting to continue the control in real time of the process, meanwhile the processor to the out of order unit will be replaced.

CONCLUSIONS

Taking into account the general conceptions of the analyzed parallel processing, a simple organization of the hardware and a high potentiality are reached, reducing at the same time the cost of software if it is compared with the criteria herein analyzed with a similar system performed in a centralized way. The design conceptions offered in this system will give the possibility of gradual increase of the number of processors giving an economical way to the performing of the principle: one task, one processor. This principle is very used in the address centralized systems and distributed control in real time using microprocessors.

BIBLIOGRAPHICAL REFERENCES

1. Barros Olivera Cándido, Guntín del Río Luis. (1979). Monitor de tiempo real para el MCS-80. <u>II Conferencia Científica del ISPJAE</u>. Cuba.

2. Cabana González Juan, Fernández de Alaiza Bertha, Teijero Páez Sergio (1981). Sistema de control digital de procesos mediante el MCS-80. <u>Control-Cibernética y Automatización</u>. No. 4. Cuba.

3. ICIDCA. (1981). Los derivados de la caña de azúcar. <u>Editorial Científico-Técnica</u>. Cuba.

4. Mc Dermott J. (1978). Distributed microprocessor systems advance process designs. <u>Electronics</u>

5. Searle B.C., Freberg D.E. (1975). Microprocessor applications in multiple processor systems. <u>Computer</u>

6. Syrbe M. (1978). Basic principles of advanced process control systems structures and a realization with distributed microcomputer. <u>Seventh Triennial World Congress of IFAC</u>. Volume 1. Helsinski.

7. Teijero Páez Sergio, Cabana González Juan. (1980). Algunas consideraciones en cuanto al sistema de programas para un SADPT utilizando microprocesadores. <u>Control Cibernética y Automatización</u>. No. 3. Cuba.

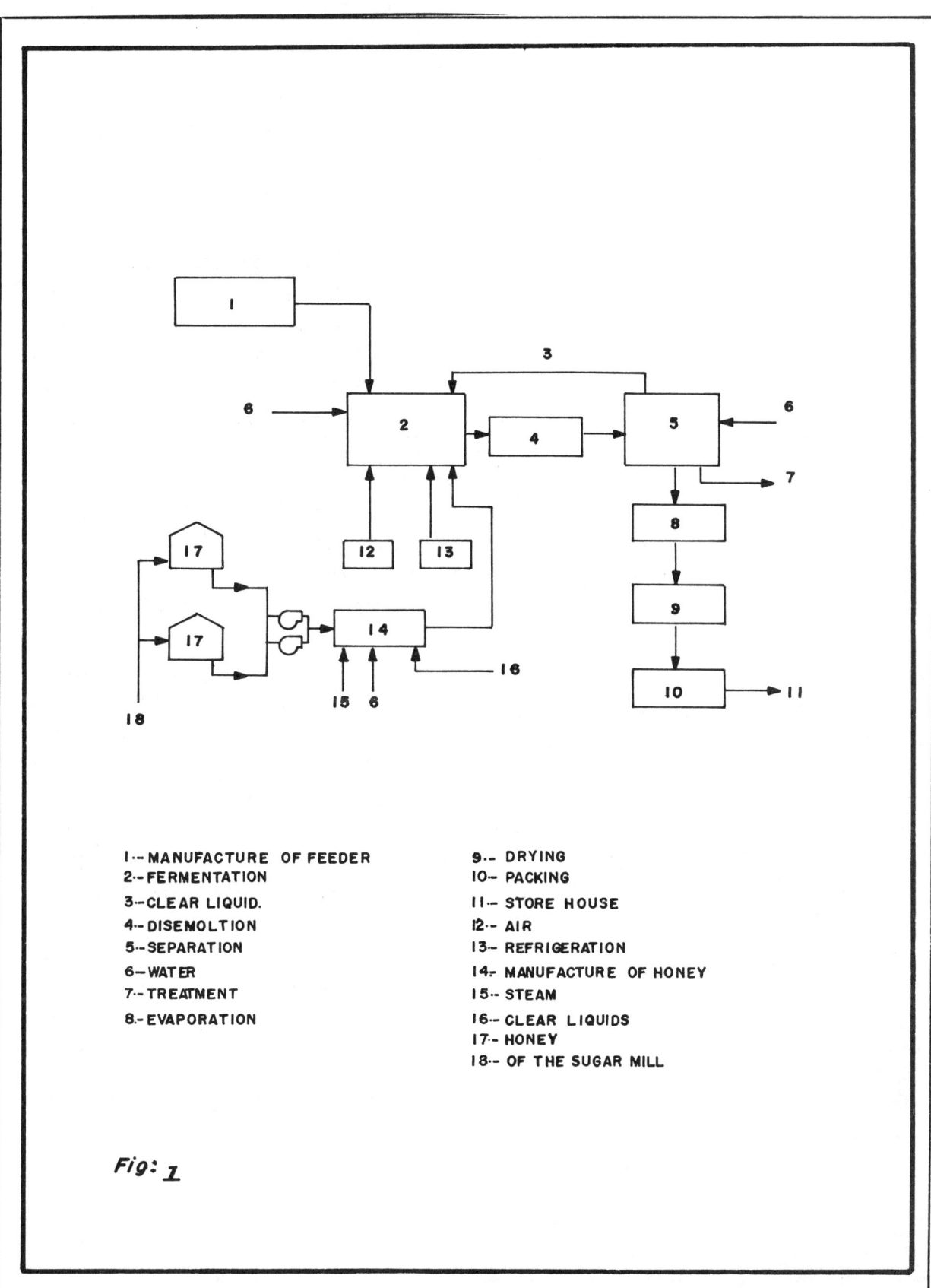

1.- MANUFACTURE OF FEEDER
2.- FERMENTATION
3.- CLEAR LIQUID.
4.- DISEMOLTION
5.- SEPARATION
6.- WATER
7.- TREATMENT
8.- EVAPORATION
9.- DRYING
10.- PACKING
11.- STORE HOUSE
12.- AIR
13.- REFRIGERATION
14.- MANUFACTURE OF HONEY
15.- STEAM
16.- CLEAR LIQUIDS
17.- HONEY
18.- OF THE SUGAR MILL

Fig: 1

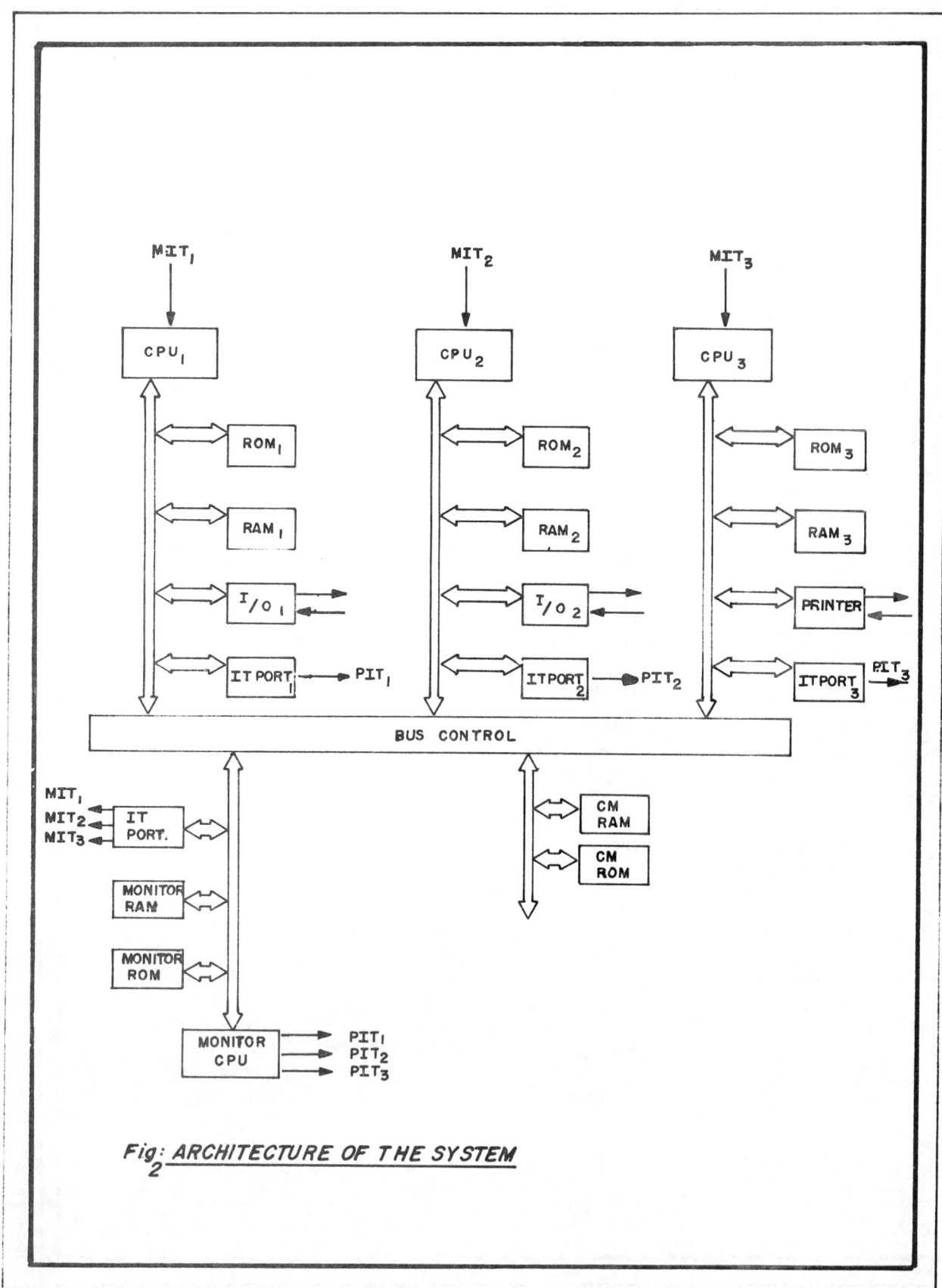

Fig. 2: *ARCHITECTURE OF THE SYSTEM*

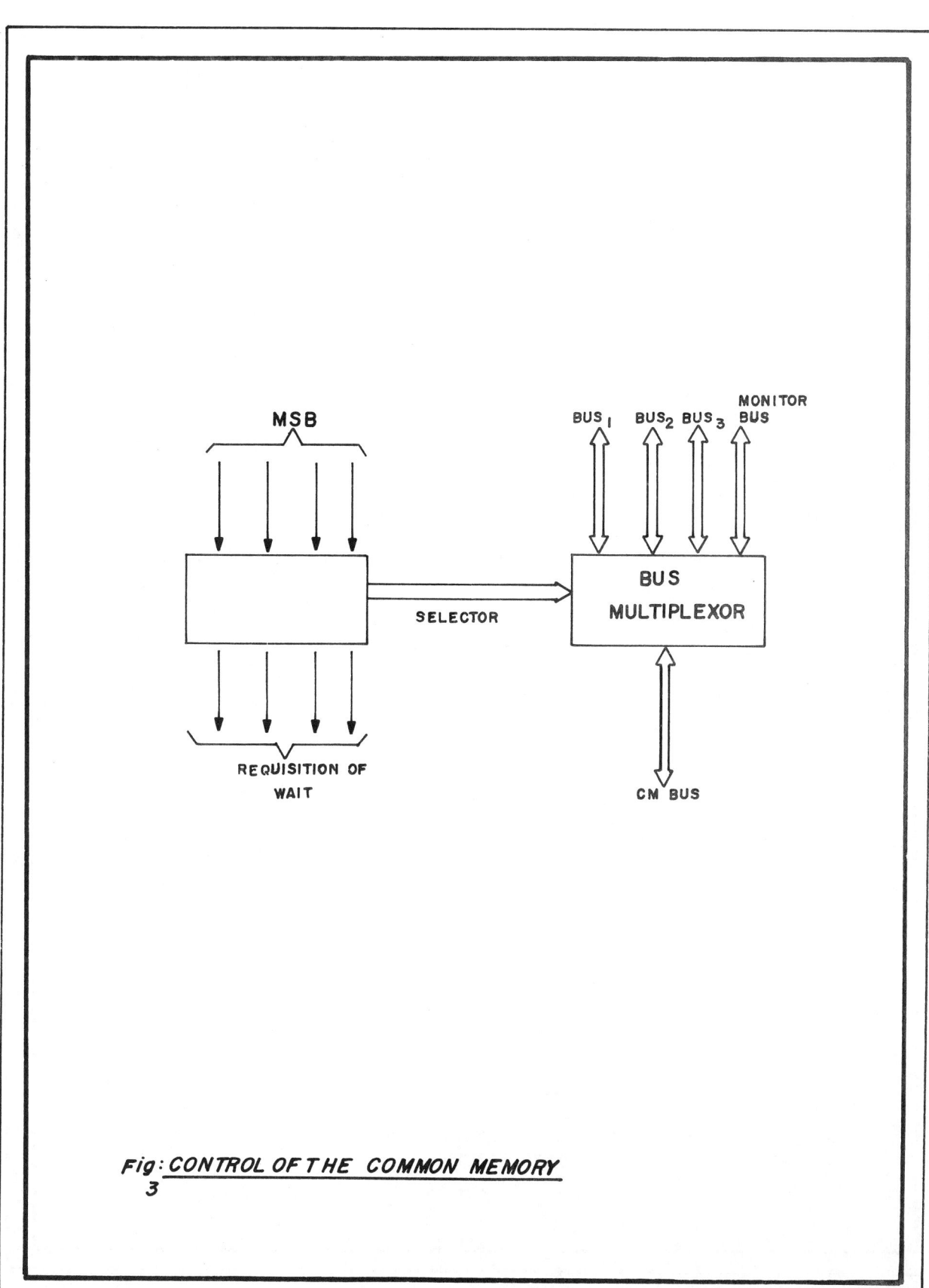

Fig: CONTROL OF THE COMMON MEMORY
3

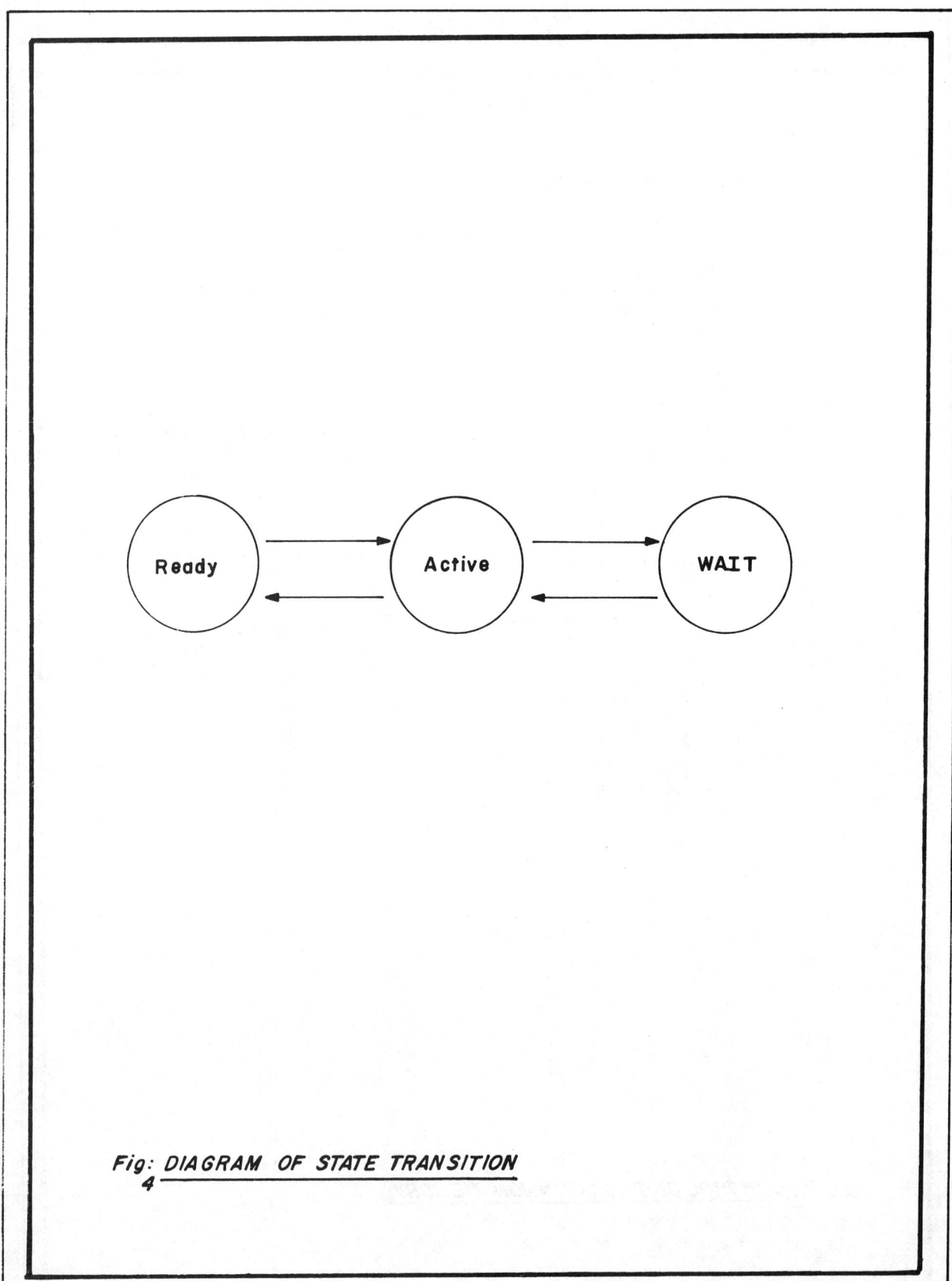

Fig: DIAGRAM OF STATE TRANSITION
4

DISCUSSION

SESSION TP 2 CHEMICAL AND BIOCHEMICAL

Paper: COMPUTER CONTROL OF SIMPLE VARIABLE FLOW PROCESSES

Authors: CP Jeffreson (Department of Chemical Engineering, University of Adelaide, Adelaide, South Australia, Australia

Discusser: Dale E. Seborg
Univ. of California
Santa Barbara,
EUA

Questions or Comments:

1) For a process with a relatively long time delay (e.g. large time delay/time constant ratio), merely keeping the product of the instantaneous process gain, $K_p(t)$, and the controller gain constant may be lead to poor closed-loop control. In such cases, it may be desirable to estimate the future process gain, $K_p(t+\Theta)$, where Θ is the time delay

Author's reply:

1) I agree, provided you are talking about a fixed (or flow insensitive) dead time. However, when dead times and process time constants all vary inversely with flowrate, the result is a simple "time sealing" which can be taken care of by the simple adaptive scheme proposed. In general for a fixed dead time which is "large enough" relative to process time constants, controler time constant adaption is not necessary (or desirable), although gain adaption usually remains necessary

Discusser: Armando B Corripio
9344 Bermuda Ave.
Baton Rouge, LA
40810
USA

Questions or Comments:

1) Could the gain variation be handled by using an equal percentage value?

2) Did you find that the dispersion effect was important in comparison with the convection effect?

Author's reply:

1) Yes. I agree that equal percentage or another characterised final control element could be used. However, I believe that this approach does not utilise one of the most valuable features of digital control, ie, the flexibility of the real time control computer. The problem with the equal percentage value is that it is rather permanent.

2) In general, studies of packed bed (percolation) processes have shown that fluid phase dispersion is not significant and can be incorporared in an "effective" heat or mass transfer coefficient.

Paper: REAL TIME DIGITAL MULTIVARIABLE CONTROL FOR A FERMENTATION SYSTEM

Authors: J Carrillo, J Alvarez, JA Gallegos (Centro de Investigación y Estudios Avanzados del IPN, México City, México)

Discusser: CP Jeffreson
Department of Chemical Engineering, University of Adelaide, Adelaide, South Australia, Australia

Questions or Comments:

1) What precautions does your control scheme take to avoid "washout" of the microorganisms? I presume that changes in dilution rate could occur much faster than the microorganisms can multiply

2) Is substrate concentration or substrate flowrate varied in your control scheme?

Author's reply:

1) The control scheme we have proposed avoids washout of the microorganisms by preventing D be greater than μ_m. Only if D is greater than μ_m you could have washout at steady state.

2) In the control scheme presented dilution rate and feed substrate concentrations are varied. They are the control variables. Biomass and substrate concentrations are the variables we try to regulate because that is very important to have a maximum productivity of the process and a certain production rate.

Paper: A PILOTO SCALE DISTILLATION FACILITY FOR DIGITAL COMPUTER CONTROL

Authors: JL Marchetti, A. Benallou, DE Seborg, DA Mellichamp, (Dept of Chemical and Nuclear Eng, University of California, Santa Barbara, California, USA)

Discusser: Prof. Daniel Tabak
Elect. & Computer Eng
Ben Gurion Univ.
Beer Sheva 84105
Israel

Questions or Comments:

1) Please elaborate on the advanced control techniques used.

Author's reply:

1) Current research projects, which will include experimental studies on the distillation column, are shown in table 4. A more complete description can be obtained from Professor Seborg or Professor Mellichamp.

Discusser: Dr. V.A. Lototsky
Inst. of Control Sci.
USSR Acad. of Sci.
Moscow, USSR, 117342
65, Profesoyusnaya str.

Questions or Comments:

1) Mr Seborg, in your presentation you have dwelled upon the pilot, education aimed distillation column. What is your opinion about the possibilities of the applicability of your results in industry and have you any experience of this kind?

Author's reply:

1) The predictive control technique has been applied in the petroleum industry (See references by Catler and Ramaker (1980) and Richalet et al (1978)). Several of our current research projects are financially supported by industry which is an indication that industry is interested in this type of process control research.

Discusser: Armando Corripio
9344 Bermuda Ave.
Baton Rouge, La
40810
USA

Questions or Comments:

1) Did you develop your own software? What real-time monitor did you use?

Author's reply:

1) The Moore Industries MI 1002 microcomputer-based process control system has a data acquisition and control package which is block oriented. We merely entered the appropriate block linkages and parameter values.

We use the standard Avanced Operating System (AOS) for the Data General real-time computers. However we did write our own FORTRAN programs to perform the control calculations and data adquisition.

Discusser: C.S. Berger
            19 Sheringhom
            Vic, 3150
            Australia

Questions or Comments:

1) Have you found linear techniques adequate for most industrial processes?

Author's reply:

1) For very nonlinear processes such as pH control, some chemical reactors and distillation columns, linear controllers with constant parameters are inadequate. Consequently, either nonlinear control laws or adaptive control schemes must be employed.

A MODEL PROGRAM FOR UNDERGRADUATE EDUCATION IN REAL-TIME COMPUTER PROCESS CONTROL

T. Olsen, R. H. Heist, H. Saltsburg and J. C. Friedly

Department of Chemical Engineering, University of Rochester, Rochester, NY 14627, USA

Abstract. Recent deveopments in the introduction of real-time computer-aided data acquisition and control into the undergraduate laboratory program at the University of Rochester will be discussed. Microcomputer technology has been utilized to provide hands-on access to real-time control for all students from the sophomore year on. Concepts of computer-experiment interfacing, digital data acquisition, data analysis and control are introduced naturally in a laboratory environment. Simple feedback loops with on-off and PID control are studied experimentally even before the students are exposed to control theory. Details of computer systems, interfaces and control experiments will be discussed.

Keywords. Education, computer control, microcomputers, control, data acquisition.

INTRODUCTION

The need for trained employees in the chemical and process industries with exposure to the use of the computers in real-time applications, including data acquisition and control, has never been greater than it is today. Computer experience among undergraduates in chemical engineering programs is quite extensive. However, experience with real-time applications remains limited to a select group of students at an even more select group of schools (in the United States and Canada, at least). There appear to be a number of sound reasons for this state of affairs in undergraduate education today. Roughly they can be categorized as 1. curricular restrictions, 2. hardware restrictions, 3. control limitations, and 4. electronics restrictions.

It is safe to say that process control education is not a strong component of the chemical engineering curriculum in the United States today (Waller (1981)). Although most schools (95%) have an undergraduate course in process control, only 75% require it of all students (Seborg (1980)). Far fewer have laboratory experiments in digital control. It is rare that more than one course is available even as an elective. A single course in an already tight curriculum tends to cover only the bare essentials of classical dynamics and control theory, with at most a quick once-over of real-time data acquisition and control. Since the control course is rarely a precursor for anything else in the curriculum, it is easy to forego curricular improvements in control for those in more fundamental areas of chemical engineering.

Hardware requirements have also limited the spread of real-time digital control education. Substantial investments in minicomputer facilities have been required in the past. Although a number of departments (including Rochester) have invested in such facilities, they have more often than not been used primarily at the graduate level for research and educational projects (Morari and Ray (1979) and Fisher and Seborg (1976)). Real-time computer control applications require almost total dedication of the computing facilities to the application at hand, and that prevents efficient use of facilities to educate large numbers of students. The departments that have used their control computers extensively in the undergraduate program, Alberta, Wisconsin, Santa Barbara, Imperial College, and Case, to name several noted examples in the USA and abroad, have involved undergraduates in control projects with good results. (See, for example, Morari and Ray (1980) and Mellichamp (1980).)

Education in real-time computer control has been limited in the undergraduate curriculum because of control theory itself. More sophisticated control applications for which digital computers are indispensible are naturally covered in advanced courses and in research. Classical control alone can be covered adequately in a typical undergraduate course and the justification for using digital control is less apparent for single loop PID control. Therefore, the average undergraduate tends to get at most a superficial exposure to computer control.

The typical chemical engineering undergraduate suffers from a lack of even simple electronics knowledge which would be helpful in understanding some of the real-time interfacing problems. Even when an electrical engineering course is required in the curriculum the education is usually (justifiably) broad and theoretical, with insufficient exposure to the practical aspects of signal amplification, sampling and conditioning. This makes it even more difficult to do justice to computer control in a single process control course.

Recognizing all of these factors which make it difficult to provide an effective, general education of all undergraduate chemical engineers in real-time computer control, the Chemical Engineering Department at the University of Rochester began a program which we called the Microcomputer Implementation Project about four years ago (Heist and others (1981), Saltsburg and others (1982a)). As it has developed we have found that nearly all of the limitations discussed above could be circumvented, and that all undergraduate students could be given a working knowledge of real-time data acquisition and control within the context of a traditional chemical engineering curriculum. Although the program still is evolving, our experience to date may be offered as one possible model for an undergraduate education in computer process control.

MICROCOMPUTER IMPLEMENTATION PROJECT

In large measure the success of the Microcomputer Implementation Project in teaching real-time computing can be attributed to its philosophy and to the independent nature of our laboratory courses. The objective was first and foremost to make the microcomputer a tool in the undergraduate laboratory. As capabilities and prices of microcomputers became comparable to traditional laboratory data collection, recording and control equipment, it was feasible to use the microcomputer as a normal part of appropriate undergraduate laboratory experiments. This emphasis gave a far greater incentive to provide the educational background to all students, divorcing that responsibility from specific courses, such as the process control course. If the microcomputer could be used as a normal adjunct to instrumentation in the laboratory, not just related to process control, time could be devoted to teaching the fundamentals to all students.

Implementation of our philosophy is simplified by the independence of the laboratory courses. The laboratory program at Rochester involves separate laboratory courses in four consecutive semesters starting with the second semester of the second year of the four year program. These laboratory courses, which have enrollments of 60 to 90 students, are formally divorced from lecture courses (although certainly coordinated with them).

A review of the laboratory program made it clear that most experiments involved slow acquisition of modest amounts of relatively low-precision data. Therefore, it was clear that the massive file storage capacity and high-speed computation capability of a large computer was not required. Given the increasing numbers of undergraduate students and the poor ability of a multiuser computer system to accommodate the real-time needs of a heavily populated laboratory, the microcomputer appeared to be the best choice for our application. We have chosen the Commodore PET as the machine that most readily meets all of our immediate educational objectives, and a significant educational discount made it a cost effective proposition in addition to any pedagogical advantage. About 30 Commodore PET and SuperPet microcomputers are currently in dedicated use in the four undergraduate laboratories.

We have tried to standardize the instrumentation in the laboratories to the extent possible by using standard interfaces for both basic instruments, such as gas chromatographs, and for data acquisition from sensors, such as thermistors and photoresistors. The microcomputer has become part of a repertoire of basic instrumentation for a variety of experiments. It has, in fact, become a universal laboratory device. Therefore, the earlier laboratory courses can be devoted primarily to teaching basic measurement techniques, including use of the microcomputer. These techniques are then used for the more engineering-oriented experiments of later courses.

REAL-TIME COMPUTING

The first laboratory course is devoted to teaching all students the basics of using the microprocessor as a laboratory tool. Instruction in high level programming language is minimal since BASIC is so widely used. On the other hand, programming techniques and program structure (e.g. structured BASIC) are discussed. Enough machine language coding is introduced so that data acquisition can be understood.

A major part of the semester is devoted to learning to use the microcomputer I/O ports which enable communication with the laboratory equipment. The first real-time I/O experiment performed uses an array of eight LED's to map the parallel port output of the PET. This elementary device, shown in Fig. 1, is sufficient to teach students how to turn on and off selected output-port bits (lights) in any manner and in any timing sequence they choose. The message is quickly understood that one can easily control anything that can be switched electrically.

Fig. 1. Photograph of the battery operated module used to map the PET data bus to eight LEDs.

Since temperature measurement is so pervasive in chemical engineering laboratory practice, the student is shown how to use the microcomputer to take readings with thermocouples and thermistors. The basic elements of the circuits necessary for signal amplification and for A/D conversion are introduced at this point. Since the student has not, and in general will not, take a full electrical engineering course but instead will have just a brief exposure to electricity and magnetism in physics, we feel it is important to provide a limited view of the basis for the conversions. However, we approach it more from the point of view of the user rather than the designer: students are taught how the electronics function but do not build circuits (Heist and others (1982)).

CONTROL EXPERIMENTS

In keeping with our philosophy, the data-acquisition function of the microcomputers was implemented first. We are just now beginning to use them for control purposes as well. Several elementary control experiments in use now and in the process of being implemented are discussed below.

The first control experiment is introduced in the sophomore laboratory course (Olsen and others (1982), Saltsburg and others (1982b)). Although chemical engineering students typically do not take a course in automatic control until their final year, this experiment so well illustrates the practical use of the concepts of interfacing the microcomputer with a laboratory experiment that it fits naturally in this first laboratory course. Figure 2 shows a photograph of a small, re-circulating air heater. It consists of a small wooden box, with a transparent front, in which a small fan continuously circulates air. The box has vents on the sides and a damper which permit manual adjustment of a throughput of room air, or load. A blackened light bulb serves as energy source and the air temperature is measured with a thermistor. The device is inexpensive enough that many replicate units can be built, permitting many students to conduct experiments with the device simultaneously. Response time is of the order of minutes so a series of experiments can be performed in a single laboratory period.

Fig. 2. Photograph of the microcomputer - air bath combination.

The students already know how to make temperature measurements with the thermistor and switch the light bulb on and off. Therefore, the progress towards combining the two procedures to control the measured temperature is easy. The student tries on-off control first and then graduates to proportional control by switching the bulb on and off at different rates to achieve a proportional fraction of the maximum power input. Students do their own programming of the control algorithm. Concepts such as continuous cycling, hysteresis, saturation and steady-state offset can be discovered experimentally as quite natural and understandable consequences of the control strategy used. Integral action is added naturally to the control algorithm to eliminate steady-state offset, and full PID control can be experimented with. Feedback loop instability can also be experienced safely and firsthand.

A control experiment used in the final laboratory course illustrates the transition from bench-top to pilot-plant scale. Figure 3 shows a Pfaudler glass-lined continuous stirred tank reactor (at the right of the photograph) controlled with a microcomputer. Rather than risk running an actual chemical reaction, we simulate an exothermic reaction by using steam injected into a water feed (Brisk (1974)). The algorithm for computing the steam injection (heat of reaction) has been programmed into the computer but the student operating the device should be oblivious to it. This simulated chemical reaction is not only safe for a student laboratory environment but also extremely versatile. Conditions can be chosen such that single or multiple steady states exist within the feasible operating space, giving rise to some very interesting dynamic behavior such as the experimentally obtained phase plane trajectories shown on Figure 4.

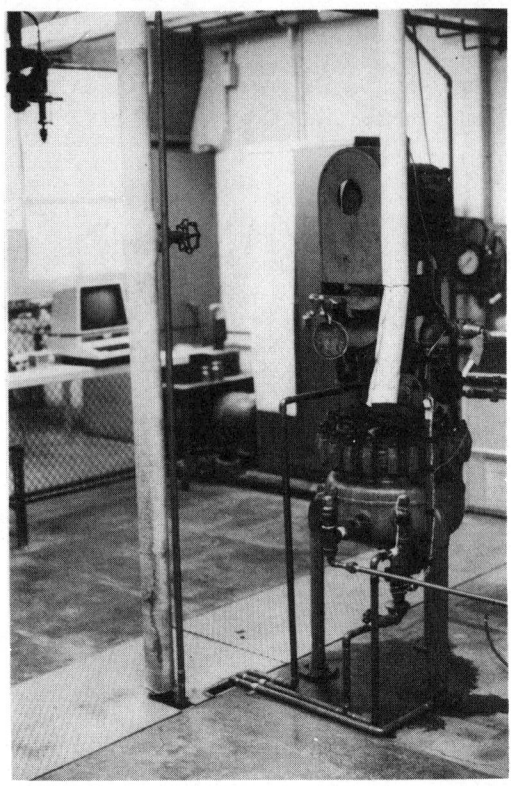

Fig. 3. Photograph of the continuous stirred tank reactor.

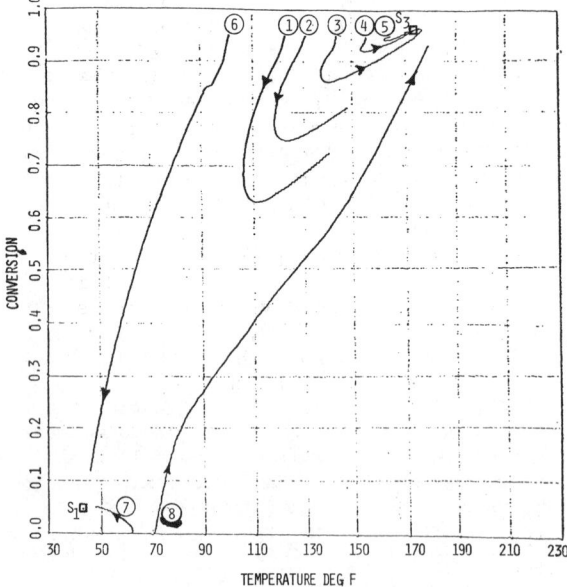

Fig. 4. Experimental phase plane trajectories for the continuous stirred tank reactor.

Three temperatures are measured: the reactor outlet, and the cooling jacket inlet and outlet. The primary variable to be controlled by the student is the cooling water flow rate, although inlet temperature and/or "concentration" can also be apparently controlled by altering the steam injection algorithm. Both flow rates are adjusted with commerical pneumatic control valves operated from the microcomputer through commercial current-to-pressure transducers. Normally, single loop DDC is implemented with the microcomputer serving as the controller for the steam injection, but the student does not control that function.

This experiment illustrates a significant advantage of the approach we have used in digital control as contrasted with a more traditional approach. The experiment was first put under computer control about five years ago when we were using a MODCOMP II minicomputer. It was used as a final year experiment (concurrent with the process control course) and in special projects. Our experience was not good, except for projects which only small numbers of students were able to complete. We found that the minicomputer was so large, powerful and versatile that it exceeded our undergraduate needs. It had to be dedicated to a single experiment for real-time control and this limited its general usefulness. More importantly, even though it has a FORTRAN compiler and rather user friendly I/O capabilities, the students took so much time learning the peculiarities of the operating system that it was not possible to generate much useful data during an experiment, or even during a single semester project. The minicomputer had to be treated as a black box controller or else few experimental results could be obtained. In contrast, the microcomputers are so inexpensive that we can have many available; they are versatile enough to be used in a number of experiments throughout the laboratory curriculum; and they are easy enough to program and use so that all students can both learn the basic notions of digital data acquisition and achieve actual laboratory implementation.

A third experiment involves a 5 cm diameter distillation column with ten trays. This column is instrumented with thermistors on alternate trays and there are ports for manual collection of samples for off-line chromatographic analysis. All temperatures are sampled essentially continuously and can be used in any fashion for control. In principle, the measured concentrations can be manually entered into the computer for updating models or state estimations. The column has a reflux splitter operated with a simple solenoid valve, an electric heater in the reboiler and a peristaltic feed pump which can all be used in a control scheme. To date we have only used the column for steady state data acquisition, but the next step is to try single loop control, using the reflux splitter and, e.g., the top temperature measurement. The facility is ideal for multivariable control experiments as well. Since the reflux splitter involves only a solenoid valve, students need use only the concepts of temperature measurement and timed on-off switching that they used in their beginning experiment (the air heater) for continuous control. In addition, through multiplexing of the temperature measurement circuits, all inputs and outputs can be handled via one 8-bit port.

All of the preceding devices serve the dual purpose of required use in the undergraduate laboratory courses and of individual use for special projects by students interested in studying more advanced topics. The final computer control experiment is sufficiently complex so that it has only been used for independent projects. Although not a traditional chemical engineering process, the model train facility shown in Figure 5 provides an opportunity to look at control problems involving scheduling of discrete events. The facility consists of six individually powered blocks of N-gauge track, modeled after the work of Mellichamp and Engleberg (1974). Each track block and four switches can be controlled directly with the microcomputer. Ten phototransistors, mounted in the track bed, provide information on the location of the trains. Since the "process" is easily understood, students can devise their own objectives and algorithms for control. The student immediately becomes aware of inadequate control. However, the experiment involves the same concept of signal sampling by computer, devising algorithms for use of the data, and outputting signals to control devices that is stressed in previous experiments. All this is done in a dramatic real-time environment.

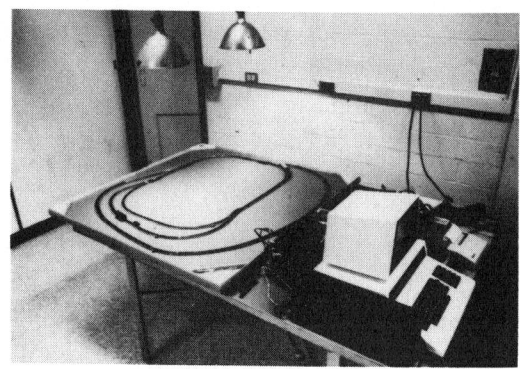

Fig. 5. Photograph of the microcomputer controlled model train layout.

DYNAMICS EXPERIMENTS

In addition to those control experiments described in some detail above, the same experimental facilities are used for studies of the process dynamics alone. Identification of process models from transient responses in the linear range of load changes is instructive for all processes. In addition to these experiments there are several others that involve process dynamics alone, with no current plans for coupling the real-time data acquisition with the feedback control. With our increased reliance on real-time data acquisition, and with students' increased familiarity with it, we find ourselves attracted more to transient experiments to illustrate traditional chemical engineering concepts rather than the more typical steady state experiments. For example, the microcomputer is used to collect and analyze the inherently transient data from gas chromatographs used extensively in our laboratories for composition measurement (Graves and others (1982)). In addition, by using the microcomputer to sample temperatures in solid spheres subjected to a step change in external temperature, the thermal diffusivity can be identified from the transient response. In both examples, laboratory productivity, as well as educational content, has been enhanced significantly by the use of the microcomputers.

Two separate experiments have been designed to measure the impulse response, and consequently fluid mixing characteristics, in "ideal" process devices. Both rely on the use of simple concentration sensing elements and the microcomputer as a real-time data acquisition device. The first experiment involves two identical well-mixed tanks in series. A pulse of red dye is injected into the feed stream of the first tank and its intensity is measured in the outlet stream of each tank. We use a very simple non-intrusive detection system consisting of a small collar containing a green LED on one side and a photoresistor on the other. This collar is mounted around a section of translucent tubing through which the outlet stream flows. The photoresistor output is ideal for sampling by the microcomputer. The measured dye intensity transients shown in Fig. 6a are a good illustration of the nearly ideal impulse response of a single first order process and two such processes in series. The second experiment involves the same ideas and techniques but the device used is 18 m of tubing. The output of eight photoresistors mounted along the tube length is shown in Fig. 6b. In contrast to the tanks-in-series experiment, this clearly shows the delays between the signals at successive locations, and, more importantly, the dispersion that occurs in the concentration signal as the fluid flows in laminar flow.

CONCLUDING REMARKS

In this paper we have discussed the approach that we are taking at the University of Rochester to teach real-time data acquisition and control to undergraduate chemical engineers. We feel that there are special circumstances in the chemical engineering field which have made it especially difficult to introduce digital control into the undergraduate curriculum. However, a meaningful exposure of all students to real-time computing is possible using microprocessor technology, coupled with a philosophy different from that traditionally adopted in chemical engineering programs.

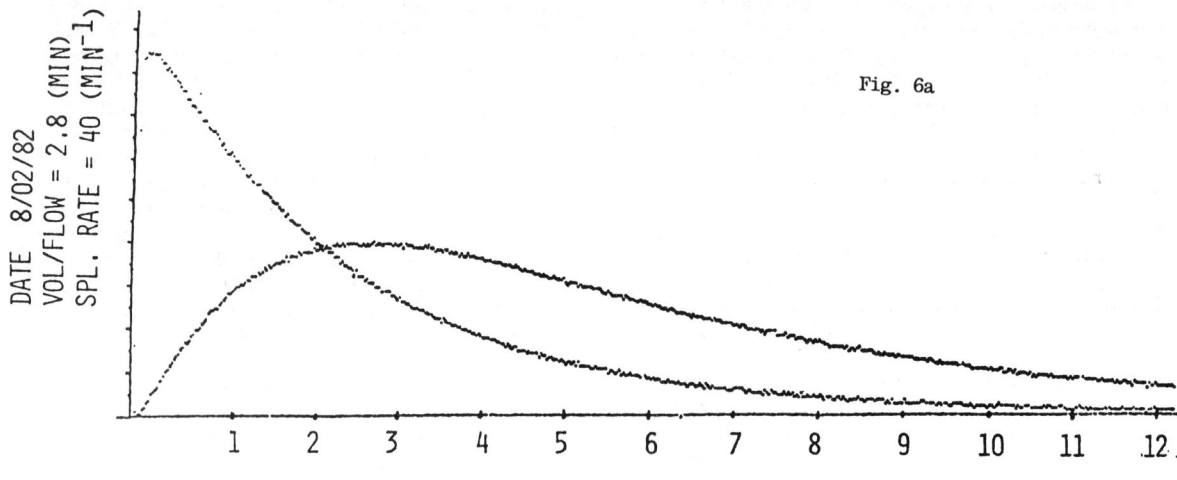

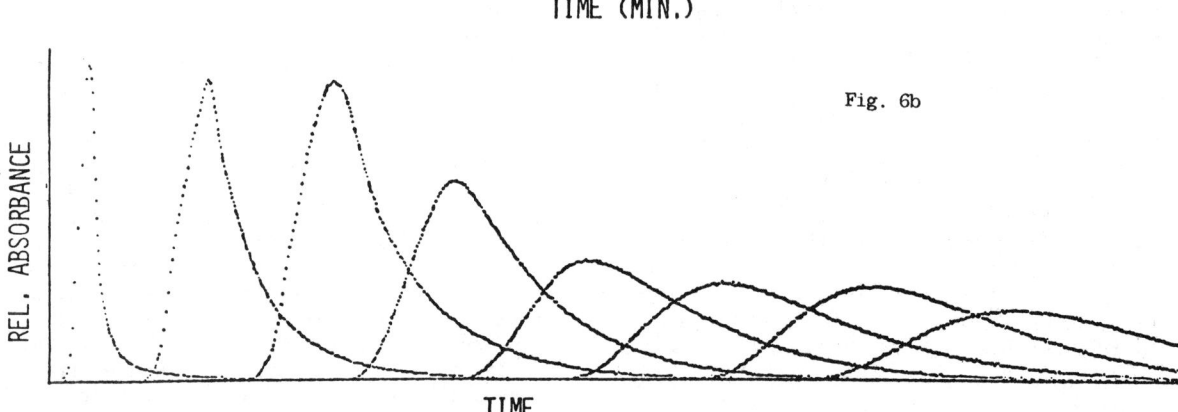

Fig. 6. Dispersion of a pulse of dye injected into the inlet of a continuous flow device. a) Output response of one and two tanks in series. b) Response at eight successive locations along a tube.

Our experience shows that the microcomputer, with suitable I/O ports and peripheral devices, is an invaluable addition to the modern engineering laboratory. It serves first and foremost as a data acquisition device, in most cases replacing the traditional strip chart recorder. That role having been established and experience in its use achieved, it is natural to close the loop with the addition of elementary digital control. We feel that this approach has rather distinct advantages:

1. The microcomputers chosen for use in our laboratories can perform the functions required for our purposes: relatively low volume, low speed data acquisition of moderate accuracy, coupled with execution of elementary control algorithms.

2. Microcomputers can be dedicated to individual experiments and can be plentiful enough so that many users can be accommodated simultaneously.

3. Students learn the general utility of digital computers in the laboratory environment rather than associating them with process control alone.

4. A working knowledge of circuits necessary for the amplification and conversion of signals in the laboratory is very useful for all chemical engineers.

5. High-level languages as simple as BASIC permit the student to learn the fundamentals of real-time programming without being excessively burdened with the operating systems and characteristics of more sophisticated machines.

6. Since the concepts are introduced early in the laboratory program and used re-

peatedly for a variety of experiments, they are rather familiar by the time more sophisticated topics such as control are discussed in a final year course.

7. A traditional control course is made more meaningful by previous exposure to some practical aspect of process control.

8. The microcomputer is an economic alternative for academic laboratory use.

A rough cost comparison for the implementation of a single loop control on our 10 tray distillation column is shown in Table 1. It can be seen that a major cost of the more traditional approach is to provide strip chart recordings for the many temperatures read continuously. An electronic controller itself costs as much as or more than a microcomputer, and we were fortunate enough to receive an educational discount of 33% on the price listed. The required interface was built based on circuits from the amateur microcomputer literature and of course was much more economical for our purposes than commercial interfaces. The economics favor the microcomputer route even if no other advantage were present. The standardization of units in the laboratory and the ability of the microprocessor to record extensive data for further analysis, on-line or off-line, cannot be duplicated for the cost.

TABLE I. Estimated costs comparison for microcomputer controlled and conventionally controlled distillation column.

Analog		Digital	
Temp. Sensing & Display	$700	PET Micromp.	$800
		Printer	$400
Multichannel Chart Recorder	$1200	Temp. Sensors & Interfacing	$25
Process Control Hardware	$\geq$ $1000	Process Control	$100
	$\geq$ $2900		$1325

The major cost involved is not in the hardware, but in the development time required of staff members. This is not circumvented by the use of microcomputers, although the machines we have used may well have required somewhat less development effort than would others. We would hope that our experience can help others compress their learning curve. Even so, considerable expenditure of time must be expected in gaining a working knowledge of real-time computer data acquisition and control. We have found this to be true even for those of our own faculty not involved directly in the microcomputer implementation project from the beginning.

There are, of course, limitations to what can be accomplished with microcomputer-based digital control experiments. Speed can be a limiting factor in some applications in which a number of signals are being sampled or several loops are being controlled. Precision of sampled data can also be a problem in some applications. Fortunately, most experiments in chemical engineering education (and many in process industry practice) do not suffer from this deficiency. More sophisticated control algorithms that require extensive on-line computation cannot be implemented on the PET microcomputer because of its low computational speed. However, these are rarely within the scope of the undergraduate program anyway. Since most industrial applications tend to load down control computers with extensive status checking and managerial bookkeeping, this limitation may well be one that students should be exposed to (although we have not yet done so).

Our experience has been rather positive to date and we feel that it may well serve as a model to be followed by other institutions in teaching real-time digital process control concepts.

REFERENCES

Brisk, M. L. (1974). In H. M. Hulburt, editor, Advances in Chemistry Series 133, Chemical Reaction Engineering II, American Chemical Society, Washington, p. 13.

Fisher, D. G. and D. E. Seborg (1976). Multivariable Computer Control: A Case Study, American Elsevier, New York.

Graves, D., R. H. Heist, T. Olsen and H. Saltsburg (1982). Microcomputers in the chemical engineering curriculum. Part four: Microcomputer-aided instrumentation. MICRO The 6502/6809 Journal, accepted for publication.

Heist, R. H., H. Saltsburg, T. Olsen and F. W. Arcuri (1981). The microcomputer in the chemical engineering laboratory. Paper presented at AIChE Annual Meeting, New Orleans

Heist, R. H., T. Olsen and H. Saltsburg (1982). Microcomputers in the chemical engineering curriculum. Part two: Analog transducers in a digital world. MICRO The 6502/6809 Journal, in press.

Mellichamp, D. A. (1980). Chemical Engineering Education, XIV, no. 1, 18.

Mellichamp, D. A. and G. P. Engelberg (1974). University of California at Santa Barbara, Department of Chemical and Nuclear Engineering Report C-74-3.

Morari, M. and W. H. Ray (1980). Chemical Engineering Education, XIV, no. 1, 32.

Morari, M. and W. H. Ray (1979). Chemical Engineering Education, XIII, no. 4, 160.

Olsen, T., H. Saltsburg, and R. H. Heist (1982). Microcomputers in the chemical engineering curriculum. Part three: Process control and the microcomputer. MICRO The 6502/6809 Journal, accepted for publication.

Saltsburg, H., R. H. Heist and T. Olsen (1982a). Microcomputers in the chemical engineering curriculum. Part one: An overview. MICRO The 6502/6809 Journal, no. 53, October, 53.

Saltsburg, H., R. H. Heist and T. Olsen (1982b). The microcomputer in the chemical engineering laboratory. Paper presented at AIChE Annual Meeting, Los Angeles.

Seborg, D. E. (1980. Chemical Engineering Education, XIV, no. 1, 42.

Waller, K. V. (1981). Chemical Engineering Education, XV, no. 1, 30.

ACKNOWLEDGMENT

Permission to publish the photographs in Figures 1,2,3 and 5 has been granted by MICRO, The 6502/6809 Journal, Robert Tripp editor-in-chief, MICRO, Inc., Chelmsford, MA, USA

A CUBAN EXPERIENCE IN THE DEVELOPMENT OF COURSES IN MICROPROCESSOR AND REAL TIME PROCESS CONTROL WITH MICROPROCESSORS

J. Olivera Reyes* and S. Teijero Páez**

Instituto Superior Politécnico "José A. Echeverría", Ministerio de Educación Superior, Central Martinez Prieto, Ciudad de la Habana, Cuba
**Empresa de Instrumentación y Control Industrial, Ministerio de la Industria Sidero Mecánica, Habana, Cuba*

ABSTRACT

The present work is related to the experience adquired in Cuba in the development of the discipline of real time process control with microprocessors. It has been divided in three courses:

a) Study and application of microprocessors.

b) Study and application of programmable interfaces in the design of real time industrial systems.

c) Applications of microporcessors in industrial process control.

With the first course it is pretended to cover the needs of an introductory level in microporcessor techniques. The MCS-80 family is the one selected for the examples. The course of programmable interfaces goes deep inside the principle features of the interface chips. Although the 8080 family interfaces are the one that are studied in more details, it saves time with 6800 and Z80 families. When the students finish this second course they must be able to a good and efficient design of a microprocessor-based system. In both courses are taught and practiced programming techniques to help the thorough building up of the new habits. Also the students practice doing different kinds of exercises varying in length and in details.

The third course deals with techniques of real time process control with particular emphasis in microprocessor based solutions. In this one, the students learn about supervisory control, digital control, sequential logic control, data collection, primary processing of the information, man-machine communication and the main conception about the hierarchy of advanced control functions and distributed control. In this paper is analyzed the general organizations of each course. Those are divided in regular, short and conferences. Each one has a different organization and evaluation system. Also, the results of four years are presented in this paper. The students came from many different fields. There were non-electrical engineers, physics, physicians, electrical and electronics engineers, computer specialists and invited non-graduated students. Finally, the books used as texts are reviewed. They were written by the authors of the paper and specially prepared to be self-studied by the whole mass of students.

KEYWORDS

Microprocessor teaching. Application of the microprocessors in the real time control. Microprocessor systems. Programmable interfaces in microprocessor systems.

INTRODUCCION

The development of the microprocessors in the last ten years constituted a great revolution in the electronics field that many specialists already compare with the development of the transistor some decades in the past. Its use bring, as a consequence, that considerable advantages were reached with respect to the previous solutions related to the size reduction of the systems, increase of the reliability, decrease of the power consumption, increase of the design flexibility, cost reduction and others.

It is important to point out that flexibility in the designs permits to make common the same structure of a microcomputer to several industrial uses or, for example, to begin a project with an electronics design relatively reduced and making it growing in function of the needs.

The use of the microprocessors in the industrial process control is a reality that has been developed since the beginning of the microprocessor in an increasing degree, continuously, in several applications. There are many industrial subprocesses in which the classical solution to the tasks is already made with microprocessors. There are the realities and many others which have caused that many specialists of our country, as universitary professors, engineers linked directly to the production are interested each day more for acknowledging deeply the engineering about the microprocessors. In such a sense the Electronics School of the Instituto Superior Politécnico "José Antonio Echeverría" has been organizing periodical courses of postgrade related to such subjects. These courses have been basically of two types: shorts and regulars; each of them has been developed in three subjects;

1) Study and application of the microprocessors

2) Study and application of the programmable interfaces and its use in the industrial system design in real time

3) Application of the microprocessors in the industrial process control

The variant of the short courses of not more than two weeks of duration has resulted of great interest, specially for the production centers whose geographical location is relatively far from the universities. The regular courses have been given, mainly, in the university branches, although, occasionally they have been given in other places.

The objective of this paper is to expose the experiences obtained by the authors and the results of these courses, as well as to present the main features of theirs textbooks used in them.

Content of the subjects

The first subject that is given is the Study and application of the microprocessors with the following main features;

Title: Study and Application of the Microprocessors

Total of Hours: 40 (regular course)

General objectives:

that the student is able to:

1. acknowledge the necessary elements of hardware and software in order to develop the design of engineered systems by the microprocessors;

2. have the necessary knowledge to understand the next study of complementary engineerings and other microprocessors.

Themes

1) Introduction to the study of the microprocessors

2) Central process unit

3) Basic hardware of the MCS-80 system

4) Programming elements

5) Applications

The selected microprocessor, as main example, is the 8080 one, although during the development of the course, in searching some generalization, are used elements and are pointed out some features of other important microprocessors, such as the 6800, Z80 and 8086. The first theme introduces the subject and is developed a historical outline, and are given the basic elements that a non specialized student in electronics should know in order to understand the course, for example, numerical systems, logical gates, flip flop, counters, register, storage, decoders, encoders, multiplexors, demultiplexors, buffers and three state logic, all of this is only seen in its most schematical way.

The theme ends by engineering them by parts the architecture of a typical microprocessor: The 8080A

In the second theme is explained in detail the functioning of the 8080 and its relation with its architecture. This theme ends with the instruction examples and its running inside the microprocessor. In the third theme is completed the CPU group of the studied system, that is, it is explained the 8224 - clock generator and the 8228 controller system. In this theme is developed the explanations of the main auxiliary circuits: the ports 8212, 8255A and 8251A, the 8259A programmable interruption controller, the interruption processes and inquiry, the engineering of addresses ports and memories, as -

well as the study of the latter. In this theme, the instructions and short programs are continuously introducing with two main objectives, the first one to teach the use of memory processing and addressed with programmable circuits and, second, to develop at the same time the indissoluble link between hardware and software.

In the forth theme is formalized the software and is completed the basic features of the programming in symbolical language, as well as the study of the assembler system of the MCS-80 family. In this theme are taught the basic techniques commonly used, such as, data block processing, bit to bit processing, excitation of the numerical indicators and whole arithmetic.

In the fifth theme is linked with a complete example, all that was previously studied. For the development of this, is followed a work methodology that permits the student to obtain a form of focusing any problem of the general features.

The second subject is the Study and Application of the Programmable Interfaces and its use in the design of industrial systems in real time. It has optional parts which depend on the particular interest of the group receiving it. Its main features are the following;

Title: Study and Application of the Programmable Interfaces and its use in the design of the Industrial Systems in real time.

Total of Hours: 40 to 50 (regular course)

General objectives:

That the student is able to

1) acknowledge the general principles in which are based the programmable interfaces and a method to perform this study.

2) acknowledge the functional and specific aspects of a popular family of programmable interfaces and the feactures of others.

3) acknowledge the way of using the programmable interfaces for the design of industrial systems in real time.

Themes

1) Introduction to programmable interfaces
2) Programmable interfaces of general interest. Parallel ports, series ports
3) Interruption subprocesses
4) Programmable timer
5) Subprocess of direct acces to memory
6) Keyboard treatment, numerical visualizers and video terminals
7) Cassette treatment and flexible discs
8) Application to an industrial process in real time

The first theme is dedicated to establish a general classification of the programmable interfaces in interfaces of general purposes, of specific purposes and universal. In this classification should be cleared the use of them and why of such classification.

In the second theme is explained the main interfaces of general purposes, divided into two groups, parallel and series interfaces. In the parallel interfaces are studying the 8255A, 6821 and Z80PID. The 8255A is studied optionally, because it was given in the first course, but the possibility exists that there are some students of this course that have not given the first one and that have studied a microprocessor by their own means which is the main requirement in order to give the second one.

The port series studied are: 8251A (optional), 6250, 6852 and Z80SIO.

At the end of this theme, the students have a general view of the main elements of the main families of microprocessors. Here are examples of their uses in the industrial processes.

In the third theme dedicated to the interruption subprocesses are generalized them. The circuits studied are the 8259A and the 6828, since the system of the Z80 microprocessor does not use aditional circuits, but each circuit of the family could ask for interruption and this is being emphasizing in the way they have been explained. In this theme is given some emphasis to the importance of the processes of interruption in the industrial systems in real time. This example constitutes here an important aspect.

The forth theme is dedicated to one of the fundamental ports of the subject: the programmable timer and are studied the 8253, 6840 and Z-80 CTC. Now the student should obtain the necessary elements to materialize the clock of real time and here may learn the hardware and software part of the problem. At the end of the theme should be mastered, by means of the example, the use of the above mentioned devices.

The subprocesses of direct acces to memory and the related devices, the 8257, 6844 and Z80DMA are treated in the fifth theme. Now it is insisted in the main cases in which these could be applied in the systems of real time. This theme could be optional depending on the features and general objectives of the course.

The sixth and seventh themes are about even more specific aspects for the ones that are interested in the processes of real time.

These themes are given depending on general objectives of the course. The experience has shown where it has been required.

A complete process of real time is developed in the eighth theme. In this process is used the main interfaces studied. This theme constitutes the element that permits to obtain a general view of all the subject.

The third subject is the Application of the Microprocessors to the Process Control. Its features are the following;

Title: Application of the Microprocessors to the Process Control.

Total of Hours: 50 hours (regular course)

General objectives:

That the student is able to

1. differentiate between the different types of SAD and its features;

2. differentiate between the different hierarchical levels which compose a SAD and the interaction among them;

3. acknowledge the different elements composing a subsystem of the supervisor, logic-sequential and digital control;

4. acknowledge the different elements composing a subsystem of data collection and alarm;

5. acknowledge the different aspects to take into account for the design of a functional structure with microprocessors.

Themes

1) Introduction
2) The SAD-PT
3) The Microprocessors and the process control
4) Hierarchical structures for SAD-PT
5) Economical essentials for SAD-PT
6) Subsystem fo the digital control
7) Subsystem of the digital control
8) Subsystem of the logic-sequential control
9) Subsystem of the data collection and alarm
10) Man-machine communication
11) Functional structure for SAD-PT with microprocessors

In the first theme are established the general criteria and unification of the subject. In the second one are given the meanings and concepts related to the SAD (SADE, SADP and SAD-PT) and is analized how the conceptions about the functional structure and the control strategy are evolved according to the development of the technical means.

In the third theme are offered the general criteria about the parameters to take into account for the selection fo the computation devices and in the specific microprocessors. Advantages and disadvantages of the utilization of these devices in the control process are given, besides, some basic aspects in the control philosophy with microprocessors.

The forth theme is about the aspects to be considered in order to divide a process in hierarchical levels.

In the fifth theme are given the most general considerations for the study of the feasibility of different industrial processes, besides, the questions more significant for the feasibility analysis using microprocessors.

The analysis of different problems related to the modelation of the stationary state, the optimization and the technical means in order to materialize a strategy of supervisory control are given in the sixth theme.

In the seventh theme are presented different methods of the algorithm design for digital control, as well as the devices of analog outlet to materialize this type of control and is given and additional reference to the adaptable control.

The analysis of the problematic related to the logic-sequential control and the technical means to materialize it constitutes the content of the eighth theme.

The ninth theme has the direct interfaces with the process form the point of view the nature (electrical or pneumatical) and the structure that is adopted in order to get the information. Also, here are offered the basic aspects related to the information processing and the techniques for detecting and treating the alarms.

In the tenth theme are analized the different aspects to be considered in the man-machine communication. Finally, in the eleventh theme is analized the design of the functional structure of SAD/PT with microprocessors, analizing the advantages and disadvantages of what was attained.

The three previous subjects have been structured in such a way that if anybody has the acknowledge constituting the basic requirements not being essential to study one in order to participate in the other.

That is why some themes are given optionally in more than one subject. These themes are those in which exist less probability of having been studied by an independent student.

Results obtained

The first of the subjects has been given between regular and short courses in 19 times up to now, with a global matricula of about 1500 students coming from, practically, all the organisms of the domestic economy. It has been given in different regions of the country, in universities, institutes, technological schools, production centers and in one occasion by correspondence that is being studied deeply.

The students have taken it have a varied occupational profile from non electrical engineers, physics, doctors, chemists, electrical and electronics engineers, specialists in computation, technical experts and invited, non graduated students. A very interesting experience is the one obtained with the last students, who in general constitute the highest marks in each course, mainly the students of the first years of the career.

The second discipline has been given for about 25% of the previous students; they are constituted about 80% of the matricula of this new course and the remaining 20% is composed by students that did not give the first of the subjects. The fifth, sixth and seventh themes have been given only 30% of the times, while the rest has been given always.

The third discipline has been of great acceptance for those dedicated to the process control. It has been given for about 200 students with a very high retention and promotion. The short courses have been highly preferred by industrial plants and industries of the country. The courses have been given in mining plants, refinery and service centers.

Bibliographical references

For the subjects 1 and 2 is used the textbook Microprocessors: Design and Application of the Systems (Study and Application of the Integrated Circuits. Part III: Microprocessors) from one of the authors of this paper. The textbook has 9 chapters, as follows;

1) Introduction to the study of the microprocessors
2) Essentials of the microprocessors
3) Central units of the processing and control
4) Port and memory addressing
5) Programmable interfaces
6) Software elements
7) Set of instructions
8) Examples of sortware
9) Microprogrammable processors

This textbook was developed linking the concepts of form, so that it could serve for the individual study, with educational purposes. Here are given the main elements of 8080, 6800 and Z-80 mircroprocessors with the rest of their respective families.

Three chapters are given to the software where are developed the standards of the assembler of the numbers mentioned. The ninth chapter is dealt with the bit-slices processors and it is developed from its essentials up to the study of the families 3000 and 2900. In the third subject is used the Process Control in real time text, using microprocessors of the other author whose general content is the following;

1) The microprocessors and the process control
2) Systems of continuous and linear controls
3) System of described and linear controls
4) System of control with logical and programmable devices
5) System of optimum and adaptable controls
6) Multiple level structures and the SAD
7) Economical essentials of the SAD-PT
8) Supervisory control
9) Digital control
10) Logic-sequential control
11) Collection and primary processing of the information
12) Man-machine communication
13) Systems of distributed control in real time using microprocessors

This textbook has permitted to develop the subjects described in the process control and it has been reviewed and enlarged the same as the frist one, taking as a base the experiences of the given courses.

CONCLUSIONS

In this paper have been given the authors' main experiences in the teaching of the microprocessor disciplines and their application to the process control. The interest shown by the different groups of students and the increasing rhythm of the matriculas and courses have shown trully that in the country is developing, the same as in the world, a clear conscience of the role that has played and is playing the microprocessors and, in particular, its extraordinary importance in the process control in real time.

Copyright © IFAC Real Time Digital Control Applications
Guadalajara, Mexico 1983

A TRAINING ABOUT REAL TIME DIGITAL CONTROL IN A FRENCH ENGINEER HIGH SCHOOL

J. P. Thomesse

Ecole Nationale Supérieure d'Electricité et de Mécanique, 2 rue de la Citadelle, B.P. 850, 54011 Nancy-Cedex, France

-I- INTRODUCTION

Since now many years, large and micro computers are used for real time digital control applications. With the development of all the concerned technologies, as well hardware as software, more and more knowledge is necessary for engineers which have to design and to realize specific applications in this field.

Specialists are necessary and every engineer in all specialities or in all the industries must have some knowledge in real time digital control.

We have therefore elaborate and carried out special teachings at the "Institut National Polytechnique de Lorraine"(INPL).

After a short presentation of this university and of its engineers school, a definition of the computer science for real time control and its objectives will be given. Then the details of such teachings will be studied with the tutorial approach used.

-II- THE "INSTITUT NATIONAL POLYTECHNIQUE DE LORRAINE"

II.1. Schools and Departments of the "I.N.P.L."

II.1.1. Schools

The "Institute" is a special university which is a federation of five National High Schools (NHS) of Engineering :
- The N.H.S. of Agronomy and Food Industries (E.N.S.A.I.A.) ;
- The N.H.S. of Electrical and Mechanical Engineering (E.N.S.E.M.) ;
- The N.H.S. of Applied Geology and Mine Prospecting (E.N.S.G.) ;
- The N.H.S. for the Chemical Industry (E.N.S.I.C.) ;
- The N.H.S. of Metallurgy and Mining Industries (E.N.S.M.I.M.).

II.1.2. Departments

Many other teaching structures have been created to meet a variety of industry's changing requirements :
- Department of Business Management and Applied Economics ;
- Department of Data Processing ;
- Department of Languages ;
- Department for the recycling of Engineers and Managerial Staff ;
- Department of International Cooperation and Exchanges.

The pluridisciplinarity of these schools and departments allows the I.N.P.L. to meet the needs of education of engineers from all countries with a combination of different teachings already existing or not.

II.1.3. Admission

Having completed his secondary studies, a student wishing to be admitted to one of the schools is faced with one of two possibilities : either enter one of the "Classes préparatoires aux Grandes Ecoles Françaises" in view of the national selective examinations, or register at a University and :
- after two years, take a special selective exam leading to admission into the first year course of studies ;
- after four years at a University, be admitted as a graduate into the second year course of studies upon examination of his university achievements.

II.2. Process control and Informatics. Trainings :

Three kinds of such trainings are available at the I.N.P.L.

II.2.1. Computer Science Specialists Training

In accordance with the Computer Science Department of the I.N.P.L. teachings of data processing for process control have been created together between both the High Schools : E.N.S.M.I.M. and E.N.S.E.M. for students of the 3rd year. These teachings are complementary to other which are appropriate to schools.

a) E.N.S.M.I.M. (N.H.S. of Metallurgy and Mining Industries)
In the field of 3rd year Computer Science section, teachings about real time languages and systems complete the training.

b) E.N.S.E.M. (N.H.S. of Electrical and Mechanical Engineering)
A 3rd year section specialized in the study of Computer Science for process control has been created. In that case, we refer to the basis electro-mechanical engineering training given during the two first years.
Teachings about automatics and electronics complete the computer science training essentially hallowed to real time, to microprocessing and to distributed systems.

II.2.2. Non-specialists training
Knowledge of computer science is also necessary for the students of all the schools. Some teachings belonging to the 1st training are given to students who are interested. Nevertheless, this training will be hourly reduced, comparatively to the specialists one.

II.2.3. Graduate specialists Education
A special 4th year receives recently graduate engineers or engineers who already possess professional experience in any speciality, in order to retrain them to computer science and process control.
This 4th year concerns engineers decided to acquire a second ability or to retrain, owing to the technological evolution.
This year can therefore easily fit engineers coming from any school in the world, who research a training in that field.

-III-
WHAT IS THE COMPUTER SCIENCE FOR PROCESS CONTROL ?

III.1. Process-control : a "cross-road" between Automatics and Informatics

This field of computer science concerns all what is related to the elaboration and the use of tools and methods of computer science, for the design and the realization of digital process control systems or applications. It is then case to use computer science in an automation purpose.
What distinguishes this computer science field from the other ones is a close dependancy towards the exterior world. The fact for computers to be used for such applications requires :
- direct link between the processors and the process itself ;
- particular time constraints ;
- particular fault tolerance constraints.
This computer science field gathers various theories, techniques and methods concerning informatics, but also automatics, electronics and some own topics too.

Our account of this "cross-road" situation between various topics, such an engineer is not only a computer science specialist but an electronics and automatics specialist too, independently from the application fields.

III.2. Automatics and Informatics

III.2.1. Automatics
Professor R. Mezencev (1) and a group of automatics (2) teachers have defined automatics and its specialists.
Automatics is the science of the systems control. In the figure n°1, Automatics is precised in three points : theories, tools and techniques, applications fields.

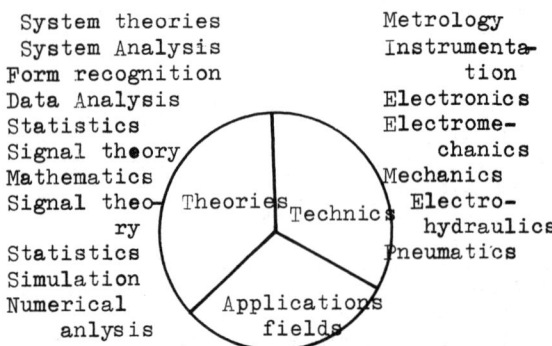

Figure 1 : Automatics Components.

III.2.2. Informatics

Informatics or computer science may be defined on the same model according to the three previous points (figure n°2).

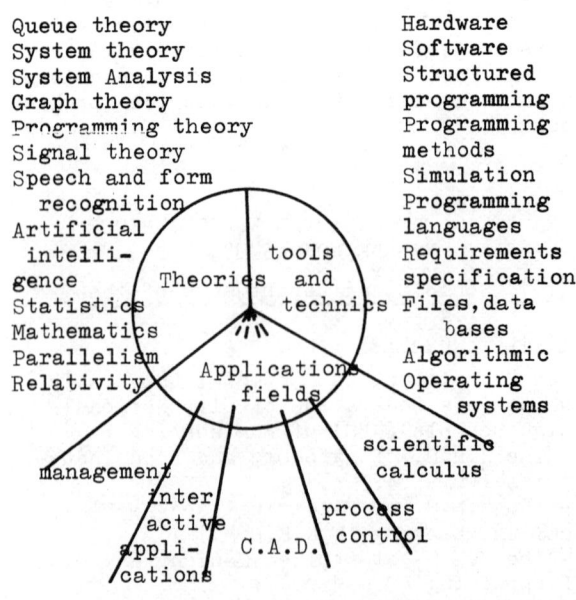

Figure 2 : Informatics components.

III.3. Intersection between Informatics and Automatics

III.3.1. Goals of these sciences

There are effectively common subjects and others which dissimulate different entities under the same term. This is well illustrated by the word "system".

Historically, automatics specialists found an interest in certain classes of systems (rules by differential equations for instance); informatics specialists were interested in other classes of systems: economic systems, information systems, ... and were therefore led to develop theories and tools in order to solve these problems. The approach is the same in both cases: there is first a system analysis step and then the control of the system after modeling.

Thus, even if the systems interesting both specialists are traditionally different, the method and purposes are similar so far as it is still the case to control a system. If informatics is the science of the information treatment, it is also the information systems control. Automatics specialists are more interested in physical systems; Informatics specialists are more interested in information systems. Both are interested in socio-economic systems.

As all the classifications, this one is unperfect. For instance does the control of a packet network belong to the activity of automatics or informatics specialists? Recapitulating back each sector theory, tools and application fields we are going to observe their intersections.

III.3.2. Theories

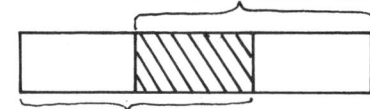

Figure 3

The hatchinged part of the figure represents all the theories related to the signal treatment, the graph theory, the Petri-Nets theory, the automata theory, and so on.
As a specific theory to automatics, we may quote the linear systems theory, the optimal control theories.

III.3.3. Tools and technics

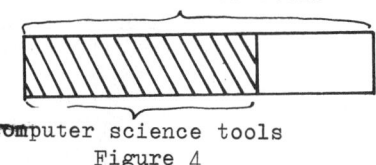

Figure 4

The whole group of the computer science tools is effectively necessary to control a process by computers.
As a consequence these tools and technics will be entirely included in the automatics technologies.

III.3.4. Application fields

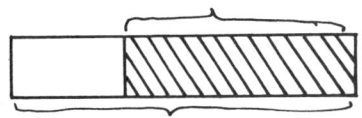

Figure 5

Refering to the fields of application, anywhere the automatics occurs, informatics will potentially do so.
The totality of the application fields related is merged with the one related to informatics.

III.3.5. Conclusion

In all cases a process control specialist must therefore possess a fairly good knowledge on all common points existing between both sciences (automatics and computer science).
But it will be not enough: neither in informatics because it will also require (for instance) knowledge on parallelism and programs proves, on C.A.D. (1) on fig 6, nor in automatics concerning for instance large systems theories, optimal control (2) on fig. 6.
A specialized formation of such specialist around the common kernel of automatics and informatics is therefore to be defined.
It will be necessary to add the knowledge of some technics about electronics and micro-electronics (3) on fig. 6.
Such a specialist will be then able to design and realize process control application, and to work as intermediate with other specialists.

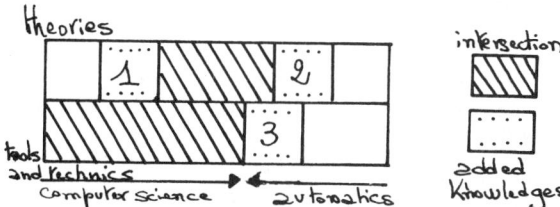

Figure n° 6

-IV- TUTORIAL APPROACH

Teaching in this section gathers 4 major points:
- theoretical teaching with exercices,
- tutorials,
- projects and training periods,
- cases studies.

In this aim, we appeal on one hand to University lecturers and researchers belonging to the concerned fields, on the other hand to manufacturers or users, to engineering firms, to computers and other components manufacturers.

IV.1. Cases studies

Technical visit with conferences in factories are organised. These visits may be the concrete and visual complement of the

case study. They occur either at the beginning of a particular case analysis in order to surround the problem concretely before studying it, or at the end as a result of the analysis already done.
Many various applications are considered as an illustration :
- flexible workshop,
- large systems in chemical industry or in metallurgy,
- traffic regulation.

IV.2. Tutorials

They will be common to all students. It has been considered fairly substantial tutorials at the beginning of the year, in order to familiarize the students with all kinds of languages and real time systems problems, before approaching theoretical teachings, assuming a good practical knowledge.

IV.3. Projects

Projects may occur in the various Laboratories of the I.N.P.L. relatively to the initial specialization of the students or not. The fact the totality of the schools belonging to the INPL covers all the industrial sectors is a considerable advantage for the development of such projects, linked to laboratories and through them to the different industries.

IV.4. Industrial training

A three months long training is foreseen in industry. As for projects, it could take place in an industry appropriate to the 4th year studebts initial speciality.
In the case of recycling engineers, the industrial training is substituted by a project extension.

Note : Complementary teachings are also foreseen at the beginning of the year in order to standardize the students level.

-V-
PROGRAM OF THE SPECIALISTS EDUCATION

The specialists in computer science and process control are formed in the 3rd year of the ENSEM (cf. § 2.1.2) or in a 4th special year (cf. § 2.3).
This program is structured in 16 modules. Some small modifications are introduced between the two above teaching structures. Indeed, some topics are already seen in the 2nd year of the ENSEM especially in automatics. Then the whole program is studied during about 600 hours. Others teachings about management, languages... are not here considered.

It is not very easy to treat all these topics in a single year. A great effort is made to link these different subjects all together and to schedule the associated all through the year.

M1 - Input/Output
. I/O programming
. Controllers
. I/O control subsystems
. Analogical and digital I/O
. Special devices programming

M2 - Operating systems
. Generalities
. Typology of operating systems
. Ressources allocation
. Scheduling
. Files management

M3 - Real time systems
. Functions of a RTS
. Synchronization and communication primitives
. Real time kernel
. Architecture of classical RTS
. Implementation of basic primitives
. Dedicated systems
. Man-machine communication

M4 - Real time Languages
. Real time Fortran
. Summary on 1970's languages (Procol - Pearl)
. New real time languages (LTR V3 "language temps réel V3" - ADA)
. Structured real time programming
= data driving modularity
= events driven   "
= actions driven   "

M5 - Parallelism
. Different primitives (semaphores, monitors...)
. Starvation and deadlocks
. Synchronization models :
= mutual exclusion
= producer/consumer model
. Communication primitives

M6 - Distributed systems
. Communication protocols
. Transport station and transport networks
. Store and forward networks
. O.S.I. Model
. Synchronization in networks
. Communication in distributed applications
. Operations on distributed data

M7 - Modeling and simulation
. Modeling by Petri nets
. Modeling by queue models
. Simulation languages
. Simulation packages

M8 - Sequential systems and automata
. Automata theory
. Graphic methods for specification
. Petri nets and variants
. Programmable automata

M9 - **Modeling - Identification Optimal control**
  . Great Industrial systems modeling
  . Optimization
  . Identification methods (correlation, statistics..)
  . Optimal control
    = application of optimality conditions
    = dynamic programming
    = different methods of optimal control
    = stochastic systems
    = Kalman filters
    = auto-adaptative systems
    = hierarchical control

M10 - **Linear systems and sample systems**
  . Continous linear systems
    = state representation
    = observability
    = controlability
    = control loops
  . Discrete linear systems
    = previous concepts in the discrete case
    = digital correctors
  . Dicrete signal treatment

M11 - **Robotics**
  . Control of displacement
  . Special sensors for Robots
  . Programming of robots
  . Flexible workshop

M12 - **Sensors - Actuators - Interfaces -**
  . Analogical - Digital converters
  . Digital - Analogical converters
  . Different sensors and actuators
  . Multiplexing
  . Intelligent sensors and special devices

M13 - **Microprocessors**
  . Functions of a microprocessor
  . Designing of systems based on microprocessors
  . Architecture of microcomputers
  . Emulators
  . Debugging of applications including microprocessors

M14 - **Data transmission**
  . Codes of transmission
  . Modulation types
  . Synchronization
  . Study of different transmission tools

M15 - **Signal theory**
  . General theorems
  . Application to the speech and picture recognition
  . Speech synthesis

M16 - **Computer aided design**
  . Functions of a CAD system
    = data-base management
    = man-machine communication
    = analysis of designed products
  . Software and hardware for CAD.

-VI-
PROGRAM FOR "NON-SPECIALISTS"

These students are specialized in different fields (food industry, mining industries, chemical industry). Computers and real time systems are tools to pilot a physical process. The aims of this education are to give some knowledge to these students in the field of real time systems. They are not specialized but they must be "acquaunted" users of all the tools.

The program is then foursed on
- real time language
- real time systems
- real time programming.

The tutorial approach in this education is based on industrial cases studies relative to the fields of the different students.

-VII-
CONCLUSION

In the 3rd year for both schools ENSMIM and ENSEM about 30 students per year are graduated. Five vacant situations are offered to each of them. The offered jobs belong to all the industrial domains. That means that specific educations in the fields of process control are reliable. The engineers are capable to work with any other ones. But it appears that the work of these engineers is fairly easier and efficient when the various process specialist engineers are also good users of process control tools. It is then necessary to give some education about these topics to all the new engineers.

Therefore, awing to the substantial needs developping in all the industries, the 4th year special section has been created. It is then possible to turn out good engineers with a double ability :

on one hand in an industrial field
on the other hand in computer science for process control.

BIBLIOGRAPHIE

(1) R. MEZENCEV : Editorial, Bulletin de l'AFCET n° 126, juin 1981, pp. 1 et 2.

(2) Club EEA : Sur une définition de l'automatique et de l'automaticien, Ile d'Yeu, avril 1981.

DISCUSSION

SESSION TP4: EDUCATION

Paper: A TRAINING ABOUT REAL TIME DIGITAL CONTROL IN A FRENCH ENGINEER HIGH SCHOOL

Authors: *JP Thomesse* (Ecole Nationale Supérieure d'Electricité et de Mécanique, 2, rue de la Citadelle, BP 850. 54011 Nancy-Cedex, France)

Discusser: *Ricardo Barrón de la Cruz*
Calle 54 # 365 S.R.
Guadalajara, Jal 44410
México

Questions or Comments:

1) The program presented in the paper is for specialists. It seems to me that it is too broad, more if you think that it is for one year period

   What is the enphasis that is given to this program?
   - Do you teach general theories and the student goes to the particular cases? or
   - Do you start from a specific case and let the student do the generalization?

   Also: What type of microprocessors or microcomputers do you use in your labs?

Author's reply:

1) About the possible approaches theory-practice or practice-theory, traditionaly in many Universities or Engineers Schools in France the first approach was choosen. But for some subjects the other one is applied. Now with the introduction of computer science and of microelectronics in many formations, the second approach is very much used.

   What I have called "case studies" is typically a type of training where some theoretical aspects are given from concrete problems and solutions. The micros used are Motorola 6800 family, and 6 8000 and Texas 99000.

Discusser: *Jesus A. Motolinía*
Atenas 89-103
Col. San Alvaro CP 0290
Delg. Atzcapotzalco
México, D.F.
México

Questions or Comments:

1) Real time digital control education is a complicated and interesting subject. How long will an educational project take? which kind of limitations do you have?. How long do you think is the efficient time for a student in order to become a designer (Real Time) engineer?

Author's reply:

1) An educational project takes us about 4 weeks during 4 or 5 months
   . One half day or a day is reserved each week
   . Two fullweeks are also reserved for the project

   It seems to us that it is enough. Obviously, the subjects must be choosen according to the allowed time.

   To be a good designer engineer, a student who has studied during 5 years after the "Baccalaureat", needs then practice in a company or in a lab. We think that all the theoretical aspects must be given in the school with some practice. But the major practice will be acquired after the studies.

GENERAL COMMENTS MADE AT THE END OF THE SESSION

Discusser: *G Gilles*
Laboratoire d'Automatique
Université de Lyon I
43 boulevard du 11 novembre 1918
69022 Villembaune Cedex, France

Comments:

Inside this session, Digital Control teaching has been related to different aspects: specialization for graduate students; initiation for undergraduate students specialized in chemical engineering; further education for upper technicians and engineers. In some cases, practical experiments follow the theoretical lectures, in others they anticipate the theoretical concepts. I think that what is important is to strongly connect theoretical lectures and practical experiments during each week of the curriculum:
practical motivations⇄theoretical concepts⇄ractical experiments.

Discusser: Dr. José de Jesús Rodríguez
             Apdo 4992 Suc "J"
             Monterrey. N.L.
             México

Comments:

I agree with Dr. Gilles about the need of a balance between theory and experience. In our country this balance is difficult due to financial problems related to the acquisition of laboratory equipment and faculty time assigned to this type of teaching job, due to the fact that the number of students participating in a lab course is small. This idea of having a good balance between theory and practice is reinforced by Piaget's theory of knowledge that says that is easy to obtain knowledge by experience and later it is possible to abstract from it and to obtain a better grasp of theory.

I think Dr. Freidly's idea about having a lab course to make chemical engineering students to use microcomputers as a tool to meassure and control early in their bachellor program and before any theoretical courses in control are taken, is a good one since this lab experience will motivate the students in the theoretical courses. We have been doing something similar at the Institute of Technology where I work, with very good results.

LOCAL OPTIMISATION FOR CORRECTING THE INPUTS IN NON-LINEAR IDENTIFICATION

M. de la Sen and M. B. Paz

Depto. de Automática e Informática, Facultad de Ciencias, Universidad del País Vasco, Leioa (Vizcaya), Spain

Abstract. In this paper, a general method to derive local <u>a posteriori</u> corrections on the inputs used for identification in a non-linear (and eventually time-varying) dynamic system over the nominal ones given <u>a priori</u>. The method is derived by using appropriate Hamiltonians (of a local nature) for the deterministic case and the Fisher information matrix for the stochastic one. The obtained corrections are used for tracking purposes of the system parameters with a delay of one step.

<u>Keywords.</u> Information matrix; nonlinear control systems; optimal inputs; tracking

INTRODUCTION

Nonperiodic sampling has been found to be very useful for improving fundamental properties of dynamic systems such as controllability, observability and identifiability. The transmission of relative measuring errors may be optimised from a numerical analysis point of view on solving systems of linear equations appearing in such problems, (De la Sen and Dormido, 1981 a, b; Dormido and De la Sen, 1979) in the linear case. Some extensions (De la Sen, 1982) have been made for the nonlinear one using this design combined with an adequiate choice of the system inputs. This paper gives a generalization of the ideas stated in De la Sen (1982), by using several Hamiltonian types, of a local nature, for optimising the inputs to achieve improvements in the identification. Each type is more convenient or not according to fast computing time possibilities leading to real-time parameter tracking algorithms. Generalizations following the ideas of Goodwin and Payne (1976) are given using information matrices for the stochastic case.

PROBLEM STATEMENT

The considered system is described by

$$\dot{x}(t) = f(x(t), q(t), u(t)), \quad z(o) = z_o,$$
$$o \leq t \leq \tau$$

$$y(t) = C x(t) + \zeta(t) \qquad (1)$$

where $x(t) \in R^n$, $y(t) \in R^p$ and $u(t) \in R^m$ are the state, output and input vectors respectively. The vector function f is assumed known, continuously differentiable with respect to its arguments and $\zeta(t)$ is the measurement noise vector.

In Matausek and Stankovic (1980), the model used to describe (1) is of the parallel type given by

$$\dot{z}^m(t) = f(x^m, q^m(t), u(t)), \quad z^m(o) = z_o, \quad o \leq t \leq \tau$$
$$y^m(t) = C x^m(t) \qquad (2)$$

In the sequel, the following features are taken into account:

(a) The parameter estimates made <u>firstly</u> are corrected <u>a posteriori</u> comparing the current system behavior with the original one.

(b) The system, if continuous, is discretized and analyzed by using a set of output samples.

(c) The input is optimised, in a local sense, from a nominal sequence of inputs defined <u>a priori</u>.

As it is well known, there is a certain freedom in the choice of the inputs in the identification problem. In the following, the objective is to locally correct the <u>a priori</u> inputs in order to improve the tracking of parameters. The corrections are not really applied to the system. They are only used to a posteriori modify the parameter vector tracking equation. So, defining for a posteriori $q_k^{m(p)}$ parameter estimate, the loss function

$$\hat{J}_{k/k+1}(q_{k-1}^{m(p)}, q_{k+1}^{m(o)}, k, u_k^{nom}, \hat{e}_k^o) =$$

$$= \hat{J}_{k/k+1}^{*}(q_{k-1}^{m(p)}, q_{k+1}^{m(o)}, k, u_k^{nom}, \hat{e}_k^o) +$$

$$+ \frac{\partial \hat{J}_{k/k+1}^{*}}{\partial u_k^T}(q_{k-1}^{m(p)}, q_{k+1}^{m(o)}, k, u_k^{nom}, \hat{e}_k^o)$$

$$\frac{1}{2}\Delta u_k^T \frac{\partial^2 J(.,.,.,.,.)}{\partial u_k \partial u_k^T} \Delta u_k \quad (3)$$

where $(\hat{\cdot})$ stands for the estimation of $(\cdot)$. The loss function $\hat{J}_{k/k+1}^{*}(.,.,.,.,.)$ has as general form a quadratic form of the error between the system and estimation model outputs plus another one of the structure distance between the a priori and a posteriori models.

Now, two Hamiltonian types are defined associated to this loss function by considering first and second - order terms around $\hat{J}^{*}(.,.,.,.,.)$ or first - order terms only.

Second-order Hamiltonian: In this case the Taylor-series expansion around the nominal values is of second order. This yields for one step-optimisation

$$H_{k/k+1} = \hat{J}_{k/k+1}^{*} + \nabla_{u_k}^T \hat{J}_{k/k+1}^{*} + \frac{1}{2}\Delta u_k^T$$

$$\nabla_{u_k u_k^T}^2 \hat{J}_{k/k+1}^{*} \Delta u_k + \mu_{k+1}^T (\phi_k \hat{e}_k^o + \psi_k$$

$$u_k^{nom} + \nabla_{u_k^T}^T (\psi_k u_k^{nom}) \Delta u_k + \nabla_{u_k^T}^T (\phi_k \hat{e}_k^o)$$

$$\Delta u_k) + \lambda_{k+1}^T (q_k^{m(o)} + \eta_k \hat{e}_k^o + \nabla_{u_k}^T (\phi_k$$

$$\hat{e}_k^o) \Delta u_k) + \lambda_{k+1}^T (q_k^{m(o)} + \eta_k \hat{e}_k^o + \nabla_{u_k}^T$$

$$(\eta_k e_k) \Delta u_k + \frac{1}{2}\Delta u_k^T \nabla_{u_k u_k^T}^2 (\mu_{k+1}^T \psi_k$$

$$u_k^{nom}) \Delta u_k + \frac{1}{2}\Delta u_k^T \nabla_{u_k u_k^T}^2 (\mu_{k+1}^T \phi_k \hat{e}_k^o)$$

$$\Delta u_k + \frac{1}{2}\Delta u_k^T \nabla_{u_k u_k^T}^2 (\lambda_{k+1}^T \eta_k \hat{e}_k^o) \Delta u_k \quad (4)$$

with $\nabla_{(.)}$ and $\nabla_{(.)}^2$, meaning the gradient operators of first and second orders and whose u_k^{nom} is the used nominal input sequence, λ_{k+1}, μ_{k+1} are Lagrange multipliers appearing with the problem constraints :

$$\hat{e}_{k+1}^o = \phi_k \hat{e}_k^o + \psi_k u_k^{nom}$$

$$\hat{q}_{k+1}^{m(o)} = \hat{q}_k^{m(o)} + \eta_k \hat{e}_k^o \quad (5)$$

$$\Delta u_k^2(i) \leq \beta_i, \text{ for each input component.}$$

The third constraint is used either for classical energy bound reasons or mathematical ones (validity, of the second-order Taylor series expansion, while the two - first ones describe dynamical equations for a simplified linearized problem given by the transition matrices $\phi_{(.)}$ (state) and $\psi_{(.)}$ (control) and the first-order estimates of the state error $(\hat{e}_{(.)}^o)$ between the nominal and estimation systems, and the parameter vector $(\hat{q}_{(.)}^{m(o)})$ of the last one. The superscript m(o) means a priori estimate in the context of De la Sen (1982). The cost function $\hat{J}^{*}(.)$ is defined in that paper to take into account a cautious control strategy according to the several optimisation problems (weighting matrices, sampling design, a posteriori corrections) which must be solved.

First-order Hamiltonian:

$$H_{k/k+1}^{(1)} = \hat{J}_{k/k+1}^{*} - \nabla_{u_k}^T \hat{J}_{k/k+1}^{*} \Delta u_k + \lambda_{k+1}^T$$

$$(\hat{q}_k^{m(o)} + \eta_k \hat{e}_k^o) + \mu_{k+1}^T (\phi_k \hat{e}_k^o + \psi_k u_k^{nom})$$

$$+ \nabla_{u_k}^T (\mu_{k+1}^T \phi_k \hat{e}_k^o + \mu_{k+1}^T \psi_k u_k^{nom}) \Delta u_k$$

$$+ \nabla_{u_k}^T (\lambda_{k+1}^T \eta_k \hat{e}_k^o) \Delta u_k \quad (6)$$

This Hamiltonian yields a simpler form than the one before and is useful in examples of a great complexity (in state or parameter orders).

Stochastic case:

Assuming a differential Wiener process available for modelling the whole system given by the initial system and the prefilter used for compensating the information lost due to the sampling, one has the incremental equation

$$d\tilde{x}(t) = A\tilde{x}(t)\,dt + Bu(t)\,dt + K\,d\mathcal{E}(t) + \bar{D}\,\Delta\hat{q}$$

$$y(t) = C^a \tilde{x}(t) \tag{7}$$

with $\tilde{x}$ being the n-state vector, y being the p-output vector and $\mathcal{E}(t)$ being the whole noise resulting of taking both noises (in the state and in the output). The estimate of the extended state vector $\tilde{x} = (\hat{e}^T, \Delta q^T)^T$ is given by

$$\hat{\tilde{x}}^o_{k+1} = \phi^a(T_k)\hat{\tilde{x}}^o_k + \psi^a(T_k)\Delta u_k \tag{8}$$

with

$$\psi^a(T_k) = \begin{bmatrix} \nabla^T_{u_k}(\phi(T_k)\hat{e}^o_k) + \nabla^T_{u_k}(\psi(T_k)\Delta\hat{q}_k) \\ --- \\ 0 \end{bmatrix}$$

and

$$\phi^a(T_k) = \begin{bmatrix} \phi(T_k) & \psi(T_k) \\ 0 & \tilde{\phi}(T_k) \end{bmatrix}$$

with $\tilde{\phi}(T_k) = (1 - k + \frac{2k}{N}) I \Delta\hat{q}_k$; $k \in (0,1)$

To solve this problem a determinant type criterion of the Fisher information matrix is used.

PROBLEM SOLUTION

The conditional covariance of $\tilde{x}_k$ satisfies the Riccati equation

$$P_{k+1} = \phi^a_k P_k \phi^{aT}_k + Q^a_k - M_k(C^a P_k C^{aT})^{-1} M^T_k \tag{9}$$

If $\lambda_k = \int_0^{T_k} \phi(T_k, \tau) K^a \, d\mathcal{E}(\tau)$

$T_k = t_{k+1} - t_k$

The sequence $\{X_k\}$ is a sequence of stochastic variables if $du(t)$ and $d\mathcal{E}(t)$ ro mean and covariance

$$E[\lambda_k \lambda^T_k] = \int_0^{T_k} \phi^a(T_k, \tau) K^a \Sigma^a K^{aT}$$

$$\phi^{aT}(T_k, \tau)\, d\tau = Q^a_k \tag{10}$$

Defining the conditional covariance of $\hat{e}_k$

as

$$S_k = C P_k C^T$$

if

$$P(y,\beta) = ([2\pi]^{rN} \prod_{k=1}^{N} \det S_k)^{-1} \exp\left\{-\frac{1}{2}\right.$$

$$\left.\sum_{k=1}^{N} (\hat{e}_k - \bar{\hat{e}}_k)^T S_k^{-1} (\hat{e}_k - \bar{\hat{e}}_k)\right\} \tag{11}$$

where the superscript $(\bar{\cdot})$ means mean value. The elements of the information matrix result to be

$$M_{ij} = E_{y/\beta} \left\{ \sum_{k=1}^{N} (\frac{\partial \bar{\hat{e}}_k}{\partial \beta_i})^T S_k^{-1} (\frac{\partial \bar{e}_k}{\partial \beta_j}) \right\}$$

$$+ \frac{1}{2} \sum_{k=1}^{N} \text{trace}\left\{S_k^{-1} \frac{\partial S_k}{\partial \beta_i}\right\} \text{trace}$$

$$\left\{S_k^{-1} \frac{\partial S_k}{\partial \beta_j}\right\} \tag{12}$$

Defining also

$$\frac{\partial \bar{\hat{e}}_k}{\partial \beta_i} = -\frac{\partial C}{\partial \beta_i} \bar{\tilde{x}}_k - C \frac{\partial \bar{\tilde{x}}_k}{\partial \beta_i} \tag{13}$$

$$\frac{\partial \bar{\tilde{x}}_{k+1}}{\partial \beta_i} = (\phi_k - M_k C) \frac{\partial \bar{x}_k}{\partial \beta_i} + (\frac{\partial \phi_k}{\partial \beta_i} - $$

$$M_k \frac{\partial C}{\partial \beta_i}) \bar{\tilde{x}}_k + \frac{\partial \psi_k}{\partial \beta_i} \Delta \bar{\hat{q}}_k +$$

$$+ \frac{\partial M_k}{\partial \beta_i} (\hat{e}_k - \bar{\hat{e}}_k)$$

$$+ \nabla_{u_k^{nom}} \phi_k \bar{\hat{e}}^o_k + \nabla_{u_k^{nom}} \psi_k \Delta \bar{\hat{q}}_k)\Delta u_k$$

which may be combined to give

$$\bar{\tilde{x}}_{k+1} = F_k \bar{\tilde{x}}_k + G_k \Delta u_k + H_k(\hat{e}_k - \bar{\hat{e}}_k)$$

$$\bar{\tilde{x}}_k = \left[\bar{\tilde{x}}^T_k, \Delta\hat{q}^T_k, \frac{\partial \bar{\tilde{x}}^T_k}{\partial \beta_1}, \ldots, \frac{\bar{\tilde{x}}^T_k}{\partial \beta}\right]^T$$

$$\frac{\partial \bar{\hat{e}}_k}{\partial \beta_i} = \Pi_i \bar{\tilde{x}}_k \tag{14}$$

After some calculation, it is possible to arrive to

$$M_{k+1} = M_k + I_{k+1} \; ; \; M_0 = 0 \qquad (15)$$

where for the mean value $\hat{x}_{(.)}$ of $\tilde{\tilde{x}}_{(.)}$ and its covariance $I_{(.)}$, one has

$$[I_{k+1}]_{ij} = (F_k \hat{x}_k + G_k \Delta u_k)^T \Lambda_i^T S_{k+1}^{-1}$$

$$\Lambda_j (F_k \hat{x}_k + G_k u_k)$$

$$+ \text{trace} \{ \Lambda_i^T S_{k+1}^{-1} \Lambda_j T_{k+1} \}$$

$$+ \frac{1}{2} \text{trace} \left\{ S_{k+1}^{-1} \frac{\partial S_{k+1}}{\partial \beta_i} \right\}$$

$$\text{trace} \left\{ S_{k+1}^{-1} \frac{\partial S_{k+1}}{\partial \beta_j} \right\} \qquad (16)$$

The remainder of the methodology is established by minimizing

$$\emptyset(M_k) = \min_{\Delta u_k \in \Lambda} \lambda_{MAX} (W^{-1} M^{-1}).$$

Second-order Hamiltonian:

If the final loss is given by

$$\partial(\hat{q}_{k+1}^{m(0)}) = 0 \qquad (17)$$

$$\partial(\hat{e}_{k+1}^o) = (\hat{e}_{k+1}^o - e_{k+1}^o)^T P_{k+1} (\hat{e}_{k+1}^o - e_{k+1}^o)$$

$$P_{k+1} > 0; \; \forall k \geq 0$$

the solution will be

$$\Delta u_k^* = -[\nabla^2_{u_k^T u_k} \hat{J}_{k/k+1}^* + \nabla^2_{u_k^T u_k} (\hat{e}_k^{oT} \emptyset_k^T)$$

$$P_{k+1} \text{diag} ((\hat{e}_{k+1}^o - e_{k+1}^o), \ldots, (\hat{e}_{k+1}^o -$$

$$e_{k+1}^o))]^{-1} [\nabla_{u_k^T} \hat{J}_{k/k+1}^x + \nabla_{u_k^T} (u_k^{nom\,T} \psi_k^T)$$

$$+ 2\nabla_{u_k} (\hat{e}_k^{oT} \emptyset_k^T) P_{k+1} (\hat{e}_{k+1}^o - e_{k+1}^o)] \text{ if }$$

$$\Delta u_k (i) < \beta_i \qquad (18)$$

or $\Delta u_k^* = \pm \beta_i$ otherwise.

where e_{k+1}^o is the error in fact obtained between the system and the estimation model at the k+1-th sampling instant. An important remark to be made is that the obtained locally optimised input provides an "<u>a posteriori</u>" correction (not applied in real time but available to correct, with one step delay, the estimates equations, and to supply a good parameter tracking which is a fundamental question in the non-linear case) of the system behaviour.

First-order Hamiltonian:

$$\Delta u_k^* = -\beta \, \text{sign} [\nabla_{u_k} \hat{J}_{k/k+1}^* + \sum_{j=1}^{n} /_{u_k}$$

$$(j^T \Lambda_k^T) \hat{e}_k^o (j) \lambda_{k+1} + \sum_{j=1}^{n} \nabla_{u_k} (j^T \emptyset_k^T)$$

$$\hat{e}_k^o (j) \mu_{k+1} + \psi_k^T \mu_{k+1} + \sum_{j=1}^{n} \nabla_{u_k} (j^T \psi_k^T)$$

$$u_k(j) \mu_{k+1}] \quad , \text{ with } \quad j^T = (0, 0, \ldots, 0, \underset{j}{1}, 0,$$

$$\ldots 0 \,) \qquad (19)$$

Stochastic case:

- for $\Delta u_k (i) < \beta_i$

$$\Delta u_k^* = -[\sum_{j=1}^{p} \sum_{j=1}^{p} G_k^T \Lambda_i^T S_{k+1}^{-1}$$

$$\Lambda_k G_k W_{ji} (k+1)]^{-1}$$

$$[\sum_{i=1}^{p} \sum_{j=1}^{p} G_k^T \Lambda_i^T S_{k+1}^{-1} \Lambda_j \hat{x}_k W_{ji}$$

$$(k+1)] \qquad (20)$$

- for $\Delta u_k (i) = \beta_i$

$$\Delta u_k^* = -\text{diag} \; (\beta_1, \beta_2, \ldots,$$

$$\beta_r) \, \text{sign} \; (\sum_{i=1}^{p} \sum_{j=1}^{p} G_k^T \Lambda_i^T S_{k+1}^{-1}$$

$$\Lambda_j F_k \hat{x}_k] \qquad (21)$$

where $\text{sign}(x) = [\text{sign}(x_1), \ldots, \text{sign}(x_n)]^T$

and S_k = conditional covariance of e_k (mean value of $\hat{e}_k$)

$\hat{x}_{k+1} = F_k \hat{x}_k + G_k \Delta u_k$ (mean value of x_{k+1})

and $\Lambda_i = \left[\dfrac{\partial C}{\partial P_i}, 0, \ldots, 0, C, 0, \ldots, 0 \right]$

and F_k (defined in a similar way) are sensitivity matrices with respect to the parameter vector.

CONCLUSION

In this paper, several techniques to improve the tracking capability of the estimation models in discrete nonlinear time-varying system have been given. These ones are based on the optimisation of local increments in the nominal inputs in both determnistic and stochastic cases.

REFERENCES

De la Sen, M. (1982). Identification algorithms in nonlinear time-varying systems. In Preprints of the IFAC Symp. on Theory and Applications of Digital Control, Delhi, 1 (5), pp. 7-11.

De la Sen, M., and S. Dormido (1981a). Aperiodic sampling and identifiability. In Preprints of the 8th triennial world Congres, Kyoto, 6, pp.71-76.

De la Sen, M. and S. Dormido (1981b). Non-periodic sampling and identifiability, Electron. Lett., 17, 922-924.

Dormido, S., and M. de la Sen (1979). A note on the transmission of relative errors in the observability problem, IEEE Trans. Autom. Control, AC-24, 634-635.

Goodwin, G.C., and R.L. Payne (1976). Choice of sampling intervals. In R.K. Mehra and D.G. Lainiotis (Eds.), System Identification. Advances and case studies, Vol. 126, Academic Press, New York, pp. 251-287.

Matausek, M.R., and S.S. Stankovic (1980). Robust real-time algorithm for identification of nonlinear time-varying systems, Int. J. Control, 31, 79-94.

A MOVING MODEL OF DISCRETE-DATA SYSTEMS AND ITS APPLICATION IN CONTROL

Wen-Teng Wu*,**, Yung-Chung Fang** and J. R. Hopper***

**Department of Chemical Engineering, National Tsing Hua University, Hsinchu, Taiwan, China*
***Department of Chemical Engineering, Lamar University, Beaumont, Texas, USA*

Abstract. A method of on-line recursive modeling of discrete-data systems is developed by the use of block pulse functions. The method is a sequential modeling. A low order model is sufficient to represent the system. With this model, on-line control is easily achieved. A direct digital control system is presented to show that the method exhibits considerable promise.

Keywords. Direct digital control; identification; recursive functions; sampled data systems; self-adjusting systems; z transforms.

INTRODUCTION

In the controller design of an industrial process, system identification is always required. Many investigations have been reported in this area (Kurz, Isermann and Schumann, 1980; Koivo, 1980; Clarke and Gawthrop, 1975; Astrom and Wittenmark, 1973). Wu and Wong (1981a) proposed a moving model and applied it to suboptimal control with inaccessible state variables. Recently Wu and Liu (1982) gave the applications of a moving model to suboptimal control of non-linear dynamic systems and obtained a satisfactory result. In the present study, a moving model of discrete-data systems is considered. The model is obtained by the use of block pulse functions.

Block pulse functions were first presented by Harmuth (1969). Chen, Tsay and Wu (1977) introduced their useful properties. Since then applications of block pulse functions to solution of time-varying systems (Rao and Ganapathy, 1979); inversion of Laplace transform (Shieh, Yates and Navarro, 1979); parameter estimation (Wu and Wong, 1980) etc. have been proposed. Wu and Fang (1981b) introduced the shift operational matrix of block pulse functions as the operator of z^{-1} in z-transformation and applied it to parameter estimation and model reduction for sampled-data systems.

In this paper a moving model of discrete-data systems is developed. The model is described in a linear form. The coefficients are determined by the use of the least-squares recursive algorithm. Then on-line control is applied.

BLOCK PULSE FUNCTIONS

The block pulse functions $\psi_i(t)$ are defined as:

$$\psi_i(t) = \begin{cases} 1, & (i-1)t_f/m \leq t < it_f/m \\ 0, & \text{otherwise} \end{cases} \quad (1)$$

where t_f is the final time and m is the number of subintervals between t=0 and $t=t_f$. A block pulse vector of dimension m is given to be:

$$\underline{\psi}(t) = \left[\psi_1(t),\ \psi_2(t),\ \ldots,\ \psi_m(t)\right]^T. \quad (2)$$

Any discrete-data function with a zero-order hold, $c_d(t)$, can be exactly represented by a block pulse series expansion if a proper sampling interval is selected.

$$c_d(t) = \underline{C}^T(m)\underline{\psi}(t) \quad (3)$$

where $\underline{C}^T(m) = \left[c_1,\ c_2,\ \ldots,\ c_m\right]$. This representation is illustrated in Fig. 1.

The shift operational matrix Q(m) is defined by Chen, Tsay and Wu (1977) as:

$$Q(m) = \begin{bmatrix} 0 & 1 & 0 & .. & 0 \\ 0 & 0 & 1 & .. & 0 \\ . & . & . & .. & . \\ 0 & 0 & 0 & .. & 1 \\ 0 & 0 & 0 & .. & 0 \end{bmatrix} = \begin{bmatrix} 0 & \vdots & & & \\ 0 & \vdots & & \underline{I}_{m-1} & \\ \vdots & \vdots & & & \\ 0 & \vdots & & & \\ \cdots & \cdots & \cdots & \cdots & \cdots \\ 0 & \vdots & 0\ 0 & .. & 0 \end{bmatrix} \quad (4)$$

where $\underline{I}_{m-1}$ is a unit matrix of (m-1)-dimensions.

A discrete-data function $c(i\theta)$ in block pulse series expansion is given as:

$$c(i\theta) \cong \underline{C}^T(m)\underline{\psi}(t),\ \ t<t_f \quad (5)$$

where the approximation denotes the discrete-data function with a zero-order hold. θ is

---
* Author to whom correspondence should be addressed.

defined as $\theta = t_f/m$. Thus $c[(i-1)\theta]$ becomes

$$c[(i-1)\theta] \simeq \underline{C}^T(m)\underline{Q}(m)\underline{\psi}(t), \quad t < t_f. \qquad (6)$$

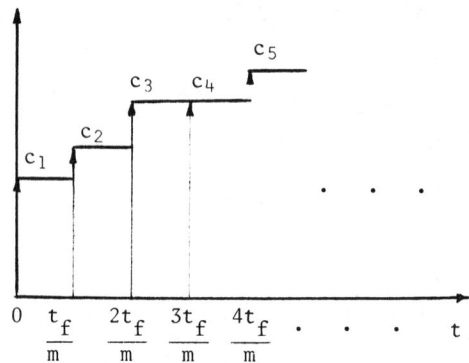

Fig. 1. Block pulse series approximation of a discrete-data function.

MOVING MODEL

(A) Parameter Estimation

Consider a first order system. The governing equation is given as

$$c(i\theta) = a_0 c[(i-1)\theta] + b_0 u[(i-1)\theta] \qquad (7)$$

where the parameters a_0 and b_0 are to be estimated.

The first step of the parameter estimation is to take block pulse series approximations of $c(i\theta)$, $c[(i-1)\theta]$ and $u[(i-1)\theta]$. Substitution of these approximations into Eq.(7) gives:

$$\underline{C}_0^T(m)\underline{\psi}(t) + a_0\underline{C}_0^T(m)\underline{Q}_0(m)\underline{\psi}(t) = b_0\underline{R}_0^T(m)\underline{Q}_0(m)\underline{\psi}(t) \qquad (8)$$

where $\underline{R}_0^T(m)$ and $\underline{C}_0^T(m)$ are the coefficient vectors of the block pulse series expansions of $u(i\theta)$ and $c(i\theta)$, respectively.

Rewriting Eq.(8), we have

$$\underline{X}_0(m)\underline{P}_0(m) = \underline{C}_0(m) \qquad (9)$$

where
$$\underline{X}_0(m) = \left[\underline{Q}_0^T(m)\underline{C}_0(m), \underline{Q}_0^T(m)\underline{R}_0(m)\right]$$
$$= \begin{bmatrix} 0 & 0 \\ c_1 & r_1 \\ c_2 & r_2 \\ \vdots & \vdots \\ c_{m-1} & r_{m-1} \end{bmatrix} = \begin{bmatrix} \underline{x}_0^T \\ \underline{x}_1^T \\ \underline{x}_2^T \\ \vdots \\ \underline{x}_{m-1}^T \end{bmatrix} \qquad (10\text{-a})$$

and $\underline{P}_0(m) = [-a_0, b_0]^T$. (10-b)

The parameter vector, $\underline{P}_0(m)$, is obtained as

$$\underline{P}_0(m) = (\underline{X}_0^T(m)\underline{X}_0(m))^{-1}\underline{X}_0^T(m)\underline{C}_0(m) \qquad (11)$$

$$= \underline{W}_0(m)\underline{X}_0^T(m)\underline{C}_0(m).$$

where $\underline{W}_0(m)$ is $(\underline{X}_0^T(m)\underline{X}_0(m))^{-1}$ provided the inverse of the matrix $(\underline{X}_0^T(m)\underline{X}_0(m))$ exists. This is called the least-squares estimator of $\underline{P}_0(m)$. Since $\underline{X}_0(m)$ and $\underline{C}_0(m)$ are given from the input-output data, the parameter vector is obtained. It is necessary to restrict m to $m \geq 2$ for the least-squares solution.

(B) Recursive Modeling

The moving model is obtained by successive parameter estimation, as illustrated in section (A), by the use of the least-squares recursive formulas. The procedure can be divided into two steps. The first step is to add new data to the last row of $\underline{X}_k(m)$ and $\underline{C}_k(m)$, where the subscript k denotes the k-th moving step. The second step is to discard the first row of the new $\underline{C}_k(m)$ and the second row of the new $\underline{X}_k(m)$.

The recursive formulas are used to avoid repeatly calculating the matrix inversion in the determination of the matrix $\underline{W}_k(m)$. In the first step, $\underline{W}_k(m+1)$ is obtained as:

$$\underline{W}_k(m+1) = \underline{W}_k(m) - \underline{W}_k(m)\underline{x}_{k+m}\underline{x}_{k+m}^T\underline{W}_k(m)Q_{kB} \qquad (12)$$

where $Q_{kB} = (1+\underline{x}_{k+m}^T\underline{W}_k(m)\underline{x}_{k+m})^{-1}$ is a scalar.

$\underline{x}_{k+m}^T$ is the last row of $\underline{X}_k(m+1)$ as

$$\underline{X}_k(m+1) = \begin{bmatrix} \underline{X}_k(m) \\ \cdots \\ \underline{x}_{k+m}^T \end{bmatrix}. \qquad (13)$$

The form of $\underline{x}_{k+m}^T$ is given in Eq.(10-a).

In the second step, $\underline{W}_{k+1}(m)$ is determined by the following Eq.(14):

$$\underline{W}_{k+1}(m) = \underline{W}_k(m+1) + \underline{W}_k(m+1)\underline{x}_{k+1}\underline{x}_{k+1}^T\underline{W}_k(m+1)Q_{kF} \qquad (14)$$

where $Q_{kF} = (1-\underline{x}_{k+1}^T\underline{W}_k(m+1)\underline{x}_{k+1})^{-1}$.

$\underline{x}_{k+1}^T$ is the second row of $\underline{X}_k(m+1)$. The form of $\underline{x}_{k+1}^T$ is also shown in Eq.(10-a). Thus the desired parameter vector is obtained to be

$$\underline{P}_{k+1}(m) = \underline{W}_{k+1}(m)\underline{X}_{k+1}^T(m)\underline{C}_{k+1}(m) \qquad (15)$$

where $\underline{X}_{k+1}(m)$ is the matrix of $\underline{X}_k(m+1)$ matrix with the second row, $\underline{x}_{k+1}^T$, discarded.

The moving model is obtained in the form of

$$c(i\theta) = a_k c[(i-1)\theta] + b_k u[(i-1)\theta] \qquad (16)$$

where a_k and b_k are given from the parameter vector of Eq.(15) in the k-th moving step.

Illustrative Example:

Consider a discrete-data system as (Shih, Wu and Chow, 1975):

$$G(z) = \frac{c(z)}{u(z)} = \frac{0.42z^7+0.279z^6-0.0525z^5+0.038z^4}{z^8-0.42z^7-0.279z^6+0.0525z^5}$$

$$\frac{-0.129z^3-0.0655z^2+0.011z-0.0015}{-0.038z^4+0.129z^3+0.0655z^2-0.011z+0.0015} \quad . \tag{17}$$

If the input, $u(z)$, is a unit-step function, the output data are obtained as follows:

t,(time)	0,	θ,	2θ,	3θ,
c(t)	0,	0.42,	0.8754,	1.131348,

	4θ,	5θ,	6θ,
	1.381853,	1/421526,	1.392267,

	7θ,	8θ,	9θ,	10θ,
	1.31236,	1.21835,	1.14919,	1.02893,

....
....

In this example m=3 is chosen. The first three input values for the unit step function are $r_1=r_2=r_3=1$ and the output data were calculated from Eq.(17) to be $c_1=0$, $c_2=0.42$ and $c_3=0.8754$. Using Eq.(11) we have

$$\underline{P}_0(3) = \begin{bmatrix} 1.085 \\ 0.42 \end{bmatrix} . \tag{18}$$

With the new data at 4θ, 5θ, .., the parameter vectors $\underline{P}_i(3)$, i=1,2,... are obtained by the use of the recursive formulas, Eqs.(12), (14) and (16).

$$\underline{P}_1(3) = \begin{bmatrix} 0.80908 \\ 0.45955 \end{bmatrix} . \tag{19}$$

$$\underline{P}_2(3) = \begin{bmatrix} 0.69447 \\ 0.56775 \end{bmatrix} . \tag{20}$$

. . . .

Simulation of the moving model with parameters obtained from $\underline{P}_i(3)$ is used to check the promise of the method. The first three unit-step response data of the model of Eq.(16) with the parameters of Eq.(18) are:

t,(time)	0	θ	2θ
c(t)	0	0.42	0.8754

The data generated by the moving model simulation are the same as that of the original system. Then using the parameter vector $\underline{P}_1(3)$ in the moving model, the response data at t= 3θ is obtained as $c(3\theta)=1.1678$. Similarly by using the parameters of $\underline{P}_2(3)$ we determine c(t) at t=4θ; using $\underline{P}_3(3)$, c(t) at t=5θ is determined; and using $\underline{P}_j(3)$, c(t) at t=$[(j+2)\theta]$ is obtained. A comparison of the unit-step responses of the original high-order system of Eq.(17) and of the moving model of Eq.(16) in the time interval of 0 to 35θ are shown in Fig. 2. Satisfactory results are obtained.

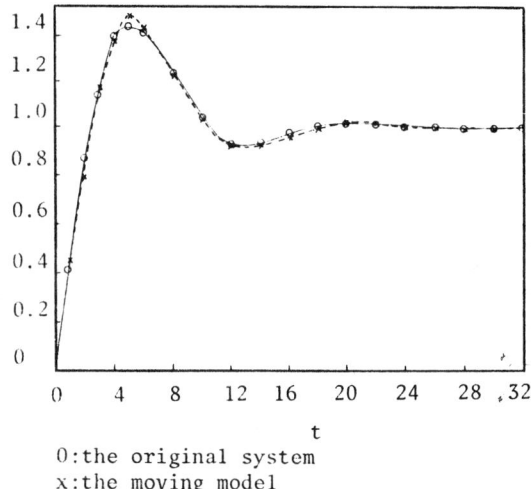

0: the original system
x: the moving model

Fig. 2. Comparison of the unit-step responses of the original high order system and the moving model.

APPLICATION OF THE MOVING MODEL TO CONTROL

Now consider a direct digital control system. The block diagram is shown in Fig. 3.

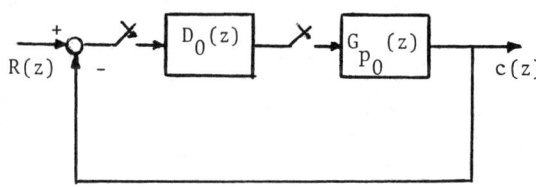

Fig. 3. A block diagram of a direct digital control system.

It is desired to find a digital compensator, $D_0(z)$, such that the output response will satisfy a given function. If the transfer function of the given function, $M_0(z)$ and the transfer function of the plant, $G_{p_0}(z)$ are known, $D_0(z)$ can be easily obtained as:

$$D_0(z) = \frac{M_0(z)}{1-M_0(z)} \cdot \frac{1}{G_{p_0}(z)} . \tag{21}$$

In the present study, the transfer function of the plant $G_{p_0}(z)$ is unknown. It is determined via a moving model $G_{p_k}(z)$.

The design of a digital compensator via a moving model may be accomplished by an equation similar to Eq.(21). The reference function and the plant are represented by the moving models, $M_k(z)$ and $G_{p_k}(z)$, respectively.

The digital compensator $D_k(z)$ is obtained to be

$$D_k(z) = \frac{M_k(z)}{1-M_k(z)} \cdot \frac{1}{G_{p_k}(z)} \quad . \quad (22)$$

Thus direct digital control of an unknown plant is achieved by the step by step calculation.

Illustrative Example:

Consider the transfer function of a plant as

$$G_p(z) = \frac{0.005z-0.0045}{z^2-1.905z+0.905} \quad (23)$$

and the transfer function of the reference model to be

$$M_0(z) = \frac{0.688z-0.0507}{z^2-0.54z+0.178} \quad . \quad (24)$$

The output $c(z)$ of the direct digital control with a unit step input is obtained as

$$c(z) = \frac{z}{z-1} \cdot M_0(z) \quad . \quad (25)$$

Assuming the plant is unknown, a moving model is used to identify the plant. The initial model of the plant is obtained by the use of the unit-step response data. Here m=3 is chosen. We have

$$[u_1, u_2, u_3] = [1, 1, 1] \quad (26\text{-a})$$

and $[c_1, c_2, c_3] = [0, 0.005, 0.010025]$. (26-b)

With the data of Eqs.(26), $G_{p_0}(z)$ is obtained by the use of Eq.(11).

$$G_{p_0}(z) = \frac{0.005}{z-1.005} \quad . \quad (27)$$

The first digital compensator is

$$D_0(z) = \frac{M_0(z)}{1-M_0(z)} \cdot \frac{1}{G_{p_0}(z)}$$

$$= \frac{0.688z^2-0.7421z+0.05095}{0.005z^2-0.00614z+0.00114} \quad . \quad (28)$$

Using $D_0(z)$ as the digital compensator, the first three control variables and output variables with the unit-step input, $R(z) = z/(z-1)$, are

$$[u_1, u_2, u_3] = [137.6, 63.476, 9.147]$$

$$[c_1, c_2, c_3] = [0, 0.688, 1.00882]$$

Similar to the procedure to determine Eq.(27), $G_{p_1}(z)$ is obtained to be

$$G_{p_1}(z) = \frac{0.00505}{z-1.01310} \quad . \quad (29)$$

where the initial value is shifted to zero.

In determination of $D_1(z)$, $M_1(z)$ is used. The data to determine $M_1(z)$ are

$$[r_4, r_5, r_6] = [1, 1, 1]$$

$$[c_4, c_5, c_6]_0 = [1.0596, 1.0299, 1.00484]$$

$M_1(z)$ is obtained by the use of Eq.(11) with the shifted initial value at zero as

$$M_1(z) = \frac{-0.02968}{z-0.84448} \quad . \quad (30)$$

Therefore $D_1(z)$ is determined to be

$$D_1(z) = \frac{-5.87342z+5.95039}{z-0.8148} \quad . \quad (31)$$

Using $D_1(z)$, we get $u_4 = -6.949$ and $c_4 = 1.0593$. With these data, $G_{p_2}(z)$ is obtained by the use of the recursive formulas Eqs.(12), (14) and (16).

$$G_{p_2}(z) = \frac{0.00551}{z-1.15967} \quad . \quad (32)$$

The reference model $M_2(z)$ is

$$M_2(z) = \frac{-0.025069}{z-0.32922} \quad . \quad (33)$$

Then $D_2(z)$ is obtained as

$$D_2(z) = \frac{-4.54651z+5.27247}{z-0.30415} \quad . \quad (34)$$

Continuing the same procedures, the desired compensators, $D_i(z)$, are obtained.

The unit-step responses of the original control system and the system via the moving model are shown in Fig. 4. The control of an unknown plant by the use of a moving model exhibits satisfactory results. If the desired output is in a vigorous oscillation, the higher order moving model should be used.

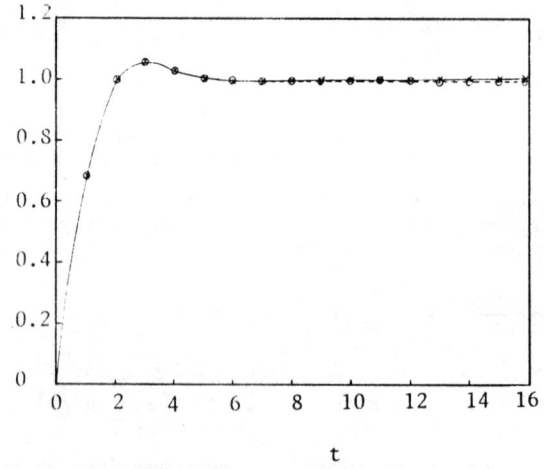

0: the desired output
x: the output via moving model control

Fig. 4. Comparison of the desired output and the output of the control system via a moving model.

CONCLUSION

A moving model of discrete-data systems is developed by the use of block pulse functions and the least-squares recursive formulas. It gives a good representation of any system.

The on-line controller design of a direct digital control system via a moving model is easily achieved. Extension to multi-variable systems is straightforward.

REFERENCES

Astrom, K.J. and B. Wittenmark (1973). On Self-tuning Regulators. Automatica, 9, 185-199.

Chen, C.F., Y.T. Tsay and T.T. Wu (1977). Walsh Operational Matrices for Fractional Calculus and Their Application to Distributed Systems. J. Franklin Inst., 303, 267-284.

Clarke, D.W. and P.J. Gawthrop (1975). Self-tuning Controller. Proc. IEE., 122, 929-934.

Harmuth, H.F. (1969). Application of Walsh Functions in Communications. IEEE. Spectrum, 6, 82-91.

Koivo, H.N. (1980). A Multivariable Self-tuning Controller. Automatica, 16, 351-366.

Kurz, H., R. Isermann and R. Schumann (1980). Experimental Comparison and Application of Various Parameter-adaptive Control Algorithms. Automatica, 16, 117-133.

Rao, B.R. and S. Ganapathy (1979). Linear Time-varying Systems-state Transition Matrix. Proc. IEE., 126, 1331-1335.

Shieh, L.S., R.E. Yates and J.M. Navarro (1979). Solving Inverse Laplace Transform, Linear and Nonlinear State Equations Using Block Pulse Functions. Computers & Elect. Engng., 6, 3-17.

Shih, Y.P., W.T. Wu and H.C. Chow (1975). Moments of Discrete Systems and Application in Model Reduction. The Chem. Engng. J., 10, 107-112.

Wu, W.T. and Y.C. Fang (1981b). Formulation of Conventional Least-squares Method for Parameter Estimation and Model Reduction of Sampled-data System via Block Pulse Functions. J.Ch.I.Ch.E., 12, 21-25.

Wu, W.T. and Y.S. Wong (1980). Recursive Parameter Estimation of Time-lag Systems via Block Pulse Function. Computers and Chem. Engng., 4, 201-203.

Wu, W.T. and Y.S. Wong (1981a). Suboptimal Feedback Control with Inaccessible State Variables via Moving Model. IFAC Kyoto Japan, 15.5 III-143.

Wu, W.T. and W.C. Liu (1982). Suboptimal Control of Nonlinear Dynamic Systems via Moving Model. Computers and Chem. Engng. In press.

AN APPROACH TO THE DESIGN OF REAL-TIME DATABASE MODELS

G. Rodriguez

Instituto de Investigaciones Electricas, Cuernavaca, Morelos, Mexico

Abstract. The implementation of a real-time database requires the design of a database conceptual model. A new approach, based on the concepts of Entities, Links, Keys and attributes (ELKA) to the design of models is presented. A small subset of the conceptual model for the National Power System Control Center in Mexico is presented as an example of a design. The concept of metamodel is introduced, which allows to automaticaly maintain documentation of a conceptual model in a computer. The use of the ELKA approach to the design of conceptual model results in a very clear statement of database requirements.

Keywords. Modeling; models; computer applications; power system control; large-scale systems; database requirements.

1. INTRODUCTION

As of March, 1981, 154 Power System Control Centers were in service or under development in the world (DyLiacco, 1981). All of these PSCC's have database needs. According to Masiello (1981) the major database categories that a modern PSCC may include are: Scan Control data, Display and Man-Machine Interface data, Real Time Applications Programs data, Study Applications Programs data, on-line Engineering data, Historical Information, Program Storage and Utility File Storage.

We currently are developing the database for the following Real-Time Applications programs: Load Forcast, Economic Dispatch, Hydro-Thermal Coordination and Interchange Scheduling that the Institute is developing for the National Energy Control Center in Mexico (Centro Nacional de Control de la Energia: CENACE).

The structured techniques for the design of a database (Atre, 1980 y Date 1977) include the Conceptual Model Design, the Logical Model Design and the Physical Model Design.

A new modeling technique (ELKA) for the design of Conceptual database models and results of the technique as applied to the CENACE database design are presented here. In Rodriguez, 1981 it was shown that the ELKA modeling technique has several theoretical advantages over the other modeling techniques published in Chen 1976, Bachman 1969, Codd 1979, and Wiederhold 1979, however, the ELKA modeling technique can be successfully used in practice as it is shown here.

In section 2 it is presented the ELKA modeling technique and its associated graphical symbolism. In section 3 an approach to build database conceptual models using the ELKA modeling technique is described.

Section 4 presents the metamodel, which is a contribution to the field of database conceptual model design since the metamodel is independent of any implementation and it is a documentation aid which can be easily automated.

2. THE ELKA MODEL AND ITS GRAPHICAL SYMBOLISM

In this section, we present informally the ELKA (Entity-Link-Key-Attribute) modeling technique. This includes a graphical language, which is useful to communicate the contents of a conceptual design of a Real-Time database to different people that will eventually interact with the actual database.

The principal descriptors utilized in the ELKA modeling technique are Entities, Links, Keys, and Attributes. With the exception of links, these descriptors have the same meaning as that ascribed to them in most of the existing literature on data models (Atre, 1980).

2.1 Entities, Attributes, and Keys. An entity is a person, place, thing, event, or concept about which information is recorded. We extend this description, defining entities as objects which are described by properties whose values can be considered to remain fixed over some interval of time. We call these properties <u>attributes</u>.

We represent an entity as a set of attributes. An attribute itself is represented as a pair composed of an attribute name and an attribute value. For example, the following set represents an entity with four associated attributes: { DAM NAME: Malpaso, MAXIMUM LEVEL: 176, MINIMUM LEVEL: 144, ACTUAL LEVEL: 155,}.

We say that the entity displays its attributes. Abbreviations are used for the attribute names, for example: DAMNAM = DAM NAME, MAXLEV = MAXIMUM LEVEL, MINLEV = MINIMUM LEVEL, ACTLEV = ACTUAL LEVEL.

Now, what is an entity class? An <u>entity class</u> is a set of entities which the user and /or the modeler have decided to group together because they all possess the same kind of properties of interest. An entity class has a unique name. This is chosen to be the common name of a tipical entity in the class.

For example, the following set is an entity class and contains two entities: {{ DAMNAM: chicoasen, MAXLEV: 392, MINLEV: 380, ACTLEV: 385}, { DAMNAM: Malpaso, MAXLEV: 176, MINLEV: 144, ACTLEV: 155}}. The name of the entity class is DAM which is the common name of each entity member of the class. Hence, when we talk about the entity class DAM, we are talking about a set of dams.

Entities are also characterized by <u>keys</u> (an entity has a key). A key of an entity belonging to an entity class is a set of non-redundant attributes which permits us to identify uniquely the entity from the other entities in that entity class.

For example, the attribute DAMNAM: Malpaso uniquely identifies the dam of Malpaso, and the attribute DAMNAM: Chicoasen uniquely identifies the dam of Chicoasen. The attribute MAXLEV: 380 may be unique when the entity class DAM has only two entities, but it may be possible that other dam has the same maximum level. Hence, MAXLEV is not a key. Note that the key in this example, is composed of one attribute, however, there are cases in which the key need two or more attributes to uniquely identify the entities in the entity class.

In the graphical symbolism of the ELKA model, we represent an entity class as a table enclosed in a box. The entity class name appears in a frame at the left lower corner of the box, the attribute names are factored as column names to the heading of the table, and the attribute values of each entity form the rows of the table. Fig. 2.1 is the table representation of our example entity class DAM.

```
DAMNAM,   MAXLEV, MINLEV, ACTLEV
Chicoasen   392     380    385
Malpaso     176     144    155
DAM
```

Fig. 2.1 Table representation,
the entity class DAM.

If attribute values are removed and only the entity class name and the attribute names are kept we have an entity class scheme. There are two alternative ways to indicate in an entity class scheme which attributes are keys: a) the attributes which form a key are enclosed in parentheses and preceded by a key name; b) the attributes which form a key are underlined. This is used when the entities have only one key. Figures 2.2(a) and 2.2(b) show the two alternatives to represent the scheme of the entity class DAM.

```
K:(DAMNAM), MAXLEV, MINLEV, ACTLEV
DAM
```
(a) Entity class scheme showing the key name and the key attributes enclosed in parenthesis.

```
DAMNAM, MAXLEV, MINLEV, ACTLEV
DAM
```
(b) Entity class scheme showing the key attribute underlined.

Fig. 2.2 Alternative representation
of the entity class scheme for DAM.

2.2 <u>Links</u>. The ELKA modeling technique utilizes a link to establish a very special relationship between two entities. A link is a reference to another entity made by an entity using a key of the refered-to entity. For example, the following two sets represent two entities:

```
valley = VALNAM:Grijalva, AREA:Oriental         (a link)
dam    = DAMNAM:Chicoasen, MAXLEV:392, MINLEV:380, VALNAM:Grijalva
where :  VALNAM=VALLEY NAME,  AREA=ELECTRICAL POWER AREA
```

The dam with key attribute DAMNAM: Chicoasen is member of the entity class DAM and is referencing another entity, the valley with key attribute VALNAM: Grijalva which is a member of the entity class VALLEY. The link causes migration, form the valley entity to the dam entity, of the attribute VALNAM: Grijalva which is a key of the valley entity. The attribute VALNAM: Grijalva is inherited by the dam entity. Also, the attribute VALNAM: Grijalva is the linking attribute.

A <u>link class</u> is a set of links which are references from all entities in one entity class to entities in another entity class. The ELKA modeling technique allows three types of link classes which may be described by functions (or mappings) from an entity class A to an entity class B with the following characteristics:

(1) <u>1-to-1 links</u>: (graphically depicted as A→B). For every a member of A there exists exactly one b member of B. For every b member of B there exists zero or one a member of A.

(2) <u>weak-m-to-1 links</u>: (graphically depicted as A◇—B). For every a member of A there exists exactly one b member of B. For every b member of B there exists zero, one or more a member of A.

(3) <u>strong-m-to-1 links</u>: (graphically depicted as A◆——B). For every a member of A theres exists exactly one b member of B. For every b member of B there exists one or more a member of A.

Every link class establishes two relationships: one from A to B and the other form B to A. We will indicate the name of the link class from A to B enclosed in parenthesis and the name of the link class from B to A enclosed in hyphens. Note however, that 1-to-1 link classes need only the name from A to B; we see an example below.

Given a link class from entity class A to entity class B, the A side of a link as the "back" and to the B side is the "front" of the link. In the case of m-to-1 link classes (weak or strong) the front is the "1" side and the back is the "m" side.

The graphical symbolism for links is as follows: a line is used to represent a link class. A 1-to-1 link class uses an arrowhead, and m-to-1 link classes use a small diamond in the back of the link to indicate the "m" side of the m-to-1 relationship. White diamonds are used for weak-m-to-1 links and black diamonds are used for strong-m-to-1 links.

The symbol for a link class is a graphical aid to help us to remember that the key attributes of one entity class have been inherited by another entity class, that is, a relationship has been established between the two entity classes.

Figure 2.3 illustrates an example of a strong-m-to-1 link class: for every entity (a dam) member of the entity class DAM there exists exactly one entity (a valley) member of the entity class VALLEY, and for every valley there exists one or more dams. We say that a valley contains many dams (at least one), and that a dam is in a valley.

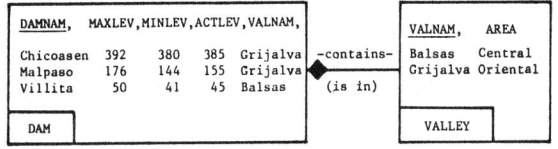

Fig. 2.3 An example of an strong-m-to-1 link class.

Figure 2.4 shows an example of a weak-m-to-1 link class: for every entity (a generator) member of the entity class GENERATOR there exists exactly one entity (a bus = electrical node) member of the entity class BUS, and for every bus there exists cero or more generators. A generator is connected to a bus and a bus has connected zero or more generators.

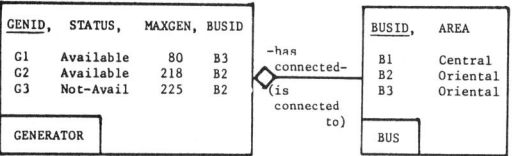

Fig. 2.4 An example of a weak-m-to-1 link class.

Fig. 2.5 shows an example of a 1-to-1 link class: for every entity (a hydro-generator) member of the entity class HYDRO-GENERATOR, there exists exactly one entity (a generator) member of the entity class GENERATOR, and for every generator there exists cero or one hydro-generator. A hydro-generator is a generator and certain generators are hydro-generators.

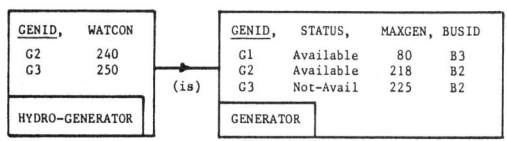

Fig. 2.5 An example of a 1-to-1 link class.

An interpretation of Fig. 2.5 is that the HYDRO-GENERATOR entity class identifies a subset of the entities in the GENERATOR class. All generators have a GENERATOR IDENTIFIER (GENID), a STATUS and a MAXIMUM GENERATION (MAXGEN), however, only the hydro-generator entities have WATER CONSUMPTION (WATCON) which is an attribute that does not apply to all members of the entity class GENERATOR.

There are no strong-1-to-1 links: if two entity classes were to be linked by a strong-1-to-1 link class, then the two entity classes can be mergerd into one.

It is part of the designing process of conceptual database models to discover the total number of attributes that an entity must include in a database. Other criteria to construct conceptual models are the topics of the next section.

3. THE ELKA APPROACH TO THE DESIGN OF CONCEPTUAL DATABASE MODELS

A conceptual database model is a set of entity class schemes interconnected through link classes.

The ELKA approach to the design of conceptual database models has two main objectives: a) to find all the entity class schemes of interest and, b) to find all the relations among the entity classes.

3.1 <u>The Entity Classes in a Conceptual Model</u>. We already saw in section 2, what is an entity class scheme. Finding the schemes of interest for a particular conceptual model is an iterative process: it is necessary that the designer interacts with the end user of the database and make a list of the attributes of interest, then, the designer groups the attributes in entity class schemes and iterates with the user as many times as necessary until both, the designer and the user are satisfied with the schemes.

3.2 The Relationships among the Entity Classes.
There are basically three types of relationships (or relations) between two entity classes: (a) one-to-one relationships, (b) one-to-many relationships, and (c) many-to-many relationships.

3.2.1 One-to-one Relationships.
To represent a one-to-one relationship between two entity classes, we use a 1-to-1 link. It is easy to discover one-to-one relationships if you remember basically two situations: (1) the subset situation and (2) the employee-spouse situation.

The subset situation is illustrated in Fig. 2.5 in which the entity class GENERATOR includes entities with a number of attributes, but certain of these attributes do not apply to all entities. Consequently, a new entity class is necessary, the entity class HYDRO-GENERATOR, to identify the subset of generators which are hydro.

The employee-spouse situation results when the entities (employees) members of the entity class EMPLOYEE are related to the entities members of the entity class SPOUSE. Since we assume that an employee either has or has not a spouse, the entity class SPOUSE must have less entities than the entity class EMPLOYEE.

In the ELKA modeling technique we represent the one-to-one relationship with a 1-to-1 link class: the entity class having more entities migrates the attributes of one key and the entity class with less entities inherits those attributes. For example, in Fig. 2.5, we say that the entity class GENERATOR migrates its key attribute GENID to the HYDRO-GENERATOR entity class, and that the HYDRO-GENERATOR inherits the attribute GENID.

3.2.2 One-to-many Relationships.
In the ELKA modeling technique we use either a weak-m-to-1 or a strong-m-to-1 link class to represent a one-to-many relationship. As for the case of a 1-to-1 link class, an m-to-1 link class (weak or strong) causes migration of attributes. The entity class at the front of the link migrates the attributes of one key and the entity class at the back of the link inherits those attributes. Two examples of one-to-many relationships are given in figures 2.3 and 2.4.

3.2.3 Many-to-many Relationships.
An example of a many-to-many relation between two entity classes follows: we have two entity classes: ISLAND whose members are electrical islands and BUS whose members are electrical buses. An island is a set of connected buses.

Fig. 3.1 shows two configurations of islands that may be formed with four buses depending on how these are connected.

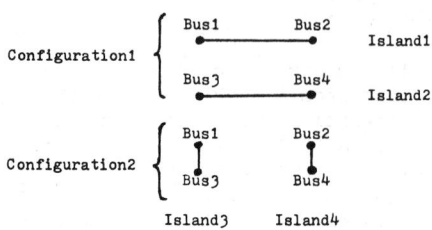

Fig. 3.1 Two configurations of islands.

The entity classes ISLAND and BUS are shown in Fig. 3.2. Only the key attribute is shown in both cases.

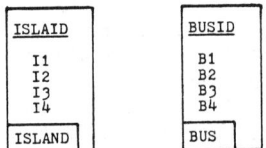

Fig. 3.2 The entity classes ISLAND and BUS.

If we construct a table to indicate which buses are in which islands, we get the table shown in Fig. 3.3.

```
ISLAND    BUSID
  I1       B1
  I1       B2
  I2       B3
  I2       B4
  I3       B1
  I3       B3
  I4       B2
  I4       B4
```

Fig. 3.3 Table representation of how buses associate with islands.

Note that a row of this table represents neither an island nor a bus, therefore, the rows of the table of Fig. 3.3 must form a new entity class whose members represent the associations between islands and buses, that is, the many-to-many relationship between ISLAND and BUS. The new entity class is ISLAND-BUS.

The entity class ISLAND-BUS has inherited the key attribute ISLAID from ISLAND and the key attribute BUSID from BUS, the relation between ISLAND and ISLAND-BUS is one-to-many and the relation between BUS and ISLAND-BUS is also one-to-many.

Figure 3.4 represents a many-to-many relation scheme between ISLAND and BUS: an island has many buses and a bus can be in many islands.

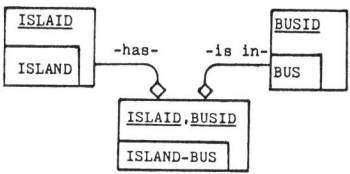

Fig. 3.4 A many-to-many relation scheme.

3.3 <u>An example of a design</u>. Figure 3.5 shows a subset of the Real-Time Conceptual Database model that the Institute is developing for CENACE.

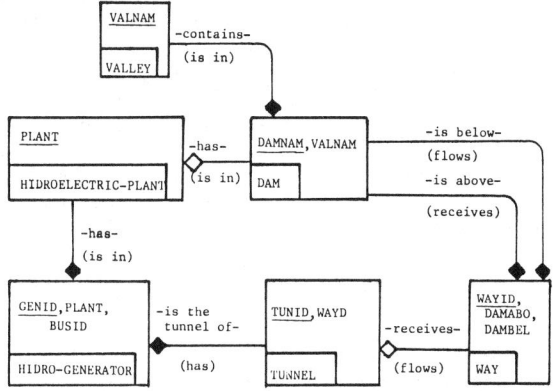

Fig. 3.5 A subset of the real-time database conceptual model for CENACE.

We present in the following paragraphs how to "read" (using the ELKA graphical symbolism conventions) the conceptual database model of Fig. 3.5. The figure shows only the attributes that are either Key attributes or linking attributes. A conceptual model does not show any attribute values, only attribute names.

We see form Fig. 3.5 that a valley-contains-one or more dams and that a dam (is in) only one valley. From the two relationships (links) between DAM and WAY, a dam - is below - one or more water ways (rivers or streams) and a water way (flows) into one dam, also, a dam - is above - one or more water ways and a water way (receives) water from one dam.

The relationship between DAM and HYDROELECTRIC PLANT is a follows: a dam - has - installed cero or more hydroelectric plants and a hydroelectric plant (is in) one dam. The entity classes HYDROELECTRIC PLANT and HYDROGENERATOR are related in the following fashion: a hydroelectric plant-has-one or more hydrogenerators and a hydrogenerator (is in) one hydroelectric plant.
TUNNEL and HYDROGENERATOR relate to each other as follows: a tail race tunnel - is the tunnel of - one or more hydrogenerators and a hydrogenerator (has) one tail race tunnel.

Finally, the relationship between the entity classes TUNNEL and WAY is: a waterway - receives - water from cero or more tail race tunnels and a tailrace tunnel (flows) into one water way.

All the entity class schemes shown in Fig. 3.5 have other attributes in addition to the ones shown, however, they are of no interest for this presentation. We have, on the average, 20 attributes per entity class in the conceptual model for CENACE and a total of 100 entity classes.

4. THE METAMODEL

The metamodel presented here is a conceptual model for a database design, in the former we store information about other databases. The idea of a metamodel is very similar to the idea of a Data Dictionary as is presented in Atre (1980). However, since the metamodel is a conceptual model, it can be implemented easily using a conventional Database management System or a system with a query language and report generator.

Figure 4.1 shows the metamodel. The entity classes: ENTITY, LINK, KEY, ATTRIBUTE, KEY-ATTRIBUTE, NON-INHERITED-ATTRIBUTED and IN-HERITE-ATTRIBUTE are necessary to represent all the entities in a conceptual model and their relationships. The entity classes: PROGRAM, MODULE and USE are used to represent how the attributes are used by the different modules that implement the algorithms of the real-time applications programs.

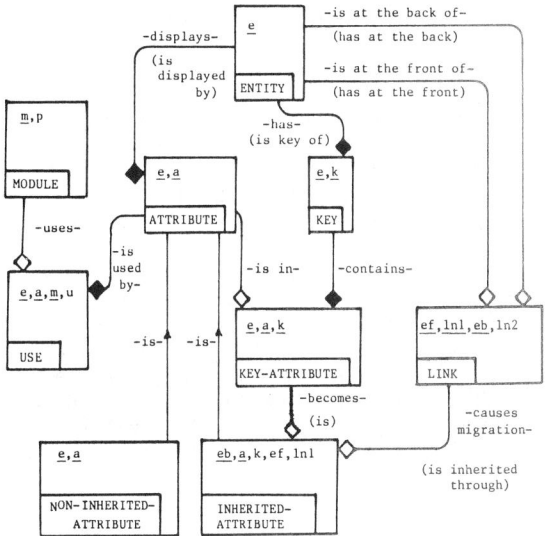

Fig. 4.1 The Metamodel.

A brief description of the metamodel follows: ENTITY represents a list of all the entity classes in a conceptual model. ENTITY and ATTRIBUTE are related as follows: an entity class - displays - one or more attributes and an attribute is (displayed by) only one entity class. The key of the entity class ATTRIBUTE is composed of two attributes: "e" and "a", this is due to the fact that an attribute name "a" can be used in different entity classes, possible with a different meaning. The entity class names (e's) must be unique and the attribute names must be unique within a given entity class.

An entity class - has - one or more keys and a key (is a key of) one entity class. The attribute values for the attribute "k" are integer numbers, for example, 1, 2, 3, etc. meaning that the entity class has 1, 2, 3, etc. keys (see the link between ENTITY and KEY).

LINK and ENTITY are related in two ways: an entity class "e" - is at the front of - cero or more links and a link (has at the front) only one entity class. On the other hand, an entity class - is at the back of - cero or more links and a link (has at the back) only one entity class.

The entity class LINK has a key composed of three attributes: the name of the entity class at the front (ef), the name of the link within hyphens (ln1) and the name of the entity class at the back of the link (eb). This is necessary since we allow the name of a link within hyphens to be used for many links (see for example the link name - is - in the metamodel itself).

The relationship between ATTRIBUTES's and KEY's is many-to-many: an attribute (e,a) - is - a key attribute in cero or more keys within the same entity class, and a key - contains - one or more key attributes. Note, the attribute "e" in the entity class KEY-ATTRIBUTE is inherited from two entity classes: ATTRIBUTE and KEY, this is because a key and its attributes belong to the same entity class.

The entity class INHERITED-ATTRIBUTE is the set of all inherited attributes in a conceptual model. The relationship between ATTRIBUTE and INHERITED-ATTRIBUTE identifies a subset: the set of pairs (eb, a) in the class INHERITED-ATTRIBUTE is a subset of the set of pairs (e,a) in the class ATTRIBUTE. Now, a key attribute "ef, a, k" in the entity class "ef" - becomes - an inherited attribute "eb, a" in the entity class "eb" and the link "ef, ln1, eb" - causes the migration - of the attribute. The relationship between LINK and INHERITED-ATTRIBUTE is strong-m-to-1 since a link-causes the migration - of one or more key attributes. On the other hand, a key attribute - becomes - inherited attribute in cero or more entity classes, consequently the relationship between KEY-ATTRIBUTE and INHERITED-ATTRIBUTE is weak-m-to-1.

The entity class NON-INHERITED-ATTRIBUTE identifies a subset of the entity class ATTRIBUTE: the set of pairs "e, a" from NON-INHERITED-ATTRIBUTE is a subset of the set of pairs "e, a" in ATTRIBUTE. An attribute is either inherited or non-inherited, consequently, the union of the two disjoint sets is the set of pairs "e, a" in ATTRIBUTE: (1) the set of pairs "eb, a" in INHERITED-ATTRIBUTE and (2) the set of pairs "e, a" in NON-INHERITED-ATTRIBUTE.

Last, we notice the many-to-many relationship between ATTRIBUTE and MODULE: an attribute - is used by - one or more modules and a module - uses - cero or more attributes. The relationships - uses - and - is used by - represent any of the relationships: reads, read by, writes, written by, modifies or modified by.

5. CONCLUSIONES

The approach presented to design conceptual models for real-time databases has been used successfully for the CENACE project. The metamodel has been implemented using a commercial query language and report generator to keep the documentation of the design automated. The use of the ELKA approach results in a uniform method to design and document database requirements. We strongly believe that the approach should be of interest to other designers of real-time databases.

An algorithm is being implemented at the Institute to find redundant attributes in a model as suggested in Rodriguez 1981.

ACKNOWLEDGMENT

I wish to thank J. Frausto of the Instituto de Investigaciones Eléctricas who has been using the ELKA modeling technique and allowed me to use a subset of the real-time database conceptual model for CENACE shown in Fig. 3.5.

REFERENCES

Atres. "Data Base: Structured Techniques for Design, Performance, and Management", John Wiley and Sons, 1980.

Bachman, C.W. "Data Structure Diagrams". Data Base 1, Vol. 2, Summer 1969.

Chen, P.P. "The Entity-Relationship Model - Toward a Unified View of Data". ACM Trans. on Database Systems, Vol. 1, No. 1, March 1976, pp. 9 - 36.

Codd, E.F. "Extending the Database Relational Model to Capture more Meaning" ACM Transactions on Database Systems. Vol. 4, No. 4, December 1979. pp. 397 - 434.

Course Notes "Modern Power System Control Centers" University of California, Berkely, May 1981.

Date, C.J., "An Introduction to Database Systems" Second Edition, Addison-Wesley Publishing Company. 1977.

DyLiacco T.E., "System Control Centers for Generation-Transmission Systems", March, 81, (see Course Notes).

Marsiello R. "Modern Power System Control Centers" May 81, (see Course Notes).

Rodríguez Ortiz, G. "The ELKA Model Approach to the Design of Database Conceptual

Models" Ph.D. Dissertation, <u>University of California</u>, Los Angeles 1981.

Wiederhold, G. and R. EL-Masri "A structural Model for Database Systems" <u>Stanford University</u>, Computer Science Department, STAN-CS-79-722, Feb. 1979.

DIGITAL IMAGE CODING BY C-MATRIX TRANSFORM[†]

R. Srinivasan and K. R. Rao

Electrical Engineering Department, The University of Texas at Arlington, P.O. Box 19016, Arlington, TX 76019, USA

Abstract: Digital image processing of NTSC Color TV images by C-Matrix transform (CMT) for bandwidth compression is investigated. The effectiveness of the CMT compared to the discrete cosine transform (DCT) for image coding is evaluated. This comparison is based on various quantitative criteria and implementation aspects. Transform coding is applied. Simulation involves quantizer-coder design and bit rate reduction.

INTRODUCTION

Transform image coding [3-7] for redundancy reduction has been attractive in view of the fast algorithms developed for efficient implementation of the discrete transforms [1,2] and also the developments in semiconductor technology. Of all the transforms, apart from the discrete Fourier transform, DCT [1,2,8-14] has been most popular in view of its near optimal performance. DCT hardware for processing audio and video has become a commercial reality. An integer version of the DCT called C-Matrix transform has recently been developed [11,15,16]. As only integer arithmetic is involved CMT is much simpler to realize and its performance is very close to the DCT. The objective of this paper is to investigate the effectiveness of CMT in image coding leading to bit rate reduction and compare with the DCT.

C-Matrix Transform

DCT can be implemented via the Walsh-Hadamard transform (WHT) through a conversion matrix which has a block diagonal structure. By integerizing the elements of this conversion matrix CMT is developed. Properties of the CMT are described elsewhere [15,16].

Data Base

The source data for the proposed system is an NTSC color TV signal sampled at $4 f_{sc}$ where f_{sc} = 3.58 MHz in the color subcarrier frequency. The sampling phase was aligned with the I signal. A window of 256 x 212 array was recorded for each frame. It means only about 15% of the visible portion of the video signal was used for similation. The following five frames were obtained from Bell Northern Research Ltd., (BNRL) Montreal, Canada (Fig. 1). Pictures A) Letters, B) special effects C) Footfall D) Beauty E) Girl.

Composite ↔ Component Transformations

In accordance with recommended sampling rate [26,]7] a $4 f_{sc}$ sampling rate has been chosen. It also eases digital processing in the demodulation process of the composite signal into the luminance (Y) and chrominance (I and Q) components using digital comb and band pass filters. As the bandwidths of I and Q are 1.5 MHz and 0.5 MHz they are subsampled to 3.58 MHz and 1.79 MHz and further subsampled in the vertical direction. Hence only alternate lines of I and Q are transmitted. At the receiver reverse processes are carried out using digital interpolators [28] to reconstruct the composite signal in a process of modulation. Details of color separation and reconstruction, decimation and interpolation of I and Q are described in [17-23]. The demodulation and modulation scheme is shown in Fig. 1.

Transform Coding

To compact most of the energy into fewer transform coefficients, CMT is applied to the Y,I,Q components in blocks of sizes 8x8, 2x4 and 1x4 respectively (Fig. 2). Hence a 2 D-CMT for Y and I and 1 D-CMT for Q is implemented. The statistics of the CMT coefficients can be developed assuming that the subpictures (Fig. 2) of all the test pictures form the ensemble. The variance distribution for Y,I and Q in the transform domain is shown in Table 1. Based on the derived statistics, bits are allocated for the Y,I and Q transform coefficients. A method suggested by Ready and Wintz [24] is based on the variance i.e.

$$b(i) = \frac{B}{N} + 2 \log_{10} \sigma^2(i) - \frac{2}{N} \sum_{j=1}^{N} \log_{10} \sigma^2(j) \quad (1)$$

where b(i) = no. of bits assigned to the ith transform coefficient, $\sigma^2(i)$ is the corresponding variance and B is the total no. of

---

[†] This paper is based on the research by Mr. R. Srinivasan as a partial requirement towards the M.S. degree from the University of Texas at Arlington, Arlington, Texas.

bits assigned to the block of N coefficients. b(i) is adjusted and integerized until

$$\sum_{i=1}^{N} b(i) = B.$$

Table 2 shows bit assignment for an average of 3 bits per pel (BPP). The coefficients are then quantized by Max quantizer that minimizes the mean square error (MSE). The performance of the CMT is compared with the DCT [23] for all the test pictures (Table 3). For all other average BPP the performance is similar. This comparison is also shown in Fig. 4. In general at all average BPP based on various quantitative criteria CMT performs very close to the DCT.

Besides having good bandwidth compression features, transform image coding also provides more tolerance to channel errors [25]. This is due to its averaging property. The probable reason for improved noise tolerance when transform coding is used, is that the message is transmitted in the frequency domain rather than in the conventional space domain. In transform coding the noise components are evenly distributed over the entire reconstructed image. Hence error propagation effects are not sutdied for the CMT system.

CONCLUSIONS

A comparison of the transform image coding using CMT and DCT as the basis vectors has shown that the CMT perform almost as good as the DCT. The integer arithmetic involved in the former lends itself more potential for implementing real time systems like broadcast TV. It is hopeful that the CMT can be a posible substitute for the DCT in signal and image processing.

REFERENCES

1. N. Ahmed and K. R. Rao, "Orthogonal transforms for digital signal processing," New York: Springer, 1975.
2. D. P. Elliott and K. R. Rao, "Fast transforms, algorithms, analyses and applications," Academic Press (to be published).
3. P. A. Wintz, "Transform picture coding," Proc. of IEEE, Vol. 60, pp. 809-820, Feb. 1970.
4. R. C. Gonzalez and P. Wintz, "Digital image processing," Addison-Wesley, 1977.
5. A. N. Netravalli and J. O. Limb, "Picture Coding: A Review," Proc. IEEE, Vol. 68, pp. 366-406, Nov. 1980.
6. A. K. Jain, "A fast Karhunen-Loeve transform for a class of stochastic processes," IEEE Trans. Commun., Vol. COM-24, pp. 1023-1029, Sept. 1976.
7. W. K. Pratt, "Digital Image Processing," New York/Chicester, John Wiley, 1978.
9. W. Chen, C. H. Smith and S. Fralick, "A fast computational algorithm for the discrete cosine transform," IEEE Trans. Commun., pp. 1004-1009, Sept. 1977.
10. J. Roese, W. K. Pratt, G. S. Robinson, "Interframe cosine transform image coding," IEEE Trans. Commun., Vol. COM-25, pp. 1329-1339, Nov. 1977.
11. H. W. Jones, D. N. Hein and S. C. Knauer, "The Karhunen-Loeve, discrete cosine and related transforms obtained via the Hadamard transform," Intl. Telemetering Conf., Nov. 1978.
12. D. Hein and N. Ahmed, "On a real-time Walsh-Hadamard cosine transform image processor," IEEE Trans. Electromag. Compat., Vol. EMC-20, pp. 353-457, Aug. 1978.
13. F. A. Kamangar and K. R. Rao, "Interfield hybrid coding of component color television signals," IEEE Trans.Commun., Vol. COM-29, pp. 1740-1753, Dec. 1981.
14. A. Jalali and K. R. Rao, "A high speed FDCT processor for real time processing of NTSC color TV signal," IEEE Trans. Electromag. Compat., Vol. EMC-24, pp. 278-286, June 1982.
15. R. Srinivasan and K. R. Rao, "An approximation to the discrete cosine transform for N=16," to be published in Signal Processing, Jan. 1983.
16. H. S. Kwak, R. Srinivasan and K. R. Rao, "C-Matrix transform," to be presented at MEXICON '82.
17. J. P. Rossi, "Color decoding a PCM NTSC television signal," SMPTE J., Vol. 83, pp. 489-495, June 1974.
18. A. G. Deczky, "Color demodulation of an NTSC television signal using digital filtering technique," ICC'75, San Francisco, CA, July 1975.
19. J. P. Rossi, "Sub-Nyquist sampled PCM NTSC color TV signal derived from four times the color subcarrier sampled signal," Intl. Broadcasting Convention, Sept. 1978.
20. N. F. Maxemchuk and D. K. Sharma, "Separation of NTSC signals by minimum mean square error filters and adaptive filters," IEEE Trans. Commun., Vol. COM-26, pp. 583-593, May 1978.
21. D. J. Goodman and M. J. Carey, "Nine digital filters for decimation and interpolation," IEEE Trans. Acoust. Speech & Signal Processing, Vol. ASSP-25, pp. 121-126, April 1977.
22. J. Y. Ouellet and E. Dubois, "Sampling and reconstruction of NTSC video signals at twice the color subcarrier frequency," IEEE Trans. Commun., Vol. COM-29, pp. 1823-1831, Dec. 1981.
23. C. E. Li, "Component coding of the NTSC color TV signal," M.S. Thesis, The University of Texas at Arlington, Arlington, Texas, 1981.
24. P. J. Ready and P. A. Wintz, "Multispectral data compression through transform coding and block quantization," Purdue University, Laboratory for Application of Remote Sensing, Information Note 050572, May 1972.
25. R. H. Stafford, "Digital television," Wiley-Interscience, 1980.
26. R. R. Green, "A world standard for digital television studios," ICC '82, Philadelphia, PA, Proc. pp. 1A.2.1-1A.2.5, 1982.

27. R. S. Hopkins, "Progress report on digital video standards," Twelfth annual SMPTE television conf., Atlanta, GA.

28. R. W. Schafer and L. R. Rabiner, "A Digital signal processing approach to interpolation," Proc. IEEE, Vol. 61, pp. 692-702, June 1973.

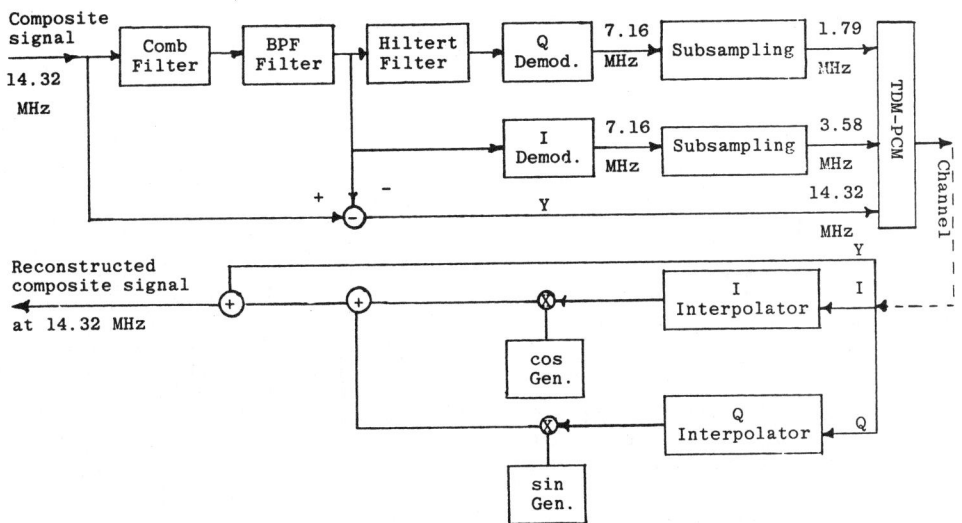

Fig. 1. Block diagram of digital color separation and reconstruction [23]

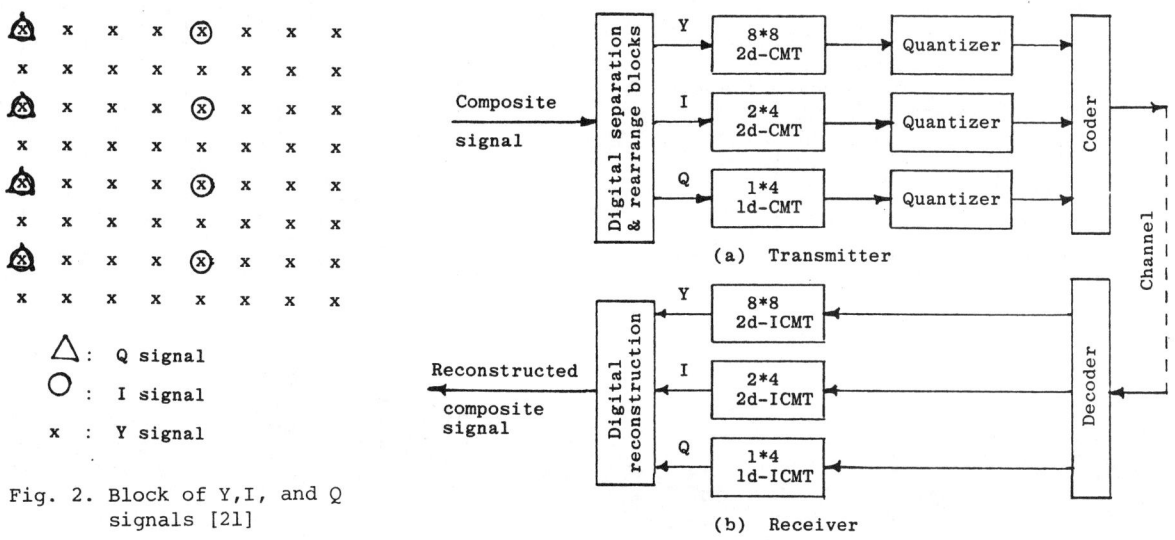

Fig. 2. Block of Y, I, and Q signals [21]

△ : Q signal
○ : I signal
x : Y signal

Fig. 3. Block schematic of the normal CMT coding system

Table 1. Variances of the CMT Coefficients

(A) Y Signal

5897.129	187.871	69.626	22.668	15.222	10.713	0.456	0.662
353.014	57.890	20.560	8.458	5.342	2.676	0.133	0.180
182.728	42.562	12.434	4.543	4.048	1.741	0.085	0.119
137.220	22.178	5.929	2.464	3.871	0.904	0.056	0.065
119.580	14.603	4.046	1.792	2.296	0.596	0.042	0.048
57.413	10.271	3.151	1.429	1.841	0.516	0.041	0.047
45.556	5.656	1.630	0.490	0.772	0.206	0.026	0.028
54.311	6.606	2.168	0.954	0.657	0.384	0.081	0.056

(B) I SIGNAL

1028.649	23.170
76.631	6.226
29.697	3.343
12.925	2.456

(C) Q SIGNAL

213.293
22.407
10.395
5.537

(A) Y Signal

```
8 7 6 5 5 4 1 2
7 6 5 4 4 3 0 1
7 5 4 3 3 3 0 0
6 5 4 3 3 2 0 0
6 4 3 3 3 2 0 0
6 4 3 2 3 2 0 0
5 4 3 2 2 1 0 0
6 4 3 2 2 1 0 0
```

(B) I Signal

```
6 3
4 2
3 2
3 1
```

(C) Q Signal

```
5
3
2
2
```

Table 2. Bit assignment matrix for the CMT system with 3 bits per pel (average bit rate) assigned to the Y, I, and Q color components.

Table 3. Performance Comparison of the CMT and DCT Systems with Average 3.5 BPP Assigned to Y,I,Q Components (Data Rate 46.24 Mbit/s)

Test Pic.	COMPONENT SIGNAL (Y,I,Q)								COMPOSITE SIGNAL			
	Mean Square Error		Mean of the Abs. Error		Max of Abs. Error		Essen. Max. 95%		NMSE (#10^{-3})		SNR (dB)	
	DCT	CMT	DCT	CMT	DCT	CMT	DCT	CMT	DCT	CMT	DCT	CMT
A	49.883	45.112	4.026	3.879	50	50	17	16	5.45	5.79	29.33	29.07
	6.644	8.716	1.523	1.812	16	16	6	7				
	12.749	11.187	2.064	1.935	17	18	9	8				
B	1.825	1.626	0.780	0.774	23	21	3	2	0.47	0.56	37.54	36.74
	2.934	4.127	0.992	1.170	10	13	4	5				
	1.667	1.524	0.875	0.834	7	8	2	2				
C	1.248	1.219	0.545	0.577	26	22	1	1	0.10	0.12	45.22	44.22
	0.592	0.727	0.519	0.609	3	4	1	2				
	0.754	0.652	0.638	0.581	2	2	2	1				
D	0.497	0.523	0.447	0.469	13	11	1	1	0.14	0.16	47.47	46.93
	0.564	0.683	0.506	0.584	2	3	1	1				
	0.807	0.757	0.678	0.623	2	2	2	2				
E	7.480	7.177	1.078	1.038	39	46	4	4	1.22	1.34	36.11	35.74
	7.208	8.218	1.191	1.330	32	31	4	4				
	1.801	2.139	0.815	0.874	16	18	2	2				

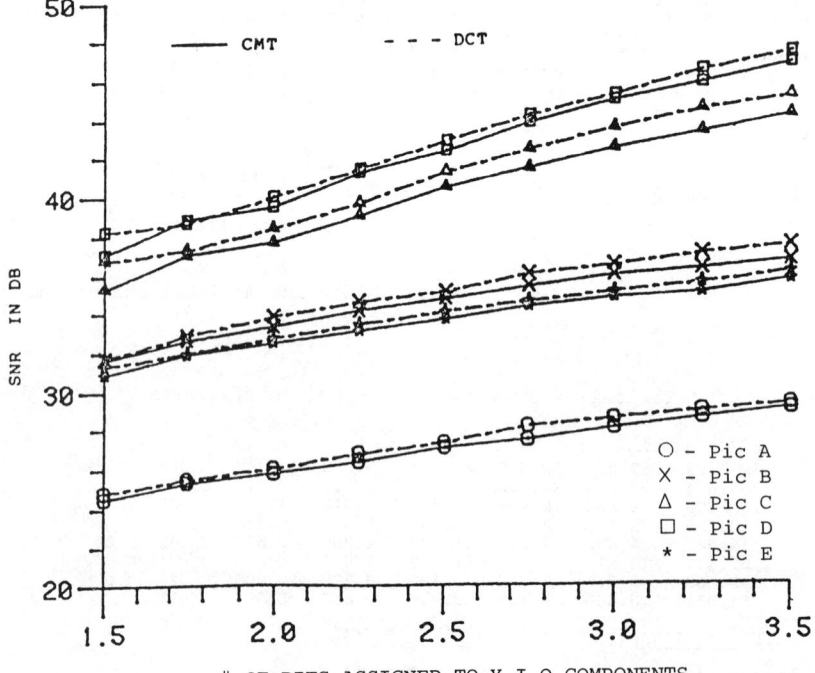

Fig. 4. SNR comparison of the DCT and CMT normal systems for the reconstructed composite signal

IDENTIFICATION AND CONTROL PROGRAMS FOR MICROPROCESSORS

J. Enríquez, A. Hormaza and A. C. Campos

Institute for Mathematics, Cybernetics and Computation (IMACC), Cuban Academy of Sciences, National Capitol, Havana, Cuba

Abstract. It is clear that the sophisticated possibilities for calculations of microprocessors are the natural technoligical support for the most developed algorithms of the modern Control Theory. A mini-library with algorithms for identification and control is proposed. With these programs the user is able to set an adaptive control configuration, according to the requirements of his plant. Some conventional algorithms are also included for standard applications. Another group of general purpose programs, as floating-point arithmetic, polynomial algebra, conversion subroutines, etc. are considered in the mini-library. Some of these programs were tested in real time digital control conditions with satisfactory results.

Keywords. Adaptive control; computer control; computer software; iterative methods; identification; PID control.

INTRODUCTION

The creation and considerable exploitation of the microprocessors has been one of the main achievements of the last decade. This technological development is a milestone in the history of Electronics. For many, it is as significant as the Industrial Revolution: the former enhanced the physical power of man, while micros are now multiplying his mental abilities.

The microprocessors have overcome the main disadvantages that the use of computers and mini-computers in the real control of industrial processes implied, i.e. the implementation costs, the volume, the unmodular design, the energy supply expenses and difficulties, etc.

Microprocessors have a very wide range of applications. Besides the military field, where technological innovations always find important applications, the micros are fundamental in space investigation, in computation and in the newly-developed domain of robotics. Other branches of human knowledge and way of life, such as education, medicine, building supervising, automobile operation, traffic control, etc. are related to them. The use of microprocessors has come to stay.

At the beginning, these devices were used as data loggers, but it is clear that their sophisticated possibilities for calculations are the natural material support for the most developed algorithms of the modern Control Theory and of the most complicated self-tuning and adaptive regulators. Full exploitation of a microcomputer in the control of plants means, not only collecting measures and controlling from the classical standpoints, but to use the best regulator and to give a satisfactory answer to many problems where old methods crumble.

Having the above in mind, the Control Department of IMACC has created a section with the aim of producing the necessary software for that kind of application. Programs were created for the INTEL's 8080 for two main reasons:
i) We have a complete system (IMACC's 8101) for general purposes which is based on this chip.
ii) It is a quite standard unit, and there are not great defficulties in translating the programs to another microprocessor machine code.

To produce such a software is not an automatical task. Even if the user has a FORTRAN version of the algorithm, its adaptation to the microcomputer will probably face some difficulties. The need to modify the versions is connected with the fact that we have to deal with shorter length computer-word (in comparison with conventional processors). The use of only two bytes for mantissa, sometimes implies the lost of certain properties of the algorithm because of the rounding errors.

We have worked in the creation of a mini-library with programs of the following type: identification, control, floating-point arithmetic with differents subroutines to make easier its exploitation, and general purposes procedures. So the user will have a minimum for a good and modern control (mainly for monovariable systems).

Some algorithms were already proved in FORTRAN. In order to avoid decimal representation errors, a FORTRAN subroutine that shows the results in mantissa-exponent format with hexadecimal notation (3 bytes) was created. This allowed us to make a first evaluation of how much we were loosing with one byte less.

We have used two different kinds of technical support in order to obtain the final versions of the programs: a special simulator of the 8080, made for the Cuban minicomputer CID-201-B, and the system IMACC 8101 we have refered previously.

At the end of this paper a brief reference is made of how some of these programs have been tested in practical conditions through real time digital control experiments in a Cuban sugar mill.

STOCK OF PROGRAMS: CURRENT STATE

Identification

Self-tuning or adaptive regulators need an on-line method in order to estimate the parameters of an established model of the plant. These methods are based on statistical considerations. They use a linear regression model:

$$y_t = \sum_{i=1}^{n} a_i y_{t-i} + \sum_{i=0}^{n} b_i u_{t-i} + e_t \qquad (1)$$

or the more general ARMA model

$$y_t = \sum_{i=1}^{n} a_i y_{t-i} + \sum_{i=0}^{n} b_i u_{t-i} + e_t + \sum_{i=1}^{n} c_i e_{t-i} \qquad (2)$$

In (2) the noise parameters are explicitly considered.

In our mini-library we have included three of these algorithms: Least Squares, Instrumental Variable and REFIL. All of them use the model (1).

The well-known difference between Least Squares and Instrumental Variable is that the first is unbiased under some condition concerning the noise, while the second is always unbiased. The user may choice, according to his experience and interest, which one of them fits better to the way he designs his regulator. The recurrent formulas for both are the following:

L.S.
$$\hat{P}_{t+1} = \hat{P}_t + \frac{1}{\sigma_{t+1}^2} g_{t+1} \hat{e}_{t+1}^T$$
$$\hat{e}_{t+1} = y_{t+1} - Z_{t+1}^T \hat{P}_t$$
$$\sigma_{t+1}^2 = 1 + Z_{t+1}^T g_{t+1} \qquad (3)$$
$$g_{t+1} = C_t Z_{t+1}$$
$$C_{t+1} = C_t - \frac{1}{\sigma_{t+1}^2} g_{t+1} g_{t+1}^T$$

For details, see Peterka (1975)

I.V.
$$\hat{P}_{t+1} = \hat{P}_t + \frac{1}{\sigma_{t+1}^2} g_{t+1} \hat{e}_{t+1}^T$$
$$\hat{e}_{t+1} = y_{t+1} - Z_{t+1}^T \hat{P}_t$$
$$\sigma_{t+1}^2 = 1 + Z_{t+1}^T g_{t+1} \qquad (4)$$
$$g_{t+1} = C_t \Theta_{t+1}$$
$$C_{t+1} = C_t - \frac{g_{t+1} Z_{t+1}^T C_t}{\sigma_{t+1}^2}$$

where Θ_{t+1} is the instrumental variable vector, $\hat{P}_t$ is the parameters estimation for the discrete time t, Z_{t+1} is a vector with the imputs/outputs of the plant and $\hat{e}_{t+1}$ is the prediction error. For a full exposition see Strejc (1977). In our versions the memory distribution is

	Subprograms	Main program	RAM memory
LS	137	1037	111 bytes
IV	137	1438	141 bytes

Although instrumental variable is a little expensive, this is a reasonable price for unbiased estimations. Both methods work quickly not expending one second in each iteration. The RAM memory increases when the order n of the plant is greater than our case (n=2) or when it is the case of a multivariable system (not included in our current versions).

According to the general experience, LS and IV may loose the convergence property when we use only two bytes for mantissa. This happens when a long running of the algorithm implies too many rounding errors. Therefore these methods are not the best for an adaptive regulator, though they may be used with a special oriented computer for automatical short running identification of previously known stationary plants.

An square root filter with double precision effect is included in our mini-library in order to overcome the above pointed difficulties. The REFIL algorithm, proposed by Peterka (1975), is particularly suited for self-adjusting or adaptive controls. Basically, it is a LS estimator —and therefore eventually biased— but instead of C_t, the upper triangular matrix G_t defined by

$$C_t = G_t G_t^T \qquad (5)$$

is propagated during the recurrent algorithm. Thus, C_t does not loose the property of being a definite positive matrix, as it happens in LS and IV methods.

The formulas for REFIL are explicitly showed below. They are a little different from the original version, in some normalizing conditions indicated also by the author:

$$\sigma_0^2 = \varphi^2 * \mathcal{X}$$
$$\mathcal{X} = 1 + \sigma_0^2$$
$$\Theta = \sqrt{\sigma_0^2 / \mathcal{X}}$$
$$\sigma_0 = \sqrt{\sigma_0^2}$$
$$j = 1, \ldots, \rho$$
$$f_j = \sum_{i=1}^{j} G_{(t)ij} Z_{(t+1)i}$$
$$a = \sigma_{j-1} / \Theta \qquad (6)$$

$$b = f_j/\sigma_{j-1}^2$$
$$\sigma_j^2 = \sigma_{j-1}^2 + f_j^2$$
$$c = a/\sigma_j$$
$$g_j = G(t)_{jj} f_j$$
$$G(t+1)_{ij} = c\, G(t)_{ij}$$
$$i = 1, 2, \ldots, j-1$$
$$d = G(t)_{ij}$$
$$G(t+1)_{ij} = c(d - b g_i)$$
$$g_i = d f_j + g_i$$

The memory distribution for REFIL in the case of monovariable systems is

	Sub-programs	Main program	RAM memory
REFIL	60	786	88 bytes

For the multivariable case we have created a second program with the following conditions:

	Sub-programs	Main program	RAM memory
REFMUV	90	752	260 bytes

REFIL has worked in practical conditions with satisfactory results, therefore it is particularly recommended for on-line identification. Nevertheless, it must be noticed that a constant linear dependence between the inputs and the outputs of the plant may cause the lack of convergence in the algorithm: this is a practical result.

Control

Though we believe that microprocessors are convinient for sophisticated control algorithms, nowadays the costs of the chips make classical control devices more expensive than microcomputers. Thinking about the possibility of a group of plants controlled by a single microcomputer, we have programmed the conventional PID algorithm. According to Smith (1972) we have prepared two versions: one of position

$$m_n = K_c \left[e_n + \frac{T}{T_I} \sum_{k=0}^{n} e_k + \frac{T_D}{T}(e_n - e_{n-1}) \right] + M_R \quad (7)$$

an another of velocity

$$\Delta m_n = K_c \left[(e_n - e_{n-1}) + \frac{T}{T_I} e_n + \frac{T_D}{T}(e_n - 2e_{n-1} + e_{n-2}) \right] \quad (8)$$

that the user can use freely.

The memory distribution is as follows:

	Sub-programs	Main program	RAM memory
Pos.	21	130	42 bytes
Vel.	21	133	48 bytes

The stock of control programs has a particular emphasis in minimum variance strategy. So, we have produced four programs upon this principle: MINVARMP, MINVAR, CONAD1, CONAD2.

MINVARMP and MINVAR are quite general regulators. In fact, the first is the minimum variace regulator for minimal phase (and therefore we can use this algorithm only when the system has been identified previously with this characteristic). The formula is

$$u(t) = -\frac{1}{b_o}[a_1 y(t-1) + b_2 u(t-1) + \ldots + a_n y(t-n) + b_n u(t-n)] \quad (9)$$
$$n \leq 10$$

The second program is rather complicated. It includes the first as a particular case, and uses three subroutines in order to solve iteratively the general polynomial equation:

$$B^* = (1-A)R + q^{-k}BS \quad (10)$$

The three subroutines may be considered as independent parts of the mini-library. So TILDE calculates the polynomial $\tilde{B}$, ASTER calculates $B^* = B^* \tilde{B}^-$ and REDCO, another square root algorithm that is used to solve a linear equation system.

The memory distribution is as follows:

	Sub-programs	Main program	RAM memory
MINVARMP	52	106	135 bytes
MINVAR	2399	788	1122 bytes

Another two very practical regulators are those named CONAD1 and CONAD2. They have reduced order (three parameters for the first and four for the second) but they can be used in a lot of situations as an aproximative solution. Both of them solve analitically equation (10) and are prepared as adaptive regulators (they use REFIL as we have already explained). Their respective formulas are

$$\Delta u(t) = -S_o y(t-1) - r_1 \Delta u(t-1) - S_1 y(t-2) \quad (11)$$

$$\Delta u(t) = -S_o y(t-1) - r_1 \Delta u(t-1) - S_1 y(t-2) \quad (12)$$
$$- r_2 \Delta u(t-2).$$

For more details, see Aguado (1982).

The memory requirements of these regulators are

	Sub-programs	Main program	RAM memory
CONAD1	52	700	60 bytes
CONAD2	52	1500	120 bytes

Floating-point arithmetic

We use the package of programs created by Chamrad (1979). The floating-point number format is as follows:

$$X = a \cdot 2^b \quad 0.5 \leq a < 1 \quad (13)$$

where a is the mantissa two-bytes long (2-complement representation with the first bit for the sign) and b is the exponent with one-byte (the first bit is an under/overflow indicator) and 64-excess code.

This arithmetic is particularly suited for control purposes with special subroutines for A/D and D/A conversors (single and double precision) but it lacks of a decimal/hexadecimal and reciprocal conversions.

In order to make possible the use of a display and the communication of the user with his computational system, we have produced two programs, CONV1 and CONV2. The first allows us to introduce decimal notation. In fact, it takes a number from a buffer of the form

$$\pm \square \square \square \ldots \square \square \quad (14)$$

where each space represents a digit or the decimal point in ASCII code (the point, of course, only once) and it produces the mantissa-exponent format of the number in the DE and B registers of INTEL's 8080. On the contrary, CONV2 takes from DE and B the mantissa and exponent and produces a buffer

$$\pm 0.xxxxx\ E \pm xx \quad (15)$$

with the decimal form of the number in ASCII code.

This arithmetic has a precision of only four decimal digits but it is quite enough for technological applications. It uses 1 Kbyte of memory.

The memory requirements for these two complementary programs are

	Sub-prog. & tables	Main program	RAM memory
CONV1	248	241	29 bytes
CONV2	290	241	29 bytes

Time considerations depend on the particular system used to display the results and it does not make sense to discuss about it, but it is clear that these programs have to be time expensive. It is clear also they do not need to be used during the whole control loop.

The arithmetic has been completed with short subroutines that make easier the work: load and store of operands, translation, zero floating-point creation, etc. (not reported in detail).

General programs

Our mini-library has been completed with a group of general purpose procedures according to the natural necessities of the user. So we have a group of subroutines for the polynomial algebra: POLADD, POLSUB, POLMUL, POLDIV, POLGCD (or POLMCD). Some of them are used in the MINVAR regulator (third and fourth), the others were programmed in order to complete the package.

We have also vector and matrix treatment programs: VECSP, VECDP, MAT and finally we we have created a subroutine for measurements treatment (MED) which allows us to produce a linear scale changing in the measurements obtained from an A/D conversor. This subroutine uses an auxiliary array with the constants needed for the calculations.

AN APPLICATION

Counting on the programs of this mini-library, it was created a configuration for the control of a vacuum-pan of a Cuban sugar mill. This result is reported in three papers (Aguado, 1982; Enríquez, 1982 and Remesar, 1982). It is well-known that this kind of process is quite unstationary, so an adaptive regulator seems to be a good solution.

The control configuration given in Fig. 1 uses the following programs: CONAD1, VECDP, CONV1, CONV2, MED, REFIL and the floating-point arithmetic.

Through out the experiments, the programs have proved to have enough precision and to give the expected results. Even when the experiments were realized under special conditions (in every step of the sampled time they were shown in the display about six parameters) it took no more than two seconds for a whole iteration.

Another special program was created in order to display the results but it is not considered in the mini-library because it is oriented to IMACC 8101 system.

CONCLUSION

A mini-library with modern algorithms for identification and control is proposed. The user of these programs can set different control configurations according to the aim pursued. The inclusion of the PID algorithm allows the substitution with real time digital control many of the present classical controls.

Many of these programs can be used for general purposes. The floating-point arithmetic has been improved with a couple of subroutines that makes easier its exploitation.

With these programs the user has the possibility to improve his control algorithm looking for better strategies toward his final goal: optimal control. Besides, he will be able to control more complicated plants that require adaptive control.

All the identification and control algorithm need the floating-point arithmetic; that means 1 Kbyte memory more.

In the future, this mini-library will be increased. The aim is to obtain a complete library with the most complicated adaptive regulators and versions for multivariable systems.

REFERENCES

Aguado, A. (1982). Reguladores de auto-ajuste simplificados. Revista CCA, No. 3, (to appear).

Aguado, A. and A. Gómez (1982). Tercer Ciclo de Conferencias Científicas ISPJAE; Controlador adaptivo para los tachos de la industria azucarera: Parte I Identificación y simulación, Cuba (to appear).

Chamrád, V. (1979). The 2nd IFAC/IFIP Symposium on Software for Computer Control; A speed-oriented floating-point subroutine package for the INTEL 8080 Microprocessors, Prague, M-1-1 - M-1-4.

Enríquez, J., J.A. Hormaza and A.C. Campos (1982). Tercer Ciclo de Conferencias Científicas ISPJAE; Controlador adaptivo para los tachos de la industria azucarera: Parte III Programas de propósito general, identificación y control para microprocesadores, Cuba, (to appear).

Peterka, V. (1975). A square root filter for real time multivariate regression, Kybernetika, Vol.11, No.1, Academia Praha, pp. 53-67.

Remesar, J.C. and J.M. Barrios (1982). Tercer Ciclo de Conferencias Científicas ISPJAE; Controlador adaptivo para los tachos de la industria azucarera: Parte II Hardware y software básico, Cuba, (to appear).

Smith, C.L. (1972). Digital Computer Process Control. Intext Educational Publishers, USA.

Strejc, V. (1977). Least squares in identification theory, Kybernetika, Vol.13, No.2, Academia Praha, pp. 83-173.

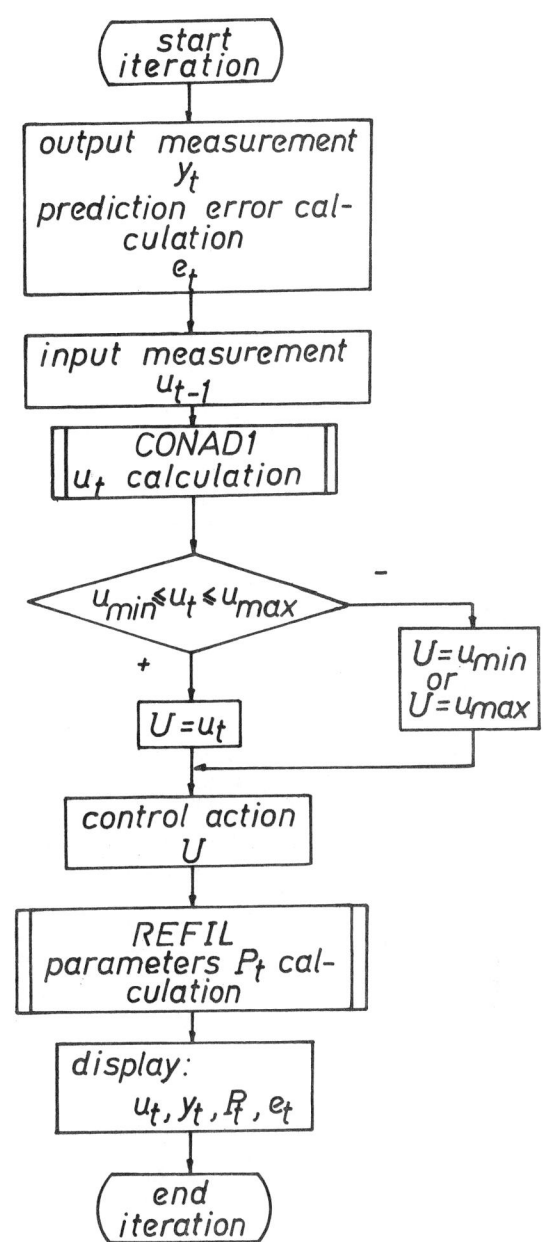

Fig.1 Adaptive control configuration for a vacuum-pan of a sugar mill.

A REAL TIME MONITOR, ITS REPRESENTATION BY PETRI NETS AND AN APPLICATION

A. Maldonado* and F. Rivera**

*Departamento de Ingeniería Eléctrica, UAM-Iztapalapa, México, DF, México
**Departamento de Electrónica, UAM-Azcapotzalco, México, DF, México

Abstract. This work is a part of a project related to the developement of a software base for applications on real-time control systems. This paper describes the design of a simplified real-time kernel oriented to its implementation in micro-computers. Also it treats the monitor primitives representation by Petri Nets. In order to show the monitor possibilities, one application of on-line monitoring is given.

Keywords. Industrial control; real-time monitor; task management; Petri Net representation; on-line operation.

INTRODUCTION

All industrial process control applications may be considered as a set of tasks interacting between them, each task related to a specific function.

Task management is the essential function of a real-time monitor. There are several design phylosofies of real-time monitors, depending on the synchronisation and communication mechanisms being utilised.

In the first part of this paper, an overview of some synchronisation and communication mechanisms implemented ordinarily in monitors is given. The second part deals on the design and implementation of the monitor proposed. In the last part, the on-line operation of a packing unit and its control in terms of cooperating tasks are described as an application example.

I STRUCTURES OF REAL TIME MONITORS

Definitions

- A real-time system is characterised by random external requests and its corresponding responses in a restraint period of time.

- A task is a dynamic entity characterised by the execution of a program. It represents a well defined activity (or function) of the specific application (industrial process control, telephone switching, etc.).

I.1 Requirements of a Real Time Monitor

In industrial process control any application may be considered as a certain number of co-existent tasks, with some degree of interdependency, and executed according to specific synchronisation rules. In addition, a priority is associated to each task according to its relative importance in the system.

Therefore, the real-time monitor must provide the following services:

- synchronisation
- communication and
- hierarchisation

Synchronisation. This function concerns the execution sequence, in time, of the tasks.

Its implementation involves one of two kinds of mechanisms: single mechanisms (semaphores, events, messages, etc.) or structured mechanisms (Hoare's monitors, path expression, etc.). The choice of the method depends not only on the objetive of the system but on its complexity.

In micro-computer applications single mechanisms seems to be the most adequate (Intel's RMX/80, based on messages). For small applications, synchronisation by events could be sufficient.

Communication. The implementation of communication among tasks is closely related to the synchronisation mechanism utilised.

Associated to semaphores and messages there exists the "mail box" system; and to events, shared memory zones. For structured synchronisation mechanisms, the mail box is also employed.

Hierarchisation. The importance of a task defines its priority. Several priority levels forms a hierarchy. If priorities are associated to interrupts, it constitues a hierarchical interrupt system, which is essential to any real-time system.

Its implementations needs a device of interrupt control (hardware) and an interrupt handler task (software).

I.2   General Organization of a Kernel

The nucleus of the real-time monitor is a specialized software module which consists of three major parts:

- primitives (P)
- scheduler (S)
- interrupt handler (IH)

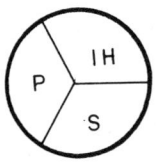

Fig. 1   Parts of a kernel

Primitives. They represent the communication and synchronisation services offered by the monitor. They are called by a task at any place in the program requiring an action concerning itself or any other task.

The primitive execution is indivisible.

Scheduler. Its essential function is the processor allocation to the task of highest priority. It also supervices the dynamical evolution of the different task states.

Generally, the scheduler is not accesible from users programs.

Interrupt Handler. It interfaces between peripheral requests (interrupts) and the scheduler.

In more sophisticated kernels, this handler takes on evoluated input/ output operations (read, write, etc.).

II   A REAL TIME CONTROL SYSTEM

This system has been designed to operate in a micro-computer kit (Zilog-80) for small control process applications.

II.1   The Real-Time Kernel

States of a Task. During its lifetime, a task can assume several states:

- The "active" state. It corresponds to the state of a task being executed.

- The "elegible" state. All necessary conditions to start running the task have been satisfied, but it does not have access to the processor.

  The elegible task with the highest priority always becomes the active task.

- The "waiting" state. It corresponds to a task that has not satisfied the conditions to run.

- The "dormant" state. The task needs a previous activation (transfer from a "dormant" state to "elegible" or "waiting" state) in order to pretend to satisfy the conditions to execution.

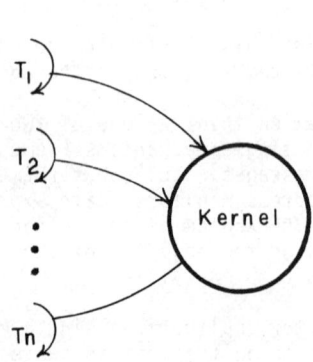

Fig. 2   Call of primitives

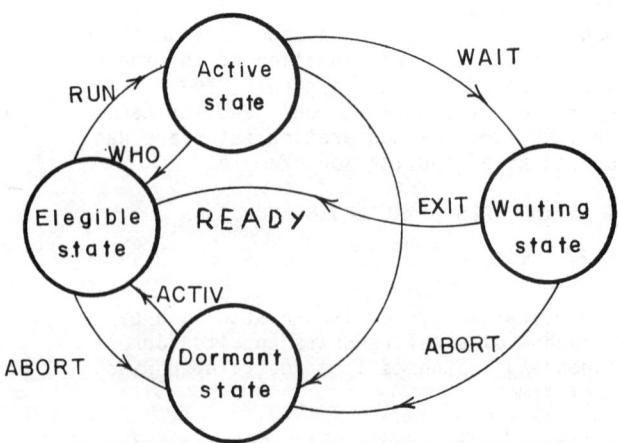

Fig. 3   States of a task

The task state transitions are carried out by the primitive calls or by inherent actions of the scheduler.

The Task Synchronisation. The synchronisation mechanism selected is based in the "event".

An <u>event</u> is a boolean variable (taking the values: "present event" and "absent event") that allows synchronisation among tasks. There exist three basic types of events:

- The interruption, coming from external devices;

- The condition, handled by tasks;

- The command, sent by the user from the keyboard.

In order to cover a wide range of applications, events must be multiples and/or combinations of the three basic tipes of events.

The communication among tasks will be implemented by common memory zones, under responsability of the user.

The Primitives. The primitives proposed are:

- SET   ev   Positionning an event

- WAIT   ev, De-   Waiting for an event and/
         lay.   or temporary hold.

- ACTIV Task   Activation of a task (iden-
         name.   tified by its name).

- STATUS Task   Request of a task state.
         name.

- EXIT   Indefinite hold of the running task.

- ABORT Task   Abortion of a task.
         name.

Petri Net Representation of the Primitives.

SET   ev

In Fig. 4 T_2 is waiting for an event; task T_1, who has the processor (p_0), sets on the event (calling the SET primitive). Once T_1 is blocked, task T_2 could take the central processor, depending on its priority.

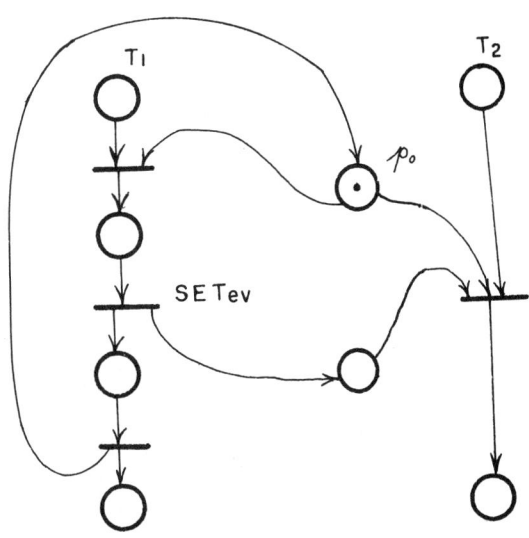

Fig. 4 Positioning an event

WAIT   ev, Delay.

Figure 5 represents the demand of task T_1 to wait for the arrival of an event and/or a temporal hold. When the event occurs and/or when the delay expires, T_1 becomes eligible.

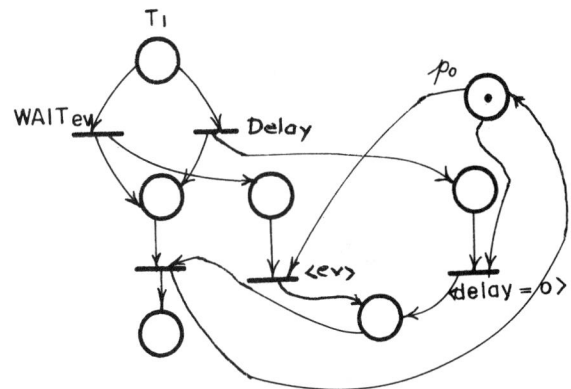

Fig. 5 Waiting for an event and/or temporary hold.

ACTIV Task name.

The task that calls ACTIV primitive, T_1 in Fig. 6, activates task T_2. This way task T_2 passes to the eligible state.

ABORT Task name.

This primitive acts in a similar way as the previous one, but task T_2 passes to the dormant state.

Both primitives are represented in Fig. 6.

III AN APPLICATION EXAMPLE

The controled system, a packing unit, is composed of three major parts: an assembling line, a bottle handler and a mechanical device to join labels.

The packing unit must perform the following operations:

- To detect the arrival of a pack;
- to label each bottle;
- to bring the bottles in the pack (six bottles by pack);
- to close the pack.

A way to resolve this problem, utilizing the real-time monitor, consists in the association of a task to each operation described above. Fig. 8 represents this solution.

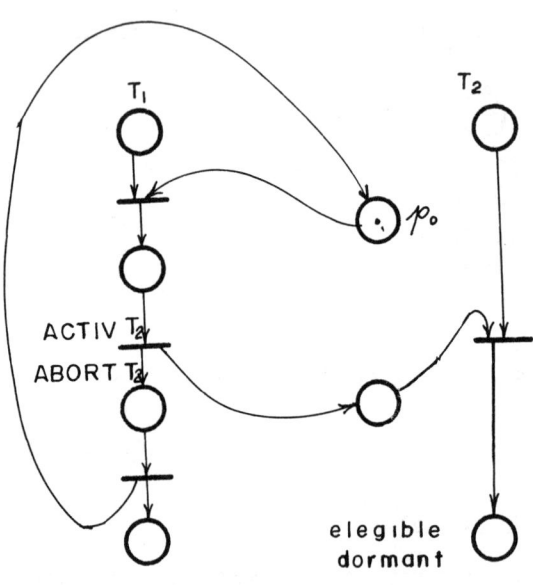

Fig. 6 ACTIV and ABORT primitives

EXIT.

A task calls EXIT primitive to release the central processor (end of a task). The task passes to the dormant state.

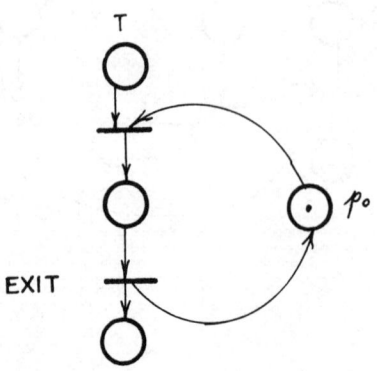

Fig. 7 EXIT primitive

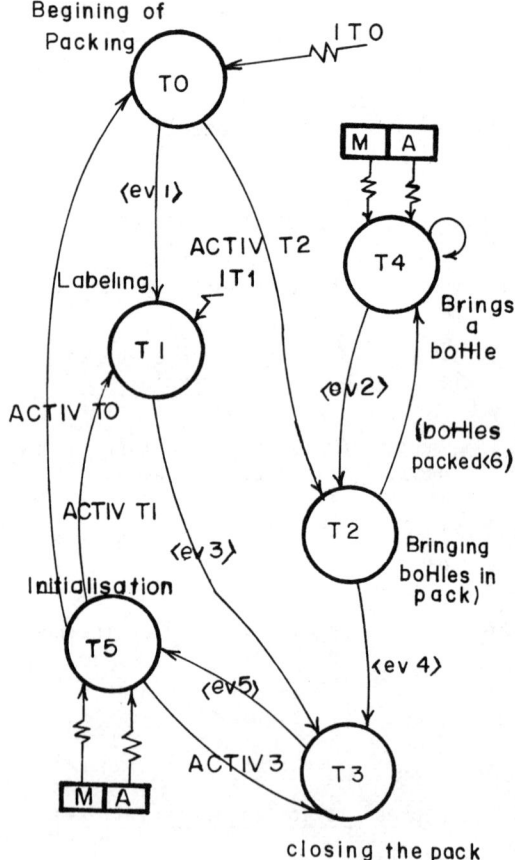

Fig. 8 Tasks of control system

IT 0 : arrival of a pack
IT 1 : begin labeling
M,A : commands from keyboard

ev 1 : pack present
ev 2 : bringed bottle
ev 3 : label joined
ev 4 : bottle bringed in
ev 5 : pack closed

This simple example shows a software modular decomposition of a problem, easy to implement, economical to maintain and to modifie.

Conclusion. The monitor described in this article was implemented in the MKE-Z80 microcomputer. Primitives were developed as MACROS, and the microcomputer hardware was additionned of a peripheral interrupt controler (8259) and a real time clock (8253).

This monitor was developed for small control process applications. Actually, it has been utilised for the automation of a reaction chemical system.

We have selected events as a synchronisation mechanism, instead of semaphores, in order to simplifie the data structure. This do not implie that semaphores could not be included in the futur to resolve more complex problems.

We want to thank Dr. Isaac Schnadower for his amicable comments.

REFERENCES

Appel O., Ayache J.M., Díaz M. (1979). Etude et developement d'un moniteur temps réel multitaches pour microordinateur. Note interne LAAS-LC 79.I.53.

Intel, RMX/80 User's Guide. (1977).

Microkit, MKE-Z80 User's Guide. (1981).

Peterson James. (1977). Petri Nets Computing Surveys, Vol. 9, No. 3. pp. 223-252.

DISCUSSION

SESSION WA2: MODELLING, IDENTIFICATION AND SOFTWARE

Paper: LOCAL OPTIMISATION FOR CORRECTING THE INPUTS IN NONLINEAR IDENTICATION

Authors: M de la Sen, MB Paz (Facultad de Ciencias, Universidad del País Vasco, Leída, Vizcaya, Spain)

Discusser: Guillermo Rodríguez
San Diego No 3
Acapantizindo
Cuernavaca, Morelos
México

Questions or Comments:

1) Did you implement the algorithms presented in your paper using a computer program?

Authors' reply:

1) The work generalizes, with four optimization procedures, a prior work of Matansek-Stankovic. At this moment we are developing a simulation program to corroborate the theoretical results.

Paper: AN APPROACH TO THE DESIGN OF REAL-TIME DATABASE MODELS

Authors: G. Rodríguez (Instituto de Investigaciones Eléctricas, Palmira, Cuernavaca, México)

Discusser: Enrique Morfín
(Comisión Federal de Electricidad)
45070 Guadalajara, Jalisco
México

Questions or Comments:

1) My question is about the conceptual data base design. How do you manage the problem to find new entities in the development of software, or in the debugging? (Physically, in the structure of your data base)

Authors' reply:

1) New entities as well as new atributes for entities already identified appear during the developing and debugging phases of implementing the software. The implementation of the metamodel helps us to keep track of the new attributes and entities. It is important to note that a good set of requirements writen in a document was of big help to design the database.

Discusser: KW Plessmann
c/o Aachen University
Templer Graben St.
5100 Aachen
Federal Republic of Germany

Questions or Comments:

1) What is the relationship of your database to real time?

2) Is time an attribute or not?

Authors' reply:

1) The database contains entities that are accessed in real-time fashion; for example we have a program that runs every 4 seconds

2) Time is an attribute of the programs and it is stored in the metamodel.

Discusser: *Juan Manuel de la Peña Serment*
Ley # 2634
Circunvalación Vallarta
Guadalajara, Jal
México

Questions or Comments:

1) Which were the main problems you found when designing and implementing your model?

Author's reply:

1) One problem we had during the data base implementation was that the relationship between entities in the conceptual model and physical files could not be one to one. So we had to "pack" several entities in a physical file. The problem arises because FORTRAN does not allow a program to use many files simultaneously.

MICROPROCESSOR-BASED CONTROL OF INDUSTRIAL SEWING MACHINES*

B. Hertzanu** and D. Tabak***

**Beta Engineering and Development Co., Beer Sheva, Israel*
***Department of Electrical Engineering, Ben Gurion University, Beer Sheva 84105, Israel*

Abstract. Microcomputer-based control of the X-Y motion of a sewing head of an industrial sewing machine is considered. A digital controller, replacing a previously used analog one, is designed. It is realized using a Motorola 6801 single-chip microcomputer. Adaptive control of the system is also discussed. In particular the model reference approach, using the Liapunov function, is adopted. A solution using this approach is proposed.

Keywords. Microcomputer-based control, industrial control, adaptive control, model-reference control, direct digital control.

1. INTRODUCTION

In the past few years there has been an accelerated growth in the application of microprocessors in a variety of control systems, such as Auslander et al. (1978); Bibbero (1977); Farrar and Eidens (1980); Glattfolder et al. (1980); Moroney et al. (1980); Neuman and Baradello (1979); Tabak and Lipovski (1979); Tao et al. (1977) and Tsuchiya (1982); (only a small sample of the existing literature is mentioned here). Microcomputers have been widely applied in control loops of different types of systems and various aspects of such implementation have been investigated. One can hardly find an application area today where microcomputers have not been applied yet. One of the technical areas where microcomputer-based control implementation has been recently initiated is the direct digital control of sewing machines (Shackil, 1981). The project, reported in this paper, falls in as a direct continuation of this effort; a direct microcomputer-based digital control system has been designed and implemented for an industrial sewing machine. To be more specific, a Beta sewing machine, with Pfaff or Adler or Durkopf industrial sewing heads, is used. The work has been further extended by applying modern control methods to this control system.

The controlled system undergoes parameter variations during its operation. The application of a PID digital controller can not handle effectively such a situation. Therefore, it was decided to apply an adaptive controller to this system. To be more specific, a model reference adaptive control approach has been adopted. (Hang and Parks, 1973; Landau, 1979; Lindorff and Carroll, 1973; Saridis, 1979; Van Amerongen and Udink ten Cate, 1975). To the best of the authors' knowledge, this is the first time where adaptive control techniques are applied to a sewing machine control system.

The description of the sewing machine control system, including some of the mechanical aspects, is given in section 2. Section 3 presents the development of the mathematical model equations of the control system. A direct, microprocessor-based, digital control implementation, using a Motorola 6801 Microcomputer, is presented in section 4. The Adaptive, Model Reference Control extension of the design is discussed in section 5. Conclusions and recommendations for further research are given in section 6.

2. THE SEWING MACHINE CONTROL SYSTEM

The system being controlled is a drive system for a computerized industrial sewing machine. The material on which the sewing operation is performed (such as boots leather), is fastened and moved in the X-Y coordinates directions, according to the specific sewing pattern. The movement of the material and the sewing operation are performed under the supervision of a microcomputer, in order to achieve the required sewing pattern. This microcomputer serves as the principal controller of the machine. It is responsible for the machine operations, such as: material movement, sewing supervision, sewing speed control, sewing parameters control, opening or closing of fasteners, and others.

The sewing machine microcomputer controls the

---
*The paper constitutes a part of an M.Sc. thesis by B. Hertzanu, advised by Prof. D. Tabak, at the Ben Gurion University

operation of the machine by a specific sewing program which contains the sewing pattern and various sewing commands. At every instant it calculates the required location (X-Y) of the sewed material in order to achieve sewing of good quality, of a desired pattern, at a desired speed.

The material movement is performed along cartesian X-Y coordinates. All necessary movement is translated into movements along the X and Y coordinates. The microcomputer translates any location requirements, obtained from the software, into an analog voltage. There is a servo loop for each of the X-Y coordinates, which controls the movement (Fig. 1). In the previous design, the loop was closed by an analog servo. The location requirement and the current location data were converted by a D/A converter. One of the main goals of this project has been to introduce a digital servo loop instead of the analog one. The loop will be closed by a special microcomputer.

We should note the difference between the microcomputer which serves as the principal controller of the sewing machine and the microcomputer which serves as a digital controller of the servo loop of X or Y position control. The principal controller processes the sewing program and establishes the desired X and Y positions. The digital servo controller receives the location specifications from the principal controller and it completes the control loop properties to achieve the desired location along the X, Y axes. Naturally, there exists a communication channel between the two microcomputers. In the present design, there is a separate microcomputer for each of the X, Y loops, although one could design it with a single microcomputer. Moreover, a design using a single microcomputer for all the above purposes could be performed. A block diagram of one of the two digital servo loops is shown in Fig. 2.

The drive system for each axis (X,Y) consists of a DC motor of 1/8 HP. The translation of the rotational motion into a linear one (along X or Y) is done by the rack and pinion (spur gearing) method. A special Thomson bearing is used for the linear motion. The gear ratio for each axis is 1:4. The pinion diameter is 0.6366". Thus, a 360° rotation of the motor axis corresponds to a 0.5" linear motion. The maximum sewing speed is:

(a) While sewing   6"/sec

(b) Slow motion (no sewing) 15"/sec

The position sensing is done with the help of an optical incremental shaft encoder, for each of the X,Y axes. The encoder is mounted on the motor axis. The basic encoder resolution is 1000 pulses/revolution. The encoder generates two pulse sequences, Channel A, Channel B. The phase shift between the two sequences is 90°. Since the encoder is incremental, there are no pulses on both channels when the motor is not turning. A block diagram of the position sensing system is shown in Fig. 3.

During the motor motion the phase difference between the pulses of channels A, B (±90°) indicates a clockwise (CW) or a counterclockwise (CCW) direction of rotation. The pulse sequences are passed through logic circuitry which generates an UP or DOWN pulse sequence, according to the phase shift between Channel A and B. The UP or DOWN sequence is counted by the UP/DOWN BINARY COUNTER. The binary number stored in the counter represents the location along the axis.

The change of position along any axis (X or Y) requires complete synchronization between the sewing needle and the material motion. It is required that the material motion be completed before the needle penetration. Material motion, while the needle is lowered, will cause defective sewing, needle breaking, unnecessary skipping, etc. An appropriate arrangement of motion is shown in Fig. 4(a). The dwell time, between two subsequent motions, is one of the control parameters. It is measured in degrees, while each stitch is considered to be represented by 360°. The dwell angle is shown in Fig. 4(b). The maximum dwell angle is 180°, the minimum 0°, and it can be changed in increments of 30° in this design. The sewing should be done with a dwell of 180°, however this will leave a very short time for the movement itself (only 10 msec for 3000 RPM). Therefore, in practice, we allow some movement even when the needle is still in the material. This is particularly possible in soft materials and it avoids the use of too high accelerations or decelerations. Other parameters which influence the motion are the sewing head RPM and Stitch Per Inch (SPI). These parameters are taken into account by the software of the principal microcomputer.

The servo controller of each axis (X,Y) receives the location requirements from the principal microcomputer, and its task is to see to it that the axis will follow the motion profile as precisely as possible. The system should have a fast reaction, attain the appropriate dwell time, without overshoots.

3. THE MATHEMATICAL MODEL EQUATIONS

A block diagram of the position control system of the motor, is shown in Fig. 5. It is a simplified model, however it was found adequate for the analysis. Ignoring second order terms, involving $\tau_a \tau_m$, the system equation in terms of the position angle θ can be written:

$$\ddot{\theta}(\tau_a+\tau_m)\frac{K_E}{A} + \dot{\theta}\left(\frac{K_e}{A} + K_g\right) + \theta K_p = \theta_d \qquad (1)$$

where θ_d is the desired position, the input to the system and τ_a, τ_m, K_g, K_p, A are system parameters. Defining the state variables to be θ and $\omega = \dot{\theta}$ we obtain the state equations:

$$\dot{\underline{x}}_p = A_p \underline{x}_p + B_p \theta_d \quad (2)$$

where:

$$\underline{x}_p = \begin{pmatrix} \theta \\ \omega \end{pmatrix}; \quad A_p = \begin{pmatrix} 0 & 1 \\ -K_1 & -K_2 \end{pmatrix}; \quad B_p = \begin{pmatrix} 0 \\ K_3 \end{pmatrix}$$

$$K_1 = K_p K_3; \quad K_2 = K_3 K_g + \frac{K_E}{A}; \quad K_3 = \frac{A}{K_E(\tau_a + \tau_m)}$$

Matrices A_p and B_p depend on the system parameters which may change in the course of the operation. Representative values of the parameters can be found from manufacturers' specifications as well as experimentally.

The system shown in Fig. 5 is the analog equivalent of the control loop. The equivalent, digitally controlled, system is shown in Fig. 6. Definite numerical values have been substituted for the K_E, τ_a, τ_m, A system parameters. The feedback gains K_g and K_p have been substituted by digital controller networks $D(z)$ and $Y(z)$, to be established by the designer.

The feedback digital controller network replaces now the position (K_p) and the velocity (K_g) feedback networks. Its output ψ should simulate the combined effect of these networks. Thus we choose at any sampling instant kT ($k=1,2,...$):

$$\psi(k) = K_p \theta(k) + K_g [\theta(k) - \theta(k-1)]$$

or:

$$\psi(z) = (K_p + K_g) \theta(z) - K_g z^{-1} \theta(z)$$

Therefore

$$Y(z) = \frac{\psi(z)}{\theta(z)} = (K_p + K_g) z^{-1} (z - \frac{K_g}{K_p + K_g}) \quad (3)$$

The exact form of the digital controller $D(z)$ can be established by various design techniques, covered in the literature (Franklin and Powell, 1980; Kuo, 1980).

Assuming $D(z) = K'$ and $Y(z) = 1$, the overall z-transform loop transfer function, including the z.o.h., for $T = 1$ msec is:

$$G(z) = 0.606 \, K' \frac{(z-0.39+j0.897)(z-0.39-j0.897)}{(z-1)(z-0.6065)(z-0.9718)} \quad (4)$$

4. DIRECT DIGITAL CONTROL IMPLEMENTATION

Before actual implementation, the digital control loop has to be established. A root locus diagram of the open loop transfer function in Eq. (4), reveals that the uncompensated system is actually unstable, as seen from Fig. 7(a). The general form of the feedback compensator $Y(z)$ is given in Eq. (3). Inspection of the root locus in Fig. 7(a) shows that the system can be stabilized by placing a zero between the two poles near $z=1$. $Y(z)$ provides a pole at $z=0$. Thus the ratio $K_g/(K_p + K_g)$ has to be smaller, but very close to unity. Choosing: $K_p = 1$, $K_g = 100$ we obtain:

$$Y(z) = 101 \frac{z - 0.99}{z} \quad (5)$$

That is, we add a zero at $z = 0.99$ (between $z = 0.9718$ and $z = 1$) and a pole at $z = 0$. The new root locus is shown in Fig. 7(b). It should be noted that the poles and the zero on the real axis near $z = 1$ are drawn out of scale, for the benefit of clearness of the root locus. The plant poles at $z = 0.6065$, $z = 0.9718$ and $z = 1$ remain the same in Figs. 7(a) and (b). The critical gain in this case is about 0.6.

The digital controller has been implemented with a Motorola 6801 single-chip microcomputer. Its word length is 1 byte (8 bits). Most of the calculations are performed using the double byte option. The cycle time is 1 μsec.

The maximum sewing speed at the moment is 3000 RPM, or 20 msec per stitch. With a dwell of 180°, the motion time per stitch is 10 msec. The minimal number of samples per stitch movement is 10. Therefore, the maximum acceptable sampling period is 1 msec, or a 1000 samples/sec sampling frequency.

The microcomputer has a 2 Kbyte EPROM memory, however, only about 1 Kbytes are utilized at the moment, the rest will be used for adaptive control and for diagnostics programs.

5. ADAPTIVE CONTROL PLANNING

The working conditions of the same sewing machine may change daily. One day it may be working with a certain type of material and holder, the other day it may work with a different type of the above. This would change the system parameters. Therefore, in order to attain good quality of sewing in all cases without the need of readjusting controller parameters, an adaptation of parameters to the existing conditions is needed.

It has been decided to plan for a future extension and further sophistication of the present control system, by turning it into an adaptive one. In fact, the model reference adaptive control approach was chosen. (Hang and Parks, 1973; Landau, 1979; Lindorff and Carroll, 1973; Van Amerongen and Udink ten Cate, 1975). A block diagram of the planned system is shown in Fig. 8. A model is formed:

$$\dot{\underline{x}}_m = A_m \underline{x}_m + B_m \theta_d \quad (6)$$

The error vector:

$$\underline{e} = \underline{x}_m - \underline{x}_p \qquad (7)$$

operates on an adaptation mechanism, which implements an algorithm minimizing a performance criterion, such as for instance:

$$J = \int_0^\infty (\underline{e}^T Q \underline{e} + P\theta_d^2) dt \qquad (8)$$

The criterion minimizes the square error $\underline{e}$ as well as the energy put into the operation.

The performance criterion is a function of the system parameters (through $\underline{e}$). As a result of the minimization procedure, the system parameters are varied (to minimize J). The updated values of the parameters are then transmitted back to the system and appropriate settings are automatically performed. The particular approach adopted in this project is that of the Liapunov function (Carroll and Lindorff, 1973; Lindorff and Carroll, 1973; Luders and Narendra, 1973). Subsequent advancement in this area has been reported by Narendra and Lin (1980), Narendra, Lin and Valavani (1980), Narendra and Valavani (1978) and Narendra and Valavani (1979) among others.

From eqs. (2), (6), (7) we obtain the error differential equation:

$$\underline{\dot{e}} = A_m \underline{e} + (A_m - A_p)\underline{x}_p + (B_m - B_p)\theta_d \qquad (9)$$

Introduce the notation:

$$A = A_m - A_p = [\underline{a}_1 \underline{a}_2] \; ; \; \underline{a}_1, \underline{a}_2 = \text{columns of } A$$

$$\underline{b} = B_m - B_p$$

Thus,

$$\underline{\dot{e}} = A_m \underline{e} + A\underline{x}_p + \underline{b}\theta_d \qquad (10)$$

Choose a Lyapunov function:

$$V = \underline{e}^T P \underline{e} + \alpha_1 \underline{a}_1^T \underline{a}_1 + \alpha_2 \underline{a}_2^T \underline{a}_2 + \beta \underline{b}^T \underline{b} \qquad (11)$$

Where P is a positive definite symmetric matrix and $\alpha_1, \alpha_2, \beta > 0$.

$$\dot{V} = \underline{e}^T (A_m^T P + PA_m)\underline{e} + 2(\underline{e}^T PA\underline{x}_p + \underline{e}^T P\underline{b}\theta_d +$$
$$+ \alpha_1 \underline{\dot{a}}_1^T \underline{a}_1 + \alpha_2 \underline{\dot{a}}_2^T \underline{a}_2 + \beta \underline{\dot{b}}^T \underline{b}) \qquad (12)$$

Denoting $Q = A_m^T P + PA_m$, the matrix Q will also be positive definite. Equating the second term in Eq. (12) to zero, we have:

$$\dot{V} = -\underline{e}^T Q \underline{e} \qquad (13)$$

and

$$\underline{e}^T PA\underline{x}_p + \underline{e}^T P\underline{b}\theta_d + \alpha_1 \underline{\dot{a}}_1^T \underline{a}_1 + \alpha_2 \underline{\dot{a}}_2^T \underline{a}_2 + \beta \underline{\dot{b}}^T \underline{b} = 0 \qquad (14)$$

The matrix $-\dot{V}$ will be positive definite, thus satisfying the asymptotic stability conditions. The condition in Eq. (14) can be satisfied by the following two equalities:

$$\underline{e}^T PA\underline{x}_p = -\alpha_1 \underline{\dot{a}}_1^T \underline{a}_1 - \alpha_2 \underline{\dot{a}}_2^T \underline{a}_2 \qquad (15)$$

$$\underline{e}^T P\underline{b}\theta_d = -\beta \underline{\dot{b}}^T \underline{b} \qquad (16)$$

From section 2 we know that

$$\underline{x}_p = \begin{pmatrix} \theta \\ \omega \end{pmatrix} = \begin{pmatrix} \theta \\ \dot{\theta} \end{pmatrix}$$

Thus, we can easily derive from Eqs. (15) and (16):

$$\underline{\dot{a}}_1^T = -\underline{e}^T P \theta / \alpha_1 \qquad (17)$$

$$\underline{\dot{a}}_2^T = -\underline{e}^T P \dot{\theta} / \alpha_2 \qquad (18)$$

$$\underline{\dot{b}}^T = -\underline{e}^T P \theta_d / \beta \qquad (19)$$

The values of α_1, α_2, β and the elements of the matrix P are chosen a priori, satisfying the positivity conditions. Matrices A_m and B_m are also chosen. The state variables θ, ω and the error vector $\underline{e}$ are measurable at each time instant. Thus, equation (17)-(19) can be easily integrated numerically and the values of the elements of A and $\underline{b}$ calculated. Keeping in mind that $A_p = A_m - A$, the components of A_p can be updated. The components of A_p are functions of the variable system parameters (such as K_p and K_g), which in turn can then be easily updated. All the above operations are performed within the "adaptation mechanism", shown in Fig. 8. This "adaptation mechanism" is in fact a procedure with the software of the control microcomputer.

6. CONCLUSION

A design of motion control of a sewing head of an industrial sewing machine has been presented in partial detail. The digital controller has been actually implemented on a set of industrial sewing machines of the Beta Co., using a variety of sewing heads. The system performed according to the specification.

An extension for the adaptive control of the system is proposed. A model-reference approach, using the Lyapunov function, is outlined in the paper. The next step in this project would be the actual implementation of the adaptive control algorithm. No doubt, the current microcomputer (M 6801) will have to be replaced by a more powerful 16-bit (IAPX86, MC68000, Z8000) or even a 32-bit (IAPX432, NS16032) microcomputer.

REFERENCES

Auslander, D.M., Takahashi, Y., and M. Tomizuka (1978). Direct digital process control: practice and algorithms for microprocessor application. Proc. IEEE, 66, 199-208.

Bibbero, R.J. (1977). Microprocessors in Instruments and Control. Wiley, N.Y.

Carroll, R.L. and D.P. Lindorff (1973). An adaptive observer for single-input single output linear systems. IEEE Trans. on Automatic Control, AC-18, 428-435.

Farrar, F.A. and R.S. Eidens (1980). Microprocessor requirements for implementing modern control logic. IEEE Trans. on Automatic Control, AC-25, 461-468.

Franklin, G.F. and J.D. Powell (1980). Digital Control of Dynamic Systems. Addison Wesley, Reading, MA.

Glattfelder, A.H., Huguenin, S. and W. Schaufelberger (1980). Microcomputer based self-tuning and self-selecting controllers. Automatica, 16, 1-8.

Hang, C. and P.C. Parks (1973). Comparative studies of model reference adaptive control systems. IEEE Trans. on Automatic Control, AC-18, 419-428.

Kuo, B.C. (1980). Digital Control Systems. Holt, Rinehart and Winston, N.Y.

Landau, I.D. (1979). Adaptive Control - The Model Reference Approach. Marcel Dekker, N.Y.

Lindorff, D.P. and R.L. Carroll (1973). Survey of adaptive control techniques using Lyapunov design. Int. J. Control, 18, 897-914.

Luders, G. and K.S. Narendra (1973). An adaptive observer and identifier for a linear system. IEEE Trans. on Automatic Control, AC-18, 496-499.

Moroney, P., Willsky, A.S. and P.K. Houpt (1980). The digital implementation of control compensators: the coefficient word length issue. IEEE Trans. on Automatic Control, AC-25, 621-630.

Narendra, K.S. and Y.H. Lin (1980). Stable discrete adaptive control. IEEE Trans. on Automatic Control, AC-25, 456-461.

Narendra, K.S., Lin, Y.H. and L.S. Valavani (1980). Stable adaptive controller design, part II: proof of stability. IEEE Trans. on Automatic Control, AC-25, 440-448.

Narendra, K.S. and L.S. Valavani (1978). Stable adaptive controller design - direct control. IEEE Trans. on Automatic Control, AC-23, 570-583.

Narendra, K.S. and L.S. Valavani (1979). Direct and indirect model reference adaptive control. Automatica, 15, 653-664.

Neuman, C.P. and C.S. Baradello (1979). Digital transfer functions for microcomputer control. IEEE Trans. on Systems, Man and Cybernetics, SMC-9, 856-860.

Saridis, G.N. (1979). Toward the realization of intelligent controls. Proc. IEEE, 67, 1115-1133.

Shackil, A.F. (1981). Design case history: Singer's electronic sewing machine. IEEE Spectrum, Feb. 1981, 40-43.

Tabak, D. and G.J. Lipovski (1979). Microcomputers in control. Invited paper in Proc. 1979 JACC, Denver, Colo., pp. 30-33.

Tao, T.F., Bar-Yehoshua, D. and R. Martinez (1977). Applications of microprocessors in control problems. In Proc. 1977, JACC San Francisco, CA, pp. 8-16.

Tsuchiya, T. (1982). Improved direct digital control algorithm for microprocessor implementation. IEEE Trans. on Automatic Control, AC-27, 295-306.

Van Amerongen, J. and A.J. Udink ten Cate (1975). Model reference adaptive auto pilots for ships. Automatica, 11, 441-449.

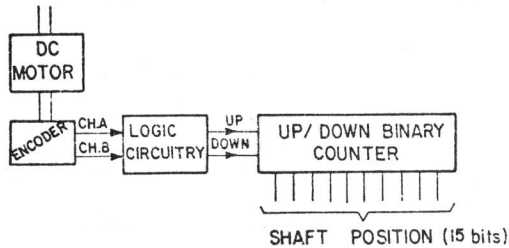

Fig.3 Position Sensing System

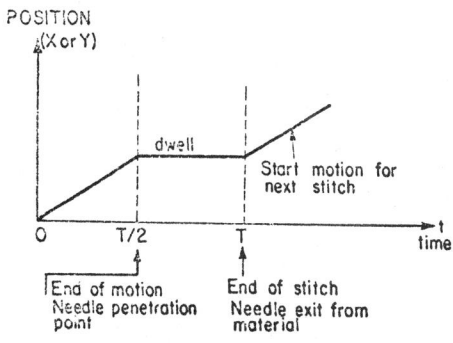

(a) Position time diagram

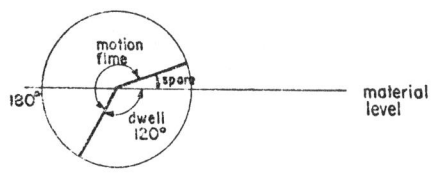

(b) Dwell diagram

Fig.4 Sewing Motion Coordination

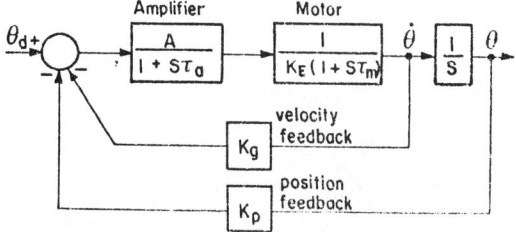

Fig.5 Control System Block Diagram

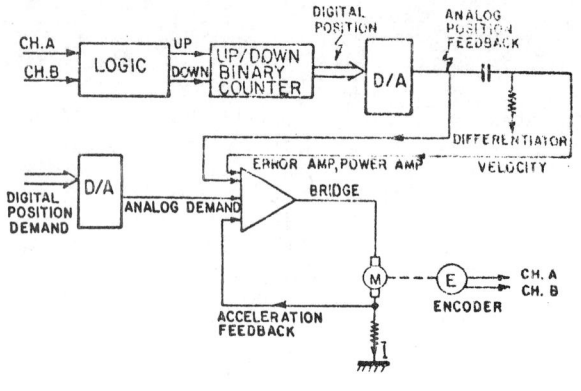
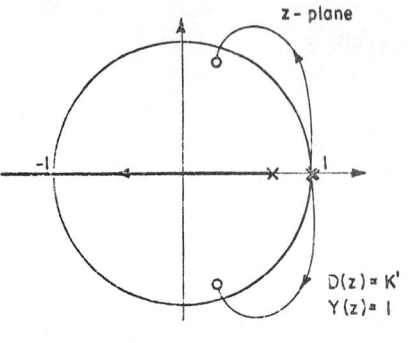

Fig.1 Analog Servo Loop

(a) uncompensated system

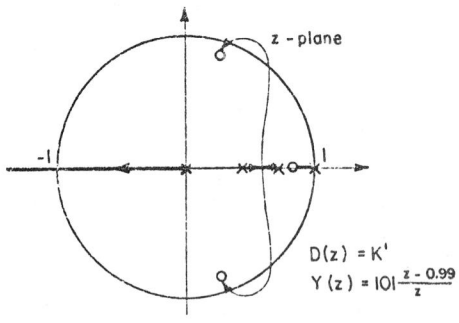

(b) compensated system

Fig. 7 Root Locus diagram of the control loop

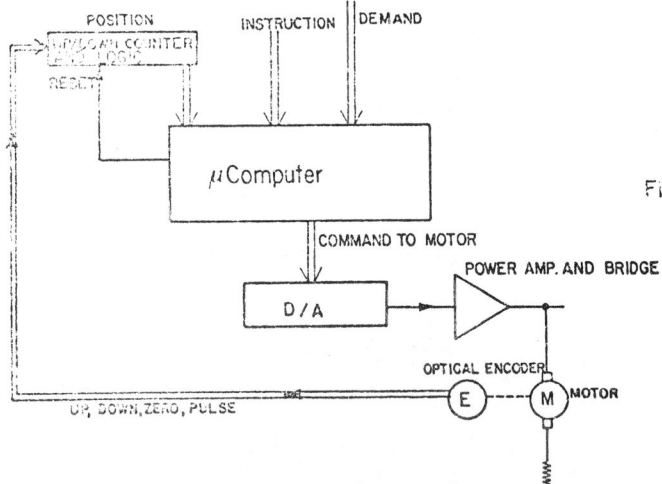

Fig. 2 Digital Servo Loop

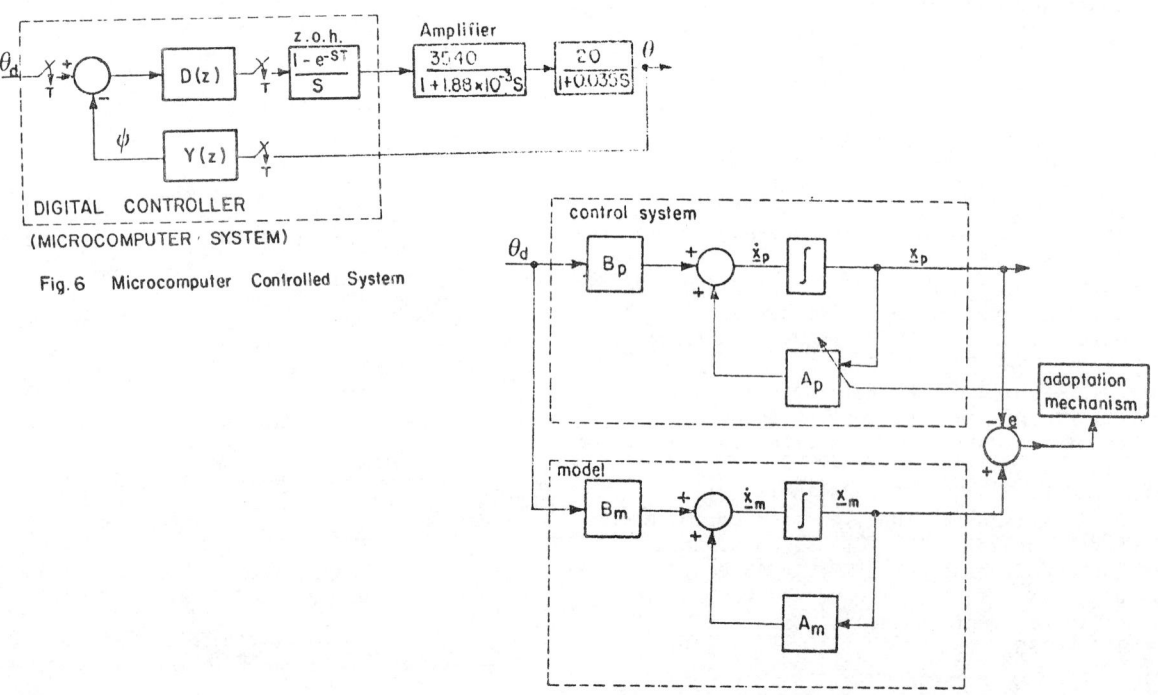

Fig. 6 Microcomputer Controlled System

Fig. 8 Model reference adaptive control system

MICROCOMPUTER DIRECT VOLTAGE CONTROL OF A PWM INVERTER

G. S. Buja and D. Longo

Istituto di Elettrotecnica e di Elettronica, Università di Padova, Padova, Italy

Abstract. A microcomputer-based closed-loop control system of the output voltage of a PWM inverter is described. The significant feature of the control system is that it demands the microcomputer to provide the dynamic control of the system output voltage in addition to the modulation of the inverter. Modelling of the overall system and optimal designing of the control algorithm are presented as well as their application to cases of practical interest. With reference to an actual case the effectiveness of the control system is verified on an experimental setup using a single-chip microcomputer as system controller. Description of the setup and some test results are given.

Keywords. Direct digital control; microcomputers; PWM inverter control; modelling; optimal control.

NOMENCLATURE

F, T frequency and period of the first harmonics of v_o, ($T=1/F$).

L_f, C_f, R_f PWM inverter output filter components.

m modulation index.

r reference input.

u disturbance.

v_i PWM inverter input signal.

v_o system output voltage.

v_{rms} rms value of the first harmonics of v_o.

V_{dc} dc voltage supplying the inverter.

Z_N nominal load.

1. INTRODUCTION

Microcomputers are being used more and more for controlling pulsewidth modulated (PWM) inverters because they make it practicable to switch the inverter output voltage as dictated by the digital modulation techniques (Abbondanti, 1978; Casteel, 1978; Murphy, 1977) and this results in a substantial improvement of the performance of the inverter-load systems (Patel, 1973; Patel 1974; Buja, 1980). Many papers have dealt with this application of the microcomputers, suggesting different methods to implement in them the digital modulation techniques. All these papers, however, refer to a use of the microcomputers as static controllers according to the following open-loop scheme: the rms value of the voltage first harmonics desired at the PWM inverter output is entered into the microcomputer and this sends to the inverter a PWM signal which is a low-level version of the inverter phase voltage, just as this has to be under rated working conditions in order to meet the requirements on the first harmonics and the constraints on the remaining harmonics, these latter being imposed by the selected modulation technique.

This paper presents a feedback control system of a PWM inverter, aimed to assign to a microcomputer the dynamic control of the inverter output voltage as well. The feedback action being formed every half cycle of the system output voltage, the study of the system is appropriately approached from the discrete system theory. Attention is focused on the analysis of the system dynamics and on the synthesis of the control algorithm with the design objective of the minimal response time. The control system is first assessed for an ideal PWM inverter-filter set delivering at its output a sinusoidal voltage and then extended to a PWM inverter followed by a single-section LC filter. With reference to

this case a control system based on a single-
-chip microcomputer has been set up and proved
in different working conditions. Application
of the control system to a PWM inverter without output filter is also discussed.

2. CONTROL SYSTEM DESCRIPTION

An overall block diagram of the control
system of the output voltage of a PWM inverter
followed by a filter is shown in Fig. 1. The
reference input is the rms value of the
voltage first harmonics desired at the system
output and the system output is the ac
voltage under control.

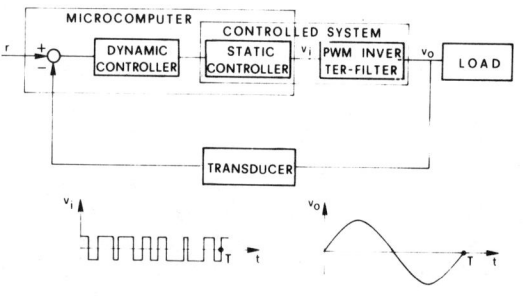

Fig. 1. Block diagram of the control system.

The controlled system is made up of the
static, microcomputer-based controller, the
PWM inverter and the filter. The static
controller acts as explained in the introduction and generates the PWM signal indicated
by v_i in Fig. 1. The PWM inverter provides
the amplification of the PWM signal, while
the filter eliminates all the higher order
harmonics from the inverter output voltage.
A filter with negligible values for both
response time and phase shift at the first
harmonics frequency is considered. Thus reference is made to an ideal PWM inverter-
-filter set which responds to v_i by delivering
a sinusoidal voltage v_o, in phase with the
first harmonics of v_i and with no delay. These
assumptions are useful to simplify in the
first instance the assessment of the control
system but they will be removed in the
sections dedicated to the applications.
However these assumptions are not so restrictive as it may appear if many switchings are
performed on the inverter output voltage and
if they are intended to eliminate the
harmonics near to the first one.

The function of the transducer is to detect
the rms value of v_o. Since the one-half cycle
integral of a sine wave is related to its rms
value by the factor $\pi/\sqrt{2}\ T$, a transducer
operating as follows is suggested: after
scaling and rectifying, v_o in entered into an
integrator circuit, which in turn is reset every half cycle of v_o. An S/H amplifier provides for
the sampling of the output of the integrator
circuit immediately before the reset, thus
obtaining a voltage proportional to the rms
value to be detected. An A/D converter
follows the S/H amplifier for interfacing
purposes. With this transducer the feedback
signal is discrete in time, the sampling
occurring once every half cycle of v_o. The
task of controlling and timing the transducer
operations may be conveniently carried out by
the same microcomputer used for the static
control of the inverter and this allows an
easy synchronization of the task with v_i. To
the purpose a routine is arranged which sends
to the transducer at the beginning of every
half cycle of v_i the following sequence of
commands: sample the output of the integrator
circuit, reset the integrator and convert the
sample data.

Finally the function of the dynamic controller is to process the error between the
reference input and the feedback value and to
supply the controlled system with a proper
manipulated rms value. Again the control
algorithm as well as the acquisition of the
reference input and the computation of the
error may be implemented in the existing
microcomputer. A keyboard or thumbwheel
switches may be used for the entry of the
reference input.

3. OVERALL SYSTEM MODELLING AND
CONTROL ALGORITHM DESIGN

By analyzing the overall system of Fig. 1 at
the beginning of every half cycle of v_i, the
model shown in Fig. 2 can be obtained. The
transfer function of the feedback loop is
z^{-1} on account of the fact that the transducer
senses the rms value of v_o at the end of each
half period. The feedback loop gain is made

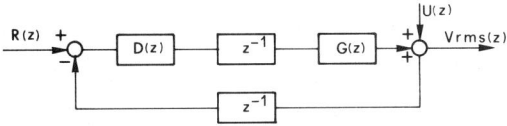

Fig. 2. Model of the control system.

equal to 1 by imposing a proper scale factor to the A/D converter. A delay is also present in the forward path. It is given by the microcomputer throughput delay due to the execution time of the following operations: command the transducer, acquire the feedback value, acquire the reference input, compute the error, compute the manipulated value and update the switching pattern. Practical experience indicates for this delay a value to some extent less than a half period of v_o, with a first harmonics frequency of 50 Hz. As a conservative choice a delay of one half period is taken. On the other hand the implementation of such a delay is advisable, thus avoiding a transient in the period of v_i and hence, of v_o, which would occur if the microcomputer updated the switching pattern before the actual half cycle of v_i had elapsed.

To model the controlled system, a discrete step is applied to its input and the discrete output is evaluated, after removing the static controller delay. The evaluation is made by computing every half cycle the rms value of v_o in the same way as the transducer and assigning the value to the time of the beginning of that half cycle. If the PWM inverter-filter set is ideal as defined above, the controlled system is linear. In addition its transfer function, indicated by $G(z)$, results in a pure gain, the value of which is set equal to 1 in a given working conditions. Note that the ratio between the rms value of the first harmonics of v_i and the value entered in the static controller, the inverter amplification and the filter gain at the first harmonics frequency determine the controlled system gain.

$D(z)$ is the control algorithm which has to be designed. Suggestion is made for a system with a minimal prototype response and zero steady-state error to a step input. This implies that $D(z)$ will include an integrator. By inspection of the system transfer function

$$W(z) = \frac{D(z) z^{-1} G(z)}{1 + D(z) z^{-2} G(z)} \quad (1)$$

and of $D(z)$ as derived from (1)

$$D(z) = \frac{W(z)}{\left[1 - z^{-1} W(z)\right] z^{-1} G(z)} \quad (2)$$

it follows that $W(z)$ must be

$$W(z) = z^{-1} \quad (3)$$

in order that $D(z)$ be realizable and stable. Substitution of (3) in (2) proves it, resulting

$$D(z) = \frac{1}{1 - z^{-2}} \quad (4)$$

with

$$G(z) = 1 \quad (5)$$

With this control algorithm the system output responds to a step input with a delay of only one half cycle and assumes from this time forward the desired value. In addition no steady-state error occurs in spite of a change of the parameters of the controlled system on account of the expected fact that $D(z)$ has a pole at $z=1$. With a control algorithm as (4), moreover, the system also exhibits a minimal prototype response to a disturbance modeled as in Fig. 2, when the disturbance is a step function. As a matter of fact, the transfer function of this disturbance is

$$W_u(z) = \frac{1}{1 + D(z) z^{-2} G(z)} \quad (6)$$

and for (4) and (5)

$$W_u(z) = 1 - z^{-2}, \quad (7)$$

which shows that the effect on the system output of the step disturbance disappears two discrete times after its occurrence.

In conclusion the system of Fig. 2 can be used effectively to track a reference rms value of a sinusoidal voltage as well as to stabilize the output sinusoidal voltage. This latter is the case of uninterruptible power supply (UPS) application.

4. APPLICATION OF THE CONTROL SYSTEM

A single-phase, half-bridge inverter modulated in such a way as to eliminate the third and the fifth harmonics from its output is considered. The modulation index for such a wave, i.e. the ratio between the peak value of the first harmonics and the peak value of the wave, ranges from 0 to 1.06. As shown in Fig. 3, the inverter output voltage is filtered with a single-section LC filter. The filter is designed to keep the total voltage distortion at its output less than 5% with m ranging from 0.4 to 1.06. From a recent paper (Bolognani and others, 1981) one obtains

$$L_f = \frac{0.6\ Z_N}{2\pi F}\ ;\quad C_f = \frac{0.8}{2\pi F Z_N} \qquad (8)$$

A resistance R_f is then inserted in series with L_f to represent the filter and inverter losses. The selected R_f gives the filter a quality factor of 10.

Nominal working conditions are: 50 Hz for the frequency of the voltage first harmonics of v_o and 220 V for its rms value, 11 Ω for the load, assumed resistive, and 1 for the gain of the controlled system at no load. Moreover a modulation index of 0.84 is taken at nominal load to compensate for a variation of -20 percent of the controlled system gain, even under this condition which is the worst as regards the filter gain. This variation may arise for instance from a drop of dc voltage supplying the inverter.

The controlled system just defined, modeled as a discrete one according to the rules given above, exibits a nonlinear behaviour, i.e. the ratio G(z) between the z-transform of the output and that of the input depends on the input itself. A reasonable approach for determining G(z) with a view to designing a proper control algorithm may be described as follows: to compute the discrete output of the system in response to a step input of the manipulated rms value from zero to that giving a steady-state output voltage rms value equal to the nominal one, to obtain the z-transform of the output and to divide it by the z-transform of the input. To simplify the determination of G(z), the following expression for the system output z-transform is taken

$$V_{rms}(z) = \frac{b_1 + b_0 z^{-1}}{(1 - z^{-1})(1 + a_0 z^{-1})} \qquad (9)$$

with the coefficients b_1, b_0, a_0 selected to obtain the inverse z-transform of (9) to assume the steady-state value of the actual output as well as its first two values. With a step input from 0 V to 220 V the discrete outputs at no load and at nominal load are analitically computed and given in Fig. 4a). As a comparison, the outputs for an ideal controlled system are given in Fig. 4b).

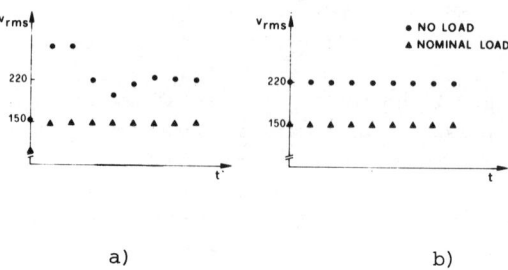

a)   b)

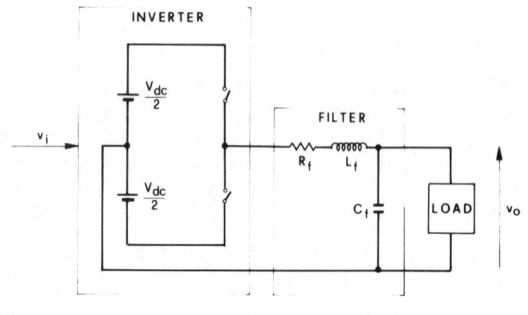

Fig. 3. Inverter-filter set under consideration.

Fig. 4. Computed controlled system responses to a step input from 0 V to 220 V: a) with the actual filter, b) with an ideal filter having the same filter gain as the actual one.

In computing (9) and, from this G(z), the output at no load is considered because this condition is the most critical for the stability of the system. The result is

$$G(z) = \frac{0.67 + 1.11\ z^{-1}}{1 + 0.78\ z^{-1}} \qquad (10)$$

An approach somewhat different from that proposed in Section 3 must be followed in this case in designing D(z). In fact, it could now occur that the output of the dynamic controller oscillates indefinitely between two values even if the discrete output of the system is constant. To avoid this, a deadbeat approach rather than the minimal prototype response is advisable. This implies to select $W(z)$ equal to

$$W(z) = z^{-1} K N_G(z) \qquad (11)$$

provided that the resulting $D(z)$ is stable, while it is surely realizable being the degree of the numerator of $G(z)$ equal to that of the denominator. In (11) $N_G(z)$ is the numerator of $G(z)$ and K a proper coefficient to give (11) a steady-state gain equal to one. Using (10), (11) becomes

$$W(z) = 0.38 \ z^{-1} + 0.62 \ z^{-2} \qquad (12)$$

K being equal to 0.56. Moreover, substitution of (12) in (2) yields

$$D(z) = \frac{0.56 \ (1+0.78 \ z^{-1})}{(1-z^{-1})(1+z^{-1}+0.62 \ z^{-2})} \qquad (13)$$

which results in a stable control algorithm. If the controlled system were linear, by using (13) as control algorithm the system would reach the steady-state condition two discrete times after the occurrence of a step input. In conclusion the root locus of the system with $G(z)$ given by (10) and that with $G(z)$ computed at nominal load and given by

$$G(z) = \frac{0.15 \ z^{-1} + 0.48}{1-0.076 \ z^{-1}} \qquad (14)$$

are sketched in Fig. 5.

5. EXPERIMENTAL SETUP AND TEST RESULTS

The method for the static control of PWM inverters described by Buja and Fiorini (1982) is referred to. It consists of storing in the microcomputer memory the switching patterns relative to a limited number of levels of the first harmonics and of demanding of the microcomputer the on-line computation of the actual pattern, by means of a linear interpolation of the stored data. This method allows both the continuous-amplitude control of the first harmonics of the inverter output voltage and the fulfilment of the constraints on the remaining harmonics, with a satisfactory trade-off between microcompu-

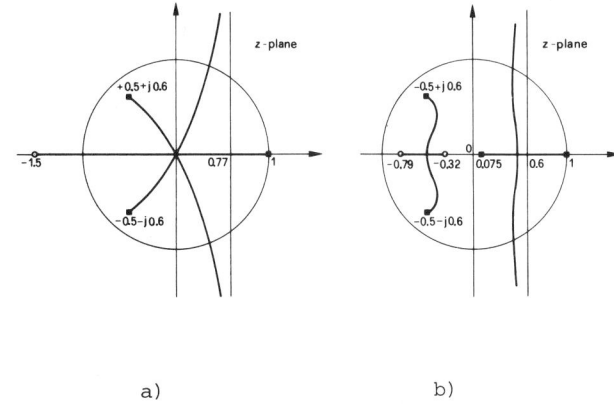

Fig. 5. Root-locus of the control system with $G(z)$ computed a) at no load and b) at nominal load.

ter throughput delay and program memory length. Consequently, a single-chip microcomputer of the MCS-48 family is effective in performing the function of static controller, as proved by Buja and others (1981). In the same reference, the implementation of the static control algorithm for the single-phase PWM inverter considered in Section 4 is detailed. Accuracy of about 1 percent both on the rms value of the first harmonics and on the elimination of the third and fifth harmonics has been achieved.

The additional implementation of the dynamic control algorithm (13) and that of the other control functions mentioned in Section 2 give rise to a total length of the program memory of about 1 Kbyte, suited to the use of the 8748 microcomputer. With a 6 MHz clock the microcomputer throughput delay is about 7 ms, but the update of the PWM signal is delayed to the beginning of its next half cycle.

The structure of the transducer is as described in Section 2. However, to get its accuracy to approach the resolution of the A/D converter (an 8 bit), two solutions are adopted. First, a filter is placed at its input to give a lower level to the harmonics of frequency higher than the fifth one, which would cause the detected rms value to be inaccurate. The additional filter does not affect appreciably the transducer dynamics. Second, to avoid the error due to the missing integration in the time interval required to reset the integrator circuit, two identical circuits operating in an alternate fashion are used.

Several tests have been performed on the experimental control system by simulating the inverter-filter set with an operational amplifier circuit. The results are satisfactory as it appears from the system responses shown in Fig. 6. Figures 6a) and 6b) refer to a step reference input from 10 V to 220 V at no load and at nominal load, respectively, while Fig. 6c) refers to a step variation of -20 percent of the controlled system gain at nominal load and output voltage of 220 V.

The difference in the settling time between the test results at no load and those expected are due both to the approximation of the actual output with (9) and to the non-linear behaviour of the controlled system. A computer simulation using the CSMP has proved this assertion. Instead the responses at nominal load coincide exactly with those expected because of the high value of the damping ratio of the controlled system.

6. FURTHER APPLICATION

The application of the control system to a PWM inverter without an output filter is now briefly discussed. As an example, this application occurs in the static supply of ac motors. The structure of the control system remains substantially that of Fig. 1, but the transducer must in addition extract the first harmonics from the system output voltage. Of course, this operation can be complex in implementation if the system operates in a wide range of frequency. A further complication arises moreover from the fact that the factor relating the half cycle integral with

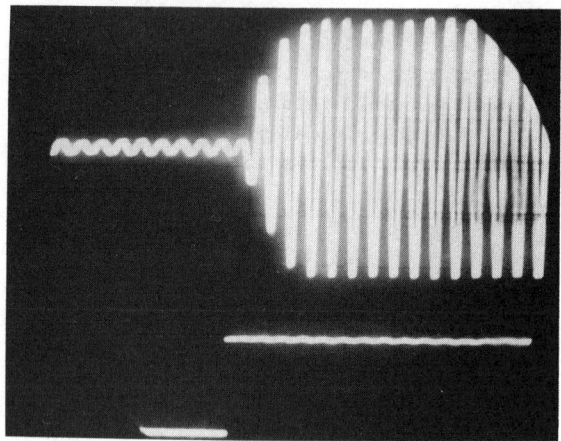

b)

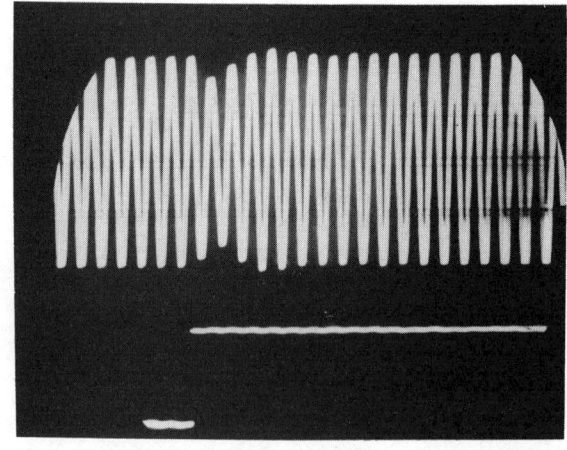

c)

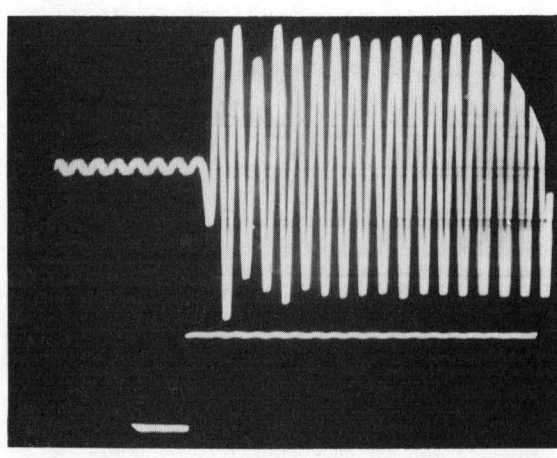

a)

Fig. 6. Experimental system responses: a) with a step input from 10 V to 220 V at no load, b) with a step input from 10 V to 220 V at nominal load, c) with a step variation of -20 percent of the controlled system gain at nominal working conditions.

the first harmonics depends on the frequency; but it can be easily overcome, if the multiplication by that factor is carried out by the microcomputer.

As regards the scheme of the control system, that of Fig. 2, with $G(z)$ and $D(z)$ given by (5) and (4) respectively, is valid if the extraction of the first harmonics in the feedback path does not affect its dynamics.

Being that his operation is performed at signal level, this condition can be satisfied with good approximation. Otherwise, the filter dynamics must be modelled as suggested in Section 4 for the controlled system and the control algorithm must be redesigned, taking into account the actual dynamics of the feedback path.

7. CONCLUSIONS

A microcomputer-based, control system of a PWM inverter was first described and was then analyzed and designed in the z domain. The case of an ideal inverter-filter set and that of an actual set have been detailed. The validity of the theoretical assessment was experimentally proved for the actual case, the more complex in investigation resulting in a nonlinear system to be controlled. The test results were found to be in tune with the expectations.

Thus the control system presented here can be effectively utilized in industrial applications of the PWM inverters as commanded or regulated voltage sources. The setup built for test purposes is also capable of industrial applications because it is very compact and inexpensive resulting from the use of a single-chip microcomputer as system controller.

8. REFERENCES

Abbondanti, A. (1978). A digital modulator circuit for PWM inverter. Proc. of 1978 IEEE Ind. Appl. Society Annual Meeting, pp. 403-501.

Murphy, J.M.D., R.G. Hoft, and L.S. Howard (1977). Controlled slip operation of an induction motor with optimum waveforms. Proc. of 1977 IEE International Conf. on Electrical Variable Speed Drives, pp. 157-160.

Casteel, J.B., and R.G. Hoft (1978). Optimum PWM waveforms of a microprocessor controlled inverter. Proc. of 1978 Power Electronic Specialists Conference, pp. 243-250.

Patel, H.S., and R.G. Hoft (1973). Generalized techniques of harmonic elimination and voltage control in thyristor inverters: Part I - Harmonic elimination. IEEE Trans. Ind. Appl., IA-9, 310-317.

Patel, H.S., and R.G. Hoft (1974). Generalized techniques of harmonic elimination and voltage control in thyristor inverters: Part II - Voltage control techniques. IEEE Trans. Ind. Appl., IA-10, 666-673.

Buja, G.S. (1980). Optimum output waveforms in PWM inverters. IEEE Trans. Ind. Appl., IA-16, 830-836.

Buja, G.S., and P. Fiorini (1982). Microcomputer control of PWM inverters. IEEE Trans. Ind. Electron., IE-8 (in press).

Buja, G.S., P. Fiorini, P. De Nardi, and M. Villa (1981). Improving the performance of microcomputer-based controllers for PWM inverters. Proc. of 1981 Powercon 8, E1-3, pp. 1-8.

Bolognani, S., G.S. Buja, and S. Ciscato (1981). Output filter design for PWM single-phase inverters. Proc. of 1981 3rd International Powerconversion Conference, pp. 246-254.

INDUSTRIAL APPLICATIONS OF VISION TECHNOLOGY

M. Ejiri

Central Research Laboratory, Hitachi Ltd., Kokubunji, Tokyo 185, Japan

Abstract. This paper reviews typical applications of vision technology in industry, and surveys its state of the art. A steady growth of applications has been seen in a number of industrial fields over the past decade. The motivation for such applications has always been the demand for more efficient production systems and product quality improvement by eliminating human error. Results so far are promising, though the barriers presented by existing limitations to binary image processing remain to be overcome.

Keywords. Image processing; pattern recognition; manufacturing processes; computer applications; artificial intelligence; assembly; inspection; classification; robots; sensors.

INTRODUCTION

To realize a more sophisticated level of automation, it has become essential to develop "intelligent" machines that can adapt to changes in external environment. Since conventional automatic production machines lack such ability, usage has been strictly limited to fixed type automation where the product changeover is relatively rare.

To cope with this limitation, and to realize modern automation systems especially for medium/small size production scale, much research has been devoted toward making machines intelligent. One such effort is computer vision research. This aims at realizing artificial eyes that can recognize three dimensional objects in an actual industrial environment. This research attempts to extract meaningful attributes of objects from their images. These attributes are thereupon used in finding, selecting or discriminating objects by comparing with a priori knowledge of the objects.

This paper first discusses functions of "intelligent" machines. Then, vision technology is focused on as the core technology of the machines, and is reviewed to reveal its state of the art from the application viewpoint. Lastly, typical industrial applications of vision technology are surveyed.

FUNCTIONS OF INTELLIGENT MACHINES

There are three major functions that future intelligent machines will require.

Understanding commands. The first is understanding commands given by a human operator. These commands are to be macroscopic preferably, in contrast to conventional microscopic instructions with one-to-one correspondence with each step of machine motion. These macroscopic commands should be as close as possible to the ones that the human is used to. Fundamentally, there are two possible forms in such commands; one is linguistic, the other is pictorial(Ejiri and others, 1972). Higher level language processing and natural language understanding have been extensively studied to date. The problem of inputting plan drawing information such as assembly drawings, logic diagrams, etc. to a computer is an example of the latter.

However, it is still very difficult for machines to understand these macroscopic commands and translate them into a meaningful set of microscopic instructions which the machines can directly handle. Therefore, instead of attempting to give the machines a thorough function of this type, the means for aiding the machines by man-machine interaction is one reasonable alternative at present.

Understanding surroundings. The second function is understanding the external surroundings of the machines. Surroundings are usually three-dimensional. The recognition of three-dimensional objects in the surroundings by scene analysis is also a very difficult problem. This field is presently being widely studied by many researchers. In these studies, TV cameras are usually hooked up to a computer with the image signals from the cameras being digitized and stored in a computer for subsequent analyses. This is why this field is often called computer vision, and every achievement in this field is now being watched with keen interest by industry.

Computer vision application in industry seeks to sense the existence of objects in the field of vision, to recognize the types of the objects, and to find the position of the objects, etc. for handling, tooling, assembling, inspecting and shipping. Geometric features such as size, peripheral length, area, moment of inertia, as well as existence of specific features such as holes, knobs, and notches are key parameters in such vision analysis.

Adaptive decision making. The third function is decision making. This is related to how machines can act upon or behave in the external surroundings by responding to human commands, and the recognition result of the external surroundings. For example, the problem of deciding an operational sequence for picking out prefered objects from among others, for properly assembling them, or for removing undesirable objects from bins, are within the scope of this function. The problem of revising the operational sequence by responding to more precise information on objects obtained in real time from tactile sensors during the course of motion is another example of the same scope.

VISION TECHNOLOGY

Of the three functions mentioned, external surroundings sensor, especially vision, is the most important from the viewpoint of industrial automation. The fact that more than 80 % of external information is obtained through the eyes in the human case attests to the relative importance of vision compared to the others.

Object Features

Generally speaking, an object has various features or attributes. These are classified into three categories as shown in Table 1. One is called geometric features which are closely related to the object geometry. Among these are shape, size, position, and orientation. These are features that can basically be measured by a scale or ruler. Shape recognition is especially important for automating classification processes, while position recognition is the key for assembly processes.

Another category of features is qualitative features that are related to human sensations such as softness, smoothness, and beauty. Machine perception of these features is essential for automating visual inspection processes on production lines. However, only processes where quantitative representation is possible, such as flaws in objects, has been automated so far.

The third category of features is those that are purposely put on an object surface. These include figures, marks, bar codes, labels and other intentional patterns. They are especially important objectives of visual recognition for automating sorting processes in physical distribution systems.

Method of Vision

When we attempt to recognize an object with an imaging device, there are basically two types of methods that can be adopted: active methods and passive methods. Details are shown in Table 2.

In the active methods, light beamed in the shape of a spot, slit, or grid is projected on the object. Reflected light image is then observed with an imaging device. The position of the light in the image plane carries information on the distance from the focal point of imaging device to the object point where the light beam is spotted. This is because the geometric relationship between the imaging device and illuminating light source is known a priori.

Among passive methods, there is one method that can obtain distance information directly. This method essentially necessitates intensive correlation calculations between two images obtained by two different cameras. Therefore, there are still great difficulties in realizing this method in industrial environments, though it is analogous, at least in appearance, to what the human does.

However, there are many cases in industrial vision applications where information on the distance does not play the main role. In these cases, recognition of projected object pattern in a two-dimensional plane is sufficient to achieve the purpose. For example, it is possible to check the existence of an object and find its location and orientation in a two dimensional plane. This is simply done by looking down on the object vertically with a TV camera. Here, only the top view of the object is utilized for analysis. Pattern matching technique is a common and basic technique that can be used in this case. Here, the input image is compared with the stored patterns in some form.

Keys for Actual Application

As mentioned, many attempts have been made to apply the two-dimensional computer vision to actual industrial use. However, successful applications are still rather limited. Some of those are briefly discussed in the later section. The reasons applications have not been widespread are summarized as follows:

1) Recognition algorithms are apt to be very complicated when reliable results are required.
2) Powerful processing hardware is usually needed as real time processing is essential.
3) Characteristics of imaging devices are not yet sufficient to always get stable and noiseless image signals.

A highly reliable algorithm is required for every application. However, as the recognition rate approaches 100 %, algorithm complexity increases exponentially. Also, hardware to process the algorithm is apt to become more and more sophisticated. A conventional computer cannot simply give us a sufficient solution. Thus, special, low-cost hardware with parallel processing capability seems to be required. Therefore, to make computer vision application a reality, the following must be taken into consideration.

1) Avoid the microscopic and sequential decision type method like edge following as much as possible to minimize confusion by noise. Adopt the homogeneous and macroscopic processing such as pattern matching and area counting methods to simplify hardware. If necessary, redesign objects to get more reliable results.
2) Use binary image from the biginning or as early as possible. Multi-level image processing usually entails lengthy processing and makes the processor scale large. Arranging the illumination condition and making the background simple are helpful for thresholding.
3) Make pre-processing good enough to cope with the possible image signal variation. Floating type thresholding is often effective for image variation.

APPLICATIONS OF VISION TECHNOLOGY

By considering the means mentioned, a few machines and systems have been put into operation in industrial field. Some will be briefly described in the following sections.

Automation of Assembly Lines

In processes where components are put together or relative position is controlled between the components, it is essential to detect the position and orientation of the components. The following machines and systems are examples of this type:

* Automatic assembly systems for semiconductor devices.
* Automatic bolting robots for concrete industry.
* A hose connecting robot for water pump testing.

In semiconductor assembly, a multiple local pattern matching technique was extensively studied and put into operation (Fig. 1) (Kashioka, Ejiri and Sakamoto, 1976). This method utilizes a plural number of standard patterns which are derived from the actual transistor image. Characteristic portions of the transistor image are selectively chosen as standard patterns so that each pattern position is uniquely determined in the image. Pattern position is first searched in a real-time mode as the image signal is fed into the image processor. By combining the detected coordinates of the two pattern portions, and by utilizing geometric a priori knowledge of each transistor pattern, the two elctrode positions, base and emitter, on which wires are to be connected, are determined.

With the automation of semiconductor assembly, many other techniques were developed, and are now being widely used for the production of ICs(Baird, 1976), LSIs(Mese and others, 1977) and hybrid ICs. Thus, the assembly of semiconductor devices has been entirely transformed recently.

The bolting robot(Uno, Ejiri and Tokunaga, 1976) detects the moving bolts on the flange of molds used for concrete pile and pole manufacture (Fig. 2). Before pouring concrete into the mold, these bolts, arranged side by side on the mold flange, should be tightened. After the concrete solidifies, these should be loosened again to remove the product. As the mold is moving at a constant speed, an image processor has to determine what is in the field of view before it passes by. A dynamic recognition method for moving objects was developed for this purpose. It descriminates the bolts, ribs, and tires of the mold. When a bolt is found, the image processor instructs a manipulator with an impact wrench on its tip to work on the bolt. However, if other objects are found, the image processor instructs the manipulator to take the rear position to avoid bumping. Thus, the image processor must have the capability of detecting the exact timing when the object passes through the vertical center line of the image field. This is realized simply by generating a few windows, each of which consist of four sub-areas. Counting of mismatched picture elements in each sub-area provides information of the degree of shape as well as position matching.

A hose connecting robot was developed (Uno and others, 1979) for the purpose of water pressure testing conducted at the end of water pump assembly lines (Fig. 3). Inlet and outlet holes are automatically detected by two solid state cameras. A pattern matching technique is again utilized here. However, only one-fourth of the circular hole pattern is matched at a time. Using simple sub-templates, one of the four quarter circular templates are dynamically selected. By this two phase matching technique, hardware size and processing time are minimized, thus giving the capability of real time visual feedback control of the arms.

Inspection of Surface Imperfection

To determine object quality, detection of abnormal portions is first attempted as possible candidates of the defects. Then, the portions are evaluated. Conventional visual inspection by the human eye lacks a quantitative description of the defect, and is highly dependent on individuals, though samples are sometimes used for comparison. Extensive efforts have been extended so far to make visual inspections more quantitative. From these efforts, machines and systems for the following processes have been developed and are now finding wider use:

* Inspection of printed circuit boards and their photomasks.
* Inspection of semiconductor wafers and their photomasks.
* Inspection of color picture tubes, shadow masks, and black matrices.

Printed circuit boards for consumer products usually have a bigger pattern width than those for computer and related products. Defect size in the former case usually differs from that of the normal conductor and insulator patterns. Therefore, the expansion and contraction method has been utilized (Ejiri and others, 1973) for detecting defects in color TV circuitry. Here, no standard pattern is used thus eliminating the registration problem between standard and input patterns.

In printed boards with finer patterns, directions of each conductor pattern is often limited to horizontal, vertical, and diagonal direction. A method using this regularity has also been developed (Minami and Sekizawa, 1976). Another method utilizes the connectivity of 3x3 local pattern features, where the rationality of vertical and horizontal combination of the features is checked (Goto and others, 1978).

For inspection of semiconductor wafers and their photomasks, the two chip comparison method is widely used (Fig. 4), as the same chip pattern is regularly formed with a constant pitch (Sandland, 1977) (Hara and others, 1980).

For color picture tube manufacture, inspections are also keys to product reliability. The dark blemish and white color uniformity are automatically checked as part of the production processes (Kinameri and Nonaka, 1979). Shape incompletion of shadow mask holes (Nakagawa and others, 1977) and of black matrices are other processes where the image processing techniques are applied.

Classification by Shape

There are processes where products are classified by their shapes into two or more categories. The following is an example where the automatic shape classification techniques are applied:

* Classification of diode chips.
* Classification of tablets and capsules.
* Classification of vegetables and fish.

Shape is one of the most important product attributes. This is especially so for electronic parts such as diode chips. If the shape differs from the normal shape, it might cause latent trouble in the future. Therefore, a machine for selecting diode chips was developed (Fig. 5) (Okamoto and others, 1975). Here, deformed chips from the normal circular shape, as well as defective chips with flaws and cracks, are checked out by simply detecting area A and peripheral length L and calculating ratio L^2/A. In this case, value 4π is expected for normal circular chips. Image is then rotated by 90 degrees on its center of gravity and overlapped with the original image to find the differences. This gives another chance to extract smaller defects. Almost the same principle was applied effectively for medicine tablet selection (Fig. 6) (Haga and others, 1979).

Classification of natural products in agriculture and fishery is also becoming important as these businesses often necessitate peak load labor for the limited periods of time. For these purposes, classification systems for cucumbers (Fig. 7) (Nakahara and others, 1979) and tomatoes are developed. The shape of a cucumber is detected as it passes under a linear solid-state sensor. The image is first thresholded into binary image and fed into an image memory. A microcomputer anlyzes the shape by finding the medial axis of the cucumber and the maximum distance of the bend from the line connecting both edges. A few other parameters are derived, and cucumbers are classified into three sizes(L,M,S) and four shape grades.

The classification system for fish (Fig. 8) (Nomura and others, 1979) also adopts the binary image analysis. Four typical fish usually caught in a mixed situation can be automatically selected by measuring such things as body length, body width, and tail height from their silhouettes.

Other Applications

Beside those mentioned above, the following machines have also been developed:

* A recognition and inspection machine for stamped and/or printed figures.
* A mark reader for packaged products.
* A Detection device for crowds.

In electronic parts, it is seldom possible to determine the part only from its shape. In these parts, the figures and the symbols stamped on the surface are the only cues for discerning the parts. Therefore, the figures and symbols on the parts can be regarded as a part of the quality that parts have. In this concern, an inspection machine for checking stamping quality was developed (Kashioka and others, 1980) and used in the stamping process of electronic parts to feed back information for ink control. The total number of picture elements that the figure consists of cannot simply give the classification boundary. In this machine, each figure is divided into several portions, and the pattern in each portion is checked independently. This is because the 20 % uniform thickness thinning of each figure is regarded as normal, but the missing 20 % of the figure in the part is serious. This machine is also used for reading figures on electronic parts placed on a circuit board to check if errorneous parts have been placed.

To automate the flow of final products in the factory, a long distance, single conveyor system is sometimes built which goes around each factory building, and finally to a warehouse. At each building, different types of products are produced and packaged in cardboard boxes, and placed on the conveyor to be carried to the warehouse. For this system, a visual device is usually needed at the entrance of the warehouse to detect product types and direct their flow to appropriate storage.

For this purpose, a vision system was developed (Fig. 9) (Ueda and others, 1977). It utilizes a special code mark designed to guard against deception by other neighboring patterns on the cardboard surfaces. The mark reader first detects an angular edge of the mark. Then, the code mark is automatically read by simultaneously tracking the upper and lower edges of the mark. As the code is formed in a two dimensinal manner, mark size becomes smaller than the conventional bar code mark. This is because the print quality of the mark is apt to be poor on a cardboard surface. The code employed here is made very similar to the arabic numerals, so as to be readable by the human eye. The code consists of 2x4 sub-areas each of which is designated as zero or one. The number of ones is plural in each numeral pattern. Thus, the Hamming distance between any two codes is at least two. This helps the system to reliably recognize the product mark.

There are a few requirements to roughly detect the number of people by visual means. In a group-controlled elevator system, for instance, information on the people waiting for the service at the elevator hall is quite important for effective traffic control of the elevator cages. Aiming at an application of this type, a visual device to detect crowds was developed (Fig. 10) (Yoda and others, 1977). The device utilizes the feature comparison method between the present and past image. Comparison is done for each block area of an image. Blocks where the image change is observed are extracted. The weighted sum of the number of blocks gives the rough number of people. Here, a direction coded image (Yoda and others, 1975) is used to minimize the effect of environmental brightness change. Furthermore, stored image data is updated each time no big image change is observed to adapt to the gradual change of the illumination condition.

FUTURE OF VISON APPLICATIONS

The state of the art of the vision applications in industry has been discussed, and the basic ideas governing each application have been briefly mentioned so far. One common feature governing these applications is that almost all eventually utilize the binary image although some floating type adaptive thresholding method is effectively used.

Another common fact is that almost all applications are "dedicated" systems, where the image processor is specially and carefully designed for a specific use. This is because the usual image processing algorithms are apt to require a longer processing time than expected in actual applications. It is true that no general purpose algorithms require shorter processing time than specific purpose algorithms.

However, there have been some attempts to make the general purpose image processor a reality. Some of these seem to have been successful, and are utilizing parameters such as area, peripheral length, moment of inertia, existence of holes, and many others detected from the top view of the objects. However, usage is still limited, as good illuminating condition required for a stable binary image is not always attainable. Furthermore, the performance of the image processor is not usually powerful enough when compared to the dedicated image processor developed for each application. It seems that the dedicated image processor will still be dominant in industrial application for some time. However, the accumulation of these application experiences, as well as the future revolution in semiconductor technology, will help in developing an efficient general purpose image processor in future.

CONCLUSION

The state of the art of the vision technology has been discussed through examples of actual industrial application. Some applications were quite effective in revolutionizing the production method used until then. This was especially so in the semiconductor industry.

To make wide use of vision systems in many other industrial environments, the more effective and efficient handling of the gray level and the color information is essential. Speeding up processing time with a unique architecture is another key.

Fortunately, this field is regarded as a very important and promising one, as many possibilities exist in conjunction with robotics and factory automation. Steady progress can be expected in the future, though this will depend largely on the future revolution in semiconductor technology.

REFERENCES

Baird, M. (1976). An application of computer vision to automated IC chip manufacture, Proc. 3rd Int. Joint Conf. Pattern Recognition, P. 3.

Ejiri,M., T.Uno, H.Yoda, T.Goto and K.Takeyasu (1972). A prototype intelligent robot that assembles objects from plan drawings, IEEE Trans. Comput., C-21, pp. 161-170.

Ejiri,M., T.Uno, M.Mese and S.Ikeda (1973). A process for detecting defects in complicated patterns, Comput. Graph. and Image Process., 2, pp. 326-339.

Goto, N., K.Ichikawa, T.Kondou and M.Kanemoto (1978). Automatic inspection system for mask patterns, Toshiba Review, 33, pp. 615-618. (in Japanese).

Haga, K., K.Nakamura, Y.Sano, N.Miyoshi and A.Komuro (1979). Tablet checker, Fuji Jiho, 52, pp. 294-298. (in Japanese).

Hara, Y., K.Okamoto, T.Hamada and N.Akiyama (1980). Automatic visual inspection of LSI photomasks, Proc. 5th Int. Joint Conf. Pattern Recognition, pp. 273-279.

Kashioka,S., M.Ejiri and Y.Sakamoto (1976). A transistor wire-bonding system utilizing multiple local pattern matching techniques, IEEE Trans. Syst., Man and Cybern., SMC-6, pp. 562-570.

Kashioka, S., T.Yasue and Y.Shima (1980). Quality measurement technique for stamped figures, Proc. Nat. Conf., Inst. Electr. and Comm. Engrs. of Japan (in Japanese).

Kinameri, K. and M.Nonaka (1979). White uniformity measurements on color TV picture tubes, Appl. Opt., 18, p. 135.

Mese,M., T.Miyatake, S.Kashioka, M.Ejiri, I.Yamazaki and T.Hamada (1977). An automatic position recognition technique for LSI assembly, Proc. 5th Int. Joint Conf. Artificial Intelligence, pp. 685-693.

Minami, M. and H.Sekizawa (1976). A new mask defect inspecting device, Kodak Photoresist Symp. p. 67.

Nakagawa, Y., H.Makihara, N.Akiyama, T.Numakura and T.Nakagawa (1977). Study on automatic visual inspection of shadow-mask master patterns, Proc. IFAC Symp. Information-Control Problems in Manufacturing Technology, pp. 63-70.

Nakahara, S., A.Maeda and Y.Nomura (1979). Automatic cucumber sorter using pattern recognition technique, Denshi Tokyo, IEEE Tokyo Section, No. 18, pp. 46-48.

Nomura, Y., S.Ito, M.Naemura (1979). Development of "MELSORT" pattern classification system and its application, Mitsubishi Denki Giho, 53, pp. 899-903. (in Japanese).

Okamoto, K., and others (1975). High speed inspection for circular shape components. Proc. 14th SICE Conf., No.3606 (in Japanese)

Sandland, P. (1977). Automatic inspection of mask defects, SPIE Seminar on Developments in Semiconductor Microlithography II.

Ueda, H., T.Uno, M.Ejiri and K.Nakamura (1977). Development of a mark reader for physical distribution systems. Proc. Nat. Conf., Inst. TV Engrs of Japan, No.11-6,(in Japanese).

Uno, T., M.Ejiri and T.Tokunaga (1976). A method of real-time recognition of moving objects and its application, Pattern Recognition, 8, pp. 201-208.

Uno, T., S.Ikeda, H.Ueda and M.Ejiri (1979). An industrial eye that recognizes hole positions in a water pump testing process, In G.G.Dodd and L.Rossol (Ed.), Computer Vision and Sensor-Based Robots, Plenum Press., New York. pp. 101-116.

Yoda, H., J.Motoike and M.Ejiri (1975). Direction coding method and its application to scene analysis, Proc. 4th Int. Joint Conf. Artificial Intelligence, pp. 620-627.

Yoda, H., J.Motoike, M.Ejiri and H.Kinoshita (1977). Development of image processor for detecting the number of persons. Proc. Nat. Conf. Information, Inst. Electr. and Comm. Engrs. of Japan, p.162 (in Japanese).

TABLE 1 Object Features

Feature category		Main usage	Typical methods	Examples
Geometric features	Shape / Size	Classification	Parameter extraction	Tablet, Capsule, Cucumber, Fish, Diode, etc.
	Position / Orientation	Assembly	Pattern matching / Area balancing / Mark finding	Semiconductor, Bolting robot, Pump-hose, etc.
Qualitative features	Flaws and cracks (Blur) (Color tone) (Texture)	Inspection	Expansion and contraction / Feature matching / Size measurement / Two chip comparison	Printed board, IC, LSI, Photomask, Black matrix, etc.
Supplemental features	Figures / Symbols / Marks	Sorting	Pattern matching	Character reader, Mark reader.

TABLE 2 Method of Vision

Method	Passive method				Active method			
Information utilized	Brightness (gray level)	Color (R,G,B; IR)	Distance					
			Stereo	Spot	Slit	Grid		
Typical processing sequence	Thresholding → Pattern matching	Edge finding → Parameter extraction	Region finding → Structure analysis	Distance finding	Edge finding → Structure analysis	Distance finding	Moire → Region finding	Fourier analysis
	Comparison/Evaluation				Comparison/Evaluation			

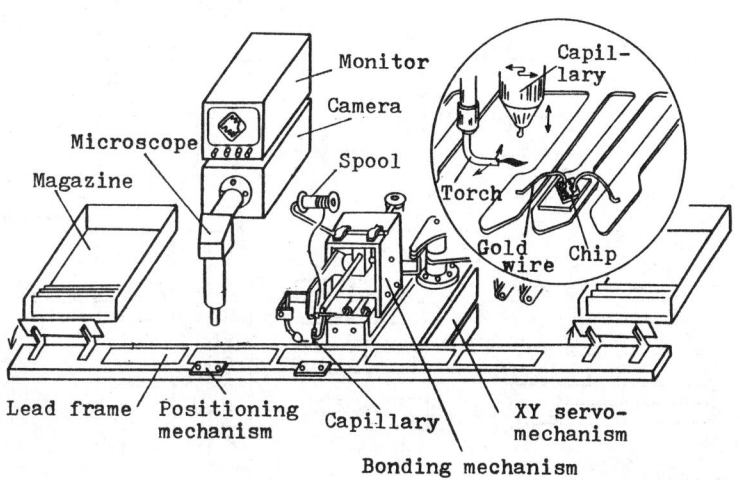

Fig. 1. Transistor assembly.

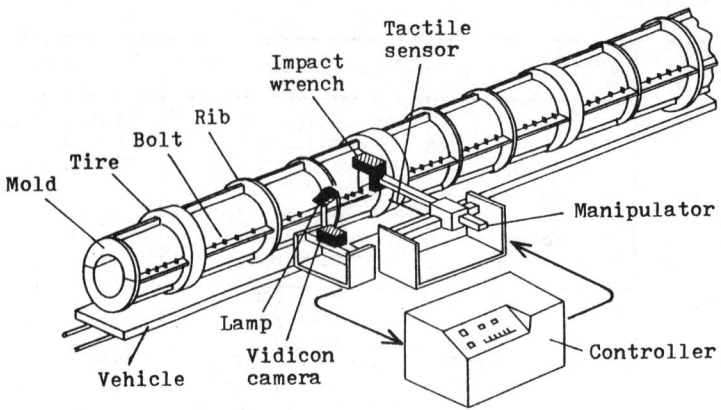

Fig. 2. Bolting robot for concrete pile and pole manufacture.

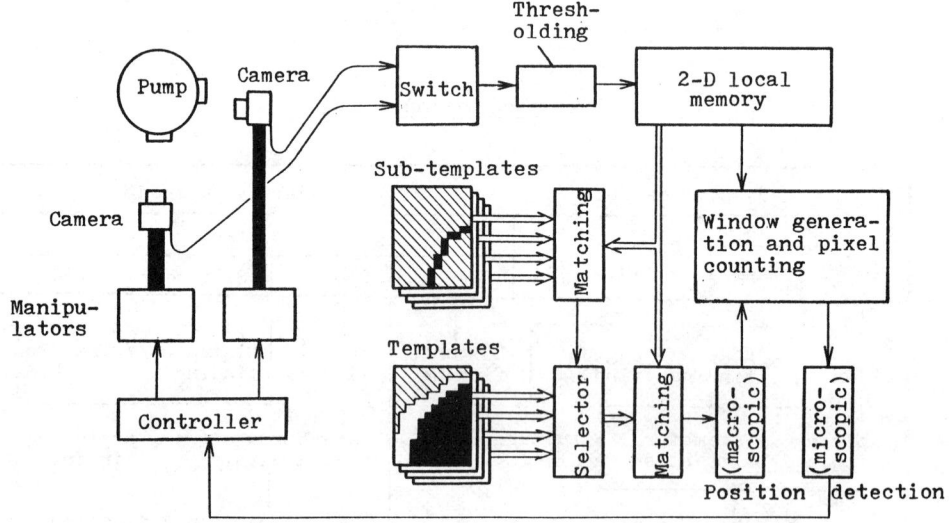

Fig. 3. Image processor for hose connecting robot.

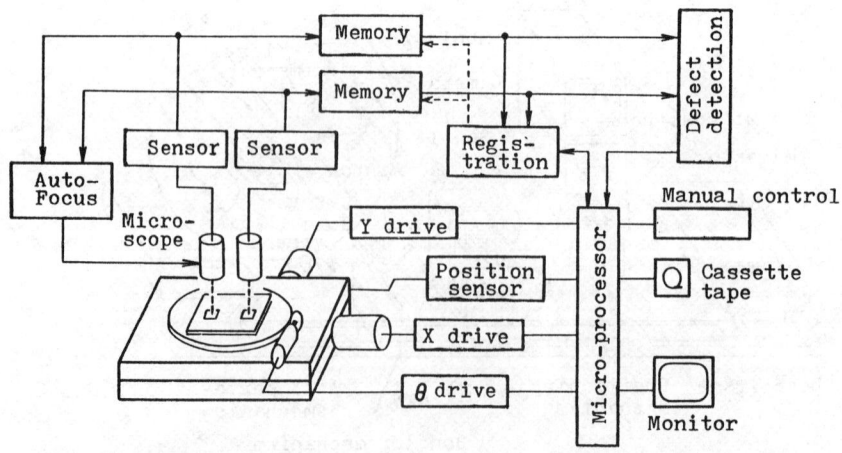

Fig. 4. Two chip comparison method for semiconductor photomask inspection.

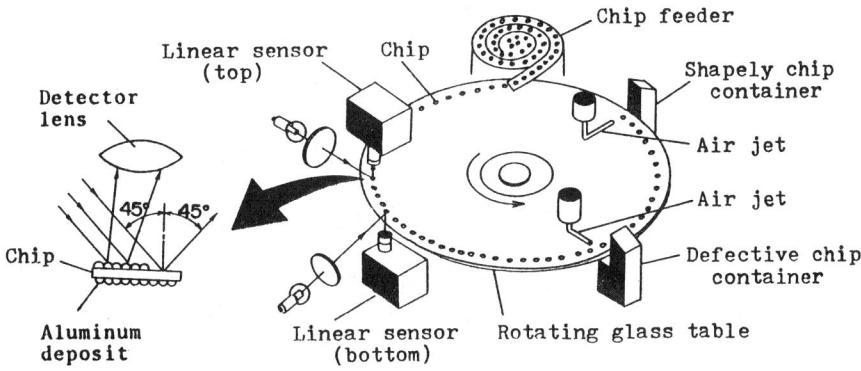

Fig. 5. Schematic diagram of diode chip classification machine.

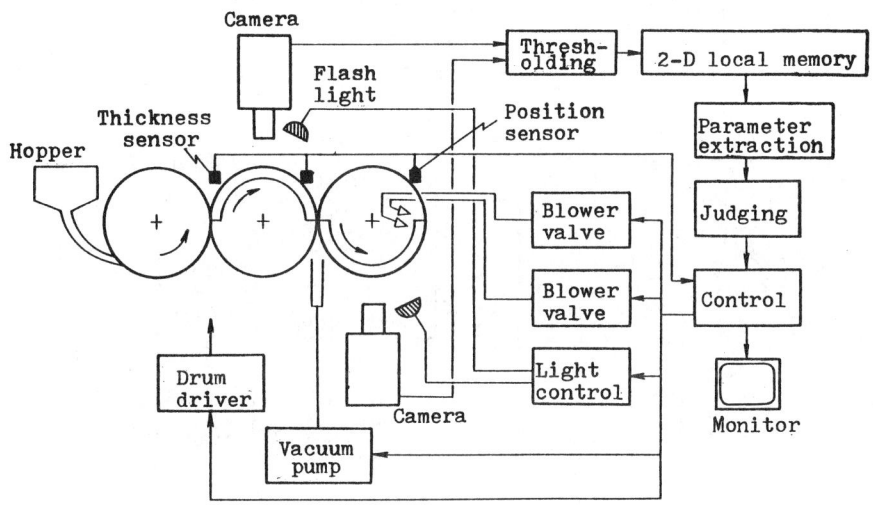

Fig. 6. Tablet checker.

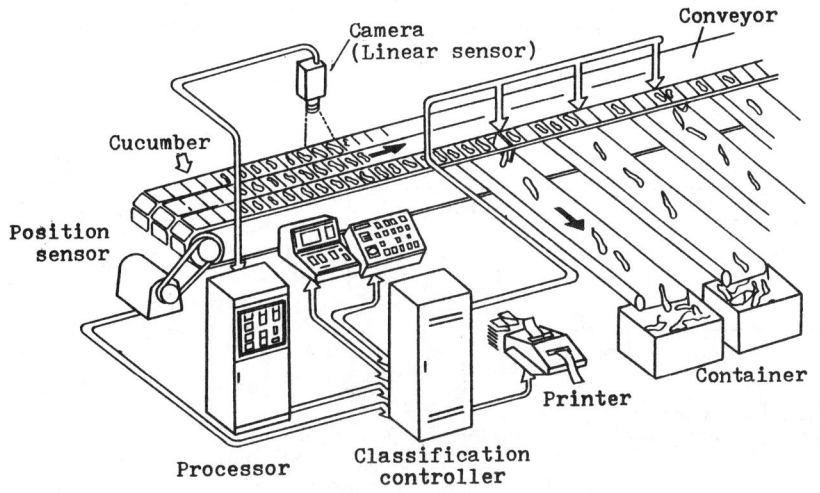

Fig. 7. Cucumber classification system.

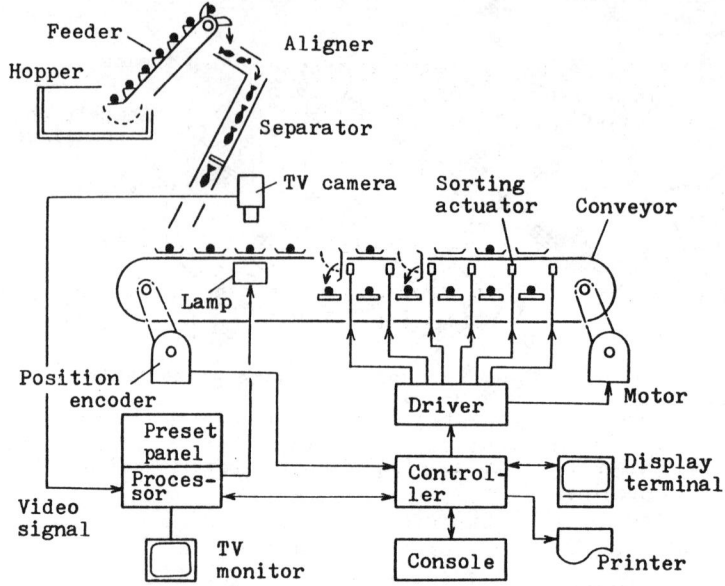

Fig. 8. Fish classification system.

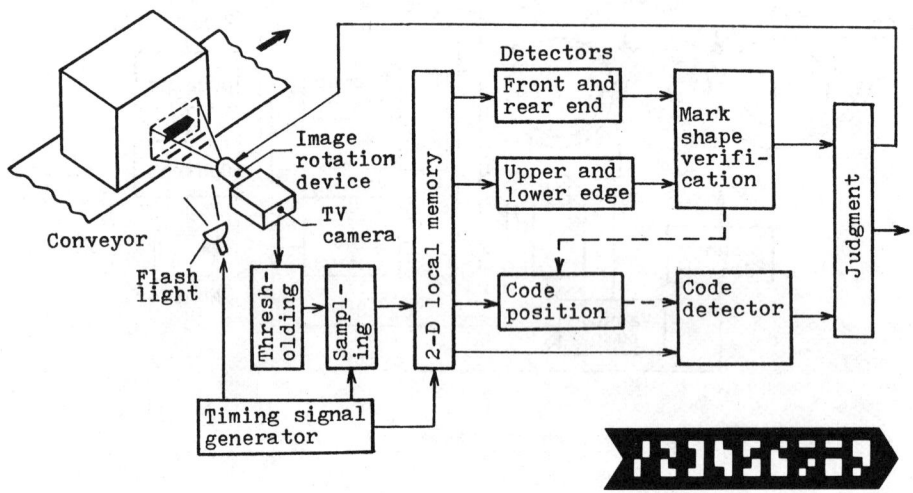

Fig. 9. Mark reader for physical distribution system.

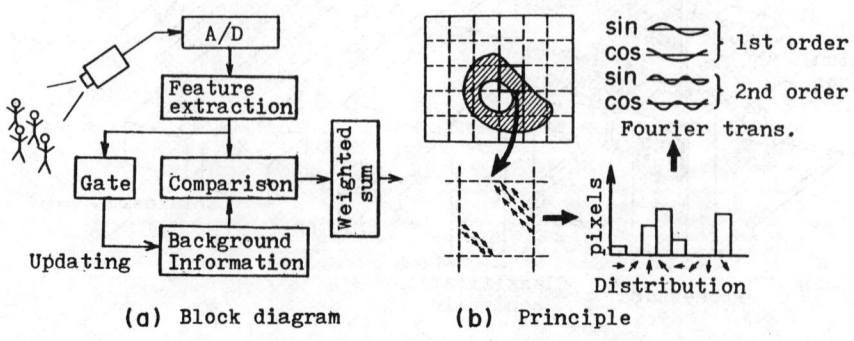

Fig. 10. Visual device for detecting rough number of people.

DIRECT DIGITAL CONTROL OF ELECTRICAL DRIVE SYSTEM BASED ON IMPROVED OPTIMAL REGULATOR THEORY

T. Tsuchiya

Department of Electrical Engineering, Faculty of Engineering, Hokkaido University, Sapporo, 060, Japan

Abstract An improved optimal regualtor theory is applied to synthesis of speed control system of an electrical drive system and the experimental results are shown. Whole procedures for design and experiment of the control system are described: Development of an improved optimal regulator theory, Derivation of a mathematical model of the controlled object based on the physical considerations, Identification of the parameters of the mathematical model from the step response of the experimental setup, Design of the control system by means of the improved optimal regulator theory, Implementation of the control algorithm with a microprocessor, Execution of the experiments and comparison the results with the simulation results.

Keywords Direct digital control, Electric drives, Microprocessors, Optimal control

1. INTRODUCTION

Application of modern control theory is one of the most interesting subject for control engineer since, with the advent of microprocessor technique, implementation of the control algorithm derived from modern control theory such as time optimal control, finite settling time control, state observer,etc. becomes very prospective (Farrar & Eidens 1980, Reeds & Meyler 1977, Seirah,Malik & Hope 1982). Above all, optimal regulator theory has been most applied to a real control problem until now. From the application of original optimal regulator control theory (Sage & White 1977) and a microprocessor as a controller of the control system point of view, however, the following difficulties should be overcome: deteriolation of the control system performance with parameter variations of the controlled object, low processing speed of the current microprocessor, constraints imposed on the control input variables, or the state variables. An improved optimal regulator theory based on original optimal regulator theory is proposed to overcome these difficulties (Tsuchiya 1981). A design method based on the improved optimal regulator theory and application of the method to an electrical drive system using a microprocessor system are presented in this paper. Electrical drive system under consideration in this paper is a static Scherbius induction motor system which makes the induction motor speed change as desired, by means of cycloconverter consisting of 18 thyristors. This system is adopted as an example for revealing the effectiveness of the method since this system is a suitable example for dealing with the above difficulties. That is, the magnitude of the control input variable which corresponds to gate signals of thyristor cycloconverter used is restricted to within certain values so as to keep normal operation of the cycloconverter. It represents the input variable constraints. Moreover, the processing time of the employed control algorithm is about 10ms in our experiment using a usual microprocessor system. This processing time cannot be disregarded because it is comparable with one-tenth of the electrical time constant of the controlled object. Input delay time due to the processing time, therefore, should be taken into account. In addition, the values of the parameters of the controlled object, of course, change with load change, disturbance, etc. A mathematical model of the controlled object is derived on the basis of the physical considerations, which results in the nonlinear second order differential equations with five unknown parameters. Values of the unknown parameters are determined by choosing the values so that the output response of the mathematical model for step change in the control input variable may approach to the output of the experimental setup for same change. It is, however, very difficult task to obtain exact values since effects of five parameters on the steady state values and the transient response of the original controlled object are very complicated. Experiment using a microprocessor has been made. The steady state error of the control system remains zero in spite of changes in the desired signal and the load. Moreover, the transient response is fairly fast with a little overshoot. Fairly close agreement between the experimental results and simulation results is obtained.

2. DESIGN METHOD BASED ON IMPROVED OPTIMAL REGULATOR THEORY

Usually, optimal regulator system designed by original optimal regulator theory has the following defects.
1. The steady state errors are not necessarily zero for any desired signals and disturbances.
2. Sensitivity for parameter variations of the controled object is not low, then, the steady state characteristic deteriolates with change of parameters.
3. Constraints imposed on control input variables are not directly considered although it could be indirectly taken into account by defining the quadratic performance index including input variables.

In addition, effective design method should be developed for considering input delay time due to the processing time required to execute the emplyed control algorithm as well as considering above points. The following design method is derived on the basis of optimal regulator theory to fulfill these requirements.

Consider the following discrete generalized original controlled object with an input delay time being equal to one sampling period. It means that the employed control algorithm is executed in a little bit shorter than one sampling period, which is accepted in real control problem in our case. Generalization to a control problem with longer input delay times than one sampling period is very easy. The sampling period T(s) will be omitted in the following equations.

$$y(k+1) = Ay(k) + G_b(k-1) + B_y d_y(k) \quad (1)$$

Output vector:
$$\bar{W}(k) = Cy(k) + B_w d_y(k)$$

Error vector:
$$e(k) = R(k) - \bar{W}(k) \quad (2)$$

where
- y (nx1) : state vector
- v (mx1) : control input vector
- R (rx1) : desired signal vector
- d_y (qx1) : disturbance vector

In electrical drive system there are two kinds of loads; one is constant torque load (conveyer, pump, etc.), the other is generator type load (generator, winder, etc.). The disturbance caused by the former case is represented by the disturbance vector d_y shown in (1), the disturbance caused by the latter case, however, is represented by parameter variations of system matrix A (Krishnan & Ramaswani 1974). Each matrix has appropriate dimensions. The original controlled object represented by (1) is assumed to be controllable and m=r is assumed for the simplicity of discussion.
A new matrix M is introduced as follows:

$$\begin{bmatrix} \bar{W}(k) \\ w(k) \end{bmatrix} = \begin{bmatrix} C \\ M \end{bmatrix} y(k) + \begin{bmatrix} B_w \\ 0 \end{bmatrix} d_y(k)$$

$$= Ty(k) + T_y d_y(k) \quad (3)$$

The new matrix M is chosen so that the matrix T is nonsingular. The vector $w(k)$ is reconstructed in terms of the state vector $y(k)$ of the controlled object. Equation (1) is transformed into (5) by using (3), where

$$TAT^{-1} = \Psi = \begin{bmatrix} \Psi_{11} & \Psi_{12} \\ \Psi_{21} & \Psi_{22} \end{bmatrix}, \quad TG_b = G_T = \begin{bmatrix} G_{T1} \\ G_{T2} \end{bmatrix},$$

$$TB_y = G_y = \begin{bmatrix} G_{y1} \\ G_{y2} \end{bmatrix} \quad (4)$$

$$\begin{bmatrix} \bar{W}(k+1) \\ w(k+1) \end{bmatrix} = \begin{bmatrix} \Psi_{11} & \Psi_{12} \\ \Psi_{21} & \Psi_{22} \end{bmatrix} \begin{bmatrix} \bar{W}(k) \\ w(k) \end{bmatrix} + \begin{bmatrix} G_{T1} \\ G_{T2} \end{bmatrix} [v(k-1)]$$

$$+ \begin{bmatrix} B_w \\ 0 \end{bmatrix} [d_y(k+1)] - \begin{bmatrix} \Psi_{11} B_w \\ \Psi_{11} B_w \end{bmatrix} [d_y(k)]$$

$$+ \begin{bmatrix} G_{y1} \\ G_{y2} \end{bmatrix} [d_y(k)] \quad (5)$$

The error system is derived as follows:

$$\begin{bmatrix} X_1(k+1) \\ X_2(k+1) \\ X_3(k+1) \\ u(k) \end{bmatrix} = \begin{bmatrix} I & I & 0 & 0 \\ 0 & \Psi_{11} & -\Psi_{21} & -G_{T1} \\ 0 & -\Psi_{21} & \Psi_{22} & G_{T2} \\ 0 & 0 & 0 & 0 \end{bmatrix} \begin{bmatrix} X_1(k) \\ X_2(k) \\ X_3(k) \\ u(k-1) \end{bmatrix}$$

$$+ \begin{bmatrix} 0 \\ 0 \\ 0 \\ I \end{bmatrix} [u(k)] \quad (6)$$

where
$$X_1(k) = e(k-1)$$
$$X_2(k) = \Delta e(k) = e(k) - e(k-1)$$
$$X_3(k) = \Delta w(k) = w(k) - w(k-1) \quad (7)$$
$$u(k) = \Delta v(k) = v(k) - v(k-1)$$

and where the assumption that the desired signal and disturbance signal are step functions was introduced. That is,

$$\Delta R(k) = 0 \quad \text{and} \quad \Delta d_y(k) = 0.$$

Equation (6) is condensed to (8).

$$X_u(k+1) = \Phi_u X_u(k) + G_u u(k) \quad (8)$$

The elements of each matrix and each vector are easily known by comparing (6) with (8). The error vector is related with new state vectors as follows:

$$e(k) = X_1(k) + X_2(k) \quad (9)$$

It is clear by observing (6) and (9) that it is necessary to regulate the state vectors $X_1(k)$ and $X_2(k)$ in order to keep the steady state error zero. The steady state error can be zero, provided that the system shown in (6) can be controlled to be stable by the appropriate control input vector $u(k)$. Moreover, it is clear that the steady state errors are kept to zero in spite of variations of parameters of the controlled object (including disturbance change due to generator type load change) as long as the system is controlled to be stable. One effective method for the determination of such control inputs is optimal regulator technique. The performance index is defined for (6) as follows:

$$J_u = \sum_{j=1}^{\infty} [e^T(j)\Gamma e(j) + u^T(j-1)H_u u(j-1)] \quad (10)$$

Equation (10) is transformed into (11) by using (9).

$$J_u = \sum_{j=1}^{\infty} [X_u^T(j)Q_u X_u(j) + u^T(j-1)H_u u(j-1)]$$

where
$$Q_u = \begin{bmatrix} \Gamma & 2\Gamma & \Gamma & 0 \\ 2\Gamma & 4\Gamma & 2\Gamma & 0 \\ \Gamma & 2\Gamma & \Gamma & 0 \\ 0 & 0 & 0 & 0 \end{bmatrix} \quad (11)$$

Solving the optimal regulator problem represented by (6) and (11), the feedback matrix F_{xu} is obtained as follows:

$$u^o(k) = F_{xu} X_u(k) = -\bar{B}_u G_u^T \bar{S}_u \Phi_u X_u(k) \quad (12)$$

where
$$S_u(n-k) = \Phi_u^T S_u(N-\overline{k+1})[I - G_u B_u(N-k)G_u^T$$
$$\cdot S_u(N-\overline{k+1})] \Phi_u + Q_u$$
$$B_u(N-k) = [G_u^T S_u(N-\overline{k+1}) G_u + H_u]^{-1}$$
$$S_u(0) = 0, \quad B_u(1) = [G_u^T S_u(0) G_u + H_u]^{-1}$$

$\bar{S}_u$ and $\bar{B}_u$ are the steady state solutions of the above discrete Riccati equation. In a real control problem, the control input variable constraints must be considered. In such a case, the above control system synthesis method is arranged as follows: As in (6) and (11), the fictious control input vector $u(k)$, which was derived by taking the first difference of the real control input vector $v(k)$ was used. It is basically impossible to deal with the problem with input variable constraints by means of the above formulation because the real control input vector $v(k)$ does not appear in the error system and performance index. Then, using the relation $u(k) = v(k) - v(k-1)$, (6) is rearranged as shown in (13). And the following performance index is defined for the system (13) at every sampling instant. (Equation (14))
The value of P is chosen to be equal to the steady state solutions of the discrete Riccati

$$\begin{bmatrix} X_1(k+1) \\ X_2(k+1) \\ X_3(k+1) \\ v(k-1) \\ v(k) \end{bmatrix} = \begin{bmatrix} I & I & 0 & 0 & 0 \\ 0 & \Psi_{11} & -\Psi_{21} & -G_{T1} & G_{T1} \\ 0 & -\Psi_{21} & \Psi_{22} & G_{T2} & -G_{T2} \\ 0 & 0 & 0 & 0 & I \\ 0 & 0 & 0 & 0 & 0 \end{bmatrix} \begin{bmatrix} X_1(k) \\ X_2(k) \\ X_3(k) \\ v(k-2) \\ v(k-1) \end{bmatrix}$$

$$+ \begin{bmatrix} 0 \\ 0 \\ 0 \\ 0 \\ I \end{bmatrix} [v(k)] \quad (13)$$

or $X(k+1) = \Phi X(k) + Gv(k)$

$$J(k) = X^T(k+N)PX(k+N)$$
$$\sum_{j=k}^{k+N} [X^T(j)QX(j) + v^T(j-1)Hv(j-1)]$$

where $Q = \begin{bmatrix} Q_u & 0 \\ 0 & 0 \end{bmatrix}$. (14)

equation under the assumption that N is infinity. The reason is not explained here for want of the space. This performance index is different from the previous performance index (11) and those used in Calovic & Cuk 1974, Pullman & Hogg 1979 and Tomizuka & Rosenthal 1979, etc. A similar performance index to those cannot be defined becuase in this case the value of the control input vector $v(k)$ does not necessarily converge to zero after infinite control interval, and a performance index with an infinite interval has no meaning. The performance index with a finite control interval must be defined as shown in (14). It is different from the others on the following points. The performance index used in the references includes the derivatives of the error vector of the real control input vector in order to introduce integral actions into the designed control system. On the other hand, the performance index (14) consists of the error and the real control input vector only. Such a performance index is suitable for the control problem with input variable constraints because the error system and the performance index include the real control input vector. If the performance index (11) consisting of the error vector and the difference of the real control input vector is defined, it is not possible to include hard constraints on control input vector through the performance index, although it is stated in Pullman & Hogg 1979 that the soft constraints can be imposed by suitable weightings in the H_u matrix. The integral actions are introduced into the control system as well, when (14) is defined (Tsuchiya 1981).
Solving the optimal regulator problem shown in (13) and (14), the optimal control input

vector is obtained as

$$v^0(k) = F_X X(k) = -\bar{B}G^T\bar{S}\Phi X(k)$$
$$= f_1 x_1(k) + f_2 x_2(k) + f_3 x_3(k)$$
$$+ f_4 v(k-2) + f_5 v(k-1) \quad (15)$$

The boundary conditions for the Riccati equation are

$$S(0) = P, \quad B(1) = [G^T PG + H]^{-1}. \quad (16)$$

$\bar{S}$ and $\bar{B}$ are the steady state solutions of the Riccati equation. In order to treat the constraints imposed on the control input variable the performance index (14) is modified as in (17). In the following, constraints for only one control input variable is considered for our experiemnt stated later. The constraints are shown as

$$V_{min} \leq v \leq V_{max}.$$

$$J(k) = X^T(k+N)PX(k+N) + \sum_{j=k}^{k+N}[X^T(j)QX(j)$$
$$+ h_0^2 H(h_0) + h_1^2 H(h_1) + h_2^2 H(h_2)]. \quad (17)$$

where

$$h_0 = v, \quad h_1 = -v + V_{max}, \quad h_2 = v - V_{max},$$

$$H(h_0) = \begin{cases} 0 & \text{for } h_1 < 0 \text{ or } h_2 < 0 \\ K_{s0} > 0 & \text{for } h_1, h_2 \geq 0 \end{cases},$$

and

$$H(h_i) = \begin{cases} 0 & \text{for } h_i \geq 0 \\ K_s > 0 & \text{for } h_i < 0 \end{cases} \quad (i=1,2)$$

When the value of the control input variable violates the value of either V_{max} or V_{min}, the penalty (represented by $H(h_1)$ or $H(h_2)$) imposed on the control input variable increases tremendously due to very large value of K_s. On the other hand, the two terms $h_1^2 H(h_1)$ and $h_2^2 H(h_2)$ have no effect on the value of the performance index, provided that the value of the control input variable does not violate its limits.

3. CONTROLLED OBJECT, CONTROL SYSTEM DESIGN AND CONTROLLER

To reveal the effectiveness of the method described, a speed control system of a wound rotor induction motor system (static Scherbius induction motor system) is designed by the method. This controlled object has both an input variable constraints due to the restricted range of the controlling angle of the cycloconverter and an input delay time due to the processing time required to execute the employed control algorithm in a microprocessor used as the controller. The system under consideration is shown in Fig.1. The processing time in the microprocessor, as stated before, cannot be disregarded when it is compared with the fast transient motion of the controlled object. This processing time should be taken into account as an input delay time. In our experiment, stated later, input delay time (which is taken to be equal to one sampling period) is taken to be 20ms, although real processing time is about 10ms. Because the processing time is sure to be longer for more sophisticated control algorithm which will be employed in our experiment in near future. In this system, the cycloconverter consisting 18 thyristors makes the induction motor speed change as desired. The range of the controlling angle, however, is limited to within certain vlaues in order to ensure normal commutations between thyristors. This represents the input variable constraints.

A mathematical model for this system is derived in the reference (Tsuchiya 1981). The following nonlinear differential equations are obtained.

$$\dot{y}_1(t) = \frac{4q}{\tau_m} y_2(t) - \frac{4q^2}{\tau_m} y_2^2(t) - \frac{B}{\tau_m} y_1(t)$$

$$\dot{y}_2(t) = -\frac{1}{\tau_e} y_1(t) - \frac{(1+q)}{\tau_e} y_2(t) + \frac{q}{\tau_e} y_1(t) y_2(t)$$
$$+ \frac{A}{\tau_e} v(t-T_A) + \frac{1}{\tau_e} \quad (18)$$

where $y_1(k)$: normalized motor rotational angular frequency

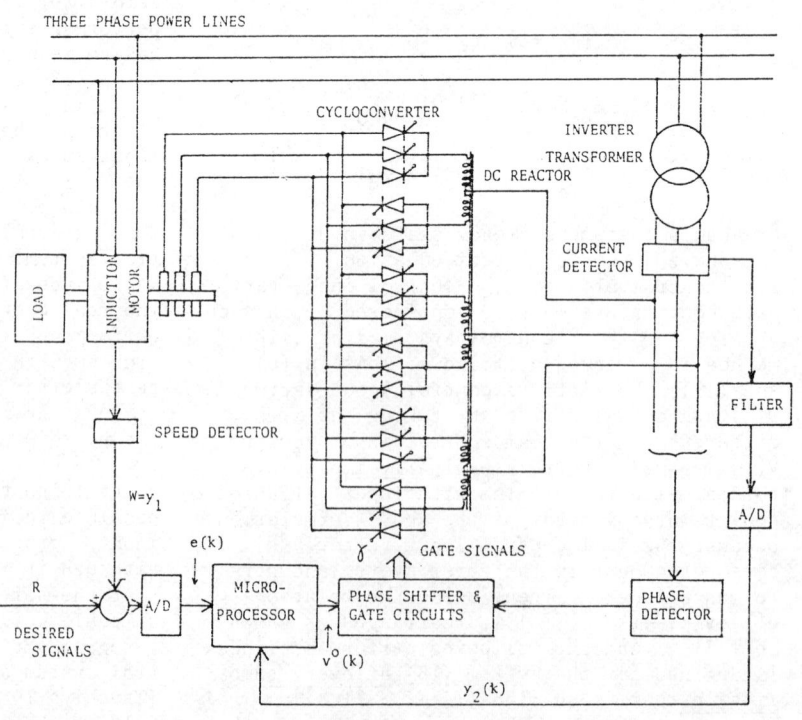

Fig.1 Controlled object

$y_2(k)$: normalized averaged rotor circuit current

τ_m, τ_e : mechanical and electrical time constants of the controlled object

B : load constant (generator type load)

q, A : system constants

T_A : input delay time (in our case $T_A = T$)

The system equations represented in continuous time form is made discrete form with a sampling period T after neglecting all the nonlinear terms $q^2 y_2^2(t)$ and $q y_1(t) y_2(t)$. Neglecting these terms is valid since the magnitude of $q y_2(t)$ is smaller than 0.2 at almost all operating conditions in this experimental setup. Then, (18) is transformed into the discrete form as shown in (19) which corresponds to (1).

$$\begin{bmatrix} y_1(k+1) \\ y_2(k+1) \end{bmatrix} = \begin{bmatrix} a_{11} & a_{12} \\ a_{21} & a_{22} \end{bmatrix} \begin{bmatrix} y_1(k) \\ y_2(k) \end{bmatrix} + \begin{bmatrix} g_{b11} \\ g_{b21} \end{bmatrix} [v(k-1)] \quad (19)$$

Disturbance term is not shown in (19), because the generator type load is applied in present experiment. Then, system parameters in (19) change with load change. Input variable constraint is $-0.5 \leq v \leq 0.0$ which corresponds to the value between 30° and 90° for the controlling angle of the cycloconverter. According to the design routine stated above, the following error system is derived.

$$\begin{bmatrix} X_1(k+1) \\ X_2(k+1) \\ X_3(k+1) \\ v(k-1) \\ v(k) \end{bmatrix} = \begin{bmatrix} 1 & 1 & 0 & 0 & 0 \\ 0 & a_{11} & -a_{12} & -g_{b11} & g_{b11} \\ 0 & -a_{21} & a_{22} & g_{b21} & -g_{b21} \\ 0 & 0 & 0 & 0 & 1 \\ 0 & 0 & 0 & 0 & 0 \end{bmatrix} \begin{bmatrix} X_1(k) \\ X_2(k) \\ X_3(k) \\ v(k-2) \\ v(k-1) \end{bmatrix}$$

$$+ \begin{bmatrix} 0 \\ 0 \\ 0 \\ 0 \\ 1 \end{bmatrix} [v(k)] \quad (20)$$

or $X(k+1) = \Phi X(k) + G v(k)$

Equations (20) and (17) constitute the optimal regulator problem. This problem, however, can only be solved by time-consuming repeated computations which cannot be executed by one microprocessor in a very short sampling period. Hence, this optimal regulator problem is transformed into a sub-optimal control problem as follows, in order to make it possible to implement it with one microprocessor.

1) $h_1 \geq 0$ and $h_2 \geq 0$

$$v(k) = FX(k) = f_1 e(k-1) + f_2[e(k)-e(k-1)]$$
$$+ f_3[y_2(k)-y_2(k-1)] + f_4 v(k-2) + f_5 v(k-1)$$
$$(21)$$

2) $h_1 < 0$ and $h_2 \geq 0$

$$v(k) = -F_h X(k) + V_{max} \quad (22)$$

F_h is obtained by defining the following performance index.

$$J_h(k) = X^T(k+N) P_h X(k+N) + \Sigma [X^T(j) Q X(j) + K_s[-v(j-1)+V_{max}]^2] \quad (23)$$

3) $h_1 \geq 0$ and $h_2 < 0$

$$v(k) = F_\ell X(k) + V_{max} \quad (24)$$

F_ℓ is obtained by defining the following performance index.

$$J_\ell(k) = X^T(k+N) P_\ell X(k+N) + \Sigma [X^T(j) Q X(j) + K_s[v(j-1)-V_{min}]^2] \quad (25)$$

This algorithm is executed in about 10ms by one microprocessor. There are five multiplications in this algorithm, whereas the multiplication is the most time-consuming calculation for the microprocessor. In our previous experiment, a multiplier-accumulator element TDC1003J was connected with the usual microprocessor system in order to execute the multiplications and accumulations in high speed, but the calculations were performed in the fixed point values. The scaling for the values treated in the microprocessor so as not to overflow were very troublesome and, nevertheless, troubles sometimes occured. In present experiment, a high speed arithmetic unit AM9511 is added in the microprocessor system to make it possible to execute the floating point calculations in high speed. Fig.2 shows the microprocessor system.

4. PARAMETER IDENTIFICATION

Obtaining a mathematical model of the controlled object, the values of the parameters are to be determined experimentally. In Fig.3 a solid line shows an experimental output response of the controlled object for step change in the control input variable v(k) and a dotted line in the same figure is a corresponding output response of the mathematical model with the most reasonable values of the parameters to be determined. The values were chosen as follows:

$\tau_m = 0.50s$, $\tau_e = 0.10s$, $q = 0.619$,
$A = 0.40$, & $B = 0.10$.

These values were chosen by the method of trial and error so that an output response of the mathematical model approaches to the

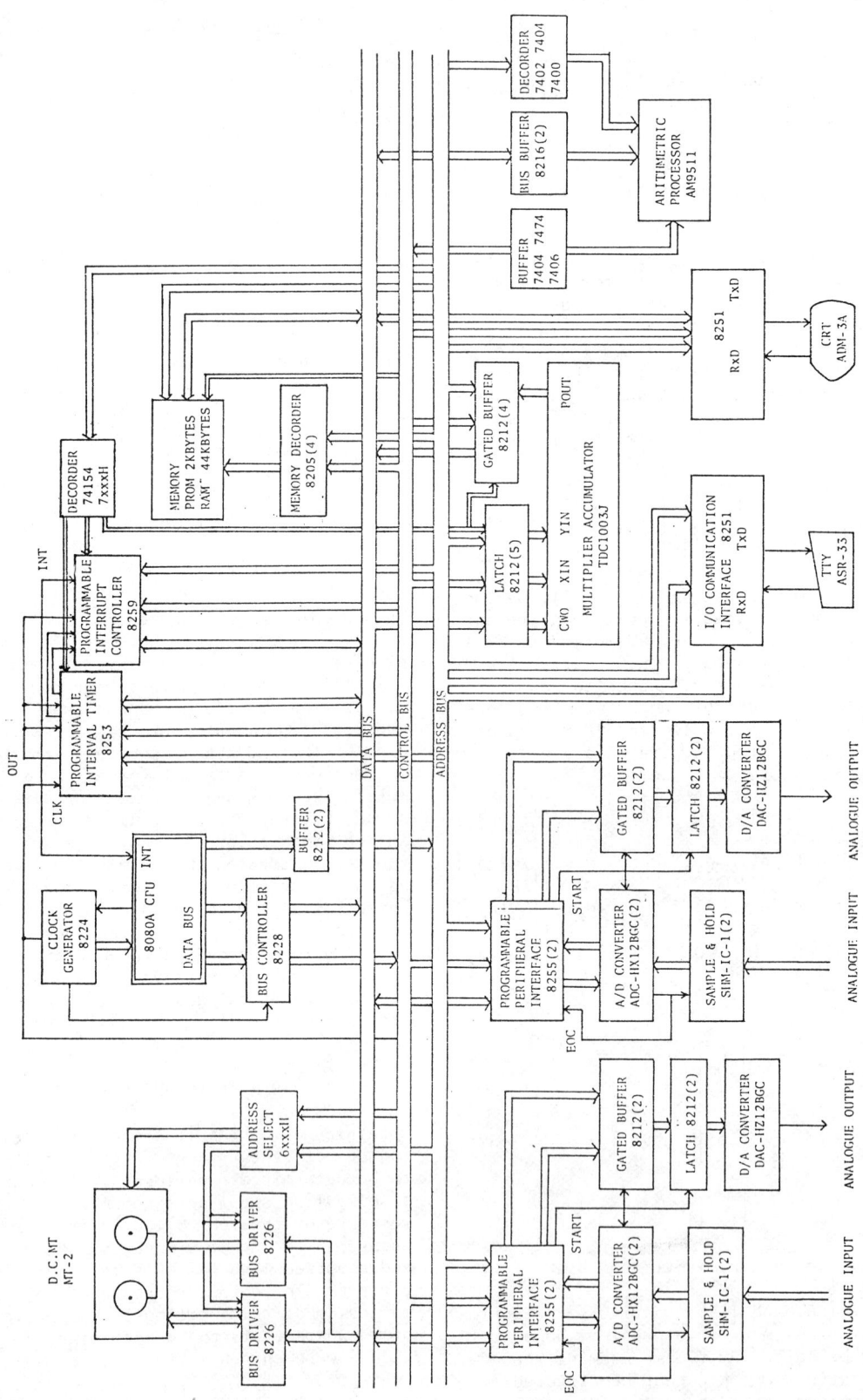

Fig.2 Microprocessor system employed in this experiment

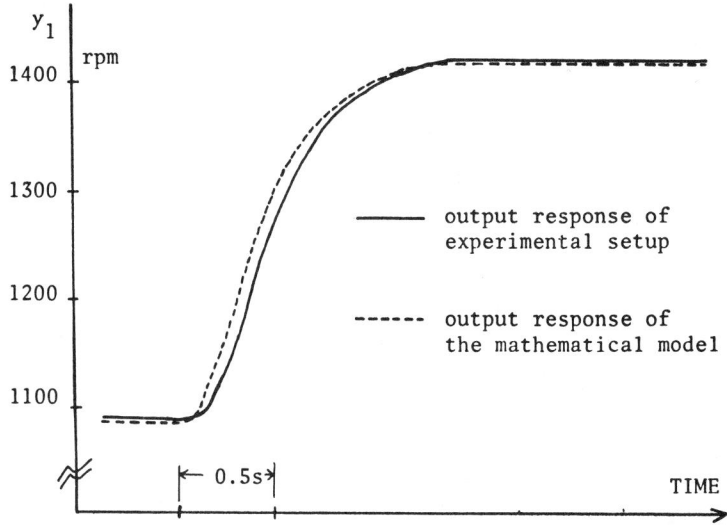

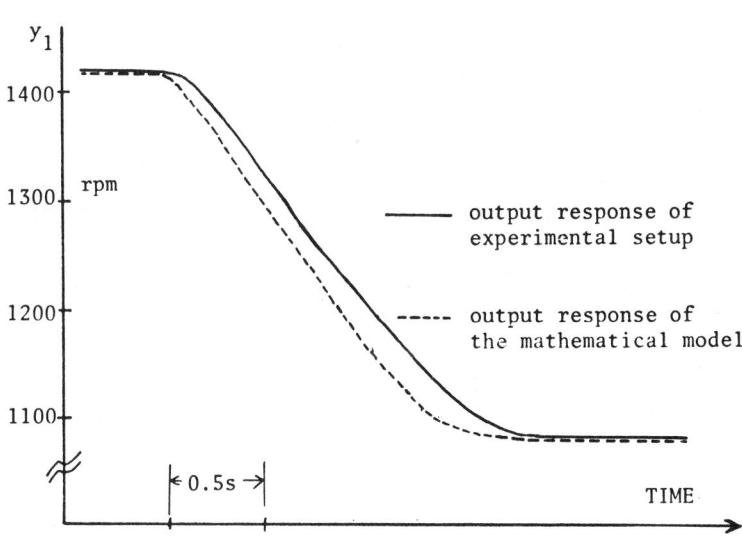

Fig.3 Step responses of the system and the model

experimental result, as close as possible, since there was no effective method to determine the parameter values because of the nonlinear mathematical representation of the controlled object.

5. EXPERIMENTAL AND SIMULATION RESULTS

Feedback gains in (21), (22) and (24) which are used in the experiment and simulation studies are obtained by the procedures stated in section 2 as follows:

$$f_1 = 7.4, \quad f_2 = 47.0, \quad f_3 = -5.5,$$
$$f_4 = 0.6, \quad f_5 = 0.4$$

for the feedback matrix F in (21). That is,

$$F = [\, f_1 \quad f_2 \quad f_3 \quad f_4 \quad f_5 \,].$$

And the elements of F_h and F_ℓ in (22) and (24) are so small that these are not needed to mention here, since the value of K_s is chosen to be very large (10^5).

Fig.4 shows an experimental result for step change in the desired signal $R(k)$. The steady state error is recognized to be zero and the fast transient response of the controlled variable $y_1(k)$ is obtained with a little overshoot. The other state variable $y_2(k)$ does not converge to the steady state value becasue of non constant control input value $v(k)$ which is the output signal of the microprocessor system. Theoretically, both the values of $v(k)$ and $y_2(k)$ must be remained constant at steady state. The reason for this are seemed to be the following points.
1) The microprocessor system is under influence of noises generated by thyristor converter used.
2) Fluctuations of the state variable $y_2(k)$ because of the harmonics generated by thyristors even though a filter to eliminate the harmonics to some extent is inserted, although in theoretical and simulation studies such a situation does not occur.

Fig.5 shows a reponse of the control system for step change in the load. As mentioned before, as the load is generator type load, then the load change results in the step change of the parameter B in (18). The system recovers zero steady state error for the step change of the load with fairly fast transient response. Magnitude of the load change in this case corresponds to about one-third of the full load. Judging from Fig.4 and Fig.5, the control system designed by the proposed method is seemed to have good characteristic for changes in the desired signal and the load. Then it may be concluded that the proposed method is useful method to design the control system and to implement the control algorithm with the microprocessor. In order to make sure the exactness of the experiment and make firm basis for extending the technique to the other field rather than power electronics field, we have made a comparison between the experimental result and the simulation result. Fig.6-Fig.8 are the simulation results for comparison with the corresponding experimental results in Fig.4. Fairly close agreements between these results are obtained except the last part of the figures.

6. CONCLUSION

Control system design method based on an improved optimal regulator theory is applied to synthesis of speed control system of an electrical drive system and the experiment using a microprocessor as a controller is carried out. Fairly good results and pretty close agreements between the experimental and

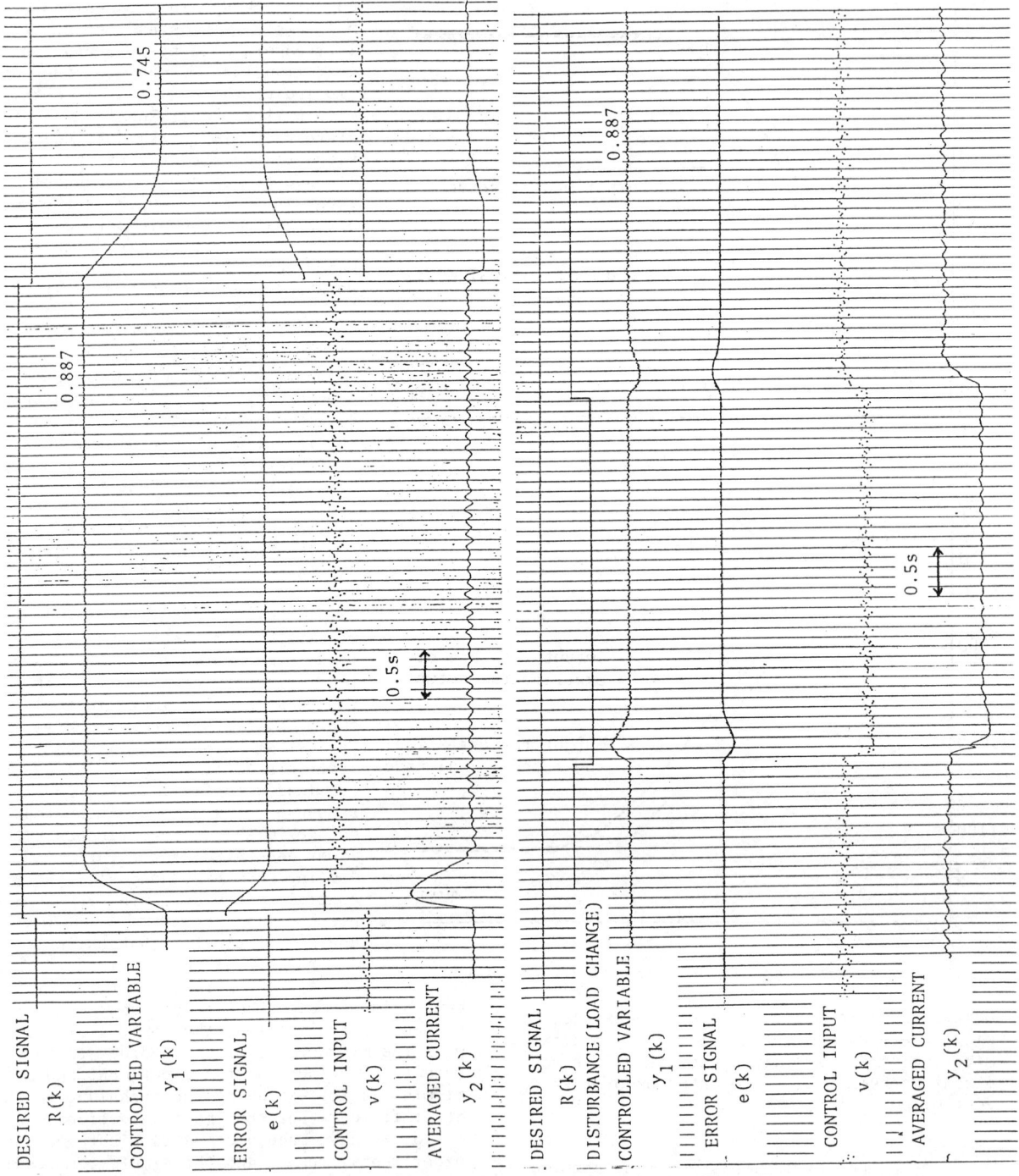

Fig.4 Experimental result for step change in desired signal

Fig.5 Experimental result for step change in the load

and the simulation results are obtained. Throughout the study, this method is considered to have reached the stage for practical use, although the success depends on mostly the results of the modelling and the parameter identification of the controlled object under consideration. The method will be applied to the case that only output variables of the controlled object are accessible and the multi-input multi-output system.

REFERENCES

Calovic,M.S. and Cuk,N.M.(1974) Proportional integral derivative realization of optimal linear quadratic regulator. Proc.IEE 121, 1441-1443

Farrar,F.A. and Eidens,R.S.(1980) Microprocessor requirements for implementing modern control logic IEEE Trans. Autom. Control 25, 461-468

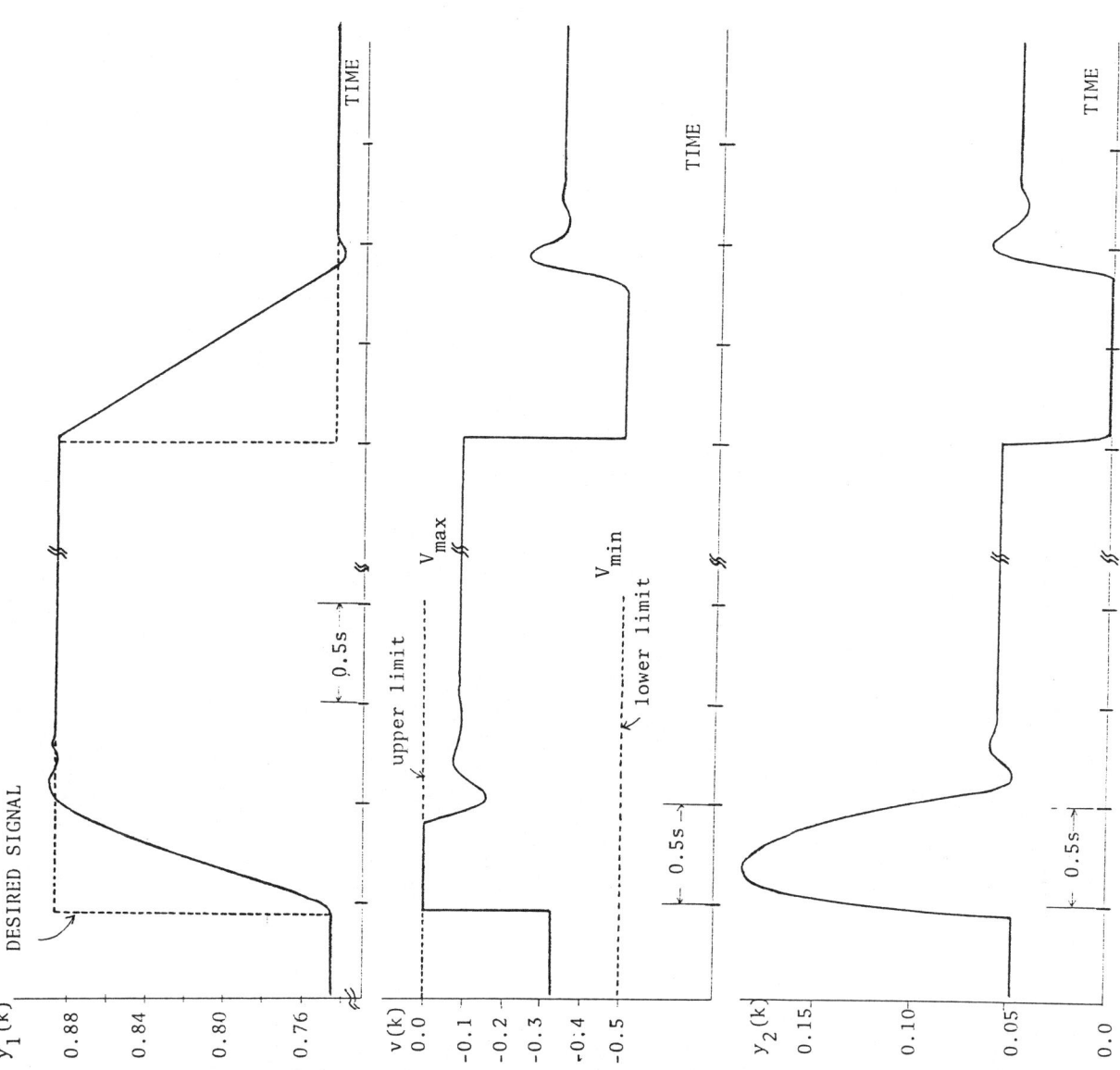

Fig.6 Simulation result of output variable $y_1(k)$ for step change in desired signal

Fig.7 Simulation result of control input variable $V_{min} \leq v(k) \leq V_{max}$

Fig.8 Simulation result of the state variable $y_2(k)$ for step change in desired signal

Krishnan,T. and Ramaswami,B.(1974) A fast response dc motor speed control system. IEEE Trans. Ind. Appl. IA-10 643-651

Pullman,R.T. and Hogg,B.W.(1979) Discrete state space controller for a turbogenerator. Proc.IEE 126, 87-92

Reed,M. and Meyler,H.W.(1977) A microprocessor based control system. IEEE Trans. Ind. Electron. & Control Instrum. IECI-24, 253-257

Sage,A.P. and White,C.C.(1977) Optimum system control, Prentice-Hall

Sheirah,M.A.,Malik,O.P. and Hope,G.S.(1982) Self-tuning microprocessor universal controller. IEEE Trans. Ind. Electronics IE-29 31-37

Tomizuka,M. and Rosenthal,D.E.(1979) On the optimal digital state vector feedback controller with integral and preview actions. Trans.ASME 101, 172-178

Tsuchiya,T.(1981) State equations of electrical drives. IEEE Trans. Ind. Electron. & Control Instrum. IECI-28, 194-200

Tsuchiya,T.(1981) Application of improved optimal regulator technique to power electronics. Optimal Control Applications and Methods 2, 209-224

STEPPING MOTOR CONTROL

R. Canales-Ruiz and L. Alvarez-Icaza

Instituto de Ingeniería, Automatización, UNAM, México

Abstract. An algorithm for computing robust sets of pulse timing for stepping motors that will follow arbitrary but predetermined trajectories is presented. The algorithm uses simulated feedback control and pole assigment techniques to compute the desired torque. One pulse per step scheme is adopted to keep small the number of commands and consequently memory requirements to store the control sequence in the computer. With the simulated controller the required torque switching angle and pulse timing are computed.

Results of the described technique applied to a second order system are presented. The guaranteed stability region for the switching angle is determined analytically. The robustness of the scheme is illustrated through large parameter variations.

Keywords. Stepping motors; torque control, digital control, non-linear control systems, pole placement.

INTRODUCTION

Stepping motors are electromechanical transducers that convert sequences of digital pulses into angular displacements. Stepping motors were originally designed to rotate a fixed angular displacements each time a pulse is applied to their controllers, and have been succesfully used to follow arbitrary positions with low speed and small accelerations. The way to obtain the timing (in the start-stop motion at every step) is to supply the pulses as a direct replica of the desired position-time relationship.

Due to the fact that stepping motors have certain advantages, such as being inherently pulse driven devices, easily interfaced with pulse generating digital systems and that under mild conditions, the shaft position is independent of load, reliable positioning can be achieved with open loop control. Stepping motors have been widely used for incremental motion in computer peripherals (printers, diskettes, drivers, plotters), wirewrapping machines, scientific instruments, testing equipment and numerically controlled machine tools.

Stepping motors are presently used only to position small to medium size loads and for a certain type of motion because appropriate algorithms are missing. For inertial loads it is not a trivial matter to compute the pulse timing that guarantee that no skipping or gainning of pulses are presented. For point to point minimum time motion, some ad hoc algorithms for pulse timing have been developed.

The contribution of this paper is to present and test a general scheme to produce the pulse timing for high speed arbitrary trajectories and loads.

SIMPLE STEPPING MOTORS CONTROL

Kuo (1979) and Leenhouts (1980) give detailed descriptions of the construction and characteristics of four winding permanent magnet stepping motors. For the purpose of this paper it would be enough to describe them as elements in which the relationship between torque (τ) and position (θ) is:

$$\tau = T \sin(N_p \theta - N_p k\pi/2) \qquad (1)$$

where $k=\{1,2,3,4\}$, depending on the energized winding, N_p number of steps per revolution to number of phases ratio and T the maximum available torque.

In Fig. 1 the available torque for each winding and the corresponding equilibrium points are shown.

The sequence of winding switchings is implemented through a controller (simple voltage, dual votage, chopping, etc) and it is common to all controllers to have clockwise and counterclockwise ports for the corresponding motion of the motor, determined by the switching times, that is, by the function $k(t) = \{1, 2, 3, 4\}$. Only in very special cases is possible to select without detailed simulations the function $k(t)$ to follow a specific trajectory. One of these occurs when the motor load is essentially viscous.

When an inertial load should be taken from one rest point to another in minimum time without caring for the specific trajectory involved (speed vs. time), is common to divide the trajectory in three parts: acceleration, sliding and braking. During the first part, to maximize acceleration, switchings occur at a fixed lead angle. This scheme is mantained until the design velocity is reached. Switching times are computed through digital simulations. In Fig. 2 a typical trajectory is shown.

Non-inertial loads

When the load of a stepping motor is viscous, the equation θ and τ

$$B\dot{\theta} = \tau$$
$$\tau = T\cos\theta \quad (2)$$

(θ here indicates electrical degrees; 1 step equals $\pi/2$ elec. deg.).

When the applied torque is zero, the resulting velocity is also zero and therefore the motor stops. This way, in each switching and for slow speeds, at every step the system reaches an unique equilibrium point.

The desired trajectory to follow $\theta_D(t)$ is divided into equal angular increments of motor resolution size. The time between two consecutive steps determines the switching times. The motion will be clockwise if $\dot{\theta} > 0$ and counter-clockwise if $\dot{\theta} < 0$.

The precise trajectory that will be followed for one step is the solution of equation 2 given by

$$\theta(t) = 2\arctg(e^{\frac{T}{B}t}) - \pi/2 \quad (3)$$

The asymptotic value for one step is $\pi/2$ and in order to reach 5% of the final value it is necessary to wait Δt seconds given by

$$\Delta t = \frac{3.25}{T/B} \quad (4)$$

This determines the maximum reacheable speed.

Start-stop motion of inertial loads

A technique to generate the switching times for a start-stop motion of inertial loads is described by Leenhouts (1979). Basicaly the velocity profiles are as shown in Fig. 2. Inertia and viscosity effects on the motion are considered. Switching times are obtained by assuming uniformly accelerated tion at each step.

The maximun acceleration and braking allowed in this method are limited by an utilization factor (lead angle) that constrains the zones for switchings in the torque vs. angle curves.

The complete procedure is as follows. At the aceleration phase, for each step, the switching time is obtained by computing the time the motor requires to move to the lead angle assuming uniformly accelerated motion. Switching times are computed this way until the desired velocity is reached.

During sliding, switchings occur at the angle in which available torque and friction torque are equal. Switching times are uniformly distributed.

In braking, the reverse procedure takes place. The transitions between the three zones are solved with asumption of uniformly accelerated motion.

INERTIAL LOADS

When load has inertia, that is when the equation of motion is given by

$$J\ddot{\theta} + B\dot{\theta} = \tau \quad (5)$$

it is no possible to use the same technique described above. The main difficulty is that is impossible to predict the equilibrium point that will be reached after a pulse train is applied.

To follow closely an arbitrary trajectory it is necessary to use detailed simulation to find the switching times. However, it is also required to guarantee the robustness of the switchings, since in practice it will be impossible to know exactly the parameters of the system value.

Robustness of switching times

In order to have a robust set of switching times, it is necessary to assure that small deviations from the nominal parameters will lead essentially to the same desired trajectory.

Consider a switching law $k(t)$ that makes a second order system follow the trajectory $\theta_1(t)$. The relation between parameters and variables is given by

$$J_1\ddot{\theta}_1(t) + B_1\dot{\theta}_1(t) = T_1\sin(\alpha\theta_1(t) + \beta k(t)) \quad (6)$$

On the other hand, if the parameters are changed to J_2, B_2 and T_2, the switchings will produce the trajectory $\theta_2(t)$.

The conditions under which $k(t)$ makes the deviation

$$\varepsilon(t) = \theta_2(t) - \theta_1(t) \quad (7)$$

small for all t, should be found.

It will be assumed that the differences

$$\begin{aligned} \delta J &= J_2 - J_1 \\ \delta B &= B_2 - B_1 \\ \delta T &= T_2 - T_1 \end{aligned} \quad (8)$$

are small.

ε satisfies the equation

$$(J_1 + \delta J)(\ddot{\theta}_1(t) + \ddot{\varepsilon}(t)) + (B_1 + \delta B)(\dot{\theta}_1(t) + \dot{\varepsilon}(t)) =$$
$$= (T_1 + \delta T) \sin\{\alpha(\theta_1(t) + \varepsilon(t)) + \beta k(t)\} \quad (9)$$

Ignoring the second order effects, the following linear time-varying differential equation is obtained

$$J_1 \ddot{\varepsilon}(t) + B_1 \dot{\varepsilon}(t) + \varepsilon(t)\cos\{\alpha\theta_1(t) + \beta k(t)\} =$$
$$= \delta J \ddot{\theta}_1(t) - \delta B \dot{\theta}_1(t) + T_1 \sin\{\alpha\theta_1(t) + \beta k(t)\} \quad (10)$$

Following the circle criterium (Willems, 1970), bounded solutions for ε are obtained when

$$-1 \leq \cos\{\alpha\theta_1(t) + \beta k(t)\} \leq -(1-2\zeta)^2 \quad (11)$$

where ζ is the relative damping coefficient of the system, that is

$$\zeta = \frac{B}{2\sqrt{TJ}} \quad (12)$$

Therefore, the switchings should only occur between torque-angle curves with negative slope as despicted in Fig. 3. It can easily be shown that the minimum damping coefficient necessary for a robust switching is $\zeta = 0.08$. The larger ζ the more room is available for placing the switching angle as is shown in Fig. 4.

SWITCHING ANGLE GENERATION

The interface between the stepping motor and the drived mechanical system is through the available torque, which is a highly non-linear function of θ angle. However, for high speed motion of inertial loads, the instantaneous torque can be substituted by the average torque between two succesive zero torque points. In this section the switching angle that matches any desired torque with the average one and that meet the robustness conditions mentioned above is derived.

In order to reduce the number of switchings involved in the motor control, only one conmutation per step is allowed.

Supose that the motor shaft is at point θ_c al time t_{θ_c} and that average torque τ is required. Calling $t_{\theta_c + \pi/2}$ the time at point $\theta_c + \pi/2$, an angle α must be found such that

$$\theta_c \leq \alpha \leq \theta_c + \pi/2 \quad (13)$$

and

$$\tau = \int_{t_{\theta_c}}^{t_{\theta_c + \pi/2}} T(\sigma) d\sigma \quad (14)$$

or

$$\tau = \int_{\alpha - \pi/2}^{\alpha} T \sin\theta \, d\theta \quad (15)$$

which leads to

$$\alpha = \pi/4 - \sin^{-1}\left(\frac{\tau \pi}{2T}\right) \quad (16)$$

When the required torque es larger in absolute value than the maximum average torque α is made equal to 0 or $\pi/2$. Moreover, should be within the allowed region determined in the previous section.

FEED BACK LAW

In this section a feedback law is used to compute through simulation the desired torque at every step, thus allow to determine the $k(t)$ function.

The dynamics of the system is described by the equation:

$$\ddot{\theta} + f(\dot{\theta}, \theta) = \tau \quad (17)$$

and the desired path is represented by $\ddot{\theta}_D$, $\dot{\theta}_D$ and θ_D. The control problem consists in finding the torque function τ_D that generates the desired path.

If the desired torque is computed as:

$$\tau_D = \ddot{\theta}_D - g_D(\dot{\theta}_D - \dot{\theta}, \theta_D - \theta) + f(\dot{\theta}, \theta) \quad (18)$$

the difference between the desired and real position satisfy the differential equation:

$$\ddot{\theta}_D - \ddot{\theta} - g_D(\dot{\theta}_D - \dot{\theta}, \theta_D - \theta) = 0 \quad (19)$$

or

$$\ddot{\varepsilon} - g_D(\dot{\varepsilon}, \varepsilon) = 0 \quad (20)$$

where $\varepsilon = \theta_D - \theta$.

In particular it is common to make linear the function g_D, that is

$$g_D(\dot{\varepsilon}, \varepsilon) = \alpha_1 \dot{\varepsilon} + \alpha_2 \varepsilon \quad (21)$$

in order to locate arbitrarily the poles by choosing the values of the α's.

Therefore the desired torque τ_D is given by:

$$\tau_D = \ddot{\theta} + \alpha_1(\dot{\theta}_D - \dot{\theta}) + \alpha_2(\theta_D - \theta) + f(\dot{\theta}, \theta) \quad (22)$$

SWITCHING TIME COMPUTATION

For computing the switching times three considerations are made:

- the system is at rest at the beginning of the motion
- equation 5 is solved numericaly
- switching angle will be computed every zero torque point.

The procedure is a follows:

a) τ_D is calculated from equation 22, where θ is the simulated angular position.
b) From τ_D the switching angle is calculated using equation 16
c) Equation 5 is integrated one step
d) After a given number of integration steps, the angle θ is tested to verify if the switching angle has been reached. If so, switching, takes place. If not, the time elapsed since the last switching angle was computed is checked. If exceeds certain value, return to step (a).
e) If θ is a zero torque point the switching angle is computed.
f) Return to step (c).

APPLICATIONS

Second order System

The procedure mentioned was applied to move an inertial load (Alvarez y Canales, 1982). The trajectory to follow is shown if Fig. 5. The results are shown in Fig. 6, where it is clear that the maximum errors occur precisely when the speed value is null, because at this point the hypothesis for switching angle computation is not fulfilled. In Fig. 7 a torque vs. θ is given.

Testing of robustness

To verify the robustness of the control scheme, the effects of modifying the parameters of the system (inertia, friction, torque) on the trajectory was tested. In Figs. 8 to 11 the results are shown.

CONCLUSSIONS

It has been shown that it is possible to use stepping motors to move arbitrary loads. Once the switching times are computed with the algorithm described in this paper, an open loop control scheme can be used. The digital computer will time the pulses. The real time close loop generation of the switching times is now been studied. The testing of robustness through simulation indicates satisfactory results.

REFERENCES

Alvarez-Icaza, L., Canales Ruiz, R. (1982). Empujador electro-electrónico de envases. Internal Report, Instituto de Ingeniería, UNAM. 67-97.

Kuo, B. C. (1979). Mathematical Modelling of Permanent Magnet Step Motors. In Kuo, B.C. (Ed.), Incremental Motion Systems. SRL Publishing Company, Champaign, Illinois, U.S.A. pp. 87-114.

Leenhouts, A. C. and Wilson C. S. (1979). Torque control of P.M. step motors in high performance applications. Proceedings of the Symposium on Incremental Motion Systems.

Leenhouts, A. C. (1980). Stepping motors in Industrial Motion Control. Proceedings of the Joint Automatic Control Conference. WP10A.

Willems, J. L. (1970). The stability of Systems with a nonlinear time-varying feedback element. In Brockett, R. W. (Ed.), Stability Theory of Dynamical Systems. Nelson, London. pp. 145-169.

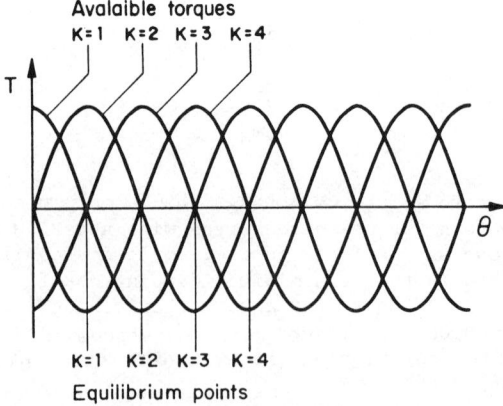

Fig 1. Avalaible torque for each winding

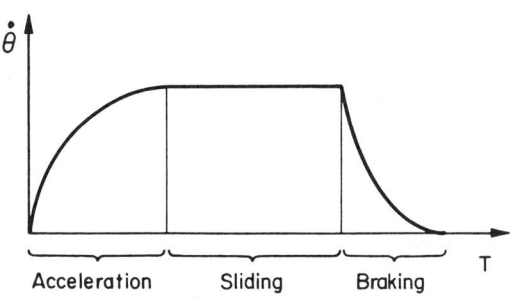

Fig 2. Typical trajectory in simple stepping motor control

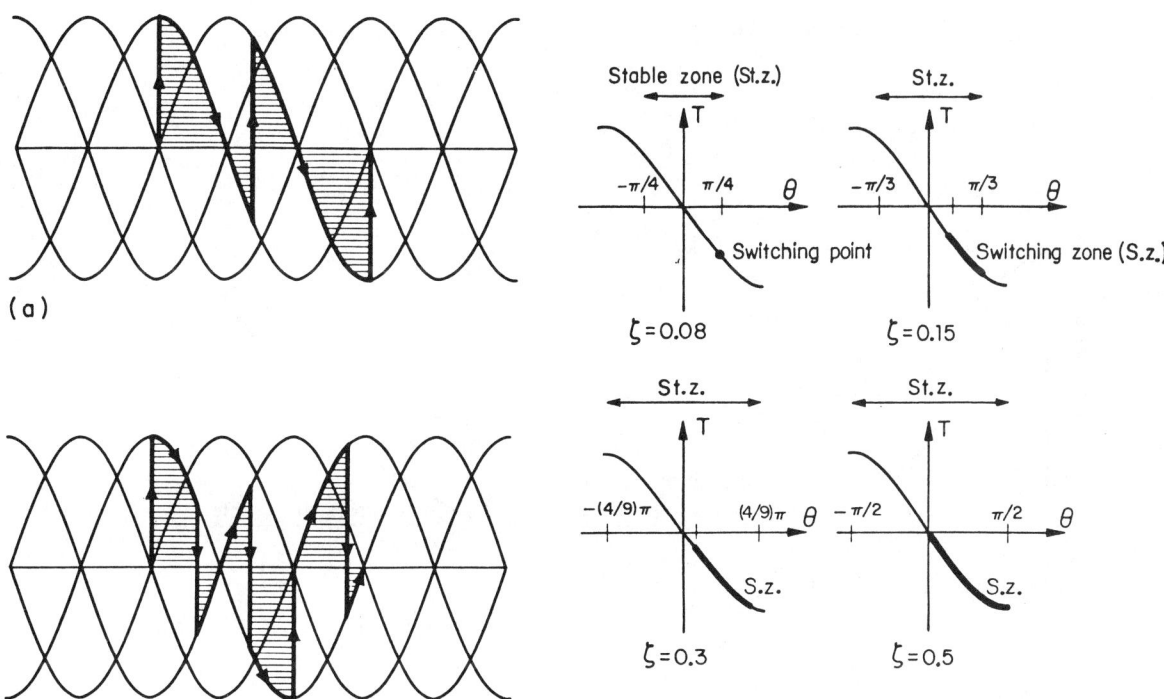

Fig 3. (a) Sequence of robust switchings
(b) Sequence of no robust switchings

Fig 4 Zones for stable switchings vs relative damping coefficient

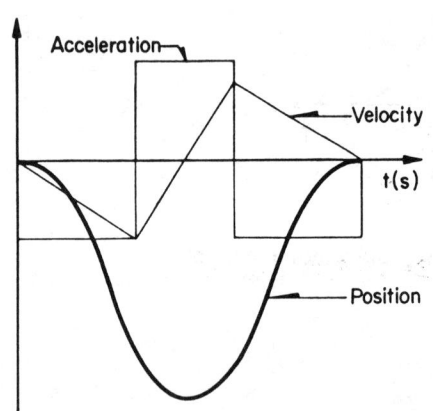

Fig 5. Trajectory to follow

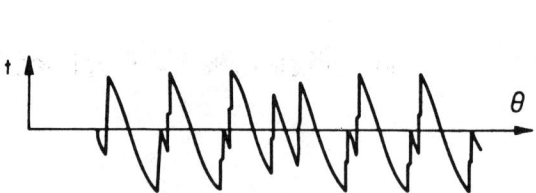

Fig 7 Simulated torque vs angle curve

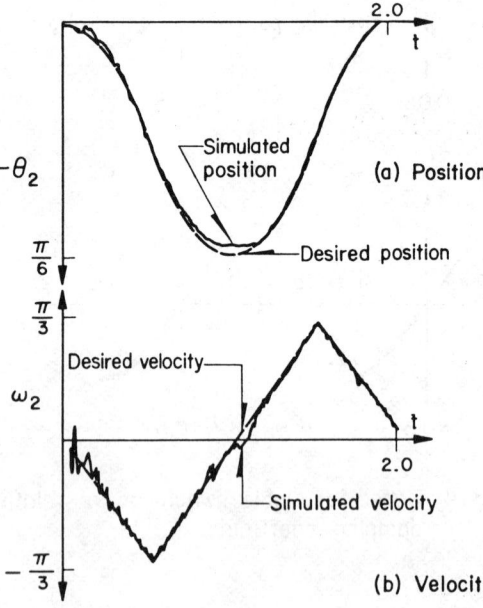

Fig 6 Simulation results

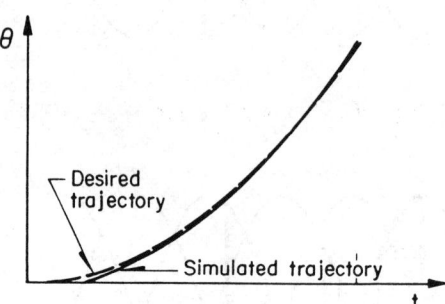

Fig 8 Simulation with $\Delta J = 0\%$, $\Delta B = 0\%$ (position)

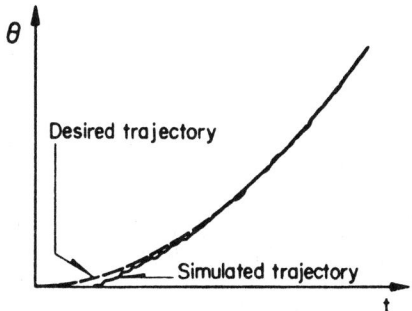

Fig 9. Simulation with △J=-60%, △B=-36.8% (position)

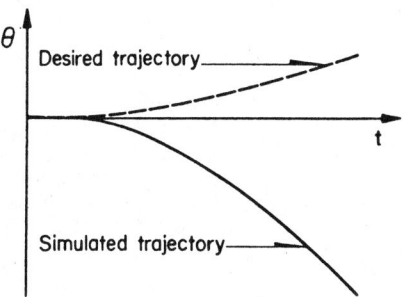

Fig 10. Simulation with △J=-80%, △B=-55.2% (position)

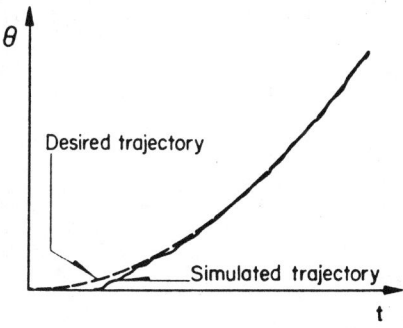

Fig 11. Simulation with △J=+100%, △B=+41.4% (position)

DISCUSSION

SESSION WA3: INDUSTRIAL APPLICATIONS II

Paper: MICROPROCESSOR-BASED CONTROL OF INDUSTRIAL SEWING MACHINES

Authors: *Benjamin Hertzanu* (Beta Engineering and Development Co., Beer Sheva, Israel)

Daniel Tabak (Department of Electrical Engineering, Ben Gurion University, Beer Sheva 84105, Israel)

Discusser: *Don R. Bristol*
2415 Manzanita Dr
Oakland, Ca 94611
USA

Questions or Comments:

1) My question is on the proposed self tuning controller. Since you are using only one feedback transducer, do you expect to tune on this transducer alone or will you add other transducers for feedback?

Authors' reply:

1) We expect to use only the existing transducers. We feel that this will allow us to vary the different types of materials without having to change the program.

Discusser: *José F. Albarrán*
Apdo Postal 475
Cuernavaca, Mor 62000

Questions or Comments:

1) In the analog servo an acceleration feedback is shown. This is not so in the digital servo loop. Please comment on that.

Author's reply:

1) I didn't participate on the analog servo loop design, therefore I don't know the reasons for including an acceleration feedback. The actual system doesn't use it.

Paper: MICROCOMPUTER DIRECT VOLTAGE CONTROL OF A PWM INVERTER

Authors: *GS Buja, D Longo* (Instituto di Elettrotecnica e di Elettronica, Universidad de Padova, Padova, Italy)

Discusser: *Roberto Ramírez C*
Motolinía 591
Col. San Francisco
Zapopan, Jal

Questions or Comments:

1) Could you be more specific on the dynamic and static control actions of the controllers included in the microcomputer system?

Authors' reply:

1) The microprocessor obtains a peak-amplitud related voltage every half-cycle. It performs a PID algorithm to generate the timing of the PWM output. Every half-cycle output-update is performed to reduce transients.

Paper: INDUSTRIAL APPLICATIONS OF VISION TECHNOLOGY

Authors: *M Ejiri* (Central Research Laboratory Hitachi Ltd. Kokubunji, Tokyo 185, Japan

Discusser: JF Albarrán
Apdo Postal 475
Curnavaca, Mor 62000
México

Questions or Comments:

1) Are you aware of work by Dr Paul Suciu (at Berkeley, Calif) on direct filtering of CCD signals to find contours on low contrast images?

Author's reply:

1) No, I don't know that particular work. However, I must point out that filtering alone doesn't solve the problem in most cases of these type of application.

Discusser: Don R Bristol
300 Lakeside Dr.
Oakland, Co. 94643
USA

Questions or Comments:

1) Have you ever used your vision equipment and patterns recognition techniques to read identification codes on metal ingots and metal coils?

Author's reply:

1) No. We have trouble with getting good contrast on metal so that it is very hard to recognize the patterns.

Paper: DIRECT DIGITAL CONTROL OF ELECTRIC DRIVE SYSTEM BASED ON IMPROVED OPTIMAL REGULATOR THEORY

Authors: T Tsuchiya (Department of Electrical Engineering, Faculty of Engineering, Hokkaido University, Sapporo, 060, Japan

Discusser: R. Schulz (Department of Electrical Engineering, Fachhochschule Bielefeld, Bielefeld, German Federal Republic)

Questions or Comments:

1) The shortest sample time by the used hardware (value to see)?

2) The constants of the PID-algorithms optimized on the process?

Author's reply:

1) We use 20 msec sampling time. The processor takes 10 msec to process the information, but we plan to use a more sofisticated algorithm. I don't remember offhand the values of the PID constants.

Paper: STEPPING MOTORS CONTROL

Authors: R. Canales, L. Alvarez (Instituto de Ingeniería, Apdo Polstal 70-472, Coyoacan 04510, Mexico, DF)

Discusser: Gustavo Oropeza G.
Federico Ibarra 782
Jard. Alcalde
Guadalajara, Jalisco,
México

Questions or Comments:

1) Are you going to use this techniques of stepping motors control for some special purpose?

Author's reply:

1) We intend to use these techniques in positioners powered with stepping motors in several of the industrial applications' projects being developed at the Instituto de Ingeniería, UNAM

DIGITAL CONTROL APPLICATION ON A CLASS GOB FEEDER

F. Saldaña and R. Solís

Vitro Tec, Monterrey, Neuvo Leon, México

Abstract. The adjustments in a glass gob feeder, to maintain a constant gob weight are made at present manually. This paper deals with the application of a digital system to control said weight, taking as a base the gob temperature. Several tests were performed in a "green" glass container production line. Correlation methods were used to determine the behavior of the variables that influence the gob weight. It was found that these variables are linearly related with a -0.9 correlation factor. Taking into consideration these results, a digital system will be used to control the process.

INTRODUCTION

In the Glass Industry the quality of the containers depends to a certain extent, in maintaining a stable weight of the gob used to form the article. The weight variation has motived several researchers (Hamilton 1980) to look for certain techniques that will permit the control of such variations. Several of these studies have been focused, mainly in trying to find the influence that may exist directly of the various factors such as: shearing velocity, glass temperature, etc., on the gob weight.

The first part of this paper has been focused on the search of the laws that regulate the behavior between the temperature variations and the position of the refractory tube that controls the glass flow (Fig. 1). For this purpose, several statistical tools have been used, and sampling methods have been designed. Both will ensure that the amount of observations have the necessary validity to draw the best conclusions.

The second part has been focused on the utilization of a digital control that will allow adjustment of the position of the tube when the temperature of the gob is known. The purpose of this control will be to stabilize the gob to the desirable weight.

DATA SAMPLING SYSTEM

In the measurement of the parameters involved in this experiment, an infrared optical pyrometer was initially used to determine the gob temperature. This pyrometer was placed at a distance of three feet from the objective; but high interference was present due to smoke and vapors in the environment. Later, an optical fiber temperature sensor was used at a distance of eight inches from the objective. In this manner, the problems of the environmental vapors and humidity were eliminated. On the other hand, to measure the adjustments of the tube mechanism, a displacement transducer of the LVDT type was used. Both variables were recorded in a two channel recorder, as shown in Fig. 1.

VARIABLE BEHAVIOR IN THE PROCESS

In order to ascertain the behavior of the variables involved, two types of analysis were used: Qualitative and Quantitative.

Qualitative Analysis: For this analysis - the data was recorded with the purpose of detecting some process characteristics (Fig. 2), that would enable a selection of the proper correlation method that would help make a quantitative analysis with greater depth and validity. A high correlation was found between the temperature and the tube position.

To estimate this correlation, it was necessary to purify the data obtained, discarding what was not in accordance with normal process operation. The following adjustment classification was made:

Those of Short Duration: Lasting up to 15 minutes (stables).

Those of Long Duration : Lasting more than 15 minutes (stables).

It could be assumed that the short duration adjustments were due to the operator incorrect handling of such adjustments; and so, they were discarded.

Quantitative Analysis: This analysis consists of making a model of the formulation taking into account the aforementioned step, where a linear process behavior is observed. The minimum squares method was used to correlate the variables, and it was found that it is governed by the following equation:

$$Y = \alpha_1 + \alpha_2$$

The correlation factor found is -0.9 and is shown in Fig. 3.

CONTROL

Once the system behavior was identified and determined, the following digital control was designed with the next characteristics:

* Microprocesor
   8 Bits.
   16 K Bytes.

* Converter A/D and D/A.
* Servo Motor.
* Positioning Tube Mechanism.
* Encoder.
* Temperature Sensor.

The system design is aimed at obtaining a modular control that can be applied to any production line, and is shown in Fig. 4. The system operating sequence is started by detecting the radiation temperature of the gob thru the optical fiber, which transmits the signal to the electronic unit. In this unit the radiation detected is changed to an analog signal so that the microprocesor can interpret the signal and use it in the control algorithm. Then it is converted into a digital signal by means of the A/D Converter.

Once the microprocesor determines the action to follow, the signal is changed again to analog, which causes the servo motor to move the Tube Mechanism, thus making the appropiate correction. The feedback is carried out by the encoder, which also indicates the tube displacements.

CONCLUSIONS

The process variables behave in a linear pattern. The benefits obtained by the application of a digital control, are the reduction in the variation of the article weight, a better quality product and an increase in the packing percentage.

REFERENCES

- Acheson J.D. (1965) "Quality Control and Industrial Statistics" Irwin.

- Hamilton B. (1980) "Control Measurements on the Glass Container Forming Machine" Glass Tech. 249-253

- Saldaña F.S. & Solís R.H. (1980) "Proyecto Mejoras al Dosificador" Diseño Experimental, Vitro Tec Monterrey.

- Saldaña F.S. & Solís R.H. (1981) "Estudio sobre la Posición del Tubo y Temperatura de la Gota de Vidrio" Vitro Tec Monterrey

- Stanley G.M. & T.F. Ensor (1977) "Some Observations Made During Studies of Glass Conditions in White Flint Forehearths". Glass Tech. 105-108

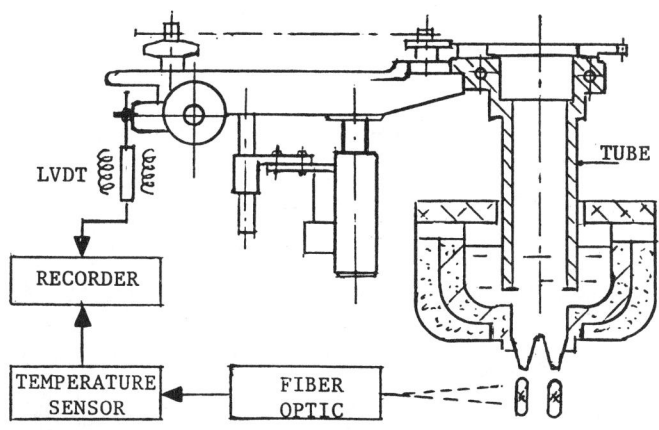

Fig. 1 Measuring Instruments

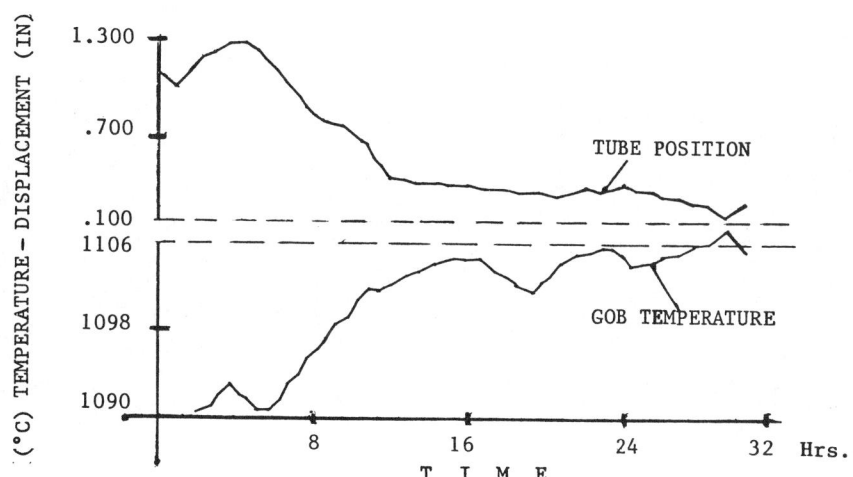

Fig. 2 Shows the tendency in the Tube Position and Gob Temperature in the time.

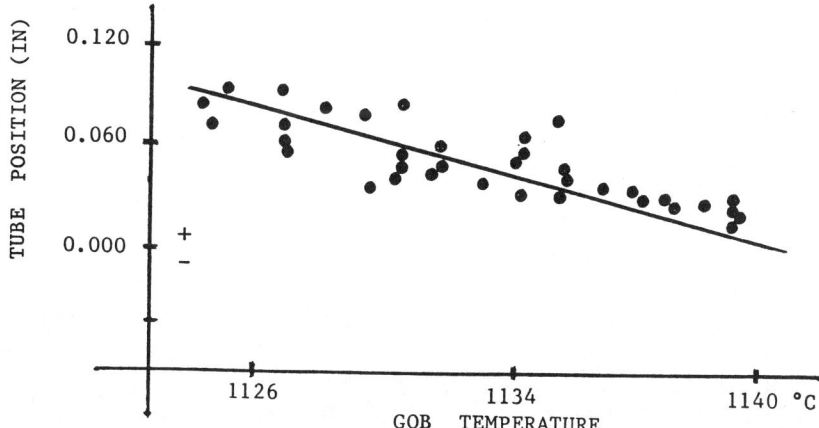

Fig. 3 Shows the correlation existing in the long-duration adjustments during a period of three continuos days.

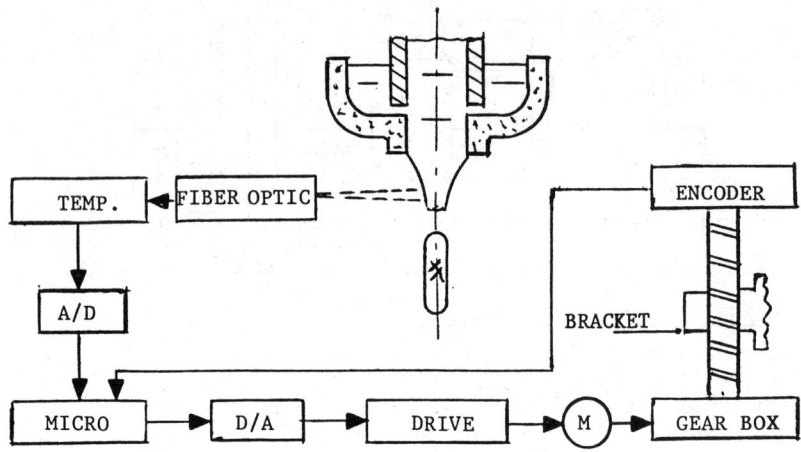

Fig. 4   Control Block Diagram

A DIGITAL CONTROLLER FOR A GLASS MACHINE WITH PRESS-BLOW ON COATED MOLDS PROCESS

M. A. González

Vitro Tec Fideicomiso, Monterrey, Nuevo Leon CP6440, Mexico

Abstract. A digital controller is presented for a glass machine using the press-blow formation process on coated molds. The controller is capable of governing the synchronization of each of the mechanism and operates in real time using a 6800 microprocessor. The objetives of the controller are to reduce the number of manually operated variables present in the ope ration of the formation process and offer the operator advantages in selecting the operations of the machine withing a single electronic system. Furthermore, interfaces with feedback sen sors render the controller the option of optimizing the process.

In this paper we present the analysis and development of the controller. Results obtained over the first year of operation in an actual production environment are also presented.

Keywords. Computer control, digital controller, digital computer, feedback, microprocessor.

INTRODUCTION

The usual way to generate sequential set points of mechanisms, is using rotating cams, at a constant velocity. These cams move a serie of buttons, each button producing a condition for opening or closing air valves or microswitches.

This way of controlling the set points produ ce problems due variations in the velocity of the motor used for the movement, and the usual wear tear in the mechanicals parts. An other important point of view is the poor re solution in the adjustments obtained using the buttons. If this adjust is made in any sequence, can generate calshes between mechanism, producing loss times or delays.

To procure a better control over these set points in the mechanisms, we developed a di gital controller, with capacity for genera ting the triggers or set points on the me chanisms. Furthermore, the digital contro ller was designed to increase the easy of operation, obtain a high level of security both for the operator and the equipment and mantain a consistent quality standard in the glass article. The digital controller pre sented in this paper, controls a machine with two independent production sections, both of which receive the gob of glass from the same feeder.

DEFINITION OF NECESSITIES

Based on the necessities of the process, we found a serie of important specifications to be considered, same of them are shown next:

Division of the mechanisms into differents group, in this way when it is necessary to make adjustment, it can be made by group. In this way clashes between mechanisms are avoided and the constant times among the members in each group is not altered.

For example in Fig. 1, we have four mechanisms. The mechanism B and C form a group, we move the time were B starts by 0.5 s., the results is shown in Fig. 2, where we can see the new start point for the mechanism B is at 1.5 s. . The difference between finish of B and start of C must be constant at 0.5 s. and the duration of the pulses of B and C remains constants also.

Some times a group can include some subgroups, in these cases the procedure is the same.

— The duration of the pulses must be less or equal to the cycle time of the process and any group or sequence of groups cannot exceed the cycle time.

— A real time operation is needed for a good control of the process, and the resolution in the system should be on the order 10 milliseconds.

— Capacity to turn on or off the operation of some mechanisms.

— Comunication between the digital computer and the operator.

— Comunication between the operator an the digital computer.

— Systems of feed back.

HARDWARE CONFIGURATION

Due to the detected necesities, we decided that a microprocessor should be able to generate the required digital control functions. We select the 6800 microprocessor, suported by RAM, EPROMS, I/O, TTL and CMOS Logics, and also interfaces and sensors. This system also contains the power supplies of +5, +12, -12 and 24 volts dc. All this equipment was installed into a control panel which functions as the operator panel, in which are colocated, thumbwheels, displays, LED'S, and switches, all of them estabilish the comunication between the operator and the digital controller.

The digital controller receives signals from the sensor, located on the machine and in the feeder. These signals help the controller optimize the process, increasing quality at the article and the sincronization between the feeder and the machine. The digital computer uses 72 output signals, 40 inputs and 5 interruptions. The interfaces with the mechanisms are made by using neumatic electrovalves, all of them located on the machine and with 24 volts dc.

SOFTWARE ORGANIZATION

One of the necesities in the proyect, is to control a machine with two sections and each of them must work independently from the other, nevertheless they have a common place, this is the part where they receive the gob of glass from the same feeder, here the operation must be made in a secuential order. Due to this requirement the software was structured into two sections, the first one "start and control" and the second one "two great working sections".

In the section of "start and control", we have the instructions for defining of the input and output variables, and the control instructions, for selecting which of the two working sections is to be executed next.

Each of the working sections generate signals for the different mechanisms corresponding to that sections, and can stop, start, eliminate movement of some mechanisms, receive signals from the feedback sensors, etc. and at no time affect or interfer with the operation of the other section.

We talk before about a 10 milliseconds resolution, for assure this resolution, we taking care to avoid problems in delays due to many operations in any part of the process, making the subroutines shorted as possible.

Another important point was to form groups for the generation of signals for the mechanisms, this results in an easy way to control the process and the time in which the changes are generated. The operator at any point at the process, can solicit an order to increase or decrese, the duration of the pulses, by moving the point of start or finish in any mechanism. At this moment the software is utilized as the tool for analyzing the orders from the operator. The software checks if the changes don't generate clashes between mechanisms, or if it exceeds the cycle time process. If it's agree with the changes, it assigns the new time and continues operating. If it are not agree the operator is inform the type of error and continues working with the original or previous set points.

During the time utilized by the software for analyses the change, the function of the machine is not altered.

CONCLUSION

The digital controller have been conected to the machine about 14 month, producing glass articles, the ambient temperature had changes from 5°C to 70°C, and the digital controller works about 85 millions of cycles in more than 10,000 hours.

By now the efficiency is 99.6%, the 0.4% lost time, was due to the operators errors during the first week's, and also, due to time used for debuging and correction in the hardware and software.

In these months, the digital controller made the process more controllable and increased the quality in the product, the operation of the machine is safety and the human effort is less.

Digital Controller for Glass Machine 543

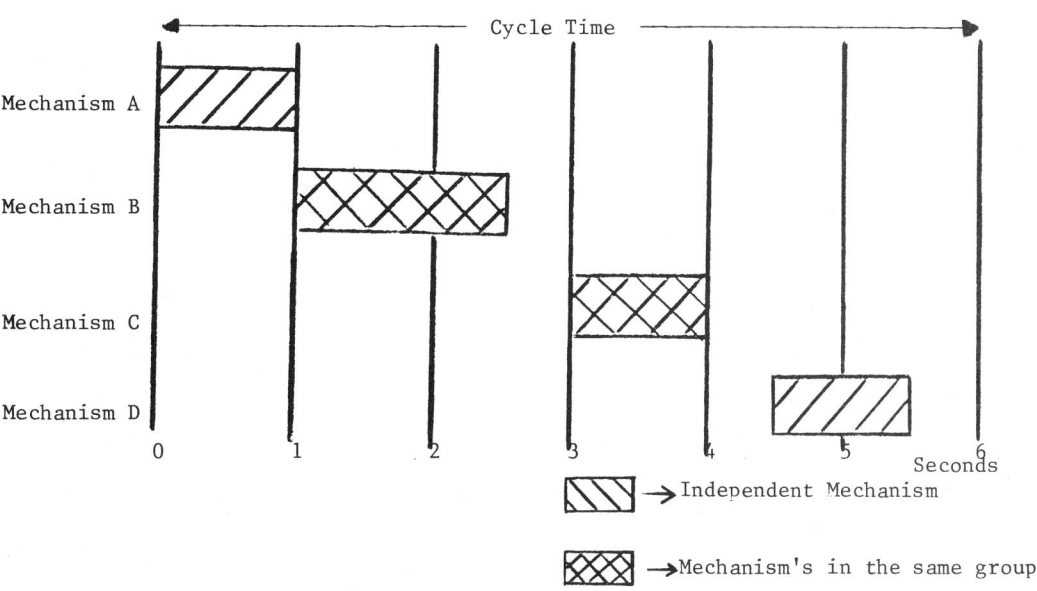

FIG. 1. ORIGINAL SEQUENCE OF MECHANISMS

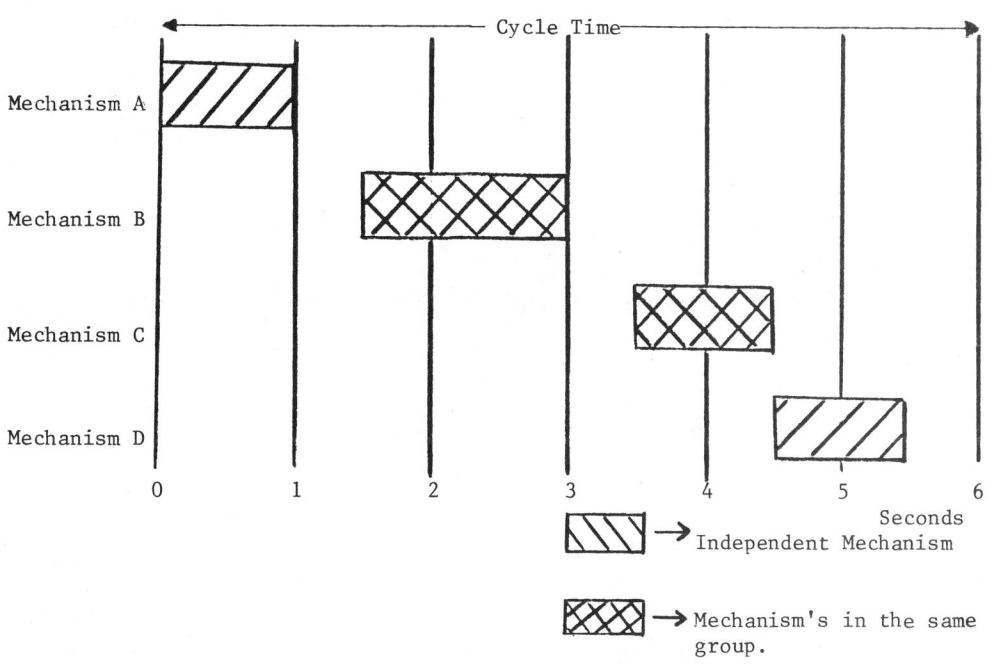

FIG. 2 SEQUENCE AFTER 0.5 S. OF DELAY APLIED TO MECHANISM B

ON THE DESIGN OF A DISCRETE FEEDFORWARD CONTROL FOR COOLING MOLDS IN THE GLASS INDUSTRY

F. L. Elizalde

Vitro Tec Fideicomiso Monterrey, Neuvo Leon, CP 64410, Mexico

Abstract. Data measurements obtained in-line on a glass forming machine under a plant environment and consisting of cooling air pressure P_{at}, ambient air temperature T_{at}, and the external mold surface temperature T_{bt}, were analyzed using discrete time series based on the Box-Jenkins method. From the analysis, a dynamic model is given for predicting the values of the mold surface temperature T_{bt}. The accuracy of the forecast model is corraborated with actual data. A control equation is derived for making the necessary adjustments on P_{at}, and compensations for variations in T_{at}, to ensure the regulation or minimum deviations, of the output T_{bt}, from the target set point.

The configuration of the control equation in a microprocessor system is presented along with the feedforward implementation. Results of in-line measurements are discussed.

Keywords. Discrete time series, feedforward-control, cooling molds, glass industry, microprocessor, dynamic models.

INTRODUCTION

The analyzed data were taken at discrete time intervals of 2 minutes under normal operation conditions. The total sampling time being 9 hours and the production container 7146044-VITA-12 OZ, at an I.S. double cavity machine (57 bpm) located at our plant, VIQUESA, in Querétaro, México.

Since T_{at} is an observable variable and not being able to be manipulated, perturbations were induced in P_{at} in order to know the dynamic response of the system (T_{bt}). A layout of the data acquisition system and asso ciated instrumentation has already been given.[7]

A preliminary analysis of the raw data showed that T_{bt} could be well represented by an ARMA (p, q)[2] model whose spectral density function had two well defined spikes at values of about 6 minutes and 2 hours. The first one being due to the proper opening-closing function and the second, not clearly identified, is belived to come probably from batch charging operation, inherent viscosity variations, gob weight variations or some action taken in the heating/cooling system during the stages of melting, refining or conditioning.

In order to get rid of the complex arma structure, the data was smoothed by taking averages (each 3 observations) and the generated new data given in figures 1 (a, b and c) was used to build the dynamic model. A sketch of the inspected system is outlined in figure 2.

DYNAMIC MODEL BUILDING

Using the smoothed data the identification and estimation stages were made and conducted to several possible models. Among them three were selected for further study since their residual sum of squares, noise autocorrelation and transfer function weights showed that the data were well represented by them. [1,5,6].

Two of the models have the general form[2]:

$$\dot{T}_{b_t} = F_2(B) P_{a_t} + F_1(B) T_{a_t} + N_t \quad \ldots (1)$$

Where:

$\dot{T}_{bt}$ = is the deviation of T_{bt} from its mean value

$F_2(B)$ is a ratio of polynomials in the back-order operator B and represents the transfer function due to P_{at}

$F_1(B)$ is the transfer function due to T_{at} and

N_t is the noise given by an ARMA (p,q) model.

The third model have the same basic structure as (1) but the variables are expressed as deviations of their values at time t from those at time t-1, and the noise model having a MA (q) structure, however in factorizing this noise model in order to have deviations in N_t, the deviations will be cancelled and arrive to equation (1).

SIMULATION

Control equation 2 was used in order to simulate the required pressure action that has to be taken to compensate aire temperature variations.

The values of T_{at} corresponded to data taken in-line at intervals of 21.49 minutes. The first value was at time 18:28 and the data covered a sampling interval of 47.3 hours supposing the same pattern each 24 hours.

These data and the resulting action for two of the selected models are shown in figure 9. Again the difference between the models is due to the difference in the estimated parameters, nevertheless what is important to see is that even with model 3 the minimum pressure when T_{at} is maximum is about 25"H_2O, that is, since the maximum pressure in the system is about 30"H_2O, there is a potential of 5"H_2O that normally is not used.

The control equation has already been implemented in a microprocessor and its adequancy is being checked in-line.

CONCLUSIONS

The discrete feedforward control system already described can compensate air temperature variations by manipulating the air pressure through a value in the air feeding system assuring the minimum deviation of the external mold surface temperatur from a given set point.

Since the system is dynamic, it can use past plant history weighed by the memory in order to predict the control actions.

The controlling of the T_{at} variations can lead to savings in energy due to the adequate used of fans and compressors when implemented in a whole machine or in a whole plant.

The control equation is relatively easy to implement in a microprocessor due to the linear structure of the model.

The evaluation of such a control system can be measured from two different points of view. The first one is its potential in reducing energy consumption and the second is the potential to reduced defective items caused by the heating/cooling process. This second point is important since it gives the operating limits that assure the production of good articles, thus increasing the plant output [4].

Another potential area of research is in making the air temperature a manipulatable variable. Some research is already being done in this area.

The basic difference among them is in the values of the transfer function and noise model parameters, which were estimated by using [3] and are given in [5] at the 95% confidence limits. Equation (1) is represented in form of a block diagram in figure 3.

FORECASTING

With the estimated parameters equation (1) was checked by making forecasts at lead times 20, 37 and 55 and assuming new observations of P_{at} and T_{at} became available at time origin 95, where the values of P_{at}, T_{at} and T_{bt} were 5.32 volts, 3.137 volts and 281.4°C respectively. Figures 4 thru 7 show some of the analyzed situations.

There are two cases that are worth while to mention since they included data that was taken in situ. Those cases are given in figures 4 and 7. Figure 4 shows the effect of a step in pressure giving an exponential response in T_{bt} reaching a stable value of 312°C, this value was also observe in plant when producing such a step.

Figure 7 gives the forecast on T_{bt} based on real P_{at} and T_{at} data. Again, comparing these results with the observed ones, it was concluded that the proposed models were able to represent real operation behavior.

The other cases are hipotetical situations however, in analyzing them it is possible to visualize new control strategics either in making adjustments in P_{at} or making T_{at} a manipulatable variable.

DISCRETE CONTROL SYSTEM

Refering to figure 3 it is seen that if P_{at} is fixed at a value of zero, the deviation from the output or total error will be given by

$$T_{bt} = F_1 T_{at} + N_t$$

The total effect of T_{at} being F_1 and the one of P_{at}, $F_2 P_{at+}$, where P_{at+} is the change of level in P_{at} during the interval $(t, t+1)$. The effect of the observable perturbation T_{at} will be cancel if

$$F_2 P_{at+} = - F_1 T_{at} \quad \ldots (2)$$

Equation (2) establishes the control action that has to be made on P_{at} in order to compensate T_{at} variations yielding at point A in figure 8, $T_{bt} = N_t$ that is, white noise with mean zero and constant variance.

BIBLIOGRAPHY

1.- Statistics for Experimenters
   G. Box, Hynter, Hunter
   Wiley Interscience, 1978

2.- Time Series Analysis: Forecasting and Control
   G.E.P. Box & G.M. Jenking
   Holden Day 1976

3.- A Computer Program for the Analysis of Time Series Models
   D.O. Pack
   AFS, 1977

4.- Análisis de Defectos y sus Probables Causas para la línea 72 de VIMOSA.
   Parte 1: Estadística
   Fernando L. Elizalde V.
   G.E.C. - VITRO TEC 1981

5.- Parámetros del Modelo Dinámico de Enfriamiento de Molduras
   Fernando L. Elizalde V.
   Reporte Interno G.E.C.- VITRO TEC 1982

6.- Aplicación del Análisis de Series Discretas de Tiempo: Modelamiento de Enfriamiento de molduras y control de peso en artículos de vidrio.
   Fernando L. Elizalde V.
   G.E.C. - VITRO TEC 1982

7.- Enfriamiento de Molduras: Set-up experimental, software y hardware.
   R. Serrano, V. García, J.A. López
   Reporte Interno G.E.C.- VITRO TEC, 1981.

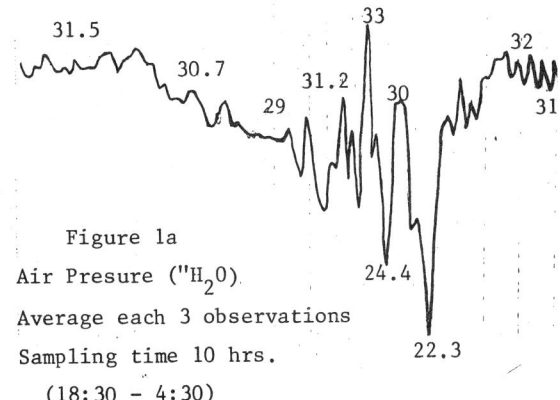

Figure 1a
Air Presure ($"H_2O$)
Average each 3 observations
Sampling time 10 hrs.
(18:30 - 4:30)

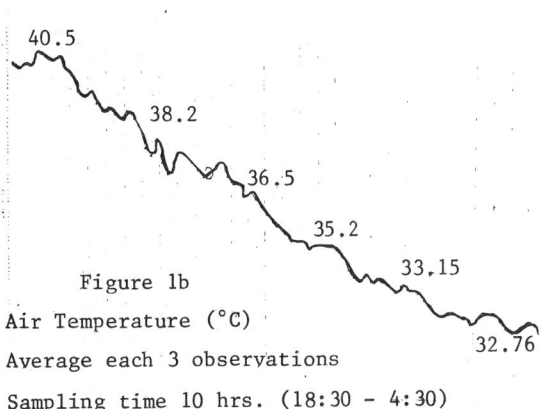

Figure 1b
Air Temperature (°C)
Average each 3 observations
Sampling time 10 hrs. (18:30 - 4:30)

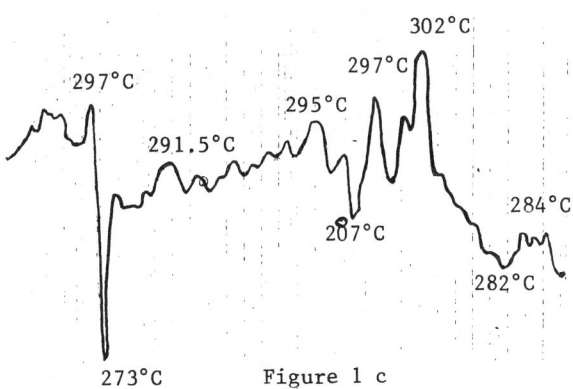

Figure 1 c
External mold temperature
Average each 3 observations
Sampling time 10 hrs. (18:30 - 4:30)

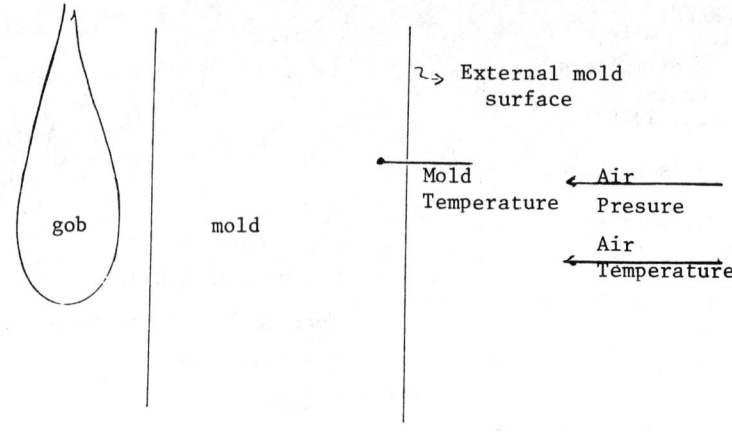

Figure 2

Sketch of the system

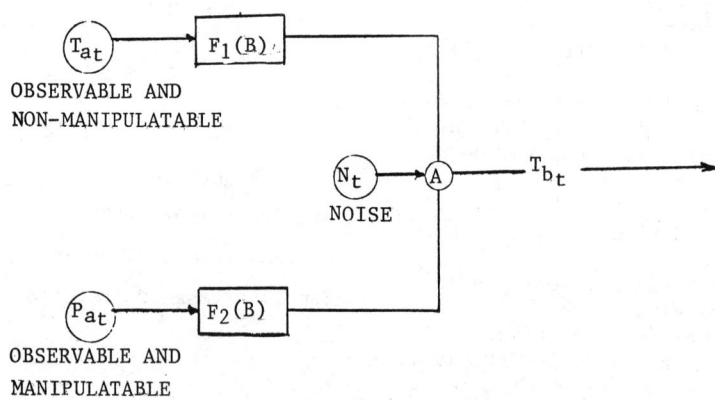

Figure 3

Block diagram of the system

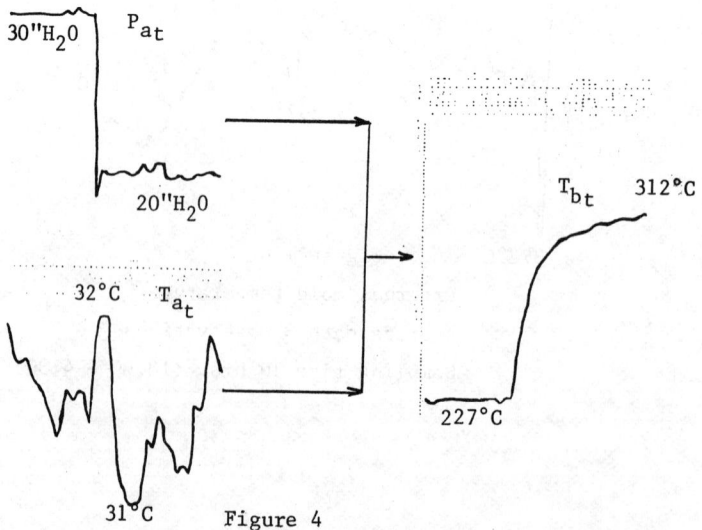

Figure 4

Step response with T_{a_t} variable
These values correspond to Data obtained in-line.

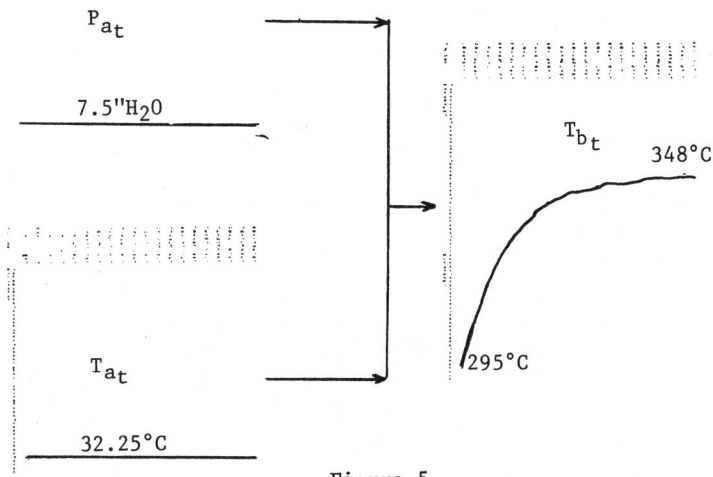

Figure 5

Step response for P_{a_t} with T_{a_t} constant

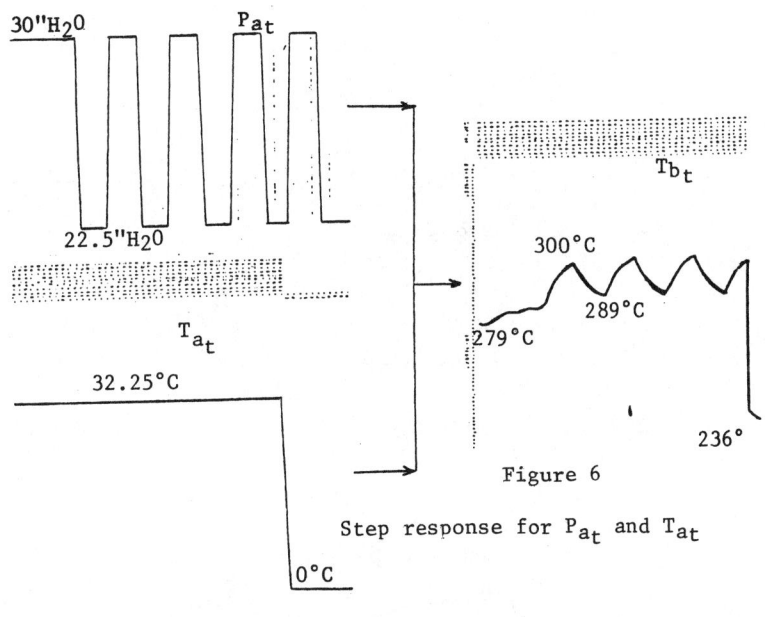

Figure 6

Step response for P_{a_t} and T_{a_t}

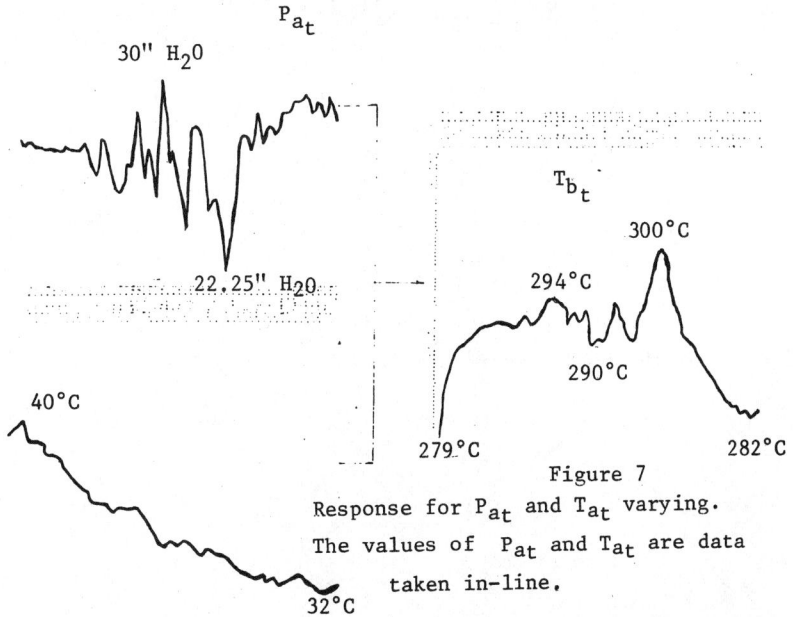

Figure 7

Response for P_{a_t} and T_{a_t} varying. The values of P_{a_t} and T_{a_t} are data taken in-line.

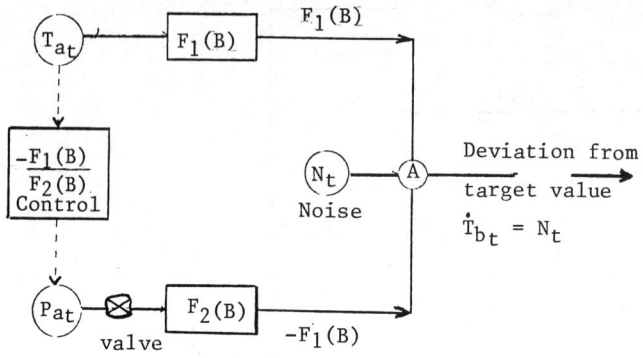

Figure 8

Block diagram of feed-forward control

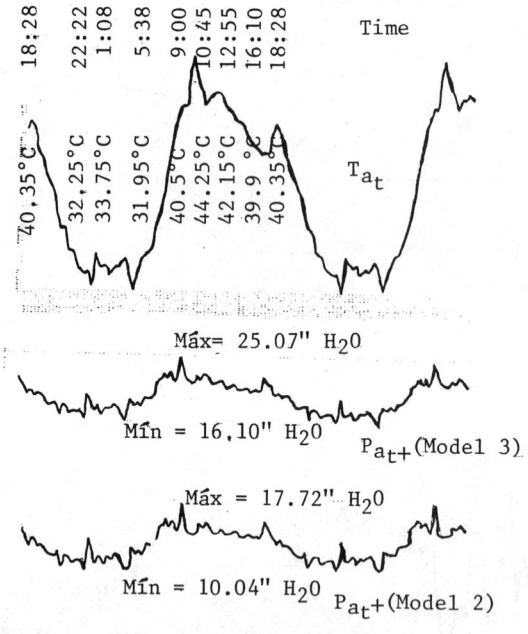

Figure 9

Corrective action of air pressure to compnesate air temperature variations allowing minimum deviation of the mold temperature from its mean value.

Copyright © IFAC Real Time Digital Control Applications
Guadalajara, Mexico 1983

AN INTELLIGENT DIGITAL CONTROLLER IN THE FORMATION PROCESS OF GLASS BOTTLES

H. Rodríguez and S. Rodríguez

Vitro Tec Fideicomiso, Monterrey, Neuvo Leon, CP 64410, Mexico

Abstract. An automatic digital controller was developed for a glass forming machine with the purpose of maintaining bottle consistency by regulating the formation process cycle period. Regulation was achieved by using a microprocessor with instrumentation interfaces to external sensors and actuatores and the software logic for predicting estimates of machine cycle periods based on present and past measurements. The controller is capable of compensating up to 12% variational changes in the cycle periods without mechanical failure. Furthermote, the software contains safeguards which continue to predict estimates of machine cycle periods in the event that the reference signal is lost. Results from on-line production measurements indica te a consistency in the quality of bottles produced during times when the cycle signal is lost or is varying.

Keywords. Real time control, microprocessors, digital computer, process control, prediction, parameter estimation industrial control.

INTRODUCTION

The I.S. machine is used extensively in the glass forming process, the machine consists of over 19 mechanisms. A necessary element for the mechanisms sincronization is a mechanical drum. This drum has many problems. For example; difficulty in adjusting the time of each mechanism, and repeatability, Kwiatkowski (1973). We designed an electronic controller to have a better process control with the following characteristics; higher reliability, flexibility to changes and all under the control of a microprocessor. We divided the proyect into 4 steps: (1) Substitution of the mechanical drum by an electronic controller, (2) Facilitate the comunication between the controller and the operator, (3) Automatic compensation for variations in cycle time by generating predictions based on previous measurements, including cases of missing sensor information and (4) Record the information and formulate a statistical analysis to improve the process controller.

In this paper we will discuss only the activities of step (3). Included in the following sections are: The technics used in the prediction algoritm, its essential functions and the analysis of the measurements which have been taken in the plant to date.

PROBLEM DESCRIPTION

After the controller was instaled and operating in the plant, we observed the hours that the controller was making bottles. We found that under certain condition it was necessary to suspend the operation of the automatic controller from the production li nebecause of improper sincronization of the controller with the intermittent gob signal and time missadjustments when the gob signal was altered beyond certain limits.

Based on these reasons we designed a predictor, capable of estimating the gob signal within certain limits and to detect changes in the cycle time of the gob signal and adjust the trigger of each mechanism in accordance to the new cycle time.

We show 6 machine cycles in Fig.-1, in the first two cycles the gob signal occur within the allowable time however in the third cycle the gob signal is missing. When this happens, the logic in the software wait un til the synchronization signal Δt after generate the predicted value for the gob signal. The selection of Δt is crucial because the proper functioning of the mechanism within the forming machine depends on preestablished duration time which may not over lapped, otherwise collisions may occur resulting in serious damage to the machine.

Should the gob signal continue to be missing the controller will automatic continue to generate the prediction synchronization signal to the forming machine based on previous measurement of the gob signal, in the case when the gob signal is beyond the limits allowed, the logic will compensate by a gradual change in the selection of Δt and a moving average prediction of the gob signal until proper phasing is achieved between the received gob signal CT and the predicted CT. An example in generating predicted synchronization signals is shown in Fig. 2.

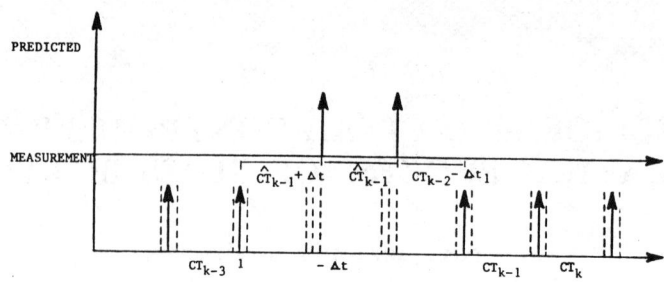

FIG. 1 GOB SIGNAL MISSING OR INTERMITTENT VARIATIONS

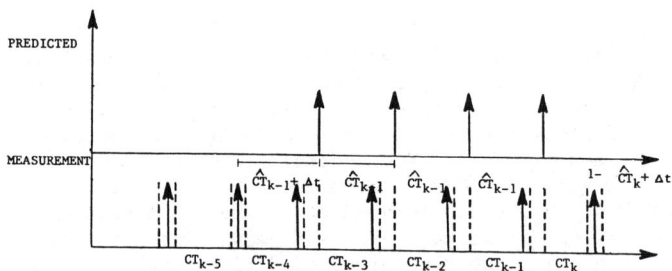

FIG. 2 IS AN EXAMPLE OF THE CASE WHEN GOB SIGNAL CYCLE TIME
CHANGES BEYOND THE LIMITS, WE SHOWN A MECHANISMS
ADJUSTMENT AFTER THREE MEASUREMENTS BEYOND THE LIMITS
AND ACTUALIZATION OF THE ESTIMATED GOB SIGNAL DURING
THE FOURTH MEASUREMENT WHICH CHANGED THE LIMITS, TO
PERMIT EXACT SINCRONIZATION WITH AUTOMATIC CONTROL AND
GOB DISTRIBUTOR.

IMPLEMENTATION

The block diagram is shown in Fig. 3, there we have shown just one microcomputer for one I.S. section, as the real time processor we are using the 6800, the interfaces consist of peripheral interface adaptors, buffers, solid state relays and electropneumatic valves. In the feedback loop we use two sensor to generate the gob signal and the machine cycle signal, they are conected to signal conditioning interfaces.

The block diagram of the control algorithm with the two main sections, software and hardware, is shown in Fig. 4.

The hardware has two input signals; gob signal and machine cycle signal. The first signal increment a counter whose output is demultiplexed and sent in proper sequenced to each I.S. section that will receive and form the glass gob into bottles. The second signal is used to reset the counters on every nachine cycle for these cases in which a gob signal is missing and eliminate time shifts in the sequencing of the sections of the I.S. machine.

The software has three main loops: (1) Control, (2) Filtering and adjusting and (3) gob signal prediction, in this paper we will discuss only the points (2) and (3) in the following sections.

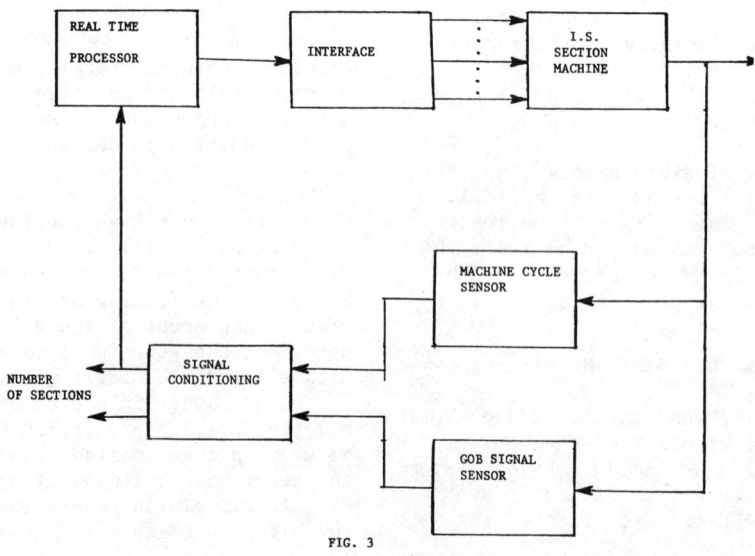

FIG. 3

SYSTEM BLOCK DIAGRAM

An Intelligent Digital Controller

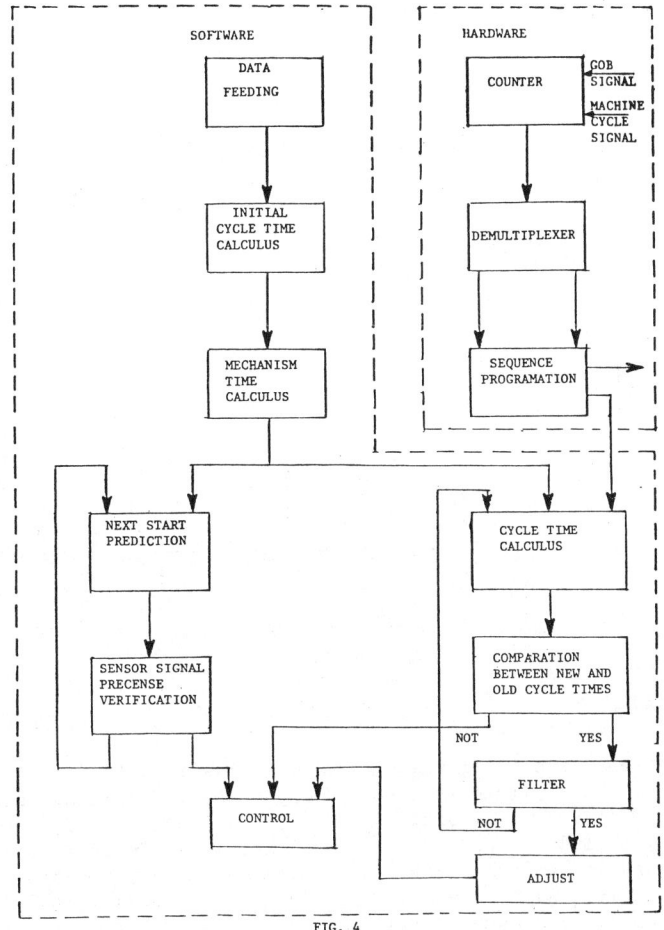

FIG. 4
BLOCK DIAGRAM OF THE CONTROL ALGORITHM

GOB SIGNAL PREDICTION

Equation (1) shows how we estimate the gob signal

$$\hat{CT}_{k-1} = \frac{1}{3} \sum CT_{k-i} \qquad (1)$$

provideed that the gob signal is present and:

$$\hat{CT}_{k-1} - \Delta t_{k-1} < CT_k < \hat{CT}_{k-1} + \Delta t_{k-1} \qquad (2)$$

Where

$\hat{CT}_k$; gob signal estimate

CT_k ; gob signal measurement

Δt_{k-1} (CT_{k-1}); time compensation depend on predicted and measure gob signals.

FILTERING AND ADJUSTING

The filtering section consist of equations (2) and (3)

$$CT - \Delta t' \quad CT_k \quad CT + \Delta t' \qquad (3)$$

Where:

t; time compensation depend on data.

CT; data cycle time

In equation (3) we have limits for the adjust, because we are unable to adjust the mechanisms in accordance with the variation of the gob signal cycle for the whole range. For the adjustment of the trigger of each mechanism we used equation (4).

$$\Delta Adj = \hat{CT}_{k-1} - CT_k \qquad (4)$$

Where:

ΔAdj; adjust compensation

CT_k; gob signal measurement

CT_{k-1}; gob signal estimate

RESULTS

In Fig. 5 we shown the bottles production increment by 7% after we did the implementation.

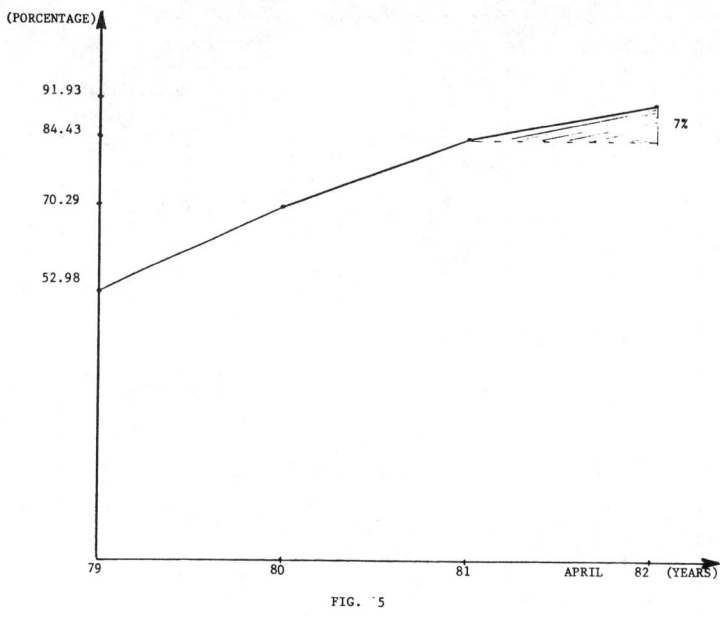

FIG. 5

CONTROLLER TIME PERCENTAGE ON BOTTLES PRODUCTION LINE

CONCLUSIONS

In this paper we presented a predictor based on previous estimations, the preliminary results, have shown improvement on reliability and bottles consistency including the cases of missing gob signal and mechanical adjustment. With protection limits we have eliminated mechanism over lapping and have guaranteed mold protection, in its present configuration the predictor is unable to estimate and make adjustments over the whole range of gob signal variation however we are currently working to derive the method to minimize impact of changes in gob signal cycles on the sequencial triggering of the mechanims on the I.S. machine and to be able to estimate and make adjustments over the whole range of gob signal variation.

REFERENCES

Kwiatkowski J.A. (1973). Electronic timing of automatic glass blowing machines. In. R.J. Mouly (Ed.), Automatic Control In Glass, IFAC, Instrument Society of America, Pittsburgh, pp. 58-65.

DIGITAL CONTROL FOR A PNEUMATIC SYSTEM

D. Figueroa and J. Heredia

Vitro Tec Fideicomiso, Monterrey, Nuevo Leon, CP 64410, Mexico

ABSTRACT.-

This paper discusses the design and operation of a digital controller for a pneumatic actuator, using a 6502 microcomputer. In essence, the controller tracks a reference velocity profile for the inversion mechanism in glass container forming machines. The velocity profile is selected a priori to minimize the stresses applied on the glass articles, since at this phase of the process, such articles are in a viscoelastic state and are highly vulnerable to accelerated motions. Furthermore, the controller should be able to move the mechanism in times varying from 500 to 1300 ms. This is due to the fact that diverse articles require different inversion times in order to minimize stress.

Key Words.- Control Equipment, Digital Control, Real Time System, Glass Industry, I.S. Machine, Glass Forming Process.

INTRODUCTION.-

One of the most critical process in the forming of glass containers is the transfer motion that takes place from the premold to the final formation phase. This invertion motion is found throughout machines of the I.S., (Independent Section) type introduced in 1925 by Hartford Empire Industries [1]. The schematic diagram of I.S. processes is shown in Fig. 1. The inversion and the one of take out, require conveying the article by the nech. Obviously, this could produce irreparable defects in the containers, since such containers are at this time, in a viscoelastic state. The application of an inapropiate velocity or acceleration is the reason for the appearance of defects.

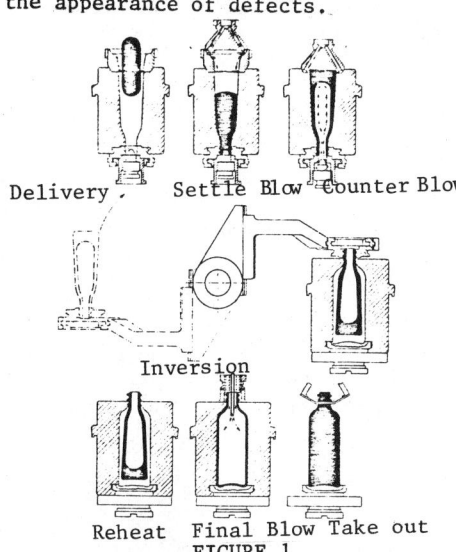

Delivery, Settle Blow Counter Blow

Inversion

Reheat Final Blow Take out
FIGURE 1

CONTAINER FORMING PHASES DIAGRAM

Considering the above, it was necessary to develop a more precise control of inversion mechanism, replacing the present mechanical control. A velocity controller was developes for the inversion transfer motion. This controller tracks a reference velocity profile during the entire inversion motion. In doing so the extensional forces on the container are below the maximam allowable predicted forces by the viscoelastic model [2].

The system description and the manner by means of which the control was developed is presented in the following sections. The comparative results are given between the present system in the last section the new digital control systems is also given. It is not the scope of the present work to describe the mathematical modeling of the mechanical system. In references [3, 4] it is fully discussed.

SYSTEM DESCRIPTION.-

The inversion mechanism consists of a pneumatic actuator which rotates an arms that holds the Parison. This rotation is done by means of a rack and pinion, as shown in Fig. 2.

The velocity in the present mechanism is

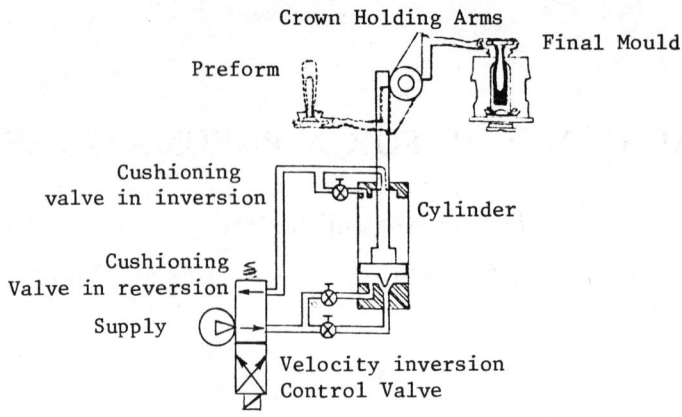

FIGURE 2

SCHEMATIC DIAGRAM, SHOWING THE PRESENT CONTROL SYSTEM.

controlled by the use of needle valves that regulate the inlet air into a pneumatic cylinder. Decceleration is achieved by reducing the exhaust flow by a pivot that closes the air discharge when it reaches a certain position. Instead of using the pivot, a hydraulic shock absorber could also be used. Both decceleration systems must dissipate large amounts of knetic energy in a very short portion of the total stroke, which produces very strong accelerations (in the order of 10 times the gravitational acceleration g). In addition to this, the adjustment of the needle valves in the present mechanical control is a difficult and arduous process. Both mechanical and process problems urged the developed of a Digital Control.

The new system was developed by using a 6502 microcomputer with 4K bytes of EPROM, 2K bytes of RAM, 2 PIAS (peripheral interface adapter) and 1 VIA (Versatile interface adapter). The microcomputer control signals, after passing through an interface of solid state relays, operate the ten (10) solenoids that form the bidirectional digital valve. This valve was designed to allow a controlled in let flow in to the chamber during the inversion, and also the return of the plunger during the reversal. (Fig. 3). Fig. 4 shows a photograph of the inversion mechanism, on the right is the control panel that was used in the plant prototype.

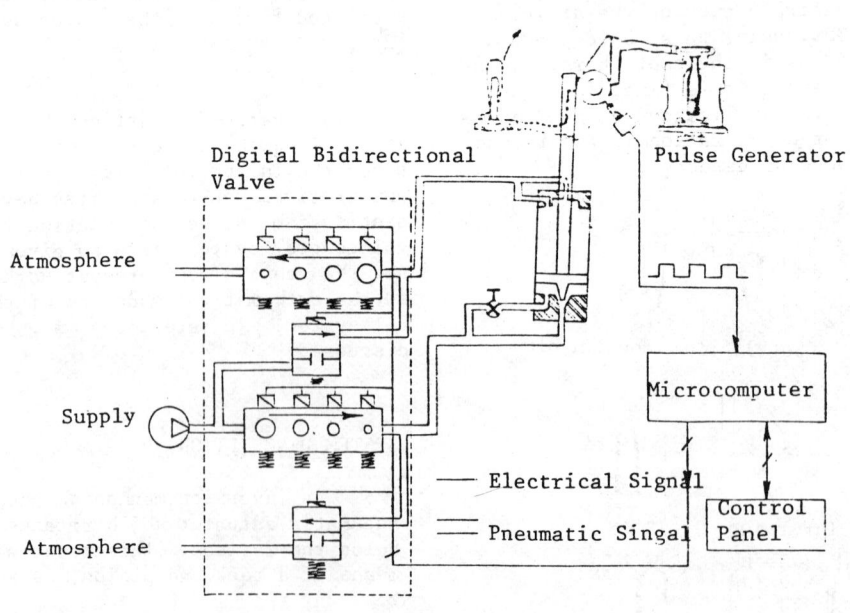

FIGURE 3

PROTOTYPE CONTROL DIAGRAM

Considering the large variation of the parameters involved in the dynamics of the system, and since the change of parameters affect the reliability of the open loop system it was therefore necessary to incorporate into the mechanism a position sensor, by using an optical sensor that detects the signals coming from a grooved disc, and that allows closing the control loop. (Fig. 5).

It is worth while to mention that not only the velocity profile must operate, within inversion times of between 400 to 1300 ms, but it must also assure that the mechanism always reaches its final position, with zero error in position and minimum error in velocity.

CONTROL METHODOLOGY.-

The control outline that was used is shown in Fig. 6. The calculations were made in development computer, and its results were programed in table form, 3, 4 . These table contain the information necessary to actuate the digital valve. The actuator converts the digital signal that is received into an air flow which controls the thrust force and the decceleration of the mechanism. The movement is discretizied by means of the pulse generator, which actuates the position feedback signal. It is mandatory to make certain corrections during the movement, to account for variations in the dynamics of the system. It should be noted that since it is required to control both the position and the velocity, it is necessary to carry out measurements in these two parameters. The measurements are made in four positions of the movement, and once they have been made they are compared with the reference measurements, so that the error is then evaluated. The same error is used to take two possible actions: if it is positive, the valves are closed during a proportional time to the error; while if negative, the control feeds more flow by opening all the valves during a time that is also proportional to the error. This corrective action takes place during each change of value in the digital valve (Fig. 7).

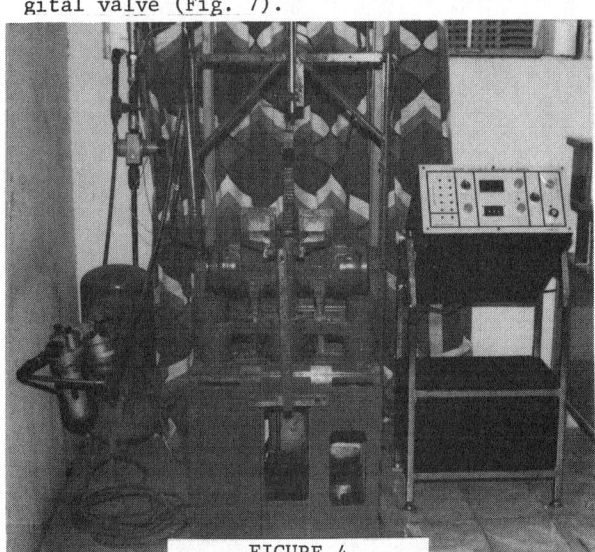

FIGURE 4
LABORATORY PROTOTYPE

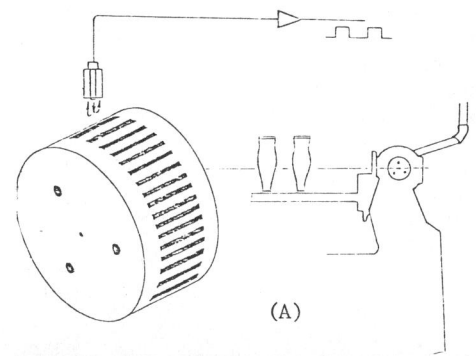

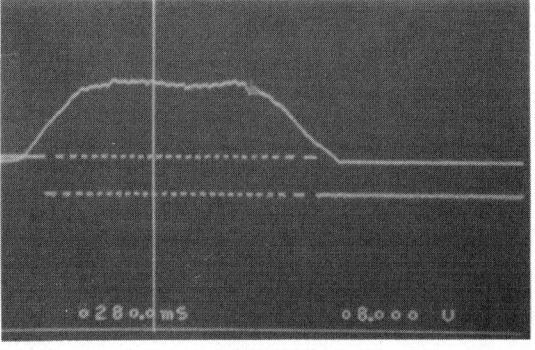

FIGURE 5
A) PULSE GENERATOR
B) TYPICAL OUTPUT

The superposition of both effects has as a goal the assurance that the mechanism always reaches its final position, and also follows a velocity profile that minimizes the flaws in the glass article.

RESULTS AND CONCLUSIONS.-

The results obtained are very satisfactory, as seen by the qualitative results in Fig. 8. These graphs were obtained before and after the digital control was implemented. It can be observed that the accelerations have been tremendously reduced, since after controlling the movement, it is possible to minimize the impulse accelerations of the present systems. Quantitatively speaking, it has been possible to reduce the acceleration from a value of 10 times the gravitational acceleration g to less than 2 g.

In Fig. 9, the result of velocity profiles can be seen for inversion times of 550, 700, 950 and 1100 milliseconds. It may be observed that for measured slower velocity profile the deviation increases with respect to the ideal cycloidal profile. This is due to the fact that a compromise was established between the form of the profile and the number of changes that take place in the valve during each cycle, in order to comply with the useful life requirements of the digital valve.

At the time of this writing, the project was still in its trial phase under plant environment. It is hoped to verify the positive experimental results and also report improvements in the quality and efficiency of the forming process of glass conatiners. Concurrent, work is being done towards improving the system stability, since there are certain aspects to which our control system does not responde with adequate stability. Lastly, it could be added that the utilization of a velocity control system for pneumatic cylinders, may have a great number of applications, as much within the glass industry, as in many other industries where such pneumatic cylinders are used as actuators.

REFERENCES.-

1.- Tooley, Fay V., "The Handbook of Glass Manufacture," Vol. II, p.p. 609-612, books for Industry Inc., 1974

2.- Rhi-Sausi J., "The implications of the extensional flows in the glass viscocity," Vol. II, Technical Symposium on Glass Fabrication, June 1981, ALAPROVI.

3.- Figueroa, D. and Heredia, J., "Mathematical Model of a Pneumatic System for Transfer Control," ALAPROVI, June 1981.

4.- Figueroa, D. and Cárdenas L. "Application of Microprocessors in Data Acquisition Systems," ANIAC, 1979.

5.- Franklin, G. and Powell, D. Digital Control of Dynamic Systems. p.p. 87, Addison Wesley, 1980.

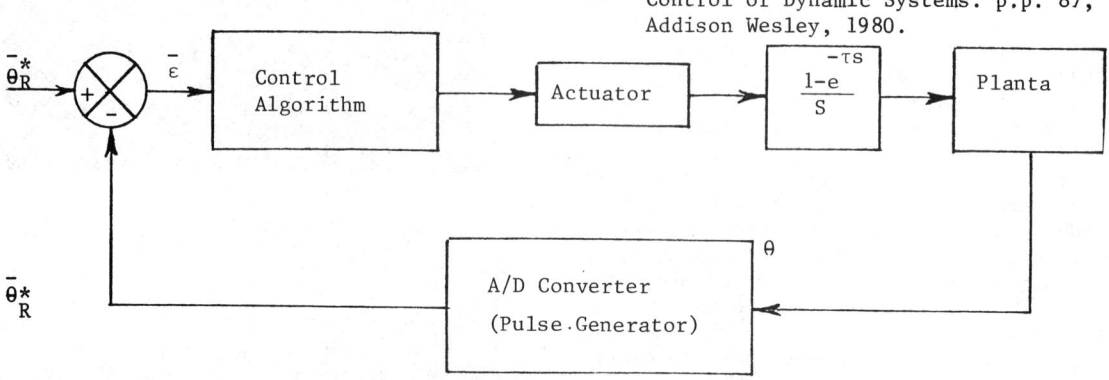

FIGURE 6
BLOCK DIAGRAM OF THE CONTROL SYSTEM

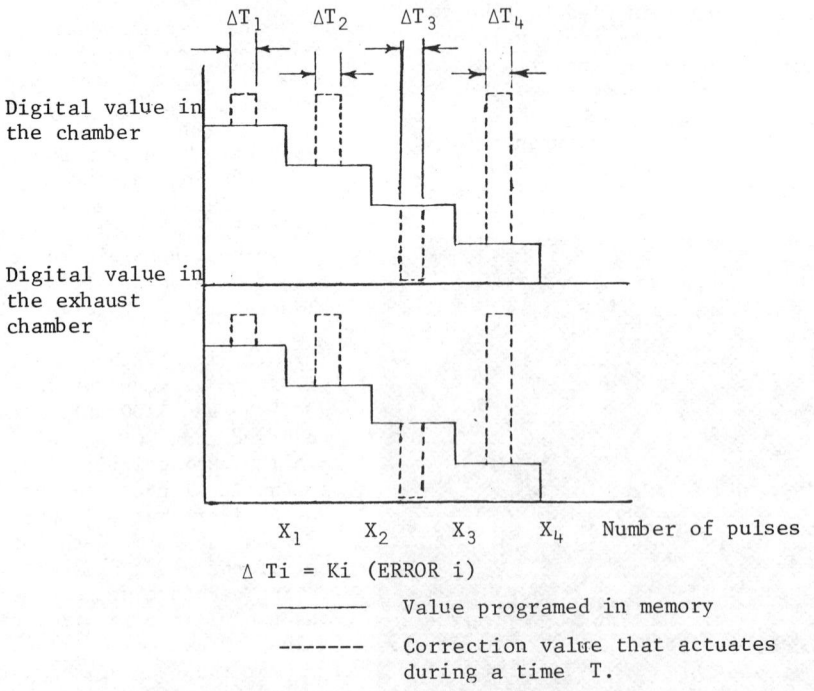

$\Delta T_i = K_i$ (ERROR i)

——— Value programed in memory

- - - - - - Correction value that actuates during a time T.

FIGURE 7. GRAPH SHOWING THE OPENING VALUES IN THE DIGITAL VALVE

Digital Control for a Pneumatic System

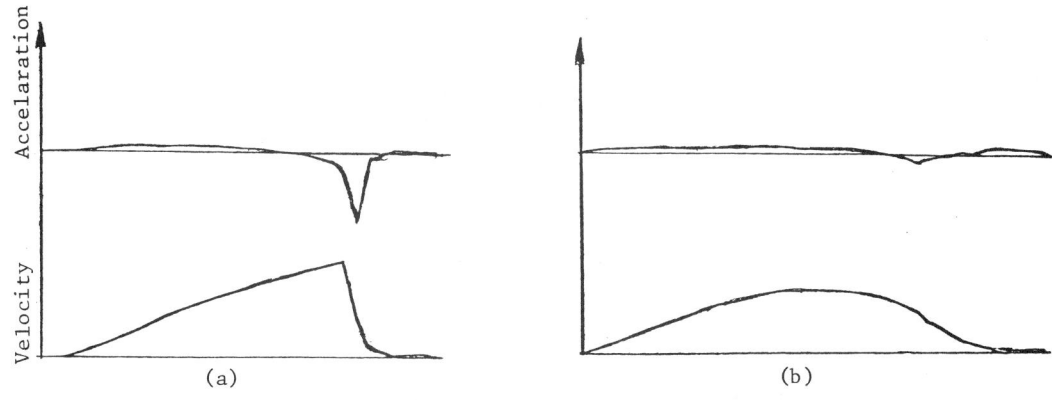

FIGURE 8. VELOCITY AND ACCELERATION GRAPHS OF THE MECHANISM
    (a) WITH PRESENT CONTROL
    (b) WITH DIGITAL CONTROL

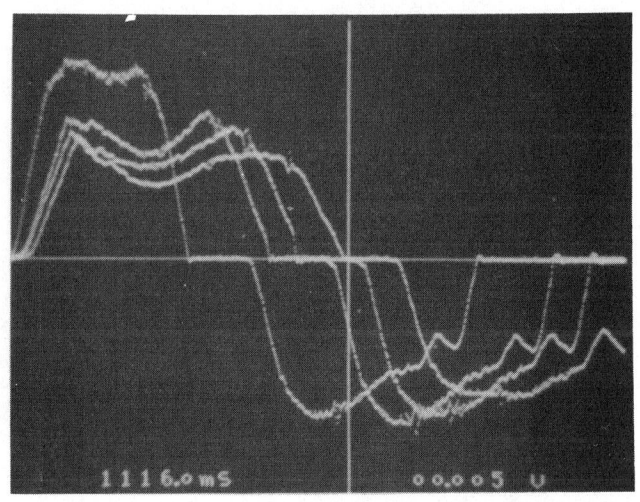

FIGURE 9

VELOCITY PROFILES FOR DIFERENT INVERSION TIME

DISCUSSION

SESSION WA4: GLASS INDUSTRY (INVITED SESSION)

Paper: DIGITAL CONTROL APPLICATION IN A GLASS GOB FEEDER

Authors: F *Saldaña*, R *Solís* (Vitro Tec Monterrey, NL. México)

Discusser: *Dr. José de Jesús Rodríguez*
Apdo 4992 Suc "J"
Monterrey, N.L.
México

Questions or Comments:

1) In the slide of the temperature you slowed there seems to be noisy variations. How are you going to process this measurement for control purposes?

Authors' reply:

1) Using the mean value. In the graph there seems to be some noise present. The temperature sensor did not have the "peak hold" included, so all variations ocurring in the formation were registered. Using a "peak-hold" what seems to be noise would be eliminated.

Discusser: *J.L. Amarillas*
Av. López Mateos 964 Nte
Guadalajara, Jal.
México

Questions or Comments:

1) How does the system you want to implement influence the output control?

Authors' reply:

1) Giving a more exact weight. At present many bottles are rejected when subjected to quality control due to weight excess or defect.

Discusser: *Roman Lamas*
Apartado Postal 6-725
Guadalajara, Jalisco 44620
México

Questions or Comments:

1) Does the system presented can be used to determine the temperature for any chemical composition of the glass?

Authors' reply:

1) Yes it can. The only thing that would change would be the control algorithm, which is a function of the type of glass used.

Paper: ON THE DESIGN OF A DISCRETE FEEDFORWARD CONTROL FOR COOLING MODELS IN THE GLASS INDUSTRY

Authors: FL *Elizalde* (VitroTec, Monterrey, Nuevo León, México)

Discusser: *Dr. José de Jesús Rodríguez*
Apdo 4992 Suc "J"
Monterrey, N.L.

Questions or Comments:

1) The application of feed-forward control for known disturbances in a slow process is appropriate. However, have you thought on using feed-back control simultaneously with the feed-forward control?

Author's reply:

1) Of course. Even in this application the gob characteristics will be included, thus closing the loop.

Paper: AN INTELLIGENT DIGITAL CONTROLLER IN
THE FORMATION PROCESS OF GLASS
BOTTLES

Authors: H Rodríguez, S. Rodríguez (VitroTec,
Monterrey, Nuevo León México)

Discusser: Abraham Cárdenas Castillo
Manuel Acuña 1025
Guadalajara, Jalisco,
México

Questions or Comments:

1) If the gob falls before it is due, is the temperature loss significant?

Authors' reply:

1) The period of time is very small; besides, the controller has a sensor at the end of the canal to detect the falling gob.

Questions or Comments:

1) Why is the process sequence controlled from the machine and not from the gob cutting process?

Authors' reply:

1) Becuase in general the variations are caused by the operator and are due to restrictions of heat extraction. Thus the control must adjust to the process.

REAL-TIME COMPUTER SYSTEM FOR THE NATIONAL ENERGY CONTROL CENTER OF MEXICO

J. L. Calderon and M. A. Avila

National Energy Control Center (CENACE), CFE, Mexico

Abstract. The Mexican Interconnected Electrical System of Comisión Federal de Electricidad (CFE) has an installed capacity of 17600 MW and a main transmission grid of 400 and 230 KV spreaded around the country, covering an approximated area of 2000 000 sq. km. The annual rate of grow is about 10%.

System Control is organized in six area control centers under the coordination of the National Control Center (CENACE). This paper makes a brief description of the Multi-level configuration of the computer system for the National Control Center of Comisión Federal de Electricidad, expected to be in service by 1983.

Software structure is described with special interest in advanced network analysis programs, such as: Network Configuration (NC), State Estimator (SE), Real-Time Equivalents (RTEQ), Real-Time Security Analysis (RTSA) and On-line Load Flow (OLF).

Multi-Area Generation algorithm based in a constrained economic dispatch and a real time area interchange calculation is also briefly described.

1.0 SYSTEM DESCRIPTION

1.1 Power system description.

At the present time, CFE has a variety of plant types having a total installed capacity of 17600 MW [1-2]. Hydroelectric plants comprise 50% of the installed capacity, 40% are thermal and the remaining 10% is made of small units (e.g., gas turbines) used for peaking power. Eighty percent of the installed capacity is currently being dispatched and is concentrated in plants larger than 70 MW.

The CFE power system consists of eight operational areas; two of them (Mexicali and Peninsular) being independent. As a rule, the principal power generating facilities are not located near large demand centers: Therefore, long transmission lines are required. Each operating area will have an Area Control Center (ACC) located as follows:

Central   (Area I)   México, D.F.
Oriental   (Area II)   Puebla, Pue.
Occidental (Area III) Guadalajara, Jal.
Noroeste   (Area IV)   Hermosillo, Son.
Norte   (Area V)   Gómez Palacio, Dgo.
Noreste   (Area VI)   Monterrey, N.L.
Mexicali   (Area VII) Mexicali, B.C.
Peninsular (Area VIII) Mérida, Yuc.

Figure 1 shows the CFE Power System.

1.2 Harris Controls System Description

Harris Controls will provide a National Control Center (NCC) M9400 System in México City and all six ACC's (M9200) [3]. In addition, the system is configured to support two future ACC's (Mexicali and Peninsular). The whole system will include 200 Remote Terminal Units (RTU) from different manufacturers:

Harris Controls	84 RTU
Leeds & Northrup	12 RTU
SCASA*	104 RTU

* National Manufacturer.

Communications between NCC-ACC and most of the RTU's will be supported by satellite (INTELSAT IV). HLDC x .25 will be used for the NCC-ACC data link. Detailed definitions and requirements are contained in the statement of work [5].

2.0 NCC APPLICATION FUNCTIONS

2.1 National control center philosophy.

An overall system philosophy was developed for CFE's National Load Dispatch and Control System which is manifested in the NCC and ACC's. The salient aspects of this philosophy are summarized below [4-5].

1. Functions at the ACCs are generally confined to Supervisory Control and Data Acquisition (SCADA), Energy Management System (EMS), and acquisition and transmission of area data pertinent to the NCC.

2. Functions at the NCC are generally confined to system-wide calculations, power flow analysis, system security assessment, dispatch of the CFE generation resources and management and transmission of information for the ACC's.

3. Emergency or abnormal conditions are to be handled by the area operators at the ACC's. The NCC will lend advice to the area operators as requested under such conditions.

4. Potentially dangerous or undesirable operating conditions for each area and the system as a whole are to be avoided through monitoring and analysis by the NCC.

5. The ACC is concerned with immediate corrective action of existing disturbances, while the NCC is more concerned with corrective advisories regarding potential or impeding abnormal system conditions.

6. The NCC is responsible for scheduling interarea power transfers, frequency and establishing operating points for generation resources. The ACCs are responsible for carrying out these schedules.

2.2 Description of the advanced subsystem at the NCC.

The advanced application subsystem (APPS) at the NCC is comprised of a number of program modules which collectively define APPS [6]

1. Network analysis Program Schedule (NAPS) [7]

2. Real-Time Security Analysis (RTSA) [8]

3. Study Network Analysis (STNA)[9]

These functions include the following programs:

a. Network Configurator (NC)[10]

b. System State Estimator (SE)[11]

c. Security Analysis (SA)[12]

d. Network Reduction (NR)[13]

e. Dispatchers Load Flow (DLF)[14]

Additionally the following programs will be implemented by Harris:

f. Production Costing Calculations

g. Reserves Assessment

h. System Data Reporting

The Instituto de Investigaciones Electricas (IIE) will develope and integrate into the system:

1. Constrained Economic Dispatch[15] as part of the Network Analysis Program Scheduling (NAPS) function.

2. Automatic Interchange Calculation [16-17]

3. Short-Term Hydro-Thermal Scheduling.

4. Short-Term Load Forecasting.

   Sections 3. and 4. will not be described in the present article.

3.0 ADVANCED NETWORK ANALYSIS PROGRAM.

3.1 Network Analysis Program Schedule

The Network Analysis Program Scheduler (NAPS) is initiated whenever a request to execute either the RTSA or the study network analysis (STNA) program is received. It is responsible for scheduling and coordinating the execution of the RTSA and STNA programs.

The NAPS program support the following features [fig. 2]:

1. Four STNA cases are to be supported.

2. The termination of the current execution of the RTSA program upon a breaker operation. Whenever a breaker action occurs, for which an event trigger to NAPS is defined, the RTSA program will be terminated if it is in execution. However, a dispatcher selectable flag is to be provided on the RTSA control page to activate/deactivate this feature.

3. The selection of a request for STNA to reconfigure a given study without solving the system.

4. The initiation of the RTSA program by the Reserve Monitoring Program if an area's reserve is insufficient (falls below a minimum value).

Figure 2 shows the interaction diagram of APPS analysis functions.

3.2 Real-Time Security Analysis.

The Real-Time Security Analysis Program is responsible for coordinating the solution of the state of the Real Time Power System. It performs the following functions [fig. 3].

1. It receives and decodes initiation requests from NAPS. The request could have been initiated through automatic triggers originating from the Real-Time System or through a dispatcher's request for execution

2. It checks that the RTSA user files are not being used by other APPS modules. If the files are busy it keeps retrying until the files are free. Once the files are available, it locks out all other programs from accessing the files and starts calling the various analysis functions.

3. RTSA calls the network configurator (NC) function to configure the system and builds the Bus Oriented Model (BOM). The State Estimator (SE) function is then called to provide a load flow solution which is used as the base for contingency simulations by security analysis (SA); if needed it calls the Real-Time Equivalence (RTEQ) program.

4. RTSA checks at various logical breakpoints whether a cancel execution request has been issued by the dispatcher. If so, it performs all processing to terminate execution of the function.

5. Various abnormal conditions requiring dispatcher attention which are detected by the various analysis functions are alarmed by Alarms Event Handler Program (AEHD).

It is expected that this program will execute approximately every 15 minutes.

3.3 Study Network Analysis.

The Study Network Analysis program is responsible for the coordination of the solution of a given study case. These study cases (four) are initialized totally by the dispatcher, to correspond to a given system configuration that he is interested in studying. STNA performs then the following functions [fig. 4].

1. It receives and decodes initiation requests from NAPS to configure a study case either from the Real-Time data base or from any of the thirty stored cases. Once a study case is defined the following options are available.

    1.a Reconfigure the system without deriving a base solution or performing contingency evaluation. (Execute the program up to the NC function only).

    1.b Reconfigure the system and derive a base solution without performing contingency evaluation. (Execute the program up to the DLF function only).

    1.c Reconfigure the system, derive a base solution and perform contingency evaluation using the full base system (Execute the program up to the SA function without calling NR).

    1.d Reconfigure the system, derive a reduced base system and perform contingency evaluation. (Execute all the functions including NR to reduce the base system before SA).

2. STNA checks at various logical breakpoints whether a cancel execution request has been issued by the dispatcher. If so, it performs the required processing to terminate the request.

3. Various abnormal conditions requiring dispatcher attention which are detected are logged and also reflected in the appropiate function pseudo status points in the Real-Time Data Base.

It is expected that this program will execute approximately every 30 minutes.

3.4 Network configuration and Modelling.

The network configuration routine is implemented as a part of either the STNA or RTSA programs. As a part of the Real Time Security Analysis (RTSA) package, the modelling function maintains the Real-Time model stored in the user files and configures the network model based on the real-time status of switching devices. It maps the measurements into the network model for use by SE. The Bus Oriented Model produced is eventually used by SA and RTEQ functions. In the process of building the model, NC detects the following abnormal topological conditions:

1. Islanding conditions, and associates network elements and measurements with in each island.

2. Open ended lines or transformers.

3. Split buses.

4. Bus outages due to opening of switching devices.

5. Generation and/or load outage.

NC also detects the following abnormal conditions in the telemetered data:

1. Point out of scan

2. Dispatcher entered data

3. Non-updated data

4. Out of range data

5. Questionable data

As a part of the study network analysis (STNA) package, the modelling function provide the dispatcher the capability of controlling the conditions of the system being analyzed. A bus oriented model for the conditions specified by the dispatcher is prepared for use by the DLF, SA and NR functions.

Figures 3 and 4 show how the network configuration and modelling function fits into the RTSA and STNA packages respectively.

3.5 Dispatcher Load Flow.

The purpose of the on-line load flow module is to calculate the network solution for the particular network model under study. The resulting solution can then be used as the starting point for the execution of other application modules, such as security analysis (SA) and network reduction (NR). As a part of the Real Time Network Analysis, the on-line load flow is used if the state estimator function is not available.

The dispatcher load flow module uses the fast decoupled method [18] This function will consider and respect:

1. All switching specified at the breaker or switch level.

2. All switch and breaker positions, specified through input, will be processed to obtain the bus oriented network model, detecting islanding if such occurs.

3. All islands with sufficient generation, resulting from breaker or switch operations will be processed by the load flow algorithm.

4. Dispatcher specified values of net interchange for each of the six areas.

5. Local and Remote bus voltage control using MVAR sources with no voltages allowed to go outside limits.

6. Local and remote bus voltage control using MVAR sources with no MVAR generation allowed to go outside limits.

7. Local and remote bus voltage control using load tap changing transformers.

8. MVAR flow control using load tap changing transformers.

9. Voltage adjustment is specified on a system wide-basis.

The interaction diagram for DLF function is shown in figure 5.

3.6 System state estimator.

The state estimator function is called as a part of the RTSA program to determine the state of the Real-Time power system. The prime purpose of this function is to create an improved data base in real-time to be used by various other functions in the NCC, the state variables are chosen to be the voltage magnitude and phase angle at each bus represented in the network model. It is understood that sufficient measurement redundancy exists to allow statistical correlation to be performed. The following additional functions are satisfied by SE:

1. Provide a historical record of measurement anomalies for use in the assessment of the reliability and accuracy of the data acquisition system.

2. Define unobservable regions in the system so that real-time equivalents may be derived for use by security analysis.

3. Provide an estimate of non-telemetered or missing data wich are within observable regions.

The state estimator is based on the classical static weighted least squares (WLS) [19] algorithm and uses the fast decoupled method for solution [20]. The program accept measurement in the form of:

1. Bus voltage magnitudes and angles.

2. Active and reactive line or transformer flows.

3. Active and reactive power injections

4. Pseudo measurements of quantities not actually measured.

Before attempting a solution, a statistical CHI square test will be done. If the current measurement fails to pass it, an observability check is made. Observability check will verify that a given measurement set covers all the nodes of the network. It will also verify that there is at least one voltage measurement for each island of the system in case of electrical separation.

Figure 6 shows the interaction and data flow pertinent to SE

3.7 Security analysis.

The purpose of the security analysis program is to evaluate the security of the power system under existing (real-time) or anticipated (study) network configuration and operating state. The security of the system is evaluated on the basis of the limit violations and their potential effect upon steady state operation which can result from the occurrence of a simulated set of contingencies. Each contingency consists of outages of power system equipment such as transmission lines, transformers, generators and loads or a combination of them.

It is important to reduce the list of contingencies to be analyzed in detail to contain the least meaningful number to be processed. Certain dispatcher preselected contingencies will be processed as a matter of normal operation: the question remains as to what other contingencies should be processed. Because there can be no assurance that an abbreviated list of additional preselected contingencies will encompass the "worst" contingencies. Automatic contingency selection (ACS) [21] is used to reduce the execution time of SA.
The following features are to be supported as a part of ACS:

1. All branches must be checked for possible security violations.

2. A maximum of 200 branch outages are to be supported. Only single branch outages must be supported.

3. The ranking shall be at least a function of potential branch overloads.

4. Contingency entries can be assigned to any of the following dispatcher alterable types:

    4.1 Contingency always to be fully executed.

    4.2 Contingency to be analyzed by ACS

    4.3 Contingency deactivated.

5. The ACS function can be readily activated/deactivated from the RTSA and STNA. If ACS is deactivated, then all contingencies shall be simulated.

The interaction of the security analysis with the rest of the network analysis programs under both real-time and study modes is shown in fig. 7.

3.8 Network reduction.

The network reduction function is implemented by means of two routines in the system. In the Real-Time mode, as a part of RTSA, the Real-Time equivalents routine is to provide a solution for the unobservable parts of the power system [22-23], as determined by the SE. As a part of the dispatcher load flow (STNA), the network reduction routine is used to derive an equivalent electrical network [24] for an area or zone to be reduced, as an input to the SA routine. The NR program is a sparsity oriented program. The Algorithm works internally, one area at a time. However the dispatcher has the option to select a zone (a sub set of an area) for reduction. All areas or zones being eliminated are reduced to their boundary buses monitoring the effect of the change on the retrieved system.

The boundary buses are the buses which connect different areas or zones. These buses are identified and retained in the reduced equivalent by the NR program. In addition to the boundary buses. any branch or bus in the system to be reduced can be specified to be retained in the equivalent network.

The data flow diagram for the NR program is shown in figure 8.

4.0 PROGRAMS TO BE WRITTEN BY IIE.

4.1 Constrained economic dispatch.

The purpose of the constrained economic dispatch function is to calculate at the NCC level, the optimum operating base point and economic constrained participation factors for every operating unit that is represented by a production cost curve. Additionally the constrained economic dispatch function calculates:

1. The constrained area interchange sensitivity factor for each area to be used by the automatic interchange calculation function.

2. The constrained external system interchange sensitivity factor for each area.

3. The base value of desired scheduled net interchange for each area.

The dispatch must be performed respecting the following constraints:

1. Real power flow limits on selected interarea and/or intraarea transmission lines or transformers or lumped equivalents thereof.

2. Real power flow limits on selected interarea or intraarea groups of transmission lines or transformers of lumped equivalents thereof.

3. Real power net interchange of each area must fall between specified limits.

4. Area spinning reserve and regulating margin.

5. Group of plants spinning reserve requirements must be met.

The constrained economic dispatch determines the generation dispatch at minimum cost while satisfying generation, transmission and reserve constraints. The penalty factors to account for incremental transmission losses must be calculated automatically each time a dispatch is executed recognizing the current state of the network.

The algorithm to be used in solving the constrained economic dispatch is based on linear programming using the dantzing-wolfe decomposition principle [15].

Figure 9 shows the interaction diagram of the constrained economic dispatch function with other functions.

4.2 Automatic interchange calculations.

The purpose of the automatic interchange calculation function (AIC) [4, 16-17] is to calculate the scheduled net interchange for each area once every four seconds, maintaining system frequency and system net (external) interchange within specified tolerances. These calculated net area interchange values are transmitted to the area control centers (ACC) every four seconds. The ACC's use these scheduled values in performing automatic generation control (AGC).

The objectives of the AIC function are summarized below:

1. The CFE system must be controlled in a way to meet the scheduled interchange of power between CFE and other companies

2. Changes in CFE load are to be followed by adjustments in the output of CFE generators.

3. Changes in CFE system load are to be economically allocated among the generation of the six areas in a manner determined on a total system basis.

4. System control error between CFE and other companies, and area interchange errors, are to effect emergency assist action by the six areas if such errors exceed threshold values.

5. CFE will operate in a way to keep a reserve in each area greater than a specified regulating margin; should an area become deficient in generation, the interchanges must be modified accordingly to maintain the regulating margin.

6. Minimize the inadvertent external interchange.

7. The AGC function in each ACC operates in a conventional manner with area control error formed on the basis of scheduled net interchange, actual tie line flow and frequency deviation from schedule.

5.0 CONCLUSIONS

This paper presents an overview of the functional aspects and interrelationship of the advanced applications software system. This design represents a new generation of energy control centers where the network and energy scheduling model reliably reflect the current conditions of the power system, and study conditions are generated from actual system data rather than hypotetical off-line data.

The subsystem is designed and intended as an aid to the dispatcher, to relieve him of some tasks and to supply him with new information which is presently not available to him. A fundamental principle of the design is to keep the dispatcher in the loop whenever practical, using his experience and basic system knowledge in concert with the application programs. The programs in no way replace the dispatcher. In fact the programs expand his responsability by providing new capabilities to operate the power system more reliably and economically.

6.0 REFERENCES

[1] "Resutados de Explotación 1980-1981" Comisión Federal de Electricidad.

[2] "Estadísticas 1965-1981" Sector Eléctrico Nacional. Comisión Federal de Electricidad.

[3] L. Rancé. "El proyecto SICTRE en el Centro Nacional de Control de Energía de la CFE" Colegio de Ingenieros Mecánicos y Electricistas [CIME] Sept. 82

[4] J.L. Calderón "Desarrollo en México de Programas de Aplicación para el Sistema de Información y Control en Tiempo Real de Comisión Federal de Electricidad". III Simposio Latinoamericano sobre Centros de Control para Sistemas Eléctricos de Potencia. Cancún, Q.R. México Junio 1981.

[5] "Sistema de Información y Control en Tiempo Real del Centro Nacional de Control de Energía". Gerencia General de Operación CFE. Octubre 1979.

[6] "APPS Network Analysis Description" Document 888-3473

[7] "APPS network Analysis Program Scheduler" Document 888-3476

[8] "Real-Time Security Analysis Executive" Document 888-3485.

[9] "Study network Analysis Executive" Document 888-3477

[10] "Network Configuration Detail Design" Document 888-3482

[11] "State Estimator Detail Design" Document 888-3490

[12] "Security Analysis Detail Design" Document 888-3474

[13] "Network Reduction Detail Design" Document 888-3488

[14] "On-line Load Flow Detail Design" Document 888-3483

[15] R. Romano, V.H. Quintana, R. López V. Valadez. "Constrained Economic

Dispatch of Multi-Area System Usin the Dantzing-Wolfe Decomposition Principle" IEEE TRANS. PAS. Vol.100 No.4 pp.2127-2137. 1981

[16] J.M. Martinez, M. Santos, D. Burciaga, V.H. Quintana R.Nieva. "Modelo General de Simulación de Control de Potencia en las líneas de enlace de un sistema Multi-Areas" México 81. IEEE. Sección México. Guadalajara, Jal. 1981.

[17] V.H. Quintana, R.Nieva, M. Santos, J.M. Martínez V.Gerez. "Automatic Tie-Line Power Control of Multi-Area Power Systems". Proc. of 22th. Midwest Symposium on circuits ans systems. Alburquerque, N.M. 1981.

[18] B.Stott, O. Alsac. "Fast-Decoupled Load Flow". IEEE Trans. Pas. Vol. 93. No.3 pp. 859-869 1974.

[19] R. Deutsch "Estimation Theory" Prentice-Hall Englewood, Cliffs. N.J. 1965.

[20] H.P. Horisberger, J.C. Richard C.Rossier. "A fast Decoupled Static State Estimator for Electric Power Systems" IEEE. PES. Summer Meeting. San Francisco, Calif. July 20-25 1975. paper F75 428-3

[21] S.Vemuri, R.E. Usher. "On-line Automatic Contingency Selection Algorithm" IEEE, PES. Summer Meeting. San Francisco Calif. July 18-23. 1982 Paper 82 SM 413-3

[22] F.L. Alvarado, E.H. Elkonyaly. "Reduction in Power Systems". IEEE. PES. Summer Meeting. México City. July 1977 Paper A77 507-7.

[23] E.H. Elkonyaly, F.L.Alvarado. "External System Static Equivalent for On-line Implementation IEEE. PES. Winter Meeting. New York, N.Y. Jan/Feb. 1978. Paper A78 060-6

[24] S.Deckmann, A. Pizzolante, A. Monticelli, B.Stott, O. Alsac. "Studies on Power System Load Flow Equivalencing" IEEE. Trans. Pas. Vol. 99. pp. 2310. 1980.

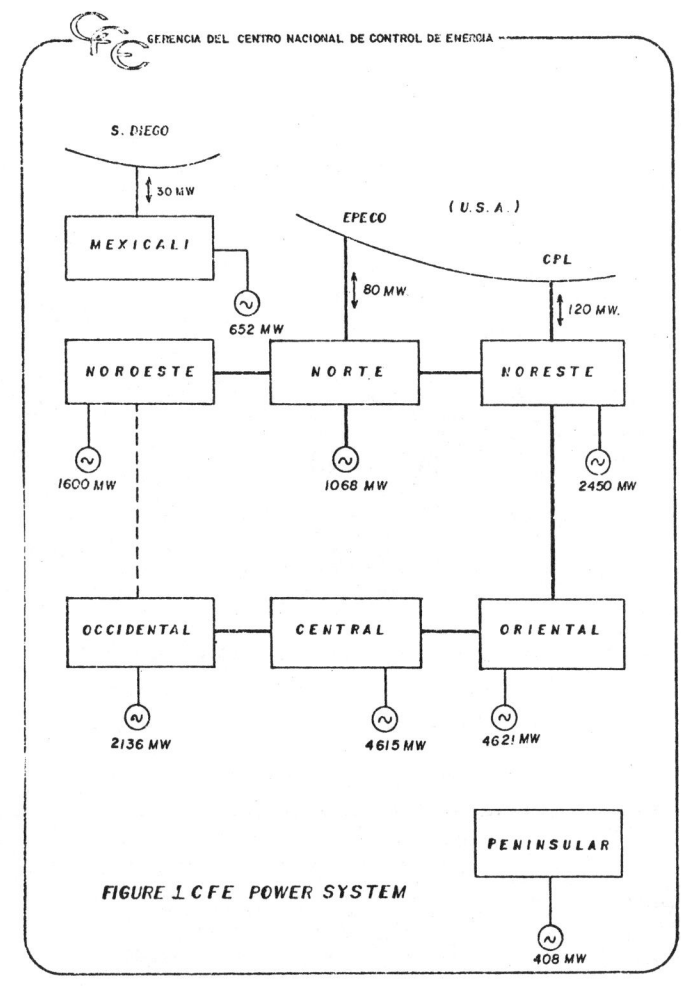

FIGURE 1 CFE POWER SYSTEM

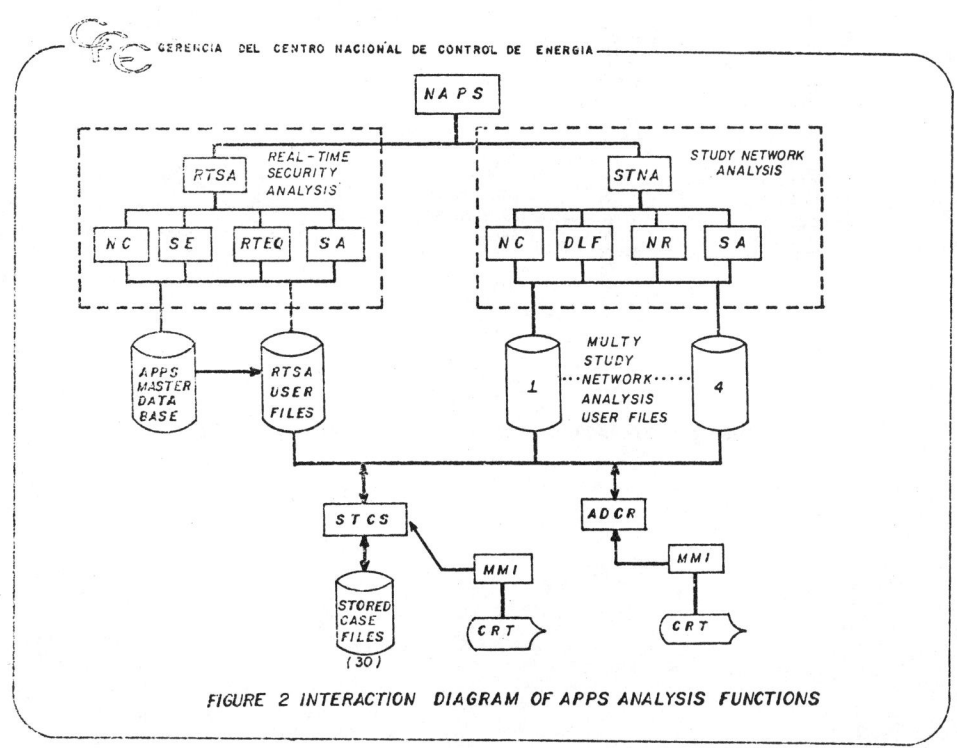

FIGURE 2 INTERACTION DIAGRAM OF APPS ANALYSIS FUNCTIONS

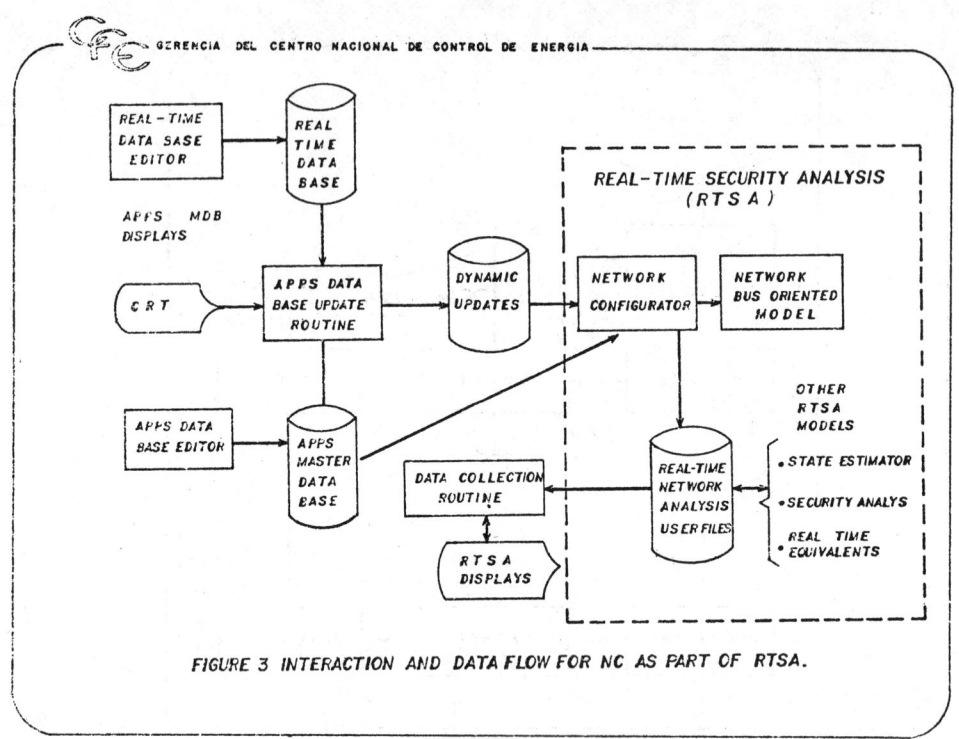

FIGURE 3 INTERACTION AND DATA FLOW FOR NC AS PART OF RTSA.

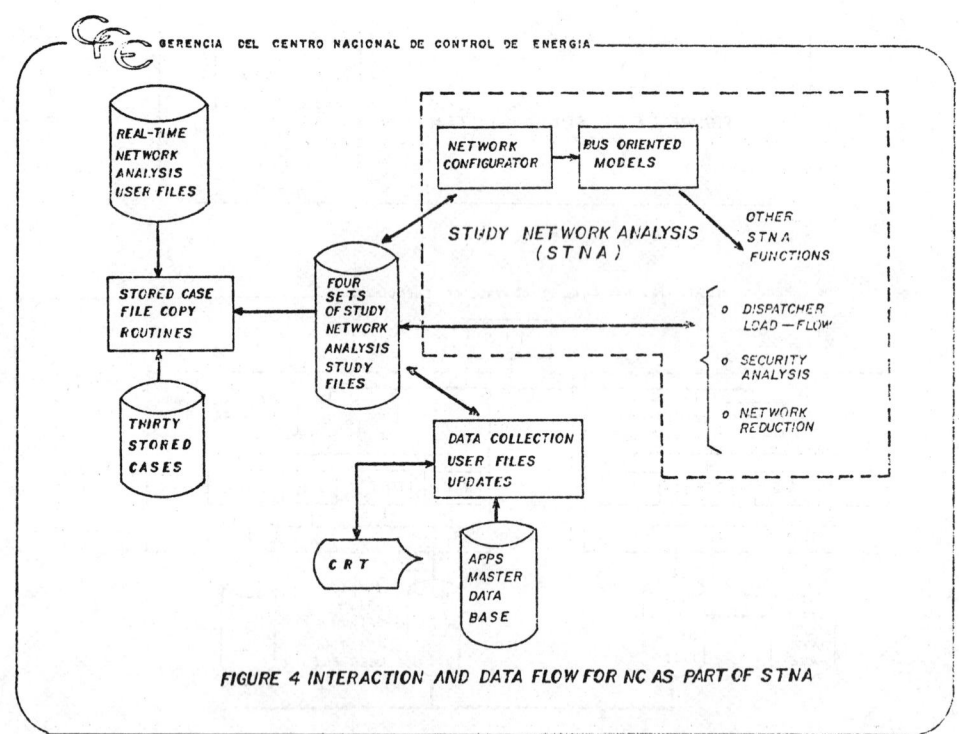

FIGURE 4 INTERACTION AND DATA FLOW FOR NC AS PART OF STNA

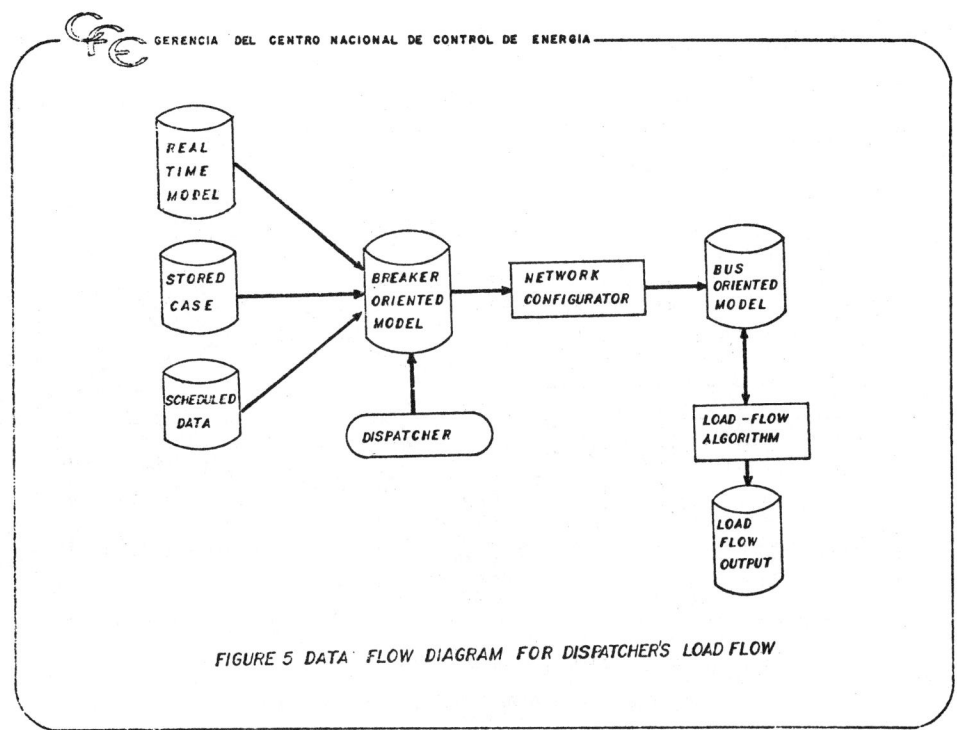

FIGURE 5 DATA FLOW DIAGRAM FOR DISPATCHER'S LOAD FLOW

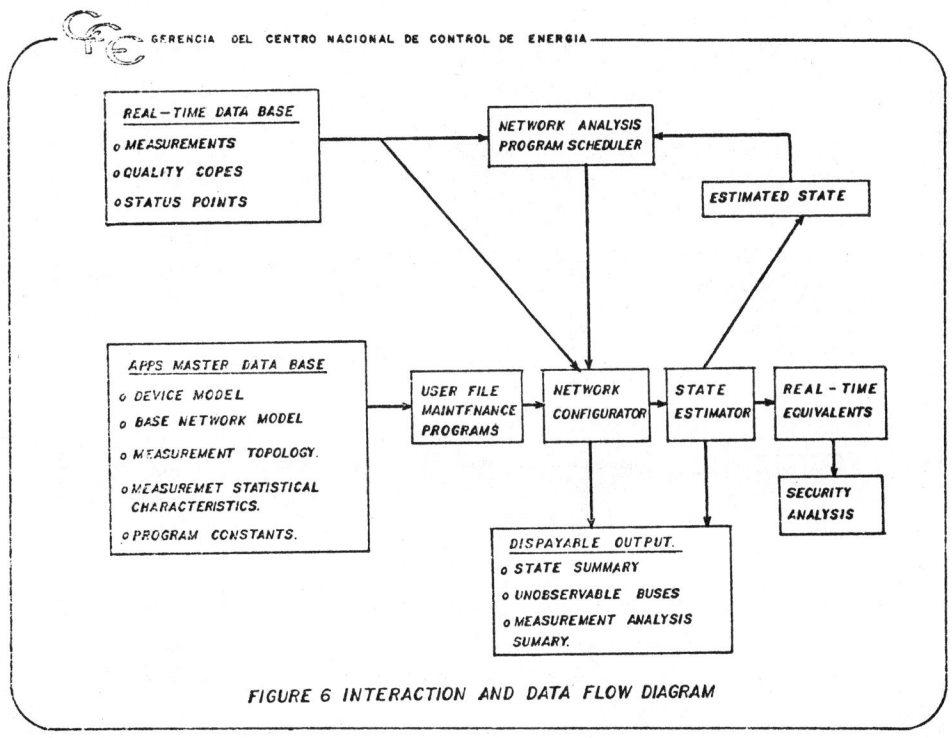

FIGURE 6 INTERACTION AND DATA FLOW DIAGRAM

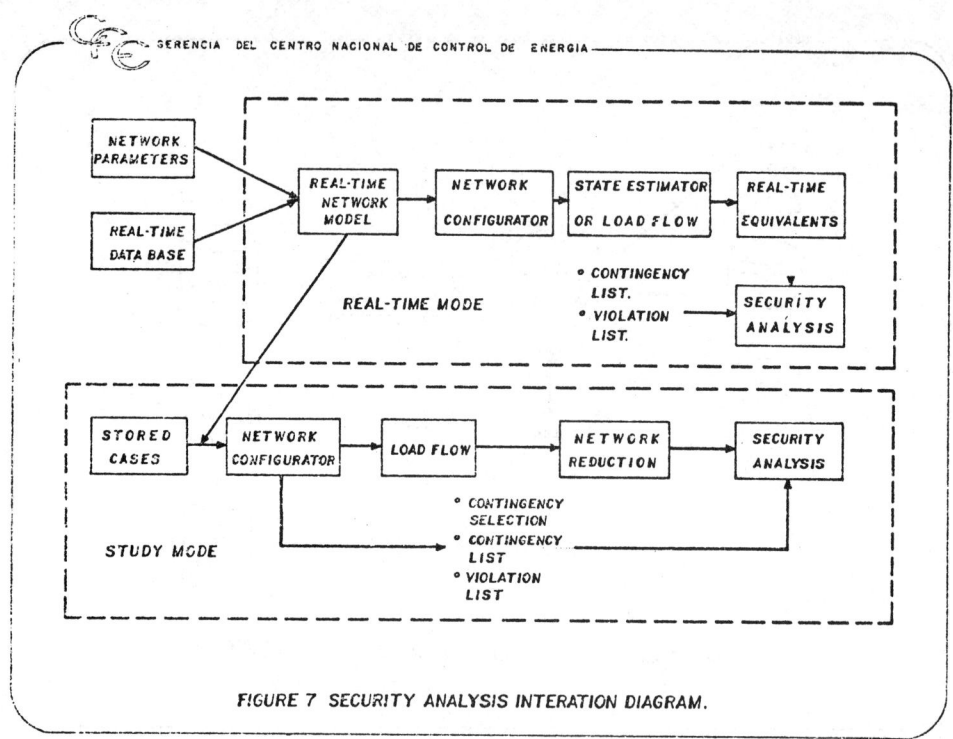

FIGURE 7 SECURITY ANALYSIS INTERATION DIAGRAM.

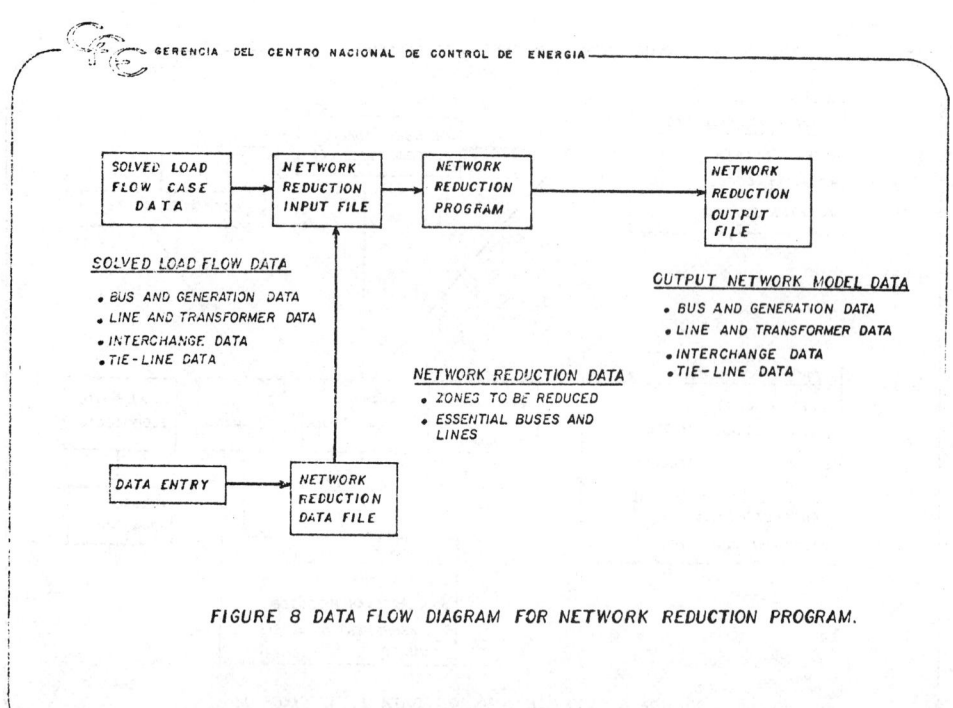

FIGURE 8 DATA FLOW DIAGRAM FOR NETWORK REDUCTION PROGRAM.

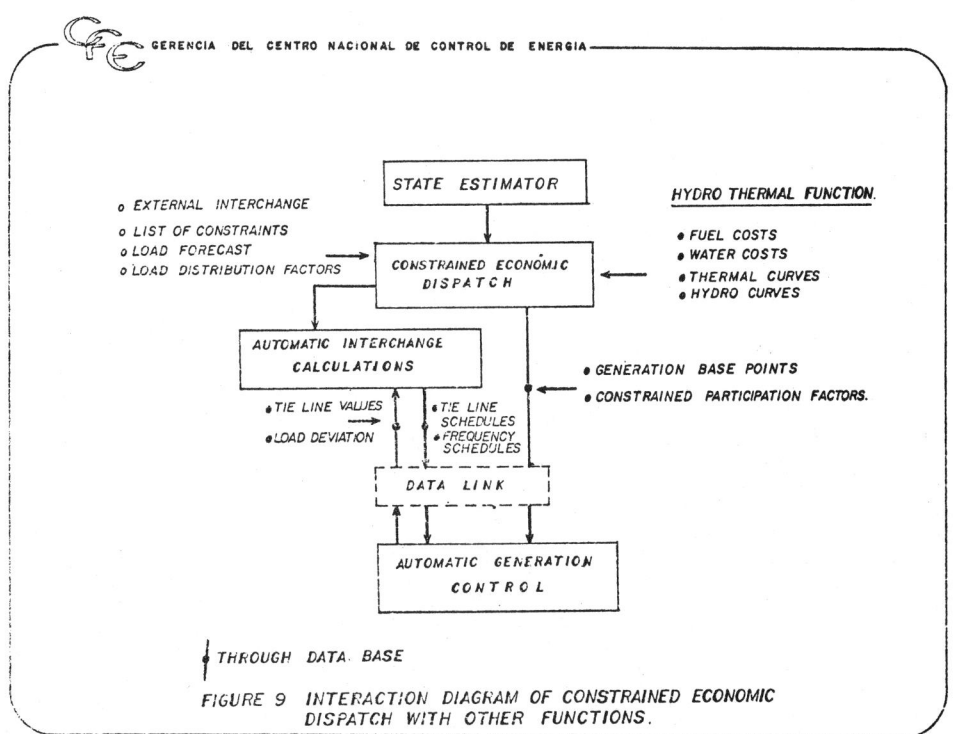

FIGURE 9 INTERACTION DIAGRAM OF CONSTRAINED ECONOMIC DISPATCH WITH OTHER FUNCTIONS.

MANAGEMENT AND IMPLEMENTATION OF THE SICTRE PROJECT

L. Rance

National Energy Control Center (CENACE), CFE, Mexico

Abstract. The increasing complexity of the Mexican Power System, growing about 10% per year, and the operation diffculties associated in the same proportion, have been the reason of CFE (Comisión Federal de Electricidad) decision to implement a real time computerized system at a national level called SICTRE (Real Time Information and Control System).

The project is one of the most important in the world for this purpose. A four computers system at the national level will be linking six Area Control Centers with dual computer systems at each one. A total of 200 remote terminal units are being installed to gather data from the electrical network. Advanced application programs will be performed basically at the National Control Center. A dedicated communications network via satellite will be used, and new buildings for the Area Control Centers have been built.

To manage and implement the whole project is very complex. This paper deals briefly with some of the major problems and activities such as planning, specifications, contract, statement of work, manufacture, schedule, training, data base and displays definition, site preparation, factory aceptance test and installation.

1. JUSTIFICATION AND PLANNING

The mexican interconnected power system is one of the biggest in the world from the geographic area point of wiew. It covers about 2 million sq. km. and it has an installed capacity of 17600 MW, using more than 35000 km of 400, 230 and 115 kv transmission lines.

CFE as many of the most important electrical utilities in the world is mainly concerned about security and reliability aspects in operation as well as quality and economy.

Nowadays many electrical utilities are in troubles to operate their network with lower production margins due to the very high expansion investments and construction delays and also due to very long transmission lines from power plants to consuming areas.

By the end of 1976 the electrical "island" in the central part of the country was eliminated (Mexico City and sourroundings) by changing the frequency from 50 to 60 hz. Then the fully interconnection with the rest of the country was made possible. CFE took all the advantages of this interconnection but it had to face the operational diffculties.

1.1. Operation Philosophy

The first step was to define a national operation philosophy in three levels:

1. National Control Center
2. Area Control Centers
3. Power Plants and Substations

The country was divided in 8 Control Areas:

1. Central (México city)
2. Oriental (Puebla)
3. Occidental (Guadalajara)
4. Noroeste (Hermosillo)
5. Norte (Torreón)
6. Noreste (Monterrey)
7. Baja California (Mexicali)
8. Peninsular (Mérida)

It was delegated at the Area level the direct supervision and control over the power plants and network corresponding to each Area.

The National Control Center in México City is in charge to coordinate the Area Control Centers with strategies and criteria looking for the optimization and security as a national goal.

1.2. The SICTRE project

At the beginning of 1977 and knowing the recent experiences of other utilities it was taken the decision to start the SICTRE project in order to provide the most updated tools available in the world oriented to operate electrical systems in the best conditions of security, quality and economy.

The multi-level structure in equipment and functions to be implemented was defined with the cooperation of Power Technologies Inc. as consultant, the participation of experts in operation, and some of the future project leaders, incorporating all the elements considered in any modern power control center in the world such as:

- Digital computers
- Real time operating system
- Video - color consoles
- Loggers
- Dynamic mimic boards
- Remote terminal units
- Advanced application programs
- Communications network (voice and data)
- Computer data links
- Ad-hoc buildings and auxiliary equipment

1.3. The Project team

The experience has shown that it is more convenient to keep the project responsability, supervision and coordination under the same Control Center organization rather than to give it to an external company or to another utility department or departments.

The CENACE (National Energy Control Center) organization has had to modify its structure adding a new branch oriented to provide and maintain all the tools related with communications, computers and software.

Usually in any utility it takes time to accept those new functions delegated in a traditional operative area, and for several months or even years there are missunderstandings, arguments and professional jealousy in the internal utility organization, mainly because the sources of personnel are different departments such as: Telecommunications, Controls, Protections, Systematization, etc.

2. SPECIFICATIONS, BID AND EVALUATION

This phase of the project took almost two years (1977-1978) which is considered normal in this kind of projects. We knew from other utilities references the high degree of difficulty and complexity because those projects have a lot of customization particularities and, in spite of simple SCADA functions, can not be considered as a standard product.

2.1. Functional specifications

During 1977 the first project team, a small group of 4 specialists, worked full time writing the specifications. They partially had the collaboration of other 3 CFE experts, some ideas and opinions of operation managers, and PTI consults.

The participation of operation experts is very important in order to clearly identify:

- The real and very specific dispatching problems.
- The operational practices and traditions
- The expansion of electrical system.

As long as they will be the ultimate users.

On the other hand to accomplish all the wishes of the operation people, it quite often results very expensive, or unpractical, or unfeasible. In this case it is possible to propose alternative solutions with the consultant help who knows the "state of the art" and the market availability.

From the beginning the relationship between the project team and the users must be very close, otherwise the skepticism and rejection of the new tools will be higher.

The specifications must be functional, not detailed, comprehensible to the potential sellers and trying not to invent anything new or too sophisticated, which sometimes consultants like to do. But it is also important to leave certain degrees of freedom in some features expecting proposals.

2.2. The Bid

Once the specifications are ready to be delivered to the sellers it is - very important to avoid from the beggining the participation of small and inexperienced companies.

It is advisable to make a survey of the vendors resources available and, if possible, to visit some other energy control centers to know their equipment and experiences.

2.3. The evaluation

By 1978 seven serious proposals were submitted to CENACE and it started another complex activity: the evaluation. At that time the project team grew up to ten people and the job was made internally without any external consultant help.

How to do the evaluation? Taking into account that it is not possible to make exact comparisons because the complexity of the problem and the seller interpretations, the project team must be very careful.

Four american manufacturers submitted their proposals: Leeds and Northrup, Control Data, TRW, and Harris, and three europeans: Brown Bovery, Siemens and Thomson Sodeteg. All of them with different approaches, different prices, different conditions and different time delivering.

There is not a standard method to get the best evaluation, some methods are more ineficient than others.

To perform a good evaluation there are some typical guide - lines:

o Comparing exactly standard items
o Assessing scorepoints in special items
o Identifying optional or indispensable items.
o Comparing breakdown prices for each major item in hardware and software.
o Determining the limitations for future expansions.
o Establishing the past performance and experience of each seller.
o Establishing the support that each seller is capable to give and how many years.

All these facts were taken into account for the evaluation of the SICTRE project proposals. The CENACE project group also developed their own method to get the conclusion.

Another very important consideration was the facilities offered to integrate mexican equipment and software developed in-house into the system.

The decision was very difficult but finally Harris was selected. The main reasons were:

o Price
o Computer technology
o Facilities for national integration

In this type of projects the seller usually does not manufacture all the equipment involved but he can provide quotation and integrate it into the system as a package.

From a practical point of view this approach is very convenient because the customer only has to negotiate with one seller. Although it can be saved a lot of money when buying directly from the manufacturers some items such as mimic boards, uninterruptible power supply, transducers, interface cabinets, and many others.

CFE decided to buy directly several items in spite of the difficulty to manage different vendors, and the project has become more complicated due to:

o The internal utility purchase procedures
o The fluctuations of foreign currencies
o The relationship between vendors handled by the customer.

At the end of the SICTRE project we will know if we have saved money or not. But there is not question about that we will have learned and acquired a lot of experience knowing the international market and motivating the possibilities for the national industry.

3. CONTRACT

The process of writing the "statement of work" and formalizing the contract, almost consumed the whole year of 1979.

The compromise with the seller was preliminarly set through a letter of intent. Then two activities started in parallel: meetings between CFE and the seller to agree the content of the statement of work, and the contract terms and conditions.

Both activities were very critical in this phase of the project. CFE was very careful to assure technically and legally its future success. The seller was strongly pushed to accept all the technical features well defined and to cover all the legal aspects avoiding unforeseens.

All the contracts in this type of

projects are very similar, but CFE/Harris contract was of a special nature.

It was itemized in 35 clauses where the most important were:

- Responsabilities of the seller
- Manufacture
- Services
- Modifications and additions
- Supervision of CFE
- System acceptance tests
- Schedules
- Price
- Payment
- Guarantee
- Transportation and insurance
- Bond
- Patents
- Penalties
- Rescission
- Divergencies
- Representatives
- Jurisdiction
- Intervention

The fact of setting penalties for delays could be dangerous or inconvenient since it develops tension and pressure between seller and customer. On the other hand a good relation between both parts significantly helps to obtain a good product. However penalties represent a way of selfprotection for the customer, since these projects are complex, intangible, and take long time to be finished.

The CFE/Harris contract was signed in October 1979 and it was agreed that the equipment had to be delivered in 30 months (April 1982).

In the last two and a half years, during the implementation of the project, CFE has requested some additions and modifications to both, software and hardware. In addition the original schedule looked very tight after the first 18 months of work, and the seller started to feel the pressure with the schedule.

CFE recently extended the dead-line by mutual agreement, conceding 17 more months until September of 1983. The reasons were:

- To adjust the schedule with the additions and modifications.
- To improve the relationship between both parts taking away some pressure.
- To think technically first
- To be more realistic with the time spent by similar projects

3.1. Statement of work

The statement of work was written in Harris facilities from January to July of 1979. During this time all the project team (10 people) lived in the factory. They had meetings with Harris' specialists to clarify ideas, to have discussions and agreements, to make adjustments and to eventually get a document with all the functional design details. This document formed part of the contract as a technical attachment. Its content defines the responsabilities, benchmarks and features that Harris have to accomplish for a good performance of the system as requested by CFE. The statement of work of the SICTRE project is a book of about 1100 pages divided in 4 volumes:

I. Introduction and system definition
II. System support functions
III. Technical items
IV. Deliverable material

3.2. IIE Contract

By the end of 1980 a contract with the Electrical Research Institute of México (IIE) was signed. The object of the contract was to develop four application programs to be integrated into the Harris' real time system.

These programs are:

- Constrained Economic Dispatch
- Automatic Interchange Calculation
- Load Forecast
- Hydro-Thermal Coordination

It almost took the whole year of 1980 to define the statement of work for these programs, due to their high complexity and sophistication.

CFE and IIE are working very close. The role of CFE is to supply:

- Knowledge of the electrical system
- Identification of the problem
- Operation know-how and experiences

The IIE assigned about 30 people to this project, most of them with master and PhD degrees with a good background in different disciplines such as:

- Computer science
- Power systems
- Mathematical algorithms

This was a good step to develop local technology, and there are not doubts about cost-efectiveness.

México as many developing countries,

still depends on foreign technology for advanced manufacturing. However, we have achieved a good engineering level to develop our own software technology.

The above is relevant considering that the cost of the software usually represents the 50% or more in any computer system project.

4. MANUFACTURING AND IMPLEMENTATION

This phase of the project has been in progress since January of 1980. All the hardware engineering and hardware integration has been completed and aproved by CFE. At the present time almost 80% of the Harris Remote Terminal Units have already been shipped to the correspondieng Area Control Center.

The software development and integration is in progress, but to trace the advance in this particular case is very difficult, since Harris is developing a new product. The documentation is constantly revised and up-dated and there are few documents released at the present time.

4.1. Modifications and additions

During the development we have had several meetings to discuss engineering details with the seller. This is a very critical phase of the project since both parts are acquiring knowledge of the total scope. CFE always fights for the best technical approaches, and the seller looks for the fastest or expensive ones.

As one gets into the details during the development process it is almost impossible to avoid some changes of the original proposal. But it is advisable to "freeze" the project at one point, otherwise it will never be finished. There is always something new or something better, mainly in this dynamic world of computers.

It also has to be considered the different position in the negotiations. Talking about cost it is not the same when the seller bids for the contract than when the customer asks for some additions. The difference is very strong and the customer loses power in the negotiations.

Anyhow some additions and modifications were decided during the SICTRE project development.

They can be summarized as follows:

o The RTU's reconfiguration and trandsducer additions.
o The addition of new masterstations for Central and Oriental Area Control Centers
o To remove Economic Dispatch functions from Area level to National level
o To remove load Forecast program from National level to Area level
o The addition of more transceivers to handle satellite communications
o To replace IMB-Bysinch (half duplex) data link communication protocol by HDLC (full duplex) protocol, to communicate Area Control Centers with National Control Center.
o To transmit State Estimator results from the National Control Center to the Areas.
o The addition of Reserve Monitor program to each Area
o The additions of video-generator and more disc capacity in the H500 computer installed in Mexico
o Other minor changes in application programs.

In order to decide these changes, 3 check points of the total project were made at different dates:

1st. December 1980
2nd. August   1981
3rd. June   1982

As a result of them, three amendments of the original contract were signed.

4.2. Information provided by CFE

The customization of this kind of projects requires a lot of information provided by the customer during the implementation phase. It sounds simple but it takes a great effort and time to accomplish these requirements.

The main tasks are located in:

o SCADA Data Base (measures, status, etc.)
o EMS Data Base (generation type, curves, etc.)
o APPS Data Base (network model parameters)
o CRT displays
o Reports
o Mimic boards

The Data Base is the heart of the system. The future success of the system highly depends on the data base definition and its quality.

The first step for the SCADA Data Base was a decentralized job: a group of dispatchers, at each Area Control Center, worked getting all the data corresponding to their own Area and filled out special formats. They followed the guide-lines and criteria provided by the project responsibles. Thousands of sheets were submitted to the factory and it started the process to debug, to edit and to integrate into the system.

At the same time, at the National Control Center, it was defined the national network model and all the necessary SCADA points to be transmitted from the Area Control Centers.

As a rough idea of the magnitude of this task, the total amount of points defined was almost 100,000 with more than 1 200 000 attributes and parameters involved.

4.3. Activities in Mexico

The SICTRE project included an extra H500 computer (same type as the ones in the systems configuration) as part of the whole purchase.

This computer was installed at the National Control Center facilities in 1981. It is a single CPU, with real time operating system, two 80 megabyte discs, 10 CRT terminals, one video-generator, and one console with a single color CRT.

The reasons to buy this extra equipment were:

- o To develop IIE application programs in the same environment where they will be integrated
- o To build about 5000 displays (one line diagrams, tabulars, etc.)
- o To debug and update the Data Base
- o To organize some training courses in Mexico
- o To practice and familiarize the future system (hardware and software)

The profit of this investment is out of the question. We are saving a lot of money by keeping in Mexico a big number of engineers and programmers.

In the future this computer could be used as a dispatcher simulator or as a spare computer.

4.4. RTU's manufactured in Mexico

By 1978 a prototype of remote terminal unit called TRIIE, was developed in Mexico by the IIE with the sponsorship of CFE.

This RTU is another development of mexican technology and it has a big proportion of national integration in hardware.

In 1979 the prototype was tested in the field with good results. It was installed in one of the biggest substations in the Mexico city area connected in parallel with a CDC remote. It simulated all the functions linked to a CDC master station for SCADA purposes at the Central Area Control Center.

This remote terminal unit is programmable, flexible, modular, small and light, and can be adapted to any master station (Harris, L&N, CDC).

By 1981 this technology was transfered to a national manufacturer for industrialization, and at the beginning of 1982 a contract between CFE and this manufacturer was signed. He will provide 100 TRIEE's in one year, starting with partial delivers by October of 1982.

The total number of RTU's in the SICTRE project is around 200, so half of them will be made in México, helping this latter to protect the mexican economy.

5. TRAINING AND SUPERVISION

From the beginning of the SICTRE project the CFE intention has been to achieve selfsuficiency for future maintenance, expansions and new developments.

CFE, as a government public utility, is always concerned to follow an institutional and nationalistic policy. Moreover, in this particular case, this policy **is necessary** in order to be able to support constantly those real time computarized systems as the users will require.

The above has meant a significant increase of the number of engineers and technicians involved in this project.

The policy mentioned means to have a group of experts in software and hardware at each Center, since the project includes 6 Area Control Centers.

By 1980 the project team grew up to 15 people, and some of them came from the Areas located in different cities of the country.

It was not easy to get people with the adequate background to be integrated into this project. With the support of the utility administration and the union collaboration, by 1981 the project team grew up to 28 people, and by 1982 to 35; out of these latter, 23 came from the Areas.

Moreover it is not easy to keep all

of them the whole duration of the project. It is impossible to avoid some resignments for personal reasons. These situations always cause some impact in the project. Our average of resignments is about 15% which can be considered normal.

The seller has provided good space facilities at the factory for CFE people including offices, telephones, terminals, furnitures and secretary services, as well as a good atmosphere and freedom to use other factory facilities. Taking into consideration that the CFE project team has to live at the factory, these facilities are very important to make easier their professional activities far away of their homes for a long time.

5.1. Courses

Any seller on this field has to offer a series of basic courses in software and hardware according to the customers. Among them CFE has selected the following:

- ASSEMBLY LANGUAGE (2 weeks) (machine language, conversion binary to octal, write source files, macros, etc.)
- REAL TIME OPERATING SYSTEM (2 weeks) (interaction with the system, create, edit, copy, map files, services, utilities, etc)
- 9000 BASE SOFTWARE (2 weeks) (Data Base software, Data acquisition, man-machine interface, diagnostics, handlers, etc.)
- 9000 EMS/PCC SOFTWARE (1 week) (Energy management system, application programs, security analysis, etc.)
- H500 CPU MAINTENANCE (3 weeks) (architecture, logic failures, diagnostics, etc).
- PERIPHERAL CONTROLLERS (1 Week) (operation, circuit diagrams, diagnostics, information flow, etc.)
- SCIENTIFIC ARITHMETIC UNIT (1 week) (floating point, architecture, failures detection, etc.)
- VIDEO DISPLAYS SYSTEMS (3 weeks) (video display generator, display editor, video trending, monitor adjustments, etc.)
- RTU MAINTENANCE (2 weeks) (operation, modules, simulator, etc.)
- 9000 MASTER STATION MAINTENANCE (3 weeks) (subassemblies, firmware, failure detection, on-line and off line diagnostics, etc.)

Some of these courses were offered several times to different people. A total of 37 courses have been taught (15 in 1980, 14 in 1981, and 8 in 1982) to 56 students (11 of them of the IIE).

Sometimes the best instructors are not available for a specific course. It is important to ask for them and to claim the best quality in those courses since most of the customer people assigned to them have a good background and they expect the best.

It is very convenient to evaluate in both directions: instructors and students. It helps to adjust deficiencies and to have some management control. Sometimes, as in our case, the language barrier makes difficult to evaluate the real performance and knowledge.

The theoretical portion of those courses must be complemented with laboratory practice. Thus the seller has to supply facilities in some of his available computer systems. He has to assign time-machine as much as he can in combination with his project schedules.

The more the seller puts into those courses in terms of quality and effort, the more it will help the customer people to feel confident and to work with enthusiasm on the project tasks.

At this stage of the project we have well trained and familiarized experts, so they are able to teach more courses in Mexico.

From now on, most of the courses related with this project will be offered in Mexico. We have already organized 2 of them with good success. Without the language barrier and with the experiences from the factory (good and bad), in-house courses have become even better. As it was mentioned before, for laboratory practices we are using the extra equipment installed in Mexico.

5.2. On-the-job-training

The first approach in order to have the customer team fully involved into the project is to work within the seller working groups with the same responsibilities and tasks as any other group memeber.

Each activity must be carefully selected and agreed with the seller. At this working level and for different reasons, it frequently happens that the customer people do not feel integrated to the group, or even rejected.

This on-the-job-training phase had a duration of one year and a half in our project. We decided to stop this approach in order to assign our people to other activities more focussed to CFE such as data base, reports, etc. and familiarize them with some functional areas not well known yet.

We have spent 40000 man-hours of our people at the factory from January 1980 to July 1982. This means that we have kept an average of 7 people constantly living at the seller's facilities.

The next phase, and as part of on-the-job-training, is the preparation of the factory aceptance tests and the tests themselves. In order to accomplish both, we are going to keep an average of 8 to 10 people at the factory facilities until the last test scheduled by September of 1983. This means to expend 22000 more man-hours there.

If we calculate the amount of money involved in courses and training we obtain a figure of about 2% of the total investment, which can be considered very reasonable.

5.3. Supervision

It is at all times convenient to have a project leader representing the customer, residing at the factory during the manufacturing and implementation. He will be the link with the utility management to keep them informed and to consult for major decission.

5.4. Dispatchers training

This part of the training is fundamental since the whole equipment and its programs are oriented to give the best tool to the dispatchers, the ultimate users.

The courses and practices should be offered at the end of the project, during the installation, and before the put in service, by the utility experts already trained.

The instructors have to do their best to teach the dispatchers all the system potentiality in order to achieve the best exploitation in the future. Furthermore the dispatchers should feel comfortable with the new tool from the beginning so the training must be very carefully conducted.

6. ACCEPTANCE TESTS

It is well known that this is probably the most critical phase of the project. Buth parts look for the same goal: to have a good system; but with different procedure or approach.

The customer wants to assure that the system will perform all its functions properly and he wishes to check everything. He will enforce the system to the worst cases, as to check its response in emergency conditions, or performing critical functions with the maximum load. The seller will have the pressure of shipping the equipment as soon as possible in order to get the payment and to vacate space floor for new projects.

The SICTRE project is getting close to this phase. By December of 1982 two Area Control Centers have to be shipped as it has been schedule, so the factory acceptance test should start, at least, in October.

According with the contract, the seller has to submit to CFE the test plan and procedures at least 3 months before the test. The proposals is going to be studied in order to make adjustments, to add some more tests, and finally to get the customer proposal, which it will be agreed with the seller.

Strong arguments can be envisioned in this phase. In order to be ready to propose the best, from the customer point of view, the project team is studying all the available documentation, analyzing other project test procedures, practicing the system functions and subsystems, and writing some preliminary documents for the final test.

The test phase is difficult for the project managers. The customer specialists, would like to check that everything is working perfectly. However perfection does not exist, mainly in this type of projects where so much software is involved. The project manager will have to recognize where is the balance point between perfection and mediocrity, looking for what is functional and practical. He has to take care of two aspects: the future dispatchers acceptance of the new equipment, and the motivation of his project team to do their best and not to feel betrayed.

The last test will be the National Control Center System. In this multi-level structure it would be desirable to test the National master station linked to all 6 Area master stations. This is very inconvenient for the seller because he can not ship any equipment in advance and he might not have space floor available for other projects. It is also inconvenient for the customer, since he can not start to install some equipment at the definitive site. How to handle this problem? One alternative is already decided: to keep two Area master stations for the final test with the national master station, and to wait for the field test to verify the performance of the whole system with the maximum load.

Another possible alternative is to install 4 Area master stations at the field, and to link them with the national master station at the factory through communication channels via satellite.

7. SITE PREPARATION AND FIELD INSTALLATIONS

New buildings were built in order to properly allocate this equipment and the people involved for maintenance and expansions.

These buildings have been designed to provide the adequate space for the equipment (computers, mimic board, consoles, etc.) as well as engineer offices, spares storage, maintenance work-shops, etc.

These buildings have been already finished. From the budget control point of view, they probably were the worst part of the total project. As usual, civil constructions have a lot of unforeseens, material cost fluctuations, and delays.

In addition, these buildings have had to be equipped with auxiliary equipment such as: air conditioning, emergency plant, uninterruptible power supply, security system, fire system, etc.

On the other hand it is not an easy task to install more than 200 remote terminal units. This job has been delegated by regions.

Installation groups have been organized under the coordination of the Area Control Center for each region. The SICTRE project management also have had to take care of budget, materials, trucks, people, etc.

Although the schedule is very tight, the buildings and RTU's can be ready before the computers are shipped.

By this time the total amount of RTU's will probably not be installed, but they will be enough for the field test and to start the operation.

8. CONCLUSIONS

Through this paper it has been framed the complexity to manage a project like the SICTRE.

In any project management there are 3 variables: time, money and people.

From an economic point of view, time and money are very important, but the people involved in the project are the most important part for the expected results.

This new technology incorporated to a modern power control center, will signify to have constantly training activities in formal courses or practices in order to train new people or to keep updated the others.

How to keep the people involved during the whole duration of the project? And how to keep them for the future maintenance of the system?

It is a very difficult part of personnel management to keep all people motivated and enthusiastic due to:

o The dynamicity of this particular environment.
o The opportunities offered by other industries or utilities.
o The frustrations lived during the long duration of the project.

It is fundamental to make appropriate decisions at the right time, specially in the adversity moments. It is also important to have the best comprehension and support from the utility executives.

Along the project development there are periods full of uncertainty and doubt about the final success. Only with a possitive and optimistic attitude and by keeping confident and satisfied all the people involved, the success can be achieved.

CFE is close to be among the most important electrical utilities in the world, using the best and more sophisticated technology oriented to operate the National Power System under the best conditions of security, quality and economy.

9. REFERENCES

1. Specifications for a computer directed on-line monitoring and analysis system for CFE.PTI/CFE March 1978.

2. Statement of work. CFE/Harris Contract. October 1979.

3. Project implementation and management. J.R. Evans IEEE. 1977.

4. Energy management system. Project implementation. R. Masiello. 1981.

5. Overview of modern power control centers. T. Dy Liacco. 1981

6. Conception et functionnement des projects informatiques "temps réel". G.C. Mapples S.D. Ward. CIGRE 1980.

7. Contrato CFE/IIE y Anexo Técnico (desarrollo de programas de aplicación del CENACE). Diciembre 1980.

8. Manufacturing reports, Harris/CFE, 1980-82.

TRANSMISSION OF DIGITAL INFORMATION VIA SATELLITE FOR THE REAL TIME CONTROL SYSTEM OF "COMISION FEDERAL DE ELECTRICIDAD", MEXICO

D. Carrasco, G. Torres and A. Vazquez

National Energy Control Center (CENACE), CFE, Mexico

Abstract.- The real time information and control system (SICTRE), of the national energy control center (CENACE), requires a communications network, covering practically the whole country, to gather information in short periods, from the most important generation plants and substations of the national electric system (SEN). This information will be processed in 6 area control centers (CCA's) to perform, among others, the functions of supervision and automatic generation control of the SEN portion pertaining to each area.

An abstract of the information compiled in each area is transmitted, also in short periods to the national control center (CENAL) wherein it will be performed in a national level, the supervision and optimization of the operation's security and economy.

This document describes the SICTRE communications problems and the features of the satellite network that is being developed to convey the neccesary links.

The impact of the path delay in the control processes is analized.

1.0 INTRODUCTION.

The communications network design for the SICTRE must be based upon the general philosophy that was developed at the beginning of the SICTRE project and consists basically of following factors:

1. Hierarchial level structure of the system.
2. Functions to be performed at each level of the system.
3. Communications between levels of the system.
4. General hardware configurations and redundancy at each level of the system.
5. Data flow between levels.

In this introduction, we describe briefly the statements of the philosophy that are important for the communications network design.

1.1 Hierarchial level structure of the SICTRE.

The SICTRE is structured in following three hierarchial levels:

1.1.1 1st. Level.- National Control Center.

The first level of the system involves the National Control Center. The NCC is primarily concerned with system wide calculations and analysis of the CFE interconnected power system. This includes economic and secure operation, establishment of power transfers between areas of the system, scheduling, operations planning and data recording.

1.1.2 2nd. Level.- Area Control Centers.

The second level of the system involves six Area Control Centers. The ACC's are concerned with and are responsible for management of information specifically related to an area and actual control of the portion of the interconnected power system in each area.

1.1.3 3rd. Level.- Remote Terminal Units.

The third level of the system, the remote terminal units (RTU's), is concerned with the acquisition of data from tranducers and implementation of control signals to end devices of the power systems.

1.2 Communications subsystems.

There are two communications subsystems to support the SICTRE:

1. NCC - ACC Subsystem
2. ACC - RTU Subsystem

1.2.1 NCC - ACC Subsystem

The first subsystem allows communications between the NCC and the six ACC's. This subsystem consists of radial redundant links to be immune to a single point of failure.

System availability requires that the loss of one channel does not interrupt communications between the NCC and an ACC. Furthermore, redundant channels may not utilize common equipment (e. g., damage to a primary channel's microwave tower may not cause loss of the redundant channel). Also, in the event of loss of a channel the performance of the system may not be degraded from the point of view of NCC to ACC or ACC to NCC data updating.

1.2.2 ACC-RTU Subsystem

The second subsystem allows communications between ACC's and their associated RTU's. Although redundancy is not always required this subsystem consists of radial redundant channels from the ACC to the RTU or RTU's on the channel.

In the design of the ACC-RTU communication subsystem, two considerations must be kept in mind. First, all RTU's on a given communications channel should operate at the same speed and second all RTU's that share a channel should share also the back-up channel.

1.3 Interlevel data flow.

1.3.1 NCC-ACC Data flow

To execute the functions of CFE system-wide nature, the NCC requires information primarily related to the bulk power transmission and generation facilities of the CFE interconnected system.

It is desirable to have the following telemetered data passed to the NCC from each ACC:

1. Watts and vars from each end of every transmission line of the network.

2. Watts and vars from every transformer connecting buses of the network in each substation.

3. Voltage at every bus represented in the network.

4. Watts and vars for generation at each bus of the network.

5. Watts and vars of each line or transformers serving a load.

6. Status indication from every switch or breaker represented in the network.

All of the foregoing data cannot be made available to the NCC from the ACC and some compromises had to be done.

The data are transmitted to the NCC on either a periodic or an aperiodic basis.

Generally speaking, analog data are transmitted on a periodic basis. Alarms and status indication data are automatically transmitted inmediatly following occurrence of an alarm or change in the status of a breaker or switch.

In addition to this data, there are other data transmitted from each ACC to the NCC. These additional data are transmitted aperiodically and generally consist of the following:

1. Parametric data from an area such as changes in values of limits, control or dispatch status of a generator unit, changes in fuel types for generators, etc.

2. Requests of the NCC by an area operator to perform services such as a load flow analysis to check the possible effects of a proposed switching action in the area.

And from the NCC to each ACC the data transmission can be:

1. Requests for data to be transmitted to the NCC from the ACC.

2. Results of the constrained eco-

nomic dispatch calculation consisting of generator unit basepoints and participation factors.

3. Input data for each area's AGC function:

4. Alarms and system security assessment messages.

5. Results of an ACC request for an NCC service.

1.3.2 ACC - RTU Data flow.

The ACC is responsible for acquiring telemeterd data from and issuing control commands to its RTU's and must accomplish this within a specific time frame on a frecuent basis.

Data acquisition deals with the following types of information:

1. Analog values (watts, vars, volts, amperes, herz)

2. Integrated values (switches, breakers, alarm contacts, miscellaneous contacts)

Controls to the RTU's deal with following types of information

1. Selection and operation of power circuit breakers, transformer tap changing mechanism, etc.

2. Control signals to generator units for raising/lowering out put under automatic generation control.

Acquisition of data from the RTU's is classified according to time categories. There are 2-seconds, 4-seconds 12-seconds, 60-seconds, and 1-hour, as follows.

2-seconds — Watts of interconnecting transmission lines between areas, for the purpose of calculating area control error for AGC.

4-seconds — Watts of generator units under AGC.
— Frecuency, for the purpose of calculating area control error.
— Report of every status change.

12-seconds — Analog data of interest to the area operator for system monitoring purposes.

60-seconds — Report of all status points

1-hour — Integrated watts and vars for tie-lines, load areas and generating stations or units

2.0 DATA REQUIREMENTS FOR THE SICTRE.

According to general philosophy a preliminary benchmark, of periodic and aperiodic information to be transmitted through NCC-ACC and ACC-RTU lines was done arriving to following conclusions:

- Information interchange NCC-ACC could be handled by a voice grade, microwave channel, 2400 bit rate, using a Bynary Sinchronous Communications (BSC) protocol.

- Information interchange ACC-RTU's in each area, could be handled by 8 to 10 voice grade, microwave or power line carrier channels, 1200 bit rate. The grouping of RTU's in a channel should be done according to traffic and geographic criteria.

However, data requirements for energy management systems - and advanced application-software have grown in the last years, making it necessary a review of the former solutions.

2.1 Data requirements for NCC-ACC's links.

Table 1 shows a breakdown of data requirements in one hour, by category. It was compiled assuming the worst case, that means that aperiodic data was considered periodic and the direction of traffic was reversed after each transmission.

Overhead bytes of common data link (CDL) and BSC protocol were added and a packet size of 128 bytes was selected.

Category	Num.of bytes (hour)
Supervisory control and data acquisition (SCADA)	581,838
Energy Management system (EMS)	157,674
Advanced Applications I (APP's)	50,286
" " II (APP's)	133,962

Miscelaneous (operators)   113,049

Table 1.   Total   1'036,809

The ideal channel capacity, that cannot be reached, at a 2400 bit rate is 1'080,000 bytes in an hour. This means that something must be changed to ensure optimal operation: reduce data requirements, increase data link capacity or both.

BSC is a half duplex protocol; so the first approach was to substitute it by a more effective, duplex, HDLC protocol.

Table 2. shows the increase of link efficiency by using an HDLC protocol over a microwave and a satellite channel. The dramatic fall of efficiency by the satellite link using BSC protocol, is due to the 0.27 second delay by each direction change.

	Microwave		
Byte/Sec	300	600	1200
NDT (BSC)	234.87	405.37	636.62
LEF % (BSC)	78.29	67.56	63.05
NDT (HDLC)	284.58	568.17	1,143.47
LEF % (HDLC)	94.86	94.69	95.29

	Satellite		
Byte/Sec	300	600	1200
NDT (BSC)	113.5	142.5	165.37
LEF % (BSC)	37.83	23.75	13.61
NDT (HDLC)	286.62	572.76	1,149.68
LEF % (HDLC)	95.54	95.46	95.8

Table 2. Link throughput analysis for 128 Byte Packet size

Where:

NDT = Net data throughput, taking into account all parameters including the data link protocol.

LEF = Data link efficiency = NDT/bit rate, presented as a percentage.

At the beginning of the SICTRE-Project (1976), it was not advisable to use bit rates larger than 2400 in the microwave public network. Now, CFE is using some 9600 bits per second links in it, but there is not enough information about their performance. Nevertheless, it seems that the second step to improve the NCC-ACC data link will be to increase the bit rate to 4800.

2.2   Data requirements for ACC - RTU's links.

Numbers of RTU's per area are as follows:

Central	50	Noroeste	32
Oriental	43	Norte	27
Occidental	42	Noreste	48

All messages interchanges between the master and RTU's used in Occidental, Noroeste, Norte and Noreste areas, consist of blocks of two or more 10-bit characters, transmitted 1200 bps. Each 10 bit character consists of a character start bit, seven bits of information, (one bit used to indicate the start of a new message), parity bit and a stop bit, as illustrated below:

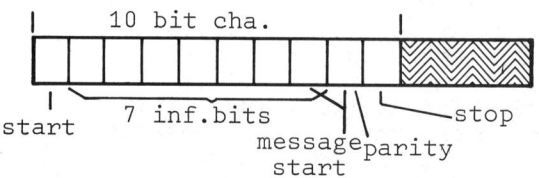

The general format of all Master to Remote and Remote to Master messages is sketched below, consisting of one 10 bit character containing the address of the RTU involved in the message interchange, additional 10-bit characters as necessary to communicate the desired intelligence of instructions, and ending always, with a 10-bit character associated with Error Control Coding (Longitudinal Redundancy Check).

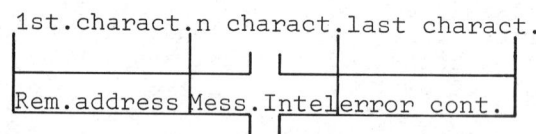

For the protocol used in Oriental and Central areas the basic message that is transmitted in either direction, is a 32-bit block consisting of three sections as shown below

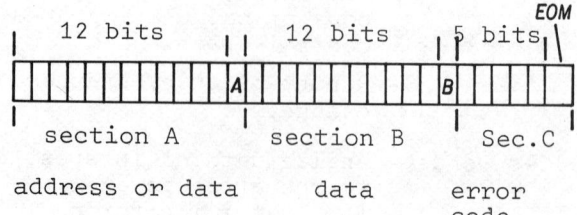

Sections A and B each contain 12 information bits an a 13th "house keeping" bit: bit A for section A and bit B for section B. Bit A is used as a flag to specify whether the preceding 12 bits contain our address or data. Section C contains a five-bit Bose-haudhuri error detecting code and an end-of-message (EOM) bit.

Both protocols used are half duplex and cannot be changed easily to duplex. This means that satellite channels throughput for ACC-RTU links is poor. The number of channels via satellite necessary to handle the information interchange ACC-RTU's without exceeding 60% percent traffic in normal conditions is:

Central *	23	Noroeste	19
Oriental	23	Norte	13
Occidental	23	Noreste	23

* Most of RTU's are closed to master station and they do not need a satellite link, nevertheless the amount of data is large enough to fill 23 microwave channels.

3.0 IMPLEMENTATION OF DATA CHANNELS.

3.1 Public Network

The public communications network does not cover many allocations of important power plants and substations of the interconnected system. Its reliability is less that the one specified for the SICTRE communications (99.98%) and many cases interruptions can last several hours.

Access to this network via microwave or UHF is complicated and expensive. For these reasons CFE developed a project to implement its own national microwave network, but it has not been authorized by the mexican ministry of communications.

3.2 C.F.E. Networks

Power line carrier reaches all points of the system, but it cannot be used as main media for the ACC-RTU links because of following reasons:

- High priority protective relaying links use an important part of the available spectra.

- Distortion by more than four repeaters at a 1200 bit rate becomes unacceptable.

- Disturbance in transmission lines can affect data messages to the ACC in those moments where they are urgently needed.

A combination of power line carrier, UHF and microwave public network will be used as back-up in the ACC - RTU links.

3.3 Intelsat satellite

In 1981, mexican government leased some transponders of an Intelsat IV satellite, that was approached to a longitude of 53° west in order to cover the whole mexican territory. Since that date, CFE decided to implement a satellite data network exclusively for the SICTRE communications requirements. This network will be commissioned during last quarter of 1982 and first of 1983.

4.0 SATELLITE NETWORK CHARACTERISTICS.

The prime communications networks NCC - ACC's (fig.1) and ACC's - RTU's (fig.2) repeater will be an Intelsat IV satellite, from 1983 to 1985, and the mexican satellite Ilhuicaua later on.

Earth stations will be installed in the NCC, in each ACC and in the allocation of the 76 most important RTU's. The information of all other RTU's will be transmitted via microwave or power line carrier to the next earth station.

4.1 Earth stations

Block diagramm in fig.3 shows primary equipment to be used in peripherals (RTU's) and ACC's earth stations. From this diagramm it is easy to infer the pieces of equipment that are necessary for the NCC-ACC's links.

4.2 Access and modulation types

Single channel per carrier (SCPC) access method and PSK modulation are used for all data channels, and SCPC-FM for voice channels that have been included to make a better use of investments. In each ACC, 3 party line/ like voice channels will be shared by all RTU's and all ACC's will share 4 voice channel to communicate to the NCC.

COMUNICACIONES ENTRE CENAL Y CCA'S

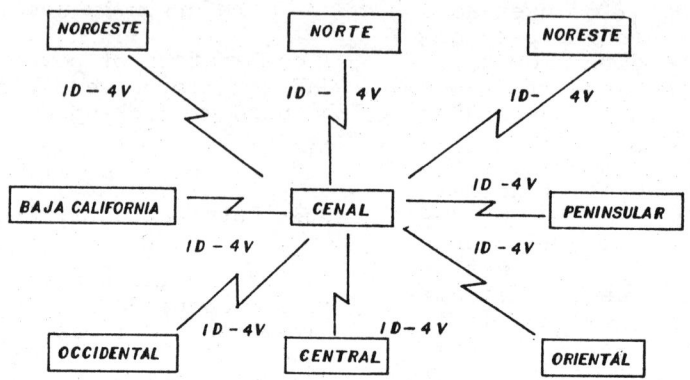

D — CANAL DE DATOS DE 2400 BAUDS BER 10^{-6}
V — CANAL DE VOZ COMPARTIDO CON PRIVACIA

Figura 1

COMUNICACIONES ENTRE CCA'S Y UTR'S

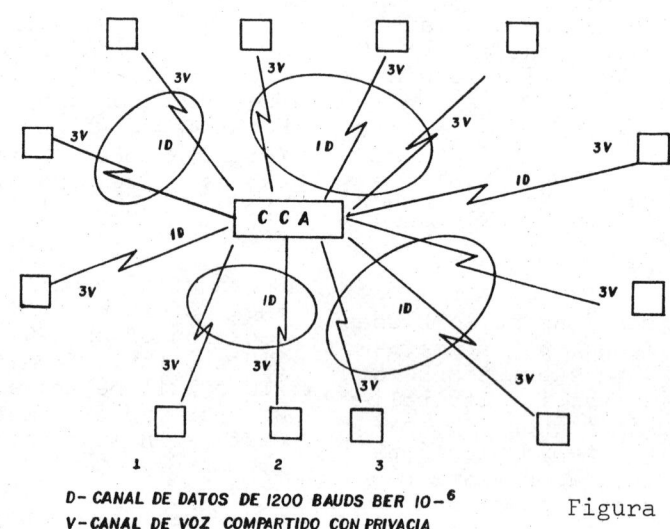

D— CANAL DE DATOS DE 1200 BAUDS BER 10^{-6}
V— CANAL DE VOZ COMPARTIDO CON PRIVACIA

Figura 2

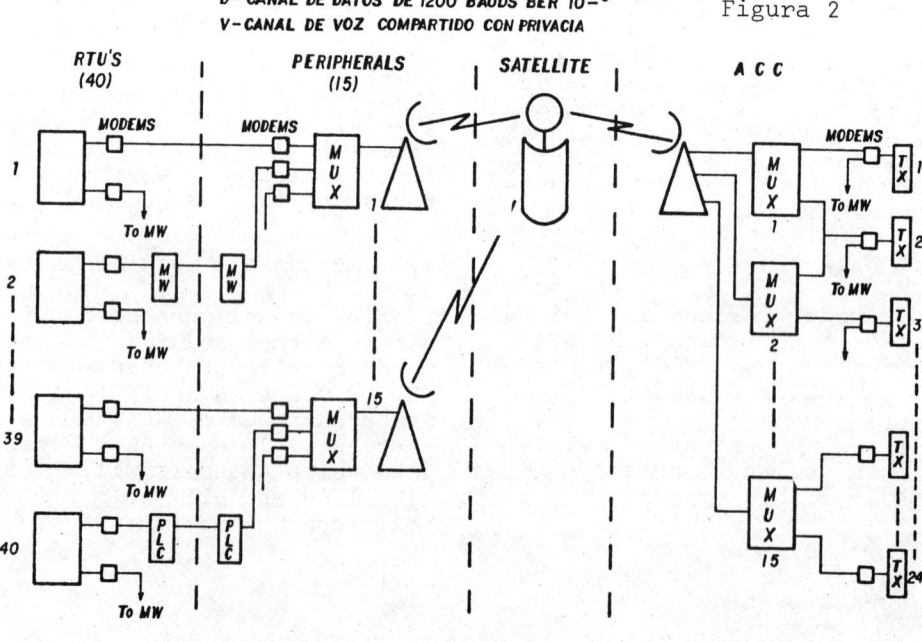

ACC-RTU'S SATELLITE LINK.

Figura 3

5.0 IMPACT DUE TO THE TRANSMISSION DELAY IN THE OVERALL PERFORMANCE.

Because of the transmission delay (aprox. 540 mseg. round trip) has been anticipated several areas of potential impact. Some of them had been corrected, like the increase of communication controllers in almost all of the area control cneters. But besides the hardware increase the most sensitive areas of research are the Data acquisition and control (DAC) subsystem in each area, and it's similar subsystem at the NCC called data acquisition and proccessing (DAP). These subsystem have several timers to control the execution of data scans and the time constants used through the process must be modified in accordance with this new delay due to satellite usage.

A more complex problem appears when the automatic switching, when a communication failure is presented to the back-up channel, (microwave) is used to carry the 100% of load. So the software must be aware of the different communication media used.

Another area of interest is the one related to the automatic generation control. Until now, because of the lack of facilities, to simulate the overall performance of this subsystem has been not possible.

It's anticipated the need to simulate during the acceptance test the closets real world conditions in order that when the field instalation occurs, to have the closets idea to get an optimal tune-up per generator per area.

At the same time also is needed the simulation through a model of the whole system in order to tune-up the CDL and another software subsystems due to the handling of message priorities and frequency of transmission.

6.0 CONCLUSIONS

There are no reports of experiences by using real time data transmission for energy control via satellite.

CFE decided to implement the satellite network to convey the necessary communications for the SICTRE, because no other option was available.

We expect to succeed with this approach.

BIBLIOGRAPHY

- Philosophy for the Load Dispatch and Control System - Power Technologies Inc. 1976.

- Statement of work of the Real Time information and control system CFE-Harris Controls 1979.

- Reporte Técnico sobre el proyecto de enlaces de datos entre CNC y CCA's. Guillermo Torres 1981.

- CFE Communications study final report Harris Controls 1982.

- Specifications for the acquisition of earth stations CFE 1982.

- Estudio de factibilidad para comunicaciones vía satélite para el SICTRE. Demetrio Carrasco 1982.

- Comunicaciones para el Centro Nacional de Control de Energía, Seminario Ejecutivo del CENACE sobre el SICTRE. Demetrio Carrasco, Arturo Vázquez 1982.

NEW CONCEPTS FOR AUTOMATIC GENERATION CONTROL IN ELECTRIC POWER SYSTEMS USING PARAMETRIC QUADRATIC PROGRAMMING

J. L. Carpentier*, G. Cotto** and P. L. Niederlander***

*Electricité de France, 2 rue Louis Murat, 75008, Paris, France
**Electricité de France, 22 avenue de Wagram, 75008, Paris, France
***EPOS, 38 rue de l'Yvette, 75016, Paris, France

Abstract. New concepts for Automatic Generation Control in Electric Power Systems are presented, where the two components of Automatic Generation Control, Load Frequency Control and Economic Dispatch are performed at the same rate, i.e. a few seconds, and where Economic Dispatch takes network security into account. This gives network security and good transients, avoiding contradictory actions of Load Frequency Control and Economic Dispatch on the generating units. The corner stone of the solution is the use of a new fast on-line Optimal Power Flow, using a new parametric quadratic programming method, which is presented in details.

Keywords. Digital control ; load dispatching ; on-line operation ; power system control ; quadratic programming ; automatic generation control.

INTRODUCTION

The Automatic Generation Control (AGC) of an electric power system has two main objectives : first adjust power generation to load, second minimize the sum of the generation costs. In the present state of the art [1] [2] [3] [4], the first objective is met through the so-called "Load Frequency Control" (LFC), the second one using an "Economic Dispatch" function (ED). Usually, these two functions are completely separated in time, LFC taking the system dynamics closely into account, with a cycle rate between 2 and 10 seconds, whereas ED only changes the set points of LFC every fifth minute. Except a few exceptions, among which New York Power Pool may be mentionned [5], no care is taken of power transmission network security in the Economic Dispatch function. In these conditions, as mentionned in [1], the usual Automatic Generation Control process has two main drawbacks :
- LFC and ED control actions may often be contradictory, which results in undesirable oscillations for the generated powers.
- the power transmission network security is not met.

Electricité de France is now studying new concepts for AGC : avoiding these two drawbacks was included in the objectives.

PRINCIPLES OF A NEW AGC SYSTEM

Respective Functions of Load Frequency Control and Economic Dispatch

The objective of the new AGC system project is to adjust generation to load in the most economical way, meeting generating unit constraints and transmission security constraints. If the latter constraints are well defined (transmission line and transformer currents within given limits under various conditions), it is not the same for all the generating unit constraints ; of course unit power as well as power ramps versus time must remain within given limits ; but also frequent and large power oscillations on a unit must be avoided, which is not so clearly defined ; if the latter constraint did not exist, the most economical AGC would only include ED : a few units, the "incremental units", in economic balance with the system, would adjust power generation to load in the most economical way. But, as load variations versus time have an oscillatory component, large oscillations might occur on the powers of these incremental units, which is undesirable. To avoid such large oscillations on a few units, the load variations are partitionned into two components : the "oscillatory" component, for which LFC shares the oscillations between a great number of units, and the "trend" component, which is the input for the ED function. Of course, to perform LFC, "regulating bandwidths" must be reserved on various units, which restrict the limits of the average available powers of these units and is costly, but this cost corresponds to sparing incremental unit wear. Unit regulating bandwidths are determined by scheduling e.g. one day in advance and incidentally adjusted every hour, defining the LFC possibilities. Then, for on-line control, the problem left is to apply LFC to the oscillatory component of load variations and ED to the trend component.

Basic Principles of the New AGC System Project

The basic typical properties of the proposed system consist of performing :
- LFC and ED functions at the same rate, a few seconds, e.g. 10 seconds (which, so far, seems sufficient for the French system).
- ED with transmission security constraints (LFC need not meet these constraints, due to the oscillatory character of phenomena and to the line thermal time lags).

In these conditions, the essential functions related with AGC include :
- Every day, power scheduling with basic determination of the unit regulating bandwidths.
- Every hour, final optimal allocation of regulating bandwidths to the various units for the next hour (associated with an optimal power flow).
- Every fifth minute, active (only) optimal power flow including electrical network security analysis and specially providing a "reduced model" of the production transmission system, to be used every 10th second.
- Every 10th second, the on-line AGC function itself, including :
1. Decomposition of input signals into oscillatory and trend components (these input signals, which are the measurements of the variations of load to provide and of the currents to adjust in critical lines, will be defined more accurately below),
2. Application of secure Economic Dispatch to the trend components of input signals : this gives variations ΔP to apply to the generating unit powers,
3. Application of Load Frequency Control to the oscillatory components of input signals: this gives variations ΔP^f to apply to the generating unit powers (of course, ΔP then $\Delta P^* = \Delta P + \Delta P^f$ meet the generating unit constraints concerning the limits of powers and power ramps),
4. Sending of the order $\Delta P^* = \Delta P + \Delta P^f$ to each generating unit.

Input Signals

It was shown in a previous paper [6] that the input signals may be taken equal to

$$\Delta b_1^* = -\alpha \int_0^t (\Delta F^* + \frac{\Delta Pe^*}{\lambda}) \, dt \qquad (1)$$

for the measurement of the variation of load to provide

$$\Delta b_L^* = -\alpha \int_0^t \frac{\Delta \Omega_L^*}{\lambda_L} \, dt \qquad (2)$$

for the measurement of the variation of a critical current to adjust, with the following notations :

t : time

Δf^* : frequency deviation

ΔP_e^* : exported power deviation

$\Delta \Omega_L^*$: current deviation in line or transformer L

$\alpha, \lambda, \lambda_L$: constants (possible to be adjusted) giving the dynamic properties of the control system.

It was demonstrated that using these input signals allows to get for the complete AGC the same transients as for the LFC integral control used now in France, which is satisfactory, specially very stable.

Load Frequency Control Function

The Load Frequency control is very simple :

$\Delta P_j^f = (B_j / \Sigma B_j) \Delta b_1^f \qquad (3)$, with

ΔP_j^f : component of ΔP^f relative to the jth unit

B_j : regulating bandwidth of the jth unit.

Secure Economic Dispatch Function

The Secure Economic Dispatch must use the trend components of the input signals $\Delta b_1 \ldots \Delta b_L \ldots$ in order to provide the corrections ΔP. This problem is not simple to solve within a very little time and need new techniques we shall call "Fast On-Line Optimal Power Flow" ("FOLOPF") and present below. The feasibility of these new techniques is the corner stone of the process and was the first study performed after the process was imagined. FOLOPF techniques are derived from the techniques of the usual Optimal Power Flow performed every fifth minute, and we shall first briefly recall the latter.

5 MINUTE RECURRENCE OPTIMAL POWER FLOW

Problem to be solved

For this program, the data are the equipment on service for generation and transmission, the fuel costs, the loads and the voltage magnitudes (approximate if not well known) at every bus. The results are the active powers P of the generating units minimizing the total fuel cost and meeting generating unit constraints and transmission security constraints, for the intact system and under contingency. Notice that the fuel costs may be taken as they are or biassed in order to take an account of constraints (as power ramp constraints) likely to appear up to one or two hours later.

Differential Injections Method

The problem is solved using the "Differential Injections" method [7][8] which belongs to the family of "compact" Optimal Power Flow Methods [9]. These compact methods are characterized by the building of a "reduced model" of the problem, valuable in a more or less wide region, where the useful constraints of the problem are expressed only versus the control variables P, i.e. the

active powers of the generating units (whereas the complete problem is usually also expressed versus the state variables θ, voltage phase angles at every bus of the network).
In the Differential Injections method, the reduced model is non-linear, the losses being a quadratic function of P, and the fuel cost may be a quadratic function of P. In these conditions, the reduced model remains valuable on a wide region. Two main steps are performed iteratively : reduced model building, reduced problem optimization, till the reduced problem is equivalent to the complete problem. The process may be summarized as follows :
1. Start : find an initial solution P^o (usually optimal but not feasible).
2. With a physical load flow (Newton method) compute the electrical system state θ = f(P).
3. Analyse and select the useful constraints; this step includes contingency analysis if security under contingency is desired ; usually the useful constraints are not numerous, less than 12 for a problem with 100000 potential constraints.
4. Build the coefficients of the reduced problem through sensitivity techniques ($d\Omega_L/dP$ for the currents, dp/dP, d^2p/dP^2, p being the system losses). At this point the reduced model is built.
5. Test if the just built reduced problem is identical to the previous one. If yes, the optimum is reached ; if no, go to 6.
6. Optimize the reduced problem : Using the reduced model, one finds : 1) a feasible solution for P, 2) a feasible and optimal solution for P. Then go to 2.

Step 1 is initialization, steps 2,3,4 reduced model building, step 5 test, step 6 reduced problem optimization.

Reduced Problem Optimization

The reduced problem optimization is performed using the Generalized Reduced Gradient (GRG) technique [10]. In this algorithm, at each step the variables are moved, when possible, in the opposite direction of the gradient projected on the linearized constraints ; the length in this direction is given by the minimum of the objective function or by the fact a new constraint is met before ; then, at each step, non-linearities of the constraints are taken into account thanks to Newton's method. A very important feature must be noticed in this method : the gradient, i.e. the <u>first</u> derivative of the objective function, only gives the steepest descent direction, but not the optimum point direction nor its position. GRG is used here to solve a quadratic program but it is a general convex programming method without special properties for quadratic programs. Usually, 3 iterations (reduced model building, reduced problem optimization) are sufficient. The numerical performances are as follows in IBM 30-81 : for a system with 500 busses, 800 lines or transformers, 65 thermal plants, it takes about 20 seconds without contingency and up to 40 seconds with contingency in the security analysis.

SECURE ECONOMIC DISPTACH TO BE SOLVED EVERY 10TH SECOND IN AUTOMATIC GENERATION CONTROL

Basic Ideas

The basic idea to get a fast solution of the secure economic dispatch to be solved every 10th second in the AGC algorithm consists of noticing that :
1) the reduced model built every fifth minute by the Differential Injections method remains valuable to describe the system operation during the 5 following minutes (except the case when a sudden event occurs : then the corresponding addition to the reduced model may be computed very quickly, which keeps this proposition valuable).
2) at the beginning of a 10 second cycle, we have the optimal solution P corresponding to the previous load : it is the solution of the previous 10 second cycle or of the Differential Injections algorithm at the beginning of a 5 minute cycle : in these conditions, it is sufficient to track the optimal solution when load changes, i.e. to perform a parametric optimization, instead of solving an optimization problem with an initial flat start.

Problem Statement

Following the above basic ideas, the problem to be solved may be written :

$$\begin{cases} \sum_j A_{1j} \Delta P_j - \frac{1}{2} \sum_{ij} \Delta P_i D_{ij}^o \Delta P_j = \Delta b_1 & (4) \\ \sum_j A_{Lj} \Delta P_j \leq b_L + \Delta b_L \quad (L \geq 2) & (5) \\ \Delta P_j^m \leq \Delta P_j \leq \Delta P_j^M \quad \forall j & (6) \\ F_Q = \sum_j (c_j \Delta P_j + \frac{e_j}{2} \Delta P^2) \text{ minimale} & (7) \end{cases} \quad (8)$$

with

ΔP_j : variation of the power P_j of the jth unit (or part of unit being defined by a quadratic cost).

Δb_1, Δb_L : trend component of the input signals defined by (1) and (2) ; they represent the variations of the conditions of the problem due to load changes since the last ten second cycle.

b_L : current margin for current in line L at the end of the last 10 second cycle ; for those lines exactly on their limits, $b_L = 0$

$A_{1j} = 1 - \frac{\partial p}{\partial P_j^o}$, with $\frac{\partial p}{\partial P_j^o}$ = 1rst differential losses (constant)

$D_{ij}^o = \frac{\partial^2 p}{\partial P_i^o \partial P_j^o}$ = 2nd differential losses (constant)

A_{Lj} : sensitivity of current in line L versus ΔP_j, incidentally after trip of a line or unit d.

ΔP_j^m, ΔP_j^M : bounds for ΔP_j, due to physical bounds and ramp limits during 10 seconds.

c_j, e_j : coefficients of the objective func-

tion ($c_j \geq 0$, $e_j \geq 0$).
(4) represents the balance between generation, load and losses,
(5) represents the transmission security constraints,
(6) represents the generating unit constraints.
This problem is a quadratic program, with a quadratic objective function (7) and one quadratic constraint (4). Moreover, we know that if $\Delta b = 0$, $\Delta P = 0$ is the optimal (and of course feasible) solution.

In linear programming, a second member parametrization would allow to track the optimal solution ΔP when Δb increases from 0 to its actual value ; in quadratic programming such a method does not exist : so, to solve our problem, we had to imagine a new parametric quadratic programming method. On the algebraic side, we were helped by the fact that the number of effective constraints is always small, as well as the number of incremental units, both less than 12 for a large scale system as France ("effective" constraint meaning a constraint on its limit, and "incremental" unit meaning $P_j^m < P_j < P_j^M$).

SECURE ECONOMIC DISPATCH SOLUTION THROUGH PARAMETRIC QUADRATIC PROGRAMMING : "FAST ON-LINE OPTIMAL POWER FLOW" ALGORITHM

Solution Philosophy

First, two ideas allow to reduce the difficulty of the problem (8) to solve : use of an intermediate quadratic program with linear constraints ; use of relaxation on constraints; in these conditions, we have to solve a parametric quadratic program with linear constraints and small dimensions, with a few additions necessary to find again the solution of (8). For the parametric quadratic program itself, we first compute the variation ΔP^O which would be optimal given Δb, assuming no new constraint is met ; then we go in the direction defined by ΔP^O as far as a new constraint makes it possible ; then we make another step taking into account the new constraint met, etc... To define ΔP^O, the techniques used in Linear Programming cannot be used; GRG cannot be used either : indeed, when the second member changes, GRG, whis is a first degree method, do not say anything about the change in the reduced gradient and the change to apply to the non-basic variables : to meet the desired variation of the second member while keeping optimal, the only way was to express the reduced gradient as a linear function of the variables and to write it remains zero : this is the basic idea of the proposed method.

Use of an Intermediate Quadratic Program with Linear Constraints

It may be shown [10] that, for the optimality properties, (8) is equivalent to (11)

$$\boxed{\begin{aligned}
&\sum_j A_{1j} \Delta P_j = \Delta b_1 &&(9)\\
&\sum_j A_{Lj} \Delta P_j \leq b_L + \Delta b_L, \quad L \geq 2 &&(5)\\
&\Delta P_j^m \leq \Delta P_j \leq \Delta P_j^M \quad \forall j &&(6)\\
&F = \sum_j (c_j \Delta P_j + \frac{e_j}{2} \Delta P_j^2) + \frac{\lambda_0}{2} \sum_{i,j} \Delta P_i D_{ij}^0 \Delta P_j &&(10)
\end{aligned}} \quad (11)$$

λ_0 being the dual variable associated with (9). So, to solve the problem (8), it is sufficient to solve (11) then to correct ΔP in order to meet (4), as performed in the GRG algorithm [10].

Use of Relaxation on Constraints

The number of constraints (5) which are effective, i.e. which work as equations (and not as strict inequalities) is always very small : so, it is advantageous to consider at each step only those relationships which are effective, dropping (or "relaxing") the others from the computation. Of course this involves two tasks :
- watching the values of the margins of the relaxed constraints : if a margin becomes zero, the constraints become effective.
- watching the effective constraints which should be relaxed : this happens when Δb_L is zero and the corresponding dual variable u_L becomes positive, or when the ΔP are not able to meet a variation Δb_L which would tend to relax the constraint.

In these conditions, (11) may be restricted to :

$$\boxed{\begin{aligned}
&\sum_j A_{1j} \Delta P_j = \Delta b_1 &&(9)\\
&\sum_j A_{Lj} \Delta P_j = \Delta b_L, \quad L \geq 2 &&(12)\\
&\Delta P_j^m \leq \Delta P_j \leq \Delta P_j^M &&(6)\\
&F = \sum_j (c_j \Delta P_j + \frac{e_j}{2} \Delta P_j^2) + \frac{\lambda_0}{2} \sum_{i,j} \Delta P_i D_{ij}^0 \Delta P_j &&(10)
\end{aligned}} \quad (13)$$

which is the problem remaining to be solved inside a step where the set of constraints (12) does not change.

Principle of Solution of a Parametric Quadratic Program Constrained by a Set of Linear Equations

The general form of the program (13) may be written :

$$\boxed{\begin{aligned}
&Ax = b &&(14)\\
&\alpha \leq x \leq \beta &&(15)\\
&F = cx + \tfrac{1}{2} x_T Dx &&(16)
\end{aligned}} \quad (17)$$

with

$$\left.\begin{aligned}
&x = \Delta P, \quad \alpha = \Delta P^m, \quad \beta = \Delta P^M\\
&b = [\Delta b_L]\\
&D_{ij} = \lambda_0 D_{ij}^0 \quad \forall j \neq i, \quad D_{ii} = e_i + \lambda_0 D_{ii}^0
\end{aligned}\right\} \quad (18)$$

We know for b = 0, x = 0 is the optimal (feasible) solution.

Reduced gradient. The variable set $\{x\}$ is partitionned into 2 subsets $\{\hat{x}, \tilde{x}\}$; following the same partition, $A = [\hat{A}, \tilde{A}]$. The partition is such that $\hat{A}$ is square non singular; $\hat{x}$ is the basic component of x, $\tilde{x}$ the non-basic component.
(16) gives $dF = (c + x_T D) dx$
(14) gives $\hat{A} d\hat{x} + \tilde{A} d\tilde{x} = 0$; both give $= (\hat{c} + x_T \hat{D}) d\hat{x} + (\tilde{c} + x_T \tilde{D}) d\tilde{x}$

$$dF = \tilde{g} \, d\tilde{x} \quad (19), \text{ with}$$

$$\tilde{g} = (\tilde{c} - \hat{c}\hat{A}^{-1}\tilde{A}) + x_T G_T^* \quad (20), \text{ with}$$

$$G_T^* = \tilde{D} - \hat{D}\hat{A}^{-1}\tilde{A} \quad (21)$$

$\tilde{g}$, row vector, is the "reduced gradient" of F versus $\tilde{x}$; (20) shows $\tilde{g}$ is a linear function of x.

If $x = x^0 + \Delta x$ (22) and $\tilde{g}^0$ is the reduced gradient at the point x^0, (20) gives

$$\tilde{g} = \tilde{g}^0 + \Delta x_T G_T^* \text{ or}$$

$$\tilde{g}_T = \tilde{g}_T^0 + G^* \Delta x \quad (23)$$

Optimality conditions. They correspond to $dF \geqslant 0$ for any possible change $d\tilde{x}$, which may be written :

$\tilde{x}_j = \alpha_j$ and $g^j \geqslant 0$ or
$\tilde{x}_j = \beta_j$ and $g^j \leqslant 0$ or
$\alpha_j < \tilde{x}_j < \beta_j$ and $g^j = 0$

At the point x^0, let us partition $\tilde{x} = \{\bar{x}, \underline{x}\}$ such that $\alpha^j < \bar{x}^j < \beta^j$ or $\bar{x}^j = \alpha^j$ and $g^j < 0$ or $\bar{x}^j = \beta^j$ and $g^j > 0$.
The optimality may be reached trying to get $\bar{g} = 0$, or

$$\bar{g}_T = \bar{g}_T^0 + \bar{G} \Delta x = 0 \quad (24)$$

$\bar{G}$ being a partition of G^* such that
$$\bar{G}_T = \bar{D} - \hat{D}\hat{A}^{-1}\bar{A} \quad (25)$$

Principle of the parametric optimization. The principle consists of trying to solve (14) and (24) at a time. If $x = x^0$, with $\bar{g} = \bar{g}^0$ and $\Delta b = b - Ax^0$, we want

$A (x^0 + \Delta x^0) = b$
$\bar{g}_T + \bar{g}_T^0 + \bar{G} \Delta x^0 = 0$, or

$$\left. \begin{array}{l} A \Delta x^0 = \Delta b \\ \bar{G} \Delta x^0 = -\bar{g}_T^0 \end{array} \right\} \quad (26)$$

We change only $\hat{x}$ and $\bar{x}$ ($\Delta \underline{x} = 0$), which gives the fundamental relationships to meet :

$$\boxed{\begin{array}{l} \hat{A} \Delta \hat{x}^0 + \bar{A} \Delta \bar{x}^0 = \Delta b \\ \hat{G} \Delta \hat{x}^0 + \bar{G} \Delta \bar{x}^0 = -\bar{g}_T^0 \end{array}} \quad (27)$$

This gives the variation $\Delta x_0 = [\Delta \hat{x}_0, \Delta \bar{x}_0, 0]$ which would be optimal and meet the complete variation Δb of the second member if no new constraint appeared. As a matter of fact, the variation Δx_0 is not always possible to perform, and the final variation for one iteration will be defined by

$\Delta x = \rho \Delta x^0$, ρ maximum such that $0 \leqslant \rho \leqslant 1$ meeting all the problem constraints.

Fast On-Line Optimal Power Flow basic algorithm

The FOLOPF algorithm derives directly from the previous analysis. It may be briefly described as follows :
1. Start from $\Delta P = 0$. Use the partition $\hat{\Delta P}$, $\bar{\Delta P}$ from the last known solution (which corresponds to $\Delta P = 0$ optimal). Compute the cost $\gamma = dF_0/dP$, the basis $\hat{A}$, its inverse $\hat{A}^{-1}$, the "dual vector" $u = \gamma \hat{A}^{-1}$, $\lambda_0 = u^1$, $\tilde{g} = \tilde{\gamma} - u\tilde{A}$.
2. Start iteration. Define $\bar{\Delta P}$ such that $\Delta P_j^m < \bar{\Delta P}_j < \Delta P_j^M$ or $\bar{\Delta P}_j = \Delta P_j^m$ and $g^j < 0$ or $\bar{\Delta P}_j = \Delta P_j^M$ and $g^j > 0$.
3. If $\bar{g} = 0$ and $\Delta b = 0$ (Δb being the residual variation to perform on the second member), the optimum is reached ; if not go to 4.
4. Compute $D(\lambda_0)$ (18), $G = \bar{G}$, $\bar{G}$ (25), ΔP^0 (27).
5. Compute ρ maximum such that $0 \leqslant \rho \leqslant 1$ and $\Delta P = \rho \Delta P^0$ remains feasible. ρ may be limited because $\bar{\Delta P}_j$ reaches one of its bounds or because a relaxed constraint becomes effective ; the definitive value of ρ must take into account a small correction to meet (4). If $\rho \neq 0$, go to 6 ; if $\rho = 0$ and a variable $\bar{\Delta P}_j$ limited ρ, go to 2 after having forbidden to $\bar{\Delta P}_j$ to belong to the next set $\bar{\Delta P}$; if $\rho = 0$ for another reason, go to 9.
6. Progression : Update $P (P^0 + \Delta P)$, Δb ($\Delta b[1-\rho]$), the relaxed constraint margins, A (the first row $A_1 \Rightarrow A_1 - \Delta P_T D^0$), γ
7. Compute $\hat{A}^{-1}$, $u = \gamma \hat{A}^{-1}$.
8. Check if a transmission line constraint must be dropped ($\Delta b_L = 0$ and $u^L < 0$ or impossibility to pivot and Δb_L tending to loosen the constraint). If so, redefine $\hat{A}$, compute again $\hat{A}^{-1}$, $u = \gamma \hat{A}^{-1}$. If $\rho < 1$ and pivoting not performed, go to 9 ; if $\rho = 1$ or $\rho < 1$ and pivoting was just performed, go to 2.
9. Pivoting, necessary because a constraint prevented ρ from being equal to 1. This is performed with the same selection criteria as in Dual Simplex method. Then go to 7.

FAST ON-LINE OPTIMAL POWER FLOW : RESULTS

Experiment organisation

Experiments were carried out for 29 cases of the French system, for typical 1980.81 hours, with about 500 busses, 800 lines, 65 thermal units, with data coming from system measurements. Starting from an actual case (about 20000 to 25000 thermal MW) loads were increased or decreased by about 1200 MW for 15 cases and about 600 MW for 14 cases. On one hand, starting from the optimal solution P of the actual case, the FOLOPF algorithm was applied ("FOLOPF solution") ; on the other hand, the optimal solution was computed directly by the usual Optimal Power Flow for the final load conditions ("direct solution"). The objective was to test how similar were the two solutions and how much computer time was required to get the FOLOPF solution.

Solution Accuracy

Experiments were carried out in two cases : with quadratic fuel costs and with linear fuel costs. With quadratic fuel costs, the differences between the FOLOPF and direct solutions P were less than 3 MW. This proves 2 things : 1) FOLOPF algorithm runs correctly and accurately ; 2) the use of the reduced model alone remains valuable even for fairly large load changes. With linear fuel costs, a difficulty appeared : the FOLOPF solution gave the same optimal cost as the direct solution, but the differences between the generated powers P could reach up to 300 MW. This is due to the existence of an infinity of equivalent optimal solutions with linear fuel costs : as the proposed AGC may have to use solutions coming from both types of computations, it was recommended to use only quadratic fuel costs.

Computation Speed

For the 1200 MW load changes, the number of FOLOPF iterations where between 2 and 12, the IBM 30-81 computation time between 8 and 90 milliseconds ; for the 600 MW load changes, the number of FOLOPF iterations were between 2 and 9, the computation times between 7 and 51 milliseconds. The number of iterations and computation times mainly depended on the number of effective security constraints.

These computations are very fast : always less than 90 ms for 1200 MW load change ; due to possible additional computations (e.g. in case of sudden system change), the maximum expected computation time for FOLOPF may be estimated between 0.2 and 0.4 second : compared with the 20 to 40 seconds necessary for a complete Optimal Power Flow, with FOLOPF the computation time is divided by a ratio 100.

Now, there exist minicomputers usable in an Energy Control Center with half the speed of the IBM 30-81 : with such a minicomputer, the maximum FOLOPF computation time for a 1200 MW load change would be 0.4 x 2 = 0.8 second ; knowing that in France 1200 MW represents a maximum load change seldom reached within a 10 second internal, FOLOPF would need at most 8 % of the cycle time, which establishes by far the feasibility of the new AGC proposed concepts.

CONCLUSION

New concepts for AGC were presented, where on one hand transmission security is met, and on the other hand LFC and ED are performed at the same rate, e.g. 10 seconds. The feasibility of these concepts is due to the use of a new tool, "Fast On-Line Optimal Power Flow" ("FOLOPF"), which divides by 100 the computer time of secure ED problems. This FOLOPF uses a new parametric quadratic programming algorithm ; it may be noticed this algorithm is general and not limited to electric power system applications ; with particular additions similar to those presented in the paper, it could even certainly in some cases be used as an approximation for more general convex programming.

REFERENCES

1. Glavitsch H. and Stoffel J., "Automatic Generation Control, a survey", International Journal of Electrical Power and Energy Systems, Vol. 2, No. 1, January 1980, pp 21-28.
2. De Mello F.P., Mills R.J. and B'Rells W.F., "Automatic Generation Control, Part I. Process Modelling", IEEE Trans. PAS 92, 1973, pp 710-715.
3. De Mello F.P., Mills R.J. and B'Rells W.F., "Automatic Generation Control, Part II. Digital Control Techniques", IEEE Trans. PAS 92, 1973, pp 716-724.
4. Schellstede G. and Wagner H., "Design aspects of a software package for Automatic Generation Control with instantaneous economic dispatch and load forecasting functions", IFAC Symposium on Automatic Control in Power Generation, Distribution and Protection, Pretoria 1980, pp 61-69.
5. Elacqua A.J. and Corey S.L., "Security constrained dispatch at the New York Power Pool", IEEE paper No 82 WM 084-2 (1982 winter power meeting).
6. Carpentier J., "Principle of a secure and economic Automatic Generation Control", IFAC Symposium on Automatic Control in Power Generation, Distribution and Protection, Pretoria 1980, pp 463-471.
7. Carpentier J., "Differential Injections method, a general method for secure and optimal load flows", Proc. PICA (1973), pp 255-262.
8. Carpentier J., "System Security in the Differential Injections method for Optimal Load Flows", Proc. PSCC 5 (1975).
9. Carpentier J., "Optimal Power Flows" (survey), International Journal of Electrical Power and Energy Systems, Vol. 1 No 1, April 1979, pp 3-15.
10. Abadie J. and Carpentier J., "Generalization of the Wolfe Reduced Gradient method to the case of non-linear constraints", in "Optimization", Academic Press, pp 37-47, 1969.

ON-LINE OBSERVABILITY DETERMINATION IN ELECTRIC POWER NETWORK

P. Albertos* C. Alvarez** and J. A. de la Puente*

*Automatic Control Dept., E.T.S.I.I., Universidad Politecnica, Valencia, Spain
**Electric Power Dept., E.T.S.I.I., Universidad Politecnica, Valencia, Spain

Abstract. The observability determination in electric power networks is discussed in the paper. A review of the different approaches is presented and the desirable features for a real-time algorithm are pointed out. Deletion of measurements as well as general topology changes ask for well adapted algorithms. The solution proposed in the paper takes advantages of both topological and numerical algorithms. The operational procedure is applied to an experimental reduced order laboratory network.

Keywords. Electric power networks, state estimation, Observability, On-line operation.

INTRODUCTION

In the base of modern electric power system monitoring and control, state estimation techniques are fundamental for security and contingency evaluation. State estimation algorithms require a good network model and the observability condition in the measurement system. That is, the available set of measurements must allow the unique determination of the system state variables, i, e, the complex bus voltages in all the buses of the network.

When a contingency happens, computer aided dispatch for electric power systems needs a fast and reliable observability algorithm before state estimation algorithms can be applied.

An observability algorithm is also needed for design purposes. In this case time saving is not so important but a good knowledge about the relevance of each measurement device in the estimated state computation is convenient. Moreover observability islands must be detected and the suggestion of additional measurements to be implemented in order to achieve complete network observability is required.

Finally, another point of view in power network observability determination is the effect of bad data measurement detection. In fact a critical or fundamental measurement, as later defined, is necessary for network observability

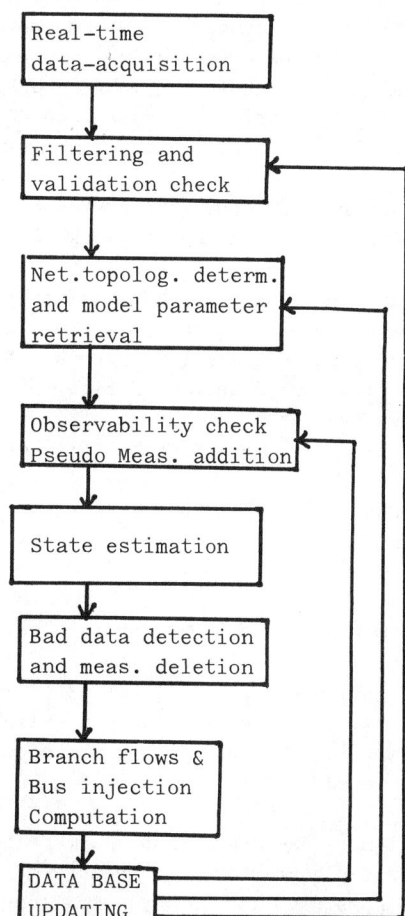

Fig.1. Real-time state estimation process

computation but state estimation algorithms cannot give us information about the correct operation of the corresponding either measurement device or telemetering station. Then at any moment, it is very important to know the set of critical measurements. So if a measurement device is turned off, the set of measurements that become critical must be known, if any.

Since this paper is mainly a discussion of the suitability of two different approaches to on-line observability determination, a general overview of the real-time measurement processing in order to perform a data base updating is previously presented.

As shown in figure 1, to update the data base a sample of the telemetered measurements is processed. A filtering and validation check is performed at first. The validity check is very simple but necessary in order to avoid a bad status data set to be processed by the state estimation algorithm. For instance, no flows are allowed to flow over open lines or to transformer presented as out of service.The status of breakers and disconnects must be consistent. The analog measurements must be between limits, and whenever possible their consistency must be checked. For instance, by comparison of measurements on the same bus or considering a set of redundant measurements.

From the information about the status of breakers and switches the actual network topology is determined. All node and branches impedances are then assumed known and retrieved from computer memory. So a complete network model is available for state estimation purpose. If changes in topology are detected or a set of measurements has been deleted, a test of observability condition must be performed. If there is some lack of observability, pseudomeasurements must be added. Usually pseudo measurements are taken from the data base as the past estimated values of the required variables.

At this moment the state estimation algorithm can be run. One of the main goals of a good state estimation algorithm is the detection of erroneous data in order to assure a reliable state estimation. So, bad data must be detected and deleted.

The deletion of any measurement must be followed by its replacement by alternative measurements, if possible and in any case, some observability computations must be done.

Once the state estimation is performed, branch flows and bus injections are computed. Also any internal variable in the network can be computed if so required. The new computed variables are the entries to update the data base.

Security monitoring, operators guidance and referency to any further computation on the network, such as cost optimization or model verification, are performed on the base of the updated data base.

Let us consider in more detail the observability problem and discuss the proposed solution. The paper is organized as follows. Next section presents a review of some required concepts and approaches to determine the observability condition of a network.Then the main characteristics of two algorithms proposed by the authors are presented. Their use in the framework of electric power system central dispatch is described in the following section. Finally some results and conclusions are presented.

OBSERVABILITY

Numerical Observability condition

Let us consider an electric power network with non linear measurement functions $h(.)$

$$z = h(x) + v \qquad (1)$$

where

- x : n-dimensional vector of state variables (magnitude and phase of bus voltages)

- z : m-dimensional measurement vector, mainly formed by injection and line flow powers and voltage magnitudes.

- v : zero-mean vandom vector representing the measurement noise.

The weighted least-squares estimator computes the optimal state x, the state estimated vector, minimizing the cost functional

$$J = [z - h(x)]^T R^{-1} [z - h(x)] \qquad (2)$$

where R is some positive definite matrix, usually assumed to be the noise covariance matrix. [1].

The minimization of (2) is performed numerically and an iterative approach is used. So, in real time operation, a linearized model of equation (1) is adopted. If we assume an operational initial point given by the state vector x_k, the measurement functions may be expressed by

$$z = H \cdot x + v \qquad (3)$$

where the state and measurement vectors are the increment with respect to the operational point and H is the jacobian matrix, obtained by

$$H = \left. \frac{\partial h(x)}{\partial x} \right|_{x = x_k} \qquad (4)$$

obviously, the jacobian matrix is numerically dependent of the operational point.

The minimization of the cost functional J, related to the linearized model (3), requires the inversión of the matrix

$$H^T R^{-1} H.$$

The observability problem in the estimation of the state of an electric power system deals with the determination of the sufficiency of the measurement set z to provide enough information to allow the computation of the state x. A power system is said to be observable with respect to a measurement set z, if all the components of the state vector can be computed from a measurement sample. Otherwise the power system is said to be unobservable. In that case some subsets of the state vector can be determined in an absolute or relative value. It is said that there are observable islands in the network.

The previons definitions are referred to an operating point. Numerical problems can arise if H-matrix inversion is performed at different states. If there are not, the power system is said to be numerically observable. This condition is more restrictive than the above one, that is named as algebraic observability.

The observability condition can be stated by the following equivalent expresions. [2]

$$\text{rank } H = n \qquad (5a)$$

$$|H^T H| \neq 0 \qquad (5b)$$

$$|H^T R^{-1} H| \neq 0 \qquad (5c)$$

$$\|A\| < \infty \qquad (5d)$$

The last inequality, where the A-matrix is the inverse matrix of $H^T R^{-1} H$ or some computed approximation to it, states that all the elements of this matrix must be upper bounded.

The check of these conditions needs the use of floating point calculations and so only off-line measurement equipment design algorithms, are based on them. Nevertheless, the last condition (5d), when considered only for a reduced number of measurement variables, may be useful in on-line observability studies, as we will show in the next section.

Topological observability condition

Power flow equations can be simplified assuming small phase angles (5) and close to nominal voltage magnitudes (v) in the bus complex voltages of the network, [3]. This simplification, when applied to the measurement functions (1), allows the system equation decoupling, grouping both active power and phase angles, P - δ, and reactive powers and voltage magnitudes, Q-V. The linearized measurement equation (3) is greatly simplified and the H-matrix becomes block diagonal with submatrices H_δ and H_v relating those groups of variables.

Let us assume a N-bus network. The dimension of the state variable vector is 2N-1 and it can be split into an X_v N-dimensional vector of bus voltage magnitudes and a X N-1 dimensional vector of bus voltage angles (the slack bus is taken as a reference for phase angles)

The power system is said to be P -δ observable if a sufficient number of active line flow and active bus injection power measurements exists. That is, if the rank of matrix H_δ is N-1. The power system is said to be Q-V observable if a sufficient number of either reactive line flow and reactive bus injection power measurements or voltage magnitud measurements exists.

Usually active and reactive powers are measured at the same points. Moreover the voltage magnitude measurements only affects the Q-V observability determination. So, network observability is assured if P -δ observability is verified. In any case, both problems may be treated in the same way, so, in the rest of the paper we will deal with the P -δ observability problem.

The measurement system decoupling allows a new approach to the observability determination and more simple observability conditions may be stated. For instance, some authors [4] have proposed algorithms based on the check of the jacobian matrix connectivity. In this way the observability condition is transformed into a connectivity one. A power network is P - δ observable if all the network nodes are related through active power measurements in such a way that all the node voltage angles are computable when a reference node is fixed. In the above framework, the logical observability condition must be stated as follows: a power system is logically observable if the jacobian matrix of the measurement function is connected.

Logical algorithms for network observability determination are based on this property. Of course, logical observability condition is a weaker condition than numerical or algebraic observability condition. But, on the other hand, on these algorithms, only logical operations are performed and so, they are suitable for on-line application. The recent paper of Bongers, Ricke and Handschin [5] is the best reference for this kind of algorithms.

Another approaches are based on the concept of topological observability. All the above mentioned methods and criteria are based on the relationship between network node voltages and measurements. In the topological approach, the observability problem is stated in the network branch frame. Instead of analysing the possibility of node voltages determination, current flow through the branches is computed. Of course, from the observability point of view the actual current values are meaningless and only the availability of enough information in the measurement vector is verified.

In order to define the topological observability condition let us introduce some preliminary concepts [6]. A network graph associated to a power system is defined as a graph whose edges are equivalent to the network branches. In order to study topological observability we must look for a spanning tree in the network, that is, a tree containing all the graph vertices.

A measurement can be assigned to a network edge if the corresponding line is related to the measurement, that is if there is a line flow measurement or a node injection measurement at any of the nodes connected by the line.

In this framework, topological observability can be stated as follows: A power system with a set of measurements is said to be topological observable if there is a spaning tree in the associated network graph in such a way that at least a measurement can be exclusively assigned to any network edge.

Recently [7] an algorithm based on topological observability dealing with a measurement graph instead of the network graph has been presented.

Logical and topological observability conditions are equivalent, but Clements, Davis and Krumpholz [6] have introduced a suficient condition to assure the observability determination of the network(theorem 2.3.4)

In the next section, two algorithms for on line observability determination are considered. The first one is based on the topological concepts and the second one is developed taking into account the algebraic observability condition stated by inequality(5d).

PROPOSED ON-LINE OBSERVABILITY ALGORITHMS

Topological approach

The first algorithm to test observability [8] can be considered as a modification of the topological method [6], incorporating the concept of redundancy level and a different way of dealing with power injection measurements.

Redundancy is defined as the ratio between the number of measurements and the number of state variables, and its value must be $\geqslant 1$ to allow the estimation of the system state. This is a global concept not related to individual measurements.

A measurement is said to be critical or fundamental if removing it from the measurement system results in an unobservabe configuration. On the other hand, a measurement is said to be redundant if its deletion does not affect the observability condition. The redundancy level of a measurement is defined as the minimum number of measurements that must be removed so that the considered measurement becomes fundamental. Obviously, the redundancy level of a fundamental measurement is zero.

The redundancy level of a measurement is a

measure of its relevance for state estimation purposes, and can be computed together with observability determination, as it will be shown.

Measurement redundancy is due to either:

- Measuring power flow at both ends of a branch.
- Measuring power flow in all the branches of a loop.
- Measuring power injection at a bus and power flow in all the branches incident to it.

or a combination of these.

The algorithm processes all types of redundancies in a systematic way and at the same time builds up a spanning forest (i.e. a set of trees [9]) for the network by selecting one branch for each flow measurement (letting aside redundant measurements). Branches are also added for injection measurements - that can be individually assigned to them, in such a way that the number of trees is reduced. If the process results in a single spanning tree, the system is observable. Otherwise each tree in the spanning forest defines an observable area (subsystem). Let us briefly present the structure of the algorithms.

In a first stage, flow measurements are processed. First of all redundancies due to both end measurements are considered. Then a spanning forest is build by means of a deep first search algorithm [9] (only branches with flow measurements are considered). At the same time loop branches are identified, so that loop redundancies can be processed. Multiple loops may produce further redundancies. Each time a redundancy is detected, the redundancy level of the related measurements is accordingly increased.

Processing of injection measurements begins with injections in nodes included in a loop of flow measurements, which form a mixed redundant configuration. Then nodes with flow measurements in all the incident branches are processed. None of these process adds branches to the forest, so the number of trees is not changed. Each tree can be considered as an equivalent node with respect to all its components.

If there is a set of r injections measurements connecting r nodes or trees (with no r-1 dimension subsets fulfilling this property), these measurements are redundant, and the connecting branches can be added to the forest. This implies the fusing of all the involved nodes into a single tree (equivalent node). This process continues until all the injection measurements are processed.

At the end of the algorithms, the system is observable if there are s equivalent nodes connected by s-1 injection measurements.

A more detailed description can be found in [8].

Numerical approach.

A second way to accomplish the study of the network observability is more convenient if an on-line state estimation has been previously performed. This method is especially well suited if only one measurement failure is to be considered.

This method 2 can be classified within the methods in which the search of the observability condition is focused on the computation of rank of the A matrix. In that way, if the i measurement is deleted, the A matrix changes and the new matrix A^i can be expressed by

$$A^i = A + \frac{1}{\beta_i} (A f_i \cdot f_i^T A) \quad (6a)$$

$$\beta_i = 1 - f_i^T A f_i \alpha_i \quad (6b)$$

where f_i: row i ot H matrix; α_i: weighting factor of measuremen z_i (usually the inverse of its variance)

β_i : <u>observability parameter</u> attached to measurement z_i.

Then, being A a n-rank matrix, the necesary condition for A^i not to be a full rank matrix is $\beta_i = 0$

Then, the observability network monitoring is equivalent to the parameter monitoring. The parameter attached to a fundamental measurement, is close to zero

In most of the current state estimation algorithms this computation is very easy to implement:

a) Owing to f_i sparsity, the computation of the row vector $f_i A$ can be simplified. Only the A rows associated to the state variables correspondieng to the non-zero elements of f_i must be computed.

b) To solve the minimization problem (2) it is usual to proceed through the triangularization of A^{-1}. So, in that process it is easy to store the changing coeficients (pivots) corresponding to the rows of A we need to compute. Then, the evaluation of each A-row only requires one back substitution process, the time required being much shorter than that used in the triangularization process.

Then, as a result of these considerations, the following algorithm for the observability coeficient determination has been proposed [2]

1) Given i determine the number and position of the non-zero elements of f_i.

2) Triangularize A^{-1}

3) Compute the rows of A we need $\theta_1...\theta_k$ taking as independent terms $c_1...c_n$, (the change vectors corresponding to the non zero state variables coeficients of f_i)

4) Compute $F_i = f_i^T \cdot A$ and $F_i \cdot f_i$

5) Compute β_i

This algorithm has the important drawback that it is not very efficient to deal with a great number of simultaneous meter devices failure (RTU loss).

It must be also noticed that due to numerical errors the β parameters never are equal to zero. Then, a measurement will be considered as fundamental when its β parameter is valued close to zero.

ON-LINE OBSERVABILITY DETERMINATION.

At the design stage, an adequate set of measurements must be chosen in order to assure power system observability, as well as good redundancy to avoid loss of the observability condition if a measurement or telemetering device failure happens.

The topological algorithm proposed in the last section fulfils all the requirements for meter-placement studies. It provides information about the network observability, relevance of any measurement in the state estimation process and in particular the set of fundamental measurements, the observable islands in the network if it is unobservable as well as possible measurements that would be required to make the entire network observable. Although only logical operations are performed, a lot of information is handled and stored for further processing.

As pointed out in the introduction, fig.1, on-line observability determination presents different requirements. Two kinds of abnormalities can be considered: a) Slight changes in the measurement system. b) drastical variations in the measurement system and/or in the network topology.

a) Slight changes in the measurement system occur when a measurement or a set of non-related measurements are wrong. Usually, as a result of the state estimation algorithm, bad-data are detected. A measurement detected as erroneous is non-critical and may be deleted. Nevertheless other measurements can become critical if they only have level-1 redundancy. So, after a bad-data detection the β-parameter attached to the measurements grouped in the same redundant group that the assumed erroneons one must be recalculated. In this way, new critical measurements, if there are, are pointed out. Only a few β-parameters must be computed and a best information about the accuracy of network state estimation is obtained.

b) If there is a change in the network topology or a group of related measurements fails, (for instance if a remote transmission unit becomes out of order), the approximative approach before described is no more useful. A complete observability test must be run. Nevertheless, most of the information provided by the topological algorithm is already known because only a few measurements or branches have been modified.

If a group of related measurements fails, the redundancy level attached to these measurements may give us information about the loss of the observability property. If it is not the case, or if the change affects the network topology, a partial application of the topological algorithm allows us the observability determination. In this way we must look for the measurements.

Let us consider, as an application, the system in fig. 2, which represents a laboratory network connected to a process computer. The application of the topological algorithm yields the redundancy levels shown in table 1, as well as the spanning tree shown in fig. 3 a.

TABLE 1. Redundancy levels for sample network.

measurement		level	
flow	3-4	4	
"	3-5	3	
"	3-6	4	
"	3-7	4	
"	4-3	4	
"	4-5	3	
"	6-7	3	
"	7-3	4	
injection	1	0	(fundamental)
"	2	0	(fundamental)
"	5	3	
"	6	3	
"	7	3	

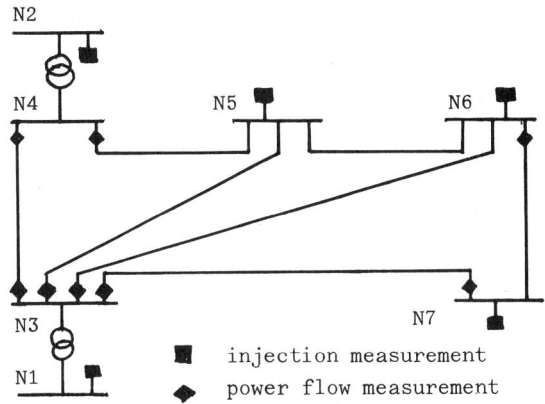

Fig.2. Sample network

As an example of slight changes in the measurement set let us consider what happens with the power flow measurement on branch 4-5 when other related measurements are deleted one by one. The applied state estimation algorithm [2] furnishes a value of the β parameter attached to this measurement of 0.23 when power injection measurement at node 5 is deleted. Then, if the flow measurement at branch 3-5 is discarded the β-parameter is highly reduced to 0.11 . Finally if the injection at node 6 fails, the β parameter value drops to 0.78×10^{-3}. This result points out that the measurement has become fundamental.

Now as a drastic change situation, let us consider in the initial measurement system a failure in a telemetering station that turns off all the measurements in node 3. The observabily test and redundancy level recomputation can be performed by a partial application of the topological algorithm to the subnetwork in the equivalent node 3* (fig. 4). As shown in the figure, a full rank spanning tree can be still built for the subnetwork, so there are no internal unobservable islands and the equivalent nodes remain connected as indicated in figure 3.c. Therefore the full system remains observable, even though the redundancy levels have decreased as shown in table 2.

CONCLUSIONS

Real time operation of electric power networks under safe conditions needs not only a reliable state estimation but the knowledge about the local reliability of the measu-

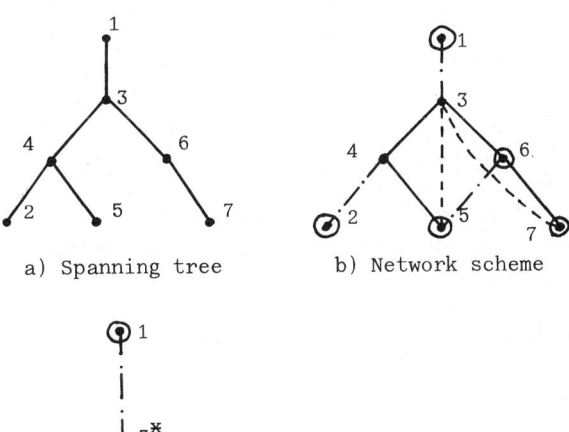

Fig.3. Spanning tree for sample network

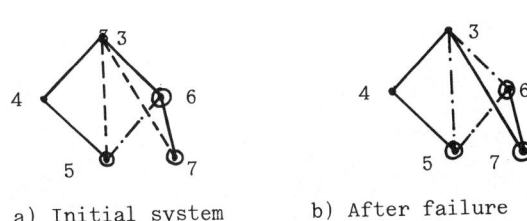

Fig.4. Subnetwork at node 3*

TABLE 2. Redundancy levels after failure at node 3.

measurement		level	
flow	4-3	2	
"	4-5	2	
"	6-7	2	
"	7-3	2	
injection	1	0	(fundamental)
"	2	0	(fundamental)
"	5	2	
"	6	2	
"	7	2	

rement equipment. In this way a review of the existent on-line observability algorithms has been presented. In order to attack different, situations generated by a single or multiple bad-data detection and/or a change in the network topology, some new concepts related to the observability condition are introduced. A general approach to deal with these situations has been proposed, based on both numerical and topological techniques. Two complementary algorithms are used in order to reduce the computation time, using the best suited according to the abnormality presented. The method has been tested on a reduced order laboratory network.

REFERENCES

1. Handschin, E. (1972). Real time control of electric power systems. Elsevier.
2. Alvarez, C. and Albertos, P. (1982). On-line observability determination as a further result of state estimation algorithms. IEEE Trans. on PAS-101.
3. Schweppe, F.G. and Wildes, J. (1970). Power system static state estimation. IEEE Trans. on PAS-89, pp 120-135.
4. Handschin, E. and Borges, C. (1972). - Theoretical and practical considerations in the design of state estimators for electric power systems. In Real time control of electric power systems. Handschin Ed. Elsevier.
5. Bongers, C., Ricke E. and Handschin E. (1981). Observability for real-time state estimation. CIGRE-meeting. Rio de Janeiro.
6. Clements K.A., Davis P.W. and Krumpholz G.R. (1981). Final report on Electric Power network measurement system design. Worcester Polytechnic Institute. Mass,
7. Quintana V.H., Simoes-Costa A. and Mandel A. (1982). Power system topological observability using a direct graph-theoretic approach. IEEE Trans on PAS-101.
8. Albertos, P., and Alvarez, C. (1982). An on line algorithm for power system observability determination. IFAC/IFIP Symposium on Software for Computer Control. Madrid.
9. Aho, Hopcroft and Ullmann. (1974) The design and analysis of computer algorithms. Addison-Wesley.

DISCUSSION

SESSION WP1: POWER SYSTEMS II

Paper: REAL-TIME COMPUTER SYSTEM FOR THE NATIONAL ENERGY CONTROL CENTER OF MEXICO

Authors: J.L. Calderón, Miguel Angel Avila R. (National Energy Control Center (CENACE) - CFE México)

Discusser: JL. Carpentier
2 rue Louis
75008 Paris,
France

Questions or Comments:

1) How are the unit participation factors determined in AGC?

2) When you carry out an AC local flow at the national level, do you represent the complete network or use equivalents for part of the network?

Authors' reply:

1) To determine the unit participation factor we make a perturbation to the solution of the constraint economic dispatch using the load forecast and the load complex distribution factors and from this new optimal base points we get the participation factors.

2) In general we use the complete network (as defined) to perform the load flow, but in some cases, when we have not enough measurements, we use real time equivalents.

Paper: MANAGEMENT AND IMPLEMENTATION OF PROJECT "SICTRE" (REAL TIME INFORMATION AND CONTROL SYSTEM) IN CFE

Authors: L Rance (National Energy Control Center (CENACE) - CFE México)

Discusser: JL Carpentier
2 rue Luois
75008 Paris
France

Questions or Comments:

1) Did you take any special disposition for the acceptance of extended real time programs such as state estimation, specifically for real time trials?

Author's reply:

1) We have planned to test the National Control Center Linked to two Area Control Centers which will remain at the factory facilities to this test. So we will simulate all the functions and application programs in the way they are considered at the National Level, taking care very specially in the data links and the performance of the programs such as state estimator. We know that this will be a very difficult job, and we expect to face it at the end of this year.

Nex year, we will have other type of problems when we test the system on the field with the difficulties to gather correct measurements and the problems to transmit them through the satellite.

Discusser: P. Albertos
ETS11
Apdo 2012
Valencia, Spain

Questions or Comments:

1) Congratulations to both institutions for this research project. I would like to know how many people (full-time) are engaged in it.

Author's reply:

1) Thank you for you comment. We (CFE=Comisión Federal de Electricidad) have about 50 people involved in this project, and the IIE (Instituto de Investigaciones Eléctricas) about 30 people. To install

and put in service the whole project we will need more people than we have now.

Taking into account that we are talking about 6 Area Control Centers and a National Control Center with all the functions described, I think we have not enough people involved in this project to cover all the activities with the desirable efficiency.

Paper: TRASMISSION OF DIGITAL INFORMATION VIA SATELLITE FOR THE REAL-TIME CONTROL SYSTEM OF "COMISION FEDERAL DE ELECTRICIDAD", MEXICO

Authors: D. Carrasco, G Torres, A. Vázquez, (CENACE, Comisión Federal de Electricidad, Mexico City, México)

Discusser: JL. Carpentier
2 rue Luois
75008 Paris
France

Questions or Comments:

1) Was there an economic study to determine the numerical size of a country necessary to make communication via satellite interesting? Would it be interesting for Europe?

Authors' reply:

1) We have not done any studies to determine the minimal size of a country to make communications via satellite interesting.

We do not think that it could be interesting for Europe, because of the large communications resources that are available to the utilities via terrestrial media.

Discusser: José de Jesús Rodríguez
Apdo 4992 Suc "J"
Monterrey, N.L.
México

Questions or Comments:

1) I assume that most of the equipment is imported. Are there any plans to have these equipments build in Mexico?

Authors' reply:

1) Part of the equipment is going to be manufactured in Mexico. At this moment we do not know what part, because the bidding to acquire the equipment has not taken place.

Paper: ON-LINE OBSERVABILITY DETERMINATION IN ELECTRIC POWER NETWORKS

Authors: P. Albertos, C Alvarez (ETSII, Universidad Politécnica de Valencia, Valencia, Spain)

Discusser: JL Carpentier
2 rue Louis Murat
75008 Paris,
France

Questions or Comments:

1) The new coefficients you propose for observability determination look very attractive. In the application of your algorithms, which values do you recommend for thresholds?

Authors' reply:

1) The β- parameters have no absolute value sense. Their value depends on measurement device variance, α_i, as well as measurement equipment configuration. We propose to consider changes in β-parameter values rather than absolute values. So, a measurement becomes fundamental if the corresponding β-parameter value decreases in more than two orders of magnitude.

Discusser: *José Calderón*
Loma Perpetua 50
México, D.F. 1620

Questions or Comments:

1) Do you perform some validation check for digital information (e.g. circuit breakers position) before running the network configurator?

Authors' reply:

1) Yes, In the general algorithm presented in the paper we propose a consistency check of logical measurements.

A SELF-TUNING CONTROLLER WITH A PID STRUCTURE

F. Cameron and D. E. Seborg

Department of Chemical and Nuclear Engineering, University of California, Santa Barbara, CA 93106, USA

Abstract. Self-tuning controllers are proposed which have the same structure as conventional PID controllers. The new self-tuning PID controllers are based on a modified version of the design method of Clarke and Gawthrop (1975, 1979). Experimental and simulation studies for a stirred-tank heating system indicate that the new self-tuners perform well and can be easily adjusted on-line.

Keywords. Adaptive control; direct digital control; heat exchangers; PID control; process control.

INTRODUCTION

During the past 20 years, major advances in control theory have occurred while inexpensive computer control systems have become readily available. However, in the process industries conventional PID control still predominates for several reasons (Clarke and Gawthrop, 1981):

1. PID control is robust and remarkably effective for a wide range of processes;

2. in comparison with aerospace control problems, industrial processes are less well understood;

3. difficult industrial control problems tend to be characterized by nonlinearities, time delays, and time-varying parameters. They also tend to change in unpredictable ways;

4. it is often difficult to establish a priori economic benefits for proposed applications of modern control. Furthermore, the control system hardware cost is usually only a small portion of the total project cost.

During the past decade, there has been considerable interest in self-tuning (or adaptive) control systems which automatically adjust controller parameters on-line in response to changes in the process or the environment. Self-tuning has considerable potential for process control problems since it provides a systematic, flexible approach for dealing with uncertainties, nonlinearities, and time-varying parameters. The current interest in self-tuning controllers was largely stimulated by the original self-tuning regulator of Aström and Wittenmark (1973) and the subsequent modifications due to Clarke and Gawthrop (1975, 1979, 1981).

Several comprehensive survey articles Aström and co-workers, 1977; Aström, 1981a, 1981b; Clarke and Gawthrop, 1979; Parks and co-workers, 1980) and tutorial papers (Isermann, 1980; Bélanger, 1982) are available. A number of experimental applications of self-tuning controllers have been reported (Aström and co-workers, 1977; Parks and co-workers, 1980; Isermann, 1980; Clarke and Gawthrop, 1981) including applications to distillation columns (Sastry and co-workers, 1977; Morris and co-workers, 1981; and Dahlquist, 1981) and other process control problems. Design methods for self-tuning controllers have been based on minimizing a selected performance index (Aström and Wittenmark, 1973; Clarke and Gawthrop, 1979) or assigning closed-loop poles and zeros (Wellstead and co-workers, 1979; Wellstead and Sanoff, 1981; Aström and Wittenmark, 1980; Vogel and Edgar, 1982).

It is important to note that the resulting self-tuners, in general, do not have a PID structure. This is one reason why industrial acceptance of self-tuners has not been very rapid.

In this paper a self-tuning controller is proposed which has the same structure as a conventional three term PID controller. The design is based on a modified version of the Clark and Gawthrop approach (1975, 1979). The resulting, self-tuning PID controller can be used in two different ways:

a) as an adaptive control system in which the PID controller parameters are adjusted continually;

b) as a means of tuning a conventional PID controller; after the PID settings have converged, the self-tuner can be tuned off.

The second type of application could be quite useful in view of the serious disadvantages associated with conventional PID tuning techniques (Yuwana and Seborg, 1982).

SELF-TUNING CONTROLLERS

In this section we review the self-tuning controller (STC) of Clarke and Gawthrop (1975, 1979). The starting point in their analysis is the single-input, single-output process model

$$A(z^{-1})y(t) = z^{-k-1}B(z^{-1})u(t) + C(z^{-1})\xi(t) \qquad (1)$$

where $y(t)$ is the measured output, $u(t)$ is the manipulated input, $\xi(t)$ is an uncorrelated sequence of random variables, and t denotes the sampling instant, $t = 0, 1, 2 \ldots$. The polynomials $A(z^{-1})$, $B(z^{-1})$ and $C(z^{-1})$ are expressed in terms of the backwards shift operator, z^{-1}:

$$A(z^{-1}) = 1 + a_1 z^{-1} + \ldots + a_n z^{-n} \qquad (2)$$

$$B(z^{-1}) = b_0 + b_1 z^{-1} + \ldots + b_m z^{-m}, \quad b_0 \neq 0 \qquad (3)$$

$$C(z^{-1}) = 1 + c_1 z^{-1} + \ldots + c_n z^{-n} \qquad (4)$$

It is assumed that all roots of $C(z^{-1})$ lie inside the unit circle.

The self-tuning controller is designed to minimize the variance of an auxilliary output $\phi(t)$,

$$\phi(t) \equiv P(z^{-1})y(t) + Q(z^{-1})u(t-k) - R(z^{-1})y_r(t-k) \qquad (5)$$

where y_r is the set point and P, Q and R are transfer functions:

$$P(z^{-1}) = \frac{P_n(z^{-1})}{P_d(z^{-1})} \quad \text{etc.} \qquad (6)$$

The selection of P, Q and R are important design decisions which play a key role in determining control system performance. (In the sequel, the z^{-1} notation will be omitted from the various polynomials and transfer functions in order to simplify the notation.)

If the process model in Eq. (1) is combined with the identity,

$$CP_n = AP_d E + z^{-k-1} F \qquad (7)$$

the predictive model in Eq. (8) is obtained

$$\phi(t+k+1) = \frac{F}{P_d C} y(t) + [\frac{EB}{C} + Q]u(t) - Ry_r(t) + E\xi(t+k+1) \qquad (8)$$

An equivalent representation of the predictive model is

$$\phi(t+k+1) = \phi^*(t+k+1|t) + \tilde{\phi}(t+k+1) \qquad (9)$$

where the (k+1)-step ahead prediction is

$$\phi^*(t+k+1|t) = \frac{F}{P_d C} y(t) + [\frac{EB}{C} + Q] u(t) - Ry_r(t) \qquad (10)$$

and $\tilde{\phi}(t+k+1)$ denotes the prediction error. If the prediction $\hat{\phi}(t+k+1 \mid t)$ is forced to zero, then rearrangement of Eq. (1) yields the control law that minimizes the variance of $\phi(t)$:

$$u(t) = \frac{CRy_r(t) - Fy(t)/P_d}{EB + CQ} \qquad (11)$$

If the model parameters (A, B, and C) were known, then E and F could be determined from the identity in Eq. (7) and the control law in Eq. (11) could be easily implemented. For the situation where these parameters are unknown, then the control law together with the predictive model in Eq. (8) provide the basis for the self-tuning controller.

Clarke and Gawthrop (1979) have discussed several alternative methods for developing recursive parameter estimation schemes based on the predictive model in Eq. (8). The predictive model used in the study is based on $\phi_y(t)$, the component of $\phi(t)$ which depends on $y(t)$ only. Then from Eq. (8):

$$\phi_y(t) \equiv Py(t) \qquad (12)$$

From Eqs. (1) and (7), it follows that

$$\phi_y(t+k) = \frac{F}{CP_d} y(t) + \frac{EB}{C} u(t) + E\xi(t+k) \qquad (13)$$

Next we set $G \equiv EB$ and define a "filtered measurement" y_f by

$$y_f(t) = y(t)/P_d \qquad (14)$$

If we assume that $C = 1$, then linear least squares estimation can be used and a suitable estimation model is given by

$$\phi_y(t+k) = Fy_f(t) + Gu(t) + \varepsilon(t+k) \qquad (15)$$

where G and F are polynomials in z^{-1} and $\varepsilon(t+k)$ is a moving average of the noise $\{\xi(t)\}$. Estimates of the elements of F and G can be obtained from the standard least squares equations shown below.

However, in both the simulation and experimental studies, the more robust U-D factorization method was used (Cameron, 1982).

Prediction:

$$\hat{\phi}_y(t) = \underline{\psi}^T(t-k-1)\, \hat{\underline{\theta}}(t-1) \qquad (16)$$

Gain:

$$\underline{K}(t) = \underline{P}(t-1)\underline{\psi}(t-k-1)[\sigma^2 +$$

$$\underline{\psi}^T(t-k-1)\underline{P}(t-1)\underline{\psi}(t-k-1)]^{-1} \qquad (17)$$

Parameter Estimation:

$$\hat{\underline{\theta}}(t) = \hat{\underline{\theta}}(t-1) + \underline{K}(t)[\hat{\phi}_y(t) - \phi_y(t)] \qquad (18)$$

Covariance:

$$\underline{P}(t) = \frac{1}{\mu}\,[\underline{I} - K(t)\underline{\psi}^T(t-k-1)]\underline{P}(t-1) \qquad (19)$$

where the parameter and data vectors are:

$$\hat{\underline{\theta}}^T(t) = [\hat{f}_0(t), \hat{f}_1(t), \ldots, \hat{f}_L(t);$$

$$\hat{g}_0(t), \hat{g}_1(t), \ldots \hat{g}_J(t)] \qquad (20)$$

$$\underline{\psi}^T(t) = [y_f(t), y_f(t-1) \ldots y_f(t-L);$$

$$u(t), u(t-1), \ldots u(t-J)] \qquad (21)$$

and the number of F and G parameters to be estimated are denoted by

$$L = n + \deg P_d \quad \text{(no. of F parameters)} \qquad (22)$$

$$J = m + k + 1 \quad \text{(no. of G parameters)} \qquad (23)$$

Then by substituting G=EB and replacing the F and G polynomials in Eq. (11) by their estimates $\hat{F}(t)$ and $\hat{G}(t)$, we obtain

$$u(t) = \frac{CRy_r(t) - \hat{F}(t)y(t)/P_d}{\hat{G}(t) + CQ} \qquad (24)$$

This certainty-equivalence control law is referred to as a <u>self-tuning controller</u> (STC) by Clarke and Gawthrop (1975, 1979).

It is convenient to introduce the notation STC-LJ to denote an STC with L parameters in the F polynomial and J parameters in the G polynomial.

A Self-Tuning Controller with a PID Structure

In general, self-tuning controllers do not necessarily have a PID structure. Consequently, it is of considerable interest to determine under what conditions a self-tuning controller will have a PID structure. As mentioned earlier, a self-tuning PID controller could be used on a continual basis as an adaptive controller, or on an intermittent basis to tune a conventional PID controller.

A number of previous studies have considered the conditions that need to be satisfied in order to obtain self-tuning controllers with a PID structure (Isermann, 1980; Wittenmark, 1979; and Wittenmark and Aström, 1980). Self-tuning PID controllers based on pole placement have recently been reported by Wittenmark (1979) and Wittenmark and Aström (1980). Banyasz and Keviczky (1982) have proposed two design methods based on prescribed overshoot to a step response and gradient search techniques. Apparently, the present paper contains the first report of a self-tuning PID controller based on the Clarke-Gawthrop design approach.

An ideal digital PID controller can be written in velocity form as (Smith, 1973):

$$\nabla u(t) = K_c\{e(t) - e(t-1) + \frac{T_s}{T_i}\,e(t) +$$

$$\frac{T_d}{T_s}\,[e(t) - 2e(t-1) + e(t-2)]\} \qquad (25)$$

where

$$\nabla u(t) \equiv u(t) - u(t-1) \qquad (26)$$

and the error signal is defined by

$$e(t) \equiv y_r(t) - y(t) \qquad (27)$$

In Eq. (25) T_s is the sampling interval and K_c, T_i and T_D denote the PID controller settings.

A number of alternatives to the ideal PID controller in Eq. (25) are available (Wittenmark, 1979). Of particular interest in this study is the form where the set point appears in only the integral term and the output y is filtered (cf. Eq. (14)):

$$\nabla u(t) = K_c\{-y_f(t) + y_f(t-1) + \frac{T_s}{T_i}\,[y_r(t)$$

$$-y_f(t)] + \frac{T_d}{T_s}\,[-y_f(t) + 2y_f(t-1) -$$

$$y_f(t-2)]\} \qquad (28)$$

This type of PID controller (without output filtering) has been considered in a number of previous studies (Bernard and Cashen, 1965; Phelan, 1977; Wittenmark, 1979; and Isermann, 1981). Rearranging Eq. (28) gives,

$$\nabla u(t) = \frac{K_c T_s}{T_i}\,y_r(t) - K_c(1+\frac{T_s}{T_i} + \frac{T_d}{T_i})y_f(t) +$$

$$K_c(1 - \frac{2T_d}{T_s})y_f(t-1) - \frac{K_c T_d}{T_s} y_f(t-2) \quad (29)$$

Next we propose a new design procedure which ensures that the STC in Eq. (24) has the PID structure of Eq. (29). This derivation is based on a new approach for incorporating integral control action into the STC plus an appropriate choice of key design parameters. As a starting point in this derivation, we rewrite the STC control law in Eq. (24) after substituting $y_f \equiv y/P_d$:

$$u(t) = \frac{CRy_r(t) - \hat{F}y_f(t)}{\hat{G} + CQ} \quad (30)$$

In order for Eq. (30) to have the PID structure of Eq. (29), we specify that the F polynomial has degree 2, or equivalently that L = 3 (cf. Eq. (22)). A reasonable choice for the output filter is either deg $P_d = 1$ (i.e. first-order filter) or deg $P_d = 0$ (no filtering). Then Eq. (22) indicates that the choice L= 3 is equivalent to choosing a second or third-order process model, i.e. n=2 or n=3. Thus in the sequel we assume that

$$P_d(z^{-1}) = 1 + p_{d1} z^{-1} \quad (31)$$

The next step in the derivation is to ensure steady-state agreement between output y and set point y_r. Since R can be freely chosen, we set

$$CR = H_o \quad (32)$$

where H_o is defined as

$$H_o \equiv (\frac{\hat{F}}{P_d})_{z=1} = \frac{\sum_{i=1}^{L} \hat{f}_i}{1 + p_{d1}} \quad (33)$$

As the final design step, integral action must be introduced into the STC control law. This can be done in a number of different ways (Clarke and Gawthrop, 1979; Aström, 1981a; Moden, 1981; Moden and Nybrant, 1981). However, some of these approaches do not guarantee offset elimination after sustained load disturbances (Moden, 1981; Morris and co-workers, 1981).

In the present paper, we introduce integral action in an alternative manner by setting

$$\hat{G} + CQ = \frac{1 - z^{-1}}{\nu} \quad (34)$$

where ν is a design parameter which will provide a convenient means of tuning the resulting STC. This approach guarantees offset elimination after both set-point and load changes (Cameron, 1982). Substituting Eqs. (31)-(34) into Eq. (30) yields a STC with the desired PID structure:

$$\nu u(t) = \nu[H_o y_r(t) - (\hat{f}_0 + \hat{f}_1 z^{-1} + \hat{f}_2 z^{-2})]y_f(t)$$

$$(35)$$

where the output filter has the form

$$y_f(t) = y(t)/(1 + p_{d1}z^{-1}), \quad -1 < p_{d1} \leq 0 \quad (36)$$

Expressions for the corresponding PID controller settings can be derived by comparing terms in Eqs. (35) and (29)

$$K_c = -\nu(2\hat{f}_2 + \hat{f}_1)/\alpha \quad (37)$$

$$T_i = \frac{T_s K_c}{\nu \hat{f}_o/\alpha - K_c - T_d/T_s} \quad (38)$$

$$T_d = T_s \nu \hat{f}_2 / \alpha K_c \quad (39)$$

where $\alpha \equiv 1 + p_{d1}$.

One of the implications of introducing Eq. (34) is that Q can no longer be specified independently, as in the Clarke-Gawthrop approach (1975, 1979). However, parameter ν can be arbitrarily specified and has an effect similar to that of conventional controller gain K_c, as demonstrated in the next section. If ν is kept constant and $\hat{G}$ is updated at each sampling instant, then Q also changes at each sampling instant in order to satisfy Eq. (34). However, this calculation is never explicitly performed since Q does not appear in the self-tuning control law of Eq. (35).

SIMULATION RESULTS

In order to evaluate the self-tuning PID controller, a simulation study was performed for a stirred-tank heating process. The objectives of the simulation study were to investigate the effect of various STC design parameters (P(z), ν etc.) and to determine whether satisfactory PID controller settings could be obtained. An experimental application to the same stirred-tank heating process is presented in the next section.

A schematic diagram of the stirred-tank heating process is shown in Fig. 1. This bench-scale process consists of two stirred tanks connected by a long pipe which introduces a time delay. The temperature of the water in the second tank T_2 is controlled by adjusting the heat duty q_1 from the electrical heater in the first tank. Two additional heaters can be used to introduce load disturbances manually. The volume of water in each tank is held constant by using overflow lines. The water flow rate through the system is controlled manually but can be adjusted to introduce a second type of disturbance. The dimensions and physical properties for the stirred-tank heating process are shown in Table 1.

TABLE 1 Typical Operating Conditions

tank volumes: $V_1 = 6514 \text{ cm}^3$, $V_2 = 3767 \text{ cm}^3$
inlet water temperature: $T_c = 18.5°C$
exit water temperature: $T_2 = 25.2°C$
ambient temperature: $T_a = 22.2°C$
control input: $q_1 = 0.502$ kW
load inputs: $q_{1L} = 0.122$ kW, $q_{2L} = 0.799$ kW
water flow rate: $w = 3.0$ kg/min

Cameron (1982) has derived a dynamic model of the stirred-tank heating system based on an energy balance for each tank and calibration curves for the temperature sensor and electrical heaters. Assuming a constant water flow rate of $w = 3.0$ kg/min, the dynamic model reduces to the following transfer function between the controlled variable T_2(°C) and the manipulated variable q_1(kW)

$$\frac{T_2(s)}{q_1(s)} = \frac{4.15 \times 10^{-3} e^{-53s}}{(119s + 1)(71s + 1)} \quad (40)$$

where the time delay of 53 sec. was determined empirically.

The simulation results in Figs. 2-4 indicate that the self-tuning PID controller can generate reasonable controller settings within a short period of time.[1] In each figure, the process is subjected to a series of set-point changes beginning at the 50th sampling instant (t=50). Figure 2 indicates that the STC results in reasonable control after the large initial transient which occurs before the first set-point change. The controller settings calculated from Eqs. (37)-(39) are shown in Fig. 3. Note that K_c and T_i converge quickly but that T_d exhibits a slow drift and eventually becomes negative. In a number of the simulation runs (Cameron, 1982) T_d became negative; however, in each case the absolute value was small and consequently the contribution of the derivative term was negligible. Alternatively, one could set $T_d = 0$ or use $|T_d|$.

---

1. In Fig. 2 and some of the subsequent figures, the initial transient response was truncated at 27.2° so that the same scale could be used in all figures.

Figure 4 demonstrates that increasing the value of design parameter ν results in more oscillatory control, as would be expected from the discussion in the previous section. Figure 5 indicates the effect of using $P_n = 1 - 0.6z^{-1}$ rather than $P_n = 1$ as was the case in Fig. 4. Now the response after the first two set-point changes is less oscillatory and exhibits little overshoot. The STC also provides satisfactory control after the load change of -0.2kw at t = 105 and the flowrate change from 3.0 kg/min to 2.56 kg/min at t=155. The calculated controller settings change only slightly after the two load changes. This is why the response to the final set-point change at t=205 is slightly different from the earlier ones.

In Figs. 2-4 the STC performance is poor during the initial transient period until the estimated parameters converge. The undesirable transients were a consequence of using zero initial parameter estimates, $\hat{\theta}(0) = \underline{0}$. The problem can be avoided by using a "commissioning period" during which the parameter estimation is performed while the process is operated using a conventional PID controller with constant settings (Wittenmark, 1973). Thus during the commissioning period, parameter estimation is performed according to Eqs. (16)-(21) but the STC control calculation in Eq. (35) is replaced by the conventional PID controller in Eq. (29). After the parameter estimates have converged (or after an arbitrary period of time), a switch is made from PID control to STC.

Figures 6 and 7 indicate that the commissioning period successfully eliminates the initial transient period. During the first 175 sampling instants, a conventional PID controller is used with conservative settings. After the STC is switched on at t = 175, the output response in Fig. 6 improves while the calculated controller constants in Fig. 7 do not change drastically. Thus an essentially "bumpless transfer" occurs. Similar results were obtained when the length of the commissioning period was varied or when different PID settings were employed during commissioning (Cameron, 1982).

EXPERIMENTAL APPLICATION

In the experimental study, the self-tuning control algorithm was implemented using a Data General Eclipse computer which is interfaced to the stirred-tank heating system. A detailed description of the real-time computing system is available elsewhere (Mellichamp, 1981). A sampling period of 60 sec and a forgetting factor of $\mu = 1$ were used in all of the experimental runs.

A preliminary STC run is shown in Figs. 8 and 9. Due to the relatively low value of

ν, the closed-loop response $T_2(t)$ is rather sluggish during the initial transient period. Note that the calculated controller parameters in Fig. 9 converge quite rapidly after the initial set-point change at t=0.

One question that arises in practical applications is how should a self-tuner be modified if the control performance is not satisfactory? That is, how does one "tune" a self-tuner? The results in Figs. 10 and 11 indicate that a response which is too oscillatory can be effectively damped by changing design parameter, $P_n(z)$. During the initial time period ($0 < t < 100$) where $\nu=1$ and $P_n=1$, the response is too oscillatory. At $t=100$ parameter $P_n(z)$ is changed from $P_n=1$ to $P_n=1-0.6z^{-1}$ and the forgetting factor is set equal to 0.1 for three sampling instants. The controller settings in Fig. 11 quickly adjust and the closed-loop response improves significantly. (An alternative approach for damping oscillatory responses is to reduce the value of ν, as indicated by the simulation results in Figs. 2 and 4.)

The results in Fig. 12 provide experimental verification that a commissioning period can be successfully employed to eliminate undesirable initial transients. During the commissioning period (0<t<35), parameter estimation is performed while a poorly tuned PID controller calculates the control action. At t=35 the STC is switched on and the control system performance improves. The calculated controller settings (not shown) do not change after the switch from PID to STC is made.

Figure 13 shows the performance of the self-tuning PID controller during a more stringent experimental test. The test consisted of two set-point changes followed by a step change in the load heater of -0.2 kW at t=110. The liquid flow rate was then changed from 3.0 kg/min to 2.56 kg/min at t=150 and a final set-point change was made at t=190. Note that the flow rate change affects both the static and dynamic behavior of the stirred-tank heating system since both the steady-state gain and the dominant time constants decrease. During the initial transient period, a large overshoot occurs due to the poor initial parameter estimate, $\hat{\theta}(0)=0$. The STC provides satisfactory control during the subsequent set point, load and flow rate changes. Note that the final set point response differs from the previous one because the process flow rate has changed.

In order to compare the self-tuning PID controller with a well-tuned conventional controller, the experimental tests of Fig. 12 were repeated using a PID controller with constant settings ($K_c=2$, $T_i=290$ sec, $T_d=50$ sec). These settings were determined empirically. The response in Fig. 14 is quite good but not appreciably better than the STC results in previous figures. Also, the STC adjustment of controller settings is done automatically and relatively fast in contrast to time-consuming empirical techniques.

CONCLUSIONS

A new method for designing self-tuning controllers with a conventional PID structure has been developed. The design method is based on a modification of the Clarke-Gawthrop approach so that the process model has a prescribed structure and integral action is incorporated in a novel manner. The resulting self-tuner eliminates offset after step changes in the set point and load variables. Experimental and simulation studies for a stirred-tank heating system have indicated that the new self-tuner performs well and can be easily adjusted on-line.

ACKNOWLEDGEMENT

Financial support from the Department of the Navy is gratefully acknowledged.

REFERENCES

Andreiev, N. (1981). A new dimension: a self-tuning controller that continually optimizes PID constants. Control Eng., 28, (8), 84-85.

Åström, K. J. (1981a). Design principles for self-tuning regulators. In K. S. Narendra and R. V. Monopoli (Eds.), Proc. Internat. Workshop on Applic. of Adaptive Control, Academic Press, New York.

Åström, K. J. (1981b). Theory and applications of adaptive control. Proc. 8th IFAC World Congress, Kyoto, Japan.

Åström, K. J., U. Borisson, L. Ljung and B. Wittenmark (1977). Theory and applications of self-tuning regulators. Automatica, 13, 457-476.

Åström, K. J., and B. Wittenmark (1973). On self-tuning regulators, Automatica, 9 185-199.

Åström, K. J., and B. Wittenmark (1980). Self-tuning controllers based on pole-zero placement. IEE Proc., 127, Pt. D, 120-130.

Banyasz, C., and L. Keviczky (1982). Direct methods for self-tuning PID regulators. 6th IFAC Sympos. on Ident. and System Parameter Estimation, Wash. D.C., pp. 1249-54.

Bélanger, P. R. (1982). A review of some adaptive control schemes for process control. In T. F. Edgar and D. E. Seborg (Eds.) Chemical Process Control 2, Engineering Found., New York, pp. 269-286.

Bernard, J. W., and J. F. Cashen (1965). Direct digital control: Instruments and Control Systems, 38, (9), 151-158.

Cameron, F. (1982), M.S. thesis, Univ. of California, Santa Barbara.

Clarke, D. W., and P. J. Gawthrop (1975). Self-tuning controller. Proc. IEE, 122, 929-934.

Clarke, D. W., and P. J. Gawthrop (1979). Self-tuning control. Proc. IEE, 126, 633-640.

Clarke, D. W., and P. J. Gawthrop (1981). Implementation and application of microprocessor-based self-tuners. Automatica, 17, 233-244.

Dahlquist, S. A. (1981). Control of a distillation column using self-tuning regulators. Can. J. Chem. Eng., 59, 118-127.

Isermann, R. (1980). Parameter adaptive control algorithms - a tutorial. 6th IFAC/ IFIP Conf. on Digital Computer Applications to Process Control, Dusseldorf.

Isermann, R. (1981). Digital Control Systems, Springer-Verlag, New York.

Mellichamp, D. A. (1981). A full-year course sequence in real-time computing. Chem. Eng. Educ., 15, 18-24.

Moden, P. E. (1981). Stochastic control with a sliding short time criterion and offset elimination. Report UPTEC 8149R, Inst. Tech., Uppsala Univ., Sweden.

Moden, P. E., and T. Nybrant (1980). Adaptive control of rotary drum driers. 6th IFAC/IFIP Conf. on Digital Computer Applic. to Process Control, Dusseldorf, pp. 355-361.

Morris, A. J., Y. Nazer, and R. K. Wood (1981). Single and multivariable application of self-tuning controllers. In C. J. Harris and S. A. Billings (Eds.), Adaptive Control: Theory and Applications, IEE Control Eng. Series 15, IEE, London.

Parks, P. C., W. Schaufelberger, C. Schmid, and H. Unbehauen (1980). Applications of adaptive control systems. In H. Unbehauen (Ed.), Methods and Applications in Adaptive Control, Springer-Verlag, New York.

Phelan, R. M. (1977). Automatic Control Systems, Cornell Univ. Press, Ithaca, NY.

Sastry, V. A., D. E. Seborg, and R. K. Wood (1977). Self-tuning regulator applied to a binary distillation column, Automatica, 13, 417-424.

Smith, C. L. (1972). Digital Computer Process Control. Intext Educ. Pub., Scranton, PA.

Vogel, E. F., and T. F. Edgar (1982). Application of an adaptive pole-zero placement controller to chemical processes with variable dead time. Proc. American Control Conf., Arlington VA.

Wellstead, P. E., J. M. Edmunds, D. Prager and P. Zanker (1979). Self-tuning pole/zero assignment regulators. Int. J. Control, 30, 1-26.

Wellstead, P. E., and S. P. Sanoff (1981). Extended self-tuning algorithm. Int. J. Control, 34, 433-455.

Wittenmark, B. (1973). A self-tuning regulator. Report 7311, Div. of Automatic Control, Lund Inst., Sweden.

Wittenmark, B. (1979). Self-tuning PID controllers based on pole placement. Report TFRT-7179, Lund Inst., Sweden.

Wittenmark, B. and K. J. Åström (1980). Simple self-tuning controllers. H. Unbehauen (Ed.), Methods and Applications in Adaptive Control, Springer-Verlag, N.Y.

Yuwana, M., and D. E. Seborg (1982). A new method for on-line controller tuning. AIChE Journal, 28, 434-440.

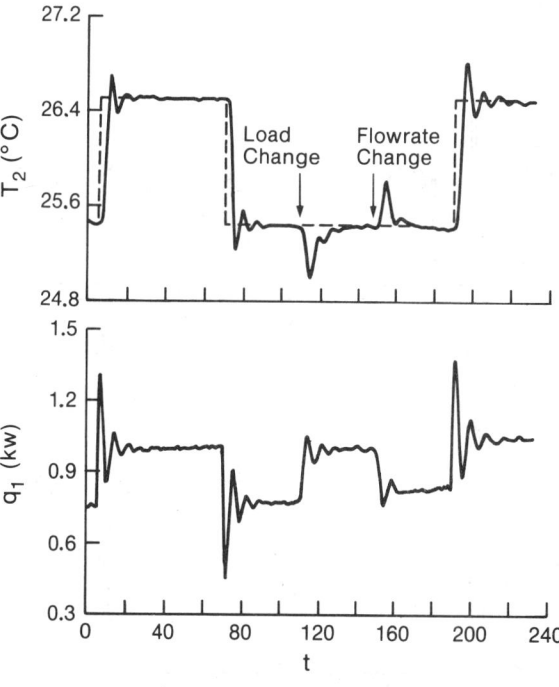

Fig. 14 Experimental response for conventional PID control.

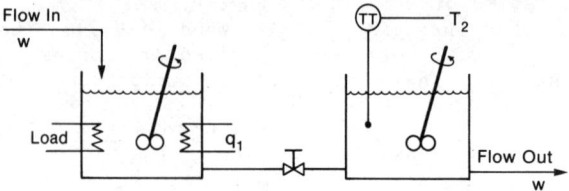

Fig. 1 Stirred-tank heating system.

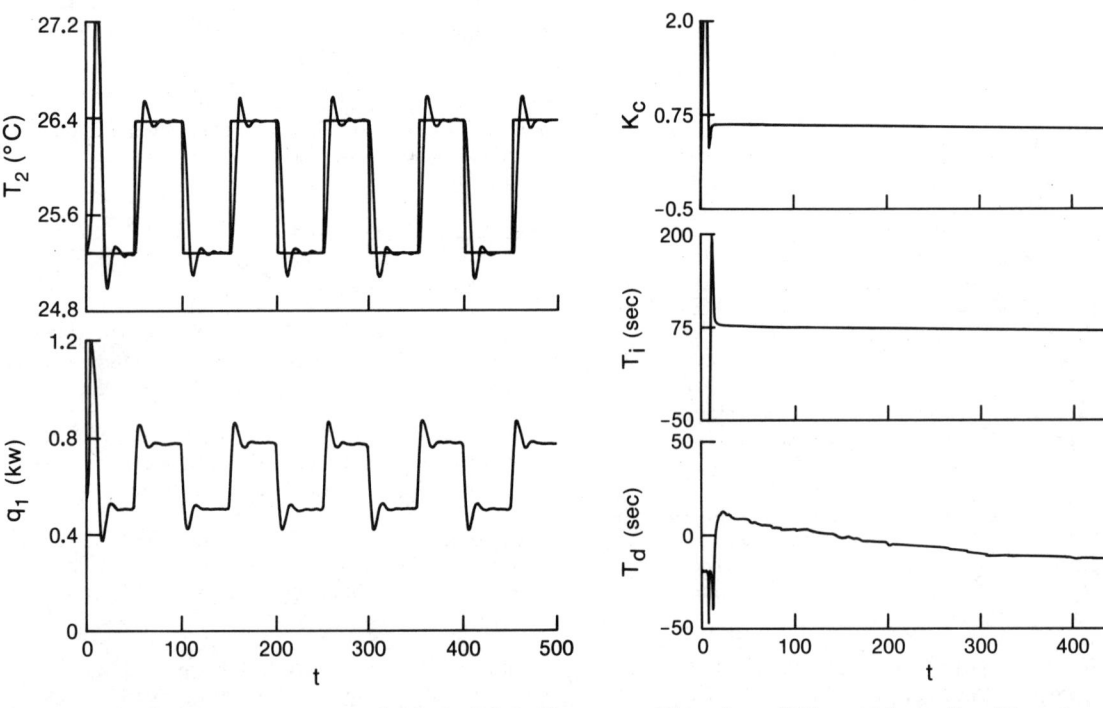

Fig. 2 Simulated response ($\nu=0.5$, $P(z)=1$).

Fig. 3 PID settings for Fig. 2

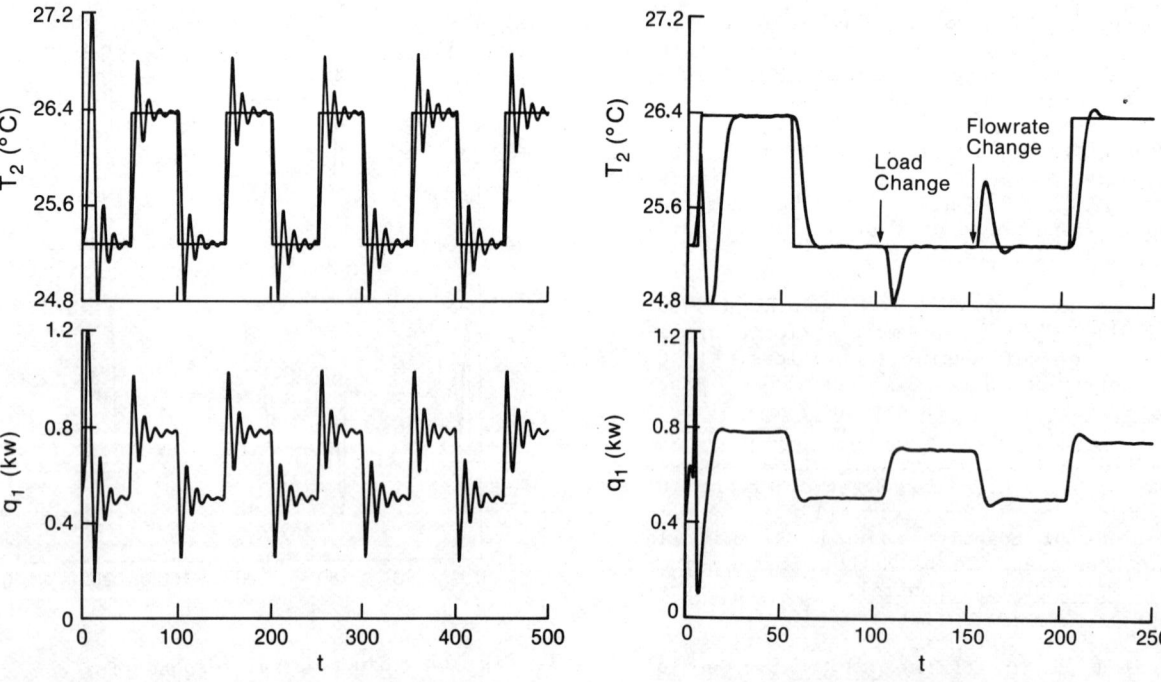

Fig. 4 Simulated response ($\nu=1$, $P(z)=1$).

Fig. 5 Simulated response ($\nu=1$, $P_n=1-0.6z^{-1}$).

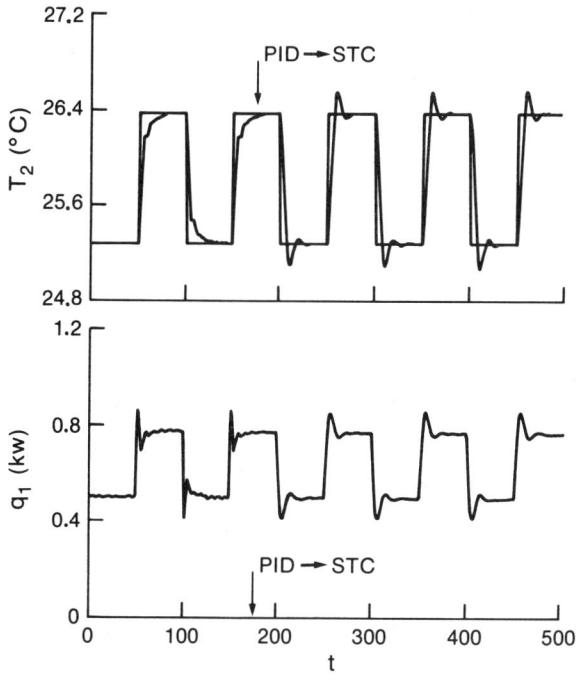

Fig. 6 Simulated response using PID control until t=175 ($\nu=0.5$, $P(z)=1$).

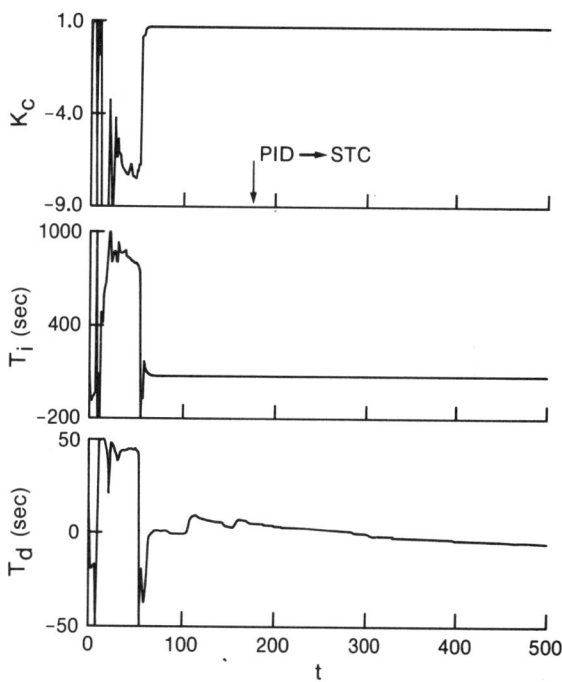

Fig. 7 PID settings for Fig. 6.

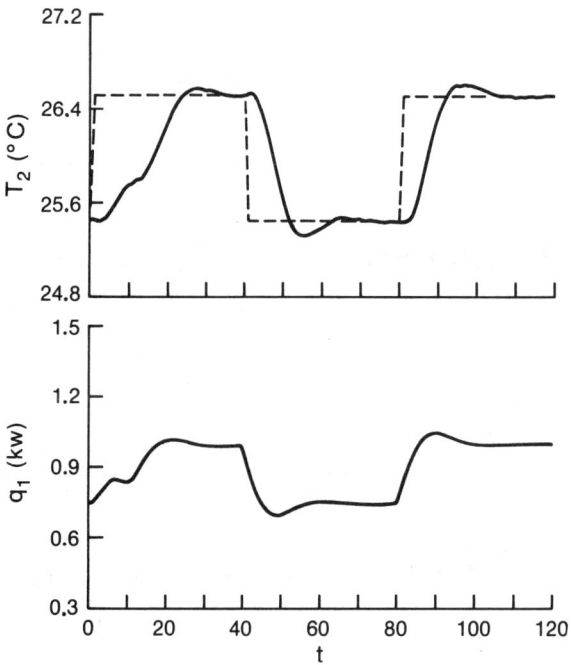

Fig. 8 Experimental response ($\nu=0.25$, $P(z)=1$).

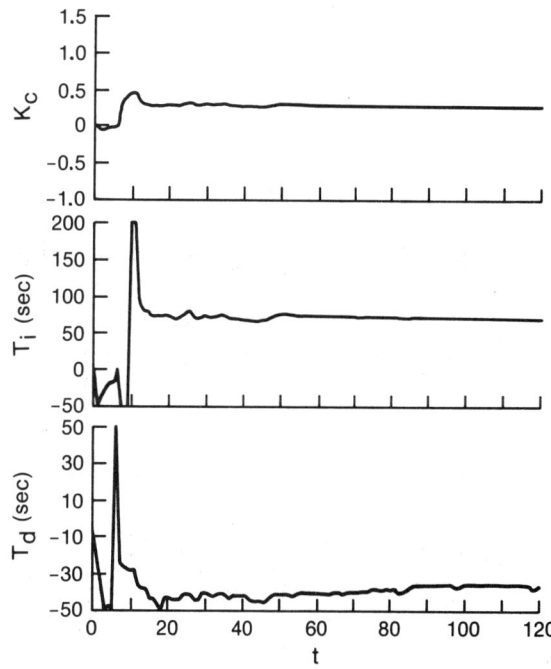

Fig. 9 PID settings for Fig. 8.

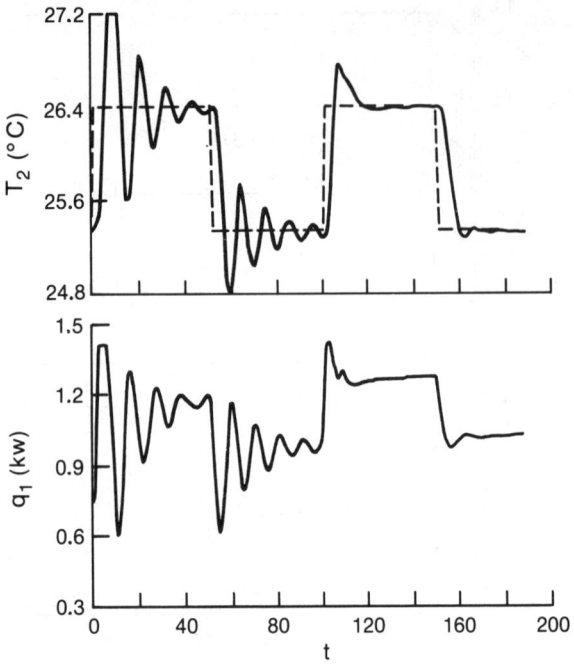

Fig. 10 Experimental response ($\nu=1$; at $t=100$, P_n change from 1 to $1-0.6z^{-1}$).

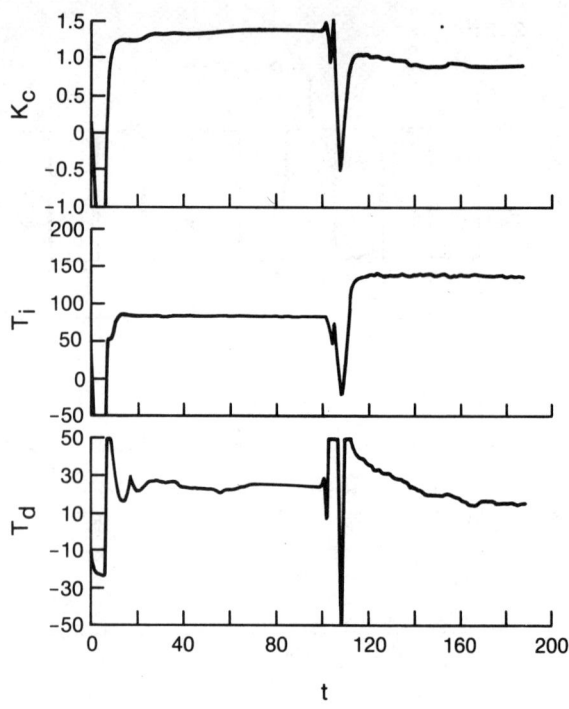

Fig. 11 PID settings for Fig. 10.

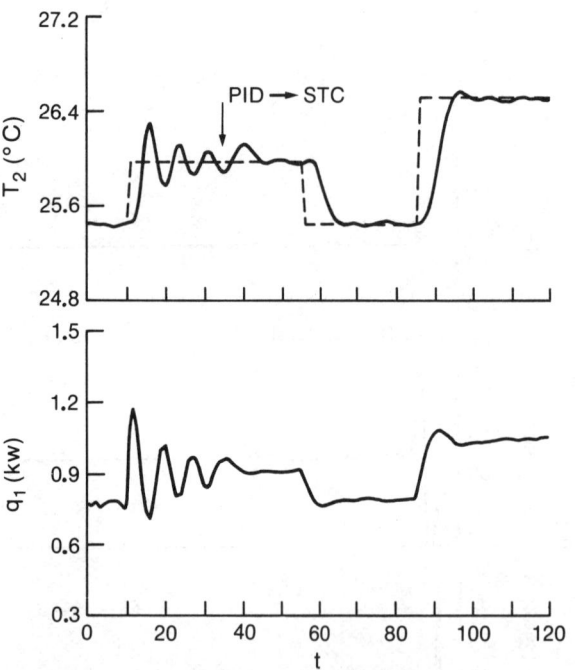

Fig. 12 Experimental response using PID control until $t=35$ ($\nu=1$, $P_n=1-0.6z^{-1}$).

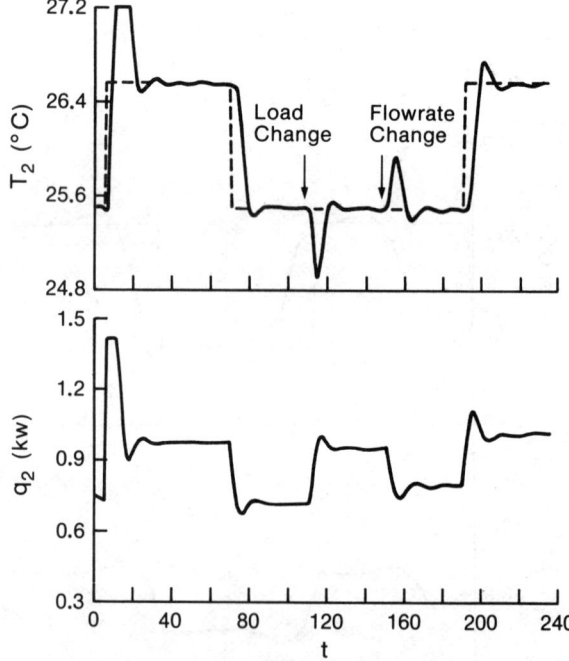

Fig. 13 Experimental response ($\nu=1$, $P_n=1-0.6z^{-1}$).

Note: Figure 14 appears immediately after the references.

Copyright © IFAC Real Time Digital Control Applications
Guadalajara, Mexico 1983

PREDICTION OF OPTIMAL DIRECT DIGITAL CONTROL SYSTEM FOR PROCESS INDUSTRIES

T. Moriyama, S. Fujii, H. Mitani, K. Achiba and T. Terada

Denka Consultant and Engineering Co., Ltd., Yuraku-cho 1-4-1, Chiyoda-ku, Tokyo 100, Japan

Abstract. The authors developed the first DDC System for Styrene Monomer Plant in 1969 and three DDC systems in 1970. Their first DDC was not profitable, and the three DDC systems were satisfied. After that, the authors experienced industries are Cement, Chemical, Detergent, Mayonase Bottle Conveying, Medicine, Petrochemical, Phono-disk making, Steel Making, Tapping Robot, Waste Water Biological Treatment, etc.

Now the authors are planning for Dual-Hierachy-distributed-batch start-scheduled optimal control system for good manufacturing fermentation proceses in a medicine industry.

The authors predict an optimal DDC system should have a flexible manufacturing control board that is an inflexible control board.

Key Words: DDC; Fermentation Control; FMS; ICB; Inflexible Control Board, Hierachy Control.

1. INTRODUCTION

The authors developed 40 sets of DDC systems for several industries.

Now they are planning for a medical industry a hierachy system: that is including schedule control, optimal fermentation control, batch auto fermentation control and sequenial batch auto control for utilities based on the demand for fermentation.

Nowaday, flexible manufacturing system (FMS) are useful for production plant of flexibility by Playback Type of Intelligent Robot.

The authors intent to recommend Inflexible Control Board for process plant of flexible process same as the above-mentioned FMS.

2. "EXTRAPORATION" OF DDC

The authors belive that an extraporation is the most important method for the prediction from the experiences.

"Taking a leaf out of a wiseman's book, Carrying knowledge into new fields" was talked by Confucius in "The Analects" in 2,000 years ago.

"Extraporation" is very famous for the authors and you by N.Wiener's "Interporation, Extraporation and Smoothing of Stationary Random Series". Prof.T.Takahashi and the late Prof.K.Izawa introduced the Confucius words, same as Extraporation to the authors.

The authors intend to introduce their experience to you in order to predict on optimal DDC system and Inflexible Control Board as the extraporation of their DDC.

Table I shows their experience list as their history of DDC. Authors developed the first DDC system for Styrene Nonomer Plant in 1969, but it was not successful and not profitable. Because, the man-machine interface is not convenient against the electronic conventional control system as 100% back up stations. cf. Picture 1.

The authors developed the first DDC System for Chloroprene Monomer plant in 1970 (reference 2) in Picture 2 and Figure 1. It was successful, because the authors planned with skillful operators as their project member for the system instead of a conventional analogue control (CAC) by the same figure of the investment cost of the CAC.

The efforts were as follows:

a) Safety control was successful (shown in Figure 1) Production was increased 6% than CAC in existing other train

b) Learning Control was introduced by POC and Monitor Typer

c) System Tie back (reference 1) was very easy by derivative type of PID control algorism

d) Turning of Time Constant of Input Filter T_f, Proportional Gain K_p, Integral Gain K_I and Derivative Gain K_D; and Nonlinear Function and Multivariable Control (AI 96, AO 48, DI 21) were effective for improvement of the control results

e) Hybride Instrumentation by Digital Control and Analogue Process Pattern Recognization was suitable

f) Yodic 500 (12 k words core only) was high reliability (MTBF 7560h, availability 100%)

The other two Yodic were adopted in the first DDC for PVC plant in Japan (reference 4) and Solvent Polymerization SBR in the same year 1970, and both were satisfied.

The authors developed 1130 Data Aquisition system for Laboratory use gas chromatography (Max. 32 sets) by connecting 1130 with Multiplexer and Autogain A/D in Picture 3.

8 k words of core memory and 512 k words of disk memory were filled by own developed program made by the authors and gas chro operators shown in Figure 2. 11 operators/day were reduced to 3 operators/day (1/shift). The authors developed SP kiln control DDC and online sampling online analyzing and one line blending SPC (ASP-SIMALTEX-U400-DIGITAL BLENDER) shown in Picture 5 in 1977 and coal Burning NSP in 1980.

64 k words of core memory and 3.2M byte min disk memory were filled be new developed program made by the authors, engineers from ONODA and operators for 2 trains of cement plant. The quality, heat efficiency, like of the brick, electrical consumption and man power were improved.

After that, engineers, managers and operators in the authors' company and their mother company are improved. Some DDC was filled the program by an manager of the plant shown in Picture 6.

By CRT and Total Distributed Control method, the authors could centralize information, sequence and turning of PID control from the many train of batch reactors as Picture 7.

The authors and Kawasaki Steel adopted DDC for co-developing test of a new pulverized coal injection (PCI) system for blast furnaces (KDD-1) (references 12,14,17) as Figure 3 and Picture 8. The operator could change and operate diskets and key-in data.

Now the authors making a bid package for a famous medicie company on the best manufacturing fermentation process with utility facilities. Host computer is adopted for time scheduling and optimal fermentation control. Slave computers are total distributed control system including sequence control for 320AI, 96AO, 448DI, 1744DO, 160PI, 144 INTERRUPT, and 145 PID loops. Basic system configurator is shown in Fig. 4. In order to reduce analogue recorders, the authors will adopt many CRT and dual host computers, double data high way, 1:N back up station for PID loops, and double sequencers as Figure 5 and Picture 9.

INFLEXIBLE CONTROL BOARD (ICB)
(Reference 10)

Nowaday, the improvement of process plant is very frequent. If we including the authors can adopt Inflexible Control Board, even if the plant is flexible manufacturing process plant, what will be happy for us as instrument and control engineers same as Flexible Manufacturing System by Playback type of Intelligent Robot as Picture 10.

The authors like making new program and developing new type of automatization always. However they think making same kind of program and routine work as debugging are very severe for young skillful engineers as same as making Mexican Piramid by stone-handling. However, the authors believe DDC and Piramid in very important for the object. Then the authors intend to predict a new type of DDC as ICB as follows:

Definition of Assumed Process for ICB

1) Type of Process:
   Continous or Batch and Continuous and Start and Stop for Shut Down

2) FMS of Process:
   Exchange of Type of Catalist. Exchange of Separation Method. Reducing of Utilities Consumption. Improvement of Plant Operation Management, etc

3) When control method and the process are changed as FMS,
   (1) No exchange of connection in Terminal Board
   (2) No exchange of Number of Measuring Signal Loops
   (3) No exchange of PID cards except turning parameters
   (4) Exchange of Sequence Control
   (5) Exchange of Input Signal and Output Signal with connection of PID cards

(6) Exchange of Signal Mode and Signal Filtering of Input and Output Signals
(7) Exchange of Feedforward Short Simple Mathmodel in DDC.
(8) Balanceless Bumpless SW for Auto/Manual and Manual/Auto and System Tie back for Output Signal Addition are possible
(9) Process Flow Sheet and Control Method is visuable easily for frequently changing of FMS
(10) Mathematical Model is changeable but is not frequently changed
(11) Shut Down of FMS is shorter and the reliability of ICB is better than existing method

Prediction of ICB assumed by the authors cf Figure 6

(1) Semigraphic Panel, Announciators, Analogue Display of Measured Valuable Valve Position, Control Status, Turning Parameter and Step of Sequence Control are Visuable in Colour CRT

(2) (3)(6)(7) are including in one chip PID Control Card

(3) Terminal Board are inflexible for wiring

(4) Input/Output Filing are done by CAD Marshalling Boards or ROM Cards

(5) Sequence Control are done by double RAM and ROM cards changing Sequence are done by Modification of Ajinomoto's Logic Chart (Cause, Effect and Interrupt Table) in Colour CRT

(6) Big Mathematical Model is filled in the host computer after debugging in out side of the computer

(7) Data High Way should be double

CONCLUSIONS

The authors intend to predict that we can easily relieve from changing the DDC program, debugging the program and cutting or removing the control board when process plant will be improved or modernized by ICB. Because, ICB will be flexible in Input/Output Filing and CRT those signal connections will be easily changed with Input/Output Filing, Feedforward Algorism and sequence control.

The inflexibility of ICB will be proved by Filing change and CRT operation. And the cost will be cheaper than programming change by skillful engineers.

Acknowledgements

The authors are grateful to Prof. the late K. Izawa, Prof. T. Takahashi and Prof. K. Tsuchiya, for many stimulating discussions.

The authors are indebted to gentlemen in Denka who did much to cooperate in their DDC.

References

1) Moriyama T.: Jan. 1967, "Electronic Control for Power Station Boiler and Low Excess Air Control", Fuel Power Generation. (in Japanese)

2) Moriyama, T.: Nov. 1971, "Direct Control for Continuous Liquid Phase Reactor", Symposium of Automatic Control Committee

3) Moriyama, T.: Dec 1971. "1130 Data Acquisition System for Laboratory Gas Chromatography", Society of Information Processing

4) Nomura, S. et al: June 1971, "DDC for PVC", Vinyl Chloride and the Polymer

5) Moriyama, T. et al: Jan./Feb., 1981, "Coal Burning in An RSP Kiln" World Cement Technology. pp 14-18 (in English)

6) Kawasaki Steel Corp.: Nov 11, 12, 1981, "Development of Pulverized Coal Flow Meter for Blast Furnace" Instrumentation Co-study Meeting/Institute of Steel Industries, Japan (in Japanese)

7) Iwamura, T. et al.: July 1981, "Cross Correlation Flow Pulverized Coal" Symposium of the Society of Instrument and Control Engineers. pp 433-434 (in Japanese)

8) Moriyama, T. et al: July, 1981, "Constant Flow Control for the High Pressure Powder Pneumatic Conveying System" Symposium of the Society of Instrument and Control Engineers pp 219-220 (in Japanese)

9) Moriyama, T. et al: June, 1981, "Industrial Safety of Coke and Coal at High Pressure Air and Oxygen for Pneumatic Conveying System", the Society of Powder Technology, Japan (in Japanese)

10) Moriyama, T and Terada T.: March, 1982, "Inflexible Control Board", Symposium of the Society of Chemical Engineers, Japan pp J305 (in Japanese)

11) Iwamura, T., Moriyama, T., et al: March 1982, "Development of a Pulverized Coal Flow Meter and Behavior of Powder Transportation", Symposium of the Society of Chemical Engineers, Japan pp J-104 (in Japanese)

12) Moriyama, T., Iwamura,T.; et al: March 1982, "Control Method of a High Pressure Powder Pneumatic Conveying System", Symposium of the Society of Chemical Engineers, Japan. pp J-105 (in Japanese)

13) Moriyama, T., Achiba, K., et al: March 1982, "Powder Properties and Transport Efficiency", Symposium of the Society of Chemical Engineers, Japan. pp J-10 (in Japanese)

14) Moriyama, T. et al.: May 1982, "A new Pulverized Coal Injection System for Blast Furnaces", Pneumatic, pp 1-22 (in English)

15) Moriyama, T. et al: July, 1982, "Behavior of Powder Particle in Turbulent Flow, Society of Powder Tech. p 1-6 (in Japanese)

16) Moriyama T.; Sept. 1982, "Measurement and Control of High Pressure Solid Gas Two Phase Flow", Symposium of the Society of Chemical Engineers, Japan pp B102-6 (in Japanese)

17) Moriyama T.: April 1983, "A New Pulverized Coal Injection System for Several Industries", MICONEX'83 (in English)

Picture 3 1130 Gas Chro DACS(cf No.5)

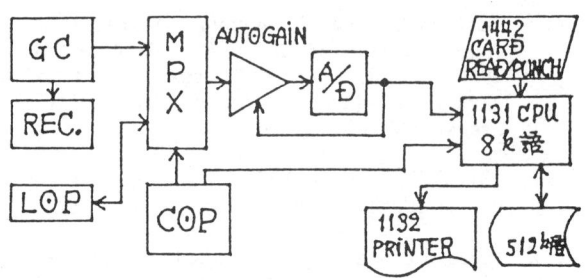

Figure 2 1130 Gas Chro DACS(cf No.5)

Picture 1 Styrene Monomer DDC(cf No.1)

Picture 2 Chloroprene Monomer DDC(cf No.4)

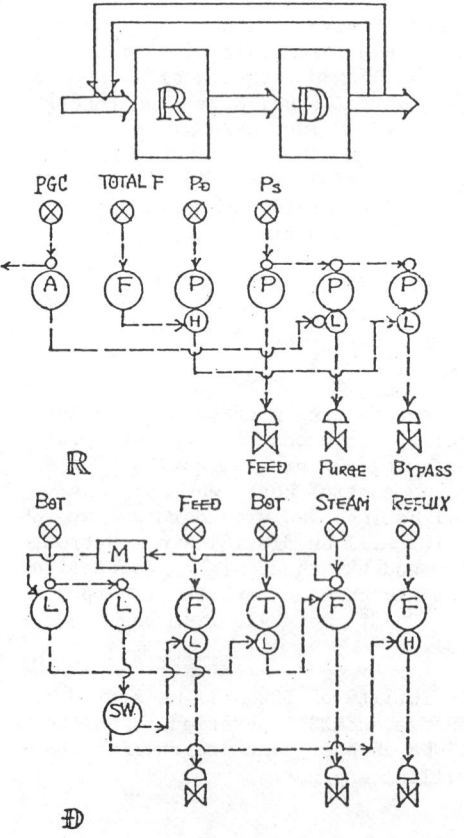

Figure 1 Chloroprene Monomer DDC(cf No.4)

Prediction of Optimal Direct Digital Control System

Table 1 Experience List for DDC by Denka. D : Denki Kagaku Kogyo K.K (Mother Company of Denka)

No.	Installed	Model	Maker	Main Specification	DKK:D or Client	Plant	DDC	SEQ	SCC	LOG	OTHERS AI	DI	PI	AO	DO
1	July '69	ICD-507	TOSHIBA	CORE 8kW DRUM 16kW	D: CHIBA	STYRENE M.	○				337				
2	March '70	YODIC-500	YOKOGAWA	CORE 12kW	D: CHIBA	PVC	○	○		○	84	171		23	201
3	May '70	ditto	ditto	ditto	D: ditto	SBR	○	○		○					
4	June '70	ditto	ditto	ditto	D: AOMI	CHLOROPRENE	○			○	96	21		48	
5	April '72	IBM-1130	IBM+SHIMAZU	CORE 8kW DISK 512kW	D: ditto	ditto					Gas Chromatography				
6	July '72	HITAC-10	Takeda+HITACHI	CORE 8kW	D: CHIBA	VAM				○	187	20	1		
7	May '75	MODEL-30	HEWLETT P.	CORE 8kW	DCE CHIBA	HIGH FLOW PNEUMA SIMULATION									
8	June '76	U-300	FUJI ELECTRIC	CORE 32kW DISK 512kW	D: CHIBA	BIOLOGICAL WATER TREAT.				○	75	21	20	1	2
9	April '77 / Mar. '81	U-400 / YEWPACK	FUJI ELECTRIC / YOKOGAWA	CORE 64kW DISK 3.2MB ROM 8kW RAM 6kW	D: AOMI	NSP CEMENT	○		○	○	192 / 12	496 / 32	0 / 0	32 / 12	272 / 92
10	July '77	CENTUM	YOKOGAWA	CORE 32kW	D: CHIBA	EVA	○	○			20	160		9	29
11	Sept '77	TOSDIC-200	TOSHIBA	IC 6kW	D: AOMI	Calcination Shaft kiln	○	○							
12	Sept '77	FUJILOG-μ	FUJI ELECTRIC	24 step	DCE CHIBA	HIGH FLOW PNEUMA TEST		○							
13	Sept '77	MEMOCON	YASUKAWA	RAM 4kW	CLIENT	HIGH FLOW PNEUMA		○							
14	Oct '77	TOSDIC-200	TOSHIBA	RAM	D: OHMUTA	ACETYLENE BLACK	○			○	152		12		
15	May '78	TOSDIC-200	TOSHIBA	RAM kW	D: AOMI	POWER STATION				○					
16	July '78	TOSDIC-200	TOSHIBA	RAM 9kW+10.5kW	ditto	CALCINATION 2 KILN	○	○			48	24		36	48
17	Nov '78	TOSDIC-200	TOSHIBA	RAM 48kW+48kW	ditto	CHLOROPRENE	○	○	○	○	48	12		18	36
18	Dec '78	TOSDIC-200	TOSHIBA	RAM 16kW+16kW	ditto	PVC MONOMER REC.	○				8		4		
19	Nov '79	MODEL 9845C	HEWLETT P.	ROM 112kB RAM 316kB	DCE CHIBA	CAD, SIMULATION ESTIMATE									
20	Nov '79	MEMOCON SC 484	YASUKAWA	RAM 4kW	CLIENT	HIGH FLOW PNEUMA		○							
21	Aug '80	MEMOCON SC 184H	YASUKAWA	RAM 8kW×5	CLIENT (USSR)	HIGH FLOW PNEUMA		○							
22	Jan '80	CENTUM	YOKOGAWA	RAM 160kW+80kW	D: CHIBA	PS	○	○		○	36	512	31	41	592
23	March '80	ST-150	CHINO	RAM 20kW	D: AOMI	CHLOROPRENE		○							
24	March '80	TOSDIC-200	TOSHIBA	RAM 4kW	D: AOMI	COKE and CARBON	○	○		○	12	6		8	16
25	May '80	YEWPACK CENTUM	YOKOGAWA	RAM 6kW ROM 8kW RAM 32kW+32kW	D: AOMI	PCP	○	○			112	64		105	96
26	Jun '80	ST-150	CHINO	RAM 20kW	D: AOMI	CHLOROPRENE		○							
27	Sept '80	YEWPACK	YOKOGAWA	RAM 12kW ROM 16kW	CLIENT STEEL	PCI (KDP-1)	○	○							
28	Nov '80	PS-80	TEAC	RAM 48kB ROM 12kB	CLIENT FOOD	BOTTLE		○		○					
29	Dec '80	YEWPACK	YOKOGAWA	RAM 6kW ROM 8kW	CLIENT STEEL	PNEUMATIC CONVEYOR	○	○							
30	Jan '80	SC-484	YASUKAWA	CORE 4kW	CLIENT SOAP	PELLET CONVEYOR		○							
31	Jan '81	SBC	YAMATAKE HONEYWELL	RAM 8kW ROM 28kW	D: AOMI	MIN REFLUX DISTILLATION	○				32		16	32	
32	Apr '81	N-2000	HITACHI	RAM 4kW	CLIENT STEEL	RADLE INJECTION		○							
33	May '81	MODEL 9845B	HEWLETT P.	ROM 112kB RAM 56kB	DCE TOKYO	ESTIMATE SALES MANAG.									
34	July '81	YEWPACK	YOKOGAWA	ROM 16kW RAM 12kW	D: OHMUTA	FUSED SiO₂	○	○		○	122		74		
35	July '81	YEWPACK	YOKOGAWA	ROM 8kW RAM 6kW	D: OHMUTA	ACETYLENE BLACK	○	○			46		31		
36	June '81	DSC	HITACHI	RAM 64kW+128kW	D: CHIBA	ABS	○	○			46	367	41	25	330
37	Aug '81	DCS	HOKUSHIN	CORE 24kW RAM 84kW DRUM 128kW	CLIENT SINGAPORE	ACETYLENE BLACK	○	○		○	85	155		28	95
38	Sept '81	CENTUM	YOKOGAWA	RAM 32kW+64kW	D: CHIBA	STYRENE M.	○	○		○	128	128	144	16	0
39	Sept '80	INTEL	INTEL	4kW+12kW	D: AOMI	ELECTRIC FURNACE	ROBOT				27	98		18	70
40	Sept '82	YEWPACK	YOKOGAWA		D: OHMUTA	FUSED SiO₂	○			○	40		40		
41	Sept '82	CENTUM	YOKOGAWA	32kW+128kW	D: OHMUTA	ACETYLENE BLACK	○			○	259		165		

Picture 4 CRT for NSP Cement (cf No.9) Picture 8 CRT for PCI (KDP-I) (cf No.27)

 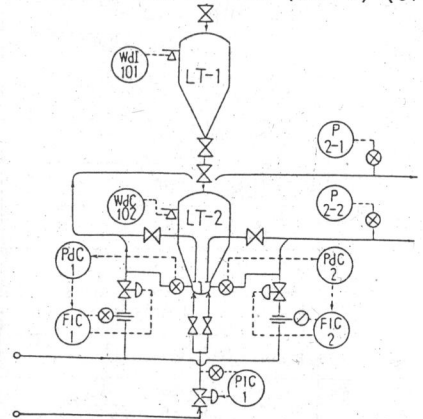

Picture 5 Computer Room for NSP Cement(cf No.9) Figure 3 PCI(KDP-I) (cf No.27)

Picture 6 PVC DDC (cf No.2) Picture 10 Tapping Robot for Electric F.(No.39)

 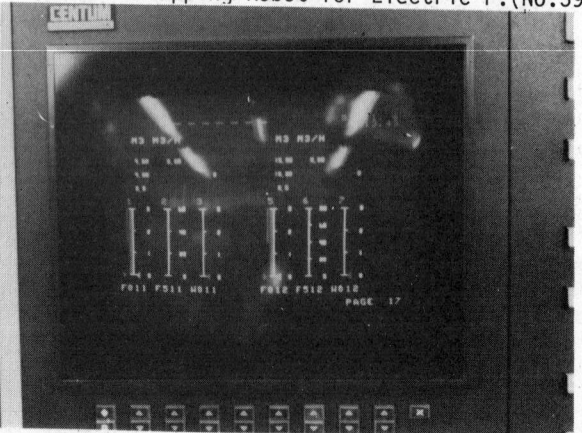

Picture 7 PS DDC (cf NO.22) Picture 7' CRT for PS DDC (cf No.27)

Prediction of Optimal Direct Digital Control System

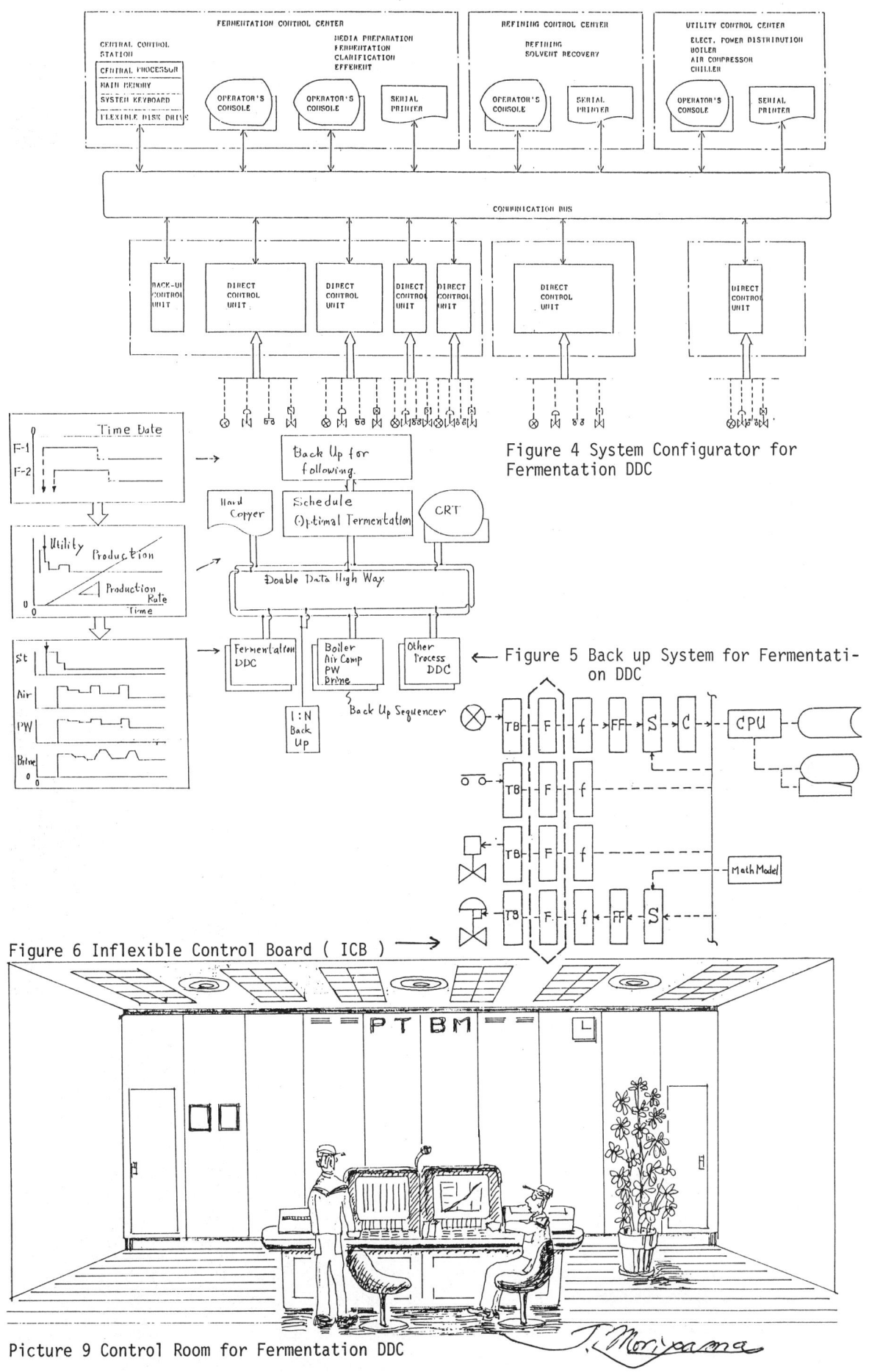

Figure 4 System Configurator for Fermentation DDC

Figure 5 Back up System for Fermentation DDC

Figure 6 Inflexible Control Board (ICB)

Picture 9 Control Room for Fermentation DDC

SELF-TUNING CONTROL OF A LIQUID-SATURATED STEAM HEAT EXCHANGER

S. Bittanti*, D. W. Clarke**, F. Romeo*, R. Scattolini*

*Dipartimento di Elettronica, Politecnico di Milano, Italy
**Department of Engineering Science, Oxford University, UK

Abstract. This paper is concerned with the self-tuning control of a simulated liquid-saturated steam heat exchanger. The problem is to control the fluid outlet temperature by adjusting the fluid flow-rate. Since the process dynamics is characterized by a number of zeros close to the unit circle, 'cancellation' controllers cannot be applied. The class of self-tuner employed are generalised-minimum variance self-tuning controllers, interpreted here as 'detuned model-reference' system. Good control performance is obtained, provided that a sensible choice of the system time delay is made.

Keywords. Adaptive Control; Heat Exchangers; Stochastic Systems; Digital Control; Process Control.

1. INTRODUCTION

This paper deals with the self-tuning control of a simulated liquid-saturated steam heat exchanger, where the fluid outlet temperature is controlled by manipulating the fluid flow-rate. Although the heat exchanger is accurately simulated as a distributed parameter nonlinear continuous-time process, the control is designed on the basis of a linear discrete-time model. This is an interesting test-case as there are several industrial processes which have similar dynamics, characterized by a number of zeros close to the unit circle. In particular, 'cancellation' controllers-which form the basis of many self-tuning control schemes-cannot be applied.

The simulated process has previously been extensively studied by Bittanti, Cividini and Scattolini (1982) to derive an identified ARMAX model, which was then used to design a variety of digital control laws (Bittanti, Scattolini, 1982). The results are used here to act as a standard against which the responses of the adaptive controllers can be measured.

The class of self-tuner employed are generalised -minimum variance self-tuning controllers (Clarke, Gawthrop, 1975 and 1979), interpreted here as 'detuned model-reference' system (Clarke, 1981), in which weighting on the control signal is used to reduce the sensitivity to zero cancellation.

The identification project discussed by Bittanti, Cividini and Scattolini (1982) led to models with more than 10 parameters, whereas most practical self-tuning excercises have shown good performance with relatively few parameters. This turns out to be confirmed in this case too, subject to a sensible choice of the model time-delay.
Heat exchanger modelling, identification and control is dealt with extensively in the literature, see e.g. the survey paper by Ray (1978). To the best knowledge of the present authors, however, this is one of the few papers where the attempt is made to use the self-tuning approach. The paper is organised as follows. In Sect. 2, the process description is given, and the control problem is stated. Then the adopted self-tuning algorithm is outlined in Sect. 3. Finally the simulation results are presented in Sect. 4.

2. PROCESS DESCRIPTION AND PROBLEM STATEMENT

In the exchanger, water is heated by pressurized saturated steam through a copper tube. In the description of the process it is assumed that there are no radial temperature gradients. Therefore the temperatures θ_1 and θ_2 of the liquid and the metal are functions only of time t and axial coordinate x: $\theta_1=\theta_1(t,x)$; $\theta_2=\theta_2(t,x)$. Since the steam is saturated, its temperature θ_s is a function of time only, $\theta_s=\theta_s(t)$. Denoting by $\dot{\theta}_i(t,x)$ and $\theta_{ix}(t,x)$ the derivatives of θ_i, i=1,2, with regard to time and space respectively, the process equations are

$$\tau_1\dot{\theta}_1(t,x)+\tau_1 v(t)\theta_{1x}(t,x)-\theta_1(t,x)-\theta_2(t,x)=0 \quad (1)$$

$$\tau_2\dot{\theta}_2(t,x)+\beta(\theta_2(t,x)-\theta_1(t,x))+\theta_2(t,x)-\theta_s(t)=0 \quad (2)$$

where v(t) is the liquid speed and τ_1, τ_2 and β depend upon the fluid and metal linear densities, the fluid and metal specific heats, the diameters of the tube and the coefficients of heat transfer liquid/metal and metal/steam. We refer the reader to (Bittanti, Scattolini, 1982) for the derivation of the above equations and the expressions for τ_1, τ_2 and β. Notice that the coefficient of heat transfer liquid/ metal is assumed to depend upon the liquid speed according to the Dittus-Boelter relation (Rohsenow,Hartnett,1973).This leads to the following relations

$$\tau_1 = \bar{\tau}_1 \left(\frac{\bar{v}}{v(t)}\right)^\nu , \quad \beta = \bar{\beta}\left(\frac{v(t)}{\bar{v}}\right)^\nu, \nu = 0.8,$$

where $\bar{v}, \bar{\tau}, \bar{\beta}$ are the nominal values of v, τ and β respectively. As τ_2 is independent of the coefficient of heat transfer, τ_2 is independent of the liquid speed.

Denoting by L the length of the tube, let $\tilde{\theta}_{\ell i}(t) = \theta_1(t,0)$ and $\tilde{\theta}_{\ell o}(t) = \theta_1(t,L)$ be the inlet and outlet liquid temperatures. The problem is to control the outlet liquid temperature by adjusting the liquid flow-rate. The process output variable is measured by a transducer simply described as a first order system whose output is denoted by $\theta_{\ell o}$

$$\dot{\theta}_{\ell o}(t) = \gamma \theta_{\ell o}(t) + \delta \tilde{\theta}_{\ell o}(t).$$

The variable $\theta_{\ell o}$ is sampled at a uniform sampling rate giving rise to a signal

$$y(k) = \theta_{\ell o}(k\Delta - \epsilon) - \bar{\theta}_{\ell o}$$

where $\bar{\theta}_{\ell o}$ is the nominal value of $\theta_{\ell o}$, $k = 0,1,2,...$ is the discrete time and $\epsilon \ll \Delta$ is introduced to allow for the computation time of the controller. As for the speed, in the time interval $t \in [k\Delta, (k+1)\Delta)$, $v(t)$ is determined by a discrete time signal $u(k)$ as follows

$$v(t) = \begin{cases} v_m & , \text{if } \bar{v}+u(k) \leq v_m \\ \bar{v} + u(k) & , \text{if } v_m < \bar{v}+u(k) < v_M \\ v_M & , \text{if } \bar{v}+u(k) \geq v_M . \end{cases}$$

Saturations v_m and v_M are introduced to model the valve nonlinearity.

The block diagram of the system is given in Fig.1, and the data and the nominal values are specified in the Appendix. The sampling interval was chosen in accordance with the system dynamics. Now the 90% settling time of the step response of the exchanger varies from about 5 s to about 20 s as the speed varies over the admissible range of the valve (Fig.2). Consequently, $\Delta = 1$ s was taken as sampling interval.

Remark 1

By linearising (1) and (2) about the nominal point, the following transfer function relating $\tilde{\theta}_{\ell o}$ to v can be derived:

$$W(s) = -6.181 \frac{(1+0.158 s)(1+6.043 s)}{s(1+0.26 s)(1+5.325 s)} \cdot \left[0.389 - e^{\{-0.944 \frac{(1+4.665 s)(1+11.123 s)}{(1+6.043 s)}\}}\right].$$

Fig.3 shows the amplitude and the phase diagrams of W. These diagrams will be useful in Sect.4 below to interprete some self-tuning results.

3. THE SELF-TUNING ALGORITHM

Referring the reader to (Clarke,1980) for details, in this section the adopted self-tuning algorithm is briefly outlined. Consider the ARMAX model

$$A(z) y(k) = B(z) u(k-d) + C(z) e(k) , \quad (3)$$

where e is a white noise process and A,B and C are polynomials in the unit backward shift operator:

$$A(z) = 1 - a_1 z^{-1} - \ldots - a_{n_A} z^{-n_A} \quad (4)$$

$$B(z) = b_1 + b_2 z^{-1} + \ldots + b_{n_B} z^{-n_B} \quad (5)$$

$$C(z) = 1 + c_1 z^{-1} + \ldots + c_{n_C} z^{-n_C}. \quad (6)$$

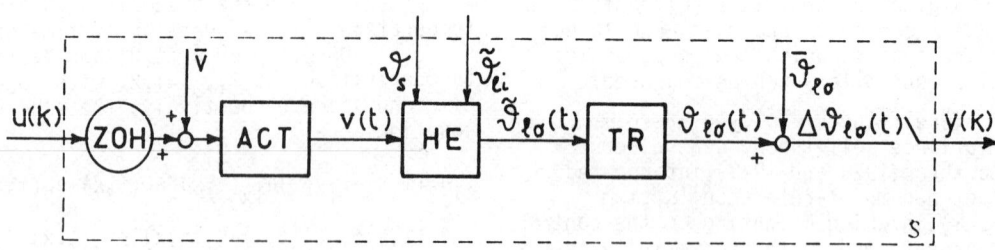

Fig. 1

The system block diagram (ZOH = zero-older-hold; ACT = actuator; HE = heat exchanger; TR = transducer).

Let $P(z)$ and $Q(z)$ be two transfer functions which will be specified by the control designer. Then, denoting by $y°$ the set point, the control objective is to minimise the variance of

$$\phi(k+d) = P(z) y(k+d) + Q(z) u(k) - y°(k). \quad (7)$$

The variable ϕ can be seen as a generalised system output.

Denoting by $P_D(z)$ the denominator of $P(z)$, let

$$\tilde{y}(k) \triangleq \frac{1}{P_D(z)} y(k) \quad (8)$$

$$\tilde{y}°(k) \triangleq y°(k) - Q(z) u(k). \quad (9)$$

Furthermore, define $E(z)$ and $F(z)$ as those polynomials such that

$$P(z) \frac{C(z)}{A(z)} = E(z) + z^{-d} \frac{F(z)}{P_D(z) A(z)}. \quad (10)$$

Then, the optimal d-step-ahead predictor $\hat{\phi}$ of ϕ is given by

$$C(z) \hat{\phi}(k+d) = F(z) \tilde{y}(k) + G(z) u(k) - C(z) \tilde{y}°(k), \quad (11)$$

where

$$G(z) = B(z) E(z). \quad (12)$$

Consequently, once the parameters of system (3) are given, the control law minimizing the variance of (7) is given by

$$F(z) \tilde{y}(k) + G(z) u(k) + H(z) \tilde{y}°(k) = 0, \quad (13)$$

where

$$H(z) \triangleq - C(z). \quad (14)$$

This leads to a control law of the following type:

$$u(k) = \frac{1}{g_0} (-f_0 \tilde{y}(k) - f_1 \tilde{y}(k-1) - \ldots - f_{n_F} \tilde{y}(k-n_F) \\ - g_1 u(k-1) - \ldots - g_{n_G} u(k-n_G) - \quad (15) \\ + h_0 \tilde{y}°(k) + h_1 \tilde{y}°(k-1) + \ldots + h_{n_H} \tilde{y}°(k-n_H))$$

In adaptive control, polynomials F,G and H have to be estimated in real time. As is well known, the estimation can be performed in two ways. The first one consists of estimating the model polynomials A,B and C and then deriving the controller polynomials F,G and H by (10),(12) and (14). In the second approach, the controller polynomials are estimated directly from the data. This also can be done in a number of ways, as e.g. the following one. Equation (11) represents a linear system the inputs of which are $\tilde{y}$, u and $\tilde{y}°$. By the superposition principle, the output can be seen as the sum of signals $-\tilde{y}$ and Ψ, Ψ being the solution of

$$C(z) \Psi(k+d) = F(z) \tilde{y}(k) + G(z) u(k). \quad (16)$$

The estimation of polynomials F,G and C can be performed by applying the Extended Least-Squares (ELS) algorithm (Ljung and Söderdtröm,1982) to equation (16). Of course, if $n_C = 0$, the Recursive Least-Squares (RLS) algorithm would suffice. The optimal control law can then be derived from (13) or, equivalently, (15).

4. DETUNED MODEL REFERENCE CONTROL OF THE HEAT EXCHANGER

The heat exchanger control system is represented in Fig.4, where S is the system of Fig.1 and controller C is given by (15) with parameters f_i, g_i and h_i replaced by their real time estimates. For conciseness, such a controller will be denoted by $C_d(n_F, n_G, n_H)$, where d is the assumed system time delay (see (3)) and n_F, n_G and n_H are the numbers of parameters of the controller polynomials (see (15)). Notice that, in view of (10), the controller parameters do depend upon d.

So far, it is apparent that the algorithm presented in the previous section enables one to derive a self-tuning controller once i) transfer functions P and Q of (7) are chosen and ii) the integers d, n_F, n_G, n_H are fixed.

As far as the choice of P and Q is concerned, note that, if Q = 0, the variance of ϕ is minimised when $y(k) = (1/P(z)) \dot{y}°(k)$. Therefore, $1/P(z)$ plays the role of a reference model for the closed loop system. In particular, P(z)=1 leads to the classical minimal variance control design (Åström,1970). A main drawback of the

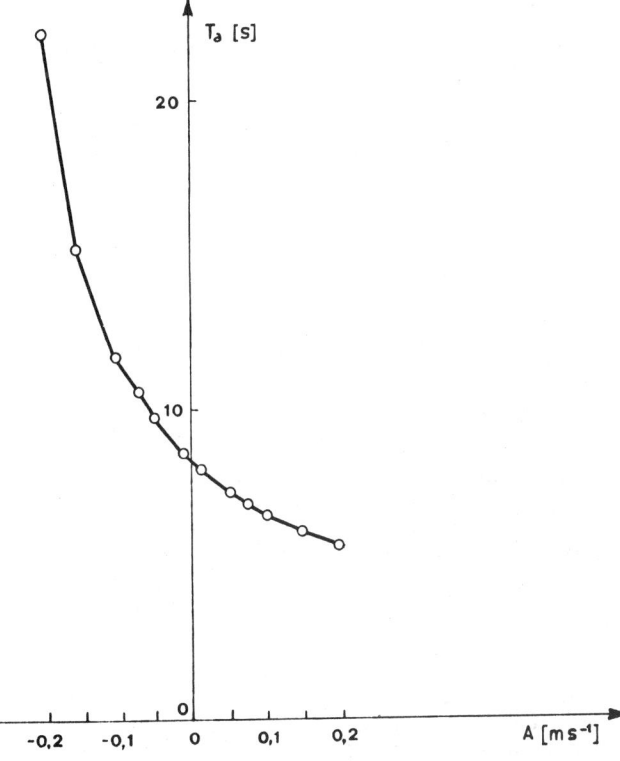

Fig. 2

90% step response settling time (T_a is the 90% settling time of $\theta_{\ell o}$ as $u(k) = 0$, $k < 0$ and $u(k) = A$, $k \geq 0$).

model reference approach is that, if one considers a linearised model of the process, the controller tends to cancel the process zeros. Hence, a poor control action is obtained when the process zeros are close to the unit circle. In the case of the heat exchanger, a number of ARMAX models have been identified off-line by Bittanti, Cividini and Scattolini (1982), any of which exhibits zeros of polynomial B close to the unit circle. This is not surprising in light of the Bode plot of Fig.3. These considerations lead to the conclusion that the model reference approach is not suited to deal with the problem at hand (unless some ad hoc trick is adopted). This is illustrated in Bittanti, Scattolini (1982), where a number of generalised minimum variance controllers were designed on the basis of one amongst the off-line identified models of (Bittanti, Cividini, Scattolini, 1982). Very poor results were also obtained by the self-tuning control system of Fig.4. They are not reported here for the sake of brevity.

In this paper attention is focused on the detuned model reference approach ($Q \neq 0$ in (7)). From (Bittanti, Scattolini, 1982), it follows that the choice

$$P(z) = 1 \quad , \quad Q(z) = 4(1 - z^{-1}) \quad , \qquad (17)$$

is a sensible one for the heat exchanger control problem based upon the off-line identified ARMAX model with $d=1, n_A=2, n_B=10$ and $n_C=6$ (see (3)-(6)). This model, which is referred to as model $M_1(2,10,6)$, was derived by the ELS algorithm and the Rissanen model selection criterion (Rissanen, 1978). Choice (17) was therefore considered for the self-tuning control too.

The corresponding control system of Fig.4 has been analysed by investigating step variations in the set-point $y°$, the steam temperature θ_s and the inlet liquid temperature $\tilde{\theta}_{\ell i}$. In detail, the following test, which will be referred to as the *three-steps test*, is considered:

$$y(k) = \begin{cases} \bar{\theta}_{\ell o} & , k<0 \\ 98°C & , k \geq 0 \end{cases} \quad ; \quad \tilde{\theta}_{\ell i}(t) = \begin{cases} \bar{\theta}_{\ell i} & , t<45 \text{ s} \\ 66.5°C & , t \geq 45 \text{ s} \end{cases} \quad ;$$

$$\theta_s(t) = \begin{cases} \bar{\theta}_s & , t<90 \text{ s} \\ 117°C & , t \geq 90 \text{ s} \end{cases}$$

Fig.5 shows the results of the three steps test for the system of Fig.4 with controllers $C_d(2,2,0)$, $d=1,2,...,6$. In all these trials, the parameters of (13) have been estimated by the RLS algorithm with exponential forgetting factor equal to 0.99. The dc term (Clarke, 1980) was also estimated with the same value of the exponential forgetting factor.

From the results of Fig.5, the conclusion can be drawn that $d=4$ leads to the best controller. $d=3$ is also acceptable, whereas $d=1$ or 2 leads to unstable systems. If d is increased over 4, then slower and slower control systems are obtained.

Remarks

2. It is interesting to compare the three-steps test diagram of Fig.5.d ($d=4$) with the

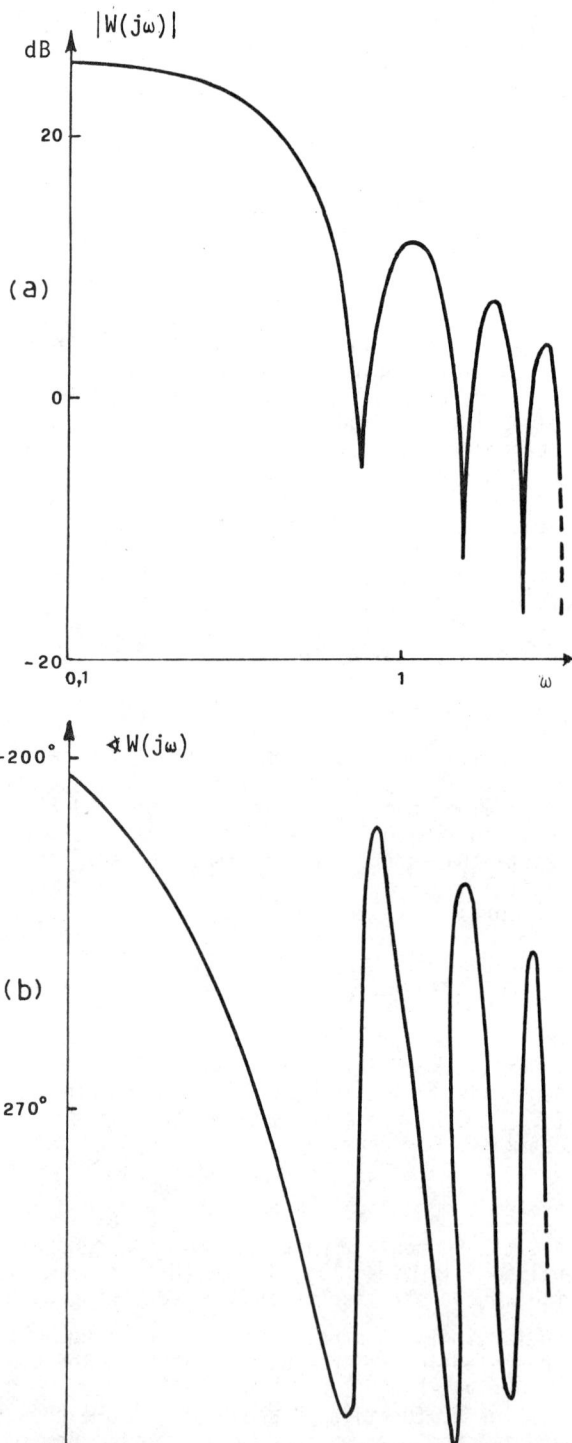

Fig. 3

Bode diagrams of transfer function relating the outlet liquid temperature to the fluid speed.

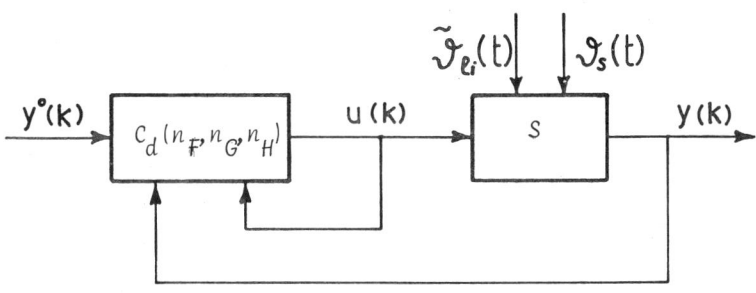

Fig. 4
Heat exchanger self-tuning control system.

best three-steps test results obtained in (Bittanti,Scattolini,1982) on the basis of model $M_1(2,10,6)$. In these latter results, the transients are quicker and smoother than the ones of Fig. 5.d. However, the self-tuning controller exhibits acceptable performance with 4 parameters only, whereas the controllers derived by Bittanti and Scattolini (1982) are defined by about 20 parameters. Moreover, the time by which the tuned controllers achieve an acceptable response was much smaller than the time necessary to identify the model $M_1(2,10,6)$.

3. In light of Fig.5, a sensible choice of d is d= 3 or 4. This can be given the following interpretation. The transfer function relating u to y in model (3) has z^{-d} as a factor. As is well known, such a delay factor has no relevance to the magnitude ratio of the system frequency response, whereas it plays a major role in its phase angle. Now, the slope of the phase angle diagram of Fig.3 about the lowest frequency at which the magnitude ratio diagram crosses the 0-dB axis is approximately -170 °s/rad. The slope of the phase angle diagram associated with the frequency response of z^{-d} is - d 180/3.14 °s/rad, i.e. - 171.9 °s/rad for d=3 and - 229.3 °s/rad for d=4. This observation suggests that, if d is equal to 3 or 4 in model (3), then it is possible to approximate the Bode phase diagram of Fig.3 around the first crossover point; in other words, the phase margin estimated by model (3) is presumably close to the true one. If d is equal to 1 or 2, the phase angle diagram slope introduced by the term z^{-d} is smaller. Therefore, the estimated phase margin is larger than the true one, so that the feedback control system is unstable. On the other hand, if d is greater than 4, the phase margin is underestimated so that the feedback control system is stable but slow.

4. Suppose that P(z)=1. Then, from (7) it follows that the following control law

$$u(k) = Q(z)^{-1}[y°(k) - y(k+d)]$$

is the one giving rise to a variance of ϕ equal to zero. Therefore, the inverse of Q can be interpreted as the controller transfer function implicitly imposed by the minimisation of the variance of (7). In this sense, (17) corresponds to an integral controller. Since a more flexible controller is the proportional-integral one, the following choice was also considered in place of (17):

$$P(z)=1 \;,\; Q(z)= 4\,\frac{1-z^{-1}}{1-0.7 z^{-1}} \,. \quad (18)$$

The specific values of the parameters of Q are suggested by the results of (Bittanti,Scattolino,1982). In Fig.6, the corresponding results of the three-steps test given by controller $C_4(2,2,0)$ show a good control action.

5. CONCLUSIONS

The application of the self-tuning approach to the control of the process at hand led to most encouraging results. Good control performance was obtained by means of simple controllers. In order to achieve this objective, a suitable choice of the assumed system time delay is important.

ACKNOWLEDGMENT

This paper has been partially supported by the Centro di Teoria dei Sistemi of the C.N.R. of Milano and by the M.P.I. Research Project 'Methods of Identification and Spectral Factorization in Multivariate Time Series Analysis'.

APPENDIX - Process data

Nominal values

$\bar{\theta}_{li}$= 65 °C ; $\bar{\theta}_{lo}$= 98.672 °C ; $\bar{\theta}_s$= 120 °C ;
$\bar{v}$ = 0.3 m/s ; $\bar{\tau}_1$= 7.2 s ; $\bar{\beta}$ = 0.193 .

Data

L = 2.44 m ; v_m = 0.1 m/s ; v_M = 0.5 m/s ;
γ = δ = 1 s^{-1} ; τ_2 = 0.31 s ; ν = 0.8 ;
ε = 0.05 s ; Δ = 1 s.

REFERENCES

Åström,K.J.,Introduction to Stochastic Control Theory,Academic Press,1970.

Bittanti,S.,A.Cividini and R.Scattolini,

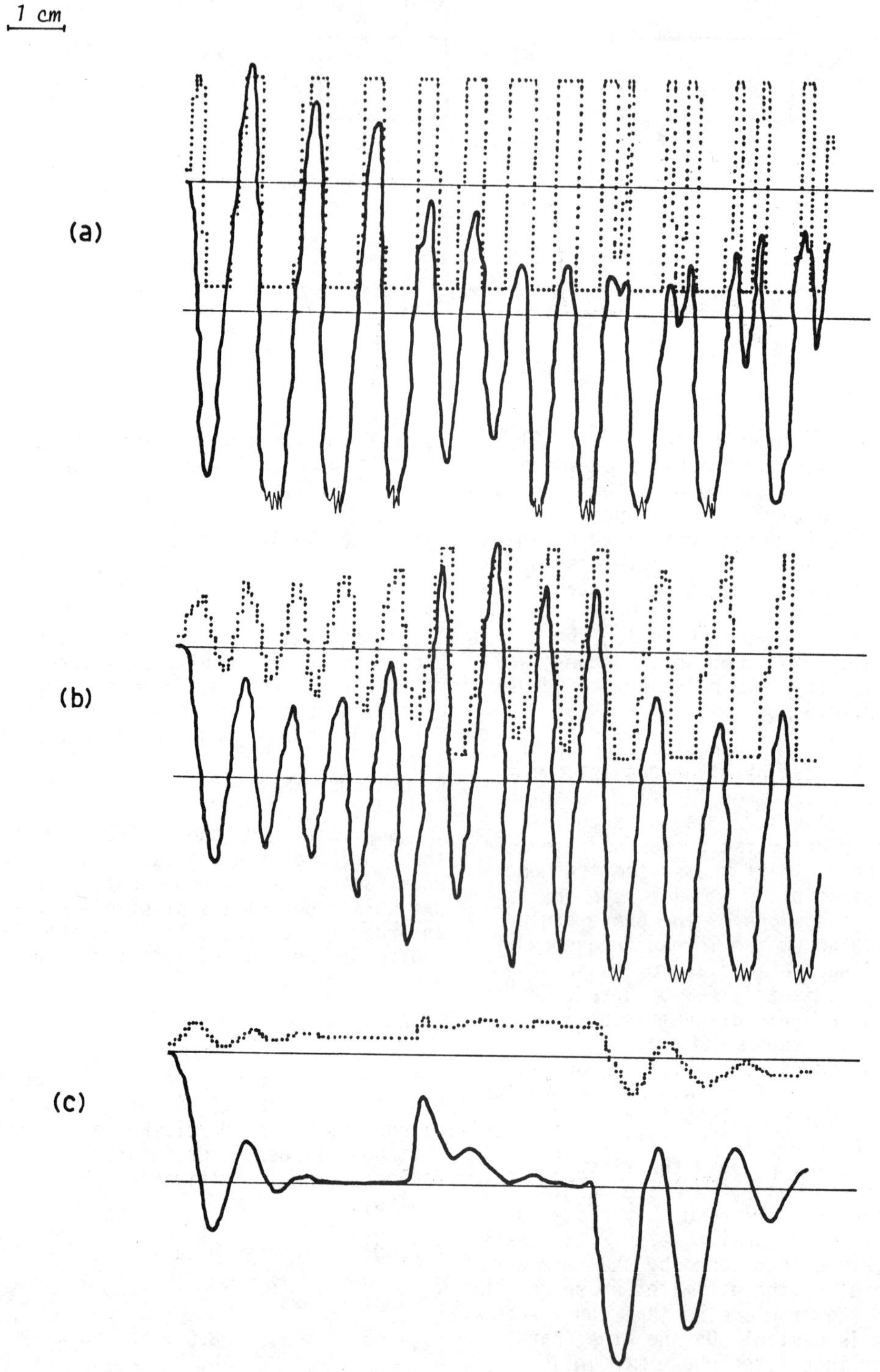

Fig. 5
Self-tuning control of the heat exchanger by controllers $C_d(2,2,0)$, $d=1$ (a), $d=2$ (b), $d=3$ (c), $d=4$ (d), $d=5$ (e), $d=6$ (f), with choice (17) of $P(z)$ and $Q(z)$. The diagrams of v (dotted lines) and of $\theta_{\ell o}$ (continuous lines) are given.

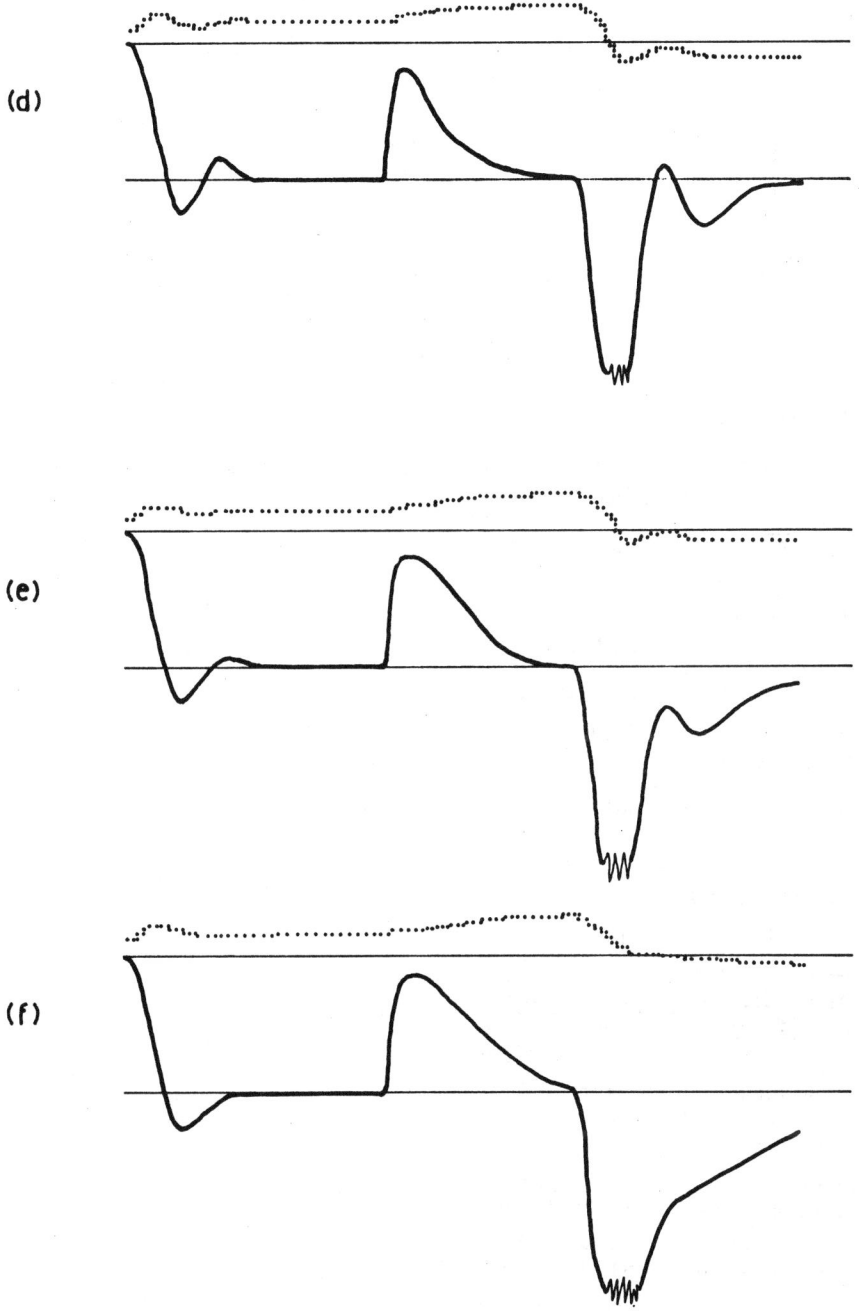

Fig. 5 (continuation)
The upper horizontal straight line corresponds to the nominal values $\bar{v}$ and $\bar{\theta}_{lo}$ of v and θ_{lo}. The lower horizontal straight line corresponds to 5 cm/s and 98 °C for v and θ_{lo} respectively. The time scale is 1 cm = 11.8 s.

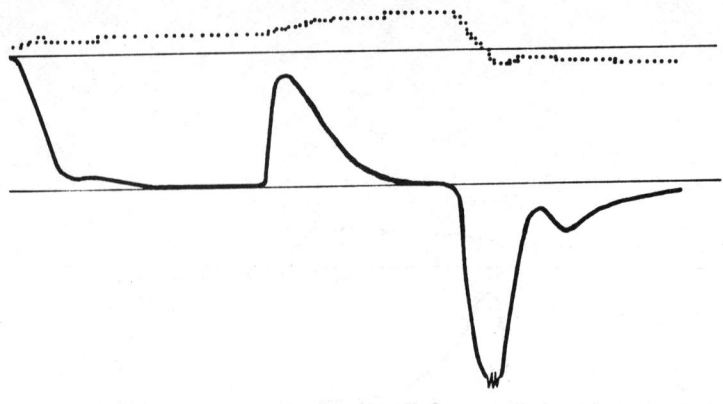

Fig. 6

Self-tuning control of the heat exchanger by controller $C_4(2,2,0)$ with choice (18) of $P(z)$ and $Q(z)$ (same scale of Fig. 5).

Identification of a Liquid-Saturated Steam Heat Exchanger, 6th IFAC Symp. on Identif. and System Par.Est.,168,1982.

Bittanti,S. and R.Scattolini,Optimal Stochastic Control of a Liquid-Saturated Steam Heat Exchanger,in Time Series Analysis:Theory and Practice 2,O.D.Anderson ed.,North-Holland, 1982.

Clarke,D.W.,Some Implementation Considerations of Self-Tuning Controllers,in Numerical Techniques for Stochastic Systems,F.Archetti and M.Cugiani eds.,North-Holland,1980.

Clarke,D.W.,Introduction to Self-Tuning Controllers,in Self-Tuning and Adaptive Control:Theory and Applications,C.J.Harris and S.A.Billings eds.,P.Peregrinus Ltd., 1981.

Clarke,D.W. and P.J.Gawthrop,Self-Tuning Controller,Proc. IEE,122,1975.

Clarke,D.W. and P.J.Gawthrop, Self-Tuning Control,Proc. IEE,126,1979.

Ljung,L. and T.Söderdtröm,Theory and Practice of Recursive Identification Methods,M.I.T. Press,1982.

Ray,G.,Some Recent Applications of Distrubuted Parameter Systems Theory - a Survey, Automatica,14,1974.

Rissanen,J.,Modelling by Shortest Data Description,Automatica,74,1978.

Rohsenow,W.M. and J.D.Hartnett,Handbook of Heat Transfer, Mc Graw Hill,1973.

DISCUSSION

SESSION WP2: CONTROLLERS

Paper: SELF-TUNING CONTROLLERS WITH A PID STRUCTURE

Authors: F. Cameron, D.E. Seborg (Department of Chemical and Nuclear Engineering, University of California, Santa Barbara, Ca 93106)

Discusser: Sergio Bittanti
Dipartimento di Elettronica,
Politecnico di Milano, Italy

Questions or Comments:

1) Can you impose that the denominator of the controller transfer function be proportional to $(1-z^{-1})$ even though polynomial $C(z)$ is different than 1?

Author's reply:

1) (Not available)

Discusser: CS Berger
Monash University
Victoria, Australia

Questions or Comments:

1) Your method assumes a 2nd order system. Doesn't that restrict its use?

2) Doesn't the use of "P" tuning parameters complicate the PID concept for engineers?

Author's reply:

1) The method assumes either a second or third-order process model depending on the selection of $P(z)$. Since second-order plus time delay models and PID controllers have been widely used in process control, this is not considered to be a serious limitation.

2) The use of P, Q and R is somewhat of a complicating factor. However, we have been successful by using simple expressions (e.g. $P_n(z) = 1 + P_{n_1} z^{-1}$)

Paper: SELF-TUNING CONTROLLERS OF A LIQUID SATURATED STEAM HEAT EXCHANGER

Authors: S Bittanti[1], D.W. Clarke[2], F. Romeo[1], R. Scattolini[1]

[1] Dipartimento di Elettronica, Politecnico di Milano, Italy
[2] Department of Engineering Science, Oxford University, England)

Discusser: R Lozano Leal
Depto de Ing. Eléctrica
Apdo Postal 14-740
México 14, D.F.

Questions or Comments:

1) Have you tried to identify the delay d on line?

Questions or Comments:

1) We have not yet tried to identify the delay d in real time. This is one of the possible extensions of the work.

Discusser: CS Berger
Department of Electrical Engineering, Monash University, Clayton, Victoria 3165, Australia

Questions or Comments:

1) As you can't choose your Q matrix a-priori to guarantee stability, wouldn't a pole placement method have been better?

Authors' reply:

1) We exploited the information derived for the known-parameter case. Notice that, in the known-parameter case, the Q was designed by combining *a-priori* plant information with a root locus analysis of the closed loop system associated with the ARMAX model of the process (a type of pole placement!)

AUTHOR INDEX

Achiba, K. 623
Aguado, A. 133
Albertos, P. 601
Alvarez, C. 601
Alvarez, J. 371
Alvarez-Icaza, L. 527
Avila, M. A. 563
Azumi, K. 43, 285

Babary, J. P. 183
Balhoff, R. A. 395
Benallou, A. 385
Benmaiza, M. 77
Berger, C. S. 109
Bittanti, S. 631
Bonilla, M. 125
Boullart, L. E. M. 297
Bouzon, B. 353
Brosilow, C. B. 407
Brown, W. A. 109
Buja, G. S. 499

Calderon, J. L. 563
Cameron, F. 613
Campos, A. C. 479
Canales-Ruiz, R. 33, 527
Carlsson, J. 227
Carpentier, J. L. 315, 595
Carrasco, D. 587
Carrillo, J. 371
Clarke, D. W. 631
Cohen, A. 377
Cohn, N. 1, 325
Corripio, A. B. 395
Cotto, G. 595

Dahhou, B. 147
Doi, K. 63
Dozortsev, V. M. 221

Ejiri, M. 507
El Hajjar, H. 183
Elizalde, F. L. 545
Enriquez, J. 479

Fang, Y-C. 461
Figueroa, D. 555
Friedly, J. C. 433
Fujii, S. 623
Fukuda, K. 43

Gallegos, J. A. 371
Gomez de Silva, J. 165
Gomez, A. 133
Gonzalez, M. A. 541
Gressel, C. 377

Haber, R. 209
Heist, R. H. 433
Heredia, J. 555
Hertzanu, B 493
Hetthessy, J. 209
Hilger, M. 209
Hirzinger, G. 243
Hopper, J. R. 461
Hormaza, J. A. 479
Hrusak, J. 69

Ienaga, Y. 285
Ishida, K. 197
Ishikawa, S. 285
Itskovich, E. L. 221

Janecek, E. 69
Jeffreson, C. P. 361
Joensuu, J. 159

Karkkainen, P. 237
Kasahara, M. 301
Kawahara, M. 257, 293
Kawasumi, T. 63
Keviczky, L. 209
Kezunovic, M. 85
Kortela, U. 159, 173
Kubo, K. 101
Kumayama, J. 63

Longo, D. 499
Lozano, R. 119,125

M'saad, M. 147
Maeda, Y. 285
Makino, T. 51
Maldonado, A. 485
Manninen, M. 237
Marchetti, J. L. 385
Matsushima, T. 345
Mellichamp, D. A. 385
Michioka, R. 51

Miki, S.   51
Mitani, H.   623
Mori, K.   43
Moriyama, T.   623
Morooka, Y.   63
Morris, E. L.   139
Mosna, J.   69

Najim, K.   147
Nakai, M.   301
Nakayama, R.   101
Nariai, Y.   51
Niederlander, P.   595
Nikiforov, I. V.   221
Noriega, A.   119

Oe, S.   93
Ohta, T.   197
Oliva, D. N.   139
Oliva, M. T.   139
Olivera, J.   415, 441
Olsen, T.   433
Ono, M.   51

Parrish, J. R.   407
Paul, R. P.   19
Paz, M. B.   455
Perel'man, I. I.   221
Plessmann, K. W.   265
Pourciel, J. B.   183
Puente, J. A. de la   601

Rance, L.   577
Rao, K. R.   475
Rivera, F.   485
Rodriguez, G.   467
Rodriguez, H.   551
Rodriguez, S.   551
Romeo, F.   631

Saito, N.   43
Saldana, F.   537
Salmelin, B.   159

Saltsburg, H.   433
Sato, K.   101
Scattolini, R.   631
Schulz, R.   203
Seborg, D. E.   385, 613
Sekine, Y.   345
Selg, H.   227
Sen, M. de la   455
Shinmura, T.   63
Simandl, M.   69
Soeda, T.   93
Solis, R.   537
Srinivasan, R.   475
Suzuki, Y.   43

Tabak, D.   493
Taguchi, J.   101
Tanifuji, S.   63
Teijero, S.   415, 441
Terada, T.   623
Thomesse, J.-P.   77, 447
Tomita, Y.   93
Torres, G.   587
Tsuchiya, T.   517

Uosaki, K.   293
Uronen, P.   309

Vajk, I.   209
Vazquez, A.   587

Wahlstrom, F.   159, 173
Wang, J. S.   191
Westerlund, T.   215
Wu, W. T.   461

Yamamoto, Y.   43, 285
Yamzaki, I.   51

Zaborszky, J.   335
Zheng, Y. F.   191